SELECTING A DESCRIPTIVE STATISTICAL PROCEDURE

Type of data	*Individual scores*	*Central tendency*	*Variability*	*Correlation coefficient*
Nominal	frequency, rel. frequency, or percent	Mode	Range	ϕ
Ordinal	frequency, rel. frequency, or percent	Median	Range	Spearman r_s
Interval or ratio (skewed distribution)	frequency, rel. frequency, or percent	Median	Range or semi-interquartile range	none
Interval or ratio (normally distributed)	frequency, rel. frequency, percent, or *z*-score	Mean	Standard deviation or variance	Pearson r and regression

The point-biserial r_{pb} is used with one interval or ratio variable and one dichotomous variable.

PARAMETRIC PROCEDURES AND THEIR NONPARAMETRIC COUNTERPARTS

Type of design	*Parametric test*	*Nonparametric test*
Two independent samples	Independent samples *t*-test	Mann-Whitney *U* or Rank Sums test
Two dependent samples	Dependent samples *t*-test	Wilcoxon *T* test
Three or more independent samples	Between-subjects ANOVA (*Post hoc* test: protected *t*-test)	Kruskal-Wallis H test (*Post hoc* test: Rank Sums test)
Three or more dependent samples	Within-subjects ANOVA (*Post hoc* test: Tukey's *HSD*)	Friedman χ^2 test (*Post hoc* test: Nemenyi's test)

UNDERSTANDING RESEARCH METHODS AND STATISTICS

AN INTEGRATED INTRODUCTION FOR PSYCHOLOGY

Gary W. Heiman
Buffalo State College

HOUGHTON MIFFLIN COMPANY Boston New York

To Karen, my wife and best friend

Sponsoring Editor: David C. Lee
Senior Associate Editor: Jane Knetzger
Packaging Services Supervisor: Charline Lake
Senior Production/Design Coordinator: Jennifer Waddell
Senior Manufacturing Coordinator: Marie Barnes
Marketing Manager: Pamela J. Laskey

Cover design: Harold Burch, Harold Burch Design, New York City
Cover image: Robert Nease Photography

Printed in the U.S.A.

Library of Congress Catalog Card Number: 97-72480

ISBN: 0-395-74590-X

23456789-HWK-01 00 99 98

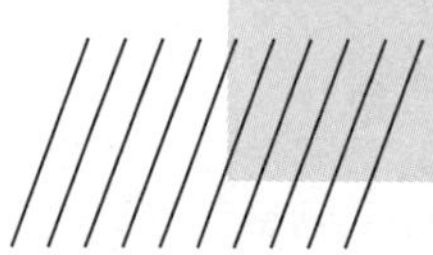

Brief Contents

Part 4

Introduction to Inferential Statistics

Part 5

Designing and Analyzing Two-Sample Experiments

Part 6

Designing and Analyzing Complex Experiments

Part 7

Alternative Approaches to Design and Analysis

Appendices

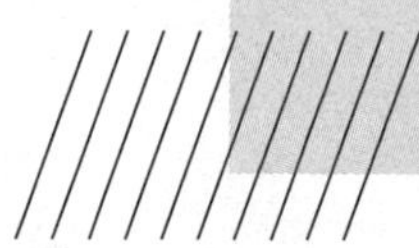

Contents

Part 2

Descriptive Statistics

6

Summarizing Data with Frequency Distributions and Percentiles 126

7

Summarizing Data with Measures of Central Tendency 154

Part 3

Correlational Research and Correlational Statistics

Part 4

Introduction to Inferential Statistics

Part 5

Designing and Analyzing Two-Sample Experiments

15

16

Part 6

Designing and Analyzing Complex Experiments

17

The One-Way Between-Subjects Experiment and the One-Way Analysis of Variance 442

18

The Two-Way Between-Subjects Experiment and the Two-Way Analysis of Variance 474

19

Within-Subjects Experiments and the Analysis of Variance 512

Appendices

A

Organizing and Communicating Research Using APA Format 603

B

Additional Statistical Formulas 645

C

Statistical Tables 679

D

Answers to Odd-Numbered Practice Problems 701

Glossary 723

References 735

Index 739

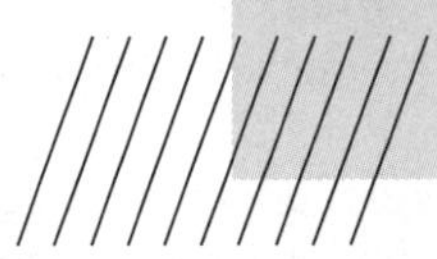

Preface

This text actively **teaches** research design and statistical procedures in an integrated fashion. It is intended primarily for an introductory two-semester psychology course that covers the material usually found in separate, one-semester courses in statistics and methods. Therefore, I have included the complete set of descriptive and experimental research methods, as well as all common primary and secondary statistical procedures. (The text can also be used in a one-semester course by assigning only some chapters.) Using one integrated text eliminates the confusion that results when separate statistics and methods texts differ in perspective, sequence, terminology, and symbols. Further, because I have previously written both a statistics textbook and a research methods textbook for use in separate courses, I am especially sensitive to the need to introduce design issues when teaching statistics, and to include statistical procedures in any discussion of research design. I am also sensitive to the times when, for clarity, it is necessary simply to teach a design principle that applies regardless of the statistics being employed, or to teach a statistical formula, without raising design concerns that only confuse the issue.

Approach

This text's approach places students in the role of researchers, focusing on the actual decisions and conclusions they make. I believe this is the way for students to develop the critical thinking skills necessary to understand and design research. At the same time, I've attempted to explain each concept clearly using a writing style that anticipates students' questions and provides many examples. In particular, statistical formulas and calculations are presented in a way that responds to students' math anxiety without pandering to them. The result should be students who successfully learn complex design issues and statistical analyses, so that they are well-prepared for advanced undergraduate courses—including independent research projects—as well as for reading the professional literature.

The central idea in this text is that the question a researcher asks leads to a particular research method, and this in turn leads to a specific statistical procedure, which is a tool the researcher uses. The researcher then considers the strengths and weaknesses of the design, and these in turn determine the interpretation of the results. The text stresses the interrelatedness of topics and continually reviews previous material while building on it. In particular, the uses of each statistic in research are discussed, and statistical and design issues are integrated in discussions about creating powerful designs. Throughout, intuitive explanations and easily grasped examples accompany more formal explanations, and all material is presented in a logical, step-by-step fashion. Overall, the approach recognizes the initial capabilities of students, and brings them up to the sophisticated level needed to understand the methods and statistics used in modern psychological research.

For the topics in methods, my goal was for students (1) to understand the terminology, logic, and procedures used in research; (2) to integrate statistical procedures with research methods; (3) to develop critical thinking skills regarding research; and (4) to learn how to design and conduct research and write APA-style reports. My goal was not to train graduate-level researchers nor to treat methods as an abstract academic discipline. Rather, I sought a middle ground of giving students a solid understanding of the basic designs and statistics found in psychological research. Therefore, the text covers all the major types of descriptive and experimental research designs and methods of control in sufficient depth that students can apply them to their research activities. For each, the impact of design decisions on reliability and validity is discussed.

For the topics in statistics, I assumed that students have a weak background in mathematics and some degree of "math phobia." My goal was that, by the end of the course, these students should understand and be able to perform the descriptive and inferential statistical procedures commonly used in psychological research. Therefore, I have included the calculations for *t*-tests, for linear correlation and regression, and for between-subjects, within-subjects, and mixed two-way factorial designs. I also discuss the logic and interpretation of three-way designs and advanced correlational techniques. In addition, the coverage of nonparametric procedures is very complete, so that the text serves as a reference source for such procedures. The importance of computing measures of effect size for each type of design is also stressed.

Throughout the text, however, the perspective is that of a psychologist, not a statistician, so I avoid dwelling on the remarkable things statisticians can do with statistics and instead discuss the things researchers commonly do. Therefore, there are no proofs or derivations, there is not a great deal of probability, and the focus is always on research and statistics, not math for math's sake. The emphasis is on the context in which statistics are used to make sense of data. Each procedure is therefore introduced in a research example, focusing on the examination of relationships between variables. Then specific procedures for describing or inferring such relationships are introduced, finally returning to the conceptual purpose and behavioral interpretation of the study. Throughout, I provide students with simplified ways to think about statistical concepts and show how these concepts translate into practical procedures for answering practical questions.

Organization

The text employs a "top-down" approach that stresses context and the interrelatedness of topics. Initial chapters give students a background in thinking scientifically and understanding research. These chapters discuss the general design and statistical issues involved in the research process. I envision that students can begin conducting their own laboratory exercises as early as Chapters 4 or 5. Later chapters then describe specific statistical procedures and fill in additional design details along with more complex designs. Appendix A, on APA research papers, can be assigned at any time.

In Chapter 1, I try to show students what we mean by "critical thinking" and how to apply the logic of the scientific method. Chapter 2 then gives an overview of how hypotheses are developed and tested, based on demonstrating a relationship between variables. Chapter 3 explains the issues of reliability and validity so that students understand how to critique and evaluate research. Chapter 4 discusses the details of conducting experiments and the ethical issues that arise. Chapter 5 presents descriptive designs—including questionnaire development—and the corresponding ethical issues. Throughout, ethical issues are presented as resulting from the design decisions that researchers make.

With this background, basic descriptive statistics are presented, including how they are used to interpret research. Chapters 6 through 8 present graphs and distributions, measures of central tendency, and measures of variability, respectively, and introduce the issues in creating powerful designs. (Procedures for computing percentiles using class intervals are presented in Appendix B.) Chapter 9 discusses z-scores, including sampling distributions and computing a z-score for a sample mean, to prepare students for understanding inferential statistics as essentially computing z-scores.

Chapter 10 presents several correlation coefficients, and Chapter 11 covers linear regression and the proportion of variance accounted for. These procedures are introduced as descriptive statistics rather than later as inferential procedures because breaking up the discussion of hypothesis testing to introduce the concept of correlation is confusing. The importance of significance testing, however, is emphasized when correlation is introduced.

The presentation of inferential statistics begins with Chapter 12, introducing probability and a preview of hypothesis testing, focusing on using the normal curve to compute probability. (Advanced probability formulas are provided in Appendix B). Chapter 13 formalizes hypothesis testing using the z-test. Chapter 14 presents the single-sample t-test and significance testing of correlation coefficients, and begins discussing the relationship between powerful designs and statistical power. Chapter 15 covers the design and analysis of two-sample, between-subject designs, and introduces measures of effect size. Chapter 16 covers the issues of controlling subject variables and the design and analysis of two-sample, within-subject designs.

Chapter 17 introduces the logic of multi-level designs, and presents the one-way, between-subjects ANOVA, including *post hoc* tests for equal and unequal *n*s and a measure of effect size. Chapter 18 discusses the logic and design of factorial experiments, and presents the two-way, between-subjects ANOVA, including *post hoc* tests for main effects and interactions. Chapter 19 presents the one-way and two-way within-subjects ANOVA, mixed-design ANOVAs, and the logic of three-way factorial designs and ANOVAs.

Chapter 20 discusses the design and analysis of various quasi-experimental designs, as well as the reasons for, and types of, single-subject designs. Chapter 21 presents the one-way and two-way chi square, as well as the nonparametric versions of all previous parametric tests (with appropriate *post hoc* tests and measures of effect size).

Appendix A provides a comprehensive look at how to search the literature and how to read and write a research report in APA format. A complete example manuscript describes a simple yet engaging two-sample experiment.

Pedagogical Features

The text is designed to be an effective learning tool during the course, and a useful reference source after the course:

- Each chapter first identifies relevant concepts from previous chapters for students to review, and then lists the learning objectives for the chapter.
- In chapters that primarily describe statistics, the "More Statistical Notation" section at the beginning of the chapter introduces new statistical notation, separate from the conceptual issues that the chapter then presents.
- Important points are emphasized by REMEMBER statements, which are summary reminders set off from the text.
- Key terms are highlighted in **bold** type, and formal definitions are accompanied by numerous mnemonics and analogies to promote retention and understanding. A complete glossary appears at the end of the text.
- All concepts, formulas, and statistical statements are presented with complete explanations. Both definitional and computational formulas are introduced in terms of what they accomplish, and computational examples of each statistic are completely worked out.
- All graphs and diagrams are explained in captions and are fully integrated into the discussion.
- At the end of each chapter, a "Putting It All Together" section provides advice, cautions, and ways to integrate material from different chapters.
- Each "Chapter Summary" provides a substantive review of the material, not merely a list of the topics covered.
- Many challenging "Practice Problems" at the end of each chapter review definitions and concepts, as well as presenting real-world application problems. Both procedural and computational statistical problems are provided. Answers to odd-numbered problems (with final and intermediate answers) are found in Appendix D. Even-numbered problems (with answers in the Instructor's Manual) can be used as assigned homework.
- A "Summary of Formulas" appears at the end of appropriate chapters.
- Appendix B provides computational formulas for computing percentiles using class intervals, linear interpolation, advanced probability, and two-way within-subjects and mixed-design ANOVAs. Practice problems are also included with these sections.
- Reference tables for selecting among designs and statistical procedures are presented on the inside front cover, and a glossary of symbols appears on the inside back cover.

Supplements

Several additional resources support the instructional and learning processes. The *Instructor's Manual and Test Bank* provides multiple-choice, short-answer, and discussion questions. The test questions are also available on disk in a program that allows instructors to edit items, add their own items, and generate exams. The *Student Work-*

book provides guided review, additional practice problems, and example test questions for each chapter of the text. Answers are included for the student. In addition, *SPSS for Windows*, by Charles Stangor of the University of Maryland, and several accompanying data sets can be shrinkwrapped with my text for students learning to use that software. Finally, I have made available a sort of capstone chapter for this book on Houghton Mifflin's Psychology Website, which can be reached at the Houghton Mifflin home page at http://www.hmco.com by going to the College Division Psychology page. This chapter provides an integrative review of the entire course by describing a number of published research studies, with prompts that guide students through the researchers' decision-making and interpretive processes.

Acknowledgments

Many people contributed to the production of this text. At Houghton Mifflin Company, I want to thank David Lee for supporting the project, and Jane Knetzger for her exceptional patience and hard work. Thanks too to all those who saw the project through to completion, including Janet Edmonds, Charline Lake, Jennifer Waddell, and Jane Judge Bonassar. Thanks also go to Barbara Muller for substantial and keen input during development and to Larissa Semanchuk for her usual invaluable assistance.

Finally, I am grateful to the following reviewers, who in evaluating the manuscript at one stage or another, provided valuable feedback:

David E. Anderson, Allegheny College
Cole Barton, Davidson College
Wendy Beller, Quincy University
Toon Cillessen, University of Connecticut
Marte Fallshore, Utah Valley State College
S. A. Fenwick, Augustana College
Gary Gillund, College of Wooster
Margaret M. Gittis, Youngstown State University
Bea Gattuso Grosh, Millersville University
Robert Keefer, Mount St. Mary's College
David S. Kreiner, Central Missouri State University
David R. Linden, West Liberty State College
Mark Masaki, Youngstown State University
Patricia A. Oswald, Iona College
P. Michael Politano, The Citadel
Gerald W. Sparkman, University of Rio Grande
L. James Tromater, University of Richmond
Toni G. Wegner, University of Virginia
Daniel B. Willingham, University of Virginia
Larry E. Wood, Brigham Young University

Gary W. Heiman

PART 1

INTRODUCTION TO PSYCHOLOGICAL RESEARCH

Okay, so you're taking a course in research methods and statistics. You probably wonder what that means. Well, essentially, you're going to learn the basics of conducting scientific psychological research: you'll learn when and how to use the methods that researchers use to conduct a study, as well as when and how to perform the different statistical procedures they use to analyze their results. To begin this journey, the following five chapters provide an overview of psychological research. The first chapter explains why students in psychology learn the scientific method and what it is. The second chapter shows you the steps involved in a psychological study. Chapter 3 then describes the major pitfalls researchers face when interpreting research. Chapters 4 and 5 then show you how psychological research is conducted.

If you're a little nervous about studying statistics, you'll be happy to see that there isn't much about statistics in these chapters. To understand how psychologists use statistics, you must first understand why and how they conduct research. As you begin to understand behavioral research, you'll see how statistics fit in and they'll become much less threatening. (Honest!)

1 Introduction to the Scientific Method

As you read this chapter, your goals are to learn:

- The assumptions and attitudes of scientists, and the goals of science.
- The difference between basic and applied research.
- The criteria for scientific hypotheses and for acceptable evidence for testing them.
- Why literal and conceptual replication are necessary.

This chapter first discusses the purpose and goals of studying research and statistics. Then it examines the scientific method and the process that leads to psychological knowledge.

INTRODUCTION (OR WHY AM I HERE?)

Students often ask the same questions about studying research methods and statistics. The answers to these questions will teach you something about psychological research and may alleviate any anxiety you have.

Why Do I Need to Learn About Research Methods and Statistics?

Psychology is the science of human and non-human behavior. Because you are a student of psychology, you are training to be a scientist, so that you can study part of "nature," just as biologists, physicists, and other scientists do. And, as in all sciences, there is only one acceptable source for your "facts": everything that psychologists think they know in the science of psychology is obtained through research. Therefore, understanding research is an essential part of being a psychologist.

But I'm Not Interested in Research, I Just Want to Help People!

Even if you're not interested in conducting research yourself, you must understand psychological research so that you can learn from other people's research. Let's say that you become a therapist and you don't consider yourself a "scientist." You hear of a new therapy that says the way to cure people of some psychological problem is to scare the living daylights out of them. This sounds crazy, but what is important is the quality of the research that does or does not support the therapy. As a responsible professional, you would evaluate the research supporting it before you would use this therapy. You could not do so without understanding research methods and statistics.

So What Am I Learning?

Research methods and statistics are the most enjoyable—and the easiest to learn—when you are actively involved, so this book's approach is to place you in the role of a researcher who wants to answer questions scientifically. Then, you'll learn how to phrase questions scientifically, how to design and conduct scientific research, and how to analyze, interpret and communicate your results. Along the way, you'll also learn how scientists operate. This involves a combination of thinking logically, being creative, and, more than anything else, applying a critical eye to all phases of a study so that you recognize its imperfections and limitations. Ultimately, you'll not only understand the research of others, but you'll be able to conduct and interpret your own research.

Where Do Statistics Fit In?

Scientists conduct research because they have a question in mind, and statistics help to answer the question. Psychological research measures the behaviors of organisms, resulting in numbers or scores. For example, to study intelligence, researchers measure the IQ scores of different individuals; to study how rats learn a maze, researchers measure the time it takes each rat to escape the maze. Such scores are **data**. (The word *data*, by the way, is plural, referring to more than one score, so researchers say "the data *are* . . .") In virtually any study, the researcher ends up with a large batch of data, which must be made manageable and meaningful. Statistics help a researcher to make sense out of the data: Statistical procedures are used to *organize*, *summarize*, and *communicate* data and then to *conclude* what the data indicate.

What If I'm Not Very Good at Statistics?

In the grand scheme of things, performing statistics is one small, though important, step in the research process. Statistics are simply a tool used in the behavioral sciences, just like a wrench is a tool used in the repair of automobile engines. A mechanic need not be an expert wrencher, and a psychologist need not be an expert "statistician." Rather, in the same way that a mechanic must understand the correct use of wrenches in order to fix an engine, you must understand the correct use of statistics: to identify which research situation calls for which statistics and to understand what a statistical result indicates about a study.

But Statistics Aren't Even Written in English!

There's no denying that statistics involve many strange symbols and unfamiliar terms. But the symbols and terms are simply the shorthand "code" for communicating statistical results and concepts and for simplifying statistical formulas. This code even has a name: It's called "statistical notation." A major part of learning statistics is merely learning the code. Once you speak this language, much of the mystery surrounding statistics evaporates.

What If I'm Not Very Good at Math?

Although statistics do involve math, it's simple math. You only need to know how to add, subtract, multiply, divide, square, find square roots, and draw simple graphs. What makes statistical procedures *appear* difficult is that they involve a sequence of mathematical operations (first you square the numbers, then you add them together, then you subtract some other number, and so on). Working through the formulas is not difficult, but because they're in code it takes a little practice.

And recognize that studying statistics is much more interesting than merely cranking out a bunch of math problems. You'll learn new ways to think, reason, and apply logic, especially when it comes to drawing conclusions from scientific research. In fact, statistics can be fun! They're challenging, there's an elegance to their logic, and you can do nifty things with them. So, keep an open mind, be prepared to do some work, and you'll be amazed by what happens. Although statistics are a little unusual, they are not incomprehensible and they do not require you to be a math wizard.

So All I Have to Do Is Learn to Compute Statistical Answers?

No! Don't get so carried away with formulas and calculations that you lose sight of the big picture. In the big picture, a statistical answer tells you something about data and therefore something about your research. Ultimately, you want to make sense of data, and to do that you must compute the appropriate statistic and correctly interpret it. You need to concentrate on learning *when* and *why* to use each procedure and how to *interpret* the answer from that procedure. Be sure to put as much effort into this as you do into learning how to perform the calculations.

All Right, So How Do I Learn Methods and Statistics?

Study. Think. Practice. Think. Practice some more. Psychological research has its own language involving very specific terms—with very specific meanings—that you must learn. The way to learn a foreign language is to speak it every day, so practice the special vocabulary of research at every opportunity. Likewise, although knowing the rules of science is important, you must practice generalizing the rules to the different types of research you'll encounter later when on your own.

Recognize that because you do not yet speak the language, you won't learn anything by simply skimming a chapter. You must learn to translate the research terminology and statistical symbols into words that you understand and this takes time and effort. Also, you cannot "cram" this material. If you try, you won't learn much (and your brain will melt). Instead, work on the material a little bit every day. Then you'll be able to digest the material in bite-sized pieces. This is the most effective—and least painful—way to learn research methods and statistics.

What's With This Book?

There are several ways this book helps you master the material. At the beginning of each chapter is a list of the major points you should learn from the chapter. After you finish the chapter, check that you understand everything in the list. Throughout each chapter, you'll see statements labeled "**REMEMBER**." These describe concepts and principles that are especially important. At the end of each chapter is a list of summary statements: Be sure you understand each one. Then there are review questions that will help you identify weak spots in your knowledge (use these as a self-test to prepare for the real test). Answers to odd-numbered problems are provided in Appendix D.

Each chapter that focuses on statistics will open with a section titled "More Statistical Notation." Here you can become familiar with new statistical symbols used in the chapter before being immersed in the concepts of the chapter. Then every new formula will be presented with some example data so that you can see how to use the formula. Master the formulas and codes at each step, because they often reappear later as part of more complicated formulas. (The examples are unrealistically simple, containing only a few scores. But, if you understand a procedure using simple numbers, you'll be able to perform it with more complex data.) For quick reference, a list of the formulas discussed in a chapter is provided at the end of the chapter.

Okay! Now that you have some idea of what this course involves, let's look at the basics of scientific research.

THE SCIENTIFIC METHOD

If you are at all curious about behavior, you're already part of the way toward becoming a researcher. All sciences are based on curiosity about nature, and psychology is based on curiosity about behavior. Your curiosity is important, because it is the basis for deciding what you want to learn about a behavior and how to go about learning it.

But being curious is not enough. Creating, conducting, and interpreting research requires mental effort, because nature is very secretive and not easily understood. On the one hand, psychological research is fun because of the challenge in devising ways to unlock the mysteries of behavior. But, on the other hand, caution is needed at every step, because it is easy to draw *incorrect* conclusions about a behavior. This is a critical problem because psychology is a source of knowledge used by society in ways that have a serious impact on the lives and well-being of others. For example, at one time, people with a criminal history were thought to have "defective" personalities, which were "remedied" by the removal of portions of their brains! Unfortunately for those undergoing the surgery, this approach was just plain wrong! Thus, because the knowledge produced by research can drastically influence the lives of others, psychology's goal is to be perfectly accurate.

To try to meet this goal, psychology uses the "scientific method." This is a rather broad term, but essentially, the **scientific method** is a set of rules consisting of certain assumptions, attitudes, goals, and procedures for creating and answering questions about nature.

Why should psychologists—including yourself—use the scientific method? We could use our intuitions and personal experiences, or logical deductions and common sense, or we could defer to the pronouncements of authority figures. But! We do not trust intuitions or personal experience because everyone has different feelings about and experiences of the world. (Whose should we believe?) Likewise, we do not trust logic because nature does not always conform to our logic. Also, we cannot rely on common sense because it is often contradictory. (Which is true: "Absence makes the heart grow fonder" or "Out of sight, out of mind"?) And, we cannot rely on what the so-called "experts" say, because there's no reason to believe that they correctly understand how nature works either. The problem with all of these sources of knowledge is that they ultimately rely on opinions or beliefs that may be created by someone who is biased or wrong or downright crazy! After all, merely because someone says something about the nature of a behavior does not make it true.

Psychology relies on the scientific method because it is the best approach for eliminating bias and opinion, for reaching a consensus about how a behavior truly operates, and for correcting errors. It does this by requiring that whenever someone makes a statement about behavior, we ask them "How do you know that?" We don't mean "believe," "feel," or "think"—we mean *know*! In science—just as in a court of law—it is the *evidence* that supports any statement that is most important. The scientific method provides the most convincing evidence because instead of being based on our own biases or intuitions, scientific evidence is based on the events as they occur in nature.

The remainder of this chapter examines the scientific approach to studying behavior. To start with, the following sections look at how scientists approach the task of science.

The Assumptions of Science

What first distinguishes scientists from nonscientists is the philosophy about nature that scientists adopt. At first glance, any aspect of nature, especially human behavior, seems to be overwhelmingly complex, verging on the chaotic. Scientists have the audacity to try to understand such a complicated topic because they do not consider nature to be chaotic. Instead, scientists make certain assumptions about nature that allow them to approach it as a regulated and consistent system. These assumptions are that nature is lawful, deterministic, and understandable.

By saying that nature is **lawful**, we mean that every event can be understood as a predictable sequence of natural causes and effects. We assume that behavior is lawful, because if it isn't (and instead is random), then we could never understand it. Thus, in the same way that the "law of gravity" governs the behavior of planets or the "laws of aerodynamics" govern the behavior of airplanes, psychologists assume there are laws of nature that govern the behavior of living organisms. Although some natural laws do not apply to all species (for example, laws dealing with nest building among birds do not apply to humans), a specific law does apply to all members of a group. Thus, when psychologists study the mating behavior of penguins, or the development of language in people, they are studying laws of nature.

Viewing behavior as lawful leads to a second, related assumption: we assume that the behavior of organisms is "determined." **Determinism** means that behavior is solely influenced by natural causes and does not depend on choice or "free will." If instead we assumed that organisms freely decide their behavior, then behavior truly would be chaotic, because the only explanation for every behavior would be "because he or she wanted to." Therefore, we reject the idea that free will plays a role. After all, you cannot walk off a cliff and "will" yourself not to fall, because the law of gravity forces you to fall. Anyone else in the same situation will also fall because that is how gravity operates. Likewise, we assume that you cannot freely choose to exhibit a particular personality or respond in a particular way in a given situation. The laws of behavior force you to have certain attributes and to behave in a certain way in a certain situation. Anyone else in the same situation will be similarly influenced, because that is how the laws of behavior operate. (Note that determinism is different from "predestination." Predestination suggests that our actions follow some grand plan that is already laid out for us. Determinism means that, while there is no overall plan, there are natural causes for every behavior.)

The third assumption is that the laws of nature are **understandable**. Regardless of how complicated nature may appear or how confused we currently are about some aspect of nature, we assume that we will eventually understand it (or there is no point in studying it). Thus, any scientific statement must logically and rationally fit with the known facts, so that it can be understood. Part of an explanation can never be that we must accept an unexplainable mystery or an unresolvable contradiction. If two statements contradict each other at present, it must be logically possible to resolve the debate eventually so that only one statement applies.

REMEMBER To be studied scientifically, any behavior must be assumed to be lawful, determined, and understandable.

Notice that the above assumptions exclude certain topics from being studied scientifically. For example, miracles cannot be studied scientifically because, by definition, miracles do not obey the laws of nature. Likewise, because of determinism we cannot study free will. (We can, however, study people's *perceptions* of miracles or free will, because their perceptions are behaviors that fit the above assumptions about nature.) Further, because nature is assumed to be understandable, any topic that requires faith cannot be studied scientifically. Faith is the acceptance of the truth of a statement without questions or needing proof. But in science it is *always* appropriate to question and to ask for proof. This means that scientists are entitled to the same religions and beliefs as anyone else. However, we cannot allow these beliefs to play a part when producing and evaluating scientific evidence. After all, if science did allow statements of faith, whose faith would we use . . . yours or mine?

The Attitudes of Scientists

Scientists also adopt specific attitudes toward the process of learning about nature. As a scientist, you should be uncertain, open-minded, skeptical, cautious, and ethical.

The starting point is to recognize that the purpose of science is to learn about nature, admitting that no one already knows everything about how nature operates. There is always some degree of **uncertainty**. For psychologists, this means that no one knows precisely what a particular behavior entails, what the factors are that influence it, or what the one correct way to study it is. All other steps in scientific research stem from this simple admission.

If no one knows for certain how nature operates, then any explanation or description of it may be just as appropriate as any other. Therefore, as a scientist you should be **open-minded**, leaving your biases and preconceptions behind. An explanation may offend your sensibilities or contradict your beliefs, but that is no reason to dismiss it. You must look in all directions, at all possible explanations, when trying to understand a behavior.

At the same time, you should be very **skeptical**: never automatically accept the truth of any scientific description. Because no one already knows how nature works, any description of a behavior or interpretation of a study may be incorrect (no one is perfect, not even psychologists). After all, the history of science is littered with descriptions that at first appeared accurate but later turned out to misrepresent nature. (The earth is not flat!) Therefore, you must skeptically and critically evaluate the evidence produced by any study: using logic and your knowledge of psychology, always question whether the factors proposed as important might actually be irrelevant (or at least not the whole story) and whether the factors proposed as irrelevant might actually be important. To aid in this process, researchers share their research findings through professional publications, meetings, and so on. Then, eventually, psychology will identify and rectify any mistakes, to arrive at the best, most accurate information.

If we assume that critical analysis will eventually produce an error-free understanding of nature, then we must also recognize that we are currently in the process of discovering which parts of our information are incorrect. Therefore, our attitude must be very **cautious** when dealing with scientific findings. Any scientific statement implicitly contains the qualifying statement "given our present knowledge and abilities." Never treat the results of any single study as a "fact" in the usual sense. Instead, a research finding is merely a piece of evidence that provides some degree of confidence in a description about nature.

Finally, there is one other attitude that scientists adopt: scientists must behave **ethically** when conducting research. Later chapters examine the specific guidelines for ethical research, but the basic principle is that neither researchers nor their research should cause harm to others.

> *REMEMBER* Scientists are uncertain, open-minded, skeptical, and cautious. They are also ethical.

Evaluating any study's conclusion means evaluating the evidence the study provides. As with the rules of evidence in a court of law, science has rules governing what evidence is admissible and how it must be gathered.

The Criteria for Scientific Evidence

When people think of scientific research, they usually think of "experiments." Although psychologists often perform experiments, they also conduct other types of research. In fact, there is an infinite number of different ways to "design" a study. The **design** of a study is the specific manner in which the study is set up and conducted. First, the design must identify the specific people or animals to study. (Note that historically, published psychological research has referred to these individuals as **subjects**. As discussed in Appendix A, however, in publications after 1994 these individuals are called **participants**. Because you'll probably be reading research articles using either term, to get you in the habit, both terms appear interchangeably throughout this text.) In addition, a design includes the specific situation or sequence of situations under which participants are studied, the way their behavior is examined, and the components of the situation and behavior that are considered.

As an example of a behavior we might design a study for, let's discuss one simple, familiar behavior: the irritating tendency of people to "channel surf"—to grab the television remote control and change the channel whenever a commercial appears. (If this seems too mundane a behavior to be "psychological," stay tuned . . .) In designing a study of this behavior, remember we seek the most convincing evidence for answering the question "How do you know that?" In science, convincing evidence is empirical, objective, systematic, and controlled.

First, the evidence must be **empirical**—meaning learned by *observation*. Psychologists study everything that a subject does, feels, thinks, wants, or remembers, from the microlevel of neurological functioning to the macrolevel of complex, lifelong behaviors. Yet ultimately, all evidence is collected and all debates are resolved by attending to observable, public behaviors. Thus, to understand channel changing, we should *observe* channel changing. Because anyone else can potentially observe this behavior in the same way, everyone then shares the same basis for determining how it operates.

Second, several people can all observe the same event and still have different personal impressions of it. Therefore, science requires **objectivity**. This means that, ideally, a researcher's personal biases, attitudes, or subjective impressions do not influence the observations or conclusions. Although people can't be perfectly objective, scientists strive for this by obtaining *measurements* that are as empirical, objective, and precise as possible. It is through empirical, objective measurement of behavior that a researcher obtains the data in a study. For example, counting the number of times a subject changes channels during a specified time period results in objective, precise data. Or, in other studies, we may use equipment that times subjects' responses or measures their physiological reactions, we may interview participants or have them perform mental or physical tests, we may observe subjects surreptitiously, and so on. In any of these approaches, we try to be as objective as possible, so that, as much as possible, the data reflect precisely what participants actually do in a given situation and not our personal interpretations of what they do.

In addition, nature is very complex, so the research situation must be simplified so that we are not confused by all that is going on. Therefore, evidence is gathered systematically. Being **systematic** means that observations are obtained in a methodical, step-by-step fashion. For example, say we think that boring commercials cause channel changing. After we've objectively measured "boring," we would then objectively measure the channel changing that occurs with very boring commercials, with less boring commercials, and again with interesting commercials. If we also think that the number of people in the room influences channel changing, we would observe subjects' responses to the above commercials first when alone, then when another person is present, then when two other people are present, and so on. By being systematic, we determine the role of each factor and combination of factors as they apply to a behavior.

Finally, evidence must be obtained under controlled conditions. **Control** is another way to simplify the situation by eliminating any factors that might influence the observed behaviors and thus create confusion. For example, while observing whether more boring commercials produce more channel changing, we would try to control how boring the television program is, so that this factor would not influence channel changing. Likewise, we control the situation by having subjects be engaged in watching television only so that other distractions won't influence their channel changing. In short, with control researchers attempt to create a clearly defined situation in which to observe only the specific behavior and the relevant factors that they're interested in.

> *REMEMBER* Acceptable scientific evidence is obtained through empirical, objective, systematic, and controlled research.

As you will see, many different research designs meet the above criteria. Which approach to take depends first upon the type of question being asked—the specific goal of the study.

THE GOALS OF PSYCHOLOGICAL RESEARCH

Overall, the goal of psychology is to understand behavior. But what does "understand" mean? Science has come to define understanding an event as being able to describe, explain, predict, and control the event.

Obviously, psychologists want to know what behavior does and does not occur in nature, so the first goal is to **describe** each behavior and the conditions under which it occurs. To describe channel changing, we would specify how frequently channels are changed, whether they are changed during all commercials, at all times of the day, and so on. We could also describe channel changing from various perspectives, in terms of the hand movements necessary to operate the remote control, or the cognitive decision making involved, or the neurological activity occurring in the brain.

But mere description of a behavior is not enough to completely understand it: we also need to know *why* the behavior occurs. Therefore, a second goal is to **explain** behaviors in terms of their specific causes. Thus, we want to explain what aspect of a commercial, either present or absent, causes channel changing and why. We want to identify the factors—the channel changer's personality, the type of program, the presence of other people in the room—that cause more or less channel changing and why. And again, there are various perspectives we can take, such as neurological, cognitive, motivational, or environmental causes.

Note that in explaining a behavior, it is important to avoid pseudo-explanations. A **pseudo-explanation** is circular, giving as the reason for an event another name for that event. For example, a pseudo-explanation of channel changing is that it is caused by the motivation to see what is on other channels—really just another way of saying that people change channels because they want to change channels. The key to avoiding a pseudo-explanation is to provide an *independent* verification of the supposed cause. If, for example, we could discover a gene that motivates people to change channels, then we would be confident that we were talking about two different things—a cause (the gene) and an effect (changing channels)—and not merely renaming one thing.

Another aspect of understanding a behavior is to know when it will occur or what will bring it about, so an additional goal of psychology is to **predict** behaviors. Thus, we want to be able to accurately predict when channel changing will and will not occur, the amount or degree of the behavior to expect from a particular person, or when and how the behavior will change as a person's physiological, cognitive, social, or environmental conditions change. In addition, notice that the accuracy with which a behavior can be predicted is an indication of how well we have explained it. If we say that a behavior has a particular cause but the presence of the cause does not allow us to accurately predict the behavior, then the explanation is wrong, or at least incomplete.

Finally, if we truly understand a behavior, we should be able to create the situation in which it occurs. Therefore, the fourth goal is to **control** behavior. Thus, in studying channel changing, we want to know how to alter the situation to produce, increase, decrease, or eliminate the behavior. And note that being able to control events is another important test of an explanation. If a cause of a behavior is identified, then **manipulating** that cause—turning it on and off or providing more or less of it—should produce changes in the behavior. If it does not, the explanation is again either wrong or incomplete.

REMEMBER To completely understand a behavior, researchers strive to describe it, explain its causes, and predict and control its occurrence.

Meeting the Goals of Science

You can now see how the science of psychology proceeds: Psychologists attempt to learn about a behavior by obtaining empirical, objective, systematic, and controlled observations that allow them to describe, explain, predict, and control the behavior. Each finding is rigorously evaluated in a skeptical yet open-minded manner, so that an accurate understanding of the laws of behavior can be developed.

You may think that this approach is massive overkill when studying a behavior as mundane as channel changing. Is it really necessary to be that fussy? Well, yes, if we want to *fully* understand the behavior. Granted, it's easier to see why we should be so fussy if, for example, we were studying something like airplane pilots who turn off their planes' engines in midflight. There is an urgency to this behavior so that understanding it in such great detail would not be overkill (pardon the pun). But recognize that channel changing is not so mundane. This behavior involves major psychological processes such as decision making, information processing, communication, neural pathways control, motivation, and social processes. Thus, an in-depth study of channel changing is worthwhile because, for example, by studying the decision making involved in channel changing, we can learn about decision making processes in general. Further, another reason for a detailed study of channel changing is that research often leads to **serendipitous** findings: In the process of studying one aspect of nature, researchers may accidentally discover another aspect, unrelated to the original research. (In studying channel changing, we may stumble upon a cure for boredom.) Thus, because we never know where an investigation will lead, we take the study of every behavior very seriously and do the best, most complete job we can.

Applied and Basic Research

There is a greater urgency for studying something like the errors an airplane pilot makes, because this behavior represents a real-life problem. Such research is called applied research. **Applied research** is conducted for the purpose of solving an existing, real-life problem. For example, the companies that pay for television commercials might conduct applied research into channel changing during their commercials so they can eliminate the resulting problem of wasted advertising money.

On the other hand, **basic research** is conducted simply for the knowledge it produces. Thus, we might study channel changing simply because it is interesting and adds to our understanding of behavior in general. Although people often have a hard time understanding why it's important to conduct basic research, such research is justified first and foremost because science seeks to understand *all* aspects of nature. Also, basic and applied research often overlap. For example, basic research into channel changing may provide information that advertising companies can apply to solve their problem (and applied research designed to eliminate channel changing will add to our basic understanding of the behavior). A third justification of basic research is that past basic research may someday be valuable in a future applied setting. Say that we learn that channel changing and turning off airplane engines in midflight share some common factor (perhaps both are caused by boredom). At that point, our basic research into channel changing would be very useful for the applied problem of preventing airplane crashes. And, finally, a justification for basic research is that it often results in serendipitous applied findings. For example, some of the most common medicinal drugs have been discovered totally by accident during the course of basic research.

In sum, the terms basic and applied are general, describing a study only in terms of its obvious, stated purposes. In reality, we never know the ultimate purpose that research will serve (which is another reason for employing very rigorous methods).

> ***REMEMBER*** The primary purpose of basic research is to obtain knowledge; the primary purpose of applied research is to solve an existing problem.

The Role of a Single Study

Regardless of whether we conduct applied or basic research, *completely* describing, explaining, predicting, and controlling a behavior constitute the *ultimate* goal of research. But, because of the extreme complexity of behaviors, no single study can fully meet this goal. Instead, a systematic and controlled study simplifies nature by examining one factor and taking one perspective at a time. Thus, one study will describe certain aspects of a behavior, another will examine an explanation, other studies will investigate ways to predict the behavior, and still others will focus on controlling it. Any specific study is therefore a momentary "snapshot" of one small portion of a behavior. (In fact, because any study has such an extremely narrow focus, most psychologists would think it too grandiose to claim that their study directly examines a law of nature.) As a result of this piecemeal approach, the discipline of psychology—and publications describing it—may appear to be disjointed and unfocused, going off in many directions at once. Yet, we assume that eventually all of these individual pieces of information will be integrated so that we can truly understand the laws of nature.

> ***REMEMBER*** Any study represents a very limited and simplified view of the complexity found in nature and contributes minutely to the goals of describing, explaining, predicting, and controlling a behavior.

Every decision researchers make depends first and foremost on whether their primary goal is to describe, explain, predict, or control a behavior. Therefore, the first step in any study is to formulate the specific question you wish to answer. That question is called a hypothesis.

SCIENTIFIC HYPOTHESES

We might ask the question, "What causes channel changing?" However, this is actually a very ambiguous question, with no hint as to the type of "snapshot" needed to answer it. Do we mean "what" in terms of the cognitive, physiological, or environmental causes? And which aspect of the cognitive, physiological, or environmental causes are we talking about? Such an ambiguous question cannot be directly answered by a study. After all, at some point we must go out and actually collect some data, so sooner or later we need to know precisely which behavior to examine and how to examine it. Therefore, we must translate any general question into a specific hypothesis that directs the research.

Creating Scientific Hypotheses

A **hypothesis** is a formally stated expectation about how a behavior operates. It is, in essence, a tentative guess about a behavior that usually relates a behavior to some other behavior or influence. Rather than asking a question beginning with "why" or "what," we phrase a hypothesis as a declarative statement or description. Then we test the hypothesis: we conduct an empirical, controlled, systematic study that provides data that help us determine if the statement is correct or not.

There are two general types of hypotheses. In keeping with the goal of explaining and controlling the causes of behavior, one type is a **causal hypothesis**: this tentatively identifies a particular cause for or influence on a behavior. For example, we might hypothesize that "channel changing is caused by the boring content of commercials." (Implicitly, we recognize there may be many other influences on channel changing, but for now, this is the one we'd study.)

On the other hand, in keeping with the goal of describing and predicting behavior, the other type is a **descriptive hypothesis**: this tentatively describes a behavior in terms of its characteristics or the situation in which it occurs, and allows us to predict when it occurs. For example, we might hypothesize that "channel changing occurs more frequently when someone is watching television alone than when other people are present." Notice that even though the number of people present might partially cause channel changing, we have not stated this. A descriptive hypothesis does *not* attempt to identify the causes of a behavior. In fact, sometimes it states simply that certain behaviors occur and can be observed and measured, giving a general goal and direction to our observations. For example, we might hypothesize that "channel changers have certain personality characteristics" and then set out to discover and describe them.

REMEMBER A causal hypothesis postulates a particular causal influence on a behavior, and a descriptive hypothesis postulates particular characteristics of the behavior or provides a goal for observations.

It is extremely important to state explicitly whether a study is examining the causes of a behavior, because, as you'll see in the next chapter, this is a critical factor in determining the design of the study. Before designing the study, however, you must be sure the hypothesis reflects our assumptions about the lawfulness and understandability of nature. If it does not, then the hypothesis is not scientific, and the evidence that supports it is not scientifically admissible. Therefore, there are specific rules for creating scientific hypotheses.

The Criteria for Scientific Hypotheses

There are five attributes that a scientific hypothesis should have: it should be testable, falsifiable, precise, rational, and parsimonious.

A hypothesis must first be testable and falsifiable. **Testable** means that it is possible to devise a test of a hypothesis. **Falsifiable** means that the test can show that the hypothesis is incorrect. Our previous channel-changing hypotheses are testable and falsifiable because we can devise a study to test them, and we may find evidence indicating they are incorrect. It is possible, however, to create hypotheses that are not testable or falsifiable. Consider the hypothesis "When people die, they see a bright light." This is not a testable hypothesis because it's not possible to study people's experience after death (they're dead!). Because it's not testable, the hypothesis is also not falsifiable. Or, consider Sigmund Freud's hypothesis that the "id" leads people to express aggression directly as well as indirectly through superficially nonaggressive behaviors. Testing people's behavior can determine whether aggression occurs, but regardless of what is observed, it isn't possible to show that the hypothesis is false: If we observe aggression, it's because of the id. If we don't observe aggression, it's still because of the id, expressing its aggression through nonaggressive behavior. Given the circular logic here, this hypothesis cannot tell us anything about the id (even Freud wasn't perfect). Note that if a hypothesis is not testable or falsifiable, it's impossible to determine its accuracy. Instead, we must take the hypothesis on faith, a non-scientific approach.

A hypothesis must also add to knowledge about the laws of nature in a meaningful and understandable way, so a hypothesis must be precise and rational. A **precise** hypothesis contains terms that are clearly defined. The use of ambiguous terms opens the hypothesis to interpretation and opinion, making it less clearly testable and falsifiable. A **rational** hypothesis logically fits what is already known about the laws of behavior. For example, our hypothesis about boring commercials causing channel changing fits with what we already know about behavior and, if shown to be correct, will mesh easily with existing knowledge. In contrast, consider the claim that some people exhibit "ESP," the ability to send and receive mental messages. Any hypothesis about this supposed ability is not rational, because it contradicts existing knowledge about the brain and physical energy already developed in psychology, biology, and physics.

Finally, a hypothesis must be parsimonious. A **parsimonious** hypothesis is one that is as simple as possible. The assumption that nature is lawful implies that many diverse events can be accounted for by an economical combination of relatively few laws. If we propose new laws or mechanisms for every situation, we are merely renaming nature without explaining it. Therefore, the rule of parsimony says that we begin with relatively simple hypotheses that apply to broad categories of behaviors. Then, only if the simple explanation fails to account for a behavior is proposing new, more complex ones justified. Thus, any hypothesis about ESP would not be parsimonious, because it would require proposing all sorts of new brain components and new energy waves, just to make it a viable hypothesis. To be parsimonious, however, there would first need to be scientific evidence for ESP that could not be explained using existing brain mechanisms that are already established by previous research. Only then would it be acceptable to propose the existence of new brain components and new energy waves.

REMEMBER Scientific hypotheses must be falsifiable, testable, precise, rational, and parsimonious.

Sources of Hypotheses

How do you come up with a hypothesis? One obvious source for generating a hypothesis is a researcher's own opinions, observations, or experiences. It is perfectly acceptable to base a hypothesis on such sources, as long as you then conduct an empirical, objective study to provide evidence for the hypothesis. A second source is existing research: When reading the results of a study that tested one hypothesis, you'll usually see the basis for several additional hypotheses. For example, if we find that channel changing increases when someone is alone, we would then want to determine why, identify the factors that modify this influence, and so on. A third source of a hypothesis, as we'll see, is the retesting of a hypothesis previously tested by another researcher.

Theories are another source of hypotheses. A **theory** is a logically organized set of proposals that defines, explains, organizes, and interrelates knowledge about many behaviors. Theoreticians may develop a theory beginning with certain ideas and concepts for which there is little scientific evidence. The theory then provides a direction for the evidence researchers will seek. Or theoreticians may develop a theory after substantial evidence has been collected, providing a way to organize diverse findings. Either way, a theory is a framework of abstract concepts that helps to explain and describe a broad range of behaviors in a parsimonious way. For example, "Freudian theory" attempted to explain and relate a vast array of normal and abnormal behaviors using such abstract concepts as id, ego, and superego. Notice, however, that a hypothesis is derived from a theory, so a study never directly tests a theory; it can only test the hypothesis. Be careful when using the word *theory*.

The most common psychological terms are actually theoretical concepts. Such terms as learning, memory, intelligence, creativity, personality, or schizophrenia are ways of combining, organizing, and summarizing many similar behaviors under one, general concept. (For example, think of all the behaviors and attributes that come to mind with the term "personality.") Such theoretical terms have an important name. Concepts such as personality or creativity are not real things (you cannot place your personality on a table). Instead, they are "hypothetical" things that we mentally "construct. " A **hypothetical construct** is an abstract concept used in a particular theoretical manner to relate different behaviors according to their underlying features or causes. A construct is an idea that allows us to describe, organize, summarize, and communicate our interpretations of concrete behaviors. When we talk about understanding behavior, we are really talking about understanding hypothetical constructs and the processes they entail. Thus, we interpret the differences among individuals' mental abilities using the construct of "intelligence," or we summarize the activities of storing and retrieving information using the construct of "memory."

REMEMBER A hypothetical construct is an abstract term used to summarize and describe various behaviors that share certain attributes.

The way to ensure that any hypothesis is rational and parsimonious is to incorporate accepted hypothetical constructs. Thus, for example, we might explain how boring commercials cause channel changing by using such hypothetical constructs as "thinking" and "motivation," and then design a study using these concepts. If we find evidence that supports the hypothesis, our description and explanation should mesh with existing knowledge in an understandable way.

A final source of hypotheses is a model. A **model** is a description that, by analogy, explains the process underlying a set of common behaviors. Whereas a theory proposes and broadly interrelates hypothetical constructs, a model provides a more concrete approach for describing and learning about specific behaviors. Thus, in the same way that testing a model airplane in a wind tunnel provides a way to learn about a real airplane, a psychological model provides a way to discuss and learn about the components of a real behavior. Psychological models usually involve a flow chart or diagram showing different stages or processes. For example, models of human memory often use a flow chart containing separate boxes that represent the flow of information through short-term and long-term memory. Although no one believes your brain actually contains little boxes labeled short-term and long-term memory, this is a useful analogy for deriving specific hypotheses about how and when information remains in memory either temporarily or permanently. Such a model is different from a theory, in that, for example, memory is one small component of Freud's theory of personality involving the id, ego, and superego.

Testing Hypotheses Through Research

Once we have created a hypothesis, we then design a study to test it. We use the logic that, if the hypothesis is true, then a specific research situation should produce certain data: The scores measuring the behavior will be high or low, or will change in a predictable manner. In other words, the test of a hypothesis is based on a prediction. Whereas a hypothesis is a general statement about how a behavior operates, a **prediction** describes the specific behaviors that will be seen in the research situation and the kinds of data that the study is expected to produce. Thus, by hypothesizing that boring commercials cause channel changing, we predict that as more boring commercials occur, subjects' scores will reflect more frequent, more rapid, or more motivated channel changing.

With a prediction in hand, we then actually conduct the study. Just how you do that, of course, is the topic of this book. Suffice it to say that there are many designs to use, depending upon whether the primary goal of the study is to describe, explain, predict, or control a behavior. Regardless of the design, we attempt to obtain objective, systematic, and controlled measurements of subject's behaviors, so that we can clearly and confidently test the hypothesis.

Then, after collecting the data, statistics come into play. There are various types of statistical procedures to use depending upon the specific design of a study. Regardless, statistics essentially make sense out of data so that it is possible to see if the scores form the predicted pattern. Thus, if we are testing whether more frequent channel changing occurs with more boring commercials, we'll use statistical procedures to determine if this pattern occurs in the data.

Finally, based upon whether the data fit the prediction, we have evidence that either does or does not support the original hypothesis. Therefore, we will have learned

FIGURE 1.1 The Flow of Scientific Research

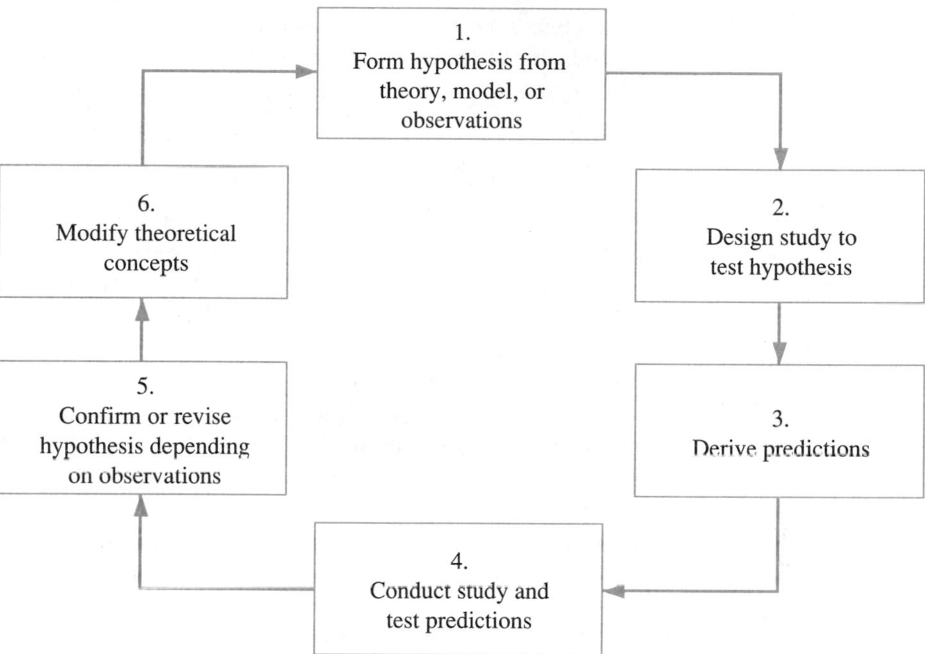

something about how the behavior operates. Then we integrate these results with existing knowledge, hypothetical constructs, and/or theories, and start the process all over again.

The "model" in Figure 1.1 summarizes the flow that most research follows. The first step is to create a rational and parsimonious hypothesis that relates to the hypothetical constructs in a theory or model so as to better describe, explain, predict, and control a behavior. The second step is to design a study that will test the hypothesis. Third, the predictions of the study must be clarified: If the hypothesis is correct, then in a particular situation, participants should exhibit the behavior in a certain way. The next step is to conduct the study and measure the behavior, and, using statistical procedures, determine if the data confirm the prediction. If they do, there is support for the hypothesis. If they do not, there is no support for the hypothesis. Then, from the conclusions about the hypothesis, we apply the outcome of the study back to the theoretical concepts or model from step 1, either adding to or correcting the description. Future research then derives from the modified theory or model, testing additional hypotheses which are used to further modify the description, and the cycle begins again. It is through this continual cycle that the science of psychology collates, organizes, and resolves the diverse "snapshots" of behaviors that individual studies provide, so that we can eventually understand the broader laws of nature.

THE FLAWS IN SCIENTIFIC RESEARCH

Recall that it's easy to make mistakes when learning about nature because nature is complex and mysterious, and it's not already clear to us how it works. Although the rules we've examined may seem sufficient to eliminate error, they do not guarantee this. There will always be many opportunities for researchers to make errors. Therefore, we must always be skeptical, considering that what we know—or think we know—depends on (1) the evidence we are attending to and (2) how we are interpreting it.

First, let's consider the evidence.

The Flaws in the Evidence

In an ideal study, we would be perfectly accurate in our measurement of the exact behavior we seek to describe, with only the relevant factors coming into play. However, no study is ideal. Rather, there are four factors that can weaken confidence in the data.

First, some behaviors cannot be studied in a completely empirical, objective, systematic, and controlled manner. For example, it is impossible to directly observe "thinking." Instead, we must observe some other behavior—such as errors in logic that subjects make—from which to draw inferences about thinking. However, the greater the inferential leap from the observed to the unseen behavior, the less confidence we have in our conclusions. Also, there is no "yardstick" for objectively measuring some behaviors, such as aggressiveness or love. Instead, we must employ more subjective measurement procedures that may include bias and error. Finally, researchers cannot always observe a behavior in a systematic and controlled fashion. For example, in studying the attitudes of women toward childbearing, we cannot separate the fact that participants have female personalities from the fact that they have female genes. Therefore, it's impossible to be sure whether it's a women's personality or her physiology (or both) that influences her attitudes. Because similar limitations are found in every study, we never have complete confidence that we *know* what the measurements reflect about the behavior or which factors were truly operating.

Second, the decisions made when designing a study may also reduce confidence in the findings. For example, if we study channel changing as it occurs in someone's living room, we cannot control such distractions as whether the phone rings in the middle of a commercial and thus prevents channel changing. But, if we study channel changing in a controlled "laboratory" setting, we create an artificial and thus biased picture of the behavior: People do not normally watch television in a laboratory! Likewise, the particular participants in a study, the way a behavior is measured, and the way factors are controlled may all result in biased evidence.

Third, often there are technical limitations that produce misleading information. For example, in the late 1800s, psychologists studied "phrenology," the idea that various personality traits are reflected by the size of bumps on the skull. Considering current techniques for studying personality and brain physiology, however, phrenology now seems silly. We must always consider the inherent limitations arising from our current technical abilities.

And finally, the results of one study can never tell the whole story. A single study is a "snapshot" that necessarily provides a biased perspective, considering certain factors and ignoring others (some of which we don't even know about). This problem is illustrated by an old fable about several blind men trying to describe an elephant (Shah, 1970). One, touching the animal's trunk, describes the elephant as like a snake. Another, touching the ear, says the elephant is like a fan. Another, touching the leg, describes the elephant as resembling a tree. And so on. In studying a behavior, researchers are like the blind men, trying to describe the entire elephant from one limited perspective at a time. For example, say our study supports the hypothesis that channel changing is caused by boring commercials. If we do not consider the television program during which the commercials appear, however, then a different study may indicate that channel changing is caused by boring programs. The problem is that either study alone does not give the complete picture if channel changing is actually caused by a combination of boring commercials and a boring program. Remember, being skeptical means not falling for the obvious explanation provided. Always consider whether your attention has been misdirected by the limited perspective of a study.

From the above, it should be clear that all research is not created equal! Individual studies can vary greatly in the extent to which they provide "good" data that accurately reflect the behavior and situation as they occur in nature. Therefore, the most important factor to consider when evaluating scientific evidence is the design of the study that produced it. As you will see in later chapters, the design and interpretation of a study are completely interrelated. So, whether you are evaluating a study performed by you or by someone else, always consider if there are flaws that suggest reasons for doubting the study's conclusions.

> *REMEMBER* A study's design determines the "snapshot" of a behavior it produces and thus ultimately the evidence for a particular hypothesis.

Even when the evidence from a study contains a minimum of flaws, you still cannot have as much confidence in the conclusions as you might think. The way the results are interpreted is also a concern, as is the intrinsic difficulty of "proving" that a hypothesis is true.

The Flaws in Testing Hypotheses

Say that we test the hypothesis that people change channels more frequently with more boring commercials. But we find that they don't, instead changing channels about the same amount regardless of how boring the commercial is. We have evidence that our hypothesis is false, so we'd say it has been **disconfirmed**. However, although we have disconfirmed the hypothesis as stated, it may contain some element of truth. Maybe more boring commercials do produce more channel changing, but only when a boring program is also present.

On the other hand, say that we find that people *do* change channels more frequently with more boring commercials. Here our finding is consistent with, or **confirms**, our hypothesis. But is confirmation the same as proving that the hypothesis is true? Absolutely not! We have not "proven" that boring commercials cause channel changing, because our observations are merely consistent with what we would expect if that hypothesis were true. But the data may *coincidentally* fit what is actually an incorrect hypothesis. There may be another, different hypothesis that actually describes this situation. Perhaps the truth is that people change channels when deprived of visual stimulation, and coincidentally (and unknown to us) our more boring commercials contain less visual stimulation. Because the data also fit this competing hypothesis, we cannot be sure whether the amount of boredom or the amount of visual stimulation actually caused channel changing in our study. Therefore, we certainly have no "proof" that our hypothesis is true. Instead, the best that we can do is to disconfirm and thus eliminate the visual stimulation hypothesis (and any other competing hypotheses.) Thus, you must fight the urge to seek only confirming evidence. The best evidence comes from disconfirming competing hypotheses while simultaneously confirming your hypothesis.

Even then, however, confirming a hypothesis is never the same as proving it. Confirming a hypothesis is actually a failure to "disprove" it. We may fail for one of two reasons: (1) The hypothesis is correct, or (2) The hypothesis is incorrect, but coincidentally the observations fit it and thus did not disprove it. We can never "prove" that our original hypothesis is correct, because no matter how many studies we conduct, someone can always argue that one more study might be the one to disprove it. For this reason, you should *never* use the terms "proof" or "proved."

Although the above is a serious objection after conducting only one study, it becomes less convincing as we observe more examples of channel changing. After enough tries we can argue that if a disproving instance were out there, we would have found it. If we continually fail to disconfirm a hypothesis, we come to believe that the hypothesis is not disprovable—that it is correct. Thus, the goal is to repeatedly confirm a hypothesis while disconfirming competing hypotheses. Eventually, with enough data, we will become confident in the truth of the hypothesis, even though, technically, we never "prove" it.

> *REMEMBER* If we confirm a hypothesis, we are merely more confident that it is true than we were before testing it.

Testing hypotheses in real research is especially difficult, because some data may confirm a hypothesis, while other data may disconfirm it. Therefore, it is essential to critically evaluate and weigh the various lines of evidence. For example, do you accept the hypothesis that dreams predict the future? If you do, it's probably because you've occasionally dreamt of events that later occurred. But consider the quality and quantity of this evidence. First, such a hypothesis is suspect because it is neither rational nor parsimonious. Second, your feelings that you accurately recall a dream and that it matches up with real events lack objectivity and thus are also suspect. Third, your conclusion is based only on confirmation: Your observation is consistent with the hypothesis that dreams predict the future. But weigh this weak evidence against the amount of disconfirmation that is available. How many times have your dreams *failed* to come true? On balance, the preponderance of evidence heavily disconfirms the hypothesis that dreams predict the future. It makes much more sense to say that those few confirming instances are nothing more than mere coincidence: Some event occurred after you had coincidentally dreamt about it.

> *REMEMBER* Our confidence in any scientific hypothesis is based on the quantity and quality of evidence that confirms and disconfirms it.

At the same time, a conclusion must accurately reflect *all* of the available evidence. For example, most scientists are not convinced that UFOs exist, because there has not been enough confirming, scientifically acceptable evidence. As in all scientific debates, additional evidence must be gathered so that, eventually, we will all be convinced one way or the other. However, notice that, in the meantime, I said that we are not convinced that UFOs exist—I did not say that UFOs don't exist. You must be open-minded and cautious, so when one hypothesis has not been confirmed sufficiently to accept with confidence, the conclusion is that the jury is still out, not that the opposite of the hypothesis is true.

Using Replication to Build Confidence in Psychological Findings

As you've seen, researchers ultimately develop confidence in a hypothesis by repeatedly confirming the hypothesis while disconfirming rival, competing hypotheses. This final component of the scientific method is called replication. **Replication** is the process of repeatedly conducting studies that test and confirm a hypothesis so that we develop confidence in its truth. The logic behind replication is that because nature is lawful, it is also consistent. Over many studies, therefore, the correct hypotheses will be consistently supported, while the erroneous, coincidental ones will not.

Researchers perform two types of replication. In a **literal replication**, the researcher tries to duplicate precisely the specific design and results of a previous study. This approach, also called *direct* or *exact* replication, is used because there are always chance factors at work—the particular subjects studied, the unique environment of the study, and so on—that may mislead us. But chance factors that appear in one study are unlikely to appear consistently in others. Therefore, literal replication demonstrates that the original results are not likely to be due to chance factors. If different researchers can repeatedly obtain the same evidence in the same situation, we are more confident that the hypothesis is accurate.

On the other hand, in a **conceptual replication**, the researcher provides additional confirmation of a hypothesis, but does so while measuring the behavior in a different way, examining different types of participants, or using a different design. Conceptual or *indirect* replication provides for greater confidence in the general applicability of the hypothesis while testing and disconfirming competing hypotheses. Thus, for example, conceptual replications of our channel-changing study might involve observing children watching Saturday-morning cartoons and adults watching late-night shows, or include various commercials for toys, automobiles, and so on. Then, by combining the findings from these different studies researchers can determine how the behavior generally operates and which factors influence it.

Thus, scientists use replication to build confidence in their "facts" in the same way that lawyers build a legal case. Literal replication is analogous to repeatedly questioning the same witnesses to make sure they keep their stories straight. Conceptual replication is akin to finding a number of witnesses who, from different vantage points, all report the same event. With enough consistent evidence from both sources, we eventually come to believe that we have discovered a law of nature.

The need for replication is often frustrating because society thinks science should quickly provide a solution to every problem. For example, a newspaper story may report the discovery of a new drug for treating cancer, but sadly, the researcher notes that it could be many years before the drug is available to the public. This reflects the recognition that consistent, convincing evidence of the effectiveness of the drug can be obtained only through time-consuming replication. After all, we accept the "law" of gravity because it always works, in every situation, from every perspective. The same logic must be used when making any other scientific claim as well.

PUTTING IT ALL TOGETHER

As should be painfully obvious by now, a study never provides unquestionable "proof." First, we can never be completely confident that the data reflect the precise behavior we wish to measure, in the precise situation we wish to observe. Further, even with good, convincing data that confirm a hypothesis, the hypothesis may still be incorrect. For that matter, even when a hypothesis is disconfirmed, it may still contain some elements of truth.

Given these problems, you may wonder why we even bother to conduct research. Well, the issues we've discussed are reasons for being skeptical about any single research finding, not for being negative about the research process. Instead of being paralyzed by such limitations, you should view dealing with them as a challenge. You simply have to recognize that testing a hypothesis involves translating a general statement about a behavior into a concrete measurable situation, and then translating the measurements back into conclusions about the general behavior. There is always room for error in the translation.

Therefore, do not automatically accept or dismiss any single study. On the one hand, researchers try to design the best study they can, providing the clearest evidence for answering the question at hand. Thus, even with flaws, a study usually tells us something about a behavior. On the other hand, what seems to describe a behavior correctly from one perspective may be incorrect when viewed from a different perspective. Look for that different perspective. And always keep in mind that a single study will not tell us everything about a behavior. A complete understanding can be gained only by replication, as we repeatedly obtain numerous and varied "snapshots" of the behavior.

CHAPTER SUMMARY

1. The *scientific method* includes certain assumptions, attitudes, goals, and procedures for creating and answering questions about nature.

2. The assumptions of psychology are that behaviors are *lawful*, *determined*, and *understandable*.

3. Scientists are *uncertain*, *open-minded*, *skeptical*, *cautious*, and *ethical*.

4. The *design* of a study is the specific manner in which the study is conducted. It should provide for *empirical*, *objective*, *systematic*, and *controlled* observations of a behavior. Control refers to the control of factors that might influence the behavior being observed.

5. Statistical procedures are used to *organize*, *summarize*, and *communicate* data and to *draw conclusions* about what the data indicate.

6. The goals of psychological research are to *describe*, *explain*, *predict*, and *control* behavior.

7. A *pseudo-explanation* is circular, explaining the causes of an event merely by renaming the event. To avoid pseudo-explanations, scientists obtain independent verification of a supposed cause.

8. The primary purpose of *basic research* is to obtain knowledge. The primary purpose of *applied research* is to solve an existing problem.
9. A *hypothesis* is a formally stated expectation about how a behavior operates. A *causal hypothesis* postulates a particular causal influence on a behavior. A *descriptive hypothesis* postulates particular characteristics or aspects of the behavior.
10. Scientific hypotheses must be *testable*, *falsifiable*, *precise*, *rational*, and *parsimonious*.
11. A *theory* is a logically organized set of proposals that defines, explains, organizes, and interrelates knowledge about many behaviors.
12. A *hypothetical construct* is an abstract concept used in a particular theoretical manner to relate different behaviors according to their common underlying features or causes.
13. A *model* is a hypothetical description that, by analogy, explains the process underlying a set of common behaviors.
14. A *prediction* is a statement about the data that are expected in a specific study if the hypothesis is correct.
15. The more a study deviates from the ideal criteria, the more likely it is to misrepresent nature, and the less confidence we have in its findings.
16. The fact that the results of a study *confirm* a hypothesis does not prove that the hypothesis is true because the results may only coincidentally fit it. *Disconfirming* a hypothesis provides the greatest confidence in a conclusion about the hypothesis.
17. *Replication* is the process of repeatedly conducting studies to build confidence in a hypothesis. *Literal replication* is the precise duplication of a previous study. *Conceptual replication* repeats the test of a hypothesis, but uses a different design.

PRACTICE PROBLEMS

(Answers for odd-numbered problems are provided in Appendix D.)

1. Why do researchers need to learn about research and statistics?
2. As a student learning statistics, what should be your goals?
3. As part of a study, a researcher measures the IQ scores of a group of college students. What four things will the researcher use statistics for?
4. What three assumptions do scientists make about nature?
5. (a) What are the attitudes that characterize scientists? (b) Why are they necessary?
6. What are the four criteria for scientific evidence, and what does each term mean?
7. What are the four goals of research?
8. What is the difference between a theory and a model?
9. What are the five criteria for a scientific hypothesis, and what does each term mean?
10. (a) What is meant by the term hypothetical construct? (b) How do hypothetical constructs simplify nature? (c) Why does incorporating hypothetical constructs help create scientifically acceptable hypotheses?

11. What is the difference between a causal hypothesis and a descriptive hypothesis?
12. What is the difference between a hypothesis and a prediction?
13. Why must we critically evaluate the design of any study?
14. Why does disconfirmation provide greater confidence than confirmation?
15. What is replication, and why does science rely on it?
16. What is the difference between literal replication and conceptual replication?
17. On a television talk show, a panelist says that listening to rock and roll music causes the listener to become a devil worshiper, a homicidal maniac, or a suicide victim. What questions would you ask this panelist before voting to ban rock and roll?
18. A theorist claims that men become homosexual when, as they are growing up, their mother either (a) tried to control them or (b) did not try to control them. Scientifically speaking, what is wrong with this hypothesis?
19. You've read some research in a developmental psychology text that contradicts what you've observed about your younger brother. Whose claim should you believe, your own or the researcher's? Why?
20. The government has announced a very large monetary grant awarded to a scientist to study the sex life of a nearly extinct butterfly. A commentator claims that this research is a waste of money. Why do you agree or disagree?
21. Researchers who accept the existence of extrasensory perception (ESP) argue that the reason others have not found convincing evidence for it is that they do not believe such mental powers exist. What rule of science is violated by this argument?
22. A researcher explains that the reason people can remember smells is because they have a memory for smells. What is wrong with this explanation?
23. There's a rumor that if you dream you're falling off a cliff and you don't wake up before you hit the ground, you will actually die. (a) What is wrong with trying to confirm this hypothesis? (b) What is the way to test this hypothesis? (c) Even if you collect the appropriate dream information, what problems remain?
24. A nurse claims that when she wore one of those silly-looking hats, patients followed instructions better than they do now that she no longer wears it. What hypothetical constructs might this situation reflect?

2

The Logic of Designing and Interpreting Research

To understand this chapter:

- Recall that we create either a descriptive or causal hypothesis about a behavior.
- Then we conduct a study that measures the behavior of participants.
- Depending on whether the data support our predictions, we have evidence that either confirms or disconfirms the hypothesis.

Then your goals in this chapter are to learn:

- How to translate hypothetical constructs into a variable and then an operational definition.
- What a relationship is and what is meant by the "strength" of a relationship.
- How a relationship in a sample of scores is used to draw inferences about the behavior of a population.
- When and why descriptive and inferential statistical procedures are used.
- What the difference between experimental and descriptive research designs is, and what true experiments, quasi-experiments, and correlational designs are.
- What the independent variable, conditions, and dependent variable are in an experiment.

You already know that in research, we translate a general, abstract hypothesis about a behavior into objective measurements and then translate the measurements back into conclusions about the behavior. In the first part of this chapter, we'll expand on this translation process, filling in some of the details so that you understand the overall flow of most research. We'll also begin to see how statistical procedures fit in (but no formulas yet). Then we'll discuss the two major types of research design and see how to draw conclusions from each.

BEGINNING THE DESIGN: ASKING THE QUESTION

It's pointless to design a study by simply grabbing some behavior out of the blue to examine: If you do, you end up with an answer in search of a question. Research proceeds well only if you first determine the question you want to answer and then design a study to answer it. For example, we've all heard the rumor that "the more you study, the more you learn." Let's say we set out to examine this proposal.

The key to developing a hypothesis lies in recognizing the hypothetical constructs that it reflects, and then fitting your hypothesis into previous research and theory in a rational and parsimonious manner. Therefore, an important first step when developing a hypothesis is to examine the psychological research literature. The term "research literature" refers to published reports of psychological research found in professional-level books and journals. This is where you will find numerous ideas for interesting studies, established procedures that you can employ, and theories, models, and previous research that ensure that your hypothesis fits with known psychological processes. (A discussion of how to "search" the literature is presented in Appendix A.)

As you read and learn from the literature, the design of the study will emerge. As you'll see, all design decisions are ultimately made simultaneously, because any one decision has an impact on all other decisions. However, a useful starting point is to identify the relevant population and sample.

Identifying the Population and Sample

By speculating that studying improves learning, we have essentially hypothesized a component of a law of nature. Any law of nature applies to a specific group of individuals (all mammals, all humans, all male white rats, whatever). The entire group to which the law applies is called the **population**. Part of designing a study is to specifically define the target population. Are we talking about young children, college students, senior citizens, or all of the above? Does the hypothesis apply to all cultures, socioeconomic classes, intelligence levels, and personality types? Say we define the population for our study as college-aged men and women who are psychology majors. Notice that now we also know the specific participants we will examine in the study (and we are most concerned with reading the literature dealing with such individuals).

The population contains all possible members of the group, so it is usually considered to be infinitely large (but it need not be). To examine the behaviors of an infinitely large population would take forever! Instead, therefore, we usually study a sample. A **sample** is a relatively small subset of a population that is intended to represent, or stand in for, the population. It is the sample or samples of subjects that are measured in a study and the scores from the sample(s) constitute the data.

Notice that although psychologists ultimately discuss the population of *individuals*, in statistics we will talk of the population of *scores*, as if we had already measured the behavior of everyone in the population in a particular situation. Thus, you can think of a population as the group of all possible scores we would obtain if we could measure the behavior of everyone of interest in a particular situation. Likewise, we discuss a sample of scores as if we had already measured the participants in a particular situation.

The definitions of a sample and a population depend on your perspective. Say that we measure the studying behavior of students in your class. If these are the only individuals we're interested in, then we've measured this small population of studying scores. On the other hand, if we are interested in the population of all college students taking this course, then we have collected a sample of scores to represent that very large population. Or, if we are interested in both the population of college males and the population of college females taking this course, then the males in the class are one sample and the females in the class are another sample, and each sample is intended to represent its respective population. And finally, scores from *one* student can be a sample that represents the population of all possible scores that the student might produce. Thus, a population is any complete group of scores found in a particular situation, and a sample is a subset of those scores that we actually measure in that situation.

The logic behind samples and populations is this: We treat a sample as rather interchangeable with any other sample we might obtain, so our participants should give scores similar to any and all other potential participants. All potential participants *are* the population. Thus, it should be true that as the sample goes, so goes the population. Therefore, we use the scores in a sample to estimate or to *infer* the scores we would expect to find if we could measure the entire population. And remember: Scores reflect behavior. By translating the scores in a sample back into the behaviors they reflect, we can infer the behavior of the population. Thus, when the nightly news predicts who will win the presidential election based on the results of a survey, researchers are using a sample to represent a population. The scores from the sample (usually containing about 1200 voters) are used to infer the voting *behavior* of the population of over 80 million voters.

Once you have identified the type of individual to be studied, completing the design involves deciding exactly what behavior to observe and how to measure it.

Identifying the Hypothetical Constructs

To investigate how studying and learning operate in nature, we must first decide what we mean by these terms. "Studying" and "learning" are actually hypothetical constructs used to describe groups of related behaviors. The term "learning" describes a person's mental activities when integrating new information, and the term "studying" describes a person's actions to acquire such information. Because these terms are intentionally general, however, they may be examined from many perspectives: Do we mean by "learning" what a rat does when learning to press a lever, or what a doctor does when learning to perform surgery? Does "studying" mean the physical rehearsal of studying to be a ballerina, or the mental activity of studying for a school exam? We cannot observe all aspects of these behaviors at once, so designing a study entails "whittling down" the complexity of each construct so that we can examine one aspect of it clearly.

Thus, we must define precisely what it is we will examine, selecting one specific aspect of "studying," of "learning," and of every other component of the situation where these behaviors occur. Then we will have a simplified, and (we hope) understandable "snapshot" of how nature works when it comes to studying and learning. The way we define and measure hypothetical constructs is by measuring variables.

Identifying the Component Variables of a Construct

In psychological research, a **variable** is any measurable aspect of a behavior or influence on a behavior that may change. A measurable aspect of a behavior may be a physical action, a mental reaction, or a physiological response. A measurable influence on behavior may be a characteristic of the participants, of the situation, or of a stimulus to which subjects respond. A few of the variables found in psychological research include your age, gender, and personality type; how anxious, angry, or aggressive you are; and how hard you will work at a task or how accurately you recall a situation.

By identifying specific variables that reflect a hypothetical construct, we define the construct in concrete and measurable terms. Then by measuring the selected variables, we obtain scores that constitute the data of a study. Thus, we "whittle away" at the complexity of a behavior while simultaneously obtaining empirical evidence that anyone else can also collect and interpret.

> *REMEMBER* We examine an aspect of a hypothetical construct by selecting a specific variable that we will measure.

Variables fall into one of two general categories. If a score indicates the amount of a variable that is present, the variable is a **quantitative** variable. A person's height, for example, is a quantitative variable because a score indicates the quantity of height that is present. Some variables, however, cannot be measured in amounts. Instead, a score classifies an individual on the basis of some characteristic. Such variables are called **qualitative**, or **classification**, variables. A person's gender, for example, is a qualitative variable, because the "score" of male or female indicates a quality, or category.

All variables are not created equal, and which you should use in a specific study depends on a number of important considerations (discussed in later chapters). Essentially, though, the variables should be a good example of the hypothetical constructs as psychologists conceptualize them, they should allow for objective and precise measurement as much as possible, and they must be compatible with other aspects of the study's design.

Thus, we could measure "studying" using such variables as how much effort is put into studying or the number of times a chapter is read, but say we select the variable of the number of hours spent studying for a college exam. We could measure "learning" by measuring how well new concepts can be explained or how quickly information can be recalled. But say we select the variable of performance on the exam.

Notice we have now translated the general hypothesis that learning involves studying into the more specific hypothesis that number of hours studying for a test is related to performance on the test. However, this is still too general.

Creating Operational Definitions

Even after selecting a variable, we are still confronted with a variety of ways in which to measure it. Ultimately, we must specifically define each variable and the way to do that is through an operational definition. An **operational definition** defines a variable by the specific operations used to measure it. Operational definitions are very important in science, because they eliminate ambiguity when discussing any aspect of a study. For example, although you and I may disagree about exactly what the concept of intelligence means, in a study I might operationally define it as a score on the XYZ intelligence test. Now at least there is no debate about what I had my participants do, how I've measured their intelligence, and what you'd need to do to replicate my study. This is far clearer than simply stating, "I determined each person's intelligence."

Thus, you must operationally define all variables in a study. In our example, we might define "hours of study" as the number of hours that students report they studied for a statistics test from Chapter 6 of this book the evening before the test. Or we might define it as the length of time we observe them studying during the afternoon before the test. Likewise, we might define "performance on a test" as the number of multiple choice questions answered correctly on the Chapter 6 test, or the number of errors made when calculating statistical answers on the test.

Notice that the variable we select is essentially an operational definition of our hypothetical construct: We have defined the construct of "studying" in terms of the variable we will use to measure it—namely, "hours of study." Then we operationally define the variable of hours of studying in terms of the specific way we will measure it. Designing the remainder of a study essentially involves operationally defining all other variables that characterize the situation in which we are examining the behavior. Thus, we must define what we mean by "studying" (is it reading the textbook, highlighting it, or outlining it?); where the studying will take place (the dorm, the library, or our laboratory); what is "performance on the test" (is it number correct or number of errors?); and so on, for every aspect of the behavior, participants, and situation that we are examining.

REMEMBER Each variable in a study must be operationally defined in terms of the procedure that is used to measure it.

By operationally defining the variables, we translate our original, general hypothesis into a specific *prediction* about the scores that will be observed: If the hypothesis is correct, then the longer that students study for the statistics test, the better their performance on the test will be. To test this prediction, we measure some students and *see* if different amounts of the learning variable do occur with different amounts of the studying variable.

TESTING A HYPOTHESIS BY DISCOVERING A RELATIONSHIP

The most basic assumption in scientific research is this: If Y is influenced by or otherwise related to X by a law of nature, then *different* amounts or categories of Y will occur when *different* amounts or categories of X occur. Thus, if nature relates those mental activities we call studying to those mental activities we call learning, then different amounts of our learning variable should occur with different amounts of our studying variable. In other words, we predict a "relationship" between our variables. A **relationship** occurs when a change in one variable is accompanied by a consistent change in another variable. Because we measure scores, a relationship is a *pattern* in which certain scores on one variable are paired with certain scores on another variable.

TABLE 2.1 Scores Showing a Relationship Between the Variables of Study Time and Test Grades

Student	*Study time in hours*	*Test grades*
Jane	1	F
Bob	1	F
Sue	2	D
Tony	3	C
Sidney	3	C
Ann	4	B
Rose	4	B
Lou	5	A

As the scores on one variable change (increase or decrease), the scores on the other variable change in a consistent manner.

What might our relationship look like? Say that we asked some students how long they studied for the test and their subsequent grades on the test. We might find the scores shown in Table 2.1.

These variables form a relationship, because as the study time scores change (increase), the test grades also change in a consistent fashion (also increase).[1] Further, when study time scores do *not* change (for example, Jane and Bob both studied for 1 hour), scores on the grade variable do not change either (they both received Fs). In research, we often use the term *association* when talking about relationships. In the same way that your shadow's movements are associated with your movements, low study times are associated with low test grades and high study times are associated with high test grades.

The simplest relationships fit either the pattern "the more you *X*, the *more* you *Y*," or the pattern "the more you *X*, the *less* you *Y*." Thus, the saying "the bigger they are, the harder they fall" describes a relationship, as does that old saying "the more you practice statistics, the less difficult they are." Relationships may also form more complicated patterns where, for example, more *X* at first leads to more *Y*, but beyond a certain point even more *X* leads to *less Y*. For example, at first, the more you exercise, the better you feel. Beyond a certain point, however, more exercise leads to feeling less well, as pain, exhaustion, or death sets in.

Although the above examples reflect relationships involving quantitative variables, relationships may also involve qualitative variables. In such a relationship, as the category or quality changes on one variable, scores on the other variable change in a consistent fashion. For example, typically men are taller than women. If you think of male and female as "scores" on the variable of gender, then this is a relationship, because as gender scores change (going from male to female), height scores change in a consistent fashion (decrease). A study may examine the relationship between any combination of qualitative and quantitative variables.

[1] The data presented in this book are a work of fiction. Any resemblance to real data is purely a coincidence.

Strength of a Relationship

In Table 2.1, there was a perfectly consistent association between study time and test grades: All those who studied the same amount received the same grade. In the real world, however, all people who study the same amount will not receive the same grade. (Life is not fair.) Consistency is not all or nothing, so a relationship can be present even if the association between scores is not perfectly consistent. There can be some *degree* of consistency so that as the scores on one variable change, the scores on the other variable *tend* to change in a consistent fashion. For example, Table 2.2 shows a relationship between the variables of number of hours spent studying and number of errors made on a test. Here higher scores on the study time variable *tend* to be associated with lower scores on the error variable. Every increase in study time is not matched perfectly with a decrease in errors, however, and sometimes the same studying score produces different error scores.

In research, the consistency found in a particular relationship is called its strength: The **strength of a relationship** is the extent to which one value of the *Y* variable is consistently associated with one and only one value of the *X* variable. It is the *degree of association* between the variables.

There are two reasons that relationships are not perfectly consistent. First, there may be external influences operating on the participants. For example, among those students who studied for two hours, perhaps for some their dorm was particularly noisy so studying was less effective. Because of such influences, different test scores occur for these students, even though they all studied the same amount. (A major purpose of studying research methods is for you to learn how to identify and then eliminate—*control*—such external influences.)

The other reason for weaker relationships comes from individual differences. The term **individual differences** refers to the fact that no two individuals are identical and that differences in genetic make-up, experience, intelligence, personality, and many other variables all influence behavior in a given situation. Because of individual differences, a particular law of nature operates in *more or less* the same way for all members

TABLE 2.2 Scores Showing a Relationship Between Study Time and Number of Errors on Test

Student	*Study time in hours*	*Number of errors on test*
Amy	1	12
Joe	1	11
Cleo	2	11
Jack	2	10
Terry	3	9
Chris	4	9
Sam	4	8
Gary	5	7

of a population. Thus, for example, our participants will exhibit individual differences in terms of their intelligence, aptitude, and motivation, so they will score differently on the test, even when they study the same amount. Because individual differences produce differences in test scores for a particular study time, scores will be only somewhat consistently associated with study times. (Later, you'll also learn ways of controlling individual differences.)

Mathematically, the scores from two variables can form a relationship of any strength, from perfectly consistent association to no association. When there is no consistent association between two variables, there is no relationship. For example, there is (I think) no relationship between the number of chocolate bars people consume each day and the number of times they blink each minute. If we measured individuals on these two variables, we might have the data shown in Table 2.3. Here there is no consistent change in the scores on one variable as the scores on the other variable change. Instead, the same blinking scores tend to show up for each chocolate bar score.

To summarize then, when there is a relationship, to some degree certain scores on one variable tend to be associated with certain scores on the other variable, and when the scores on one variable change, there tends to be a consistently different group of scores on the other variable. In fact, perfectly consistent relationships do not occur in actual psychological research: We cannot perfectly control external influences, and individual differences are always present. Therefore, it is never enough merely to say that you have observed a particular relationship. You must also describe the strength of the relationship.

> ***REMEMBER*** In research, we are concerned not only with the existence of a relationship but also with its strength.

Using Relationships to Discover Laws

Now you can see how relationships are used in scientific research. The goal of science is to discover the laws of nature, and a relationship between variables is a telltale sign that a law of nature is at work. Above, we hypothesized that the behaviors we call studying are linked with the behaviors we call learning. Therefore, we translated this hypothesis into measurable variables to see if there is this linkage—a relationship—

TABLE 2.3 Scores Showing No Relationship Between Number of Chocolate Bars Consumed per Day and Number of Eye-Blinks per Minute

Student	*Number of chocolate bars consumed per day*	*Number of eye-blinks per minute*
Mark	1	20
Ted	1	22
Ray	2	20
Denise	2	23
Maria	3	23
Irene	3	20

between the variables. We observed the relationship in our data, so we have observed one instance, at least, where it looks like studying is linked to learning, and so we've learned something about how nature operates. However, because we did not observe that eye-blink scores are related to chocolate-bar scores, we have no evidence that eating chocolate bars is linked to blinking in nature. That, more or less, is the basis for interpreting scientific data.

Graphing Relationships

Researchers often use graphs to show a relationship. Therefore, it is important that you be able to recognize a relationship and its strength when looking at a graph. In case it's been a long time since you've drawn a graph, the horizontal line across the bottom is called the X axis, and the vertical line at the left-hand side is called the Y axis. How do we decide which variable to call X or Y? Any study implicitly asks this question: For a *given* score on one variable, what scores occur on the other variable? The "given" variable is the X variable (plotted on the X axis), and the other variable is the Y variable (plotted on the Y axis). Thus, previously we asked, "For a given amount of study time, what test grade occurs?" so study time is the X variable and test grade is the Y variable.

Notice that once you have identified the X and Y variables, there is a special way of communicating the relationship between them. The general format is: "Scores on the Y variable change as **a function of** changes in the X variable." Thus, so far we have discussed relationships involving "higher test grades as a function of greater study times" and "number of eye blinks as a function of amount of chocolate consumed." Likewise, if you hear of a study titled "Differences in Career Choices as a Function of Personality Type," you know that the researcher looked at how Y scores that measure career choices changed as X scores that measure personality types changed.

> *REMEMBER* The "given" variable in a study is designated the X variable, and we describe a relationship as "changes in Y as a function of changes in X."

Figure 2.1 shows the graphs from four sets of data. Graph A plots the original test-grade and study-time data from Table 2.1. Notice that where the two axes intersect is labeled as a score of 0 on both X and Y. Then, on the X axis, scores become larger positive scores as we move to the *right*, away from zero. On the Y axis, scores become larger positive scores as we move *upward*, away from zero. To fill in the body of a graph, we plot a "dot" to represent each pair of X and Y scores. Each dot on the graph is called a **data point**. For example, for a person who studied 1 hour, by travelling vertically from 1 to the data point and then horizontally back to the Y axis, we see the corresponding test grade was F. (Notice that two people originally studied for 1 hour and received an F, so their data points are on top of one another.)

To see the overall relationship, just as you read a sentence from left to right, read the graph from left to right along the X axis, simultaneously observing the pattern of change in the Y scores. In essence you should ask, "As the scores on the X axis increase, what happens to the scores on the Y axis?" In Graph A, the pattern is such that as the X scores increase, the Y scores also increase. Further, the graph shows that there is perfectly consistent association because everyone with a particular X score obtained the same Y score.

FIGURE 2.1 Plots of Data Points from Four Sets of Data

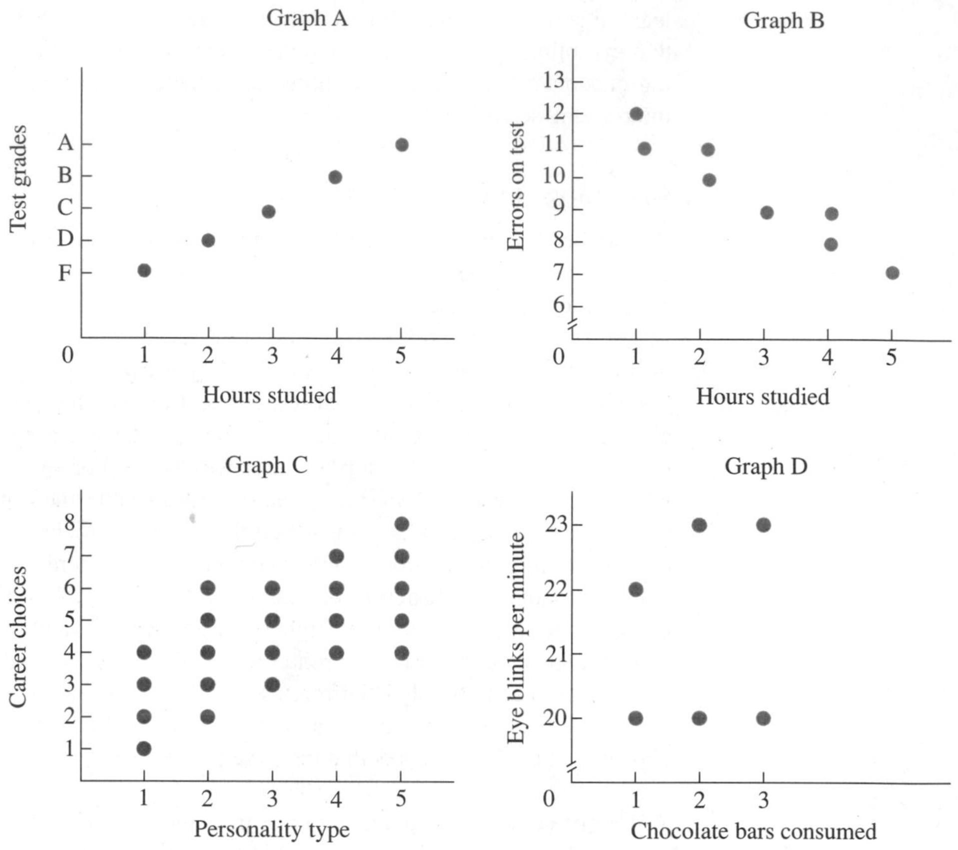

Graph B shows test errors as a function of number of hours studied, from Table 2.2. Here, increasing *X* scores are associated with decreasing values of *Y*. Further, because there are different values of *Y* at an *X* score, this relationship is weaker than in Graph A and is not perfectly consistent. (Notice the two diagonal lines (//) on the *Y* axis: Whenever there is a gap between 0 and the lowest score being plotted, the axis is compressed using this symbol. Here we "cut out" the part of the *Y* axis between 0 and 6 and slid the *Y* values down closer to 0.)

Say that Graph C shows the relationship between *Y* scores reflecting different career choices and *X* scores reflecting different personality types. Again, there is a relationship here, but a relatively wide range of different career-choice scores are paired with each personality-type score. Therefore, this is an even weaker, less-consistent relationship than in Graph B.

Graph D shows the eye-blink and chocolate-bar data from Table 2.3. There is no consistent pattern of change in *Y* scores, with more or less the same values of *Y* associated with each value of *X*. As shown here, whenever a graph shows an essentially flat pattern—so that *Y* scores tend to neither increase nor decrease as *X* scores increase—the graph reflects zero association and no relationship.

Later you'll learn when to connect the data points with lines and when to create other types of graphs. Regardless of the final form, always label the *X* and *Y* axes to indicate what the scores measure (not just *X* and *Y*), and create the graph in a way that honestly presents the data, without exaggerating or minimizing the pattern formed by the data points.

APPLYING STATISTICAL PROCEDURES

The data from real research seldom produce as clear a picture as in the previous examples. Instead, researchers are usually confronted with a mind-boggling array of different numbers that may have a relationship hidden in it. Also, we need more objective, quantitative ways of describing the strength of the relationship present. Therefore, we employ statistical procedures to help bring order to this chaos.

Descriptive Statistics

To understand the sample data and the relationship they form, we employ descriptive statistics. **Descriptive statistics** are procedures for organizing and summarizing scores so that we can describe and communicate the important characteristics of the sample data. (When you see *descriptive*, think *describe*.) Descriptive statistics have three general purposes.

First, these procedures tell us whether a relationship is present. Thus, if we are studying test errors as a function of hours studied, then we first want to determine whether there is a relationship between error scores and the amount of time studied.

Second, descriptive statistics are used to summarize the scores and describe the particular relationship we have found. Thus, we want to know how many errors are associated with a particular amount of study time, how much errors decrease with increased study time, how consistently errors decrease, and so on.

Third, descriptive statistics help us to meet the scientific goal of predicting behavior, because we can use a participant's score on one variable to predict his or her score on the other variable. Thus, once we establish the relationship in our study, we will know the typical error score that a student obtains for a given amount of study. On the basis of this relationship, we can predict the error scores of any other students, if we know how long they studied.

> *REMEMBER* Descriptive statistics are used to determine whether a relationship is present, to describe the relationship and the scores in it, and to predict the scores on one variable using the scores on another variable.

Drawing conclusions about sample data, however, is only one component of research. Combined with it is the process of drawing inferences about the population.

Representativeness and Drawing Inferences About the Population

After we describe and understand the relationship in sample data, we want to do the same thing for the population represented by the sample (because then we are describing how the variables and underlying behaviors relate for *everyone* in nature). Of course, usually we cannot measure the scores in the population, so we must *infer* the description of the population, based on the sample data. In our study-time research, inferring the relationship that would be found in the population means inferring that every time the studying variable changes, we would obtain a different *population* of error scores. Thus, say we find that a sample who studied 1 hour made between 11 and 15 errors on the test. We expect that if all students studied for 1 hour, they would produce a population consisting of error scores generally between 11 and 15. However, say that participants who studied for 2 hours produced between 7 and 10 errors. Therefore we expect that if we went back and had all students study for 2 hours, they would produce a different population consisting of scores between 7 and 10 errors. And so on, inferring that as everyone's score on one variable changes, his or her score on the other variable tends to change in a particular fashion.

But remember, we do not actually observe *everyone*! Sample data provide only one example of the larger population. The problem is that the sample may not be a "good" example. To produce accurate conclusions about a population, a sample must be "representative" of that population. In a **representative sample**, the characteristics of the participants, and thus their behaviors, accurately reflect the characteristics and behaviors of individuals in the population. Thus, a representative sample of college students will contain the same proportion of good or poor, motivated or unmotivated, and male or female students as in the population. Then, our sample's scores will be a good example of the scores we would find if we measured the entire population. Thus, if the average person in the population scores 50, then the average score in a representative sample will be around 50. If 30% of the scores in the population are 45, then around 30% of the scores in a representative sample will be 45, and so on.

The basis for creating a representative sample is random sampling. **Random sampling** is a method of selecting a sample in which: (1) All individuals in the population have the same chance of being selected for a sample, and (2) All possible samples have the same chance of being selected. Random sampling relies on the luck of the draw. Although we'll discuss the details of random sampling later, for now think of it as analogous to placing the names of everyone in the population in a large hat, mixing thoroughly, and then with your eyes closed, drawing participants for a sample. Thus, random sampling is unbiased in selecting participants and selects from all segments of the population.

Random sampling should produce a representative sample, because it allows the characteristics of the population to occur in a sample in the same ways that they occur in the population. For example, say that the population of students at your college are all standing in a field, and 60% of them are female. Random sampling is analogous to blindly wandering through that field. Out of a sample of 100 people you encounter, about 60 of them should be female, because that's how often females are out there. Or, if 30% of the individuals in the population have a score of 45, then 30% of the sample should also have a score of 45. In the same way, random sampling should produce a sample having all of the characteristics of the population so that the sample is a good example of the population.

At least we *hope* it works that way! I keep saying that a random sample "should" be representative, but this is a *very* big should! Nothing forces a sample to be representative of a particular population. The trouble is that whether a sample is representative is determined by which participants we select, and that is determined by random chance.

Therefore, whether a sample is representative depends on random chance. We can, just by the luck of the draw, obtain a very unusual sample whose characteristics do not match those of the population. Thus, it is possible to wander through that field of students described above and not encounter *any* females, or to encounter *only* females. Likewise, even though 30% of the population may have a score of 45, by luck we may not select anyone having this score, or we may select only those having this score.

If samples are unrepresentative, then the entire logic of research falls apart. When we select students who study for 1 hour, if by chance we unknowingly select only poor students, then their test scores are unrepresentative of the typical student who studies for 1 hour. Or, if we select only exceptionally good students who study for 2 hours, their scores do not represent the typical student who studies for 2 hours. And so on. If the sample of scores at each study time does not accurately represent the population of scores we'd find at that study time, then the *relationship* in the sample data does not accurately represent the relationship we'd find in the population. Then, if we could study the population of all students, we might find a relationship very different from this one, or we might find *no* relationship!

REMEMBER Never automatically assume that a sample accurately represents the population.

Thus, random sampling is a double-edged sword. Usually, random sampling works pretty well in producing a representative sample. But it can backfire, producing a very unrepresentative sample.

If as the sample goes is *not* how the population goes, then we do not have an accurate snapshot of nature, and any evidence we have is misleading and our conclusions are wrong. Therefore, whether a sample relationship accurately represents the relationship that would be found for everyone in the population is one of the most important concerns we have when interpreting research. But, the problem is that we can never *know* whether a sample is representative. We'd have to measure the entire population to see if the sample and population match, but if we could measure the population, we wouldn't need the sample to begin with!

To rescue us from this dilemma, we use additional statistical procedures.

Inferential Statistics

As we've seen, we are always talking about two things: the sample of scores we measure and the population of scores being represented. Therefore, we separate statistical procedures into two categories: descriptive statistical procedures, which deal with samples, and inferential statistical procedures, which deal with populations. **Inferential statistics** are procedures for deciding whether the sample data represent a particular relationship in the population. As the name implies, the focus of inferential procedures is to make accurate *inferences* about the scores and relationship represented by a sample.

So, after we use descriptive statistics to understand the data and relationship observed in a sample, we always stop and perform other procedures that allow us to decide whether the same relationship would be found in the population. If the sample data are deemed representative, we then ask the same things about the population we asked about the sample: What are the characteristics of the scores and of the relationship that would be found in the population, and can we use the scores in the population on one variable to predict the scores on the other variable? But, because we cannot measure the entire population, we use the descriptive statistics, computed from our sample, as the basis for *estimating* the scores that would be found in the population.

Eventually we will discuss inferential statistics in great detail. Until that time, simply think of them as ways of deciding whether sample data are "believable": Should we believe that we would find similar data, forming a similar relationship, in the population?

> *REMEMBER* Inferential statistics help us to draw inferences about the population.

SUMMARY OF THE FLOW OF A STUDY

In sum, we originally hypothesized that, in nature, studying is related to learning. If this is true, then we should find a relationship first in our sample data where, as scores that reflect more study time occur, scores that reflect higher test performance also tend to occur. Therefore, we use descriptive statistical procedures to examine the relationship between the scores in the sample. If we observe the predicted relationship here, we begin to have evidence that supports our hypothesis.

Our interpretation does not stop there, however, because we are interested in more than just the relationship observed in the sample. The goal is to draw the inference that a similar relationship would be found for *all* students in the population: That more or less for all students, higher studying scores are associated with better test performance. But, we cannot automatically assume that sample data accurately represent the relationship that occurs in the population. So, we perform inferential statistical procedures that allow us to decide whether the sample relationship is likely to be found in the population. Only then do we proceed to infer that for all students, greater amounts of study time are associated with higher test performance.

By claiming that our specific observations provide evidence for the general case, we are generalizing. To **generalize** means to apply the conclusions of a study to other individuals or situations. In interpreting research, we generalize in two ways. First, as above, we generalize the relationship between the variables in the sample to a relationship between the variables in the population. Second, we then generalize the relationship between the variables in the population to the relationship between the broader behaviors and hypothetical constructs that we originally set out to study. That is, we translate the scores back into the general behaviors and events they reflect, and argue that a similar relationship would be found with other variables and operational definitions: Ultimately, we want to say that as scores on *any* variable reflecting amount of studying increase, scores on *any* variable reflecting amount learned will also increase. If we can make this claim, then we have finally come full circle, confirming the original hypothesis that nature lawfully operates in such a way that when more of the process we call studying occurs, more of the behavior we call learning also occurs.

The above description shows the steps involved in a typical study as we translate from the general to the specific and then back to the general again. You can visualize these translations as the double funnels shown in Figure 2.2.

FIGURE 2.2 The Steps in a Typical Research Study

The flow of a study is from a general hypothesis to the specifics of the study, and then back to the general hypothesis.

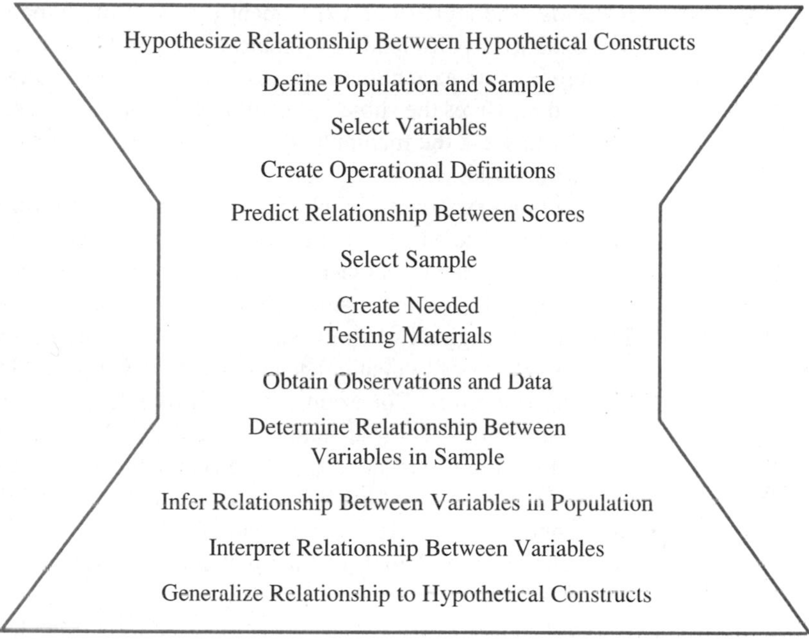

Begin with hypothetical constructs regarding a behavior. Identify the applicable population and specify the hypothesis as a relationship between variables. Next create operational definitions for the variables and predict the relationship between the scores you seek. Then obtain a random sample of participants from the population and design a method for obtaining their scores. Using descriptive statistics, determine whether there is the relationship between the scores in the sample. Then, using inferential statistics, attempt to infer that the variables are related in the population. Finally, on the basis of this inference, generalize to the broader relationship in nature involving the hypothetical constructs you began with.

Note that this flow from the general to the specific and back to the general is mirrored in the organization that psychologists use when publishing reports of research in the psychological literature. This organization was created by the American Psychological Association or "APA," the national association of psychologists in the United States, and is known as "APA-style" or "APA-format." This format is presented in Appendix A in detail. A published report will basically follow the above discussion, although the sequence may not be so obvious. The trick to reading the literature is to look for what the author is saying about each step.

When reading the literature, you will come across a variety of research designs, because there are many ways to go about demonstrating a relationship. However, research can be broken into two major types: experiments and descriptive studies.

EXPERIMENTAL RESEARCH METHODS

Recall from the previous chapter that a causal hypothesis proposes the causes of a behavior. To test such a hypothesis, we usually employ **experimental methods**. The logic of an experiment is this: If my hypothesis is correct, then if I *do* this or that to participants, I should see an influence or change in their behaviors. Therefore, in an **experiment** the researcher actively changes or *manipulates* one variable and measures the subject's resulting behavior by measuring another variable. Then we look to see if the manipulation changed the behavior so that the predicted relationship is observed.

Usually an experiment implies a laboratory setting (with images of elaborate equipment and mad scientists), but this need not be the case. The hallmark of experimental methods is that the researcher actively manipulates some aspect of a situation, and then determines whether the manipulation causes a change in the behavior being studied. Thus, if we originally hypothesized that more studying *causes* fewer test errors, we would systematically manipulate amount of study time and then measure the number of resulting test errors. For example, to compare 1, 2, 3, and 4 hours of study time, we could randomly select four samples of students. Then we'd give one sample 1 hour of study time, administer the test, and then count the number of errors each subject makes. We could give another sample 2 hours of study time, administer the test, and then count the errors, and so on. If we understand the laws of nature governing learning and test taking, then as the length of time that students study increases, the number of errors they make should decrease.

There are names for the components of any experiment, and you will use them daily in research and statistics.

The Independent Variable

An **independent variable** is a variable that is systematically changed or manipulated by the experimenter. Implicitly, it is the variable that we think causes a change in behavior. Above, we manipulate study time because we think it influences test errors, so amount of study time is our independent variable. Or, in an experiment to determine whether eating more chocolate causes people to blink more, amount of chocolate consumed would be the independent variable. You can remember the independent variable as the variable that the experimenter manipulates *independent* of what the participant wishes: Some students will study for 4 hours whether they want to or not. (An independent variable is also called a **factor**.)

We select an independent variable because it is relevant to a hypothetical construct of interest. Then we create an operational definition of the variable, stating how we will measure and manipulate it. An independent variable may be quantitative, in which case we manipulate the amount of a variable that is present in a given situation. Or an independent variable may be qualitative, in which case we manipulate a quality or attribute of the situation. (If, in a different study, we compare the test performance of students who study in either their dorm room or a library room, we are manipulating a quality of the room.)

Conditions of the Independent Variable

An independent variable is the *overall* variable a researcher examines that may have many different amounts or categories we could examine. A **condition** is a specific amount or category of the independent variable that the researcher selects to create the specific situation under which the participants are observed. Thus, although the independent variable above was amount of study time—which could be any amount—our conditions were 1, 2, 3, or 4 hours of study. Likewise, if we compare studying in a dorm room to studying in a library room, the independent variable is type of room, and the conditions are "dorm" and "library." (A condition is also known as a **level** or a **treatment**.)

We have a special name for the condition in which we present zero amount of the independent variable. A **control group** is a group of participants that is measured on the dependent variable but receives zero amount of the independent variable, or otherwise does not receive any treatment. A control group shows how subjects behave without the treatment, providing a "baseline" or starting point for evaluating the influence of the variable when it is present. Groups which receive a nonzero amount of the independent variable or otherwise do experience the treatment are called **experimental groups**. Thus, in our study-time experiment, a control group would be students who spend zero time studying for the exam. Their test scores then provide a starting point for determining, literally, whether the other amounts of study time are better than nothing.

Sometimes it is not possible to administer zero amount of the independent variable, and then the control group is tested under a "normal" or "neutral" condition. For example, if the independent variable is temperature, there must be some temperature present, so the control condition might be normal room temperature. Or, if we are presenting cheerful or sad statements to alter participants' moods, the control condition might involve emotionally neutral statements. And sometimes a control condition involves comparing subjects' performance to the result we would expect if they were guessing. For example, to study "psychic abilities," researchers "send" participants a telepathic message about a card drawn from a deck of playing cards. The number of cards someone correctly selects is then compared to the number we'd expect if the person were merely guessing. (As it turns out, "psychics" do not perform above a chance level, Hanssel, 1980.)

True Versus Quasi-Experiments

Not all experiments are true experiments involving true independent variables. In a **true experiment** the independent variable is controlled *by the researcher*, so that a **true independent variable** is something that the experimenter does *to* participants.

You can recognize a true independent variable because subjects can be randomly assigned to any condition. **Random assignment** means that the condition of the independent variable a participant will experience is determined by random chance. Random assignment is a second step that occurs after we use random sampling to select the subjects for a study. For example, we might write the names of the study-time conditions (0, 1, 2, or 3) on slips of paper and mix them in a hat. Then, when a participant we've randomly selected arrives for the study, we would select a slip to randomly assign him or her to one of the conditions.

> *REMEMBER* A true experiment, with a true independent variable, allows random assignment of participants to any condition.

A true independent variable is something that a researcher exposes participants to. However, there are many behavior-influencing variables that we cannot control, such as age, race, background, or personality. Such variables are called "quasi-independent variables," and studies that employ them are called "quasi-experiments." A **quasi-independent variable** is *not* something that the experimenter does *to* participants, and so they cannot be randomly assigned to conditions. Instead, subjects are assigned to a particular condition because they already qualify for that condition based upon some inherent characteristic. For example, say we hypothesize that growing older causes higher test scores. We can't randomly select participants and *make* some of them 20 years old and others 40 years old. Instead, we would randomly select one sample of 20-year-olds and one sample of 40-year-olds. Similarly, if we wanted to examine whether a qualitative variable such as gender was related to test scores, we would select a sample of females and a sample of males.

> *REMEMBER* A quasi-experiment does not involve random assignment of participants to conditions.

Note that regardless of whether we have a true or quasi-independent variable, the experimenter always determines a subject's "score" on the independent variable. In our examples, students in the sample that studied 1 hour have a score of 1 on the study-time variable, or people in the 20-year-olds sample have a score of 20 on the age variable. Then if a relationship exists, as we change the conditions of the independent variable, we will observe a consistent change in subjects' scores on the other variable. This other variable is called the dependent variable.

The Dependent Variable

The **dependent variable** reflects some aspect of participants or their response under each condition of the independent variable. You can identify the dependent variable as the one in which scores are presumably caused or influenced by the independent variable, so scores on the dependent variable *depend* on the conditions of the independent variable. In our studying experiment, the number of errors on the test is the dependent variable, because we believe that the number of errors depends on how long students study. Or, if we manipulate the amount of chocolate people consume and then measure their eye blinking, eye blinking is the dependent variable. (The dependent variable is also called the **dependent measure**.)

As usual, we must also operationally define a dependent variable. Most of the time, the dependent variable *quantifies* a behavior: It measures the amount or degree of a behavior—how strongly it is exhibited, or its frequency of occurrence. At other times, the variable *qualifies* the behavior, distinguishing one behavior from another in terms of a quality or characteristic. For example, if we look at the causes of different personalties, personality type is the dependent variable, even though instead of measuring more or less personality, we would just identify different ones.

> *REMEMBER* Changing the independent variable presumably causes participants' behavior to change, and the dependent variable measures that behavior.

After conducting the study-time experiment, we need to examine the results. Table 2.4 shows a useful way to diagram the design of an experiment, label the components, and organize the data.

Each column is a condition of the independent variable—study time—under which three participants were tested. Each number in a column is one student's score on the dependent variable of number of test errors. Is there a relationship here? Yes. Why? Because as scores on the variable of amount of study time change (increase), scores on the variable of number of test errors also tend to consistently change (decrease).

Notice that in an experiment, we ask, "What scores occur on the dependent variable for a *given* amount of the independent variable?" In other words, we always ask, "Are there consistent changes in the dependent variable *as a function of* changes in the independent variable?" Above, we are asking "For a given study time, what error scores occur" or "Do error scores decrease as a function of increasing study time?"

As we'll see, there are special descriptive statistics for summarizing scores and describing the relationship found in an experiment. Likewise, there are special inferential statistics for deciding whether the scores found in the conditions are likely to represent different populations of scores that would be found if we tested everyone from the population in the experiment. If so, we then proceed to generalize the results to the broader hypothetical constructs as we discussed earlier.

> *REMEMBER* In experiments, we look for a relationship such that as the conditions of the independent variable change, scores on the dependent variable tend to change in a consistent fashion.

Approaches to Conducting Experiments

Essentially, experiments involve changing a *stimulus* to which subjects react and then measuring each subject's *response*. In the research literature, you'll find there are

TABLE 2.4 Diagram of an Experiment Involving the Independent Variable of Number of Hours Spent Studying and the Dependent Variable of Number of Errors Made on a Statistics Test

Each column contains subjects' scores measured under one condition of the independent variable.

	Independent variable: number of hours spent studying			
	Condition 1: 1 hour	*Condition 2: 2 hours*	*Condition 3: 3 hours*	*Condition 4: 4 hours*
Dependent variable: number of errors made on a statistics test	13 12 11	9 8 7	7 6 5	5 3 2

numerous ways to manipulate independent variables and measure dependent variables. However, there is some standard terminology for the more common techniques that usually involve one of the following approaches.

Manipulating the independent variable A common approach is to present participants with different stimuli or change the characteristics of a stimulus. For example, in studying perception, researchers have presented simple shapes that differ along some physical dimension and then measured subjects' recognition of the stimuli (e.g., Stevens, 1975). In studying sex stereotypes, researchers have presented various toys to see whether boys or girls are more stereotypical in their manner of play (Blakemore, LaRue & Olejnik, 1979).

At other times, the independent variable is the context in which a stimulus is presented, while the stimulus itself is kept constant. For example, environmental researchers present different amounts of background noise while measuring participants' performance on a set of mental problems (Hockey, 1970). Or, in learning research we may present different amounts of rewards or punishments to influence subjects' ability to learn a task (e.g., Bower & Hilgard, 1981).

Another approach is to manipulate the instructions or information given to participants. For example, Loftus (1975) examined how different information conveyed in questions about an event affected subjects' recall of the event. Or, to study sexist attitudes, Phesterson, Kiesler, and Goldberg (1971) had people evaluate a painting they were told was painted by either a male or female.

Sometimes researchers employ confederates as a means of manipulating an independent variable. **Confederates** are people enlisted by a researcher to act as participants or "accidental" passers-by, thus creating a particular social situation in which the "real" participants are observed. For example, Darley and Latané (1968) had a confederate fake a seizure and then measured participants' willingness to help as a function of the number of other people present. Asch (1951) had confederates drastically overestimate the length of a line to see if subjects would conform and also overestimate it (they did).

A different approach is to stress or overload a psychological system and then to infer from subjects' responses how the system normally operates. For example, we have procedures for inducing speech errors—getting your "mords wixed"—to infer speech-production processes (Fromkin, 1980). Researchers have also produced emotional stress by having participants produce a behavior that conflicts with an attitude they advocate (Sherman & Gorkin, 1980).

Some researchers manipulate physiological processes. Here the conditions may involve giving participants different amounts of alcohol or other drugs (Taylor & Leonard, 1983), manipulating the amount of sleep subjects get (Horne, 1978), or varying their "sensory deprivation" (Suedfield, Ballard, Baker-Brown, & Borrie, 1986). Using animal subjects, researchers may employ surgical techniques to create different conditions in which parts of the brain are removed or altered (e.g., Tokunaga, Fukushima, Kemnitz & Bray, 1986). The dependent variables in such studies indicate whether these manipulations produce differences in motivation, memory, and so on.

And finally, researchers manipulate a variable because they believe it changes an internal psychological state, which then influences a behavior. This internal state is called an intervening variable. An **intervening variable** is influenced by the independent variable, which in turn influences the dependent variable. It "intervenes" or comes between the independent and dependent variable. For example, let's say we hypothesize that being frustrated makes people angry, and that greater anger leads to greater amounts of aggressive behavior. Then we would manipulate the independent variable of participants' frustration level to change the intervening variable of their anger, which should then influence the dependent variable of their aggressiveness.

Measuring the dependent variable Most commonly, the dependent variable directly measures the behavior of interest. For example, in studying eating behavior, Schacter (1968) measured the quantity of food eaten by participants as a function of different eating cues present. In a classic learning study, Greenspoon (1955) counted the frequency with which subjects spoke certain words after the researcher said "mmmmm hmmmm" to reinforce them. And Cunningham, Shaffer, Barbee, Wolff, and Kelley (1990) measured participants' willingness to help a confederate after hearing cheerful or sad statements.

On the other hand, the dependent variable may reflect an unseen behavior. For example, a common indirect measure is **reaction time**, the amount of time a participant takes to respond to a stimulus, with differences in reaction time used to infer differences in underlying mental processes. For example, Sternberg (1969) used reaction times to different stimulus words to infer how people search through their memory. Also, researchers take physiological measurements of breathing, heart rate, and perspiration level to infer participants' anxiety or stress levels (e.g., Zimbardo et al., 1966). The number of words correctly recalled under various conditions can be used to infer the cognitive processes of retention and forgetting (e.g., Craik & Lockhart, 1972). Or, the length of time a baby stares at a stimulus is used to infer a positive emotional reaction to it (e.g., Tronick, 1989).

Another approach is to ask participants to make judgments about a stimulus. Often, this approach involves a **forced-choice** procedure, in which subjects must select from a limited set of choices (e.g., a multiple-choice test). Or, participants may perform a **sorting task**, in which they indicate their judgments of similarity or differences by sorting stimuli into different groups (e.g., Flowers, Warner & Polansky, 1979). And participants may describe their feelings or thoughts, in which case they are providing **self-reports**. Often, self-reports involve completing **Likert-type questions**. Here participants rate a series of statements, typically using a scale of 1 to 5, where 1 indicates "strongly agree" and 5 indicates "strongly disagree" (for example, Schwarz & Clore, 1983, had participants rate statements about their mood and the causes for it). In another form of self-reports, subjects provide a running commentary of the mental steps they are performing, as when, for example, they describe their approach for identifying liars (Ekman & O'Sullivan, 1991).

DESCRIPTIVE RESEARCH METHODS

In addition to testing causal hypotheses, recall that we may also create a descriptive hypothesis that describes a behavior without identifying its causes. We test such hypotheses using descriptive or nonexperimental methods. In a **descriptive design**, we simply observe behaviors or relationships so that we may describe them, without manipulating the variables of interest. The logic behind descriptive methods is this: If my hypothesis is correct, then I should observe the predicted characteristics of the behavior, participants, or situation. To test such hypotheses, we may interview subjects, directly measure their behavior, or examine their history. We may study only one subject or conduct a survey of many people. Or we may surreptitiously watch people or animals in their natural habitat.

Don't confuse descriptive *hypotheses* and descriptive *research* with descriptive *statistics*. Descriptive statistics are used to describe *data* from any type of design, whether it be experimental or descriptive. Descriptive hypotheses and descriptive research pertain to a research approach where we seek to describe *behaviors*.

The most common descriptive method is a correlational design. In a **correlational design**, we measure participants' scores on at least two variables (which, as usual, we must operationally define) and then determine whether the scores form the predicted relationship. Originally, for example, we used a correlational design to study the relationship between amount of study time and test grades: We *asked* a random sample of students how long they studied for a test and what their grade was on the test, and then looked for a relationship. (This is different from the above experiment where we *made* students study for a particular amount of time.) Or, in a different study, we would have a correlational design if we measured subjects' career choices and their personality type, asking, "Is career choice related to personality type?" Notice, here we have no independent or dependent variable, and we simply look to see if Y scores consistently change *as a function of* changes in a "given" X variable.

Correlational procedures are especially useful in predicting behaviors. Researchers often conduct such studies in natural "field" settings, where a wide range of natural behaviors is likely. This makes the data useful for predicting future, natural behaviors. Thus, we may test the hypothesis that scores on a certain variable can be used to accurately predict a behavior, or we may set out to discover variables that are useful for making predictions.

Not all descriptive designs are correlational designs. Sometimes the hypothesis is simply that certain behaviors or situations operate in a certain way, and we merely observe participants on one or more variables, without predicting that they form a relationship. Thus, we might surreptitiously observe people while they study, simply to describe this behavior. Or we might distribute a survey of questions about various variables that we think constitute studying, so we can describe what studying is, develop theoretical constructs, or derive additional hypotheses for later study.

> *REMEMBER* Descriptive designs are used to demonstrate a relationship, predict behaviors, and describe a behavior or participant.

In later chapters, we'll see specific descriptive statistics for summarizing the data from descriptive studies, especially when they are correlational designs. Then, as in an experiment, there are inferential procedures for deciding whether similar results are likely to be found in the population. If so, then as usual, we generalize from the specific variables we observe to a more general description in terms of hypothetical constructs and broad behaviors.

Approaches to Conducting Descriptive Research

Any of the approaches for measuring variables in experiments may be used in descriptive procedures. Likewise, a correlational design may examine the relationship between virtually any variables, regardless of how they are measured. We often examine the relationship between scores on two tests or questionnaires. For example, Runco and Albert (1986) examined the relationship between participants' intelligence test scores and their creativity test scores. We may also relate test scores and a physical or physiological attribute, as when Martinez and Dukes (1987) related subjects' self-esteem scores to their gender. Or we may use subjects' records, as when Firth and Britton (1989) correlated the attendance factors relating to psychological "burnout" in nurses. Participants may create the record themselves, as in Cann and Donderi (1986), where subjects kept a record each morning of the previous night's dreams. We may also relate environmental events and behavior. For example, Anderson and Anderson (1984) related daily temperature in a large city to the incidence of crime. And, we may measure a variable in a laboratory procedure, as did Jenson (1993) when relating reaction times to intelligence test scores.

> *REMEMBER* In experimental designs, the researcher actively controls or manipulates the behavior or situation. In descriptive designs, the researcher only observes and measures the behavior or situation.

PUTTING IT ALL TOGETHER

The terms and logic introduced in this chapter are used throughout the scientific world. Psychologists thoroughly understand such terms as relationship, independent and dependent variable, condition, and descriptive statistics. These terms are a part of their everyday vocabulary, and they think in these terms. For you to understand research and apply statistical procedures (let alone understand this book), you too must learn to think in these terms. The first step is to be careful always to use the appropriate terminology.

CHAPTER SUMMARY

1. A *sample* is a relatively small subset of a *population*, which consists of all the members of a specific group. In statistics, the population is the entire group of scores being described, and a sample is a subset of those scores.
2. A hypothetical construct is studied by measuring a *variable*, which is any measurable aspect of a behavior or influence on behavior that may change. Variables may be *quantitative*, measuring a quantity or amount, or *qualitative*, measuring a quality or category.
3. An *operational definition* defines a construct or variable in terms of the operations used to measure it.
4. Most studies focus on demonstrating a *relationship*, a pattern in which, as the scores on one variable change, scores on another variable also change in a consistent fashion. On a graph, the pattern is formed by the *data points*.

5. The *strength* of a relationship is the degree of consistent association between the scores on the two variables.

6. The term *individual differences* refers to the fact that no two individuals are identical.

7. The "given" variable in any study is always designated the X variable, and a relationship is described in terms of "changes in Y *as a function of* changes in X." Experiments investigate changes in the *dependent variable* as a function of changes in the *independent variable*.

8. A relationship between the variables in a sample is used to infer a relationship between the variables in the population, which in turn is used to make inferences about hypothetical constructs.

9. *Random sampling* is the method for selecting a sample so that all individuals in the population have the same chance of being selected and all possible samples have the same chance of being selected.

10. Random sampling should produce *representative samples*, so that the characteristics of the sample match the characteristics of the population. By chance, however, a sample may be unrepresentative.

11. *Descriptive statistics* are procedures used to organize, summarize, and describe sample data. *Inferential statistics* are procedures for deciding whether sample data are likely to represent a particular relationship in the population.

12. Causal hypotheses are tested using *experimental designs*, in which the researcher demonstrates a relationship by systematically manipulating the *independent variable* and then measuring participants' scores on the *dependent variable*.

13. Each specific amount or category of the independent variable is known as a *condition*. A *control group* receives zero amount of the independent variable, as opposed to an *experimental group*, which does receive the treatment.

14. In a *true experiment*, with a *true independent variable*, the researcher controls the variable and participants can be randomly assigned to a particular condition. *Random assignment* means that the condition a participant experiences is determined in a random and unbiased manner.

15. In a *quasi-experiment*, with a *quasi-independent variable*, participants cannot be randomly assigned to any condition but, instead, are assigned to a particular condition based upon some inherent characteristic.

16. *Confederates* are people enlisted by a researcher to create a particular social situation for participants.

17. An *intervening variable* is an internal characteristic that is influenced by the independent variable, which in turn influences the dependent variable.

18. Dependent measures may include *forced-choice procedures* (including *sorting tasks*) and *self-reports* (including *Likert-type questions*).

19. Descriptive hypotheses are tested using *descriptive designs*, in which no variables are manipulated. Instead, variables are simply measured as they occur. In a *correlational study*, two or more variables are measured to determine whether the predicted relationship occurs.

PRACTICE PROBLEMS

(Answers for odd-numbered problems are provided in Appendix D.)

1. How can you recognize when a relationship exists between two variables?

2. Why do researchers look for relationships between variables?

3. (a) What's the difference between a sample and a population? (b) How are samples used in research?

4. Why can't you expect to observe a perfectly consistent relationship between variables?

5. What is a data point?

6. What is random sampling?

7. (a) Why do random samples occur that are representative of the population? (b) Why do unrepresentative samples occur?

8. What are descriptive statistics used for?

9. What are inferential statistics used for?

10. (a) What is the difference between descriptive and experimental research methods? (b) What is the primary consideration for selecting one approach over the other?

11. (a) What is the difference between the independent variable and the conditions of the independent variable? (b) What is the dependent variable?

12. (a) What is a control group? (b) What is an experimental group? (c) Why do we employ control groups?

13. What is the difference between a true independent variable and a quasi-independent variable?

14. (a) What is the difference between a correlational study and other descriptive research methods? (b) What is the difference between an experiment and a correlational study?

15. In study A, a researcher gives groups of participants various amounts of alcohol and then observes any decrease in their ability to walk. In study B, a researcher notes the various amounts of alcohol that people drink at a party, and then observes any decrease in their ability to walk. What is the name for each type of design? Why?

16. In each of the following, identify the independent variable, the conditions, and the dependent variable:

(a) A researcher studies whether scores on a final exam are influenced by whether background music is played softly, is played loudly, or is absent.

(b) A researcher compares freshmen, sophomores, juniors, and seniors with respect to how much fun they have while attending college.

(c) A researcher investigates whether being first-born, second-born, or third-born is related to intelligence.

(d) A researcher examines whether length of daily exposure to a sun lamp (15 minutes versus 60 minutes) accounts for differences in self-reported depression.

(e) A researcher investigates whether being in a room with blue walls, green walls, red walls, or beige walls influences aggressive behavior in a group of adolescents.

17. In problem 16, which are true experiments and which are quasi-experiments?

18. A student, Foofy, conducts a survey of the beverage preferences of college students on a random sample of students. Based on her findings, she concludes that most college students prefer sauerkraut juice to other beverages. What statistical argument can you give for not accepting her conclusions?

19. For the following data sets, which sample or samples have a relationship present?

Sample A		*Sample B*		*Sample C*		*Sample D*	
X	*Y*	*X*	*Y*	*X*	*Y*	*X*	*Y*
1	1	20	40	13	20	92	71
1	1	20	42	13	19	93	77
1	1	22	40	13	18	93	77
2	2	22	41	13	17	95	79
2	2	23	40	13	15	96	74
3	3	24	40	13	14	97	71
3	3	24	42	13	13	98	69

20. In which sample in problem 19 is there the strongest degree of association? How do you know?

21. Below are graphs of data from three studies. Which depict a relationship? How do you know?

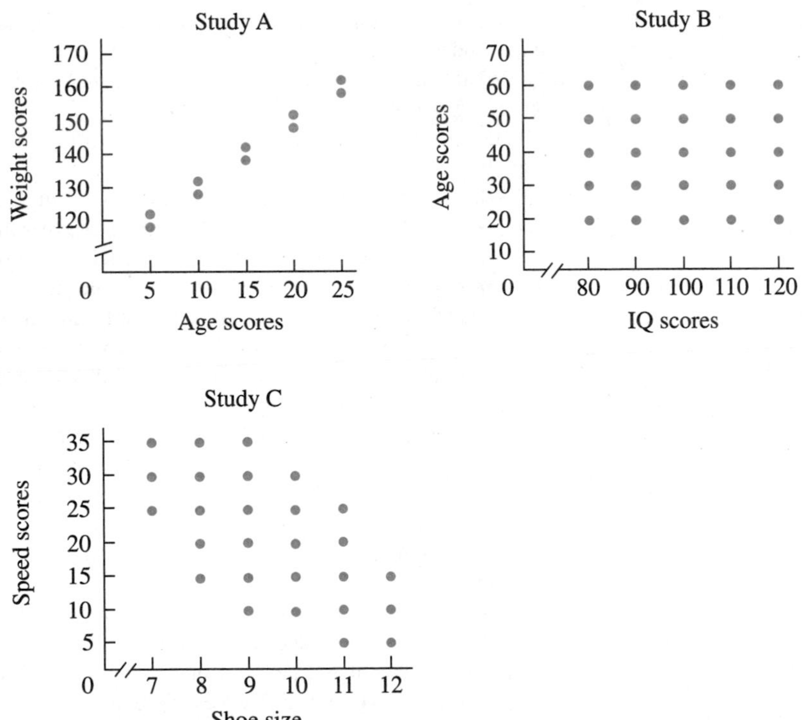

22. Which study in problem 21 demonstrates the strongest relationship between the variables? How do you know?

23. In problem 21, why is each relationship a telltale sign that a law of nature is at work?

24. (a) Another student, Poindexter, says that Study A in problem 21 examines age scores as a function of weight scores. Is he correct? (b) Poindexter also claims that in Study C, the researcher is asking, "For a given shoe size, what speed scores occur?" Is he correct? (c) Let's say the studies in problem 21 were experiments. In each, which variable is the independent (or quasi-independent) variable and which is the dependent variable?

25. What does it mean to use (a) A "forced-choice" procedure? (b) "Likert-type" questions? (c) A "sorting task?" (d) "Self-reports?"

26. A researcher proposes that happier students perform better on a test. To alter their mood when taking a test, students are tested in a room with either red, blue, or green walls. What name do researchers have for the variable of mood in this context?

3

Understanding Reliability and Validity

To understand this chapter:

- From Chapter 2, understand the flow of a study, what a relationship is, and how experimental and descriptive designs differ.

Then your goals in this chapter are to learn:

- What is meant by reliability and validity.
- What the various types of validity are.
- How subject, environmental, researcher, and measurement task variables may threaten reliability and validity.
- What a confounding is.
- What the general approaches to controlling threats to reliability and validity are.
- How experimental and descriptive designs differ in terms of reliability and validity.

In Chapter 2, we saw what is *apparently* a pretty straightforward process: We predict and then examine a relationship between variables and then use the relationship to draw conclusions about a hypothesis. You may be wondering why then, in Chapter 1, I made such a big deal of the potential flaws in research and the need for skeptical and critical evaluation of every research finding. Well, although *overall* a study is straightforward, it is in the specifics of planning and conducting a study that all sorts of errors can creep in. This is because nature is very complex, and so any situation in which a behavior occurs is loaded with fluctuating variables. Any of these variables may be a design flaw that misleads us about a behavior. Therefore, ***the*** issue when designing research is to identify and deal with these variables.

This chapter introduces the common potential flaws that "threaten" confidence in research conclusions. Then we'll see how to deal with such flaws, and how they influence our interpretations of descriptive and experimental designs. As a first step, let's examine a study that we might conduct.

DESIGNING AN EXAMPLE STUDY

Say that while in the student lounge, you observe that your friends are more successful at meeting members of the opposite sex than you are (aren't they always?). You believe this is because they are better-looking than you (aren't they always!). You decide to pursue this topic as a research project. You realize that your observation is related to the broader hypothetical constructs of "first impressions" and "physical attractiveness." In the research literature, you find many social psychology studies that show that first impressions are influenced by physical attractiveness (e.g., Eagly, Ashmore, MaKijani & Longo, 1991). Although this finding holds for *both* genders, you'll examine the descriptive hypothesis that females form more positive first impressions of males who are more physically attractive. (If this hypothesis offends you because it suggests people are shallow and insensitive, remember that scientists must be open-minded and accept nature as it is, warts and all.)

Your first step in testing this hypothesis should be to take a moment and think about the behavior.

Considering the Context of the Behavior

When designing any study, it is *extremely* important that you first consider the overall context or situation in which the behavior occurs. Using your knowledge of psychology and some common sense, try to identify all of the factors that may influence the behavior and your study of it. This approach not only provides numerous ideas for the specific study, but also allows you to foresee potential problems. So, what's involved when a woman forms a first impression and judges a man as attractive?

First, what might a female consider when determining a male's physical attractiveness? She probably considers his facial attractiveness, the color and style of his hair, his height and weight, his posture, his body's shape, his style of dress, his cleanliness, and so on. She might also consider his behavior when they meet: Whether he is silent or talkative, what he says and how he says it, whether he is friendly or condescending. Does he exhibit nervous tics or irritating mannerisms? Does he make eye contact? Does he smile? Does he make physical contact, and how? (Is his handshake firm or mushy?)

The personal characteristics of a woman can also influence her perceptions. Her height and weight determine whether she judges a male as tall or short, heavy or thin. Also, is she different from him in age, style of dress, educational level, culture, or language? Is she actively seeking to meet men, or is she happily married and thus disinterested? Further, how does she form a first impression? It may involve whether she judges him to be intelligent, creative, sexy, likable, interesting, decisive, or some combination of these qualities. How long after she meets him is she beyond first impressions and getting to know him?

You should also consider the environment for the meeting. Is she interviewing him for a job or is the meeting social? Is the meeting in a crowded room or an empty one? Does it occur at a shopping mall, a party, or a funeral? Is the environment noisy or quiet, dark or well-lit? Remember, too, that you must somehow observe her behavior. Doing so may make her nervous and interfere with her "normal" reactions. Does the gender of the researcher make her more or less self-conscious? How will you measure her impressions? Will she answer honestly?

The above provides only a partial list of the many questions to consider when you examine how a female forms a first impression and judges a male's attractiveness. Notice, however, that these questions center around four general components found in any study. The *researcher* observes the *participants* in a specific *environment* and applies a *measurement procedure*. When designing a study, variables related to each component are the variables you must consider.

1. **Subject variables:** The personal characteristics and experiences of participants that may influence their responses.
2. **Researcher variables:** The behaviors and characteristics of the researcher that may influence the reactions of subjects.
3. **Environmental variables:** The aspects of the environment that can influence scores.
4. **Measurement variables:** The aspects of the stimuli presented or the measurement procedure employed that may influence scores.

It is by dealing with these variables that we reduce a complex situation into a controlled and understandable one. We accomplish this by operationally defining each important variable in the situation.

Refining the Design

Let's say that to start the design, we first identify the population (and the sample). Physical appearance is probably irrelevant to blind people when forming first impressions, so we limit the population to sighted individuals, and further, to adult sighted females who are citizens of this country. We must locate a random sample of such women, and given your initial observations at the lounge, we decide to sample from among the women attending your college. In fact, we'll go to the most common source for participants—the current introductory psychology course.

We decide that an important component of a male's physical attractiveness is the variable of his "facial attractiveness." Of the many ways we might operationally define facial attractiveness (for example, rating it ourselves or having a panel of judges rate it), we will directly ask our female participants how attractive a man is just after they've met him. We also decide that an important component of a first impression is how much he is initially "liked" by a woman. Although we might measure the variable of his "likability" by determining whether a female agrees that he meets some definition of likability, or by recording the number of times she uses "like" when describing him, we decide to simply ask her how much she likes him. To quantify a woman's judgement of a man's likability and attractiveness, we'll use a 6-point rating scale (a "Likert scale") and have our participants answer these two questions:

How much do you like this person?
1 2 3 4 5 6
DISLIKE LIKE

How attractive is this person's face?
1 2 3 4 5 6
UNATTRACTIVE ATTRACTIVE

We can now formulate our prediction: At an initial meeting between a male and our female participants, women who produce higher attractiveness ratings for a man will also produce higher likability scores for him. Completing the design involves defining the individual components of "an initial meeting between a male and our female participants." Let's say we define "female participants" as 12 female volunteers from an introductory psychology class. We define the "meeting" as a social, one-on-one introduction, and "initial" as lasting about 2 minutes. And we define "meeting males" as bringing each woman to a student lounge and introducing her to each of 10 of our male friends—we'll call each man a *model*. After each meeting, a participant will rate the model's likability and attractiveness, answering our questions using pencil and paper.

You may think that this design is finished. However, it's not a good design, because we have not completely thought it through.

Critically Evaluating the Study

Say that you conducted the likability study and obtained data like that in Table 3.1. For simplicity, I've sorted the likability scores into two groups based on whether a model was rated as less attractive (rated a 3 or below) or as more attractive (rated a 4 or above). Sure enough, higher likability scores were given to more attractive models. It looks like we can conclude that, in general, a woman's first impressions of a man are related to his physical attractiveness.

But, recall that the goal is to *know* how a behavior operates. However, in any study, the only thing we know is that each participant obtained a particular score—a number—at a particular time. We *don't* know what these numbers actually reflect: We don't

TABLE 3.1 Summary of Data from the Likability Study

	Less Attractive Models	*More Attractive Models*
Likability Scores →	2	5
	1	6
	2	6
	1	5
	3	4
	"	"
	"	"

know that subjects are really—and only—reacting to the stimuli or variables we want them to, so we don't know what a score really means, and so we don't know what the relationship really reflects. That's why we must always critically evaluate any study. Critically evaluating a study means answering the question, "How confident are we that the scores actually reflect the hypothetical constructs, variables, and behaviors we think they reflect, and that the observed relationship actually reflects the relationship we think it reflects?"

There are many aspects of the likability study that we took for granted. Consequently, there are numerous potential flaws that reduce our confidence that the scores indicate what we think they indicate, and that the relationship reflects what we think it reflects. Here are just some of the things that could go wrong:

1. "Facial attractiveness" cannot be accurately measured by our procedure.
2. Some models are wearing a more pleasant aftershave, have more muscular bodies, and dress better than others, and some talk warmly to the females while others talk only of car engines and beer-blasts. Some models will be rated as more attractive, but *not* because of their faces.
3. Some participants barely speak English. Their answers to our questions are partially a test of their English.
4. Some meetings actually last only 30 seconds while others last over 2 minutes, so we don't always measure the same kind of first impression.
5. Some subjects already know the models. Their responses don't reflect an *initial* anything.
6. Some females cannot decide between a "4" and a "5" on the rating scales, and mentally flip a coin when responding. Others are extremely nearsighted (but don't wear glasses). They can barely see the model and instead guess at his attractiveness.
7. Sometimes the lounge is hot, sometimes it's noisy, and sometimes other people are wandering through it. Therefore, some subjects don't pay much attention to the model.
8. The researcher is a female who introduces the more attractive models in a more positive manner. Or, the researcher is a male who's jealous of the attractive models and gives negative introductions for them. Either way, likability scores reflect the tone of the introduction.
9. The lounge is usually lit by sunlight, but when clouds pass by, some participants meet the models in a darkened room. Everyone looks more attractive in a dark room.
10. Two subjects are inadvertently given a pencil with a broken point and think that their reaction to this is what is being studied. They focus not on attractiveness or likability but on where the hidden camera might be.

Although such problems may strike you as unlikely, they can and do occur. For example, May and Hamilton (1980) found that something as mundane as the type of background music being played influenced how females rated a male's attractiveness. Therefore, because we are skeptical and open-minded, always consider whether such things might be occurring in any study. In general, there are two aspects of a study to examine.

First, always critically examine the operational definitions being employed, because they are a major source of potential flaws and can produce all kinds of controversy among researchers: There are many ways to define a construct or variable, and each definition will provide a very narrow perspective and type of evidence that may produce a misleading picture of the behavior. Rely on the research literature for accepted definitions that produce the most precise and objective data possible.

Second, recall that the more systematic and controlled a study is, the greater our confidence that we are not being misled by other things that are going on. The problems in the likability study occur because of the many unintended and misleading influences that might affect the participants. Therefore, always examine a study to identify potential unintended influences. These influences are called extraneous variables.

Extraneous variables An **extraneous variable** is any variable that can potentially influence our results, but is not a variable we wish to study. Such variables come from those four components of a study: the researcher, the participants, the environment, or the measurement task. Thus, extraneous variables are operating if a researcher's rudeness or the environment's temperature influences attractiveness scores, or if subjects' nearsightedness or the difficulty of the measurement task alters likability scores.

Notice that extraneous variables are extremely important at all stages of a study. It was extraneous variables associated with the researcher, participants, the environment, and the measurement task that we actually considered originally when we considered the context of the study. Now recognize that if you missed any, these are then the variables that come back to haunt you, causing problems when interpreting the study.

Sometimes an extraneous variable changes *unsystematically*, changing with no consistent pattern. For example, if the lounge is sometimes dark and sometimes not in a random pattern, then lounge lighting changes unsystematically. Unsystematic extraneous variables produce problems, because if an influence on a behavior is inconsistent from moment to moment, then the behavior will be inconsistent. Yet, the whole logic of research is to demonstrate a relationship: We look for a consistent behavior in one situation that then changes to a different, consistent behavior when the situation is different.

Even more detrimental to a design is that sometimes extraneous variables change *systematically*, consistently changing along with our variables of interest. For example, if the researcher is more positive when introducing more attractive models, then politeness systematically changed with attractiveness. The problems created by systematic extraneous variables are so great that they have their own name.

Confounding variables A **confounding** occurs when an extraneous variable systematically changes along with a variable of interest. Then we cannot tell which variable is actually operating in the study (we are confused or "confounded" by the extraneous variable). Here's another example: Say that coincidentally our more attractive models were better dressed. Then style of dress would be a confounding variable, or we would say it is confounded with attractiveness. This confounding is shown in Table 3.2. Again, the models are in two groups, either less attractive or more attractive. However, the less attractive models can also be described as poorly-dressed, while the more attractive models can be described as well-dressed. We have the predicted relationship here, because what we are *calling* the more attractive models receive higher likability scores. The trouble is, we don't know if we've labeled the groups with the

TABLE 3.2 Diagram of a Confounded Relationship

Intended Variable →	*Less Attractive Models*	*More Attractive Models*
Confounding Variable →	*Poorly Dressed*	*Well Dressed*
Likability Scores →	2 1 2 1 " "	5 6 6 5 " "

correct variable. Maybe facial attractiveness was irrelevant in this situation. Maybe we should be calling the groups "poorly-dressed" and "well-dressed" because that was the relevant variable.

> REMEMBER If an extraneous variable changes simultaneously with a variable of interest, then the variables are confounded.

Identifying a confounding *after* the study is conducted just makes interpreting the study impossible: Above we don't know whether the participants were reacting to a model's attractiveness or to his manner of dress. Therefore, the goal is to identify and eliminate *potential* confounding variables when designing a study. This is especially so for experiments, because here we manipulate the independent variable to cause changes in the dependent scores. If a confounding is present, then we cannot know whether the independent variable is actually causing the change. For example, say that we wanted to show that higher attractiveness causes greater likability. We might create an experiment involving the above two groups, manipulating attractiveness by introducing participants to models that we've classified as either more or less attractive. Now each column in Table 3.2 represents a condition under which females are tested, showing their scores on the dependent variable of likability. As before, the problem is that there are two ways to describe these conditions: either in terms of the independent variable of a model's attractiveness or in terms of the confounding variable of his manner of dress. Because we don't know which variable is really influencing participants, we cannot confidently claim that facial attractiveness causes likability.

Confounding variables arise from any source. There's a confounding if a male researcher introduces the unattractive models and a female researcher introduces the attractive ones. Or, we have confoundings if, along with changes in the models' attractiveness, there are systematic changes in the temperature or noise level of the lounge, or in how interested participants are in meeting men. Or, say in a different study, we studied the influence of different types of background music on the models' likability, comparing a song that has lyrics with another that does not. Confoundings would be present if the volume of one song is greater than that of the other, or if one song is more familiar than the other. In each case, we will not be able to determine which variable is the one that produced differences in a model's likability.

REMEMBER Always be on the lookout for confounding variables.

Thus, in summary, whether designing a study or evaluating someone else's, critically evaluating a study involves considering its operational definitions and looking for potentially influential unsystematic and systematic—confounding—extraneous variables. The question is always whether the data reflect what we think they do. However, we have special terminology so that we can communicate different aspects of the question. These aspects fall under two general concerns called reliability and validity.

UNDERSTANDING RELIABILITY

When asking what the data actually reflect, one concern is whether the measurements are reliable. **Reliability** is the degree to which a measurement is consistent and reproducible. We ask, "Regardless of what the scores actually measure, do they measure it consistently, without introducing error, so that measuring an event produces the same score whenever we measure it?" For example, we assume a woman's liking of a particular model will not change rapidly. Thus, if she indicates she likes him today, she should like him to the same degree tomorrow. If her two scores differ, then we have incorporated some measurement error: Either today's score, tomorrow's score, or both contain error. The problem is that obtaining different scores each time we measure the behavior will lead to different conclusions each time, so we don't know which conclusion is correct. Thus, unreliable data are "untrustworthy" in the sense that they reflect error and lead to inconsistent conclusions.

REMEMBER Reliability is the degree to which measurements are consistent and do not contain measurement error.

Often, reliability is reduced by unsystematic extraneous variables, so any aspect of a study that is inconsistent threatens reliability. Thus, we have unreliable scores if, because of their eyesight or indecision, females guess when rating the attractiveness of the models. Or, we have unreliable scores if our stopwatch is inaccurate and so meetings sometimes last 30 seconds and sometimes 2 minutes. Obviously, to obtain reliable data, we seek to keep all aspects of any research situation as consistent as possible.

UNDERSTANDING VALIDITY

The other concern is whether the data accurately reflect the constructs, variables, and relationships we think they do. Any time researchers question whether they are drawing the correct inferences from results, they are concerned with validity. **Validity** is the extent to which a procedure measures what it is intended to measure. When a procedure lacks validity, it is "untrustworthy" because it reflects the "wrong" aspects of a situation to some degree, so we cannot trust our inferences about the situation.

As we see below, researchers break the issue of validity into several subparts, depending on the particular inference being drawn.

Drawing Valid Inferences About Measurements

First, we are concerned with whether a score actually reflects what we think it does, in terms of the variable and in terms of the hypothetical construct we wish to measure. For each type of inference, we have a corresponding type of validity.

Content validity is the degree to which measurements actually reflect the *variable* of interest. Here we question whether a procedure actually and only measures the behavior we seek (are we tapping the appropriate "contents" of the target behavior?). Thus, when females are supposedly rating attractiveness, we ask, "Are their scores really and only measuring the attractiveness of the model?"

A procedure lacks content validity, first, if it lacks reliability. For example, if participants are guessing somewhat when rating a model's attractiveness, then a score partly reflects his attractiveness and partly reflects the random error of guessing. Second, content validity is decreased if any systematic or unsystematic extraneous variable is also measured by a score. Thus, we lose content validity in measuring attractiveness when participants barely speak English, because their scores partially reflect the variable of language ability. Likewise, content validity is decreased if a woman's rating partially reflects her response to a model's behavior or style of dress. By reflecting these other variables, such scores can mislead us about the variable of facial attractiveness. In essence, we don't know what we're talking about when it comes to facial attractiveness, because we have not measured facial attractiveness alone.

On the other hand, **construct validity** is the extent to which a measurement reflects the *hypothetical construct* of interest. Here, we question whether the variable we are measuring actually reflects the construct as it is conceptualized from a particular theoretical viewpoint. If a procedure lacks construct validity, then any inferences drawn about the broad underlying psychological processes will be in error.

A classic question of construct validity occurs with intelligence tests, which measure people on such variables as vocabulary or problem-solving ability. While content validity is the question of whether we really measure these variables, construct validity is the question of whether these variables really measure "intelligence." Perhaps there are other variables that more accurately reflect this construct. Likewise, some females might argue that they are not so shallow as to judge a male's attractiveness based on his facial appearance. They are essentially theorizing about what should constitute the construct of attractiveness and its relationship to first impressions. For them, our study lacks construct validity because it examines the "wrong" variable—facial attractiveness—and thus will lead to an incorrect interpretation of behavior.

> *REMEMBER* Content validity refers to whether we actually and only measure an intended variable. Construct validity refers to whether a variable actually and only reflects the intended hypothetical construct.

In addition to considering the content and construct validity of any single variable, we also consider validity when drawing inferences about the relationship between the variables. Because we may consider the relationship either in terms of the sample data, or when generalizing beyond the study, we have two types of validity, internal and external.

Drawing Internally Valid Inferences About the Relationship

Internal validity is the degree to which the mathematical relationship we observe between subjects' scores actually and only reflects the relationship between the variables of interest. Thus, we have internal validity if the *X* and *Y* scores reflect only the *X* and *Y* variables we think they do, so that we draw the correct inferences about what was going on *in* ("internal" to) the study. However, we lose internal validity if the relationship between the *X* and *Y* scores reflects other, unintended variables. That is, we lose internal validity when there is a confounding variable operating. Then the relationship is "untrustworthy" in that it does not only reflect the *X* and *Y* variables that we think it does. If we are inaccurate about what occurred internal to our study, then we cannot confidently draw inferences about the relationship in nature.

Don't confuse internal validity with content or construct validity. For example, the variables of likability and attractiveness may each measure what we intended, so they have content and construct validity. But say that the researcher inadvertently gave a more positive introduction to some models, and because of the introduction, females rated these models as more attractive and more likeable. If so, the relationship that we see between likability scores and attractiveness scores actually reflects the relationship between likability and the extraneous variable of type of introduction.

Internal validity is most important when discussing causes and experiments. Internal validity in experiments is our confidence that changing the independent variable produced the observed changes in the dependent variable. If other extraneous variables systematically changed along with the independent variable, however, then our inferences about the cause of the behavior are not correct—not valid. Instead, the observed relationship actually reflects other laws of nature. Above, the data also confirm the rival hypothesis that "better introductions produce both greater attractiveness and greater likability." Because the observed relationship confirms both hypotheses, we have less confidence in our original hypothesis.

> *REMEMBER* Internal validity is the degree to which we can draw the correct inferences about the relationship occurring within a study.

Drawing Externally Valid Inferences About the Relationship

Recall that once we conclude what was going on *in* a study, we then *generalize* the conclusions beyond the study. Thus, we will use the relationship between facial attractiveness and likability to generalize to the broader conclusion that greater physical attractiveness (however it is measured) is related to a more positive first impression (however it is measured) for everyone in the population. But, are we correct in this conclusion? **External validity** is the degree to which the results accurately generalize to other individuals and other situations. In other words, external validity is the question of whether the study provides a good example of the relationship that occurs in situations "external" to our study.

External validity is threatened by any extraneous variable that makes observations unique and atypical, so that they are unrepresentative of the relationship generally found in nature. For example, our female participants are all from the same college so the sample represents the type of women who attend that college. And by testing only college women, we have reduced confidence that the relationship generalizes to the population of all women, including those who don't go to college. Likewise, the way we operationally defined the variables and the specifics of our procedure may result in scores or a relationship that would not be found in other settings. If a similar relationship cannot be found with other participants, in other settings, then we have a biased and misleading perspective: We will incorrectly describe the relationship between likability and facial attractiveness—as well as incorrectly describing the relationship between the constructs of first impressions and physical attractiveness.

REMEMBER External validity is the degree to which we can draw the correct inferences when generalizing beyond a study.

Several subparts to external validity arise depending on the researcher's perspective when generalizing. One important subpart is ecological validity.

Ecological validity A component of external validity is whether we are studying events commonly found in the real world. **Ecological validity** is the extent to which research can be generalized to common behaviors and natural situations. If a design does not have ecological validity, we end up focusing on what participants *can* do in a study instead of what they *usually* do in real life. This issue is most important in highly controlled laboratory experiments. For example, for years researchers studied "paired-associate learning," in which subjects learned pairs of nonsense syllables (for example, learning that BIM and YOB go together). But learning nonsense syllables is not an everyday, real-life behavior, and so such research lacks ecological validity: We cannot be confident that it accurately generalizes to and describes natural, everyday learning processes.

Researchers often lose ecological validity in their quest for internal validity, developing a rather unusual measurement task in order to control and simplify the behavior. (Paired-associate learning allows for much more controlled observations than, say, classroom learning.) However, where possible, a balance between internal and ecological validity is best. The challenge is to maintain control while having participants perform tasks that bear some resemblance to those found in the real world.

REMEMBER Ecological validity is the extent to which the situation and behaviors in a study are those found in the natural environment.

The preceding issues of reliability and validity are ***the*** issues in designing research, so they are summarized for you in Table 3.3. (You need to know these terms.)

TABLE 3.3 A Researcher's Terminology for Questioning the Types of Inferences Made in a Study

Inference Made	*Research Term*
Do the scores reflect error?	Reliability
Do the scores reflect the variable?	Content validity
Do the scores reflect the hypothetical construct?	Construct validity
Does the relationship reflect the variables in the study?	Internal validity
Does the relationship generalize beyond the study?	External validity
Do the results generalize to natural behaviors and situations?	Ecological validity

MINIMIZING THREATS TO VALIDITY AND RELIABILITY

All research suffers to some extent from problems of reliability and validity. The best we can do is minimize the *major* threats to reliability and validity, so that we are as confident in a conclusion as possible. The problem with the likability study was that we did not take the necessary steps to minimize these threats. However, after a study is completed, there is no way to solve such problems. The design and interpretation of a study are completely interrelated, so the appropriate time to worry about the interpretation is when you are designing the study.

Notice that although we focus on identifying threats to reliability and validity, such issues are ultimately empirical questions. That is, we rely on replication to determine whether we've made a mistake when interpreting a study. In particular, literal or direct replications—studies that duplicate a previous study—increase our confidence that a particular approach is internally valid. Conceptual replications—studies that employ somewhat different procedures—increase our confidence that the overall approach is externally valid.

Therefore, researchers resolve issues of reliability and validity by first relying on the psychological literature. There you will find operational definitions of constructs and variables that are considered to be reliable, as well as having the various types of validity. Of course, you can devise your own procedures, but you run the risk that they will have weak validity and reliability. Therefore, you must demonstrate that a procedure you've developed is reliable and valid. (Some techniques for demonstrating this are discussed in Chapter 10.) However, given the time and effort involved, whenever possible it is best to adopt procedures that are commonly accepted in the literature.

In addition, you'll make decisions about how to test a hypothesis so that in some ways your study is unique, with its own unique threats to validity and reliability. The first step in dealing with such threats is to identify potentially important extraneous variables. Use your knowledge of research and psychology to examine the general context of the behavior, as we did in our likability study. In particular, consider the four general components of a study: Researcher, participants, environment, and measurement procedure. For each, look for any extraneous variables that might reduce your confidence that the scores and relationship actually reflect what you think they reflect.

Once you identify serious threats, often you will want to control them.

Controlling Extraneous Variables

Recall that the term *control* means that we prevent fluctuating extraneous variables from influencing the results. The basic approach to creating a more controlled study is to employ a more precise operational definition of each component. More precise definitions automatically limit extraneous variables.

Typically, researchers control extraneous variables in one of the following ways.

Eliminating extraneous variables Many problems can be eliminated by eliminating extraneous variables. For example, in the likability study, we could eliminate distracting noises by moving to a lounge where they do not occur. We could refuse to admit intruders who might distract participants. We could ensure the models are all unknown to participants by obtaining them from a different school, and we could avoid the use of broken pencils.

Keeping extraneous variables constant If we cannot eliminate an extraneous variable, we may keep it at a constant level for all participants. For example, we should keep the temperature and illumination in the lounge at constant, normal levels, and we should keep the researcher's behavior constant, precisely defining how all participants are treated. Also, we could redefine our subjects, selecting females who all speak English well and have corrected vision. Likewise, we could select models with the same body type and manner of dress, and provide them with a "script" of what to say and how to behave.

Balancing extraneous variables Sometimes an extraneous variable cannot be eliminated (for example, we cannot eliminate the variable of a researcher's gender). Likewise, keeping the variable constant may not be feasible, because if it is present in one particular way it may create a rather unique situation, reducing external validity for generalizing our findings (employing only a male researcher or only a female researcher might influence participants in a particular way, and this would reduce generalizability). In such cases, we may intentionally change the variable to "balance" its biasing influence. For example, we could balance the researcher's gender as shown in Table 3.4. Each *X* represents a likability score. We would have a male researcher introduce half of the less attractive and half of the more attractive models, while a female

TABLE 3.4 Diagram of Likability Study Where Researcher's Gender Is Balanced

	Less Attractive Models	*More Attractive Models*
Likability scores with male experimenter	X X X X	X X X X
Likability scores with female experimenter	X X X X	X X X X

researcher would introduce the remaining models. In one sense, this design is consistent, because across all participants and all models, the potential positive or negative influence of the male researcher should balance out any potential effects of the female researcher. Then we show a more general relationship between attractiveness and likability that occurs despite the sex of the researcher who is present.

Likewise, if we thought the time of day participants were tested might have a biasing influence, we'd balance those tested early in the day with those tested later. Or, if we thought a particular decor in the lounge could influence first impressions, we'd balance testing some participants in each of several different lounges. (We'll see more about balancing variables in the next chapter.)

REMEMBER To control extraneous variables, we eliminate them, keep them constant, or balance their influence.

Deciding on the Controls to Employ

Ideally, of course, we seek the best possible study from all perspectives, so we always correct any flaws that we can. It is a fact of life, however, that the things that influence reliability and validity are interrelated. There are always trade-offs, so that improving one aspect of a study may have a negative impact on another aspect. (For example, the most reliable procedure may not be construct valid.) In particular, as we'll see repeatedly, those procedures that increase internal validity often decrease external validity, and vice versa.

Therefore, there is no set of rules to follow when deciding whether to control an extraneous variable. How to deal with any issue of reliability and validity depends on your particular research hypothesis and how you examine the behavior. You may simply accept a threat to reliability or a certain type of validity because the threat is not all that important to your study. In doing so, however, recognize the limitations produced by the threat and refrain from drawing inferences that are invalid because of them.

REMEMBER Whether you should deal with a threat to reliability or validity depends on whether it seriously threatens the purpose of your study.

How to deal with threats to reliability and validity is determined primarily by the type of hypothesis you are testing and thus by the type of design you employ.

ISSUES OF VALIDITY AND RELIABILITY IN DESCRIPTIVE STUDIES

Recall that in descriptive designs the researcher does not manipulate the variables of interest. The original likability study in the lounge was a descriptive (correlational) design because we simply described the reactions of females when first meeting a model, without trying to alter the his facial attractiveness or likability.

Descriptive studies are usually conducted outside of the laboratory. Few extraneous variables are controlled, and the full richness and complexity of nature is present. Therefore, descriptive studies tend to produce high external validity: They tend to be repeatable with other participants and settings. For example, even with the controls we've proposed, the likability study has much in common with the way people meet in the real world. Therefore, our results should generalize to other such meetings in the real world.

It is because of their greater external validity that descriptive methods are best for testing hypotheses about describing and predicting behaviors. With such hypotheses, the goal is to describe behaviors as they normally occur and to predict behaviors based on their occurrence in natural settings. Therefore, we use the methods that produce observations and relationships that are typical of such behaviors in the real world.

Different designs result in trade-offs, however: The cost of high external validity in a descriptive study is reduced control. Therefore, uncontrolled unsystematic extraneous variables may reduce reliability or content validity. And, more importantly, descriptive research tends to have weak internal validity because of potential confoundings. That is, numerous extraneous variables might still be fluctuating in the likability study, because we introduce males and females who can behave in all sorts of ways, and we cannot control all aspects of the environment. Therefore, it is likely that confounding variables are operating, so we have less confidence that the relationship reflects only the variables of interest.

> *REMEMBER* The strength of descriptive approaches is that they tend to have greater external validity, but their weakness is that they tend to have less internal validity.

It is because of their weak internal validity that descriptive approaches cannot be used to make valid inferences about the causes of a behavior.

Problems in Inferring the Causes of a Behavior

Recall that experimental methods are used when testing causal hypotheses and descriptive procedures are used when testing descriptive hypotheses. It's not that we can't test causal hypotheses using descriptive procedures, it's just that they provide minimal confidence in our conclusions. This is because merely showing a relationship between X and Y does not automatically mean that changes in X *cause* changes in Y. One variable can be related to another variable without causing it to change: A person's weight is related to his or her height, but greater weight does not cause greater height. The key to inferring a causal relationship lies in the *manner* in which the relationship is demonstrated.

The first requirement for demonstrating the cause of a behavior is to produce the correct temporal sequence: To say that X causes Y, we have to show that X occurs *before* Y. We have the least confidence about causality in descriptive studies because they do not establish positively which variable occurs first. In the likability study, we cannot say that greater attractiveness causes greater likability, because we are unsure of the order in which these reactions occur. Perhaps as we predicted, females first perceived a model as more attractive and this then caused greater likability. But it might be that participants first decided they liked the model, and his greater likability then caused them to rate him as more attractive.

The second requirement for confidently inferring a cause is to be sure that no extraneous variable could actually be the cause: In other words, that there are no confoundings. We have the least confidence about causality in descriptive studies because we tend not to control extraneous variables and therefore there is a great potential for confoundings. Recall that in the likability study, the models rated as more attractive might also, coincidentally, be better dressed, be better mannered, or receive a better introduction. Therefore, although we may see data that seem to show that greater likability is caused by greater attractiveness, the problem is that the data will also fit rival hypotheses that greater likability is caused by these other variables. With so many potential causes present, we have virtually no internal validity for identifying the variable that actually causes greater likability.

Thus, with descriptive designs, we only describe the relationship between the variables, without inferring that one causes the other. Changes in X might cause changes in Y, but it's also possible that: (1) Changes in Y cause changes in X, or (2) Some other, third variable causes both X and Y to change. (This latter idea is often called "the third variable problem.") Therefore, all we can say with confidence is that there is a relationship, or association, between the scores on the two variables.

REMEMBER In descriptive designs, we are unsure that changes in X cause changes in Y, because we cannot be sure that X changes first or that it is the only variable that could cause Y to change.

Recognize that statistics are not a solution to the problem of causality. There is no statistical procedure that will prove that one variable causes another variable to change. Think about it: How could some formula written on a piece of paper "know" what causes a behavior to occur? How could a mathematical result prove any statement you make about the real world? Statistics don't *prove* anything!

The way to demonstrate a relationship so that we have greater confidence about the causes of a behavior is to conduct an experiment.

ISSUES OF VALIDITY AND RELIABILITY IN EXPERIMENTS

The way to increase internal validity is through greater control. If we could control all extraneous variables so that the only variable that systematically distinguished the models was their facial attractiveness, then we could confidently claim that attractiveness causes likability. To provide this type of control, we conduct an experiment. Laboratory experiments typically yield the greatest control, because we solicit participants to come to our controlled environment and be tested under situations we determine.

For example, here are some controls we could employ in a likability experiment. We might have a panel of judges rate different models' facial attractiveness prior to the study, and then create conditions of the independent variable of attractiveness by, for example, selecting one model consistently rated as low to create a "low attractiveness" condition, one rated as medium to create a "medium attractiveness" condition, and so on. We could control extraneous subject variables by selecting English-speaking females with corrected eyesight. To eliminate extraneous variables stemming from a model's behavior, dress, or body type, we could show participants a photograph of each model's face (as in May & Hamilton, 1980). To give all models the same personality characteristics, all subjects would read the same paragraph describing the model. Finally, we can conduct the study in a controlled laboratory, keep the environment constant, keep the researcher's behavior constant, and balance the researcher's gender in each condition. Then we would hope to see the relationship where, as the conditions change so that a model's attractiveness increases, his likability score also increases. Now, however, by controlling many extraneous variables, we have a much clearer idea of what these scores and the relationship reflect.

In particular, the above controls provide much greater internal validity for concluding that the relationship actually and only reflects the variables of facial attractiveness and likability: Because participants don't experience a model's behavior or dress, it makes no sense to argue that differences in likability were due to these variables. Because subject characteristics, the environment, and the researcher are much more consistent, it's also unlikely that these variables influenced participants' ratings. And so on.

However, recall there is a trade-off between internal and external validity. The more controls we add, the more atypical and unnatural the situation becomes. For example, above we've created a rather strange situation, because females do not normally form first impressions from meetings with a photograph! By creating an unusual situation, we reduce external validity, because the results generalize poorly beyond the study and are not repeatable with other participants and settings. Thus, with experimental methods, we have greater internal validity for understanding a particular relationship, but at the cost of getting results that may be atypical of, and thus less generalizable to, other settings.

REMEMBER Experiments tend to have greater internal validity but less external validity.

There are variations in how an experiment may be conducted and all experiments are not created equal: Not all experiments are equally strong on internal validity or equally weak on external validity.

Internal Validity in True and Quasi-Experiments

Recall that a true independent variable can be controlled by the experimenter so that participants can be randomly assigned to any condition. However, a quasi-independent variable does not allow for random assignment. Random assignment is very important because it is a way of "balancing" and thus controlling extraneous subject variables.

For example, say in a quasi-experiment, we want to determine whether a model's likability changes as a function of a female's age. We select the conditions of 18 or 22 years old. We cannot randomly assign females to be a particular age, so for each condition we must select participants who are already that age. In the process, however, we may end up with two groups who also differ in terms of all sorts of other subject variables. For example, older women are likely to have had a greater amount of dating experience. Thus, we might produce the study diagrammed in Table 3.5. The quasi-independent variable of age is confounded by the extraneous subject variable of amount of dating experience. Further, the women in these two age groups may also differ in terms of their year in college, their maturity, their preferences in men, or who knows—perhaps the amount of their exposure to makeup. It might be differences due to any of these variables that actually cause the differences in likability ratings.

TABLE 3.5 Diagram of Quasi-Experiment Showing a Confounding of the Independent Variable of Age with Amount of Dating Experience

	Condition 1	*Condition 2*
Independent Variable	*18-year-olds*	*22-year-olds*
Confounding Variable	*Little dating experience*	*Much dating experience*
Likability Scores →	X X X X X	X X X X X

Therefore, we cannot confidently identify which variable is actually the cause. In other words, the absence of random assignment severely reduces internal validity.

Note that the lack of random assignment also limits the internal validity of descriptive designs. Here we simply measure participants on the variables, so we cannot prevent extraneous subject variables from producing a confounding. For example, when we measured likability and attractiveness in the lounge, those women who found certain models less attractive might coincidentally have been younger, shorter, or less experienced. These variables might have then determined likability.

On the other hand, the previous likability experiment involved the true independent variable of the attractiveness of a model that a participant sees, so we would randomly assign subjects to conditions. Then we could be confident that the above confoundings do not occur, because by luck, we should obtain various types of subjects in each condition, so that differences in subject variables should balance out. Thus, we'd expect that each condition will contain some participants who are experienced daters and some not, some older women and some younger, some taller and some shorter, some freshmen and some seniors, some who have worn more makeup and some who have worn less, and so on. Overall, each condition should contain a balanced mix of the same types of subjects as the other conditions. Therefore, we have greater confidence that the true independent variable is not confounded by extraneous subject variables.

> REMEMBER Random assignment to conditions balances subject variables, so true independent variables provide greater internal validity than quasi-independent variables.

Thus, on the one hand, a true experiment provides the greatest confidence that we have demonstrated a causal relationship. First, by presenting the conditions of the independent variable and then measuring the dependent variable, we're confident of the temporal sequence between the supposed cause and its effect. Second, random assignment balances extraneous subject variables, reducing their potential as confoundings.

And third, by controlling other extraneous variables, we hope to be able to say that the "only" variable that systematically changed was the independent variable, so that it must be the cause of changes in dependent scores.

On the other hand, such controls build our confidence only by default. If, for example, the influence of the extraneous variables of a model's behavior and dress are eliminated, then by default we are more confident that the scores and relationship reflect what we think they reflect. But, being skeptical, we recognize that we've eliminated only *some* of the many extraneous variables that might be operating, and so it's always possible that we've missed an important one. (Perhaps, for example, the researcher's expectations of how the different models should be rated was communicated to participants.) Also, we cannot be positive that the measurements are otherwise valid or perfectly reliable. (Perhaps our judges did not rate a model's attractiveness as the subjects will see him, so we may not actually manipulate a model's attractiveness in their eyes.) And so on.

Thus, although a well-controlled, true experiment provides substantial confidence that we have validly and reliably demonstrated a causal relationship, it will not be "proof." A quasi-experiment has the added problem of lacking random assignment, so we have even less confidence in conclusions about causality with these designs. Therefore, the results of any single experiment allow us to *argue* to some degree that we have identified a causal relationship. Only as researchers conduct replication studies that control different extraneous variables do we become truly confident of the cause of a behavior.

Laboratory Versus Field Experiments

A laboratory setting is not a slice of real life. The situation is artificial because participants know they are being studied, and a researcher is present who does not behave the way people in real life behave. Therefore, the results may lack ecological validity, because we study strange behaviors under strange conditions. Also, the subjects may be unrepresentative because we can study only those who will come to our laboratory. The bottom line is that although a laboratory experiment has increased internal validity, it may lack external validity, producing results found only in a very similar situation with very similar participants.

Sometimes this is an acceptable state of affairs. Typically, for basic research that seeks the underlying causes of behavior or tests hypotheses derived from theories, we are usually concerned with internal validity and thus lean toward highly controlled laboratory experiments. Such research adds to our understanding of basic processes, even though it may create a unique or artificial situation that is less externally valid. (For example, laboratory studies involving reaction-time do not generalize well to everyday behaviors, but they do tell us how basic cognitive processes operate.) Likewise, a vast amount of research has studied participants drawn from college Introductory Psychology courses. Such subjects may decrease external validity, because they primarily represent the population of Introductory Psychology students. Usually, however, such participants are acceptable, because the primary emphasis of the study is not external validity.

When we seek greater external validity, however, we may leave the laboratory and conduct the study in the real world. Then we are conducting what is generally called **field research**. When studying a causal relationship, we perform a field experiment. A

field experiment is a true or quasi-experiment conducted in a natural setting. Field experiments are common in applied research, when we seek to observe complex behaviors in a natural setting, or to determine whether theoretical explanations are supported in the real world.

The setting may be a factory, a school, a shopping mall, a street corner, or any place the behavior occurs. Yet a field experiment still incorporates the procedures of an experiment: As much as possible, we randomly assign participants to conditions, systematically manipulate the conditions of the independent variable, control extraneous variables, and measure the dependent variable in a reliable manner. Then, because we conduct a controlled experiment, we have internal validity when arguing that changing the independent variable caused the dependent scores to change. But, because we conduct the experiment in the field, we also have external validity when concluding that the results apply to other participants or situations.

> ***REMEMBER*** Field experiments are conducted to show a causal relationship in a natural setting.

One approach to field experiments is to disguise the fact that an experiment is being conducted while observing the general public. Often, the researcher manipulates an environmental condition. For example, Barefoot, Hoople, and McClay (1972) investigated violations of "personal space" by observing whether people used a public drinking fountain depending on whether someone else was sitting close by. Or, Ellsworth, Carlsmith, and Henson (1972) stared at drivers who were stopped at an intersection, to determine if staring prompted them to take flight and depart more rapidly (it did).

A researcher may also use confederates to create a particular social situation. For example, Shaffer, Rogel, and Hendrick (1975) had a confederate leave his belongings with a subject in the college library, and then another confederate searched the belongings and stole a wallet. The dependent variable was whether the subject tried to stop the thief. Or Crusco and Wetzel (1984) examined whether restaurant patrons left a larger tip depending on whether they had been touched by their (confederate) waitress.

The other approach to field experiments involves studying an existing group of participants, because a behavior occurs only in certain situations or with certain subjects. For example, Neri, Shappell, and DeJohn (1992) studied the errors made by airplane pilots on flight simulators as a function of different conditions of fatigue. Likewise, we may study police officers or nurses on the job, because they exhibit specific types of behavior, operate in a role of authority, normally wear uniforms, and so on (e.g., Lavender, 1987). And, educational field research often involves students, as in Fisher and Harris (1973), who manipulated different note-taking and review styles in college classes.

In any of the above, the more natural setting should produce more generalizable results. However, field experiments are far from perfect. They may lose ecological validity because the situation is somewhat contrived: Would a real thief steal a wallet when another person is sitting at the same table watching? Even external validity is still limited to situations and participants similar to the ones we study: Helping was studied when a nonviolent theft occurred, and subjects were the type who sit alone in the library. Most importantly, however, because of lessened control of extraneous variables, there is usually the trade-off where—you guessed it—we lose internal validity. For example, whether people stop and drink from a fountain may actually be due to the time of day and where they are going. The size of a tip may depend on how much money a subject has or the perceived quality of the waitress's service. Confederates' behaviors may be inconsistent, because they must react to a participant's uncontrolled behavior. And we have less control over environmental variables, such as wind and temperature conditions, or traffic noises and other distractions.

Thus, once again we see this general principle of research: The fewer controls in field experiments mean we study a more natural and realistic situation and although our results generalize better, we have greater confusion about what specifically influenced the behavior. On the other hand, the greater controls in laboratory experiments mean the situation is easier to understand, but it may also be an atypical situation that does not generalize well.

> *REMEMBER* Field experiments tend to have greater external validity but weakened internal validity. Laboratory experiments tend to have greater internal validity but weakened external validity.

PUTTING IT ALL TOGETHER

Although validity and reliability are always important concerns, recognize that some behaviors simply cannot be studied using certain methods, and that your options may be limited because you must always be ethical and not harm participants. Therefore, any design is always guided by the questions "What is it we wish to study?" and then "What potential flaws must we deal with?" Select a procedure for its strengths while trying to minimize its weaknesses.

The key to minimizing a study's weaknesses is to anticipate potential threats and build in controls that eliminate them. Therefore, keep these points in mind. (1) State the hypothesis clearly, asking one question about one specific behavior. (2) Be a psychologist, using your knowledge of behavior to identify potential flaws in the design. (3) Rely on the research literature, employing solutions that others have already developed. And (4) Assume that "Murphy's Law" always applies: Anything that can go wrong will go wrong. Design the study accordingly.

At the same time, recognize that you cannot control every aspect of a behavior. Therefore, focus on the *serious* threats to validity and reliability. Any study ultimately involves two major concerns: First, try to eliminate the flaws that can be dealt with practically. Second, consider any remaining flaws when interpreting the results. Leave everything else for the next study.

CHAPTER SUMMARY

1. *Extraneous variables* potentially influence a study's results, but they are not variables we wish to study. They may be subject variables, researcher variables, environmental variables, or measurement task variables, and they may change systematically or unsystematically.

2. *Reliability* is the degree to which a measurement can be consistently reproduced and avoids error.

3. *Content validity* is the degree to which a measurement reflects the variable or behavior of interest.

4. *Construct validity* is the degree to which a measurement reflects the hypothetical construct of interest.

5. *Internal validity* is the degree to which the relationship found in a study reflects only the variables of interest.

6. When an extraneous variable changes systematically with a variable of interest, then the extraneous variable is a *confounding* variable.

7. *External validity* is the degree to which the relationship found in a study generalizes to other individuals and situations.

8. *Ecological validity* is the extent to which a study's results can be generalized to natural settings and natural behaviors.

9. Extraneous variables are controlled by eliminating them, keeping them constant, or balancing their influence.

10. To demonstrate that changes in variable X cause changes in variable Y, a study must show that X occurs first and is the only changing variable.

11. Descriptive methods are best for testing descriptive hypotheses because they tend to have greater external validity. However, they also have less internal validity.

12. Experimental methods are best for testing causal hypotheses because they tend to have greater internal validity. However, they also have less external validity.

13. In a *true experiment*, with a true independent variable, participants are randomly assigned to a particular condition. This helps to balance out subject variables between the conditions and thus prevent confounding.

14. In a *quasi experiment*, with a quasi-independent variable, participants cannot be randomly assigned to conditions, and thus subject variables may confound the independent variable.

15. A *field experiment* is an experiment conducted in a natural setting. Field research tends to have greater external validity, but reduced internal validity.

PRACTICE PROBLEMS

(Answers for odd-numbered problems are provided in Appendix D.)

1. Explain what researchers mean when they express concern about: (a) Reliability, (b) Content validity, (c) Construct validity, (d) Internal validity, and (e) External validity.
2. What do we mean by the term "confounding variable"?
3. What question about a study is raised by the term ecological validity?
4. (a) What are the advantages and disadvantages of descriptive research methods? (b) How do they influence a researcher's conclusions?

5. (a) What are the advantages and disadvantages of experimental methods? (b) How do they influence a researcher's conclusions?
6. (a) What is a field experiment? (b) What is the advantage of such studies? (c) What is the disadvantage?
7. (a) What is the difference between a true independent variable and a quasi-independent variable? (b) What is the problem with using the latter to draw causal inferences?
8. After taking a test, you hear the following complaints. Each is actually about whether the test is reliable or valid. Identify the specific issue raised by each. (a) "The test is unfair because it does not reflect my knowledge of the material." (b) "The questions were 'tricky' and required that I be good at solving riddles." (c) "My essay makes the same points as my friend's, but I obtained a lower grade." (d) "Based on my grade and how I studied in this course, I believe I know how well I'll do in other courses." (e) "I doubt that what caused students to perform poorly or well on the test was how much they studied."
9. While testing the less attractive models in the likability study, you are in a nasty mood. Coincidentally, while testing the more attractive models, you are in a good mood. (a) What do we call the variable of your mood? (b) What two techniques could you use to deal with this variable?
10. You conduct a study in which you test participants' memory for a story, comparing the recall of six who read it silently to that of six others who read it out loud. (a) Will you have greater external validity if you test only males or only females, or if you test both genders? (b) Draw a diagram to show how you would balance subject gender in each condition.
11. To test the hypothesis that drinking red wine daily prevents heart disease, we select elderly subjects who have consumed either zero, one, or two glasses of red wine daily during their lives and determine the health of each subject's heart. (a) What type of design is this? (b) If we find the predicted relationship, how confident can we be in our conclusion that drinking more wine causes reduced heart disease? Why?
12. How would we conduct the above study using a correlational design?
13. Red wine contains an acid that causes headaches, so people who drink more red wine probably take more aspirin. Taking aspirin may prevent heart disease. (a) In question 11, what term refers to the variable of amount of aspirin that subjects take? (b) What problem of validity pertains to this situation, and how does it affect our conclusions?
14. In question 11, describe how you would control the variable of aspirin by: (a) Eliminating it, (b) Keeping it constant, and (c) Balancing it.
15. We hypothesize that the greater a person's fear of sexually transmitted diseases, the fewer sexual partners they've had. (a) How should we study this relationship? (b) What inherent flaw must we accept?
16. Poindexter conducted a survey. In his sample, 83% of females employed outside the home would rather be in the home raising children. After performing all the necessary statistical analysis, he concluded that "the statistical analyses prove that most working women would rather be at home." What is the problem with this conclusion?
17. A researcher reads numerous traffic accident reports and finds that red-colored automobiles are involved in the most traffic accidents. He concludes that certain colors cause more accidents. (a) What type of research method was used? (b) Are the researcher's inferences correct? Why? (c) What specific confounding variables might be operating? (HINT: Think subject variables.) (d) What design is needed to bolster his inference? Describe this study.
18. You test the effectiveness of motivational training by providing it to half of your college's football team. The remaining members form the control group, and the dependent variable is the coach's evaluation of each player. (a) What type of design is this? (b) What is the advantage to testing the football team, instead of testing introductory psychology students? (c) What specific flaws occur as a result of your design?

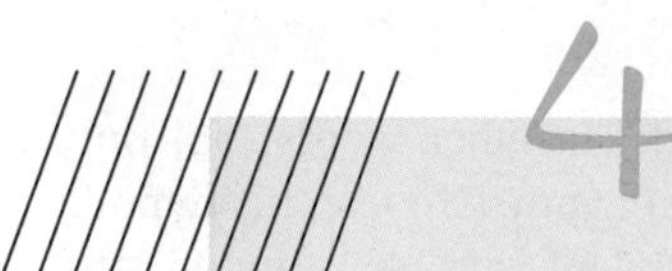

4 Design Issues and Ethical Concerns in Experiments

To understand this chapter:

- From Chapter 2, recall the components of an experiment.
- From Chapter 3, understand what we mean by extraneous variables, how they produce confoundings, and how they threaten reliability and validity.

Then your goals in this chapter are to learn:

- How to design reliable and valid independent and dependent variables and the basic mechanisms for maintaining consistency.
- What order effects are and how to deal with them.
- What demand characteristics are and how to prevent them from threatening a study.
- What research ethics are and how to design and conduct an ethical study.
- What animal research involves and the ethical issues in animal research.

In this chapter, you'll begin to learn specific techniques for dealing with threats to reliability and validity. For now, we'll focus on experiments, because they provide the greatest opportunity to control extraneous variables. We'll also see the ethical issues that arise in human and animal research and how to handle them.

TWO EXAMPLE STUDIES

The discussions in this chapter will refer to two experiments taken from the research literature. The first investigated the hypothesis that the facial muscles used for smiling provide feedback that improves mood. (In other words, not only does being happy make you smile, but smiling makes you happy.) To test this hypothesis, Strack, Martin, and Stepper (1988) measured the dependent variable of mood after participants held a pen in their mouths using their teeth (which mimics smiling) or using their puckered lips (which does not mimic smiling).

The second study tested whether people become more aggressive as room temperature increases. Bell (1980) placed participants in a laboratory under a particular temperature condition and then provoked them by being rude. Subjects later evaluated the experimenter, and the dependent variable for measuring aggressiveness was how negative the evaluation was.

The goal of these studies is to show that a person's mood improves when mimicking a smile or that aggressiveness intensifies with higher temperatures. Remember, however, that all we will *see* is a bunch of scores. Just because particular scores occur in a condition or different scores occur in each condition does not necessarily mean that these behaviors operate as we think they do. To increase confidence in our interpretation of these scores, we try to create a valid and reliable procedure. This starts with how we design the independent variable.

DESIGNING THE INDEPENDENT VARIABLE

Assuming we've selected a construct- and content-valid independent variable, then the first step is to select the specific conditions. This entails decisions about the number of conditions and the specific amount or category of the independent variable that each condition will present. We can examine almost any number of conditions: The number we select depends on the hypothesis we're testing (and some statistical issues we'll encounter later). Likewise, the specific amount of the variable present in each condition also depends on the hypothesis. In the temperature study, for example, we're investigating the influence of "hot" temperatures, so we don't want conditions where it's cold. Of course, we don't want such a high temperature that we might cause heat stroke in the participants, either. So, we might decide to compare the temperature conditions of 70 versus 90 degrees (Fahrenheit).

Controlling Threats to the Reliability and Validity of a Treatment

Once the conditions have been selected, you must precisely define how to create them. In so doing, consider the four components of an experiment: The participants, the researcher, the measurement procedure, and the environment. For each, try to anticipate and eliminate anything that might threaten reliability and validity.

At the top of our list is maintaining internal validity by eliminating any confoundings (so that no systematic differences between the conditions exist other than the independent variable). It is for this reason that participants in the smile study hold the pen in either their lips or their teeth, because then the "only" difference between conditions is whether the smile muscles are engaged. If instead, they either held a pen between their teeth (smile) or held nothing, these two conditions would differ in whether subjects held *something* in their mouths or not, as well as in whether they employed their smile muscles or not.

We are also concerned about reliability: We want the scores to be consistent and to contain no errors. Therefore, first we try to manipulate the independent variable reliably: All participants in a condition should receive the same amount or category of the variable, and when we change to another condition, all subjects there should receive the same new amount. Thus, in the temperature study, room temperature should be precisely 70 degrees for everyone in one condition and precisely 90 degrees for everyone in the other condition. If there is inconsistency in the manipulation, then we are not always presenting the condition we think we are, and our conclusions will be based on the wrong amounts of the independent variable.

Manipulation Checks

We may also "check" that a manipulation actually influenced subjects. A **manipulation check** is a measurement, in addition to the dependent variable, that determines whether the independent variable had its intended effect. Thus, if the goal was to test subjects in a hot room, we may check that they thought it really was hot. Or, if the intent was to make participants happy, we may check that they really were happy. Manipulation checks are especially important for checking on an intervening variable. Thus, if we thought that higher room temperature produced greater aggressiveness by way of the intervening variable of producing greater anger, we could also directly measure subjects' anger. Usually, such checks are made after the dependent variable has been measured. Then we may ask subjects what they experienced in a particular condition, or have them perform some task that reflects the influence of a treatment. This increases our confidence that the independent variable worked as intended, which in turn increases internal validity for explaining why dependent scores changed as they did.

> *REMEMBER* To create a valid and reliable independent variable, minimize confounding, specify the number and type of conditions, and employ a manipulation check.

DESIGNING THE DEPENDENT VARIABLE

Assuming we've selected a construct- and content-valid dependent variable, then the next step is to decide how to measure each behavior and assign it a score. **Scoring criteria** define the system for assigning scores to different responses. These criteria determine if a response is correct or not, what constitutes the beginning and end of a response, how to distinguish one response from another, and all of the other decisions necessary for consistently assigning a particular score to a particular response. Then, we'll "know" what participants did or did not do to receive a certain score, and that any other subject who received that same score did the same things.

You need scoring criteria even for straightforward variables. For example, say we're going to test participants' memory by presenting them a list of words, and then counting the number of words recalled. Is it a correct response if subjects write "bare" but the list presented "bear"? What if for "mother" they recall "mom"? What if subjects recall the correct words but in a different order than in the list?

Scoring criteria will be even more elaborate when a behavior is more difficult to quantify, as when measuring aggressiveness, sexism, or motivation, because here often you must evaluate a subject's behavior subjectively. For example, in the temperature study, we might define an aggression score as the number of aggressive acts exhibited toward the researcher. Then the question is what do we mean by an aggressive act? The strategy is always to minimize inconsistency and bias by minimizing the interpretation you must give to a behavior. Instead, look for observable, concrete behaviors that have a distinct beginning and end. Thus, we could define aggressive acts in terms of such observable behaviors as yelling, hitting, slamming things on the desk, etc.

Likewise, you must define how to assign scores when such behaviors occur. Does each word a participant yells count as one aggressive act, or is each uninterrupted string of words one aggressive act? Does a nasty look receive the same aggression score as a punch? And so on. Or, if we are observing whether someone smiles in response to a stimulus, how shortly after the stimulus must a smile occur in order to be considered a response to that stimulus? How long must a smile last for it to be counted?

Recognize that participants do not behave in an ideal way. You must try to anticipate every possible variation on the expected response, so if it happens, you know how to score it. For help, refer to the literature to find acceptable scoring criteria. Beyond this, your decisions are usually arbitrary: Simply define one way to score each behavior so that you eliminate inconsistency, sloppiness, and error.

REMEMBER Precise scoring criteria that focus on observable behaviors are necessary for measuring a behavior reliably.

One important goal for your scoring criteria is that they result in a sensitive measure.

Creating a Sensitive Dependent Measure

Recall that the goal of an experiment is to observe different behaviors as the conditions of the independent variable change. Therefore, the measurement procedure must *discriminate* or distinguish between behaviors, giving a different score each time a behavior is different. Only when two subjects exhibit the identical behavior should they receive the same score. If a slight difference in behavior produces a different score, then we have a sensitive dependent measure. A **sensitive measure** produces different scores for small differences in behavior. This lets us observe even a small influence on responses that a manipulation may produce.

Sensitivity is increased through observing responses that can differ subtly, and by precisely measuring those differences. For example, if we're judging a subject's aggressiveness, a five-point rating scale is more sensitive than a simple yes-no judgment. Or, in the smile study, measuring participants' mood using precise physiological equipment is more sensitive than asking them questions that rate their mood, which is more sensitive than asking them merely whether they're happy or sad. Always try to measure the amount or degree of a behavior precisely, as opposed to measuring a yes-no type of response.

Part of creating a sensitive measure is to avoid restricting the range.

Avoiding a Restricted Range of Scores

So that we can distinguish subtle differences in participants' behavior, it should be possible for them to obtain any of a wide range of different scores. If the design artificially limits subjects to only a few possible scores, then we have the problem of a restricted range. **Restriction of range** occurs when the range of possible scores on a variable is limited. For example, subjects can exhibit great differences in aggressiveness, so our procedure should allow scores that also reflect these differences. If we restrict the range, however, everyone scores close to or at the same score, *regardless* of the independent variable. Therefore, we are less likely to see a relationship in which there are different scores as the conditions of the independent variable change.

To avoid a restricted range, first consider the scores you will assign. If, in the aggression study, participants can obtain a score of only 1, 2, or 3, the range is restricted. But, if they can score anywhere between 0 and 100, the range is not restricted. Second, look for aspects of the testing situation that, *practically speaking*, limit the behaviors or scores that may occur.

One such aspect is when the task is too easy or otherwise biased so that all scores are likely to be near the highest possible score. In that case, the data will show **ceiling effects**: The lowest potential scores—from the worst-scoring subjects—are very high, so scores cannot differ much because would-be higher-scoring subjects cannot get much higher (everyone's scores are "hitting the ceiling"). For example, if in the temperature study we accept almost any action as being aggressive, then not-so-aggressive subjects will have a high score. Then the range is restricted, because other subjects who get seriously aggressive cannot score substantially higher, and we will fail to discriminate between differences in aggressive behavior.

The range is also restricted if the task is too hard or otherwise biased toward producing low scores. Here, the data will show **floor effects**: The highest potential scores—from the best-scoring subjects—are very low, so scores cannot differ much because would-be lower-scoring subjects cannot get much lower (everyone's scores are "hitting the floor"). For example, most laboratory subjects will not physically assault a researcher, so it's unreasonable to make high aggressiveness scores depend on such actions: Realistically, no one will obtain such scores and everyone will have low scores. Then, truly nonaggressive participants cannot score lower than more aggressive subjects and, again, we will fail to discriminate between behaviors.

Thus, a sensitive procedure is, in part, one that avoids ceiling and floor effects. To do that, the typical participant should start off with scores in the middle between very low scores and very high scores. But it should also be realistically possible for subjects to obtain higher or lower scores as the conditions of the independent variable change.

REMEMBER We want a sensitive procedure that will reflect slight differences in behavior and that will produce an unrestricted range of scores.

Observing Reliable Behaviors

Unreliability comes not only from inconsistency in measuring behavior but also from inconsistency in the behavior itself. Remember that we use scores to describe the typical behavior that is found in a condition. Therefore, each score should reflect a reliable, representative example of a subject's typical behavior.

There are two general approaches to increasing the reliability of data: Practice trials and multiple trials.

Practice trials A participant's score may be unrepresentative because he or she has not "warmed up" on the measurement task. When this is a possibility, we provide **practice trials**: We test participants as in the real study, but we ignore their scores from these trials when analyzing the results. (A "trial" is one complete measurement or observation.) Practice trials are especially useful when studying physical reactions, as in a reaction-time task. If the task is complicated or involves elaborate equipment, practice trials ensure that participants understand the task. In addition, practice trials allow subjects to get used to being observed, so they behave more naturally when we collect the real data.

Multiple trials A major component of reliability is the number of real trials a participant performs in a condition. A score that is based on only one trial may reflect all sorts of extraneous factors: A participant may be momentarily distracted or may guess well; the trial may be especially easy or difficult; or some other aspect of the situation may make the trial peculiar. In such a case, the response and score are not representative of a subject's typical response and score.

To avoid the bias from one unique trial, we observe each participant several times in a condition, observing **multiple trials**. Again, we are using the strategy of balancing out extraneous variables. We assume that differences in a participant's motivation or attention on different trials balance out, that easy trials balance with hard ones, and so on. We then compute each subject's total score or average score for all trials. When interpreting these summary scores, we are more confident that they reliably reflect the typical response, because we have balanced out the random fluctuations found in individual trials.

There is no magic number of trials you should observe per condition. Depending on the study, there may be as few as 1 or 2 trials or as many as several hundred. To prevent confounding, all conditions should involve the same number of trials. Observe a greater number of trials when each trial is more easily influenced by extraneous variables and is thus likely to be unreliable.

> ***REMEMBER*** Providing practice trials and testing multiple trials tends to increase the reliability of the data.

Although multiple trials add reliability, they also create the problem of order effects.

The problem of order effects Multiple trials introduce a new extraneous variable called order effects. **Order effects** are the influence on a particular trial that arises from its position in the sequence of trials. Order effects have two components.

Practice effects are the influence on performance that arises from practicing a task. Even after practice trials, participants may perform initial trials poorly because they are not warmed up. After more trials, performance may improve because participants become quicker or more accurate. With even more trials, however, performance may decrease again because subjects become fatigued or bored.

The other component is **carry-over effects**, the influence that a particular trial has on performance of subsequent trials. Carry-over effects may arise first from simply experiencing a trial. For example, if a trial happens to be very frustrating, this feeling may "carry over," lowering performance on subsequent trials. Trials might also be especially boring, easy, or anxiety-provoking, all of which may influence responses to later trials.

Carry-over effects also arise from a **response set**, which is a bias toward responding in a particular way because of previous responses made. This occurs whenever responding becomes more of a habit than a natural reaction to the stimulus. Thus, you have a response set when completing a multiple choice exam if, after the first few questions, you superstitiously believe the correct answer is always choice 4. You might also develop response sets from strategies that have proven successful. For example, unscramble each of the words below:

ookb

reet

oatc

oabt

You can solve the first three words quickly because of the response set that says to always place the final letter first. If you stumble on the fourth word, however, it's because this strategy no longer works.

The problem with practice and carry-over effects is that a response is somewhat unique because of where in the sequence it occurs. (Above, your solution of each word is faster or slower than it would be if the word was in a different location in the sequence.) Therefore, participants' overall performance is tied to the unique order used, and so the summary scores are an unreliable indication of the subjects' typical response, especially when compared to other orders.

Counterbalancing order effects To solve the problem of order effects we again use the strategy of balancing, here balancing the effect of any single order by including several different orders. Balancing order effects is called counterbalancing. **Counterbalancing** is systematically changing the order of trials for different subjects in a *balanced* way to *counter* the biasing influence of any one order. For example, say that in the temperature study, we measure how aggressive participants feel towards the experimenter by having them answer 10 different questions. If we number the questions 1 through 10, a simple counterbalancing scheme is to present the questions to half of the participants in each condition in the order 1 through 10, and to present the questions to the remaining subjects in the order 10 through 1. This design is shown in Table 4.1.

For each participant, we could summarize their 10 responses by computing an average aggressiveness score. We then ignore the order in which participants completed the questions and look at all aggression scores found in each condition (each column). Higher scores in the 90 degree condition will support the hypothesis that increasing temperature increases aggressiveness. Further, we are confident that the results are not biased by the order in which questions are answered, because in each condition there is not one particular order present. If we were to test additional people using other orders, we would have even more confidence that the order of trials was not biasing any conditions. (There are several variations of this technique discussed in Chapter 16.)

REMEMBER Whenever you measure multiple trials, consider counterbalancing the order of trials.

TABLE 4.1 Diagram of Temperature Experiment with the Order of Trials within Each Condition Counterbalanced

Each X Represents a Participant's Summary Score from 10 Questions.

	Independent variable of temperature	
	70 degrees	*90 degrees*
Participant's Scores Obtained with Order 1–10	X X " "	X X " "
Participant's Scores Obtained with Order 10–1	X X " "	X X " "

Some researchers use the term counterbalancing when balancing any extraneous variable, in addition to the variable of order. Thus, if a male experimenter tests half the participants in each condition and a female tests the remainder, we have counterbalanced for experimenter gender. If half the participants in each condition are male and half are female, we have counterbalanced for subject gender.

> *REMEMBER* To create a valid and reliable dependent variable, establish scoring criteria, create a sensitive procedure that avoids a restricted range (and floor/ceiling effects), include practice trials, observe multiple trials, and counterbalance the order of trials or other extraneous variables.

After designing the independent and dependent variables, the next step is to control any remaining extraneous variables.

CONTROLLING EXTRANEOUS VARIABLES

The goal of controlling extraneous variables is to prevent them from fluctuating *within a condition* (so that all participants experience the condition in the same way), and *between the conditions* (so that, except for the independent variable, everything is the same in the different conditions). You can view all fluctuating extraneous variables as essentially resulting in inconsistency either within or between conditions. Therefore, the basic strategy is to build in consistency. Thus, everyone in the smile study should hold the same type of pen in their mouths, for the same period of time, all the while maintaining the same posture and performing the same tasks. In the temperature study, all participants should be dressed the same way, seated the same distance from the heater, acclimated to the temperature to the same extent, and so on. In each case, if everything is the same for all participants except for the conditions of the independent variable, then we have greater confidence that we "know" what the scores and relationship reflect.

REMEMBER The key to designing a convincing experiment is to build in consistency within and between conditions.

Although every study will necessitate its own controls, there are several general techniques for creating consistency.

Instructions to Participants

Instructions must clearly explain what the task is and how participants should approach it: Describe the sequence of events, identify the stimuli they should attend to, and explain how they should indicate a response. The goal is to have all participants perform precisely the same intended task, without introducing extraneous stimuli or behaviors that make the task different for different participants.

Creating effective yet consistent instructions requires considerable effort. They should be clear for the least sophisticated subjects, avoiding psychological jargon and unfamiliar words. They should anticipate participants' questions. (Should people guess when responding? Should they hurry?) Instructions should also prohibit unwanted behaviors. (Subjects should not look around, fidget, or talk, so they don't miss crucial aspects of the task.) And, the same instructions should be presented to all participants. (When manipulating a variable through instructions, change only the necessary parts and avoid a confounding by keeping constant all other aspects.) Never "ad lib" instructions, because you can't reproduce them reliably. Instead, read them from a prepared script, using an easily reproduced, neutral voice, or, better yet, play a tape recording of them. To help make this all work, researchers usually dress and behave rather formally, because this elicits greater compliance from participants.

Using Automation

Another way to control extraneous variables is through **automation**: Using electronic or mechanical devices to present stimuli and to measure and record responses. Electronic timers, slide projectors, video and audio tape recorders, and computers ensure controlled and reliable stimulus presentations. Automating the data-collection process similarly ensures that the scoring system is consistently and accurately applied and provides for more reliable and sensitive measures. Automation also eliminates experimenter errors and inconsistency that may result because: (1) The experimenter is so busy directing the study that parts of a behavior are missed, and (2) The experimenter has expectations about how the study should turn out and thus inadvertently influences participants or records scores accordingly.

With automation, however, we must guard against instrumentation effects. **Instrumentation effects** are changes in the measurement materials that occur because of use, making the measurements less reliable. This occurs when, over the course of a study, slides, films, and videotapes become scratched and blurred, paper materials get mutilated, or equipment becomes worn and timers become less accurate. Part of the "instrumentation" is the experimenter, who may become more experienced, more bored, or more crazed as time passes, and who may inadvertently change the procedure. Because of such changes in the material or procedure, the measurements obtained late in a study can be different from those obtained early on, making the experiment unreliable overall.

To minimize instrumentation effects, always keep equipment in order and make copies of materials so that all participants can be presented with pristine stimuli. Be vigilant to keep the experimenter's behavior constant. Finally, test some participants from each condition during the early, middle, and late stages of the study, so that potential instrumentation effects are balanced in all conditions and thus cannot confound the study.

REMEMBER Although automation improves reliability and precision, instrumentation effects reduce reliability because the materials change through use.

Testing Participants in Groups

When building in consistency, an important issue is whether to test participants individually or in groups. Group testing is most common when the task requires subjects' written responses, such as on a questionnaire. The advantages of group testing are: (1) Collecting data is more efficient, and (2) If we can test an entire condition at one time, all participants will experience the same consistent condition. The disadvantage is that participants may make noise, block one another's view, or otherwise distract each other so that many extraneous variables are introduced. Therefore, we need to control such behaviors carefully when testing groups. Usually, we can do this through particularly explicit instructions. How successful we will be—and the wisdom of group testing in general—depends on the particular experiment and how susceptible participants will be to the presence of others.

Pilot Studies

To be sure that they have developed a reliable and valid procedure, researchers often conduct a pilot study. A **pilot study** is a miniature version of a study that researchers use to test a procedure prior to the actual study. Using participants similar to those in the actual study, we determine such matters as whether the instructions are clear, whether the task can be done given time constraints or other demands, and whether we have developed a workable, sensitive, and reliable scoring procedure. In addition, a pilot study allows the experimenter to work out any bugs in the equipment or procedure so that it runs smoothly and consistently.

Pilot studies are also used to create and validate stimuli. For example, say that an independent variable involves showing participants films that contain different amounts of violence. Our personal judgment is of little help in determining the amount of violence in a film, because we may be particularly sensitive or insensitive to violence. Therefore, we would show the films to pilot subjects and have them rate the amount of violence each contains. (Note: A pilot study is different from a manipulation check; pilot studies occur before a study, manipulation checks during the study.)

For any problems identified, we alter the stimuli, task, or instructions and conduct more pilot studies until we have the desired situation in each condition.

REMEMBER To maintain consistency in the testing procedure, create clear instructions, employ automation but limit instrumentation effects, consider group testing, and conduct pilot studies.

There is one more major threat to reliability and validity. In any study, we must deal with demand characteristics.

DEMAND CHARACTERISTICS

Imagine you are a subject in an "experiment" taking place in an elaborate "laboratory." A "psychologist" with lab coat and clipboard puts a plate of cookies in front of you and says, "Normal people crave cookies at this time of day, so eat if you want." I bet you'll eat one. On the other hand, imagine the psychologist says, "Only people who have no self-control eat at this time, but eat if you want." I'll bet you don't. In any situation, the social and physical surroundings provide cues that essentially "demand" that we behave in a certain way. In research, these cues are called demand characteristics. A **demand characteristic** is an extraneous cue that guides or biases a participant's behavior. Demand characteristics arise despite the instructions we provide, and participants don't necessarily respond to them intentionally, or even consciously.

Certain demand characteristics occur repeatedly in different research situations, so we have specific names for them.

Types of Demand Characteristics

Recognize that for many people, research procedures are mysterious, and rumor has it that psychologists do strange things to people and study only intelligence, sexual deviance, and crazy people. Therefore, participants tend to be on guard: They are sensitive to the whole idea of being "studied" and may alter their behavior accordingly. This is the demand characteristic called reactivity. **Reactivity** is the bias in responses that occurs because participants are aware they are being observed. Participants know they're under the gun, and they "react" to the mere presence of an experimenter who is observing their behavior. Therefore, they may respond in unusual ways, become nervous and giddy, or make more errors than they normally would.

A similar demand characteristic is the **Hawthorne effect**, which refers to the influence on participants' performance—usually an improvement—that occurs due to the novelty of being in a study. This effect was named for a study of worker productivity conducted at Western Electric Company's Hawthorne factory. The researchers manipulated numerous variables that should have decreased productivity, but the participants instead continuously increased their productivity. They indicated that, because their employer and the researchers had given them special attention, they felt compelled to increase productivity (Roethlisberger & Dickson, 1939). (There is, however, some controversy over this account: See Bramel & Friend, 1981.)

The Hawthorne effect is different from reactivity. If participants are unnaturally motivated to perform a boring task because of their enthusiasm for the study, we have the Hawthorne effect. If their performance is then unnatural due to nervousness about being videotaped, we have reactivity.

An additional important demand characteristic is called social desirability. **Social desirability** occurs when people provide what they consider to be the socially acceptable response. Essentially, participants "edit" their responses so that they aren't embarrassed or so they won't appear weird or abnormal. Thus, in the smile study, some participants may act happy, not so much because they are happy, but because they think it's expected. Others may want to act happy, but inhibit this because they're afraid it's inappropriate. (Notice that reactivity and social desirability may also restrict the range of scores: For example, all subjects in the temperature study may stifle their aggressiveness so that they don't look bad. If so, all the scores will be low, producing floor effects. If we were measuring "niceness," we might see ceiling effects.)

The environment and measurement task can also produce such demand characteristics. The surroundings may distract participants or cause them to react to extraneous factors. This is especially true in fancy laboratories with one-way mirrors and complex equipment that can play upon a person's fears. (Once, when using a bank of electronic timers, I had to convince subjects they would not be electrocuted by having them look under their chair to see that there were no wires!) Likewise, participants react in accordance with how they view the measurement procedure. For example, participants may interpret our 10 questions about their aggressiveness towards the experimenter as really being a personality test. Then, instead of honestly answering the questions, they may respond in ways they think will project their ideal personality.

Finally, the experimenter is an important source of demand characteristics. Participants are very sensitive to **experimenter expectancies**, which are cues the experimenter provides about the responses that subjects should give in a particular condition. These cues occur because the researcher knows the predictions of the study and may inadvertently communicate them to participants. Then the ultimate self-fulfilling prophecy is produced. In the smile study, for example, we expect people to be happier when mimicking a smile. Subtle actions on our part can register on participants and, sure enough, they'll respond with the predicted mood.

As these examples illustrate, demand characteristics cause people to play a role, being good (or not so good) participants who perform on cue. The problem is that they are responding to these cues, instead of to our variables. Then we lose internal validity because it is the cues that cause a particular response, instead of the independent variable. We lose reliability, because not all participants react to the same cues in the same way, so scores will be inconsistent. And, we lose external validity because the results will not generalize to other situations where such demand characteristics are not present.

REMEMBER Demand characteristics are cues that bias subjects, resulting in responses that are not valid, reliable reactions to our variables.

Fortunately, we have several techniques for controlling demand characteristics.

General Controls for Demand Characteristics

Our first line of defense against demand characteristics is to provide participants with as few cues as possible. If they have no cues, their only recourse is to act naturally. Thus, instructions should not divulge the specific purpose, manipulation, or predictions of a study. Anytime we keep participants in the dark regarding the specifics about a condition they receive, we are using a **single-blind procedure**. In addition, we should not include extraneous, distracting information (such as placing an unnecessary time limit on a task). Also, we hide threatening equipment, and we avoid threatening actions by the experimenter.

It is especially important to limit the cues from the experimenter. A prime reason for using automation is that equipment will not communicate experimenter expectations. When a potentially biasing experimenter must be present, however, we employ a double blind. In a **double-blind procedure**, both the participants and the researcher who interacts with them are unaware of the specific condition being presented. The original researcher trains others to actually conduct the study, but they do not know the conditions or predictions—they are "blind" to them. Such procedures are especially common when testing the effect of a drug or other medical treatment. If the researcher knows when a particular drug is being administered, his or her expectations about its effects may be communicated to patients. These expectations *alone* can produce the expected physical reaction or recovery from an illness. If the experimenter is blind to such information, however, no expectations can be communicated.

> *REMEMBER* A double-blind procedure keeps both the researcher and the participants blind to the conditions, so that their expectations do not influence the results.

Our second defense against demand characteristics is to make those cues that must be present as neutral as possible. Thus, the researcher tries to be rather bland, being neither overly friendly nor unfriendly. In the instructions, we try to neutralize participants' fears and suspicions by presenting the task without implying that it is difficult or easy and without indicating what the "normal" or expected response is. Also, we try to provide a response format that subjects are comfortable with: We have children act out responses using toys or we give college students a paper-and-pencil test. And we try to select a researcher with whom participants will be comfortable performing the task (in terms of gender, age, etc.). We also make participants comfortable by allowing them to "habituate" to the procedure. With **habituation**, we familiarize participants with a procedure before beginning actual data collection. For example, we may first play with children until our presence is no longer disruptive. If we are videotaping responses, we allow participants to become comfortable with being recorded before the study begins.

Our final defense is to have participants ignore demand characteristics by creating experimental realism. **Experimental realism** is the extent to which the measurement task engages participants. The goal is to create a measurement task that subjects find so interesting or engrossing that they "forget" about demand characteristics. Thus, in the smile study, if the task is so engrossing that participants forget about the pen in their mouth, the results will be less influenced by demand characteristics. Experimental realism does not mean, however, that the task is like real life, so experimental realism is different from ecological validity. With experimental realism, the task may be very strange and unreal, but participants' responses are actual, honest responses to it.

You should apply the preceding strategies to virtually all designs. If a procedure will still produce very strong demand characteristics, however, then you may need to employ unobtrusive measures or deception.

Using Unobtrusive Measures and Deception

With an **unobtrusive measure**, we measure participants' behavior without making them aware the measurement is being made. Thus, an unobtrusive measurement of aggressiveness might involve observing participants through a one-way mirror. Other unobtrusive measures may include the use of hidden cameras and recorders. Or we may observe telltale evidence left by participants (for example, to measure how far people sit from each other, we could measure the distance separating their chairs after they've left the room). In each case, participants cannot be overly influenced by demand characteristics if they are unaware of the measurement being made.

Sometimes an unobtrusive measure is used in conjunction with deception. **Deception** involves the creation of an artificial situation or "cover story" that disguises a study. Participants are then unaware of the manipulation or the behavior actually being studied, so they do not feel pressured to respond in a certain way. For example, in the smile study, imagine what you would do if a researcher simply said, "Here, hold this pen in your lips." You'd probably feel very self-conscious and behave unnaturally. The original researchers eliminated such problems by telling participants that the study investigated how physically impaired people use their mouths to do tasks that others do by hand. Then, while holding the pen in either their lips or teeth, participants used the pen for various tasks, including marking various stimuli. Among the stimuli were several cartoons, and participants' rating of how humorous they found the cartoons was the dependent variable for measuring their mood. Thus, what might have been a bizarre task was transformed into a rational, engaging task in which participants were less self-conscious and responded in a more natural way.

Likewise, Schacter, Goldman, and Gordon (1968) sought to measure how much food people would eat under different conditions. But blatantly watching participants as they eat is likely to make them highly reactive. Therefore, the researchers told their subjects that they were in a taste study in which they were to rate the taste of different crackers, eating as many crackers as necessary. The real dependent variable was the number of crackers eaten.

A special type of deception is often used in conjunction with a control condition (a condition that receives zero amount of the independent variable). Because only the experimental group receives the treatment, it also only experiences the associated demand characteristics. Therefore, we have a confounding: The experimental group behaves differently from the control group either because of the treatment or because of the accompanying demand characteristics. For example, let's say we give an experimental group a drink of alcohol while a control group receives nothing. Any impairment the experimental group exhibits may be due to the alcohol, or it may arise because giving people alcohol implies that we expect them to act drunkenly, so they do.

To keep such demand characteristics constant, control groups are given a placebo. A **placebo** provides the demand characteristics of a treatment. Thus, we would give the above control group something that smells and tastes like alcohol, but that is not alcohol. Then we communicate to both groups the same demand characteristics for acting drunkenly, so any differences in their behavior are due to the real alcohol given to the experimental group. Similarly, in studies that require the experimental group to perform an involved task prior to making a response, we have the control group perform a similar, placebo task to eliminate differences between the groups in terms of motivation or fatigue.

> ***REMEMBER*** A placebo is given to control groups so that they experience the same demand characteristics as experimental groups.

Deception is not always necessary or wise (as with everything in research, you must balance the pros and cons). In the temperature study, we might off-handedly tell subjects that the room's thermostat is broken. However, Bell and Baron (1976) found that participants in a hot room often guessed that the study dealt with temperature anyway. Catching the researcher in a lie will worsen the demand characteristics, because now subjects *know* they should be on guard.

Concealing the Experiment

When we are especially concerned about reactivity and social desirability, the final approach is to conceal the entire experiment. Many studies, for example, have been performed while subjects sit in a waiting room, supposedly waiting to be taken into the experiment. Mathews and Cannon (1975) studied how the noise level in a room influenced a subject's willingness to help a confederate who dropped some books. Doing this in a formal laboratory with the experimenter watching might communicate that helping behavior was being studied and that the expected behavior was to help. Instead, a confederate dropped the books while walking past the subject in the "waiting room," and the subject was unobtrusively observed.

The most extreme way to conceal an experiment is to move out of the laboratory and conduct a field experiment with the general public. One reason that field experiments have greater external validity is that with them we can disguise the fact that an experiment is being conducted. For example, Isen and Levin (1972) discreetly manipulated mood by allowing some subjects the pleasant experience of finding money in the change return of a pay phone. Then a passing confederate dropped a manila folder, and the dependent variable was whether subjects helped the confederate. Similarly, Mathews and Cannon (1975) conceptually replicated their above study of the influence of background noise on a subject's helping behavior. They had a confederate drop books on a sidewalk in front of random passersby, while another confederate operated a noisy lawn mower. Such field studies overcome the reactivity or social desirability that occurs with a laboratory experiment, and the inability of laboratory subjects to forget they are in a study because of its poor experimental realism.

> ***REMEMBER*** We reduce demand characteristics through minimizing cues, habituation, experimental realism, unobtrusive measures, deception, and field experiments.

Believe it or not, with the information you've learned from this chapter, you are ready to begin designing experiments. (There are a few more issues to consider, but they're best left until you've gotten some statistics under your belt.) To summarize the discussions so far, Table 4.2 presents a checklist of the major issues for you to consider when designing an experiment.

Similar techniques are used when conducting research involving animals.

RESEARCH INVOLVING ANIMALS

Psychological research is not limited to the study of humans. Psychologists study other animals in part simply because they demonstrate interesting behaviors: For example Patterson (1978) showed that a gorilla will use American Sign Language to tell lies. Researchers also study animals to test models of behavior that then generalize to all

TABLE 4.2 Checklist of Issues to Consider When Designing an Experiment

Create valid and reliable independent variable
- Minimize confounding
- Specify type and number of conditions
- Include manipulation check

Create valid and reliable dependent variable
- Establish scoring criteria
- Create a sensitive procedure
- Avoid restricted range and floor/ceiling effects
- Include practice trials
- Observe multiple trials
- Counterbalance order of trials

Maintain consistency of testing procedure
- Create clear instructions
- Employ automation but limit instrumentation effects
- Consider group testing
- Conduct a pilot study

Control demand characteristics
- Minimize reactivity, social desirability, experimenter expectancies
- Use experimental realism, deception, double blinds, habituation, unobtrusive measures, field experiments

species, including humans. For example, much of what we know about basic brain functioning is based on models that originated from animal research. A common design is to surgically alter an area of the brain and then determine how a behavior is different relative to the behavior in unaltered animals. For example, researchers have observed the eating behavior of white rats in which various parts of the hypothalamus had been surgically damaged (e.g., Tokunaga, Fukushima, Kemnitz & Bray, 1986).

Researchers also test animal models without using surgical procedures. For example, Ebert and Hyde (1976) selected mice that exhibited high or low aggression. The level of aggression in subsequent offspring indicated that this trait is genetically transmitted. Similarly, animal research has been the basis for many developments in learning and conditioning: Ivan Pavlov's principles of classical conditioning were based on the behavior of dogs, and B. F. Skinner's work on operant conditioning was based on the behavior of rats and pigeons.

Though some people are incensed by the comparisons, animal research often has substantial external validity, generalizing well to many aspects of human behavior, such as education, clinical therapy, and the workplace. Humans are animals too, and some laws of nature apply to all animals in the same ways. For example, a hypothalamus is a hypothalamus, and the model of how a rat's hypothalamus influences eating behavior has generalized well to humans (see Schachter, 1971, for an entertaining review). Likewise, animal research is often the first step in the development of a new drug or physical treatment. When the treatment works with animals, it often works with humans too.

Controls Used with Animal Research

Many animal studies are true experiments conducted in a laboratory: We obtain a random sample of animals (sometimes trapped in the wild but usually purchased from commercial suppliers), randomly assign them to conditions, and apply all of the previous controls for reliably and validly manipulating the independent variable. For internal validity, we keep constant the extraneous variables that might produce a confounding, so we maintain the environment consistently for all subjects, test subjects in the same manner, and so on. Likewise, for a valid and reliable dependent variable, we define the scoring criteria, provide practice trials, observe multiple trials and counterbalance order effects.

Believe it or not, experimenter expectancies and demand characteristics are a problem even in animal studies. A researcher can inadvertently make errors in measuring or recording scores that are biased toward confirming the research hypothesis. And a researcher's expectations can produce subtle differences in the way that animals are handled and tested, biasing their behavior so that they confirm the hypothesis. These problems even occur when dealing with something as simple as a rat learning a response (Rosenthal, 1976). Researchers minimize such biases by handling all animals in the same way, automating data collection, and employing double-blinds.

Further, control groups are often used to identify the influences of merely handling and testing the animals. For example, when testing a drug or surgical procedures, control animals are injected with a placebo or undergo the anesthesia and surgery without receiving the actual treatment. As a result, they experience the same trauma that experimental animals experience.

> *REMEMBER* In research involving animals, we control extraneous variables in the same ways as in research involving humans.

As we'll see in the final sections of this chapter, whether studying animals or humans, every experiment must be conducted in an ethical manner.

RESEARCH ETHICS

We've seen that psychological research can be rather tricky and devious. As such, it gives rise to a dilemma: On the one hand, we want a well-controlled study, even though we may need to be deceptive, or to elicit responses that participants want to keep private or that cause them discomfort. We justify these actions on the grounds that scientific knowledge benefits humanity. On the other hand, we should treat participants properly because they have basic rights to privacy, to respect, and to safety. Therefore, the issue of **research ethics** can be summed up as a concern for balancing a researcher's right to study a behavior with the right of participants to be protected from abuse.

The Cooperativeness of Participants

Why do people allow themselves to be mistreated by a researcher? First, researchers are often viewed as authority figures and, as with all authority figures, people tend to think that they are benevolent and honest. Second, participants assume that research is valuable for society, so they believe their participation is important. The result is that participants respect the goals of research and trust the researcher and so they are open to abuse.

An example of how motivated participants are was demonstrated by Orne (1962) who tried to give subjects a task they would refuse. He gave each person 2,000 sheets of paper, each of which contained 224 addition problems. No justification for the task was given, and subjects were merely told that the researcher would return "sometime." Five and one-half hours later, the participants were still working and the experimenter gave up! In the next attempt, participants were told that, after completing each sheet, they should tear it up into a minimum of 32 pieces and then continue with the next sheet. Participants performed this task for several hours until, again, the experimenter gave up. Subjects later reported that they viewed this meaningless task as some sort of important psychological endurance test.

Thus, an overriding demand characteristic in any study is for participants to be cooperative. In fact, people cooperate even to their own detriment. The classic example of this is Milgram (1963), who convinced participants that they were assisting him to train a "learner" confederate to learn verbal stimuli. Each time the learner made an error, the participants pressed a switch that he or she believed administered an increasingly larger electrical shock to the learner. Despite protests from the learner (who could be heard but not seen, and who eventually emitted deathly silence), and despite the fact that the electrical switches were labeled "DANGER: SEVERE SHOCK," a majority of participants delivered what they believed was as much as 450 volts of electricity! (The electrical outlets in your home deliver at least 110 volts.)

Milgram applied no coercion other than to tell participants to continue, using only the authority that they had implicitly given him. Yet they complied, even though they believed they were harming another person, and in many cases became *very* emotionally and physically distressed themselves. Further, these were not young, impressionable freshman college students, but adults of various ages and backgrounds!

Milgram was soundly criticized for his tactics, but there are many studies suggesting that a researcher can probably get participants to do almost anything: They are willing to suffer great mental or physical discomfort, and they are reluctant to protest or to protect themselves. However, merely because people volunteer for a study does not mean that we have the right to take advantage of them. In essence, there is an implicit contract between participants and researchers. Their side of the contract is to help in our study and to trust us. Our side of the contract is to not abuse their trust. Being ethical means living up to our part of the contract.

> ***REMEMBER*** We must treat participants ethically, because they are helping us and trust us not to harm them.

The APA Principles of Ethical Conduct

To assist researchers when dealing with ethical issues, the American Psychological Association adopted the Ethical Principles of Psychologists and Code of Conduct (1992). These principles govern the full range of a psychologist's activities. In particular, they deal with the care of human and nonhuman research participants, and apply to any type of study (not just experiments). These principles can be summarized as follows.

Identify potential risks First identify potential *physical* risks: Is there anything in the study that could physically endanger subjects? Are we manipulating a potentially dangerous independent variable or could measuring the dependent variable be harmful?

Is all equipment working properly and safely, and will presenting stimuli cause pain or physical damage? Are we adhering to all accepted procedures when injecting drugs, drawing blood, and the like?

Second, identify potential *psychological* risks: Will participants experience undue anxiety, depression, or other unpleasantness, because we are invading their privacy, producing negative emotions, or lowering their self-esteem? Distress can occur directly as the result of a manipulation, as when we intentionally cause people to become sad or depressed. Distress may also result indirectly from a procedure. For example, you might think Milgram's study was not so unethical, because no one actually got hurt. But if Milgram had not disconnected the electrical wires, the participants would have killed—murdered—the learner! Think about how they felt when *that* dawned on them.

When identifying potential risks, consider not only the particular procedure, but also the participants. Having healthy teenagers perform strenuous exercise may not be risky, but it is risky for the elderly and for people with heart conditions. Likewise, films containing sex, violence, and mayhem are standard fare for some adult moviegoers, but they may be upsetting for others, or for children.

Dealing with deception is a particularly difficult issue. On the one hand, Christensen (1988) found that among people who had participated in deceptive and nondeceptive experiments, those in the deceptive studies enjoyed the experience more, became better educated about psychological research, and did not mind being deceived. On the other hand, this does not mean we can freely deceive participants on a whim. Deception may be harmful because, when subjects learn of it, they may feel foolish, depressed, or angry. Therefore, the APA guidelines explicitly require that deception be used *only* when it is a necessary component of a design.

Thus, in evaluating the ethics of deception, first consider whether it is necessary for producing the desired study. (It is doubtful, for example, that Milgram could have elicited such extreme obedience without employing deception.) Second, consider the amount of deception involved. The greater the deception—the bigger the lie—the more objectionable it is. Finally, and most important, consider how severe the psychological impact of the deception will be. The extreme emotional reactions resulting from Milgram's deception are a serious ethical concern, while the more minor reactions arising from the deception in our smile study are much less objectionable.

Protect subjects from physical and psychological harm Once the potential risks in a study are identified, try to eliminate or at least minimize them. If participants will be stressed or embarrassed, can we alter the design to eliminate these feelings and still get at the intended behaviors? Can we "tone down" a manipulation so that it affects participants but still is not too extreme? Thus, if we want to make one group more depressed than another, we don't need to make the first group suicidal! In fact, we might consider whether the study will work if, instead, we make one group happier than another. We can also minimize risks by screening out high-risk individuals. If the study involves exercise, for example, we screen out people having a heart condition. If we're manipulating depression, we screen out people who are already clinically depressed.

One important rule is to minimize participants' anxiety by keeping all information about them confidential. They are never identified in publications or discussed in casual conversations. (Often we assign participants a number instead of recording their names so they remain anonymous.)

Justify remaining risks For any risks to participants that remain in a design, fairly and honestly decide whether they are justified by the study's scientific worth. That is, the knowledge to be gained from a study must clearly and convincingly justify the risk to subjects. Recognize that as the potential for physical or psychological harm becomes greater, the scientific justification for the study becomes less convincing. As we've seen, one study never definitively "proves" a hypothesis. Therefore, conducting an unethical or dangerous study is just not worth it.

> *REMEMBER* The primary ethical concern in any study is to minimize potential harm to participants and to ensure that any remaining risk is justified.

Ethical issues are not always clear-cut, so seek the advice of others. In fact, to conform to the APA's Ethical Principles (as well as to federal and state regulations), colleges and research institutions maintain a **Human Subjects Review Committee** (also called an Institutional Review Board, or IRB). This committee consists of individuals from various disciplines, so a broad perspective is represented. The committee's job is to review a study's procedures to ensure the ethical treatment of participants. All researchers must obtain approval from their IRB *before* conducting any study.

One thing a review committee always looks at is how you'll deal with the issue of informed consent.

Obtain informed consent Out of respect for the participants' right to control what happens to them, inform them about the study *prior* to their participation, and then let them decide whether they wish to participate. That is, obtain **informed consent**. The APA's Ethical Principles state that informed consent is required unless there is minimal risk, such as when we merely observe anonymous people in a field setting. But always obtain informed consent when conducting laboratory experiments. The usual procedure is to provide participants with a written description that contains four components.

First, describe the purpose and procedures of the study. Sometimes you must withhold specific details because of demand characteristics, but usually you can provide general information without biasing subjects. Tell them as much as you can.

Second, explicitly warn participants of any physical or psychological risks associated with the procedure, describing any details that could reasonably be expected to influence a subject's decision to participate in the study. Even if you cannot divulge all aspects of a procedure, you must warn participants of any negative consequences of it. (For example, Milgram should have at least warned his subjects that they might learn some unpleasant things about themselves.)

Third, inform participants that they are free to discontinue their participation at any time during the study, *without penalty*. Volunteering for a study produces such strong demand characteristics to be cooperative that this option *does not* occur to people. Be sure that withdrawing is a realistic option, without any hidden coercion because participants are enrolled in a class or have a job where the study is being conducted.

Finally, obtain participants' signatures as their explicit consent to participate. In the case of minors and others who are not capable of making this decision, obtain consent from their parents or guardians.

Take care of participants after the study Finally, after a study, any negative consequences of the procedure must be removed so that subjects feel as good about themselves as they did when they first entered the study. First, we alleviate any concerns they have by providing a **debriefing**. That is, we fully inform participants about all aspects of the study, including about the manipulation and any deception used and why. Second, we remove any adverse physical or emotional reactions in participants that we may have created. Thus, if we tested the effects of alcohol, we care for subjects until they are sober. If we created anxiety or depression, we try to reverse these feelings, explaining why we think they are normal reactions to our manipulation. If follow-up counseling or check-ups may be needed, we provide qualified, professional help. Finally, we give participants a means of contacting us later in case unforeseen problems arise.

> ***REMEMBER*** Always obtain informed consent, debrief subjects, and care for them after the study.

Regardless of the specific design of a study, we approach the ethical issues in the same manner. To help you remember them all, they are listed in Table 4.3.

The Ethics of Unobtrusive Measures and Field Experiments

Ethical issues get particularly complicated with unobtrusive and deceptive procedures, especially in terms of whether we're violating a participant's right to informed consent. In a laboratory setting, however, participants are aware that they will be observed even if they cannot see the observer, so using one-way mirrors and such is usually acceptable. Likewise, in a waiting-room situation, participants have given their tacit agreement to be observed by showing up for the study. However, if an unobtrusive procedure might embarrass, victimize, or otherwise harm a person, then explicit prior informed consent is needed.

When subjects know they are participating in a field experiment, we deal with informed consent and debriefing as described above. However, the most difficult ethical situation arises with unobtrusive or hidden field experiments. After all, they involve the ultimate deception, because participants are not even aware a study is being conducted! Therefore, they have not formally volunteered, nor have they been given a chance to provide informed consent. As usual, our approach is to first minimize the risks and then be sure the remaining risks are scientifically justified.

The issue of risk is especially important in field experiments, because they allow us to manipulate all sorts of real-life situations. However, we are not free to abuse the unsuspecting public in the name of "science." (We don't have the right to yell "fire" in

TABLE 4.3 Checklist of Ethical Issues to Consider When Designing an Experiment

- Identify physical risks to participants
- Identify psychological risks to participants
- Eliminate or minimize risks
- Justify risks in terms of scientific value of results
- Obtain informed consent
- Provide debriefing and care for subjects after the study
- Submit planned procedure to IRB for review

a crowded theater just to see what happens!) In a laboratory setting, informed consent and the lack of experimental realism are protection for participants: Because they have volunteered to experience our artificial situation, it has less of a real impact on them. In field experiments, however, this is not the case. Our deceptions and pranks can cause people to become really frightened, really angry, or really dangerous! Therefore, researchers have an even greater responsibility to respect and protect participants. In short, there are limits to our right to conduct field studies that impose upon others. And, as usual, after resolving these ethical issues for ourselves, we obtain approval from the Human Subjects Review Committee.

Role Playing and Simulations

One possible solution when a laboratory or field procedure is just too risky is to have people simulate being in the experiment through **role playing**: Participants pretend they are in a particular situation and we either observe their behavior or have them describe how they would behave. Because the situation is not real, physical or psychological harm is unlikely. However, caution must still be exercised. For example, Haney, Banks, and Zimbardo (1973) created a prison simulation that unexpectedly turned sinister: College men pretending to be guards or prisoners exhibited the worst, most dangerous behaviors associated with a real prison!

In practice, role playing is used infrequently because it produces very limited validity and reliability. First, demand characteristics can run rampant: Participants may alter their reactions or descriptions to conform to perceived expectations or to keep their real behaviors private. (Would subjects simulating the Milgram, 1963, electric-shock study actually admit that they'd electrocute someone?) Second, subjects often cannot accurately predict how they would respond. For example, in studies of personal space, participants have given verbal descriptions, manipulated dolls, or drawn lines on paper to indicate how far they would stand from someone else. Yet such predictions seldom match the person's actual behavior when observed under real conditions (Hayduk, 1983).

Ethics and Animal Research

Scientists and the general public continue to debate the ethics of conducting laboratory experiments with animals. It is true that such research often exposes animals to unpleasant and harmful manipulations, such as surgical procedures, electric shock, or food or water deprivation. Further, the way to conduct a manipulation check of a surgical procedure is to perform an autopsy. And, even with nonsurgical procedures, animals may be physically or psychologically altered by the treatments, so they usually cannot be studied again and are destroyed.

On the one hand, "animal rights" advocates say that animal experiments are unethical because they violate the rights of animals to live free and unharmed. They argue that even though humans have the ability to exploit other animals, they do not have the right to do so. Some make the more radical argument that laboratory studies of animals do not even provide useful information, so there is no justification for what is seen as animal abuse. From these perspectives, the only ethical way to study animals is through descriptive studies conducted in natural settings.

On the other hand, animal researchers argue that such experiments are justified by the knowledge they produce: The previous criticism is wrong in that animal research has most definitely added substantially to the well-being of humans (and other animals) in important ways. Animal research has been the basis for virtually all modern drugs and surgical techniques, for the identification of numerous toxins and carcinogens, and for many psychological principles. From this perspective, it would be unethical if researchers did not conduct animal research to benefit society. Thus, the argument here is that researchers have the right—and the responsibility—to pursue any useful scientific information.

This issue boils down to whether you think the goal of benefitting humans takes precedence over the rights of other animals. If you think it does, then animal research is justified, because there is no other way to obtain the data. It would be more unethical to perform experimental surgical or medical procedures on humans: We cannot undo surgical alterations, and, when first testing a drug, we may have no idea of the harmful side-effects that can occur. Likewise, we cannot control the breeding practices of humans in order to study genetics, nor can we administer to humans many of the aversive conditions that have led to important discoveries with animals.

In addition, there are scientific and practical reasons for conducting animal laboratory studies. Descriptive research limits the variables and controls we can employ, so this approach is an inadequate substitute. Also, laboratory research with animals can be conducted quickly and efficiently: Animals are easily obtained and housed, their environment can be controlled and manipulated easily, and, for genetic studies, they have a short gestation period.

Regardless of where your personal feelings fall in this debate, it is wrong to think of animal research as involving the mindless torture of abused animals. As with all people, some researchers are less than ethical and may mistreat their animals. However, for the vast majority of researchers, laboratory animals are valued subjects in whom much time, energy, and expense have been invested. It is in the researchers' interest to treat them well, because animals who are abused make poor subjects for a reliable and valid study. Furthermore, the APA's Ethical Principles (1992) provide guidelines for the treatment of research animals, and there are federal, state, and local regulations for the housing and care of animals as well. Because of such rules, animals are well cared for, undergo surgery in sterile settings with anesthesia, and are disposed of in a humane manner.

Finally, APA's ethical guidelines require that we evaluate animal research in the same way we do human research. First, the harm caused to an animal must be minimized. Thus, we prefer designs that provide positive events as opposed to aversive events, we prefer mildly aversive events to drastic ones, and we prefer temporary physical alterations to permanent surgical ones. Second, we are not frivolous in the treatment of animals, so every aspect of a procedure must be necessary. As usual, the key issue is whether the procedure is justified by the scientific importance of the information that may be learned. Finally, every research institution must have an institutional review board that ensures the ethical treatment of animals.

REMEMBER Acceptable animal research minimizes the harm done to subjects and must be justified as scientifically important.

Scientific Fraud

There is one more aspect of ethics to consider. Unfortunately, one reason that scientists must always be skeptical about research findings is that other scientists are sometimes guilty of fraud. They may report data from a study inaccurately or they may publish data when no research was conducted. Often their motivation is to provide further support for their previous conclusions: They needed a replication that failed to materialize, so they "faked" the data. At other times, they are responding to professional pressures to be productive researchers.

It is unethical to perpetrate any form of scientific fraud. This includes falsifying results, as well as keeping secret a result that contradicts one's views. Such fraud not only violates every rule of science, but causes enormous harm: Given the extent to which researchers share and integrate research findings, a fraudulent finding can undermine many areas of psychological knowledge. There is no justification for research fraud.

PUTTING IT ALL TOGETHER

With this chapter, you're a little more familiar with the issues in designing a good study (actually, you're a lot more familiar). However, don't develop a false sense of security about the "proof" that a controlled experiment provides. We are still only collecting evidence. The techniques we've discussed are necessary for producing "good" convincing evidence.

CHAPTER SUMMARY

1. A *manipulation check* is used to check that participants were influenced as intended.
2. *Scoring criteria* define the system for assigning a score to a response. They should produce a *sensitive measure*, producing different scores for small differences in behavior.
3. *Restriction of range* occurs when the range of possible scores on a variable is limited. With *ceiling effects*, all scores tend to be high, so scores cannot get much higher. With *floor effects*, all scores tend to be low, so scores cannot get much lower.
4. Reliability is improved with *practice trials* and by observing participants on *multiple trials* within each condition.
5. *Order effects* are the influence on a particular trial due to its position in a sequence of trials. *Practice effects* are the influence due to practicing a task, and *carry-over effects* are the influence that a trial has on performance of subsequent trials. A *response set* is a bias in responding because of previous responses made.
6. *Counterbalancing* controls for order effects by presenting different orders of trials within each condition.

7. The key to designing a convincing experiment is to build in *consistency*, both *within* and *between* conditions. Consider *automation*, *instrumentation effects* (changes in equipment and materials that occur through use), *instructions*, *group testing*, and *pilot studies*.
8. *Demand characteristics* are extraneous cues that guide or bias a participant's behavior.
9. *Reactivity* is the demand characteristic occurring because participants are aware that they are being observed.
10. The *Hawthorne effect* is the influence due to the novelty of being in a study.
11. *Social desirability* is the demand characteristic occurring because participants want to behave in a socially acceptable manner.
12. *Experimenter expectancies* is the demand characteristic from cues the researcher provides about the responses participants should give.
13. Demand characteristics are reduced through *habituation*, *unobtrusive measures*, *deception*, and *unobtrusive field experiments*.
14. With *experimental realism* participants are engaged by the task and thus are less concerned with demand characteristics.
15. A *placebo* provides the demand characteristics of a treatment.
16. In a *single-blind* procedure, participants are unaware of the nature of the treatment. In a *double-blind* procedure, the researcher who tests participants and the participants are unaware of the nature of the treatment.
17. *Research ethics* deal with balancing the right of a researcher to study a behavior with the right of participants to be protected from abuse.
18. The *APA's Ethical Principles of Psychologists and Code of Conduct* require that animal and human participants be protected from physical or psychological harm and that potential harm is scientifically justified.
19. All research is reviewed by the appropriate institutional review committee.
20. Researchers should obtain *informed consent*, should *debrief* participants, and should take care of them after the study.
21. In *role playing*, participants pretend they are in a particular situation.

PRACTICE PROBLEMS

(Answers for odd-numbered problems are provided in Appendix D.)

1. (a) What is a manipulation check? (b) What is the difference between a pilot study and a manipulation check?
2. (a) Why do we include practice trials? (b) Why are multiple trials generally better than a single observation? (c) What problem arises with multiple trials?

3. (a) What are practice effects? (b) What are carry-over effects? (c) What is a response set?
4. What is counterbalancing?
5. What is a sensitive measurement procedure?
6. What are the pros and cons of testing participants in groups?
7. (a) What do we mean by automation? (b) Why can automation be good for a study? (c) Why can automation be bad for a study?
8. (a) What do we mean by demand characteristics? (b) In terms of reliability, internal validity, and external validity, how do demand characteristics harm a study?
9. (a) What is experimental realism and why do we seek it? (b) How is experimental realism different from ecological validity?
10. (a) What are unobtrusive measures? (b) Why do researchers employ unobtrusive measures and/or deception?
11. What is meant by research ethics?
12. What are the three major issues about risks you must resolve to *design* an ethical study?
13. What are the two major steps you must include to *conduct* an ethical study?
14. (a) Why is informed consent needed? (b) What four things must you do with participants to obtain "informed consent?"
15. (a) What is role playing? (b) What is the advantage of this approach? (c) What is the disadvantage of this approach?
16. In the smile study discussed in this chapter, should participants in a condition be tested individually or in groups? Why?
17. In a study of memory, you read out loud one list of either similar or dissimilar words and then measure participants' memory for the list. (a) What problems with reliability may occur because you read the lists? (b) In terms of demand characteristics, what is the problem with you reading the lists? (c) How can you eliminate the problems in (a) and (b) above? (d) What problem may then arise over the course of testing many subjects?
18. In question 17, you ask subjects to write down the list of words. (a) To score recall reliably, what decisions should you make? (b) How could you ensure that the person scoring the responses was unbiased?
19. In question 17, (a) What problem affects the reliability of the memory *behavior* you are observing? (b) What preliminary task can you add to the design to improve reliability? (c) How can you expand your observations of each subject to improve reliability? (d) What problem have you created? (e) Precisely describe how you would deal with the problem.
20. You wish to test the proposal that women become more sexually aroused by erotic films as a function of whether the plot has a weak or strong theme of love and romance. After showing subjects one type of film, you measure the dependent variable using a questionnaire about their arousal. (a) What demand characteristics are a major problem? (b) How would you reduce these demand characteristics? (c) What ethical problems might arise with this study?
21. (a) In problem 20, why should you test each participant in a condition with more than one film? (b) What possible bias must you then eliminate? (c) Describe how you would accomplish this.
22. (a) What are two major criticisms of laboratory animal research? (b) How would you answer these criticisms?
23. When conducting a study, a researcher wears a white lab coat and carries a clipboard and stopwatch. How might these details influence the internal and external validity of the study?
24. Consider the hypothesis that greater exposure to violence on television results in more aggressive behavior. (a) For ethical reasons, what may be the best design for determining whether this relationship exists? (b) What is the trade-off involved in being ethical?

25. You deliver different speeches to your participants to make them more or less sexist. Then you examine how sexism influences whether they help a confederate of the opposite sex. (a) Why could your manipulation appear to work even though it does not really alter subjects' views? (b) Why could your manipulation appear not to have worked although it really did alter subject's views? (c) What can you do to check that your speech altered sexism?

26. In the previous chapter, we discussed having students study for different amounts of time before taking a statistics exam in their research methods course. Ethically speaking, what's wrong with this design?

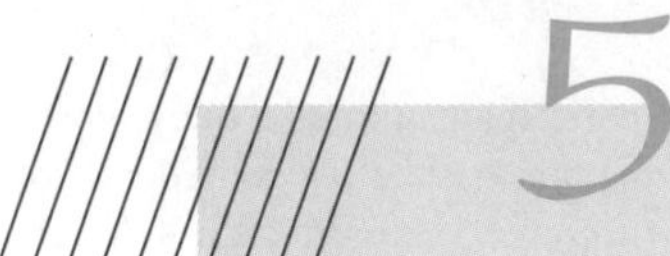

5 Design Issues and Ethical Concerns in Descriptive Research

To understand this chapter:

- From Chapter 2, be sure you understand the idea of a representative sample.
- From Chapter 3, understand the nature of descriptive research.
- From Chapter 4, understand how to deal with threats to reliability and validity, especially counterbalancing order effects.

Then your goals in this chapter are to learn:

- The different approaches to observational research.
- How field surveys are conducted.
- The different sampling procedures researchers use.
- How to design questionnaires and interviews.
- The ethical issues in descriptive research.

This chapter addresses the basics of descriptive research. In it, we discuss procedures for observing participants and for asking them questions through questionnaires and interviews, as well as different sampling techniques and, as usual, ethical concerns. Although the focus here is description, recognize that many of these procedures also can be used in experiments.

THE USES OF DESCRIPTIVE RESEARCH

As you know, the term **descriptive research** refers to the observation and description of a behavior, the situation it occurs in, or the individuals exhibiting it. We've seen that one type of descriptive procedure is a correlational design. This is a descriptive study where we examine the mathematical relationship between two or more variables. (A relationship between variables is essentially a statistical issue discussed in Chapter 10.) Other types of descriptive studies may not look for a specific relationship and therefore may measure only one variable or involve only one participant. The goal here is simply to describe a certain behavior or type of individual. Therefore, descriptive research is usually conducted as field research in order to minimize demand characteristics and increase external validity.

On the one hand, the disadvantage of such descriptions is that we examine a behavior without much control, so there is great potential for confoundings, and we may miss hidden influences or misinterpret what we see. Therefore, with descriptive research we can only speculate on the causes of a behavior. On the other hand, there are three advantages of descriptive research:

1. Descriptions are informative. For example, we might describe the mating rituals of frogs in the wild or the actions of drivers in a large city, because these are interesting behaviors. For applied research, we may describe consumer attitudes or the behaviors of drug addicts.
2. Descriptions are the starting point for identifying variables and building hypothetical constructs that can be tested later using other methods. For example, much research in clinical psychology developed from the constructs of Sigmund Freud, even though he only observed and described behaviors. Likewise, descriptions can provide an indirect test of a theory or model. (A researcher might ask, for example, "Are the predictions from a Freudian model confirmed by a description of a schizophrenic?")
3. Description is sometimes the only way to study a behavior or situation, because it is either practically or ethically impossible to produce it in an experiment. For example, the only way to learn about the migratory behaviors of whales or the childhood experiences of a serial killer is by observing and describing them.

Thus, descriptive research is a legitimate approach, even though we cannot use it with confidence to infer the causes of behavior. In addition to correlational designs, descriptive research falls into two general categories: Observational studies and field surveys.

OBSERVATIONAL STUDIES

In **observational research**, we do not directly ask participants to respond. Rather, we observe them in an unobtrusive manner. There are three general observational methods.

In **naturalistic observation**, the researcher observes a wide range of behaviors in an unobtrusive manner. Naturalistic observation usually implies a rather unstructured approach in which we have not identified a specific, limited behavior or situation to study. To be unobtrusive, we may use hidden cameras, observe from camouflaged hiding places, or simply blend in with the crowd in a public place. If unobtrusive techniques are not possible, then at a minimum we habituate participants to our presence before beginning the study. For example, the famous studies performed by Jane

Goodall (1986, 1990) involved naturalistic observation of chimpanzees in which she observed their general lifestyles in the wild.

More commonly, researchers perform **systematic naturalistic observation**. We are again unobtrusive, but here we identify a particular behavior to observe, and are more "systematic" in our observations. For example, Heslin and Boss (1980) observed people being met at an airport, categorizing their various nonverbal actions, or Boesch-Acherman and Boesch (1993) observed the use of natural tools by wild chimpanzees.

When the behavior of interest involves private interactions among members of a group, we may perform **participant observation**. Here the researcher is a "participating" member of the group being observed. Usually, the researcher's activities are "disguised" or hidden from subjects. In a classic example, Rosenhan (1973) arranged for "normal" people to be admitted to a psychiatric hospital to observe how patients were treated. Less frequently, the participant observer is "undisguised."

> *REMEMBER* The basic approaches to observational research are naturalistic observation, systematic naturalistic observation, and participant observation.

The Pros and Cons of Observational Designs

The overriding advantage of observational designs is that, through unobtrusive observation, we are describing natural behaviors that are not influenced by reactivity or other demand characteristics. There are, however, several disadvantages:

1. Descriptions are highly susceptible to experimenter expectations, so that we may see only what we expect to see. Further, with participant observation, the researcher may inadvertently cause participants to behave in the expected way.
2. We may be very limited in the individuals that we can find to observe and so we may have an unrepresentative sample.
3. We often have only our verbal descriptions of a behavior. Such *qualitative* data may lack precision and accuracy, and they are not very sensitive for identifying subtle differences in behavior.

To minimize the above problems, we can try to build in more control. For example, we try to develop specific quantitative scoring criteria, looking for certain movements, speech patterns, facial expressions, and so on. Then we may count the frequency of each behavior or measure its duration, or determine whether it occurs during a given time period. To facilitate accurate data recording and to minimize the time spent not observing, we may produce structured scoring sheets so that we can simply check off categories of behavior, or we may automate by tape-recording subjects for later scoring.

An especially important control is to eliminate experimenter biases through the use of multiple raters.

The Problem of Multiple Raters and Inter-rater Reliability

Special reliability concerns occur when, in descriptive or experimental designs, a score is based on a researcher's *judgment* about a participant's behavior. Because of experimenter expectancies, a researcher may not produce objective, valid, and reliable judgments. Therefore, we enlist the aid of others, called *raters*, who are usually "blind" to the hypothesis and are trained (often in a pilot study) to use our scoring criteria. For example, in research into whether chimpanzees can learn American Sign Language, the researcher might cue the chimp as to the correct sign or erroneously give the chimp credit for a sign because it is "close enough" to the desired sign. To eliminate these biases, one researcher shows the object, and another "rater," who cannot see the object, observes and records the chimp's sign. Raters are also common in studies that involve scoring such behaviors as creativity or aggressiveness.

Of course, a rater may not reliably score a behavior. The rater might miss or forget part of a subject's action, or, over the course of a study, become more attuned and sensitive, or become fatigued and less motivated. Also, one rater might judge a behavior differently from another.

Any rater may introduce such errors, and we usually cannot eliminate them. But, recall that if an extraneous influence cannot be eliminated, we balance it out. The approach with raters is to employ **multiple raters**, having more than one rater judge each participant. Then we combine their ratings, usually computing the average rating per subject. This average rating should balance out the biases of individual raters, giving us a more reliable measure of each subject's behavior. Further, multiple raters form a sample of observers from which we can infer that any observer would judge the behavior in roughly the same way, so we can generalize results with greater confidence.

To be convinced that raters are consistent, we determine their inter-rater reliability. **Inter-rater reliability** is the extent to which raters agree in the scores they assign to a participant's behavior. First, to test that we'll have high inter-rater reliability we may conduct a pilot study. Then the inter-rater reliability is usually determined in the actual, completed study. Inter-rater reliability may be computed as the percentage of agreements between raters—for example, the percentage of times that two raters agreed on the sign a chimp produced. (Better than 90% agreement is usually considered reliable.) Also, as we'll see in Chapter 10, there are statistics for describing inter-rater reliability.

> ***REMEMBER*** Using multiple raters who have high inter-rater reliability is a major concern in **any** design where data are based on subjectively evaluating participants' responses.

Additional Sources of Data in Descriptive Approaches

The term observational research usually implies that a researcher was physically present to observe participants. The research literature, however, contains three common terms that indicate that other procedures were employed: Archival research, *ex post facto* research, and case studies.

Archival research In **archival research** the source of the data is written records. Typically, the records come from schools, hospitals, government agencies, or police. For example, Faustman and White (1989) examined medical and psychiatric records to describe stress disorders found in some war veterans. Archival research is also used to describe social trends and events, as in Connors and Alpher's (1989) analysis of alcohol-related themes in country-western music. (They concluded that alcohol is usually used for drowning one's sorrows in such songs.)

The advantage of archival research is that it allows access to behaviors that would otherwise be unobservable. It also allows us to verify participants' self-reports (e.g., we can compare actual college-grade records to participants' reports). Archival research also presents some disadvantages, however. First, obtaining access to records may be very difficult. Second, the accuracy of the data depends entirely on the people who created the records. Usually, the records are not made with a researcher's question in mind, so they may not address our variables. And they may contain verbose, indirect descriptions, requiring much subjective interpretation on our part. Often, too, there are few controls in place to prevent the inclusion of the record-keeper's personal biases or errors. Thus, we may have considerably less confidence in archival data than in data derived from our own direct observations.

***Ex post facto* research** In ***ex post facto* research**, the study is conducted after the events of interest have occurred (*ex post facto* means "after the fact"). Usually, this research involves examining archival records. For example, in an *ex post facto* (correlational) study, Anderson and Anderson (1984) examined police records and meteorological records to determine whether the number of criminal assaults increased with increased daily temperature. Less commonly, *ex post facto* research may involve having participants report a past event, such as completing a questionnaire regarding their stress levels before and after an earthquake.

The problem with *ex post facto* designs is again that we obtain potentially unreliable data. Often, we cannot precisely quantify the variables and events that occurred nor ensure that they reliably occurred for all participants. Likewise, written records are only as accurate as the people keeping them, and subjects' self-reports may be biased and erroneous.

Case studies A **case study** is an in-depth study of one situation or "case." Usually, the case is a person, and such studies are frequently found in clinical research, providing an in-depth description of a particular patient's clinical symptoms and reactions to therapy (e.g., Martorano, 1991; Stagray & Truitt, 1992). However, a case may also involve a specific event or organization, as in Anderson's (1983) study of the government's decision-making procedures during the "case" of the Cuban missile crisis. The advantage of case studies is that they provide an in-depth description of an individual or event. The disadvantage is that they may yield data with poor reliability and validity. Further, their generalizability may be poor because any one case probably does not typify other cases.

FIELD SURVEYS

Another common approach in descriptive research is the field survey. In a **field survey**, people complete a questionnaire or interview in a natural setting so that we may infer the responses we would see if we could poll the population. Field surveys may apply to a narrow population, as when, for example, surveying nurses treating AIDS patients to gauge their burn-out rates (George, Reed, Ballard, Colin, & Fielding, 1993). They are also used to describe the attitudes of the general population, as when, for example, surveying the public's reactions to crime (Harrison & Gfroerer, 1992), or describing consumer attitudes (Fornell, 1992). Field surveys may also describe common experiences: The Roper Organization (1992) polled Americans to determine how often they experience ghosts and UFOs (more than 30% reported they had!).

Administering a Survey

Developing reliable and valid survey questions is of paramount importance, and, as we'll see later in this chapter, requires considerable work. In addition to asking good questions, however, the key to field surveys is obtaining a representative sample. Therefore, we typically survey a large sample involving hundreds of participants. The survey may be conducted at a shopping mall or other public place, because members of the target population frequent that location. If we are unable to reach a representative sample at any one place, we can try to reach a broader segment of the population either by mailing the survey or by conducting it over the telephone.

A mailed survey is most useful when a large sample is needed and/or a lengthy questionnaire is being used—and when the researcher is not in a hurry to get the data. A telephone survey is typically shorter and can be completed quickly. Telephone surveys are conducted, for example, overnight during election campaigns to get the momentary "pulse" of the voters. The major difficulty with mailed or telephone surveys is in getting people to participate. Many will not bother mailing a survey back (even with a stamped, addressed return envelope), so the return rate may be only 10% or 20%. People may also refuse to participate in a telephone survey, especially because "psychological surveys" have frequently been used as a cover for obscene phone calls or telemarketing scams.

There are many techniques for improving the return rates of mailed surveys (see Kanuk & Berenson, 1975), and for enlisting subjects in a telephone survey (see Lavrakas, 1993). Essentially, the approach is to introduce the survey as an above-board, scientifically important, and professional project. Participants are informed about the purpose of the research, who the researchers are, and how they can be contacted. Then brief questions are asked that can be accurately and rapidly completed. Throughout, the goal is to convince participants of the legitimacy of the research and to engage their interest and cooperation.

The Flaws in Field Surveys

Recognize first that field surveys reflect how people feel at the time of the survey: They probably do *not* describe how people will feel later. Also, a survey can be biased if there are hidden criteria that eliminate certain types of participants. For example, telephoning during weekdays will reach only a certain type of person, or distributing surveys at a shopping mall in the evening will encounter only certain types, and thus produce a biased sample.

Another problem is that survey results will be prone to the volunteer bias. The **volunteer bias** is the bias that arises from the particular people who participate in a study. This is a problem in both descriptive and experimental studies, because there are considerable differences between people who volunteer for a study and those who do not (Rosenthal & Rosnow, 1975). Among other things, volunteers tend to have a higher social status and intelligence and to exhibit a greater need for social approval. An especially important factor is how strongly people feel about the issues being raised in a

TABLE 5.1 Summary of Terminology Used in Descriptive Research

Observational research:	Observation of participants by
Naturalistic observation:	Unobtrusive, rather unstructured observation
Systematic observation:	Unobtrusive but rather structured observations
Participant observation:	Unobtrusive, systematic observations, with researcher a member of group being observed
Archival research:	Any type of design in which data are collected from formal records
Ex post facto research:	Any type of design in which data are collected after events have occurred
Case study:	In-depth description of one individual, group, or event
Field survey:	Polling people in the field

survey. For example, a mailed survey about abortion in the United States is most likely to be completed by those who very strongly favor abortion or very strongly oppose it. This is the danger with radio and television call-in surveys: The callers are probably a biased sample, consisting of those people who are especially motivated to make the call. In field surveys, always consider whether the "silent majority" has been missed.

REMEMBER Field surveys can be influenced by low return rates, biased sampling, and the volunteer bias.

So that you can remember the previous terms used in different descriptive research, a summary is presented in Table 5.1 above.

TYPES OF SAMPLING TECHNIQUES

Recall that an aspect of external validity is how well our conclusions generalize to the broader population, so we are always concerned with obtaining a representative sample. Although external validity is an issue in experiments, we are especially concerned about it in descriptive field research. By taking the study to the participants, we can be more selective in defining the target population and in the techniques used to select participants.

Researchers often identify potential participants from governmental or commercial mailing lists, or from membership lists of social or civic groups. On the one hand, such lists should not include people from outside the target population. A survey of voter attitudes, for example, should include only people who vote, so we may use a list of people who voted in the last election (although this does not mean they will vote in the next election). On the other hand, we should not exclude any important segments of the population—a list of voters from the last election does not include people who are voting for the first time. Likewise, a survey about gun control should not include only registered gun owners—but it should not totally exclude them either. Thus, to obtain a representative sample, the goal is to include all of the important subgroups in the target population.

How we go about obtaining a sample is, in research lingo, referred to as our "sampling technique." There are two general types of sampling techniques: Probability sampling and nonprobability sampling.

Probability Sampling Techniques

Probability sampling relies on random sampling to select participants. The most common type of probability sampling is simple random sampling. **Simple random sampling** is the selection of participants so that all members of the population have an equal chance of being selected and all samples have an equal chance of being selected. Ideally, by giving everyone an equal chance, the type of subjects found in the population should occur in the same ways in the sample. One way to randomly sample is to use books containing tables of random numbers: We close our eyes and point to a number in the table, and then select the subject having the same identification number. We may also program a computer to select participants randomly.

Another probability technique is **systematic random sampling**, where we select every *n*th person from a list of the population. For example, after randomly selecting a starting point in the list, we might then select every third, or every tenth, name in the list. A variation of this approach used in field research is, for example, to survey every tenth person entering a shopping mall. We must be careful, however, that there is no hidden bias here: If, for example, people are listed by age, then we may fill the sample with young people before we get to older ones further down the line.

Both simple and systematic random sampling rely on chance, so we may *not* contact all types of individuals found in the population. To ensure that the various subgroups of the population are represented in a study, we may employ stratified sampling. In **stratified random sampling**, we randomly select from the important subgroups so that their representation in the sample is proportional to their representation in the population. For example, if government records reveal that 60% of the target population is male, then 60% of a sample should be male. If the sample will contain 100 people, from the pool of identified men, we randomly select 60. If someone declines to participate, we select another male to replace him. We may also combine selection criteria: For example, if 5% of the population are females who own guns, we randomly select 5 participants from the pool of female gun owners. Other "strata" that are typically included in field research are socio-economic level, race, and geographic location.

Sometimes it is too expensive or too difficult to contact the individuals in a population. In such cases, we might use an alternative technique called cluster sampling. In **cluster sampling** certain groups or "clusters" are randomly selected and then all members of each group are observed. To study homeless people, for example, we might randomly select a few areas in a city where homeless people are found and then study all subjects in each area. Similarly, to study workers at a large factory, we might randomly select a few departments and study the workers in each. Because the clusters are randomly selected, there should be no bias in subject selection, so the sample should be representative.

Nonprobability Sampling Techniques

Nonprobability sampling does not rely on random sampling, so every member of the population does not have an equal opportunity to be selected. Therefore, we are likely to miss certain types of subjects and obtain a less representative sample.

A common form of this approach is **convenience sampling**, in which we study participants who are conveniently available. Studying the students sitting in the student union or the people riding on a bus involves convenience samples. These are not random samples, because only those people who are present at the one place and time of the study have any chance of being selected, and the reason they are present is usually not a random event. Therefore, a convenience sample is representative of a very limited population.

To at least some extent, most "random" samples are also convenience samples: Given a researcher's limitations in terms of travel, time, and cost, there will always be some members of the population who have no chance of being selected. In reality, then, a truly random sample is an ideal that is seldom attained. In particular, the vast majority of psychological research—both descriptive and experimental—is based on convenience sampling of college-level, introductory psychology students. Therefore, participants are technically only representative of the population of introductory psychology students.

Another type of nonprobability sampling that may produce a more representative sample is **quota sampling**. As in stratified sampling, we ensure that the sample has the same percentage—the same "quota"—of each subgroup as in the population. Unlike stratified sampling, however, in quota sampling we do not randomly sample from each subgroup. Instead, we rely on convenience samples to fill each quota. For example, say that we want 20 six-year-olds and 20 seven-year-olds in our sample. If we obtain these subjects by testing a convenient class of first-graders and a convenient class of second-graders, we are using quota sampling.

Finally, sometimes we seek to study a "hidden" population, as when studying drug addicts or prostitutes. Then we may use **snowball sampling**. Here we identify one participant, and from him or her we identify other participants, and from them we identify others, so that the sample tends to build or "snowball." However, such a sample is probably not very representative, because only those people within our network of acquaintances have any chance of being selected, and they may be different from others in the population.

For a summary of the above sampling techniques, consult Table 5.2.

TABLE 5.2 Summary of Sampling Techniques

Probability Sampling	
Simple	Randomly select participants from the population.
Systematic	Select every *n*th individual from the population.
Stratified	Randomly select from subgroups, proportionate to each group's representation in population.
Cluster	Randomly select clusters and test all members per cluster.
Nonprobability Sampling	
Convenience	Select subjects who are conveniently available.
Quota	Obtain convenience samples to represent subgroups, proportionate to each group's representation in the population.
Snowball	Locate participants through other participants.

DESIGNING INTERVIEWS AND QUESTIONNAIRES

As we've seen, a common method of gathering data in descriptive research is to ask people questions using interviews and questionnaires. However, don't make the mistake of thinking that questionnaires and interviews are used solely in descriptive designs. Such self-reports are often used in experiments to measure the dependent variable and as a manipulation check.

When you design an interview or questionnaire, your first step should be to search the psychological literature for existing questionnaires and tests rather than creating your own questions. The advantage of using existing procedures is that their reliability and validity have already been established. Detailed descriptions of various psychological tests can be found in research reports, as well as in specific reference books (e.g., Robinson, Shaver & Wrightsman, 1991).

Creating your own questions involves a lot of work, and then you must demonstrate their reliability and validity. If you choose to develop your own questions, you have a number of decisions to make. As an example study we'll discuss, say that from industrial-organizational psychology you learn about the construct of "job satisfaction," the degree to which workers find working a satisfying experience. How would you go about measuring job satisfaction?

The first step is to consider the type of questions to ask.

Using Closed-Ended Versus Open-Ended Questions

In a **closed-ended question**, the researcher provides the alternatives from which a participant can select. Multiple-choice, true-false, yes-no, and rating-scale questions are all closed-ended questions. Some closed-ended questions that might reflect job satisfaction are:

(a) At work, my favorite activity is
1. Working with my hands.
2. Solving mental problems.
3. Supervising others.
4. Completing paperwork.

(b) At work I become angry
1. Never.
2. Once in a while.
3. Frequently.
4. Most of the time.

(c) Mark all the words that describe your co-workers:
— Stimulating
— Stupid
— Helpful
— Boring

(d) On most days, do you look forward to going to work?
1. Yes
2. No

Closed-ended questions are also called "objective" questions, and that is their overwhelming strength: A response can be assigned a score objectively and reliably, with a minimum of subjective interpretation or error on the researcher's part. For example, in question (a) above, when participants select choice 1, we can assign them a score of 1 on that question, reliably assigning the same score to all people who gave the same response.

The disadvantage of closed-ended questions is that they may yield limited information. First, they measure only the variable(s) we've selected. In question (b) above, we ask about anger at work, but participants may feel that happiness is the more relevant emotion. Second, participants can select only from the choices provided, even if they would like to give a different response. In question (c) above, perhaps a worker considers co-workers to be "intelligent," a response that is not available. For these reasons, closed-ended questions are used when reliability is a major concern but we are not interested in discovering new variables that may be relevant.

Conversely, in an **open-ended question**, the participant determines both the alternatives to choose from and the response. Any question equivalent to an essay question, whether written or oral, is open-ended. Thus, we might ask the open-ended questions "Describe your favorite activities at work" and "How often do you become angry at work?" The advantages of open-ended questions are the opposite of those of closed-ended questions. Open-ended questions allow subjects to provide a wide range of responses, so researchers may discover new relevant variables. Further, participants can respond in their own words, so they are not limited to just the one perspective or phrasing that is present in a closed-ended question. On the other hand, the disadvantage of open-ended questions is that scoring each response requires subjective interpretation by the researcher, so scores may not be reliable. Two people who should be given the same score because their behaviors are the same may obtain different scores because of differences in the wording of their responses. (How would you score the responses "The paperwork is easy" versus "The paperwork is very easy"?) Further, the scoring of open-ended questions is highly susceptible to experimenter biases and expectations.

As usual, we counteract problems of subjective scoring by using double blinds and multiple scorers who demonstrate high inter-rater reliability. The crucial component, however, is the scoring criteria. One way to score open-ended questions is the technique called content analysis. In **content analysis**, we score a participant's written or spoken answer by counting specified types of responses. We may assign a score based on the number of times a certain word, a certain feeling, or a particular perspective occurs. For example, a worker's score might simply reflect the number of positive references made to paperwork, regardless of whether the word "very" occurs.

Content analysis is also applied in other contexts, such as when scoring physical actions in observational research, or when scoring subjects' diaries or conversations. (See Holsti, 1969, or Krippendorf, 1980, for further information on content analysis.)

Even with strict scoring criteria, open-ended questions tend to provide less reliable and objective data, so they are used at those less frequent times when reliability is not a major concern. In particular, open-ended questions are used with initial research into some unexplored behavior so that we can identify potentially relevant variables, or when each participant's response is likely to be unique.

Of course, you may employ both open-ended and closed-ended questions in the same questionnaire. This mix has the advantage of providing both reliable questions that are narrow in scope and less reliable but wider-ranging questions.

Using Interviews Versus Questionnaires

You must also decide whether to use an interviewer to ask the questions or to provide participants with a questionnaire to complete. The advantage of interviewers is that they ensure that participants complete the questions as instructed. In addition, an interviewer can react to the information provided by a participant, either requesting clarifying information or exploring additional topics that arise. The drawback to interviewers, however, is that they may inadvertently heighten the demand characteristics of reactivity and social desirability, or they may communicate expectancies about the desired response.

Just the opposite is true of questionnaires: The interaction between the researcher and participant is minimal, so there is less risk of biasing participants with experimenter expectations. Reactivity and social desirability also may be reduced, because completing a questionnaire anonymously can be much less threatening than talking to another person. Questionnaires also provide more efficient data collection, because many people can be tested at one time. The disadvantages of questionnaires, however, are that subjects may not complete them as instructed and the information obtained is limited to the inflexible questions presented.

The role of an interviewer is not all-or-nothing. At one extreme is the structured interview. In a **structured interview**, participants are asked specific, predetermined questions in a controlled manner. The most structured interview is when the interviewer simply reads closed-ended questions to participants and records their responses. The interviewer reads each question in a neutral manner with no additional comments or hints, and responds to a subject's comments by merely repeating the question. (This approach is common, for example, in telephone surveys or when testing young children.) A less structured interview may involve asking open-ended questions, but the interviewer follows a script to ensure that all participants are treated in a consistent fashion. (This approach is commonly used in intelligence tests.)

At the opposite extreme is an unstructured interview. In an **unstructured interview**, the researcher has a general idea of the open-ended questions that will be asked, but there is freedom of discussion and interaction between participant and interviewer. Such interviews are commonly used when a researcher begins studying a behavior or is developing a complete description of an individual, such as during a clinical diagnosis. The lack of structure allows us to explore a wide range of issues, but the trade-off is that we lose reliability, and the interviewer may lead participants into saying things they do not mean.

REMEMBER Interviews are preferred when the researcher must react to participant responses, but questionnaires are more reliable and less susceptible to demand characteristics.

Constructing Questions

The principles for creating questions in interviews and questionnaires are largely the same. Any question constitutes a *trial* where what we ask is a stimulus and a participant's answer is his or her response. Therefore, we exercise the same concerns when designing a question that we did when presenting an independent variable and measuring a dependent variable.

First, recognize that participants must interpret the meaning of a question and that we must interpret the meaning of their response. For valid inferences, you should minimize the extent to which subjects must decide what it is we are asking. For example, the question "Do you experience job satisfaction with your job?" is a problem because most workers will not know what we mean by job satisfaction. Instead, we should ask participants to report specific behaviors. If job satisfaction is related to being paid enough, we should ask workers whether they are paid enough. Then *we* translate their responses back into our variables and constructs.

Second, design questions that are sensitive to the subtle differences that exist between people. We create a question to *discriminate* or differentiate between participants on the variable being measured. We assume that there are differences between people on the variable because, otherwise, why bother to measure it? Therefore, you should design sensitive and precise questions that subjects will answer differently, to reflect such differences.

Finally, recall that a score should reflect a participant's "typical" behavior. However, a participant's response may be atypical because of a question's unique wording or perspective, or because the participant experiences a momentary distraction or misinterprets the question. As usual, we counter such problems by increasing reliability through multiple trials: We create a number of different questions designed to measure the same variable or behavior. By varying the wording and perspective across different questions, we should balance out the unique aspects of any one question. Then we have a more reliable estimate of a person's typical response to such items.

> *REMEMBER* The goal of question construction is to reliably and validly discriminate between participants on the variable being studied.

When generating questions, try to create many examples that reflect a variable. Then select those questions that are best suited to your purposes. Select questions using the following criteria.

Wording the questions Phrase questions so that you are confident you know what participants are communicating by their response and so that you can discriminate between different subjects. To meet these goals, avoid the following types of questions.

First, avoid **double-barreled questions**. These are questions that have more than one component. Consider the question "Should you be given more flexibility and less supervision on your job?" What if a person agrees to one part but not the other? The meaning of any response here will not be clear to you, so instead, ask two separate questions. Always phrase a question so that it states just one idea.

Second, avoid **leading questions**. These are questions that are so loaded with social desirability or experimenter expectancies that there is one obvious response. Consider the question "Should very bad workers who are always late receive low pay?" This question won't discriminate between subjects, because everyone knows what the correct answer "should be." Always phrase questions in a neutral manner, avoiding biased or inflammatory statements.

Third, avoid **Barnum statements**. These are questions that are so global and vague that everyone would agree with them or select the same response for them. (They are named after P. T. Barnum, who was famous for such statements.) For example, asking "Do you sometimes worry?" or "Have you had difficulty in some college courses?" will elicit the same answer from virtually everyone. This is the problem with horoscopes and palm readings: They are so general that people can always think of personal experiences that seem to fit. Always phrase a question so that it targets a specific behavior.

Finally, avoid questions that contain **undefined terms**. For example, asking "Should workers who are always late receive low pay?" will leave you wondering how participants interpret "always" and "low pay." Instead, either define such terms in the question, or have participants provide the definition in their response. Thus, you might ask "What pay should a worker receive who is late for work an average of twice a week?" In this version you define "always late" and you allow participants to define "low pay" in their response.

REMEMBER A question should be a clear, precise, and unbiased statement of a single idea, to which different participants are likely to respond differently.

Creating the responses for closed-ended questions The above guidelines also apply to the wording of the response alternatives in closed-ended questions. Each alternative should be constructed to maximize our confidence that we know what participants wish to communicate when they select it. Therefore, the choices should be worded in a precise and unbiased manner, should convey one idea each, and should be mutually exclusive.

We must also determine the **response scale**, which is the number and type of choices to provide for each question. For example, we may ask yes-no (or true-false) questions, such as "Do you deserve a raise in pay?" Then we can assign a score of 1 for "yes" and a score of 2 for "no" (or arbitrarily pick any other two numbers). With only two scores, however, we have the problem of a restricted range, so that we cannot finely discriminate between subjects and instead we gloss over differences between those who firmly believe a statement is true and those who think it is only somewhat or sometimes true. Also, if we are scoring a test for correct answers, someone's *apparent* correct response may actually be a lucky guess.

To alleviate problems of restricted range, sensitivity, and guessing, increase the number of response choices. Multiple-choice questions are most appropriate for measuring factual information or discrete responses. For example, we might ask:

Do you deserve a raise in pay?

1. No
2. Yes, a $1 raise
3. Yes, a $2 raise
4. Yes, more than a $2 raise

Four choices allow us to discriminate more precisely. However, the fourth choice does not distinguish between those who seek a $3 raise and those who seek a $4 or $5 raise. To obtain finer discriminations, we'd provide additional choices.

When measuring responses that fall along a continuum, we often employ Likert-type questions. These consist of a declarative statement accompanied by a rating scale. Most often, the scale is "anchored" at each end by the words *agree* and *disagree*. Thus, we might ask:

My hourly salary is sufficient for me.

1 2 3 4 5

STRONGLY AGREE — STRONGLY DISAGREE

We can also change the wording of the question to measure other experiences and attitudes, using such anchors as *seldom/frequently* or *like/dislike*.

Notice that creating a Likert-type question involves three decisions. First, consider the wording of the anchors. Above, including the word "strongly" implies extreme feelings. Because of social desirability, participants may be less likely to select the extreme positions of 1 or 5. Labeling the anchors with only *agree/disagree* would imply less extreme feelings, and thus would be more likely to get a wider range of responses. However, we'd also have a less clear definition of what participants were communicating. The way to resolve this issue depends on how threatening a particular question is. Usually, we define the anchors clearly and then attempt to minimize demand characteristics by our wording of the statement being rated.

Second, you must select the number of response alternatives. The above rating scale allows us to discriminate among only five levels of agreement. When greater sensitivity is needed, include more alternatives. (Do not allow subjects to place responses between the points on the scale, because such responses cannot be scored reliably.) How large a scale you should select depends on participants' ability to differentiate their feelings. On a scale of 1 to 20, for instance, people probably cannot distinguish between a 16 and a 17. Instead, they are likely to guess between the two, and then our interpretation of what a 16 or 17 indicates is in error. Resolving this issue depends on the particular question, but we usually employ scales with between 5 and 7 choices.

Finally, note that with five, seven, or any odd number of choices, there is one neutral "middle of the road" choice. The more threatening an issue, the more likely it is that people will play it safe and choose the midpoint. If so, they defeat our primary purpose of discriminating among participants. The solution is to use an even number of choices. With six options, for example, there is no middle ground, so participants must commit one way or the other. In general, we use an odd-numbered scale when we assume that people can be legitimately neutral on an issue, and we use an even-numbered scale to counter strong demand characteristics or to force participants to take a stand.

Once you have created the basic questions to ask, you can often generate additional, comparable questions for multiple trials merely by changing the wording and perspective. For example, to measure how interesting workers find their job, we can ask them to answer these two questions:

My job is interesting.
1 2 3 4 5
STRONGLY AGREE — STRONGLY DISAGREE

My job is boring.
1 2 3 4 5
STRONGLY AGREE — STRONGLY DISAGREE

Then we "code" similar responses across related questions with the same score. That is, strongly agreeing with "My job is interesting" is equivalent to strongly disagreeing with "My job is boring." Therefore, we can record a response of 5 on the "boring" question as a score of 1, a 4 as a 2, and so on. Then for both questions, the lower the score, the more interesting the job. Likewise, with multiple-choice questions we score the choices so that each score reflects the same response (e.g., a 1 is assigned to the choice describing minimum job satisfaction).

Note that pilot studies are extremely valuable when developing questionnaires and interviews, because you can ask participants questions about your questions. Thus, you can check that the questions convey the intended meaning, that participants can differentiate among the choices, or that minimal demand characteristics are present. Alter problem questions and conduct additional pilot studies until you have created the desired questions.

The Problem of Order Effects

Because multiple questions are essentially multiple trials, we again face the problem of order effects. Practice effects, for example, occur if participants first find the questions to be novel or they feel great reactivity; but with more questions, they become comfortable, or later, they become fatigued or bored. Likewise, carry-over effects occur if participants respond in a biased fashion to later questions because of earlier questions. (Have you ever found that the way one question is worded on an exam provides the answer to a later question?) Participants may also develop response sets, especially over repeated closed-ended questions. If, for example, initial multiple-choice questions consistently call for choice 1, participants may superstitiously select choice 1 for subsequent questions. Or if people select the strongly agree option on initial Likert-type questions, they may continue to make this response automatically.

There are several techniques for dealing with order effects:

1. **Provide practice questions** By providing practice questions prior to presenting the questions of interest, we allow participants to warm up to the questions and to habituate to their content, without contaminating the data.
2. **Counter-balance order effects** We can balance out the effects of one particular order by creating different orders of questions for different participants so that questions that appear early in some questionnaires appear later in others, and vice versa. Across all participants completing the different versions, the total sample will not be biased by one unique order of questions.
3. **Prevent response sets** To prevent rote responding, we vary the question format to try to force participants to read and think about each question. Thus, in multiple-choice questions, we randomly vary which choice is correct. In Likert-type questions, we present both positive and negative statements to be rated, and we vary the scale by mixing agree/disagree with frequently/infrequently, and so on. We may also intermix multiple-choice with Likert-type questions. Note, however, that we do not drastically change the format from question to question. Rather, we generally present a block of one type of question (containing, say, 10 questions) before changing to a different format for the next block of questions. In this way, we avoid confusing subjects and increasing their errors, while still minimizing response sets.
4. **Use alternate forms** **Alternate forms** are different versions of the same questionnaire given to different subjects. Here we change the order, wording, and perspective of questions so that the questionnaires appear to be different, yet still measure the same variables. Alternate forms are especially useful when we must measure the same participants repeatedly. For example, say we wanted to measure workers' job satisfaction immediately before and after giving them a raise. We would not use the identical questionnaire both times, because people might duplicate their previous responses on the second testing in order to appear consistent, or they might intentionally change their responses because they think we expect them to. Ideally, the alternate forms will hide the similarity of participants' past and present responses so that they answer the second version

honestly. (Alternate forms involve different questionnaires, so we must ensure that they are comparable in terms of validity and reliability.)

Creating Catch Trials

Participants may not follow instructions when completing questionnaires and interviews. Some people may give no thought to the questions and select answers randomly, just so they can be finished. Others may be untruthful, responding solely to demand characteristics. And still others may make errors when responding. We can incorporate specific questions to "catch" such participants.

To identify people who may be answering questions randomly, we can include a specific question several times throughout the questionnaire, but reorder the choices. Consider these examples:

When working I prefer to be	When working I prefer to be
1. Left alone.	1. Supervised occasionally.
2. Supervised occasionally.	2. Supervised frequently.
3. Supervised frequently.	3. Left alone.

A person's preference should be the same on both questions. Anyone who fails to be consistent is either completing questions randomly or recording responses erroneously.

To identify participants who may be responding to demand characteristics, we can create questions for which we know the truthful response. Say that when questioning teenagers about the extent of their drug use, we are concerned that peer pressure may cause some subjects to overstate their drug involvement. To identify these people, we might ask the following:

I have taken the pill known as a "watermelon"

1. Never.
2. Between 1 and 5 times.
3. Between 5 and 10 times.
4. More than 10 times.

There is no pill known as a watermelon. All participants should select response 1, unless they are untruthful or made an error when responding.

With such questions, we can estimate the frequency with which participants were untruthful or made errors when responding to all other questions. Also, we may use such questions to eliminate untrustworthy subjects from the data.

Administering the Questionnaire or Interview

Administering a questionnaire or interview requires the same controls that are found in experiments. You should control the environment so that there are no extraneous distractions. You should provide unbiased instructions for completing the questions (even if they seem self-explanatory). And you should keep the behaviors of the researcher neutral and consistent.

To minimize demand characteristics, be careful when creating a title for your questionnaire. Ask yourself if a title is really necessary. Does it bias participants? (Think about how you would respond to a questionnaire titled "Survey of Deviant Sexual Fantasies." What if it were titled "Survey of Common Sexual Fantasies"?) Second,

TABLE 5.3 Checklist for Question Construction

- Use closed- or open-ended questions?
  - Is reliability or breadth of information most needed?
- Use interview or questionnaire format?
  - How strong are demand characteristics?
  - Must responding be structured for participants?
  - What breadth of information is needed?
- Are questions worded correctly?
  - Avoid double-barreled questions?
  - Avoid leading questions?
  - Avoid Barnum statements?
  - Avoid undefined terms?
- What is response scale?
  - Are measurements sensitive to subtle differences?
  - Use multiple-choice or Likert scale?
  - What description should anchor Likert scales?
  - How many choices in each scale?
  - Is odd or even number of points needed?
- How to administer questioning?
  - Avoid order effects, especially response sets?
  - Are alternate forms needed?
  - Are catch trials needed?
  - Include clear instructions?
  - Is pilot study needed?

consider whether deception is needed, in the form of "filler" or "distracter" questions. These are not included in the data, but they alter the overall appearance of the questionnaire and disguise its actual purpose. (For example, you might include filler questions about nonsexual fantasies to reduce reactivity to sexually oriented questions, and title the questionnaire "Survey of Common Fantasies.")

Last, but certainly not least, you are usually obligated to use the techniques discussed in Chapter 10 to demonstrate that the questions have a minimum level of reliability and validity. After all, without this you are not measuring what you seek to measure, so you're just wasting time.

To help you remember the various considerations when constructing questions, Table 5.3 above presents a summary of the previous discussion.

ETHICAL ISSUES IN DESCRIPTIVE RESEARCH

The ethical issues in descriptive research are the same as in experiments: To minimize potential harm to participants, and to justify the risks by the knowledge we will gain. The risks arise because being observed by a researcher or completing a survey may be embarrassing, stressful, or unpleasant for participants. As with all research, your procedure should first be approved by your institution's Human Subjects Review Committee.

If the participants are aware a study is being conducted, we follow the usual rules: All responses are kept confidential, and we alleviate participants' fears about what the data will divulge about them or what it will be used for. As always, we obtain explicit informed consent. The fact that participants complete a questionnaire or interview is *not* informed consent, because they may feel coerced to do so. Upon completion of testing, provide a debriefing.

The principal ethical dilemma arises with unobtrusive research. From one perspective, observational techniques are another name for spying on people. With participant observation, we are present under false pretenses and we violate a subject's expectation of privacy. With archival studies, we examine private records without obtaining informed consent. In fact, in any form of unobtrusive field research, participants are not aware that a study is being conducted, we do not obtain their informed consent, and so we may be violating their rights.

A classic example of this dilemma is a study by Middlemist, Knowles, and Matter (1976). They wanted to eliminate demand characteristics while measuring whether an invasion of one's personal space created physical tension. Their solution was to observe males as they visited the urinal in a public restroom! They invaded personal space by having a confederate use the adjacent urinal, and the measure of subjects' resulting tension was the amount of time each took to urinate. To be unobtrusive, a researcher occupied one of the stalls and used a periscope to observe each participant, timing the interval between unzipping and rezipping.

Although technically this was a field experiment, the ethical problem is with the secret observation of participants. We might justify this study by claiming that this is "scientific research" for the "good of humanity." But, some would argue, this is no different from when a government agency or the police spy on citizens, claiming that it helps catch criminals. After all, spying is spying, and it is wrong to invade people's privacy and violate their rights, regardless of whether it is for scientific advancement and the good of humanity, or for national security and rooting out evil.

Others would argue that a public behavior is just that—public—and so it's open to anyone's observation. Thus, a male who uses a public restroom has tacitly agreed to be observed by other males. Conversely, if a male wishes to keep his urinal behavior private, he should not use a public restroom. From this perspective, some researchers claim that it is unethical for scientists *not* to conduct unobtrusive field research, because then they would miss potentially valuable information.

There is no easy resolution to this debate. You might suggest that we obtain informed consent after the study, but by then subjects' rights are already violated (and telling participants afterwards might be more upsetting than not informing them at all). Instead, it is again the researcher's responsibility to weigh the violation of subjects' rights against the potential scientific information to be gained. Therefore, first decide just how "public" participants consider a behavior to be. Are you invading their expected privacy? How strenuously would they object if you asked their permission? How upset would they be if they found out about your spying after the fact? (If you are unsure of the answers, conduct a pilot study in which you ask people these questions.)

Then weigh the invasion of privacy against the potential scientific benefits. For example, Koocher (1977) argued that the above urinal study needlessly invaded participants' privacy, because it replicated findings already demonstrated by other, less questionable techniques. (But see the reply of Middlemist et al., 1977.) Also consider whether the procedure really needs to be conducted as an unobtrusive field study. Remember that the APA's Ethical Principles (1992) state that deception must be necessary and that informed consent is needed unless the risk to participants is minimal. The more the behavior being studied is an innocuous, mundane public behavior, and the greater the necessity for an unobtrusive field study, the more the study can be justified ethically.

REMEMBER You must be particularly sensitive to the ethics of unobtrusive field research.

PUTTING IT ALL TOGETHER

Because we always want valid and reliable data, there are actually few differences in the mechanics of how we conduct descriptive or experimental research. The issues are largely the same, so any of the concerns you learned in the previous chapter apply to descriptive research discussed in this chapter and vice versa. Thus, for example, in any research, you should create clear instructions or consider automating. Likewise, you should always consider multiple raters and interrater reliability, the volunteer bias, and using the appropriate sampling technique to obtain the most representative sample. Essentially, from these chapters you've been developing a collection of techniques that, depending upon the specific hypothesis, you may mix and match as needed.

CHAPTER SUMMARY

1. In *descriptive research*, the goal is to describe a behavior, the situation it occurs in, or the individuals exhibiting it. The three major types of descriptive designs are correlational studies, observational studies, and field surveys.
2. In *naturalistic observation*, the researcher observes participants' behaviors in an unobtrusive and unsystematic manner. In *systematic naturalistic observation*, the researcher unobtrusively observes a behavior in a systematic manner. In *participant observation*, the researcher is a member of the group being observed.
3. *Ex post facto* research is conducted after a phenomenon has occurred. *Archival research* is conducted using subjects' existing records. A *case study* is an in-depth description of one individual, organization, or event.
4. When a researcher subjectively evaluates a participant's behavior and assigns a score to it, it is important to use *multiple raters*. They should have high *inter-rater reliability*, the extent to which raters agree in the scores they assign a particular behavior.
5. A *field survey* involves having participants complete a questionnaire or interview in the field, either in person, by mail, or over the telephone.
6. The *volunteer bias* is the bias that arises because people who will participate in a study are different from those who will not.
7. In *probability sampling techniques*, every member of the population has an equal likelihood of being selected. In *simple random sampling*, participants are selected in a random fashion. In *systematic random sampling*, every *n*th individual is selected from a list of the population. With *stratified random sampling*,

participants are randomly selected proportionately from within each important subgroup in the population. With *cluster sampling*, certain groups are randomly selected and then the members of each group are observed.

8. In *nonprobability sampling*, every member of the population does not have an equal likelihood of being selected. In *convenience sampling*, participants are those who are available. In *quota sampling*, the population is proportionately sampled, but convenience samples fill each quota. With *snowball sampling*, potential subjects are identified by other subjects.

9. The goal of psychological questions is to reliably and validly discriminate among participants on the variable of interest. With *closed-ended questions*, participants select from alternatives provided by the researcher. With *open-ended questions*, participants determine the alternatives to choose from.

10. *Content analysis* is the procedure for scoring open-ended questions by looking for specific words, themes, or actions.

11. In a *structured interview*, participants are asked predetermined questions. In an *unstructured interview*, the questions are less rigidly predetermined.

12. Researchers should avoid (a) *double-barreled questions*, which have more than one component; (b) *leading questions*, which are biased so that there is only one obvious response; (c) *Barnum statements*, which are global truisms to which everyone responds in the same way; and (d) questions that contain *undefined terms*.

13. A *response scale* is the number and type of choices provided for each question. When using Likert-type questions, the rating scale should (a) Have anchors that do not overly bias participants, (b) Contain choices that subjects can discriminate among, and (c) Force participants to indicate a preference.

14. Order effects in questionnaires can be controlled by (a) Including practice questions, (b) Counterbalancing with different orders so the influence of any one order is balanced out, (c) Alternating between different question formats and (d) Using *alternate forms* of a questionnaire that contain questions that are worded differently but measure the same behaviors.

15. The important ethical issue in unobtrusive field research is that by not obtaining informed consent, the researcher may violate participants' rights.

PRACTICE PROBLEMS

(Answers for odd-numbered problems are provided in Appendix D.)

1. What does the term descriptive research convey?
2. What is the difference between naturalistic observation, systematic observation, and participant observation?
3. (a) What are the strengths of observational designs? (b) What are their weaknesses?
4. (a) What is archival research? (b) What is *ex post facto* research? (c) What are the major weaknesses of these designs? (d) Why do researchers employ them?

5. (a) What is a case study? (b) What is the strength of this approach? (c) What is its weakness?
6. (a) What important ethical issue pertains to unobtrusive observational research? (b) According to the APA's Ethical Principles, when can we not obtain informed consent with these approaches?
7. (a) What is the difference between simple and systematic random sampling? (b) What is stratified random sampling? (c) What is cluster sampling?
8. (a) What is the difference between probability and nonprobability sampling techniques? (b) Why are probability techniques more likely to produce a representative sample? (c) Which probability technique should produce the most representative sample?
9. (a) What is quota sampling? (b) What is snowball sampling?
10. (a) What is convenience sampling? (b) Why are all samples somewhat convenience samples? (c) How do such samples influence external validity?
11. (a) When do researchers mail surveys? (b) When do they employ telephone surveys? (c) Why is it important to ensure high subject-participation rates in both types of surveys?
12. For each of the following, indicate whether you should use a written questionnaire, a structured interview, or an unstructured interview: (a) When measuring the attitudes of first-graders (b) When measuring the contents of people's daydreams (c) When measuring people's attitudes toward researchers.
13. (a) What are the advantages and disadvantages of open-ended questions? (b) Of closed-ended questions?
14. I ask students in my class to rate their agreement with the following statements. Give the term that identifies what is wrong with the wording of each statement. (a) The material in this textbook is sometimes difficult. (b) I like reading this book, but I dislike the statistics. (c) A good student will like this book. (d) With this book I can get an acceptable grade.
15. (a) What are alternate forms? (b) What is the design concern about them? (c) When are they necessary?
16. On a personality test, the question "Do you prefer cooked carrots or raw carrots?" occurs several times. Why?
17. (a) What is the volunteer bias? (b) How might it decrease external validity?
18. A student complains that a college exam was unfair because it contained some questions that very few students could answer correctly. How would a researcher justify the inclusion of such questions?
19. A researcher wants to observe children unobtrusively at a day-care center, judging how aggressively they behave after watching an adult behave aggressively. (a) What problem arises with this scoring technique? (b) In terms of the scoring criteria, how would you insure reliability? (c) In terms of the scorer, how would you ensure reliability? (d) What must you then do to show you have reliability?
20. In problem 19, what are the major ethical issues in this design and how should you handle them?

PART 2

DESCRIPTIVE STATISTICS

Now that you understand the basics of descriptive and experimental research, it's time to see how statistical procedures fit in. Our starting point is descriptive statistics that summarize the important characteristics of data. What do we mean by important characteristics? Essentially, they involve answering the following five questions about the data:

1. *Which scores occurred?* We answer this question by presenting the scores in tables and graphs.
2. *Are the scores generally high scores or generally low scores?* We can describe the scores with one number that represents the "typical" score.
3. *Are the scores very different from each other, or are they close together?* There are mathematical ways of describing "close."
4. *How does any one particular score compare to all other scores?* We have a system for evaluating any score relative to the other scores.
5. *What is the nature of the relationship we have found?* We can summarize a relationship and then use it to predict scores on one variable if we know a score on another variable.

The following four chapters show how to answer the first four questions. In Part 3, we answer the final question.

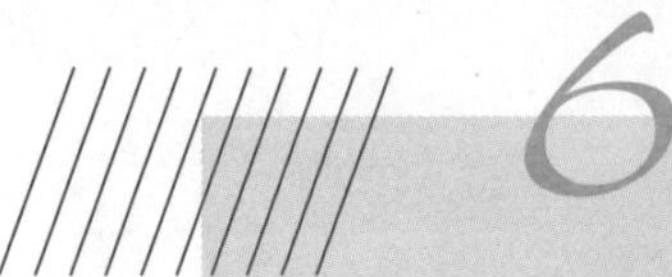

Summarizing Data with Frequency Distributions and Percentiles

To understand this chapter, from Chapter 2 remember the difference between a sample and population of scores, how a sample is used to draw inferences about a population, what a data point is and what the phrase "as a function of" means. Then your goals in this chapter are to learn:

- The scales of measurement that researchers use.
- How simple frequency, relative frequency, cumulative frequency, and percentile are computed, and what each tells you.
- How the different types of frequency tables, bar graphs, histograms, and polygons are created.
- What normal, skewed, bimodal, and rectangular distributions are.
- How the proportion of the total area under the normal curve corresponds to the relative frequency of scores.

Recall that with statistics we make sense out of data. This chapter shows how to do this by displaying data in graphs and tables. A basic rule is that you should always create a table or graph. As the saying goes, "A picture is worth a thousand words," and nowhere is this more appropriate than when trying to make sense out of a large group of scores. Also, statistics allow you to communicate your results to others, and a table or graph is often the most efficient way to do this. Finally, a table or graph often makes it easier to see the relationship hidden in data.

Before we examine the relationship between the scores of two variables, however, we first summarize the scores on *each* individual variable, asking "Which scores occurred?" In fact, buried in any batch of scores are two important questions: Which scores occurred, and how often did each score occur? We can answer both questions simultaneously with certain types of tables and graphs. But first let's look at some basic statistical rules and notation.

MORE STATISTICAL NOTATION

The scores we measure in a study are the *raw scores*: They are "uncooked" and not yet "digestible." Descriptive statistics help us cook down the raw scores into an organized and interpretable form. One way to do this is to create a **distribution**, which is the general name for any organized set of data. Then we can see the pattern the scores form, or in statistical language, see how the scores are *distributed*. From the sample, we then infer how the scores in the population are distributed.

Another way descriptive statistics cook down the raw scores is through transformations. A **transformation** is a mathematical procedure for systematically converting a set of scores into a different set of scores. One reason to transform data is to make different kinds of scores comparable. For example, it's difficult to compare your performance if you obtain 8 out of 10 on a math quiz and 75 out of 100 on an English quiz. However, by transforming each score to a percent of the total, we are no longer comparing apples to oranges. Second, we transform data because it makes scores easier to work with. For example, if the scores each contain one decimal place, multiplying every score by 10 eliminates the decimals.

Speaking of decimals, "close counts" in statistics, so carry out calculations to the appropriate number of decimal places before you "round off" an answer. The convention is to carry out calculations so that the *final* answer after rounding has two more decimal places than the original scores. For example, if you have whole-number scores, the final answer should contain two decimal places. However, do not round off at each intermediate step in the calculations: *Round off only at the end*! Therefore, if the final answer will contain two decimals, carry out the intermediate calculations to at least three decimal places, and then round off the final answer. Use the following rules when rounding:

- If the number in the next decimal place is 5 or greater than 5, round up. For example, if rounding to two decimal places, 2.366 is rounded to 2.370, which becomes 2.37.
- If the number in the next decimal place is less than 5, round down: 3.524 is rounded to 3.520, which becomes 3.52.

Recognize that we add zeroes to the right of the decimal point to indicate our level of precision. Rounding 4.996 to two decimal places produces 5, but to show the precision of two decimal places, we report it as 5.00.

Finally, in this chapter we'll be counting scores. When we count how *many* scores there are, we use the symbol N to stand for the total number of scores in a set of data. (When you see N, think *Number*.) An N of 10 means that we have 10 scores, or $N = 43$

means that we have 43 scores. In statistical terminology, N is called the *sample size* because it indicates how big a sample is. Note however, that N stands for the total number of scores, *not* the number of different scores. If, for example, the 43 scores in a sample are all the same number, N still equals 43. (When we have one score for each subject, N also corresponds to the number of subjects in the sample.) Get in the habit of treating N as a quantity itself so that you understand such statements as "the N participants in the sample" or "this sample's N is larger than that sample's N."

We also count how *often* each individual score occurs. How often a score occurs is the score's **frequency**, symbolized by f. Also learn to treat f as a quantity: One score's f may be larger than another score's f, we can add the fs of different scores, and so on. As we'll see, there are several ways to describe a score's frequency, so we often combine the terms frequency and f with other terms and symbols.

TYPES OF MEASUREMENT SCALES

Recall that a big part of your job in this course is to learn when to use a particular statistical procedure. Which procedure to employ in a study is determined by three things. First, you must decide what it is you want to know—what question about the characteristics of the sample or population do you want to answer? Then, your choice of procedures depends on the specific research design being employed, because different designs are analyzed differently. Finally, within a particular design, you'll measure a variable in such a way that the scores have certain underlying mathematical characteristics. The particular mathematical characteristics of scores also determine which particular statistical procedure to use. Therefore, part of your job is to learn to recognize the different types of scores researchers obtain. This means being able to recognize the type of measurement scale involved and whether the scale is continuous or discrete.

The Four Scales of Measurement

Numbers mean different things in different contexts. The meaning of the number 1 on a license plate is different from that of the number 1 in a race, which is different still from the meaning of the number 1 in a hockey score. The kind of information that scores convey depends on the *scale of measurement* that is used to measure the variable. There are four types of measurement scales: Nominal, ordinal, interval, and ratio.

With a **nominal scale**, each score does not actually indicate an amount; rather, it is used simply for identification, as a name. (When you see *nominal*, think *name*.) License plate numbers and the numbers on the uniforms of football players reflect a nominal scale of measurement. In research, a nominal scale is used to identify the categories of a qualitative, or classification, variable. For example, we cannot perform any statistical operations on the words *male* and *female*. Therefore, we might assign each male participant a 1 and each female a 2. However, because this scale is nominal, we could just as easily assign males a 2 and females a 1, or we could use any other two numbers. Because we assign numbers arbitrarily in a nominal scale, they do not have the mathematical properties normally associated with numbers. For example, giving a male the number 1 does not indicate more than 0 yet less than 2, as it usually does.

When a variable is measured using an **ordinal scale**, the scores indicate rank order: The score of 1 means the most or least of the variable, 2 means the second most or least, and so on. (For *ordinal*, think *ordered*.) In psychological research, ordinal scales are used, for example, to rank participants in terms of their aggressiveness, or to have subjects rank the importance of certain attributes in their friends. Each score indicates an amount of sorts, but it is a relative amount. For example, relative to everyone else being ranked, you may be the number 1 student, but we do not know how good a student you actually are. Further, with an ordinal scale there is not necessarily an equal unit of measurement separating each score. In a race, for example, first may be only slightly ahead of second, but second may be miles ahead of third. Also, there is no number 0 in ranks (no one can be "zero-th").

When a variable is measured using an **interval scale**, each score indicates an actual amount, and there *is* an equal unit of measurement separating any two scores: The difference between 2 and 3 is the same as the difference between 3 and 4. (For *interval*, think *equal* interval.) Interval scales include the number 0, but it is not a "true" zero. It does not mean zero amount; it is just another point on the scale. Because of this, an interval scale allows negative numbers. Temperature when measured in Celsius or Fahrenheit is an interval scale, because zero degrees does not mean that zero amount of heat is present—it means only that there is less than 1 degree and more than -1 degree. Interval scales are often used with psychological tests, such as intelligence or personality tests. Although a score of zero may be possible, it does not mean zero intelligence or zero personality. Note that with an interval scale, it is incorrect to make "ratio statements" that compare the amount of a variable at one score relative to the amount at another score. For example, at first glance it seems that 4 degrees Celsius has twice as much heat as 2 degrees. However, if we measure these temperatures using the Fahrenheit scale, 2 and 4 degrees Celsius are about 35 and 39 degrees Fahrenheit, respectively, so now the one amount is not twice that of the other.

Only with a **ratio scale** do the scores reflect the true amount of a variable that is present, because the scores measure an actual amount, there is an equal unit of measurement, *and* 0 truly means that zero amount of the variable is present. Therefore, ratio scales do not allow negative numbers. Further, only with ratio scales can we make ratio statements, such as "4 is twice as much as 2." (So for *ratio*, think *ratio*!) In psychological research, ratio scales are used to measure quantitative variables, such as the number of errors made on a test, the number of friends someone has, or the number of calories consumed in a day.

To help you remember the four scales of measurement, Table 6.1 summarizes their characteristics.

Discrete and Continuous Scales

The other important attribute of a measurement scale is whether it is continuous or discrete. A **continuous scale** allows for fractional amounts; It "continues" between the whole-number amounts. With a continuous scale, decimals make sense. Age is a continuous variable because it is perfectly intelligent to say that someone is 19.6879 years old. To be continuous, a variable must be at least theoretically continuous. For example, intelligence tests produce whole-number scores. You cannot obtain an IQ score of 95.6. But theoretically, an IQ of 95.6 makes sense, so intelligence is a theoretically continuous interval variable.

TABLE 6.1 Summary of Types of Measurement Scales

Each column describes the characteristics of the scale.

	Type of measurement scale			
	Nominal	***Ordinal***	***Interval***	***Ratio***
What does the scale indicate?	Quality	Relative quantity	Quantity	Quantity
Is there an equal unit of measurement?	No	No	Yes	Yes
Is there a true zero?	No	No	No	Yes
How might the scale be used in research?	To identify males and females as 1 and 2	To judge who is 1st, 2nd, etc., in aggressiveness	To convey the results of intelligence and personality tests	To state the number of correct answers on a test

On the other hand, a **discrete scale** can be measured only in whole-number amounts. Here, decimals do not make sense. Usually, nominal and ordinal variables are discrete. In addition, some interval and ratio variables are discrete. For example, the number of cars someone owns and the number of children someone has are discrete ratio variables. It sounds strange when the government reports that the average family has 2.4 children and owns 1.78 cars, because these are discrete variables being treated as if they were continuous. (Imagine a .4 child driving a .78 car!)

There is a special type of a discrete variable. When there can be only two amounts or categories of the variable, it is a **dichotomous** variable: Pass/fail, male/female, and living/dead are examples of dichotomous variables.

The Impact of a Particular Scale

There are two important reasons to pay attention to the scale of measurement being employed. First, certain mathematical and statistical procedures make sense only with certain types of scores. It makes no sense to think that the football player with number 12 is bigger, faster, or has more of anything relative to player number 10, nor does calculating the average player's number make any sense.

Second, different scales provide different degrees of precision and sensitivity. Recall that when developing scoring criteria or selecting the response scale for questionnaires, you should seek the most sensitive, most discriminating measure. Nominal scales discriminate only grossly among participants, because scores essentially reflect only a yes/no classification. Ordinal scales, by measuring relative amounts, are more sensitive, but by being discrete and having no constant amount between scores, they still lack precision. Interval or ratio scales are the most sensitive measures, especially when they are continuous: Here scores can reflect very small, subtle differences in behaviors. Thus, if a behavior can be precisely quantified, it is best to measure it using a continuous, interval, or ratio scale.

REMEMBER The sensitivity of measurements and which statistical procedure to apply to them depend on the type of measurement scale employed.

CREATING SIMPLE FREQUENCY DISTRIBUTIONS

There are several ways to answer the question "Which scores occurred and how often did each occur?" The most common way is to create a simple frequency distribution. A **simple frequency distribution** shows the number of times each score occurs in a set of data. As we saw, the symbol for a score's frequency is f. To find f for a score, count how many times that score occurs in the data. If three subjects scored 66, there are three scores of 66, so the frequency of 66 (its f) is 3. Creating a simple frequency distribution involves counting how often every score in the data occurs.

Presenting Simple Frequency in a Table

To see how we present a simple frequency distribution in a table, let's begin with the following sample of raw scores. (They measure some deep psychological trait.)

14 14 13 15 11 15 13 10 12
13 14 13 14 15 17 14 14 15

There are only 18 scores here, but in this disorganized arrangement, it's difficult to make sense out of them. Watch what happens, though, when we arrange them into a simple frequency table, as shown in Table 6.2. The table consists of a score column that shows each score, and an f column that shows the number of times the corresponding score occurred. Notice that the score column begins with the highest score in the data at

TABLE 6.2 Simple Frequency Distribution Table

The left-hand column identifies each score, and the right-hand column contains the frequency with which the score occurred.

Score	f
17	1
16	0
15	4
14	6
13	4
12	1
11	1
10	1
	Total: $18 = N$

the *top* of the column. Below that in the column are all *possible* whole-number scores in decreasing order, down to the lowest score that occurred. Thus, although no subject obtained a score of 16, we still include it.

Now we can easily see the frequency of each score, as well as the overall distribution. We can also determine the combined frequency of different scores by adding together their individual *f*s. For example, the score of 13 has an *f* of 4 and the score of 14 has an *f* of 6, so the frequency of 13 and 14 is 4 + 6, or 10.

Note that even though there are 8 scores in the score column, *N* is not 8. There are 18 scores in the original sample, so *N* equals 18. Likewise, if we add together all the values in the *f* column, the sum will equal 18: The 1 participant who obtained 17, plus the 0 participants scoring 16, plus the 4 who scored 15, and so on, equals the *N* of 18.

> *REMEMBER* The total of all individual frequencies always equals ***N***.

As a check on any frequency distribution you create, add up the frequencies. If the sum of the frequencies does not equal *N*, you made a mistake.

That's how to create a simple frequency distribution. Such a distribution is also called a **regular frequency distribution** or a plain old **frequency distribution**.

Graphing a Simple Frequency Distribution

Graphs provide an easy way to communicate the overall distribution of a set of scores. Essentially, they show the relationship between each score and the frequency with which it occurs. We ask, "For a given score, what is its corresponding frequency?" Therefore, in creating the graph we place the scores on the *X* axis and the frequency of the scores on the *Y* axis. Then we look for changes in frequency *as a function of* changes in the scores, looking at how the frequencies change as the scores increase.

> *REMEMBER* A graph of a frequency distribution always shows the scores on the *X* axis.

A variable will involve one of the previous four types of measurement scales. The type of scale is important, because it determines whether we graph a frequency distribution as either a bar graph, a histogram, or a polygon.

Bar graphs A frequency distribution of nominal or ordinal scores is graphed as a bar graph. A **bar graph** is a graph in which we draw a vertical bar centered over each score on the *X* axis. *In a bar graph, adjacent bars do not touch.*

Figure 6.1 shows two bar graphs of simple frequency distributions. The corresponding frequency tables are included so that you can see the data being plotted, but usually a bar graph does not include the table. The upper graph shows the nominal variable of political affiliation, where a score of 1 indicates Republican, a 2 indicates Democrat, a 3 indicates Socialist, and a 4 indicates Communist. The lower graph shows ordinal data reflecting a baseball team being ranked first, second, and so on during the last 20 years. In both graphs, the height of each bar corresponds to the score's frequency.

The reason we create bar graphs here is that both nominal and ordinal scales are usually discrete scales. The space between the bars indicates this fact. Interval and ratio

FIGURE 6.1 Simple Frequency Bar Graph for Nominal and Ordinal Data

The height of each bar indicates the frequency of the corresponding score on the X axis.

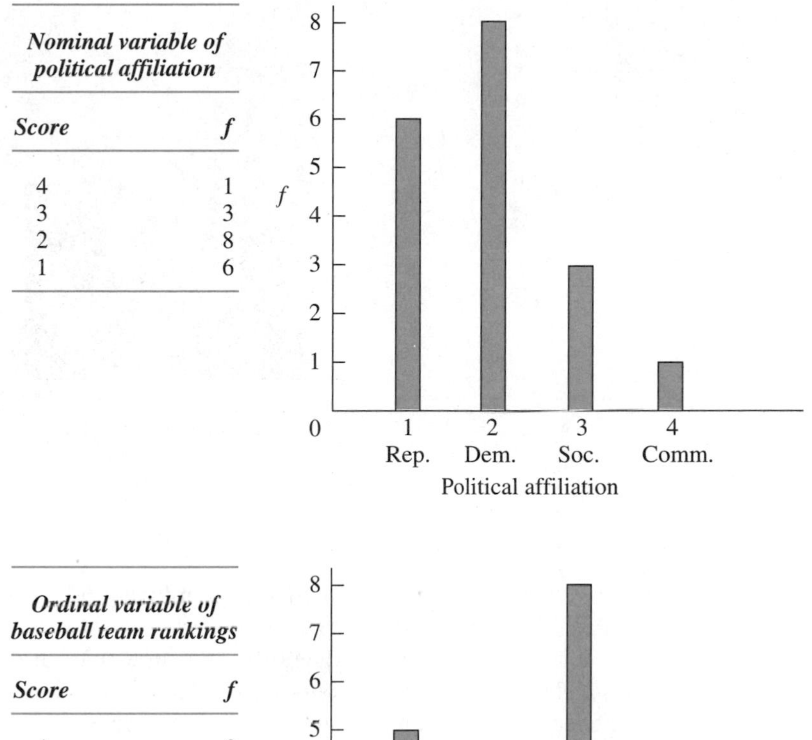

Nominal variable of political affiliation	
Score	***f***
4	1
3	3
2	8
1	6

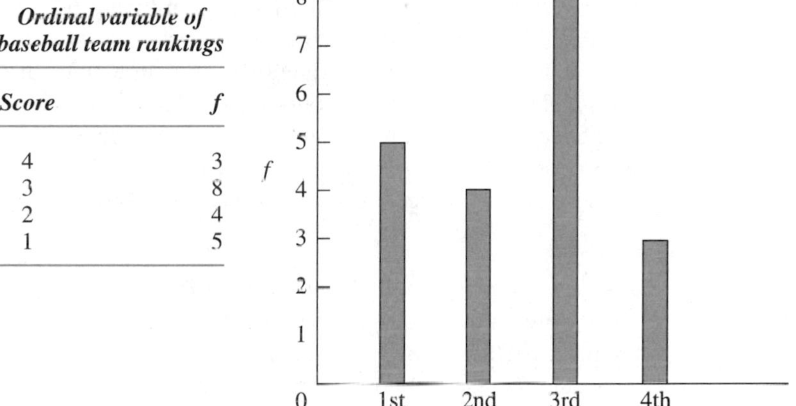

Ordinal variable of baseball team rankings	
Score	***f***
4	3
3	8
2	4
1	5

scales are usually at least theoretically continuous, so these scales are not usually plotted using bar graphs. Instead, we have two ways of graphing such scores, depending upon how many *different* scores the data include.

Histograms Create a histogram when plotting a frequency distribution that contains a *small* range of interval or ratio scores. A **histogram** is similar to a bar graph except that *in a histogram adjacent bars touch*. The absence of a space between the bars in a histogram signals that the scale continues between scores. For example, say that we measured a sample of students on the ratio variable of the number of college

FIGURE 6.2 Histogram Showing the Simple Frequency of College Courses Taken in a Sample

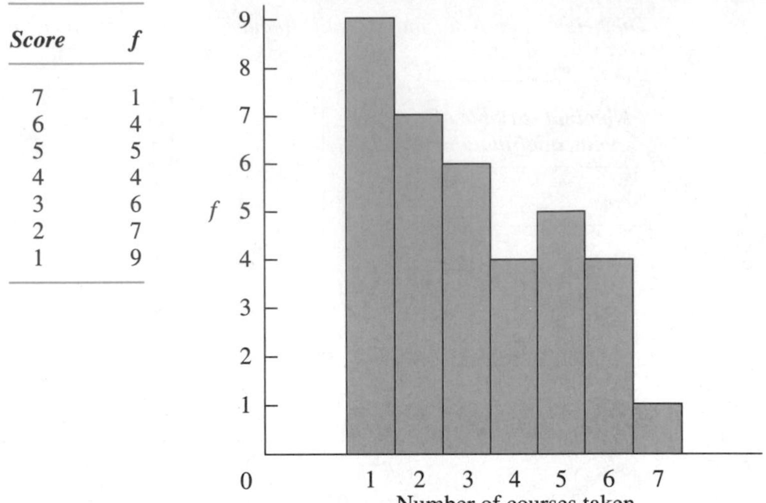

Score	*f*
7	1
6	4
5	5
4	4
3	6
2	7
1	9

courses they've taken, obtaining the data in Figure 6.2. Again, the height of each bar indicates the corresponding score's frequency.

Don't create a histogram when there is a large range of different scores (say if subjects had taken from 1 to 50 courses). The 50 bars would need to be very skinny, so the graph would be difficult to read. Likewise, sometimes we plot more than one sample of scores on the same graph, and even with a small range of scores, overlapping histograms are hard to read. Instead, in such situations create a frequency polygon.

Frequency polygons To construct a **frequency polygon**, place a data point over each score on the *X* axis at the height on the *Y* axis that corresponds to the appropriate frequency. Then connect the data points using straight lines. Figure 6.3 shows the previous data course plotted as a frequency polygon.

Notice that, unlike a bar graph or histogram, a simple frequency polygon includes on the *X* axis the next score *above* the highest score in the data and the next score *below* the lowest score (in Figure 6.3, scores of 0 and 8 are included). These added scores have a frequency of 0, so the polygon touches the *X* axis. In this way we create a complete geometric figure—a polygon—with the *X* axis as its base.

Often in statistics you must read the frequency of a score directly from the polygon. To do this, locate the score on the *X* axis and then move upward until you reach the line forming the polygon. Then, moving horizontally, locate the frequency of the score. For example, as shown by the dashed line in Figure 6.3, the score of 4 has an *f* equal to 4.

REMEMBER The height of the polygon above any score corresponds to the frequency of that score.

FIGURE 6.3 Simple Frequency Polygon Showing the Frequencies of College Courses Taken in a Sample

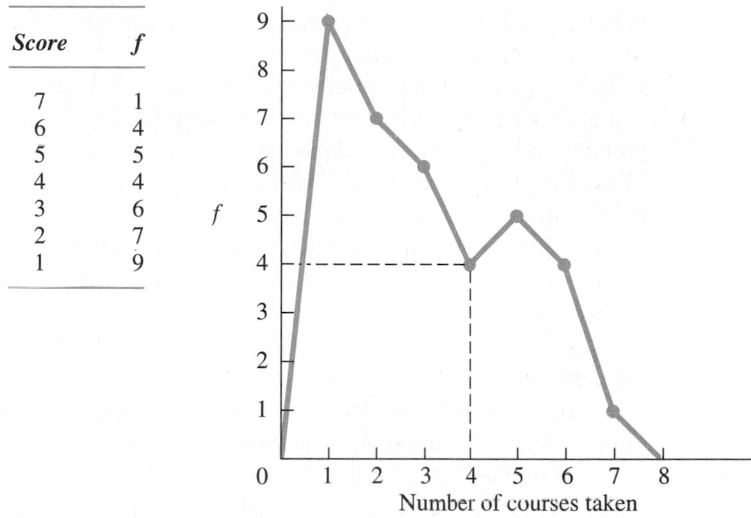

Score	f
7	1
6	4
5	5
4	4
3	6
2	7
1	9

TYPES OF SIMPLE FREQUENCY DISTRIBUTIONS

There are special names for polygons that have certain characteristic shapes. Each shape comes from an idealized frequency distribution of an infinite population of scores. By far the most important frequency distribution is the normal distribution. (This is the big one, folks.)

The Normal Distribution

Figure 6.4 shows the polygon of the ideal theoretical normal distribution.

FIGURE 6.4 The Ideal Normal Curve

Scores farther above and below the middle scores occur with progressively lower frequencies.

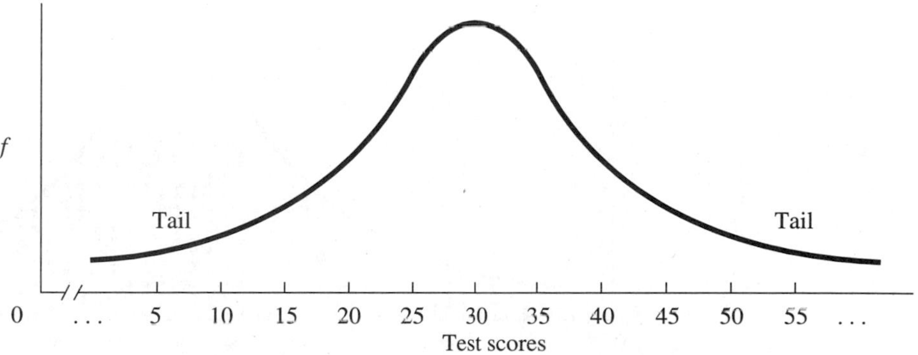

For reference, assume that these are test scores from a population of college students. Although specific mathematical properties define such a polygon, in general it is a bell-shaped curve. But don't call it a bell curve! Because this shape occurs so often, the polygon itself is called the *normal distribution* or the *normal curve*, or we say that the scores are *normally distributed.*

To help you interpret the normal curve (or any polygon for that matter), imagine that you are flying in a helicopter over a parking lot. The X and Y axes are laid out on the ground, and the entire population of subjects is present (it's a very big parking lot). Those people who received a particular score stand in line in front of the marker for their score on the X axis. The lines of people are packed so tightly together that, from the air, all you see is a dark mass formed by the tops of many heads. If you painted a line that went behind the last person in line at each score, you would have the outline of the normal curve. This "parking lot view" is shown in Figure 6.5.

Thus, you can think of the normal curve as a solid geometric figure made up of all the participants at their different scores. The height of the curve above any score gives the corresponding f of the score on the Y axis, which is the same as counting the number of people in line at the score. Likewise, we might, for example, read off the simple frequencies on the Y axis for the scores between 30 and 35 and, by adding them together, obtain the f of scores between 30 and 35. We'd get the same answer by counting the number of people in line above each of these scores and adding them together. And if we read off the simple frequencies on the Y axis for all scores and added them together, we would have the total number of scores (our N). This is the same as counting the number of people in all lines in the parking lot.

As you can see from Figures 6.4 and 6.5, the normal distribution has the following characteristics. The score with the highest frequency is the middle score between the highest and lowest scores (the longest line of people in the parking lot is at the score of 30). The normal curve is *symmetrical*, meaning that the half below the middle score is a mirror image of the half above the middle score. As we proceed away from the middle score toward the higher or lower scores, the frequencies at first decrease only slightly. As we proceed farther from the middle score, the frequencies decrease more drastically, with the highest and lowest scores having relatively low frequency.

FIGURE 6.5 Parking Lot View of the Ideal Normal Curve

The height of the curve above any score reflects the number of subjects obtaining that score.

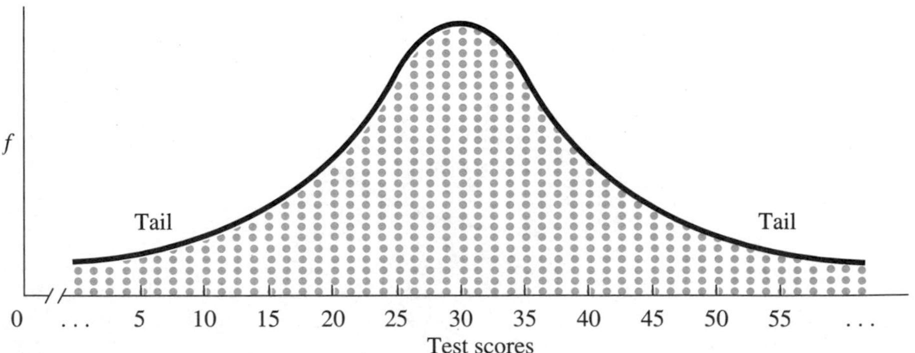

In statistics, the scores that are far above or below the middle score of any distribution are called the **extreme scores**. In a normal distribution, the extreme scores have a relatively low frequency. In the language of statistics, the far left and right portions of a normal curve containing the relatively low frequency, extreme scores are called the **tails** of the distribution. In Figures 6.4 and 6.5, the tails are roughly below the score of 15 and above the score of 45.

Because the ideal normal curve represents a theoretical infinite population of scores, it has several characteristics that are not found with polygons created from actual data. First, with an infinite number of scores, we cannot label the Y axis with specific values of f. (Simply remember that the higher the curve, the higher the frequency.) Second, the theoretical normal curve is a smooth curved line. There are so many different whole-number and decimal scores that we do not need to connect the data points with straight lines. The individual data points form a solid curved line. Finally, regardless of how extreme a score might be, theoretically such a score will sometimes occur. Therefore, although the frequency of scores decreases as we proceed into the tails of the distribution, there is never a frequency of zero so the theoretical normal curve approaches but never actually touches the X axis.

Before you proceed, be sure that you are comfortable reading the normal curve. Can you see in Figure 6.4 that the most frequent scores are between 25 and 35? Do you see that a score of 15 has a relatively low frequency and a score of 45 has the same low frequency? Do you see that there are relatively few scores in the tail above 50 or in the tail below 10? Above all, you must be able to see this in your sleep:

> **On a normal distribution, the farther a score is from the central score of the distribution, the less frequently the score occurs.**

The normal curve model The pattern reflected in a normal distribution is very common in psychological research: Most individuals tend to score at or slightly above or below a middle score, with progressively fewer individuals scoring at more extreme low or high scores. Of course, scores from the real world never form the precise mathematical normal curve. Nonetheless, they are often close enough to forming the ideal normal curve for us to use it as a model. That is, we apply the **normal curve model**, meaning that we assume a population of scores comes close enough to forming the normal curve that we treat it as if it does form a normal curve and is normally distributed. Then, because there are precise mathematical procedures for describing the normal curve, we can apply these procedures to describe the distribution of actual scores.

Notice what the normal curve model enables us to do. Recall that the goal of research is to understand how everyone in a population will behave in a particular situation. In statistical terms, this means describing the population of scores we would find in a particular situation. Saying that the population forms a normal distribution is a major part of describing the population, so we are well on the way to our goal.

Variations in the normal curve Different samples of data, however, will produce variations in the shape of a normal distribution, and we have a statistical term for describing such differences. Consider the three curves in Figure 6.6. The word *kurtosis* refers to how flat or peaked—how skinny or fat—a distribution is. Curve A is generally what we think of when we discuss the ideal normal distribution, and such a curve is called mesokurtic (*meso* means middle). Curve B is skinny relative to the ideal normal

FIGURE 6.6 Variations of Bell-Shaped Curves

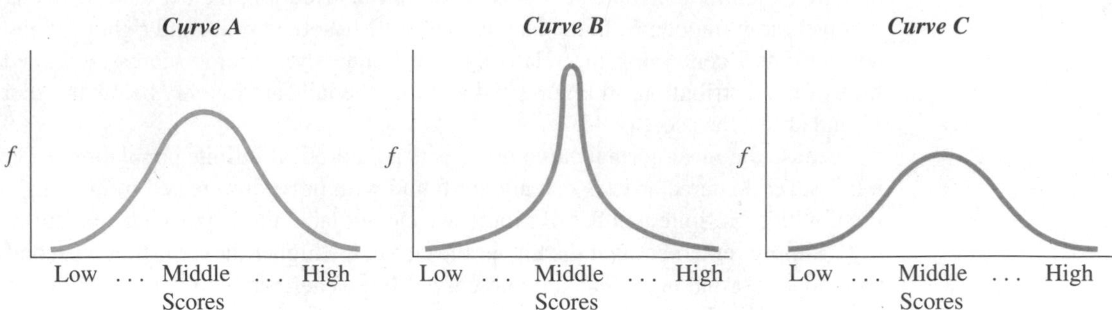

curve, and it is called leptokurtic (*lepto* means thin). In leptokurtic distributions, only a few scores around the middle score have a relatively high frequency. On the other hand, Curve C is fat relative to the ideal normal curve. And such a curve is called platykurtic (*platy* means broad or flat). Here there is a wide range of different scores around the middle score that each have a relatively high frequency.

These terms help us to describe various normal distributions. For statistical purposes, however, as long as we have a reasonably close approximation to the normal curve, such differences in shape are not all that critical.

Other Common Frequency Polygons

The distribution from every variable does not form a normal distribution. When a distribution does not fit the normal curve, it is a *nonnormal* distribution. The three most common nonnormal distributions are *skewed*, *bimodal*, and *rectangular* distributions.

Skewed distributions A **skewed distribution** is similar in shape to a normal distribution except that it is not symmetrical: The left half of the polygon is not a mirror image of the right half. *Instead a skewed distribution has only one pronounced tail.* As shown in Figure 6.7, a distribution may be either negatively skewed or positively skewed.

FIGURE 6.7 Idealized Skewed Distributions

The direction in which the distinctive tail is located indicates whether the skew is positive or negative.

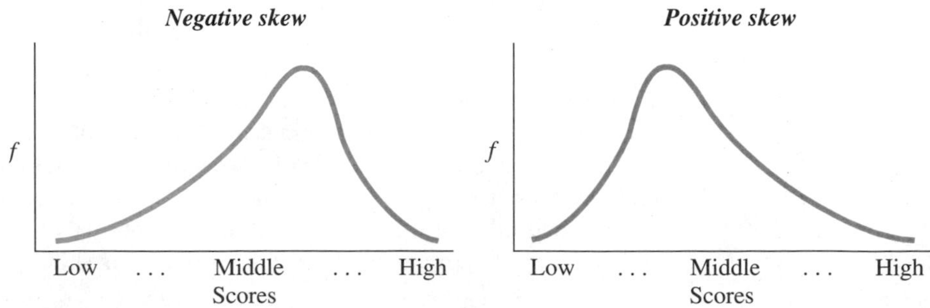

A **negatively skewed distribution** contains extreme low scores that have low frequency, but does not contain extreme high scores that have low frequency. The left-hand polygon in Figure 6.7 shows an idealized negatively skewed distribution. This pattern might be found, for example, when measuring the running speed of professional football players. Most would tend to run at higher speeds, with a relatively few linemen lumbering in at the slower speeds. To remember that such a curve is negatively skewed, remember that the one pronounced tail is over the lower scores, sloping toward zero, toward where *negative* scores would be.

On the other hand, a **positively skewed distribution** contains extreme high scores that have low frequency, but does not contain extreme low scores that have low frequency. The right-hand polygon in Figure 6.7 shows a positively skewed distribution. This pattern is common, for example, when measuring reaction time. Most frequently the scores will tend to be rather low (fast), but every once in while a participant "falls asleep at the switch," requiring a large amount of time and thus producing a high score. To remember that such a curve is positively skewed, remember that the one tail is on the side away from zero, toward where the higher, *positive* scores are located.

> *REMEMBER* Whether a skewed distribution is negative or positive corresponds to whether the distinct tail is toward or away from where the negative scores would be located.

Bimodal and rectangular distributions An idealized bimodal distribution is shown in the left-hand side of Figure 6.8. A **bimodal distribution** is a symmetrical distribution that contains two distinct humps reflecting relatively high frequency scores. At the center of each hump is a score that occurs more frequently than the surrounding scores (and technically the center scores in each hump have the same frequency). Such a distribution would occur with test scores, for example, if most students scored around 60 or 80, with fewer students failing or scoring in the 70's or 90's.

The right-hand side of Figure 6.8 shows a rectangular distribution. A **rectangular distribution** is a symmetrical distribution shaped like a rectangle. There are no discernible tails, because the extreme scores do not have relatively low frequencies. Such a distribution occurs whenever the frequency of all scores is the same.

Distributions of Real Data Versus Ideal Distributions

If you're wondering why you need to know the names of the previous distributions, it's because we use descriptive statistics to describe the important characteristics of a

FIGURE 6.8 Idealized Bimodal and Rectangular Distributions

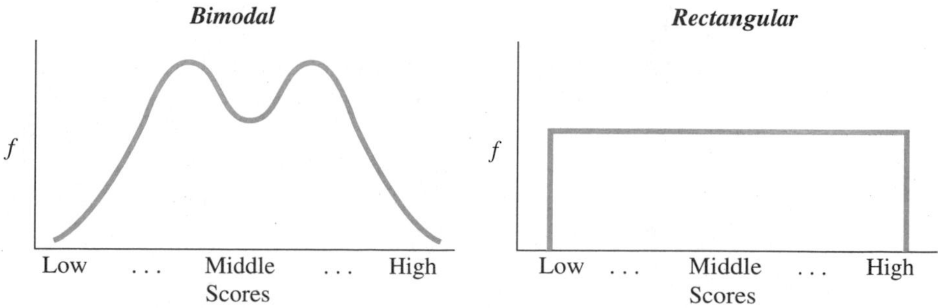

FIGURE 6.9 Simple Frequency Distributions of Sample Data with Appropriate Labels

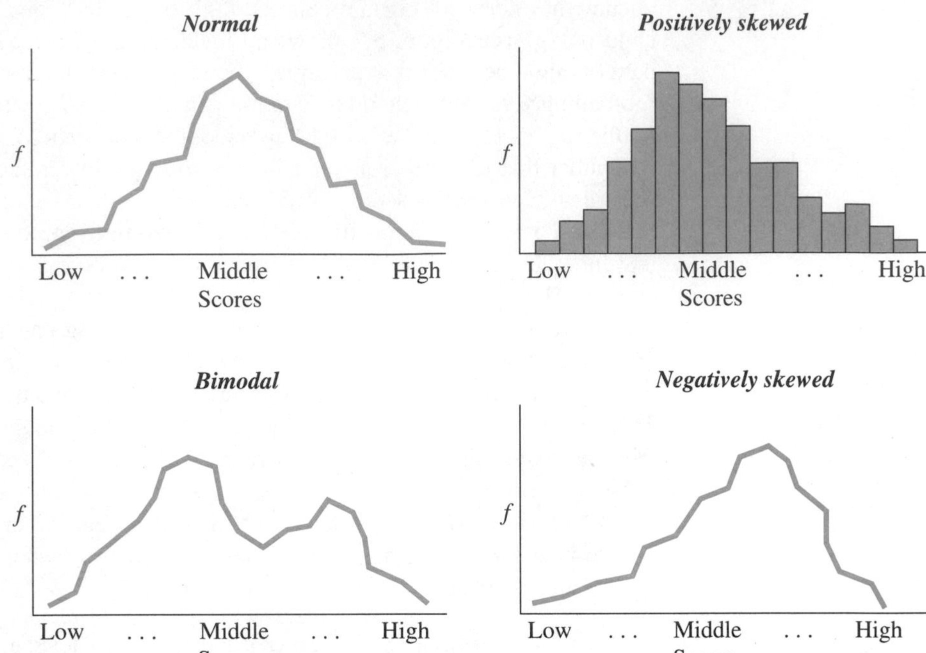

sample of data. One important characteristic is the shape of the frequency distribution which the data form, so we apply the names of the previous distributions to sample data as well. Real data are never pretty, however, and the distribution in a sample will tend to be a bumpy, rough approximation of the smooth idealized curves we've discussed. For example, Figure 6.9 above shows several frequency distributions of sample data, as well as the corresponding labels we might use. Notice that we can apply these names even to choppy histograms or bar graphs.

We generally assume that the sample represents a population that more closely fits the corresponding ideal polygon: We expect that if we measured the population, the additional scores and their corresponding frequencies would "fill in" the sample curve, smoothing it out to form the ideal curve.

We'll return to simple frequency distributions throughout the remainder of this book. However, counting each score's simple frequency is not the only thing we do in statistics.

CREATING RELATIVE FREQUENCY DISTRIBUTIONS

Another way to organize scores is to transform each score's simple frequency into relative frequency. **Relative frequency** is the proportion of the total N made up by a score's simple frequency. While simple frequency is the *number* of times a score occurs in the data, relative frequency is the *proportion* of times the score occurs. (A proportion is a decimal number between 0 and 1 that indicates a fraction of the total.) The symbol for relative frequency is *rel. f.*

Why compute relative frequency? We are again asking how often certain scores occurred, but relative frequency is often easier to interpret than simple frequency. For example, the finding that a score has a simple frequency of 60 is difficult to interpret, because we have no frame of reference. However, we can easily interpret the finding that a score has a relative frequency of .20, because this means that the score constitutes .20 of all scores in the sample.

Here is your first statistical formula.

THE FORMULA FOR COMPUTING A SCORE'S RELATIVE FREQUENCY IS:

$$Rel.\,f = \frac{f}{N}$$

To compute the relative frequency of a score, divide that score's frequency by the total *N*. For example, say that out of an *N* of 10 scores, the score of 7 has a simple frequency of 4. What is the relative frequency of 7? Using the formula, we have:

$$Rel.\,f = \frac{f}{N} = \frac{4}{10} = .40$$

The score of 7 has a relative frequency of .40, meaning that 7 occurred .40 of the time in the sample.

Conversely, to compute the simple frequency that corresponds to a certain relative frequency, multiply the relative frequency times *N*. Thus, to find what simple frequency constitutes .4 out of an *N* of 10, multiply .4 times 10, and voilà, 4 is .4 of 10.

Presenting Relative Frequency in a Table

A distribution based on the relative frequency of the scores is called a **relative frequency distribution**. To create a table showing relative frequency, first create a simple frequency table as we did previously. Then show the relative frequency of each score by adding a third column labeled "*rel. f.*"

As an example, say that we asked a sample of mothers how many children they each have, obtaining the results shown in Table 6.3. To compute *rel. f*, we need *N*, which is the total number of scores in the sample. Then the relative frequency for each score is the *f* for that score divided by *N*. The score of 1, for example, has $f = 4$, so the relative frequency of 1 is 4/20, or .20: In this sample, .20 of the participants have 1 child.

We can also determine the combined relative frequency of several scores by adding the individual relative frequencies together. For example, a score of 1 has a relative frequency of .20, and a score of 2 has a relative frequency of .50, so the relative frequency of 1 and 2 is .20 + .50, or .70: Mothers having 1 or 2 children compose .70 of this sample.

TABLE 6.3 Relative Frequency Distribution of Number of Children

The left-hand column identifies the score, the middle column shows each score's frequency, and the right-hand column shows each score's relative frequency.

Score	*f*	*rel. f*
6	1	.05
5	0	.00
4	2	.10
3	3	.15
2	10	.50
1	4	.20
	Total: 20	1.00 = 100%

You may find that working with relative frequency is easier if you transform it to a percentage. (Remember that officially relative frequency is a proportion.) By converting relative frequency to a percentage, you have the percent of the time that a score or scores occurred. To transform a proportion to percent, multiply the proportion times 100. Above, .20 of the scores were the score of 1, so (.20)(100) = 20%: 20% of the scores were 1. To transform a percent back to a relative frequency, divide the percent by 100.

To check your work, remember that the sum of all the relative frequencies in a distribution should equal 1.0: All scores together should constitute 1.0, or 100%, of the sample.

> *REMEMBER* Relative frequency is interpreted as the proportion of the time that certain scores occur in a set of data.

Graphing a Relative Frequency Distribution

As with simple frequency, relative frequency is graphed as a bar graph if the scores are from a nominal or ordinal scale, and as a histogram or polygon if the scores are from an interval or ratio scale. Figure 6.10 presents examples using the relative frequency distribution we created back in Table 6.3. These graphs are drawn in the same way as graphs of simple frequency except that here the *Y* axis reflects relative frequency, so it is labeled in increments between 0 and 1.0.

Finding Relative Frequency Using the Normal Curve

When data form a normal distribution, an extremely valuable procedure is to determine relative frequency directly from the normal curve. One reason for visualizing the normal curve as the outline of a parking lot full of people is so that you think of the normal curve as forming a solid geometric figure having an area underneath the curve. What we consider to be the total space occupied by people in the parking lot is, in statistical terms, the **total area under the curve**. This area represents the total frequency of all scores.

FIGURE 6.10 Examples of Relative Frequency Distributions (Using Data in Table 6.3)

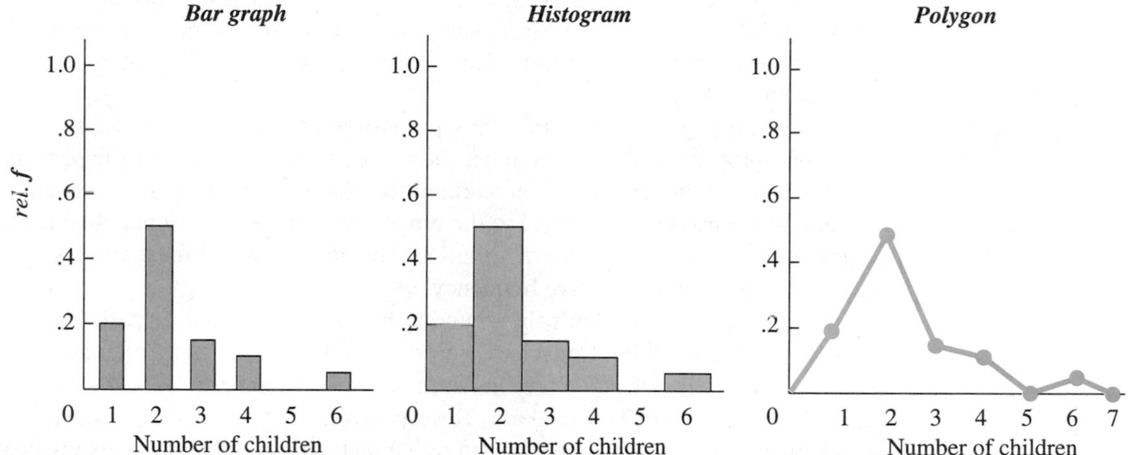

To find the relative frequency of scores in any portion of the distribution, we first find the area of that portion of the normal curve. We take a vertical slice of the polygon above the scores and the area of this portion of the curve is the space occupied by the people obtaining those particular scores. Therefore, this area represents the combined frequency for these scores. We then compare this area to the total area, to determine the "proportion of the total area under the curve." Now here's the important part: The proportion of the total area under the normal curve at certain scores corresponds to the relative frequency of those scores.

For example, Figure 6.11 shows a normal curve with the parking lot view. A vertical line is drawn through the middle score of 30, and so .50 of the parking lot is to the left of the line. Because the complete parking lot contains all participants, a slice that is .50 of it contains 50% of the participants. (We can ignore those relatively few people who

FIGURE 6.11 Normal Curve Showing .50 of the Area Under the Normal Curve

The vertical line is through the middle score, so 50% of the distribution is to the left of the line and 50% is to the right of the line.

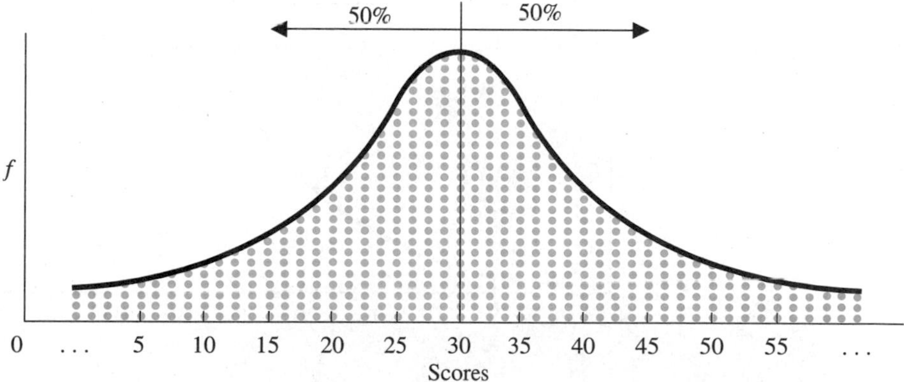

are straddling the line.) People are standing in the left-hand part of the lot because they received scores of 29, 28, and so on, so in total, 50% of the participants obtained scores below 30. Fifty percent of the participants had these scores so, in other words, scores below 30 occurred 50% of the time. Thus, the scores below 30 have, altogether, a relative frequency of .50.

In the same way, we can identify any proportion of the total area under the curve. Think of a proportion of the area under the curve as the proportion of the participants who are standing in that part of the parking lot. The proportion of people standing in that part of the parking lot is equal to the proportion of time that people obtained those scores out of all the scores in the distribution. The proportion of time that certain scores occur out of all scores *is* relative frequency.

Of course, statisticians don't fly around in helicopters, eyeballing parking lots. But the same principle applies. Consider Figure 6.12. Say that by using a ruler and protractor, we determine that the total area under the curve—the entire polygon—occupies an area of 6 square inches. This total area corresponds to the total of all frequencies for all scores, which is N. Say that we also determine that the area under the curve between a score of 30 and 35 covers 2 square inches. This area corresponds to the frequencies of the scores found there. Therefore, the total frequency of scores between 30 and 35 constitutes 2 out of the 6 square inches created by all scores, so these scores constitute two-sixths, or 33%, of the total distribution. In other words, the scores between 30 and 35 occur 33% of the time, so they have a relative frequency of .33.

Because area corresponds to frequency, we would obtain the same answer if we used the formula for *rel. f.* First, we would add together the simple frequencies of every score between 30 and 35. Then, dividing the sum by N, we would again find that the relative frequency is .33. The advantage of using area under the curve, however, is that we can get the answer without knowing the N or the simple frequencies of any scores. Whatever the N might be and whatever the actual frequency of each score is, we know that the area comprised by these scores is 33% of the total area, and that's all we need to know to determine their relative frequency. This is especially advantageous because, as we'll see, statisticians have a system for finding the proportion of the total area under

FIGURE 6.12 Finding the Proportion of the Total Area Under the Curve

The complete curve occupies 6 square inches, with scores between 30 and 35 occupying 2 square inches.

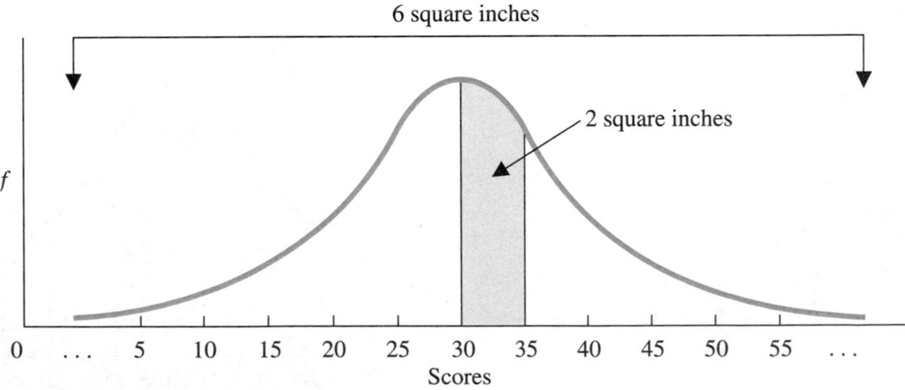

any part of the normal curve, which means we can easily determine the relative frequency for any group of scores within a normal distribution. (No, you won't need a ruler and a protractor.) Until that time, simply remember this:

> *REMEMBER* Total area under the normal curve corresponds to the times that all scores occur, so a proportion of the total area corresponds to the proportion of time certain scores occur, which is their relative frequency.

CREATING CUMULATIVE FREQUENCY DISTRIBUTIONS

Sometimes we want to know not only how often a particular score occurred, but also its standing relative to other scores in the data. Knowing that 10 people received an exam grade of 80 may not be as informative as knowing that 30 people scored above 80 or that 60 people scored below 80. When we seek such information, the convention is to count from *lower* scores, computing cumulative frequency. **Cumulative frequency** is the frequency of all scores at or below a particular score. The symbol for cumulative frequency is *cf*. The word *cumulative* implies accumulating: To compute a score's cumulative frequency, we accumulate, or add, the simple frequencies for scores below and at that score.

Presenting Cumulative Frequencies in a Table

To create a cumulative frequency table, first create a simple frequency table. Then add a third column labeled "*cf*." As an example, say that we've measured the ages of a sample of adolescents. To summarize these scores, we create Table 6.4.

TABLE 6.4 Cumulative Frequency Distribution of Age Scores

The left-hand column identifies the scores, the center column contains the simple frequency of each score, and the right-hand column contains the cumulative frequency of each score.

Score	*f*	*cf*
17	1	19
16	2	18
15	4	16
14	5	12
13	4	7
12	0	3
11	2	3
10	1	1
	Total: 19	

To compute cumulative frequency, begin with the *lowest* score. No one scored below 10 and 1 person scored 10, so we put 1 in the *cf* column opposite 10: This indicates one person is 10 years of age or younger. Next, there were two scores of 11. We add this *f* to the *cf* for 10, so the *cf* for 11 is 3: 3 participants are at or below the age of 11. Next, no one scored 12 and three children scored below 12, so the *cf* for 12 is also 3: There are 3 people at age 12 or below. In the same way, the *cf* of each score is the frequency for that score plus the cumulative frequency for the score immediately below it.

As a check on any cumulative frequency distribution you create, the *cf* for the highest score must equal *N*: All of the *N* participants obtained either the highest score or a score below it.

Graphing a Cumulative Frequency Distribution

Usually, it makes sense to compute cumulative frequency only for interval or ratio data, and the convention is to create a polygon. Figure 6.13 shows the cumulative frequency polygon of the data from Table 6.4. In a cumulative frequency distribution, the *Y* axis is labeled "*cf*." As with previous polygons, include on the *X* axis the next score *below* the lowest score in the data: 10 was the lowest age score above, but 9 is included on the graph. Unlike previous polygons, however, with cumulative frequency do *not* include the next score above the highest score in the data. Also, notice that cumulative frequencies never decrease, because there cannot be fewer subjects having a score of 13 *or below* than received a score of 12 (or below). Therefore, as the scores on the *X* axis increase, the height of the polygon must either remain constant or increase.

FIGURE 6.13 Cumulative Frequency Polygon Showing the Cumulative Frequencies of the Scores in Table 6.4.

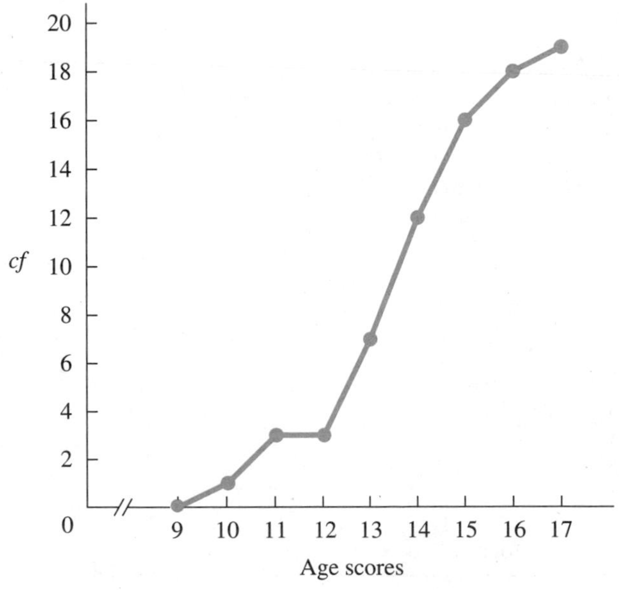

COMPUTING PERCENTILE

There is one other way to answer the question of how often various scores occurred. Previously, we transformed simple frequency into relative frequency, because the proportion (or percent) of the time a score occurs is easier to interpret than the number of times it occurs. Likewise, cumulative frequency can be difficult to interpret because it is only the number of times scores at or below a score occur. Therefore, our final approach is to transform cumulative frequency into a percent of the total. A **percentile** is the percent of all scores in the data that are at or below a certain score. While cumulative frequency indicates the *number* of participants who obtained a particular score or below, a percentile indicates the *percentage* of participants who obtained a particular score or below. Thus, for example, if a person scores at the 25th percentile, we know that 25% of all participants scored at or below that person's score.

One way to compute percentile is to again examine the area under the normal curve. On a graph of a normal distribution, lower scores occur to the left of a particular score. Therefore, a percentile for a given score corresponds to the percent of the total curve that is to the *left* of the score. For example, on the distribution of test scores in Figure 6.14, 50% of the curve is to the left of the middle score of 30. Because scores to the left of 30 are below it, 50% of the distribution is below 30 (50% of the people are standing to the left of the line in the parking lot, and all their scores are below 30). Thus, the score of 30 is at the 50th percentile. Likewise, to find the percentile for a score of 20, we would find the percent of the total curve that is to the left of 20. In Figure 6.14, 15% of the distribution is to the left of 20, so 20 is at the 15th percentile.

We can also work the other way, finding the score at a given percentile: If we seek the score at the 85th percentile, we measure over until 85% of the area under the curve is to the left of a certain point. If, as in Figure 6.14, the score of 45 is at that point, then 45 is at the 85th percentile.

Notice that we make a slight change in the definition of percentile when using the normal curve. Technically, a percentile is the percent of scores *at* or below a certain score. However, we should use the normal curve to describe large samples or populations, and then we can treat those participants scoring *at* the score as a negligible portion of the total (remember we ignored those relatively few subjects who were

FIGURE 6.14 Normal Distribution Showing the Area Under the Curve to the Left of Selected Scores

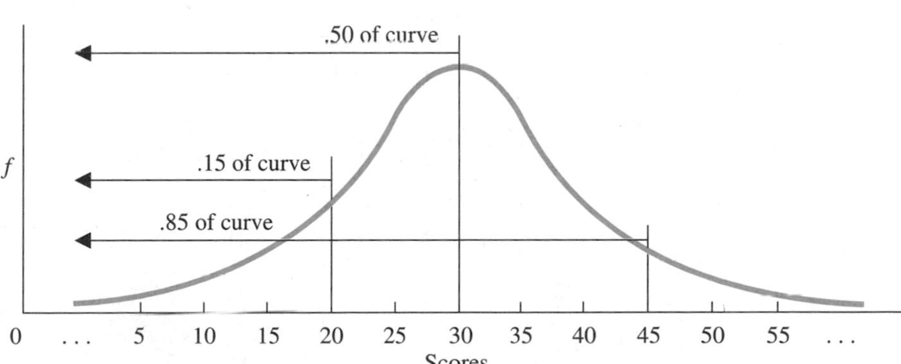

straddling the line). Then a percentile is the percent of all scores that are *below* a certain score. Thus, in Figure 6.14, the score of 30 is at the 50th percentile, so this means that 50% of the scores are below 30 and 50% are above it.

When describing a small sample, however, we should *not* say that 50% of the scores are above the 50th percentile and 50% are below it; those participants scoring *at* the 50th percentile might actually constitute a relatively sizeable proportion of the sample, which we cannot ignore. And, if we say that 50% are below, 50% are above, but, for example, 10% are at the score, we have the impossible total of 110% of the sample! Therefore, with small samples, percentile is defined and calculated as the percent of scores at or below a particular score.

Thus using the normal curve to compute percentile will be reasonably accurate only if we have a rather large sample or a population that closely fits the normal curve. With a small sample, or if the data form a nonnormal distribution, we instead calculate percentile using a computer program or using the formulas presented in Part I of Appendix B.

A WORD ABOUT GROUPED FREQUENCY DISTRIBUTIONS

A rule of thumb for any type of frequency table is that there should be between about 8 and 18 rows in the table. Fewer than 8 scores tends to produce a very small, often unnecessary table, while more than 18 scores tends to produce a very large, inefficient table. In the previous examples, we examined each score individually, so we created **ungrouped distributions**. When there are too many individual scores to produce a manageable ungrouped distribution, we create a grouped distribution. In a **grouped distribution**, different scores are combined to form small groups and then we report the total *f*, *rel. f*, or *cf* in each group.

For example, say that we measure how anxious 25 people each get when making a speech by giving them an anxiety questionnaire. Their anxiety scores span a wide range, so we create the grouped distribution shown in Table 6.5.

TABLE 6.5 Grouped Distribution Showing *f*, *rel. f*, and *cf* for Each Group of Scores.

The left-hand column identifies the lowest and highest score in each class interval.

Scores	***f***	***rel. f***	***cf***
40–44	2	.08	25
35–39	2	.08	23
30–34	0	.00	21
25–29	3	.12	21
20–24	2	.08	18
15–19	4	.16	16
10–14	1	.04	12
5– 9	4	.16	11
0– 4	7	.28	7
	Total: 25	1.00	

The group labeled 0–4 contains the scores 0, 1, 2, 3, and 4, while 5–9 contains scores 5 through 9, and so on. Although we don't know the frequencies of each individual score, we do see that scores between 0 and 4 have a total f of 7, while scores between 5 and 9 have a total f of 4. Likewise, the relative frequency of scores between 0 and 4 is .28, while for 5–9, it is .16. The cumulative frequency for each interval indicates the number of scores that are at or below the *highest* score in the interval. Thus, there are 7 scores at 4 or below, with 11 scores at 9 or below. (For details on how to create and graph grouped distributions, consult Part I of Appendix B.)

PUTTING IT ALL TOGETHER

All of the procedures you've learned in this chapter indicate how often certain scores occur, but each provides a slightly different perspective that allows you to interpret the data in a slightly different way. Which procedure you should use in a particular setting is determined simply by which provides the most useful information. Regardless of whether you've conducted an observational study or survey, or you are summarizing the scores on one variable from a correlational study or the dependent scores from an experiment, choose the technique that most accurately and efficiently summarizes the data.

As an aid to learning statistics, start drawing the normal curve. When you are working problems or taking tests, draw the normal curve and indicate where the low, middle, and high scores are located. Being able to see the frequencies of the different scores will greatly simplify your task.

CHAPTER SUMMARY

1. A *transformation* is a mathematical procedure for systematically converting one set of scores into a different set of scores. Transformations make scores easier to work with and make different kinds of scores comparable.
2. Round off the final answer in a calculation to two more decimal places than are in the original scores. If the digit in the next decimal place is equal to or greater than 5, round up; if the digit is less than 5, round down.
3. The total number of scores in a sample is symbolized by N.
4. Which statistical procedure to use in a particular study depends on the *scale of measurement*: (1) In a *nominal scale*, numbers name or identify a quality or characteristic; (2) In an *ordinal scale*, numbers indicate rank order; (3) In an *interval scale*, numbers measure a specific amount, but with no true zero; and (4) In a *ratio scale*, numbers measure a specific amount and 0 indicates truly zero amount.
5. In a *continuous* variable, decimals make sense. In a *discrete variable*, decimals do not make sense. A *dichotomous variable* is a discrete variable that has only two amounts or categories.
6. A *simple frequency distribution* shows the frequency with which each score occurred. The symbol for simple frequency is f.

7. When graphing a simple frequency distribution, if the variable involves a nominal or ordinal scale, create a *bar graph*. If the variable involves an interval or ratio scale and there are relatively few different scores, create a *histogram*. If there is a wide range of scores from an interval or ratio variable or there is more than one sample of scores, create a *polygon*.

8. In a *normal distribution* forming a *normal curve*, extreme high and low scores occur relatively infrequently, scores closer to the middle score occur more frequently, and the middle score occurs most frequently. The low frequency, extreme low and extreme high scores are in the *tails* of the distribution.

9. The *normal curve model* assumes that a population of scores generally forms a normal distribution.

10. A *negatively skewed distribution* is a nonsymmetrical distribution containing low frequency, extreme low scores, but not containing low frequency, extreme high scores. A *positively skewed distribution* is a nonsymmetrical distribution containing low frequency, extreme high scores, but not containing low frequency, extreme low scores.

11. A *bimodal distribution* is a symmetrical distribution containing two areas where there are relatively high frequency scores. A *rectangular distribution* is a symmetrical distribution in which the extreme scores do not have relatively low frequencies.

12. The *relative frequency* of a score, symbolized by *rel. f*, is the proportion of time that the score occurred in a distribution.

13. The *proportion of the total area under the normal curve* above a score or scores equals the relative frequency of the score or scores.

14. The *cumulative frequency* of a score, symbolized by *cf*, is the frequency of all scores at or below the score.

15. *Percentile* indicates the percent of all scores at or below a given score. On the normal curve, the percentile of a score is the percent of the curve to the left of the score.

16. In an *ungrouped distribution*, individual scores are examined. In a *grouped distribution*, different scores are grouped together and then the total *f*, *rel. f*, or *cf* for each group is computed.

PRACTICE PROBLEMS

(Answers for odd-numbered problems are provided in Appendix D.)

1. (a) What is a transformation? (b) Why do we transform data?

2. (a) To how many places do we round off a final answer? (b) What are the rules for rounding up or down when rounding to two decimal places?

3. What are the three things to consider when deciding on the particular statistical procedure you should employ?

4. (a) Define the four scales of measurement. (b) Rank order the four scales of measurement from most sensitive to least sensitive and explain your answer.
5. What do each of the following mean? (a) *N* (b) *f* (c) *rel. f* (d) *cf.*
6. What does a rectangular distribution show about the relationship between frequency and the different scores?
7. What type of frequency graph should you create when counting each of the following? (a) The males and females at a college. (b) The different body weights reported in a statewide survey. (c) The number of sergeants, lieutenants, captains, and majors in an army battalion. (d) The people falling into one of eight salary ranges.
8. What is the difference between graphing a relationship as we did in Chapter 2 and graphing a frequency distribution?
9. What does it mean when a score is in one of the tails of the normal distribution?
10. (a) How is percentile defined in a small sample? (b) How is percentile defined for a large sample or population when calculated using the normal curve?
11. (a) What is the difference between a score's simple frequency and its relative frequency? (b) What is the difference between a score's cumulative frequency and its percentile?
12. What is the difference between how we use the proportion of the area under the normal curve to determine a score's relative frequency and how we use it to determine a score's percentile?
13. For each of the following calculations, to how many places will you round off your final answer? (a) When measuring the number of questions students answered correctly on a test. (b) When measuring what proportion of the total possible points students have earned in a course. (c) When counting the number of people having various blood types. (d) When measuring the number of dollar bills possessed by each person in a group.
14. The intermediate answers from some calculations based on whole number scores are $X =$ 4.3467892 and $Y = 3.3333$. We now want to find $X^2 + Y^2$. What values of X and Y do we use?
15. Round off the following numbers to two decimal places: (a) 13.7462 (b) 10.043 (c) 10.047 (d) .079 (e) 1.004
16. A professor observes that a distribution of test scores is positively skewed. What does this indicate about the difficulty of the test?
17. In reading psychological research you encounter the following statements. Interpret each one. (a) "The IQ scores were approximately normally distributed." (b) "A bimodal distribution of physical agility scores was observed." (c) "The distribution of the patients' memory scores was severely negatively skewed."
18. In the chart below, identify the characteristics of each variable.

Variable	*Qualitative or quantitative*	*Continuous, discrete, or dichotomous*	*Type of measurement scale*
gender	________	________	________
academic major	________	________	________
time	________	________	________
restaurant ratings	________	________	________
speed	________	________	________
money	________	________	________
position in line	________	________	________
change in weight	________	________	________

19. From the data 1, 4, 5, 3, 2, 5, 7, 3, 4, 5, Poindexter created the following frequency table. What five things did he do wrong?

Score	*f*	*cf*
1	1	0
2	1	1
3	2	3
4	2	5
5	3	8
7	1	9
	$N = 6$	

20. (a) On a normally distributed set of exam scores, Poindexter scored at the 10th percentile, so he claims that he outperformed 90% of his class. Why is he correct or incorrect?
(b) Because Foofy's score had a relative frequency of .02, she claims she had one of the highest scores on the exam. Why is she correct or incorrect?

21. The following shows the distribution of final exam scores in a large introductory psychology class. The proportion of the total area under the curve is given for two segments.

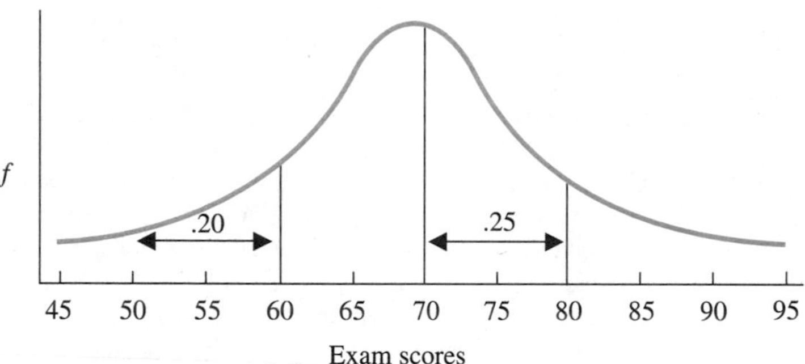

(a) Order the scores 45, 60, 70, 72, and 85 from most frequent to least frequent.
(b) What is the percentile of a score of 60?
(c) What proportion of students scored between 60 and 70?
(d) What is the percentile of a score of 80?

22. Organize the scores below in a table showing simple frequency, relative frequency, and cumulative frequency.

49	52	47	52	52	47	49	47	50
51	50	49	50	50	50	53	51	49

23. Organize the scores below in a table showing simple frequency, cumulative frequency, and relative frequency.

16	11	13	12	11	16	12	16	15
16	11	13	16	12	11			

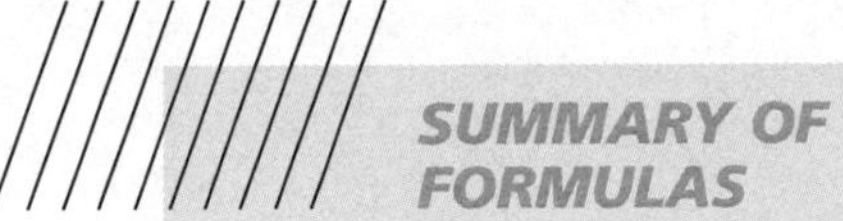

SUMMARY OF FORMULAS

1. *THE FORMULA FOR COMPUTING A SCORE'S RELATIVE FREQUENCY IS:*

$$Rel.\,f = \frac{f}{N}$$

where f is the score's simple frequency and N is the number of scores in the sample.

7

Summarizing Data with Measures of Central Tendency

To understand this chapter:

- From Chapter 2, understand what a relationship is, what independent and dependent variables are, and what the logic of an experiment is.
- From Chapter 6, understand the four measurement scales and using the area under the curve to calculate percentile.

Then your goals in this chapter are to learn:

- How measures of central tendency describe data.
- What the mean, median, or mode indicates and when each is appropriate.
- How a sample mean is used to describe both individual scores and the population of scores.
- What is meant by "deviations around the mean" and what they convey about each score's location and frequency in a normal distribution.
- How to interpret and graph the results of an experiment.

Recall that descriptive statistics describe the important characteristics of a set of data. The graphs and tables discussed in Chapter 6 are important because the type of distribution we find is one important characteristic. However, graphs and tables are not the most efficient way to summarize data. Instead, we can compute individual numbers—*statistics*—that each communicate an important characteristic of the scores. Which statistic we compute depends on what it is we want to know about the data. This chapter discusses the important characteristic called central tendency. A measure of central tendency allows us to answer the question "Are the scores generally high scores or generally low scores?" But first . . .

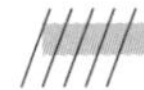

MORE STATISTICAL NOTATION

Recall that the data in a sample is used to estimate the scores we expect to find in the entire population. So that we know when we're describing a sample and when we're describing a population, we use the following system. A number that describes a characteristic of a sample of scores is called a **statistic**, and the symbols for different statistics are letters from the English alphabet. A number that describes a characteristic of a population of scores is called a **parameter**, and the symbols for different parameters are letters from the Greek alphabet. Thus, your bowling average from a sample of your games is a statistic, and the symbol for a sample average is from the English alphabet. If we use that average to estimate the average in your population of all bowling scores we are estimating a population parameter. The symbol for a population average is from the Greek alphabet.

Whether calculating statistics or parameters, formulas are written so they can be applied to any data. Usually, the symbol X stands for each individual raw score in a study. When a formula says to do something to X, it means to do it to all of the scores you're calling X scores.

A new symbol you'll see in conjunction with X is Σ, the Greek capital letter S, called sigma. Sigma means to sum, so ΣX is pronounced "**sum of X**" and literally means to add up all the X scores. For example, the sum of the scores 1, 2, and 3 is 6, so $\Sigma X = 6$. Notice that we do not care whether each X is a different score. If the scores are 4, 4, and 4, then $\Sigma X = 12$.

Finally, formulas often call for a series of mathematical steps, and you must identify the order in which to perform the steps. Sometimes, the steps are set apart by parentheses. Parentheses mean "the quantity," so always find the quantity inside the parentheses first and then perform the operations outside of the parentheses on that quantity. For example, $(2)(4 + 3)$ indicates to multiply 2 times "the quantity 4 plus 3." So first add, which gives $(2)(7)$, and then multiply to get 14.

Now, on to central tendency.

UNDERSTANDING CENTRAL TENDENCY

To understand central tendency, you need to first alter your perspective of what a score indicates. For example, if I am 70 inches tall, don't think of my score as indicating that I have 70 inches of height. Instead, think of a score as indicating a *location* on a variable: Think of me as being located on the variable of height at the point marked 70 inches. Think of any variable as an infinite continuum—a straight line—with each score indicating an individual's location on that line. Look at Figure 7.1: My score locates me at the address labeled 70 inches.

FIGURE 7.1 Locations of Individual Scores on the Variable of Height

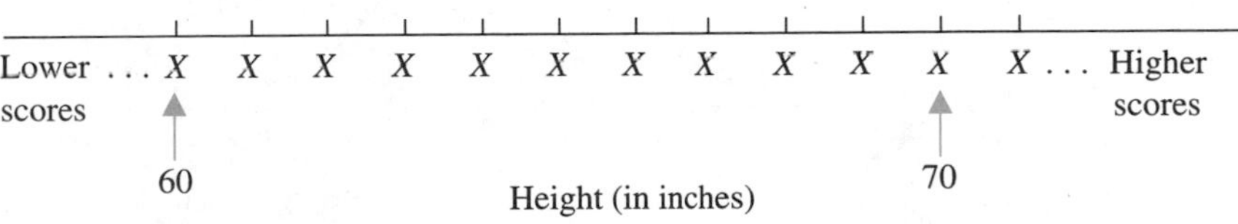

If my brother is 60 inches tall, then he is located at the point marked 60 on the height variable. The idea is not so much that he is 10 inches shorter than I am, but rather that we are separated by a *distance* of 10 units—in this case, 10 "inch" units. In statistics, scores are locations, and the difference between any two scores is the distance between them.

From this perspective, a frequency polygon shows the location of all the scores in a sample or population. For example, Figure 7.2 shows the polygons for two samples of height scores, one consisting of low scores and the other consisting of higher scores. Visualize the "parking lot view" of the normal curve described in Chapter 6, where participants' scores determine *where* they stand. A high score puts them on the right side of the lot, a low score puts them on the left side, and a middle score puts them in the crowd in the middle. Further, when two distributions contain different scores, then the *distributions* have different locations on the variable.

We began this chapter by asking, "Are the scores generally high scores or generally low scores?" Now you can see that we are actually asking, "*Where* on the variable is the distribution located?" A **measure of central tendency** is a statistic that summarizes the location of a distribution on a variable. Listen to its name: It is the score that indicates where the "center" of the distribution "tends" to be located. Essentially, it describes *around* where most of the scores are located and provides an "address" for the distribution. This means that ideally most of the scores are close to this score. Thus, in Sample A in Figure 7.2, the most frequent scores are in the neighborhood of 59, 60, and 61 inches, so the measure of central tendency will indicate that most scores are located around 60 inches. In Sample B, the scores tend to center around 70 inches.

Notice that the above example illustrates how we use descriptive statistics: From them we get an idea of what's in the data and we can "envision" the important aspects of the distribution *without* looking at every individual score. If a researcher said only that one normal distribution is centered at 60 and the other is centered around 70, we can envision the essence of the information presented in Figure 7.2. Although we lose some detail, we get an idea of whether the scores in each sample are generally high or low scores. We'll see other statistics that add more to our mental picture of a distribution, but measures of central tendency are at the core of summarizing a distribution.

> ***REMEMBER*** The first step in summarizing any set of data is to compute the appropriate measure of central tendency.

FIGURE 7.2 Two Sample Polygons on the Variable of Height

Each polygon indicates the locations of the scores and their frequencies.

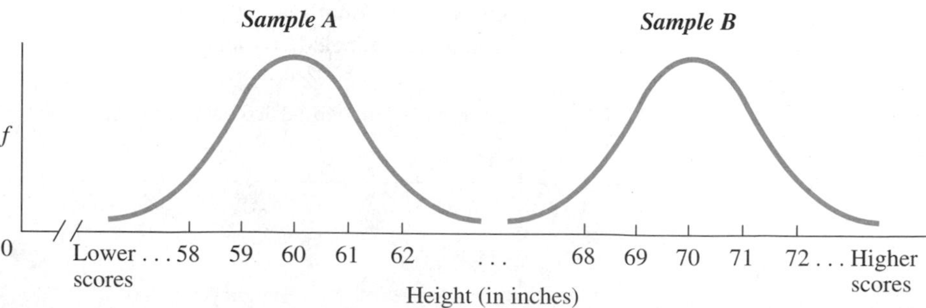

The trick is to compute the correct measure of central tendency so that you *accurately* envision where most scores in the data are actually located. Recall that one of your goals is to learn when to use a particular statistical procedure. Which measure of central tendency you should calculate depends on two factors:

1. The scale of measurement used, so that the summary makes sense given the nature of the scores.
2. The shape of the frequency distribution the scores produce, to ensure that the measure of central tendency accurately summarizes the distribution.

There are three common measures of central tendency. In the following sections we first discuss the mode, then the median, and finally the mean. With this background, we'll then see how measures of central tendency are used in research.

THE MODE

One way to describe where *most* of the scores are located in a distribution is to find the one score that occurs most. The most frequently occurring score is called the **modal score** or simply the **mode**. Consider the test scores 2, 3, 3, 4, 4, 4, 4, 5, 5, and 6. The score of 4 is the mode, because it occurs more frequently than any other score in the sample. You can see how the mode summarizes this distribution from the polygon in Figure 7.3. Most of the scores are at or around 4. Also, notice that the distribution is roughly a normal curve, with the highest point over the mode. When a polygon has one hump, such as on a normal curve, the distribution is called **unimodal**, indicating that one score qualifies as the mode.

There may not always be a single mode in a set of data. For example, consider the scores: 2, 3, 4, 5, 5, 5, 6, 7, 8, 9, 9, 9, 10, 11, and 12. Here 5 and 9 are tied for the most frequently occurring score. This sample is plotted in Figure 7.4. In Chapter 6, we saw that such a distribution is called **bimodal**, meaning that it has two modes. Describing

FIGURE 7.3 A Unimodal Distribution

The vertical line marks the highest point on the distribution, thus indicating the most frequent score, which is the mode.

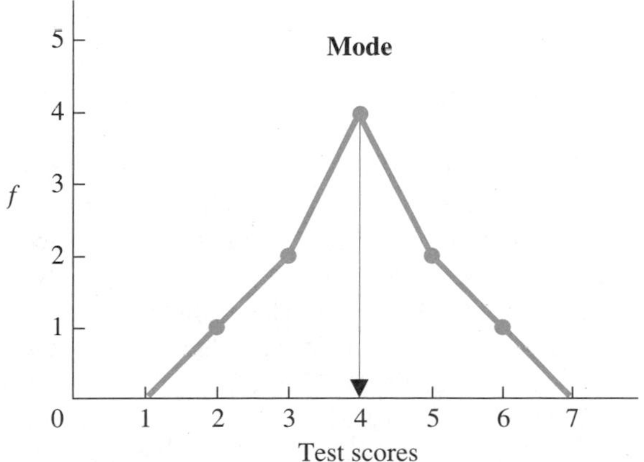

FIGURE 7.4 A Bimodal Distribution

Each vertical line marks one of the two equally high points on the distribution, indicating the location of one of the two modes.

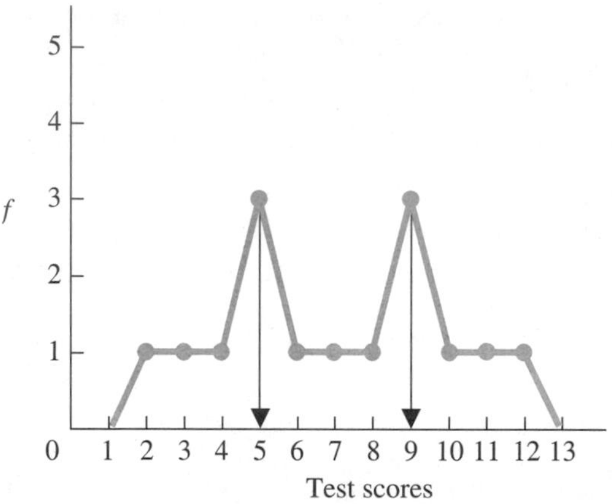

this distribution as bimodal and identifying the two modes summarizes where most of the scores tend to be located, because most scores are either around 5 or around 9.

Uses of the Mode

The mode is typically used to describe scores from a nominal scale of measurement (when participants are classified using a qualitative variable). For example, say that we asked subjects their favorite flavor of ice cream and counted the number of responses in each category. Such data might produce the bar graph shown in Figure 7.5. It makes sense to summarize such data by indicating the most frequently occurring category. Reporting that the modal response was a preference for category 5, "Goopy Chocolate," is very informative. However, with data that are ordinal, interval, or ratio, although it is useful to know the mode, we can usually compute a better measure of central tendency. This is because there are two potential problems with the mode.

First, depending on the distribution, several scores may be tied for the highest frequency, and identifying many modes does not summarize the data. In the most extreme case, we might obtain a rectangular distribution with scores such as 4, 4, 5, 5, 6, 6, 7, and 7. Either there is no mode, or all scores are the mode. Either way, the mode should not be determined.

A second problem is that the mode does not take into account any scores other than the most frequent score, so it ignores much of the information in the data. An *accurate* summary should reflect all of the scores so that we can accurately envision the distribution. For example, the mode is 7 in the skewed distribution containing the scores: 7, 7, 7, 20, 20, 21, 22, 22, 23, and 24. The problem is that this gives you the wrong idea about the scores, because the majority of the data is not located *around* 7 (most scores are up there between 20 and 24). Thus, the mode may not accurately summarize where *most* scores in a distribution are located.

FIGURE 7.5 Bar Graph Showing the Frequencies of Preferred Ice Cream Flavors

The mode is flavor 5, "Goopy Chocolate."

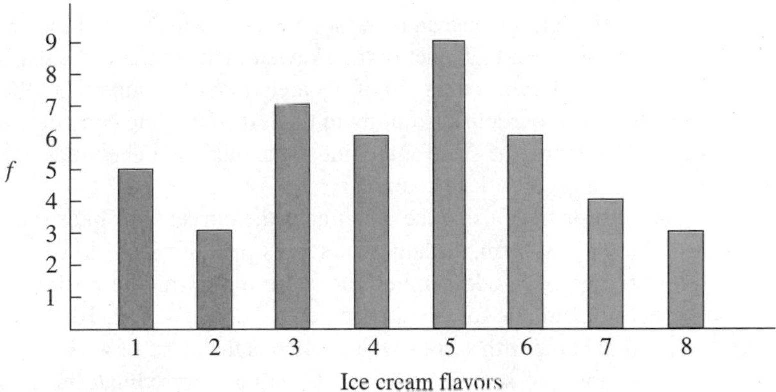

THE MEDIAN

Often a better measure of central tendency is the median. The **median** is simply another name for the score at the 50th percentile. Recall that 50% of a distribution is at or below the 50th percentile. Thus, if the median is 10, 50% of the scores are at or below 10. The median is typically a better measure of central tendency than the mode because (1) only one score can be the median, and (2) the median will usually be around where most of the distribution is located. The symbol for the median is its abbreviation, *Mdn*.

As we saw in Chapter 6, with a large sample or population, the 50th percentile is the score that separates the lower 50% of the distribution from the upper 50% of the distribution. For example, look at Graph A of Figure 7.6.

FIGURE 7.6 Location of the Median in a Normal Distribution (A) and in a Skewed Distribution (B)

The vertical line indicates the location of the median, with one half of the distribution on each side of it.

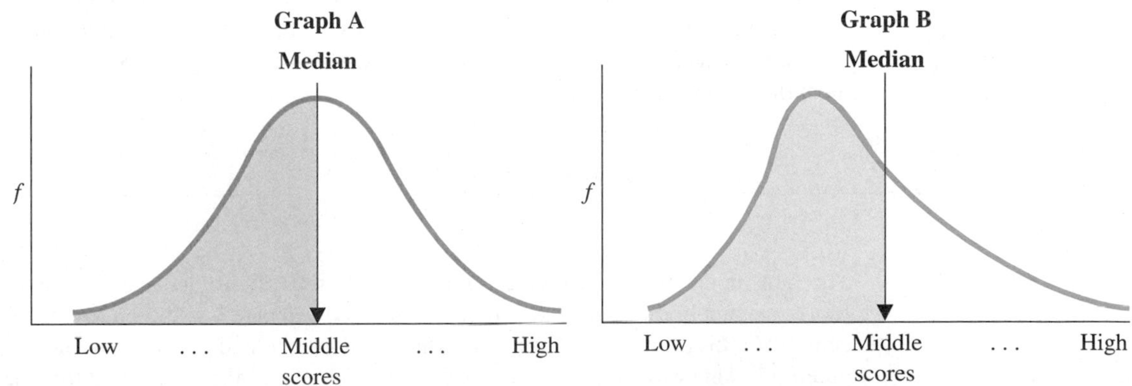

Because 50% of the scores are to the left of the line, they are below the score at the line. Therefore, the score at the line is the 50th percentile or the median. (In a normal distribution the median is also the most frequently occurring score, so it is the same score as the mode.)

In fact, the median is the score below which .50 of the area of *any* shaped polygon is located. For example, in the skewed distribution in Graph B of Figure 7.6, the vertical line is drawn so that .50 of the area under the curve is to the left of the line. The scores of those participants falling to the left of the line constitute the lower 50% of all scores. Therefore, the score at the line is the median. (The median will not always be one of the actual scores that occurred.)

Instead of using the area under the curve, we can estimate the median using the following system. Arrange the scores in order from lowest to highest. If the number of scores is an odd number, the score in the middle position is the approximate median. For example, for the nine scores 1, 2, 3, 3, 4, 7, 9, 10, 11, the score in the middle position is the fifth score so the median is the score of 4. If N is an even number, the average of the two scores in the middle is the approximate median. For example, for the ten scores 3, 8, 11, 11, 12, 13, 24, 24, 35, 46, the middle scores are at position 5 (the score of 12) and position 6 (the score of 13). The average of 12 and 13 is 12.5, so the median is approximately 12.5.

To calculate the median precisely, use the formula discussed in Part B.1 of Appendix B to find the score at the 50th percentile. (Most computer programs employ this formula, providing the easiest solution.)

Uses of the Median

The median is not used to describe nominal data. For example, to say that 50% of a survey preferred "Goopy Chocolate" *or below* is confusing. On the other hand, the median is often the preferred measure of central tendency when the data are ordinal (rank-ordered) scores. For example, say that a group of students ranked how well a college professor teaches. If the professor's median ranking was 3, we know that 50% of the students rated the professor as number 1, 2, or 3. Also, as we shall see, the median is appropriate when interval or ratio scores form a very skewed distribution. In addition, for any distribution, the median may be one of the measures of central tendency we compute, because it's nice to know the score at the 50th percentile.

Computing the median still ignores some information in the data, however, because it reflects only the frequency of scores in the lower 50% of the distribution. It does not consider the mathematical values of these scores or the scores in the upper 50%. Therefore, the median is usually not our first choice for describing the central tendency of most distributions.

THE MEAN

The most common measure of central tendency in psychological research is the mean score, or simply the mean. The **mean** is the score located at the exact mathematical center of a distribution. The mean is what most people call the average and it is computed the same way you compute an average: Add up all the scores and then divide by the number of scores you added. Unlike the mode or the median, the mean includes every score so it does not ignore any information in the data.

The symbol for a *sample* mean is $\overline{X}$. It is pronounced "the sample mean" (not "bar X," which sounds like the name of a ranch!). As with other symbols, get in the habit of thinking of $\overline{X}$ as a quantity itself, so that you understand statements such as "the size of $\overline{X}$" or "this $\overline{X}$ is larger than that $\overline{X}$."

To compute $\overline{X}$, recall that the symbol meaning "add up all the scores" is ΣX and the symbol for the number of scores is N. Then,

THE FORMULA FOR COMPUTING A SAMPLE MEAN IS

$$\overline{X} = \frac{\Sigma X}{N}$$

As an example, take the scores 3, 4, 6, and 7. Adding the scores together produces $\Sigma X = 20$, and N is 4. Thus, $\overline{X} = 20/4 = 5$. Saying that the mean of these scores is 5 indicates that the exact mathematical center of this distribution is at the score of 5. (As in this example, the center of the distribution may be a score that does not actually occur in the data.)

What is the mathematical center of a distribution? The center of a distribution is its balance point. Visualize a polygon as a teeter-totter on a playground. A score's location on the teeter-totter corresponds to its location on the X axis. The left side of Figure 7.7 shows the scores 3, 4, 6, and 7 sitting on the teeter-totter. The mean score of 5 is the point that balances the distribution. The right side of Figure 7.7 shows how the mean is the balance point even when all scores do not have the same frequency (the score of 1 has an f of 2). Here the mean is 4 and it balances the distribution.

Uses of the Mean

You should compute the mean whenever getting the "average" score makes sense. Therefore, do not use the mean when describing nominal data. For example, say we are studying political affiliation and assign a 1 if a person is a Democrat, a 2 if a Republican, and so on. It is meaningless to say that the average political affiliation was 1.3. The mode or percentages would be much more informative. Likewise, we prefer the median when describing ordinal data (because it's strange to say, for example, that, on average,

FIGURE 7.7 The Mean as the Balance Point of a Distribution

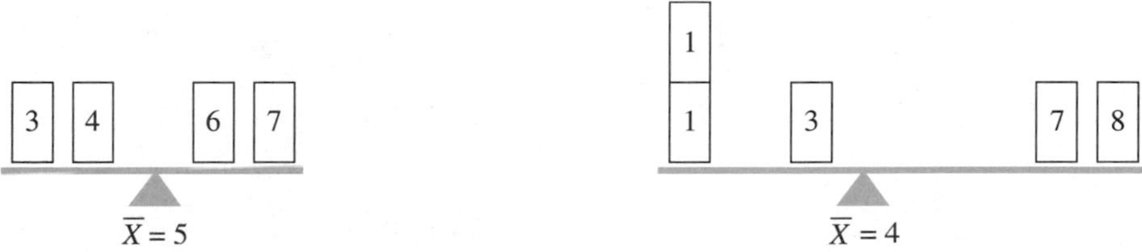

runners came in 5.7th in a race). This leaves the mean to describe interval or ratio data. And, the mean makes the most sense when the variable is at least theoretically continuous (when decimals make sense).

In addition to considering the type of scores, however, you must consider the shape of the distribution. The goal is to identify the point around which most of the scores tend to be located. The mean is simply the mathematical center of a distribution. Therefore, the mathematical center of the distribution must also be the point around which *most* of the scores are located. This will be the case when the distribution is *symmetrical* and *unimodal*. For example, say that we are studying the intelligence of cats by timing how long it takes them to escape from a maze. The data are 1, 2, 3, 3, 4, 4, 4, 5, 5, 6, and 7 minutes, which form the roughly normal distribution shown in Figure 7.8. The mean is appropriate here because it *is* that point around which most of the scores are located: Most of the cats did take around 4 minutes to escape.

The ultimate symmetrical distribution is the normal distribution, so we compute the mean whenever we have a normal or approximately normal distribution. Notably, on a perfect normal distribution all three measures of central tendency are located at the same score: In Figure 7.8, the score of 4 is the mean, it is the median, and it is the mode. If a distribution is only roughly normal, then the mean, median, and mode will be close to the same score. You might think that in such cases, any of the measures of central tendency would be good enough. Not true. Because the mean uses all of the information in the data, the mean is the preferred measure of central tendency. Therefore, the rule for you as a researcher is that the mean is the preferred statistic to use with interval or ratio data unless it clearly provides an inaccurate summary of the distribution.

> *REMEMBER* Describe the central tendency of a normal distribution of interval or ratio scores by computing the mean.

The mean provides an inaccurate summary when a distribution is highly skewed. For example, say that our previous cat subjects produced the escape-time scores of 1, 2, 2, 2, 3, and 14, forming the positively skewed distribution shown in Figure 7.9.

FIGURE 7.8 Location of the Mean on a Distribution Formed by the Escape Times 1, 2, 3, 3, 4, 4, 4, 5, 5, 6, and 7

The vertical line indicates the location of the mean score, which is the balance point of the distribution.

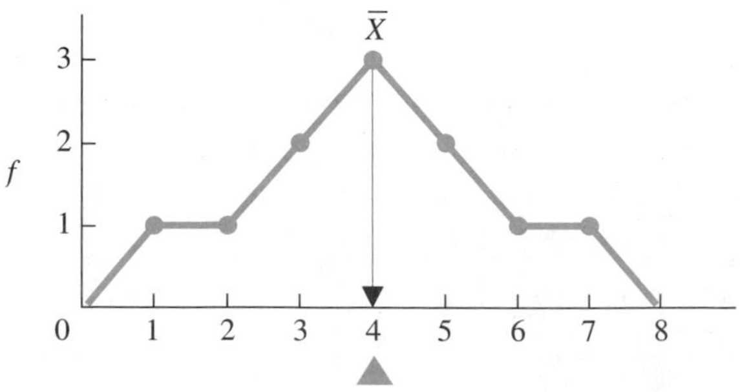

FIGURE 7.9 Location of the Mean on a Skewed Distribution Formed by the Time Scores 1, 2, 2, 2, 3, 14

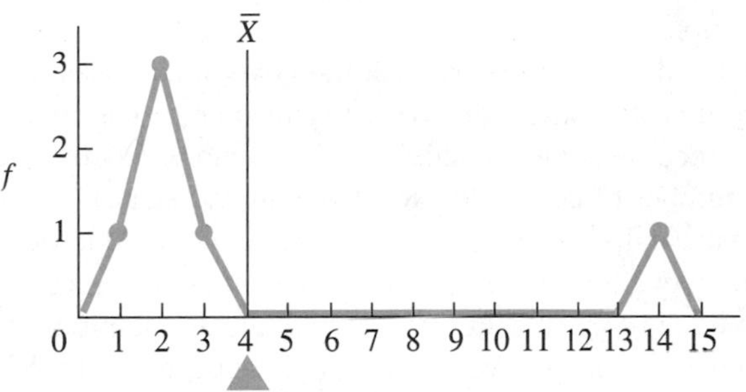

Without the 14, the scores would form a symmetrical distribution with a mean of 2. However, because that one exceptionally slow cat has the extreme score of 14, the mean is pulled away from the low scores so that it can balance the distribution. Picture the distribution as a teeter-totter with a bunch of little guys—1's, 2's, and 3's—sitting at one end, trying to balance the big guy—the 14—sitting way at the other end. To keep the teeter-totter level, the balance point must be located at the mean of 4. The problem is that a measure of central tendency is supposed to describe where most of the scores tend to be located. But in Figure 7.9, most scores are not around 4. As this illustrates, the mean is where the mathematical center is, but in a skewed distribution the mathematical center is *not* where most of the scores are located.

The solution is to use the median to summarize very skewed distributions. Figure 7.10 shows the relative positions of the mean, median, and mode in such distributions. When the distribution is positively skewed, the mean is larger than the median. When the distribution is negatively skewed, the mean is less than the median. In both cases, the mean is pulled toward an extreme tail and does not accurately summarize the distribution. Likewise, the mode tends to be toward the side away from the extreme tail, so most of the distribution is not centered around the mode either. However, the median is

FIGURE 7.10 Measures of Central Tendency for Skewed Distributions

The vertical line shows the relative positions of the mean, median, and mode.

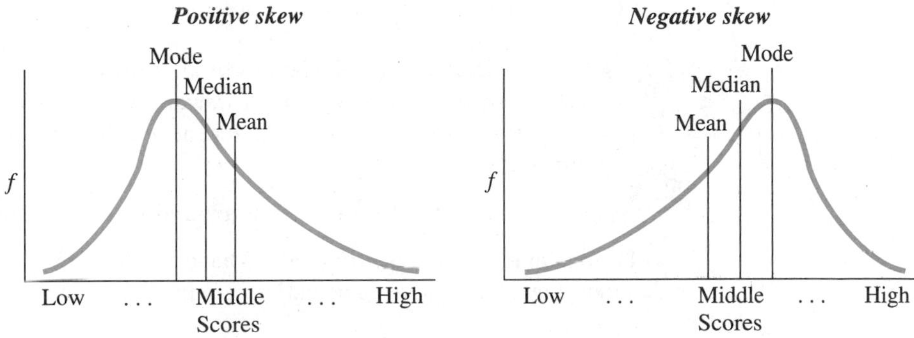

not thrown off by extreme scores occurring in only one tail, because it does not take into account the actual values of the scores. Thus, of the three measures, the median most accurately reflects *around* where most of the scores tend to be located in a skewed distribution.

It is for the above reasons that the government uses the median to summarize such skewed distributions as that of yearly income or the price of houses. For example, the median income in the United States is around $35,000 a year. But there are a relatively small number of corporate executives, movie stars, professional athletes, and the like who make millions! Averaging in these extreme high incomes would produce a mean income over $50,000. However, because most incomes are not at or around $50,000, the median is a much better summary of the distribution.

Believe it or not, we've now covered measures of central tendency. In summary:

1. Use the mode with nominal (categorical) data or with a distinctly bimodal distribution of any type of scores.

2. Use the median with ordinal (ranked) scores or with a very skewed distribution of interval/ratio scores.

3. Use the mean with a symmetrical, unimodal distribution of interval/ratio scores.

Most often in psychological research, data are summarized using the mean. This is because most often psychologists measure variables using interval or ratio scores that, *simply by coincidence*, form a roughly normal distribution. Because the mean is used so extensively, we'll delve further into its characteristics and uses in the following sections.

Deviations Around the Mean

To understand why the mean is the mathematical center or balance point of a distribution, you must understand that *in total*, the mean is just as far from the scores above as it is from the scores below it. The distance separating each score from the mean is called the score's **deviation**, indicating the amount the score **deviates** from the mean. A score's deviation is equal to the score minus the mean. In symbols, a deviation is the quantity $(X - \overline{X})$. We read from left to right, so $(X - \overline{X})$ says "X minus the mean." Thus, if the sample mean is 47, a score of 50 deviates by $+3$, because $50 - 47$ is $+3$. A score of 40 deviates from a mean of 47 by -7, because $40 - 47 = -7$.

> *REMEMBER* Always subtract the mean *from the raw score* when computing a score's deviation.

When we determine the deviations of all scores in a sample, we find the *deviations around the mean.* Then the *sum of the deviations around the mean* is the sum of all differences between the scores and the mean. And here's why the mean is the center of any distribution:

The sum of the deviations around the mean always equals zero.

For example, the scores 3, 4, 6, and 7 have a mean of 5. The upper portion of Table 7.1 shows how to compute the deviations around the mean for these scores.

TABLE 7.1 Computing Deviations Around the Mean

The mean is subtracted from each score, resulting in the score's deviation.

Score	*minus*	*Mean score*	*equals*	*Deviation*
3	−	5	=	−2
4	−	5	=	−1
6	−	5	=	+1
7	−	5	=	+2
			Sum =	0

Score	*minus*	*Mean score*	*equals*	*Deviation*
1	−	4	=	−3
2	−	4	=	−2
2	−	4	=	−2
2	−	4	=	−2
3	−	4	=	−1
14	−	4	=	+10
			Sum =	0

The lower portion of the table shows the deviations for the skewed distribution containing the scores 1, 2, 2, 2, 3, and 14, which has a mean of 4. In each sample, the sum of the deviations is zero. In fact, for *any* distribution of *any* shape, the sum of the deviations around the mean will always be zero. Every time, the sum of the positive deviations equals the sum of the negative deviations, so the sum of all deviations is zero.

Many of the formulas we'll encounter involve something like the sum of the deviations around the mean. The code for finding the sum of the deviations around the mean is $\Sigma(X \quad \overline{X})$. Always start inside parentheses, so first find the deviation for each score: $(X - \overline{X})$. The Σ indicates to then sum these deviations. Thus, as in the upper portion of Table 7.1, $\Sigma(X - X) = -2 + -1 + 1 + 2$, which equals zero.

USING THE MEAN IN RESEARCH

It is important to understand that the sum of the deviations equals zero, because then you understand that only the mean score is literally "more or less" the score that all participants obtained: Individual scores may be higher or lower than the mean, but those that are higher balance out with those that are lower. Because of this, the mean provides researchers with a very useful tool. As we see in the following sections, a sample mean is used in three ways: To predict scores of individuals, to describe a score's relative location within a distribution, and to draw inferences about the population.

Using the Sample Mean to Predict Scores

Because the mean is the central score in a distribution, it is the *typical* score in the distribution. This implies that if all scores in the sample were the same, they would all be the mean score. Therefore, the mean is the best description of any individual score. For example, if your friends have a B average in college, they may not always get B's, but you operate as if they do. If asked what you think they received in a particular course, you'd estimate B. For every other course, you'd also estimate B. Likewise, if the class average on an exam was 80, the best estimate for every student in the class is 80.

Notice that above you are using the mean to *predict* an individual's score in the sample. We can also use the mean to predict any additional scores that are *not* in the original sample. Because the mean is typical of the scores already observed in a particular situation, it is our best prediction of any other scores we might observe in that situation. Thus, your best guess is that your friends will continue to be B students, so for any future course you'd predict they'll get a B. Likewise, because the mean is the typical score of our participants, we assume that it also typifies the score of any similar individuals who might have participated. Thus, we'd predict that, like those students who averaged 80 on the exam, any similar students who missed the exam would have scored 80.

> *REMEMBER* We use the sample mean to predict the scores in the sample, or to predict additional scores we'd expect to find in the sample.

Of course, these predictions will sometimes be wrong. The amount of error equals the difference between the mean and the actual score someone obtains. In symbols, this prediction error equals $(X - \overline{X})$. We've already seen that $(X - \overline{X})$ is the amount a score deviates from the mean. The reason we use the mean to predict scores is that then the *total* prediction error is the sum of these deviations, $\Sigma(X - \overline{X})$, and this *always* equals zero. For example, the exam scores of 70, 75, 85, and 90 have a $\overline{X}$ of 80. One student, Quasimodo, scored the 70. We would predict he scored 80, so we'd be wrong by -10. But another student, Attila, scored the 90. By estimating an 80 for him, we'd be off by $+10$. In the same way, our errors will cancel out so that the total error is zero. Likewise, we expect the differences between the mean and the scores that other students *would* have obtained to be about the same as the differences between the mean and the scores that the students in the sample *did* obtain. Thus, using the mean as the predicted score for anyone else should also result in a total prediction error equal to zero.

Using any single score other than the mean produces a total error *greater* than 0. If, for example, we described the above scores using a score of 75 or 85, the sum of the deviations would be $+20$ or -20, respectively. In statistics, a total prediction error of zero is best because it means that, *over the long run*, we overestimate by the same amount that we underestimate. (There is an old joke about two statisticians shooting targets. One hits 1 foot to the left of the target, and the other hits 1 foot to the right. "Congratulations," one says. "We got it!") If we cannot perfectly describe every score, then we want our errors—the over- and underestimates—to cancel out to zero. Only the mean provides that capability.

Of course, this isn't the whole story. Although the *total* error in predictions will equal zero, any individual prediction may be off by a country mile. Later chapters will discuss how to reduce these errors in prediction. For now, simply remember that *unless you have additional information about the scores*, the mean is the best score to use when predicting or describing scores. This is because the over- and underestimates over all such predictions will cancel out to zero.

Using the Sample Mean to Describe a Score's Location

Because the mean is at the center of a distribution, it also provides a frame of reference for describing the location of any individual score. That is, we can transform a raw score by computing the amount it deviates from the mean. Then the deviation score communicates the score's location relative to the mean and, indirectly, the score's location relative to the rest of the distribution. For example, say that we give participants a test of creativity and obtain the raw scores 1, 2, 3, 3, 4, 4, 4, 5, 5, 6, and 7. Look at the approximately normal distribution these scores form, shown in Figure 7.11. Below each score is its deviation, computed as $X - \overline{X}$. A deviation consists of a number and a sign. A positive deviation indicates that the score is *greater* than the mean, and located to the *right* of the mean on the distribution. Conversely, a negative deviation indicates that the score is *less* than and thus falls to the *left* of the mean. The size of the deviation (regardless of its sign) indicates the *distance* the score lies from the mean. A deviation of 0 indicates that the score is equal to and at the mean. The larger the deviation, the farther the score is above or below the mean.

Deviation scores in a normal distribution also communicate the *frequencies* of the corresponding raw scores. As you know, scores lying farther in the tail of a normal distribution occur less frequently. Therefore, the larger a deviation (whether positive or negative), the farther into the tail the corresponding raw score lies, and so the less frequently the raw score occurs. Also, the larger a deviation, the less frequently the *deviation* occurs. Above, the raw score of 7 produces the deviation of +3. Not only is 7 an extreme score that occurs only once, its deviation is also extreme and occurs only once. In fact, the frequency of any score's deviation will equal the frequency of that score. As in Figure 7.11, whether we label the X axis using raw scores or deviations, we have exactly the same frequency polygon.

The advantage of using deviation scores, however, comes when interpreting scores. A problem we will encounter repeatedly is that psychologists usually do not know how

FIGURE 7.11 Frequency Polygon Showing Deviations from the Mean

The first row under the X axis indicates the original creativity scores, and the second row indicates the amounts the raw scores deviate from the mean.

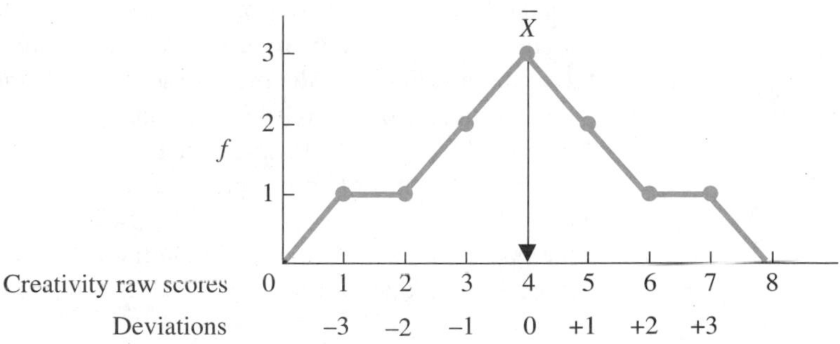

to evaluate an individual score in the grand scheme of things in nature. If, for example, I tell you that your raw score on a creativity test is 6, you won't know if your score is good, bad, or indifferent, because you have no frame of reference. However, if I say that your score produced a deviation score of +2, you can then at least interpret your score *relative* to the rest of the distribution. By envisioning the previous graph in Figure 7.11, you know that a positive deviation of +2 indicates that you are above average (which with creativity is good), that you are to the right of and therefore above the 50th percentile (which is also good), and that you are out there in the direction of the less common scores, where the most creative people are (and that's good too). Conversely, if these scores reflected the number of blunders on a test, you'd know you were out there among those who made the highest number of blunders (and that's not so good). But either way, at least you'd have a better idea of how to interpret—*make sense of*—the score, and that is the purpose of statistics.

We'll elaborate on using deviations around the mean in subsequent chapters. For now, remember that on the normal curve, the larger the deviation (whether positive or negative), the farther the raw score is from the mean and thus the less frequently the score and its deviation occur.

Using the Sample Mean to Describe the Population Mean

Recall that ultimately we seek to describe the entire population of scores we would find in a given situation. Populations are unwieldy, so we also summarize them using measures of central tendency. Here is what we want to know: If we examined the population, around which score would most of the scores be located?

Because we usually have interval or ratio scores that form at least an approximately normal distribution, we usually describe the population using the mean. The mean score of a population is a parameter, symbolized by the Greek letter Mu whose symbol is μ. Thus, to indicate that the population mean is 143, we'd say $\mu = 143$. A mean is a mean, however, so a population mean has the same characteristic as a sample mean:

1. μ is the arithmetic average of all scores in the population.
2. μ is the score at the mathematical center of the distribution.
3. The sum of the deviations around μ is zero.

We saw earlier that these characteristics make the mean the best score to use when predicting any individual's score in a sample. For the same reasons, μ is the best score to use when predicting any individual's score in the population.

How do we determine the value of μ? If we know all of the scores in the population, then we compute μ using the same formula used to compute $\overline{X}$: $\mu = \Sigma X/N$. Of course, usually a population is infinitely large, so we cannot directly compute μ. Instead we estimate the value of μ based on the mean of a random sample. If, for example, a sample's mean in a particular situation is 99, then our best guess is that the population μ in that situation would also be 99. We make such an inference because it is a population with a mean of 99 that is most likely to produce a sample with a mean of 99. That is, we are very likely to obtain a sample of participants who score around 99 when most individuals in the population score around 99. On the other hand, we are very unlikely to obtain an entire sample that scores around 99 if, for example, the population mean is

4,000. Here few participants should score around 99, but many should score around 4,000. Thus, wherever most scores in a sample are located should be where most scores in the population are located, so a random sample mean should be a good estimate of the population μ (assuming the sample is representative).

REMEMBER A sample mean is the best summary score to use when thinking about the scores in either a sample or in the corresponding population.

SUMMARIZING A STUDY USING CENTRAL TENDENCY

Now you can understand how means are used in research. In descriptive research, we compute the mean anytime we have a sample of normally distributed scores to describe (or we compute other measures of central tendency when appropriate). Thus, we might compute the mean number of times participants exhibit a particular behavior in an observational study, or compute the mean response in a survey. Based on such sample means, we can describe the typical participant and predict the scores of other individuals, including those of the entire population.

We perform similar steps when summarizing the results of an experiment. For example, say that in a study of human memory, we predict that people will make more mistakes when recalling a long list of words than when recalling a short list. We conduct an overly simplistic experiment involving three conditions. In one condition participants read a list of 5 words and then recall it. In another condition, participants read a 10-item list and recall it, and in a third condition, they read a 15-item list and recall it. If the predicted relationship exists, then as the independent variable of list-length increases, scores on the dependent variable of recall errors also will tend to increase.

Say we test 3 participants per condition (an unrealistically small number) and obtain the scores shown in Table 7.2. It appears there is a relationship here, because a different and higher set of error scores tends to be associated with each condition. Most experiments involve many more participants, however, and with many scores it is often difficult to detect a relationship by looking at the raw scores. But that's what measures of central tendency are for: To summarize the scores and simplify the relationship. Thus, your first step is *always* to compute a measure of central tendency for the scores in each condition of an experiment.

TABLE 7.2 Numbers of Errors Made by Subjects Recalling a 5-, 10-, or 15-Item List

Independent variable: Length of list

Condition 1: 5-item list	*Condition 2: 10-item list*	*Condition 3: 15-item list*
3	6	9
4	5	11
2	7	7

Summarizing a Relationship

As we've seen, the measure of central tendency to compute depends on the characteristics of the scores. In an experiment, the scores reflect the *dependent variable*. Therefore, you should compute the mean, median, or mode depending upon (1) the scale of measurement used to measure the dependent variable, and (2) for interval or ratio scores, the shape of the distribution they form.

> REMEMBER The measure of central tendency used in an experiment is determined by the type of *dependent* scores.

In determining the shape of the distribution, consider how the scores are assumed to be distributed in the *population*. Obtain this information from the research literature, seeing how other researchers treat such scores. In our memory experiment, recall errors is a ratio variable that forms an approximately normal distribution, so we will compute the mean score for each condition (computing the mean in each column in Table 7.2). Examining these means, we quickly recognize evidence of a relationship:

Condition 1: 5-item list	*Condition 2: 10-item list*	*Condition 3: 15-item list*
$\overline{X} = 3$	$\overline{X} = 6$	$\overline{X} = 9$

To interpret these data, envision the kinds of distributions that produced each mean. Thus, recalling a 5-item list resulted in one distribution of scores located around the mean of 3. Recalling a 10-item list resulted in a different distribution located around 6 errors, and recalling a 15-item list resulted in a different distribution, located around 9 errors. A relationship is present here, because as the independent variable changes (from 5- to 10- to 15-item lists), the scores on the dependent variable also tend to change (from around 3, to around 6, to around 9, respectively).

This experiment "worked" because it demonstrates that list length is a variable that literally makes a *difference* in recall scores. Researchers often communicate that they have found a relationship by saying that they have found a difference (between the means). If no difference is found, then they have not found a relationship. Recognize, however, that not all of the means must differ for a relationship to be present. For example, we might find that only the mean (and raw scores) in the 5-item condition differs from the mean (and scores) in the 15-item condition. We still see a relationship if, at least sometimes, as the conditions of the independent variable change, the dependent scores also tend to change.

> REMEMBER A relationship is present when the mean scores on the dependent variable change as a function of changes in the independent variable.

The above logic also applies to the median or mode. For example, say that we study the dependent variable of political party affiliation, to see if it changes with a person's year in college. Political parties are categories involving nominal scores, so the mode is the appropriate measure of central tendency. We might see that freshmen most often claim to be Republican, but the mode for sophomores is Democrat, for juniors

Socialist, and for seniors Communist. These data reflect a relationship because as college level changes, political affiliation tends to change. Or say we learn that the median income for freshmen is lower than the median income for sophomores, which is lower than that for juniors or seniors. This tells us that the location of the corresponding distribution on the variable of income is different for each class, so we know that the income "scores" of individuals must be changing as their year in college changes.

Finally, recall that for a reliable study, we often test participants over multiple trials within each condition. To summarize the trials, we may compute a measure of central tendency for each participant's raw scores. For example, we would have had a better memory experiment by testing each participants' recall on several lists in a condition. Here each person's "score" would actually be the *mean* number of errors he or she made over all lists in the condition. Then we would again compute an overall mean for each condition, computing the mean of the participants' means.

To easily see a relationship when it is present, we summarize experiments by creating graphs.

Graphing the Results of an Experiment

Recall that in creating a graph, the "given" variable is placed on the X axis, and in an experiment, the given variable is the independent variable. Therefore, we always place the independent variable on the X axis and the dependent variable on the Y axis. Instead of plotting the raw dependent scores on the Y axis, however, we plot the measure of central tendency we've computed. Thus, each data point reflects either the mean, median, or modal dependent score for a condition of the independent variable.

The measure of central tendency to compute depends on the characteristics of the *dependent* variable. We must then decide on the type of graph to create, however, and this depends on the characteristics of the *independent variable*. As we see below, we create either a line graph or a bar graph.

Line graphs When the independent variable is an interval or ratio variable, we connect the data points with straight lines. This type of graph is called a **line graph**. In our memory experiment, list length is a ratio variable, so we create the line graph shown on the left in Figure 7.12. First, we label the X and Y axes with the specific variables. The label on the Y axis is *mean* recall errors, and the X axis is labeled with the conditions of the independent variable. Thus, for example, the mean error score for the 5-item list was 3.

We connect the adjacent data points with straight lines because we assume that, with interval or ratio data, the relationship continues in a straight line between the points shown on the X axis. Thus, if there had been a 6-item list, we assume its mean score would fall on the line connecting the means for the 5- and 10-item lists.

The line graph conveys the same information as the sample means did previously. For each mean, we envision the distribution that would produce it. As shown on the right in Figure 7.12, we envision a sample of scores and their corresponding data points that would occur *around*—above and below—the mean's data point. The different vertical locations of the means indicate that they have different values on the Y axis, and thus that there are different scores in each condition.

Notice that you can easily spot such a relationship, because the changing positions of the means on the Y axis produce a line graph that is *not* horizontal. On the other hand,

FIGURE 7.12 Line Graphs Showing (A) The Relationship for Mean Errors in Recall as a Function of List Length and (B) The Data Points We Envision Around Each Mean

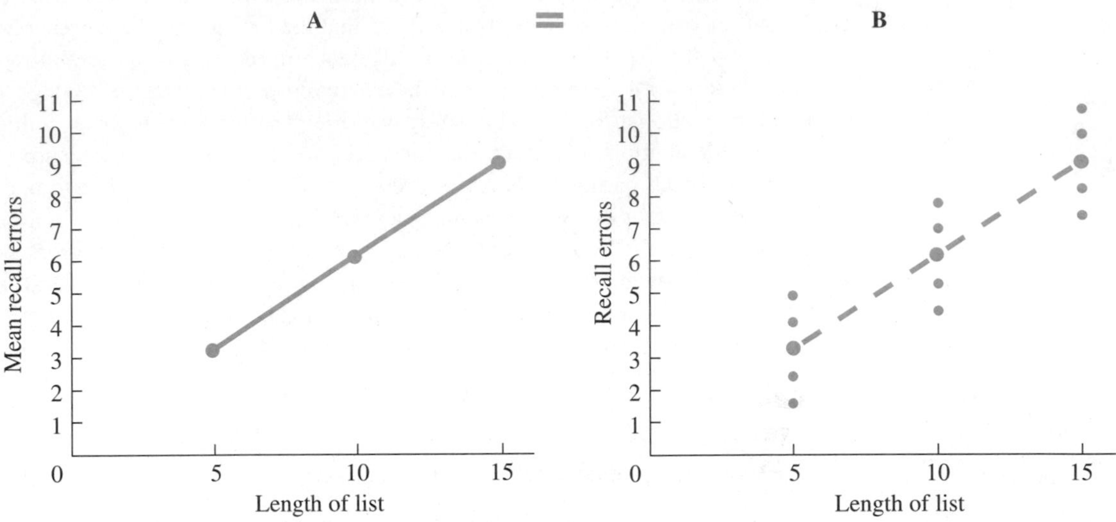

say that each condition had produced a mean of 5. As shown on the left in Figure 7.13, these results produce a horizontal (flat) line. Here as list length changes, the mean stays the same. This implies that (as on the right of the figure) the individual scores stay the same, regardless of the condition. Because the dependent scores stay the same when the independent variable changes, no relationship is present.

> REMEMBER On a graph, if the summary data points form a line that is not horizontal, then the individual Y scores are changing as the X scores change and a relationship is present.

FIGURE 7.13 Line Graphs Showing (A) No Relationship for Mean Errors in Recall as a Function of List Length and (B) The Kind of Data Points We Envision Around Each Mean

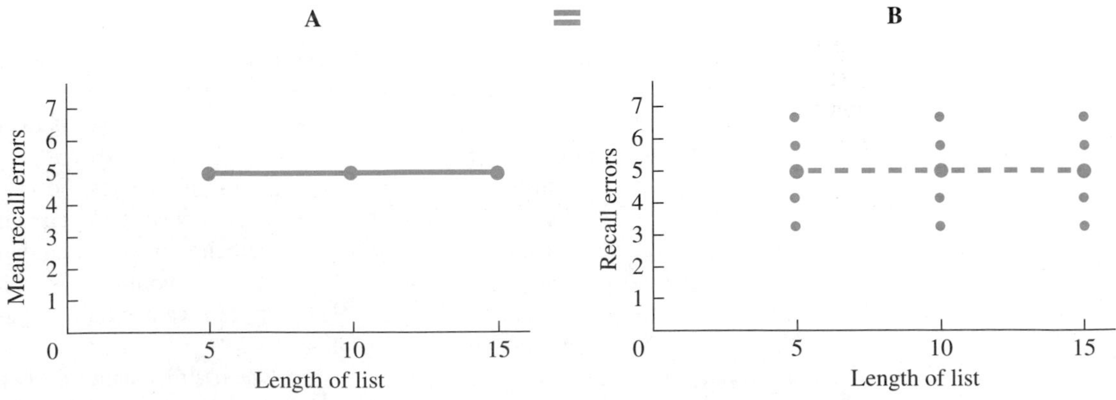

Bar graphs Create a bar graph when the independent variable is a nominal or ordinal variable. The rule here is the same as in Chapter 6: Create a bar graph *whenever* the X axis reflects a nominal or ordinal variable. Each bar is centered over a condition on the X axis, and the height of the bar corresponds to the mean score for the condition.

For example, say that we conducted another experiment, comparing the recall errors of psychology majors, English majors, and physics majors. Here, the independent variable of college major is a nominal variable, so we have the bar graph shown in Figure 7.14. This too shows a relationship: The tops of the bars do not form a horizontal line, so there are different means and thus different scores in each condition. The bars communicate that we arbitrarily placed psychology to the left of English. Therefore, if we inserted the additional category of sociology between psychology and English, we could not assume that the mean for sociology majors would fall on a line running between the means for psychology and English majors.

> *REMEMBER* The type of dependent variable determines the measure of central tendency to calculate and the type of independent variable determines the type of graph to create.

Inferring the Relationship in the Population

In our original memory experiment we predicted that recall errors would increase as list length increased, and the sample data confirmed this. Although interpreting the sample data is always the first step, it is only part of the story. The big question remains: Is this how nature works? Do longer lists produce more errors in recall for everyone in the population?

Recall that before making *any* inferences about the population, we must perform inferential statistics. For the moment, assume that the data from the memory

FIGURE 7.14 Bar Graphs Showing Mean Errors in Recall as a Function of College Major

The height of each bar corresponds to the mean score for the condition.

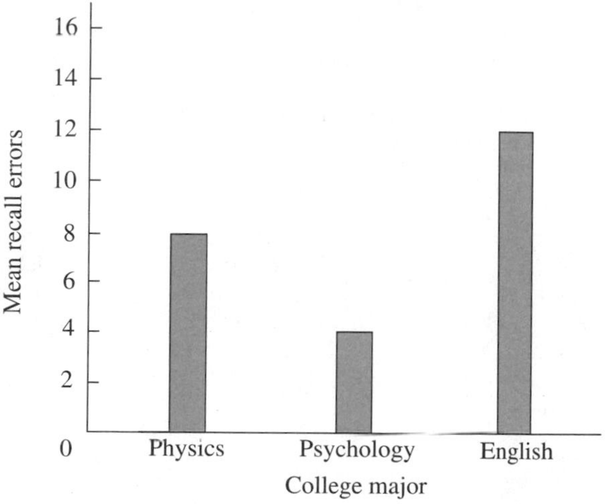

experiment passed the inferential test. Then, we use each sample mean to estimate the population mean that would be found for that condition. The mean score for the 5-item condition was 3, so we infer that if the population were to recall a 5-item list, the mean of the population of scores would be 3. In essence, we expect that everyone in this situation makes around 3 errors. Similarly, we infer that if the population recalled a 10-item list, μ would equal the sample mean of 6, and if the population recalled a 15-item list, μ would be 9.

We conceptualize the populations of error scores in the following way. We've assumed that recall errors are normally distributed in the population, and we've estimated the location of the distribution of dependent scores for each condition. Thus, based on our assumptions and data, we can envision the population of recall errors we would expect for each condition as the frequency polygons shown in Figure 7.15. Now we have frequency distributions with the dependent scores on the *X* axis. Because these distributions have different values of μ, we can see the relationship that we think exists in the population: As the conditions of the independent variable change, the scores on the dependent variable tend to change so that there is a different population of scores for each condition. Essentially, for every 5 items in a list, everyone's score tends to increase by about 3 errors, and so the distribution slides 3 units to the right for each condition. (The overlap among the distributions simply shows that some people in one condition make the same number of errors as other people in an adjacent condition.)

Remember that the population of scores reflects everyone's behavior. If, as an independent variable changes, *everyone's* behavior tends to change, then we have learned about a law of nature involving that behavior. Above, everyone's recall behavior tends to change as list length changes, so we have evidence of how human memory generally works in this situation. Now we interpret the results "psychologically," generalizing to the theories, models, and hypothetical constructs that originally led to the study. At the same time, we try to understand and explain the relationship, answering such questions as why do scores increase at this rate, what cognitive or physiological mechanism is responsible for such scores, and so on. And of course, don't forget that our confidence in these conclusions must be tempered by the usual concerns about construct and

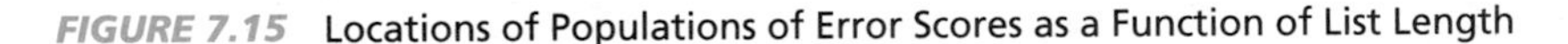

FIGURE 7.15 Locations of Populations of Error Scores as a Function of List Length

Each distribution contains the recall scores we would expect to find if the population were tested under each condition.

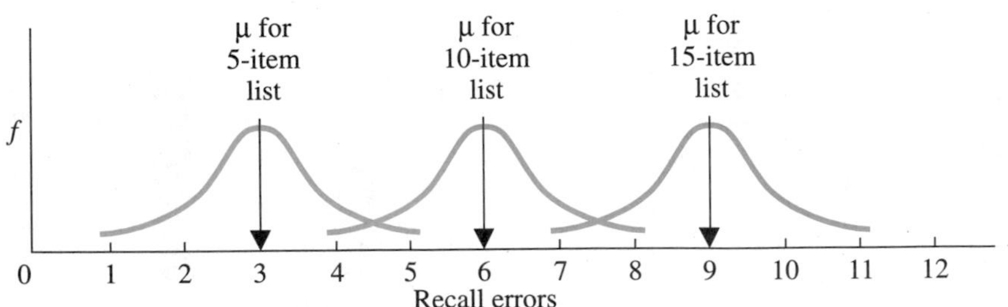

content validity, reliable and sensitive measurements, potential biases from demand characteristics, possible confoundings, and so on.

DESIGNING A POWERFUL EXPERIMENT

As we've seen, our conclusions about a hypothesis hinge on first finding a relationship in the sample data. However, a relationship is not an all-or-nothing thing. It's possible, therefore, to *not* find a clear and convincing relationship in the sample data, even when the variables are related in nature. For example, say that in the memory experiment above, even though the variables are related in nature, we obtained these results:

5-item list	*10-item list*	*15-item list*
$\overline{X} = 3.11$	$\overline{X} = 3.12$	$\overline{X} = 3.13$

These means barely differ from each other, so the individual raw scores are barely changing as the conditions change, and the line graph will form close to a flat line. This pattern is close to the pattern we would see when there is no relationship. Therefore, we—and our statistical procedures—might misinterpret the data, deciding that the safest conclusion is that these variables are not related. Then we would miss the relationship that really exists.

There is little we can do if the relationship in nature is really as weak as in these data. What would really be a shame, however, is to miss a substantial relationship that exists in nature because we obtain overly weak sample data. That is, we may obtain a poor example of the relationship in the population. Through some flaw in the design, strange participants, fluctuating variables or just plain luck, we may obtain some weird scores that form little or no relationship, even though there is a strong relationship in nature.

Remember that data are always tied to the design. Therefore, we design a study so that if there is a relationship between the variables in nature, the data are likely to clearly and convincingly show it. The term that applies to this strategy is that we seek to maximize our "power." Power is actually an aspect of inferential statistics, so we'll wait until Chapter 13 for a precise statistical definition. For now, just remember that a **powerful design** is more likely to show a convincing sample relationship.

Thus, your goal is create a powerful design. A powerful design might produce these results.

5-item list	*10-item list*	*15-item list*
$\overline{X} = 0$	$\overline{X} = 6$	$\overline{X} = 12$

Here, each change in the independent variable makes a relatively large difference in scores, and when graphed, these means will produce a steeply slanted line graph. Clearly, these data present convincing evidence for a relationship and so we are unlikely to miss the relationship in nature here.

There are a number of things that researchers do to maximize power. For example, you know to seek sensitive dependent variables that reflect subtle differences in behavior. This increases power because then you are more likely to see differences in participants' scores when a variable actually produces differences in their behavior. Another way to increase power is create a strong manipulation.

Selecting a Strong Manipulation

As you've just seen, large differences in means between the conditions produces a convincing relationship. We assume that manipulating the conditions of the independent variable cause differences in the means (and raw scores), so we want to select conditions that will produce large differences. A **strong manipulation** involves conditions that will produce large differences in scores between the conditions. You can create strong manipulations by using two strategies.

First, select amounts or categories of the independent variable that are substantially different from one another. For example, using list lengths of 5, 10, and 15 items is a reasonably strong manipulation of list length: If longer lists do influence memory, adding 5 items in each condition should show this. Conversely, comparing list lengths of 5, 6, and 7 items would not be a strong manipulation: Even if list length does influence memory, we might not see much of an effect in our study. In a different study, say we are presenting happy or sad words to influence participants' mood. For a strong manipulation, the happy words should be *very* happy and the sad ones should be *very* sad.

A second aspect of a strong manipulation is to have participants experience a condition sufficiently for it to dramatically influence their behavior. For example, in a previous chapter we discussed having participants hold pens in their mouths to mimic smiling and influence their mood. A strong manipulation would have them hold the pen for a long period of time, so that we dramatically alter their mood. Or, say we are studying how different speaking styles influence the persuasiveness of a message. Rather than presenting a short message with only one example of the style, we'd present a longer message with many aspects conveying that style.

> *REMEMBER* For power, create a strong manipulation so that the conditions are likely to produce large differences in dependent scores.

In later chapters, we'll discuss additional techniques for increasing power.

PUTTING IT ALL TOGETHER

As you may have noticed, the mean is *the* measure of central tendency in psychological research, whether you are conducting a descriptive or experimental design. To be literate in research and statistics, you should understand the mode and the median, but the truly important topics in this chapter involve the mean and its characteristics, especially when applied to the normal distribution.

We will eventually discuss inferential statistical procedures, and they will tend to occupy all of your attention. Despite the emphasis such procedures receive, however, remember that the mean (or another central tendency measure) forms the basis for interpreting any study. You always want to say something like "the participants scored around 3" for a particular situation, because then you are describing their typical *behavior* in that situation. Thus, regardless of what other fancy procedures we discuss, remember to ultimately focus on the measure of central tendency to identify *around* where the scores in each condition are located.

CHAPTER SUMMARY

1. A *statistic* describes a characteristic of a sample of scores, and is symbolized by a letter from the English alphabet. A *parameter* describes a characteristic of a population of scores and is symbolized by a letter from the Greek alphabet.

2. *Measures of central tendency* summarize the location of a distribution of scores on a variable, indicating around where the center of the distribution tends to be located. Which measure to compute depends on (a) the scale used to measure the variable, and (b) the shape of the distribution.

3. The *mode* is the most frequently occurring score(s) in a distribution, and is used primarily to summarize nominal data.

4. The *median*, symbolized by *Mdn*, is the score located at the 50th percentile. It is used primarily with ordinal data and with highly skewed interval or ratio data.

5. The *mean* is the average score, located at the mathematical center of a distribution. It is used with interval or ratio data that form a symmetrical, unimodal distribution. The symbol for a sample mean is $\overline{X}$, and the symbol for a population mean is μ.

6. The amount a score *deviates* from the mean is computed as $X - \overline{X}$. A deviation indicates the location of the raw score relative to the mean and relative to the distribution. In a normal distribution, the larger the deviation, the less frequently the score and the deviation occur.

7. The *sum of the deviations around the mean*, $\Sigma(X - \overline{X})$, always equals zero. In the absence of any other information about the scores, the mean is the best score to use when describing or predicting an individual score, because the total error across all such estimates will be the sum of the deviations around the mean, which will equal zero.

8. In graphing the results of an experiment, the independent variable is plotted on the X axis and the dependent variable is on the Y axis. A *line graph* is created when the independent variable involves a ratio or interval scale. A *bar graph* is created when the independent variable involves a nominal or ordinal scale.

9. On any graph, if the summary data points form a line that is not horizontal, then the individual Y scores change as a function of changes in the X scores and a relationship is present. If the data points form a horizontal line, then a relationship is not present.

10. The sample mean in each condition of an experiment is the estimate of the population μ for that condition. When a relationship in the population is present, there will be different values of μ for two or more conditions of the independent variable.

11. A *powerful design* is more likely to show a clear and convincing sample relationship.

12. A *strong manipulation* of the independent variable will produce large differences in dependent scores between the conditions.

PRACTICE PROBLEMS

(Answers for odd-numbered problems are provided in Appendix D.)

1. (a) What is the difference between a statistic and a parameter? (b) What types of symbols are used for statistics and parameters?

2. (a) What does a measure of central tendency indicate? (b) What two aspects of the data determine which measure of central tendency to use?

3. (a) What is the mode, and with what type of data is it most appropriate? (b) What is the median, and with what type of data is it most appropriate? (c) What is the mean, and with what type of data is it most appropriate?

4. Why is it best to use the mean as the measure of central tendency for a normal distribution?

5. Which measure of central tendency is used most often in psychological research? Why?

6. Why is it inappropriate to use the mean as the measure of central tendency in a very skewed distribution?

7. For the following data, compute (a) the mean, and (b) the mode:

55 57 59 58 60 57 56 58 61 58 59

8. (a) In problem 7, what is your best estimate of the median (without computing it)? (b) Explain why you think your answer is correct. (c) Calculate the approximate median using the approach presented in this chapter.

9. After receiving many high grades, Foofy receives one low grade, resulting in an embarrassingly low grade average. (a) What has happened to her grade distribution to make it produce such a mean? (b) How could she summarize her grades to minimize the impact of that one low grade?

10. (a) For the data below, compute the mean:

18	16	19	20	18	19	23	54	20	16
18	19	18	19	18	40	30	19	18	38

(b) Why should we use the mean to predict any score in the sample?

11. (a) For the data in question 10, what is the mode? (b) By comparing the mean and the mode, determine the shape of the distribution. How do you know what it is?

12. What two pieces of information about the location of a score does a deviation score convey?

13. A scientist collected the following sets of data. For each, indicate which measure of central tendency she should compute.

(a) The following IQ scores:
60, 72, 63, 83, 68, 74, 90, 86, 74, 80

(b) The following anxiety scores:
10, 15, 18, 15, 14, 13, 42, 15, 12, 14, 42

(c) The following blood types:
A−, A−, O, A+, AB−, A+, O, O, O, AB+

(d) The following course grades:
B, D, C, A, B, F, C, B, C, D, D

14. On a normal distribution, four participants obtained the following deviation scores: −5, 0, +3, and +1.

(a) Which participant obtained the lowest raw score? How do you know?
(b) Which participant's raw score had the lowest frequency? How do you know?
(c) Which participant's raw score had the highest frequency? How do you know?
(d) Which participant obtained the highest raw score? How do you know?

15. In a normal distribution of scores, five people obtained the following deviation scores: $+1$, -2, $+5$, and -10.
    (a) Which person obtained the highest raw score?
    (b) Which participant obtained the lowest raw score?
    (c) Rank-order the deviation scores in terms of their frequency, putting the score with the lowest frequency first.
16. Foofy says a deviation of $+5$ is always better than a deviation of -5. Why is she correct or incorrect?
17. What is μ, and how do we usually determine its value?
18. (a) What is a powerful design? (b) What is a strong manipulation? (c) Why do we seek strong manipulations?
19. For the following experimental results, specifically interpret the relationship between the independent and dependent variables:

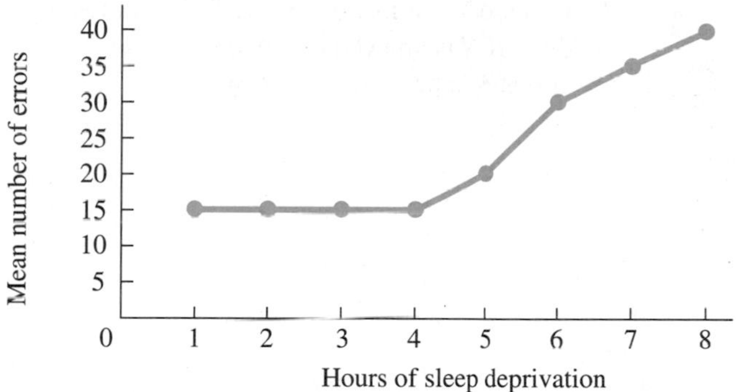

20. (a) In problem 19, give a title to the graph, using "as a function of." (b) If you participated in the above study and had been deprived of 5 hours of sleep, how many errors do we expect you would make? (c) If we tested all people in the world after 5 hours of sleep deprivation, how many errors would you expect each to make? (d) What symbol stands for your prediction in part c? (e) What issue of validity is raised by part c?
21. For each of the following experiments, determine which variable should be plotted on the Y axis and which on the X axis, whether a line graph or bar graph should be created, and how to summarize the dependent scores.
    (a) A study of income as a function of age
    (b) A study of politicians' number of positive votes on environmental issues as a function of the presence or absence of a wildlife refuge in their political district
    (c) A study of running speed as a function of carbohydrates consumed
    (d) A study of rates of alcohol abuse as a function of ethnic group
22. Using the words "statistic" and "parameter," how do we learn about a relationship in a population?
23. Dr. Grumpy tested people on the Grumpy Emotionality Test as a function of increases in the amount of sunlight present when they were tested. The resulting line graph slants downward. What does this tell you about: (a) The means for the conditions? (b) The raw scores for each condition? (c) The population μs? (d) The relationship between emotionality and sunlight in nature?
24. In problem 23, what would constitute a strong manipulation?

SUMMARY OF FORMULAS

1. *THE FORMULA FOR COMPUTING THE SAMPLE MEAN IS*

$$\bar{X} = \frac{\Sigma X}{N}$$

where ΣX stands for the sum of the scores and N is the number of scores.

2. To estimate the median, arrange the scores in rank order. If the number of scores (N) is an odd number, the score in the middle position is approximately the median. If N is an even number, the average of the two scores in the middle positions is approximately the median.

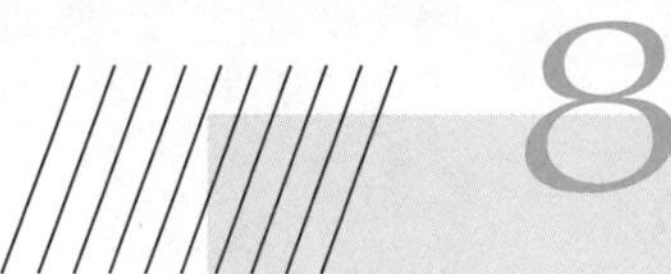

8

Summarizing Data with Measures of Variability

To understand this chapter:

- From Chapter 2, understand what is meant by the strength of a relationship.
- From Chapter 7, recall that the mean is the center of a distribution, the difference between $\overline{X}$ and μ, and why the sum of the deviations around the mean is zero.

Then your goals in this chapter are to learn:

- What is meant by variability.
- When the range is used and how to interpret it.
- When the standard deviation and variance are used and how to interpret them.
- How to compute the variance and standard deviation for a sample, for a population, and as an estimate of the population based on a sample.
- How variance is used to measure errors in prediction.
- How for power, we control variability.

The previous chapter introduced the first statistic to compute when describing *any* set of data: Compute the appropriate measure of central tendency. This information simplifies the distribution and allows you to envision its typical score. But! All participants will not behave in the same way, and so many scores may be very different from this typical score. Therefore, to completely describe a distribution, you must also answer the question "Are there large differences between the scores, or are there small differences between the scores?" In this chapter, we discuss procedures for describing the differences between scores.

First, though, here are some more symbols and terms.

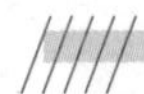

MORE STATISTICAL NOTATION

In this chapter, we'll see formulas involving several steps, so you must identify the quantity on which to perform the next operation. For example, a square root sign operates on "the quantity," so first compute the quantity inside the square root sign. Thus, $\sqrt{2 + 7}$ becomes $\sqrt{9}$, which is 3. Likewise, the length of the dividing line in a fraction determines the quantity that is in the numerator (the number above the line) and in the denominator (the number below the line). First, complete fractions involving a short line. Thus if we have:

$$\frac{6 + \frac{12}{2}}{4}$$

First divide 12/2, then add 6, and then divide by 4 (the answer is 3). And note: If you become confused in the midst of a formula, there is an order of precedence to mathematical operations. Unless otherwise indicated, first perform squaring or finding a square root, then multiplication or division, and then addition or subtraction. Thus, $(2)(4) + 5$ is $8 + 5$ or 13. Or, $2^2 + 3^2$ becomes $4 + 9$, which is 13. However, $(2 + 3)^2$ first produces 5^2, which is 25.

A new symbol you'll see is ΣX^2, which is called the **sum of the squared *X*'s**. It indicates to first square each score and then add up the squared scores. Thus, ΣX^2 for the scores 2, 2, and 3 first becomes $2^2 + 2^2 + 3^2$, which is $4 + 4 + 9$, which equals 17.

Learn right here to avoid confusing ΣX^2 with an operation that looks similar, but is in fact very different. The symbol $(\Sigma X)^2$ stands for the **squared sum of *X***. Because ΣX is inside the parentheses, first find the sum of the X scores and then square that sum. Thus, $(\Sigma X)^2$ for the scores 2, 2, and 3, is $(2 + 2 + 3)^2$, which is $(7)^2$, or 49. Notice that above, ΣX^2 gives 17 but $(\Sigma X)^2$ gives the very different answer of 49. Be careful with these terms.

This chapter also introduces *subscripts*, a letter placed below and to the right of a statistical symbol that identifies the scores used in computing the statistic. Pay attention to subscripts, because they are part of the symbols for certain statistics.

Finally, understand that many statistics will have two different formulas. A "definitional formula" defines a statistic and is important because it shows you where the answer comes from when you compute the statistic. However, actually using definitional formulas tends to be very time-consuming. Therefore, you'll also see computational formulas. As the name implies, computational formulas are used when actually computing a statistic. Trust me, computational formulas give exactly the same answers as definitional formulas, but they are much easier to use.

THE CONCEPT OF VARIABILITY

Recall that a measure of central tendency describes the center of a distribution and is used as the typical score to describe or predict all of the scores in a distribution. By themselves, however, such measures provide an incomplete description of any distribution. The mean, for example, tells us the central score and often where the most frequently occurring scores are, but it tells us little about scores that are *not* at the center of the distribution and/or that occur infrequently. Consider the three samples of data shown in Table 8.1. As you can see, each sample has a mean of 6. If we could not see the raw scores, we might believe that the three samples were identical. Obviously, they are not. Sample A contains relatively large differences between many of the scores, Sample B contains smaller differences between the scores, and Sample C contains no differences between the scores.

Thus, to describe a set of data accurately, we need to know how much the individual scores differ from each other. The type of statistic we need is called a measure of variability. **Measures of variability** summarize and describe the extent to which scores in a distribution differ from each other. Thus, when we ask whether there are large or small differences between the scores, we are asking the statistical question "How much variability is there in the data?" When there are many relatively large differences between the scores, the data are said to be relatively *variable* or to contain a large amount of *variability.*

In Chapter 7, you saw that a score indicates a participant's location on a variable and that the difference between two scores is the distance that separates them. From this perspective, measures of variability tell us how *spread out* the scores are. For example, Figure 8.1 visually presents the distances separating the scores in the previous samples. There are relatively large differences among the scores in Sample A, so the distribution is spread out. There are smaller differences between the scores in Sample B, so this distribution is not as spread out. There are no differences between the scores in Sample C, so there is no spread in this distribution. Thus, Sample A has the largest differences, so the scores in Sample A are spread out the most. In statistical terms, the scores in Sample A show the greatest variability.

There is nothing magical about what produces variability. Recall that participants will differ because of individual differences: Differences in genetics, physiology, experience, mental ability, and so on, that make individuals behave differently when in the same situation. Also, there are always fluctuations in subject, researcher, environmental, and measurement variables that cause participants to behave differently from moment to moment. These differences in behaviors in a particular situation result in differences—variability—in scores.

You should always describe the variability of a distribution, because it shows two important and related aspects of the data. First, the opposite of variability is how *consistent* the scores are. Small variability indicates that the scores do not differ greatly, so they must be rather similar and consistently close to the same value (and thus participants responded in a similar fashion). Conversely, larger variability indicates that scores and behaviors were inconsistent, each being rather different from the next.

TABLE 8.1 Three Different Distributions Having the Same Mean Score

Sample A	*Sample B*	*Sample C*
0	8	6
2	7	6
6	6	6
10	5	6
12	4	6
$\bar{X} = 6$	$\bar{X} = 6$	$\bar{X} = 6$

FIGURE 8.1 Distance Between the Locations of Scores in Three Distributions

An X over a score indicates a subject who obtained that score. Each arrow indicates how spread out the scores in the sample are.

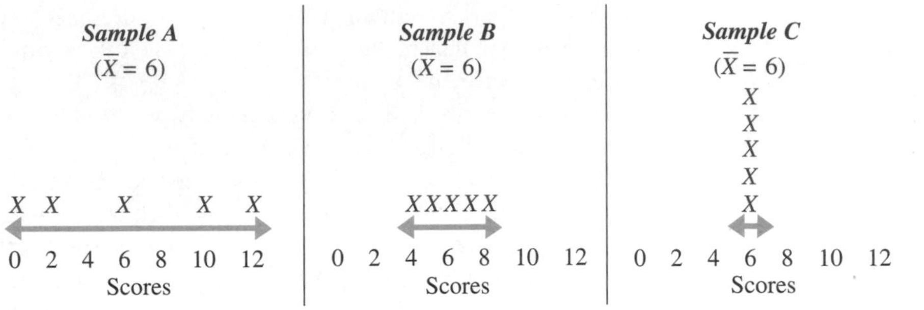

Sample A ($\overline{X} = 6$)	*Sample B* ($\overline{X} = 6$)	*Sample C* ($\overline{X} = 6$)
X X X X X over 0 2 6 10 12	XXXXX over 4 to 8	five X stacked over 6
0 2 4 6 8 10 12 Scores	0 2 4 6 8 10 12 Scores	0 2 4 6 8 10 12 Scores

Second, a measure of variability indicates how accurately a measure of central tendency describes the distribution: The greater the variability, the more the scores are spread out, and so the less accurately they are represented by *one* central score. Conversely, the smaller the variability, the closer the scores are to each other and to the one central score. Thus, by knowing the amount of variability in each of the above samples, we know Sample C contains consistent scores (and so 6 very accurately represents it), Sample B contains less consistent scores (and so 6 is not so accurate a summary), and Sample A contains very inconsistent scores (and so 6 is not even close to most scores).

REMEMBER Measures of variability indicate how spread out the scores are.

As with all statistics, the specific measure of variability to compute depends first on what it is you want to know and second on the scale of measurement involved and the shape of the distribution obtained. In the following sections we discuss three common measures of variability: The range, the variance, and the standard deviation.

The Range

One way to describe variability is to determine how far the lowest score is from the highest score. Sample A above is spread out the most because the lowest and highest scores are farther apart than in the other two groups. The descriptive statistic that indicates the distance between the two most extreme scores in a set of data is called the **range**.

THE FORMULA FOR COMPUTING THE RANGE IS

$$\text{Range} = \text{highest score} - \text{lowest score}$$

In Sample A, the highest score is 12 and the lowest score is 0, so the range is $12 - 0$, which equals 12. Sample B is less variable, because its range is $8 - 4 = 4$. The range in Sample C is $6 - 6 = 0$, so it has no variability.

Although the range communicates the spread in a sample, it is a rather crude measure. Because it involves only the two most extreme scores, the range is based on the least typical and often least frequent scores. Therefore, we usually compute the range as the *sole* measure of variability only with nominal or ordinal data. (For interval or ratio data forming a highly skewed distribution, we may compute a variation of the range called the "semi-interquartile range." As discussed in Part 2 of Appendix B, this statistic reflects the range between the scores at the 25th and 75th percentiles.)

For normally distributed interval or ratio scores, we calculate the variance and standard deviation.

Understanding the Variance and Standard Deviation

Most of the time psychological research involves interval or ratio data that more or less fit a normal or at least a symmetrical distribution, so the mean is the best measure of central tendency. When the mean is the appropriate measure of central tendency, the variance and the standard deviation are also appropriate. These are two very similar ways to describe variability. They are the best measures of variability, because they consider every score in the distribution.

We *use* the variance and the standard deviation to describe and communicate how different the scores are from each other. We *calculate* them, however, by measuring how much the scores differ from the mean. The mean is our reference point because the mean is always the center of a distribution, and when the scores are spread out from each other they are also spread out from the center. Thus, if a high score is far from a low score, both scores are relatively distant from the mean. If all scores are close to each other, they are also close to the mean.

The variance and standard deviation are appropriate with normal or other symmetrical distributions, because with them the mean is the point *around* which the distribution is located. *The variance and standard deviation allow us to quantify "around."* For example, if the grades in a statistics class form a normal distribution with a mean of 80, then you know that most people have a score *around* 80. But are most scores between 79 and 81, or between 60 and 100? By computing the variance and standard deviation, you can define "around." Here's how.

The distance between a score and the mean is the numerical difference between the score and the mean. Recall that this difference is symbolized as $(X - \overline{X})$ and is the amount the score deviates from the mean. Thus, a score's deviation indicates how much it is spread out from the mean. Of course, some scores will deviate from the mean by more than others. Because we want to summarize the variability of many scores that deviate by different amounts, we should determine the average amount that the scores deviate from the mean. We could call this the "average of the deviations." The larger the average of the deviations, the greater the variability between all of the scores and the mean.

To compute an average, we sum the scores and divide by N. We *might* find the average of the deviations by first computing $(X - \overline{X})$ for each participant, then summing these deviations to find $\Sigma(X - \overline{X})$, and finally dividing by N, the number of deviations. Altogether, the formula for the average of the deviations[1] would be:

$$\text{Average of the deviations} = \frac{\Sigma(X - \overline{X})}{N}$$

[1] In advanced statistics, there is a very real statistic called the average deviation. This isn't it.

We *might* compute the average of the deviations using this formula, except for a *big* problem. Recall that the sum of the deviations around the mean always equals zero, because the positive deviations cancel out the negative deviations. This means that the numerator in the above formula will always be zero, so the average of the deviations will always be zero. So much for the average of the deviations!

But remember our purpose here: We want a statistic that tells us something like the average of the deviations, so that we know the average amount the scores are spread out around the mean. The trouble is, mathematically, the average of the deviations is always zero. Therefore, we will calculate slightly more complicated statistics called the variance and standard deviation. But think of the variance and standard deviation as producing a number that, like an average, indicates the typical amount that the scores differ from the mean.

DESCRIBING THE SAMPLE VARIANCE

So how do we compute something like the average of the deviations? If the problem is the positive and negative deviations, then the solution is to *square* the deviations: First find the difference between each score and the mean, and then square that difference. This eliminates all negative deviations, so the sum of the squared deviations is not necessarily zero and neither is the average of the squared deviations. (As you'll see, we also choose this solution because it results in statistics that have very useful characteristics.)

By finding the average of the squared deviations, we are computing the variance. The **variance** is the average of the squared deviations of scores around the mean. When we calculate this statistic for a sample of scores, we are computing the *sample variance*. The symbol for sample variance is S_X^2. Always include the squared sign (^{2}), because it is part of the symbol for variance. The capital S indicates that we are describing a sample, and the subscript X indicates that the variance is computed for a sample of X scores.

THE DEFINITIONAL FORMULA FOR THE SAMPLE VARIANCE IS

$$S_X^2 = \frac{\Sigma(X - \overline{X})^2}{N}$$

Use the above formula *only* when describing a sample of data (as opposed to the population).

REMEMBER The symbol S_X^2 stands for the variance of the scores in a sample.

As an example, say that we measure the ages of a few children and find scores of 2, 3, 4, 5, 6, 7, and 8, with a mean age of 5. To compute S_X^2 using the above formula, arrange the data as shown in Table 8.2. First, compute each deviation $(X - \overline{X})$, by subtracting the mean from each score. Next, as shown in the far right column, square each deviation to get $(X - \overline{X})^2$. Then add the squared deviations to find $\Sigma(X - \overline{X})^2$, which here is 28. The number of scores, N, is 7. Filling in the formula for S_X^2 gives

$$S_X^2 = \frac{\Sigma(X - \overline{X})^2}{N} = \frac{28}{7} = 4.0$$

Thus, in this sample, the variance, S_X^2, equals 4.0. In other words, the average squared deviation of the age scores around their mean is 4.0.

Computational Formula for the Sample Variance

To simplify the preceding definitional formula, we have the following computational formula. Again, this formula is used only when describing the variance in a *sample*.

THE COMPUTATIONAL FORMULA FOR THE SAMPLE VARIANCE IS

$$S_X^2 = \frac{\Sigma X^2 - \frac{(\Sigma X)^2}{N}}{N}$$

This formula says to first find the sum of the X's or ΣX, then to square that sum, and divide the squared sum by N. Then subtract that result from the sum of the squared X's, ΣX^2. Finally, divide that quantity by N.

For example, arrange the previous age scores as shown in Table 8.3. The ΣX is 35, ΣX^2 is 203, and N is 7. Putting these quantities into the computational formula, we have

$$S_X^2 = \frac{\Sigma X^2 - \frac{(\Sigma X)^2}{N}}{N} = \frac{203 - \frac{(35)^2}{7}}{7}$$

The squared sum of X, $(\Sigma X)^2$, is 35^2, which is 1225, so

$$S_X^2 = \frac{203 - \frac{1225}{7}}{7}$$

TABLE 8.2 Calculation of Variance Using the Definitional Formula

Subject	*Age Score*	$-$	$\overline{X}$	$=$	$(X - \overline{X})$	$(X - \overline{X})^2$
1	2	$-$	5	$=$	-3	9
2	3	$-$	5	$=$	-2	4
3	4	$-$	5	$=$	-1	1
4	5	$-$	5	$=$	0	0
5	6	$-$	5	$=$	1	1
6	7	$-$	5	$=$	2	4
7	8	$-$	5	$=$	3	9
	$N = 7$					$\Sigma(X - \overline{X})^2 = 28$

TABLE 8.3 Calculation of Variance Using the Computational Formula

X score	X^2
2	4
3	9
4	16
5	25
6	36
7	49
8	64
$\Sigma X = 35$	$\Sigma X^2 = 203$

Now, 1225 divided by 7 equals 175, so

$$S_X^2 = \frac{203 - 175}{7}$$

Because 203 minus 175 equals 28, we have

$$S_X^2 = \frac{28}{7}$$

Finally, after dividing,

$$S_X^2 = 4.0$$

Thus, as we found using the definitional formula, the sample variance for these age scores is 4.0.

Do not read any further until you understand how to work this formula!

Interpreting Variance

The good news is that the variance is a legitimate measure of variability. Ideally, though, we want the average of the deviations, and the bad news is that the variance does not make much sense as an average deviation. There are two problems. First, because the variance is the average of the squared deviations, it is always an unrealistically large number. For the age scores of 2, 3, 4, 5, 6, 7, and 8, the $\overline{X}$ is 5 and S_X^2 is 4. To say that these scores differ from the mean of 5 by an *average* of 4 is plain silly! Not one score actually deviates from the mean by as much as 4, so this is certainly not the average deviation. The second problem is that variance is bizarre because it measures in squared units: Above we measured ages, so the variance indicates that the scores deviate from the mean by 4 *squared* years! (Whatever that means.)

Thus, it is difficult to interpret the variance as the "average" deviation in the data. Does this mean that computing the variance is a waste of time? No, because variance is used extensively in the statistics we will discuss later. Also, variance does communicate the *relative* variability of scores. If someone reports that one sample has $S_X^2 = 1$ and another sample has $S_X^2 = 3$, you know that the second sample is more variable, because it has a larger average squared deviation. This tells you that the scores are relatively inconsistent and less accurately described by their mean score. Thus, think of variance as a number that generally communicates how variable the scores are: The larger the variance, the more the scores are spread out.

The measure of variability that more directly communicates the average deviation is the standard deviation.

DESCRIBING THE SAMPLE STANDARD DEVIATION

The variance is always an unrealistically large number because we square each deviation. To solve this problem, we take the square root of the variance. The answer is called the standard deviation. The **standard deviation** is the square root of the variance, or the square root of the average squared deviation of scores around the mean. To create the definitional formula for the standard deviation, we simply add the square root sign to the previous definitional formula for variance.

THE DEFINITIONAL FORMULA FOR THE SAMPLE STANDARD DEVIATION IS

$$S_X = \sqrt{\frac{\Sigma(X - \overline{X})^2}{N}}$$

Notice that the symbol for the sample standard deviation is S_X, which is also the square root of the symbol for the sample variance ($\sqrt{S_X^2}$ is S_X.) Conversely, squaring the standard deviation produces the variance.

REMEMBER The symbol S_X stands for the standard deviation of the scores in a sample.

To compute S_X using this formula, first compute everything inside the square root sign to get the variance: Square each score's deviation, sum the squared deviations, and then divide that sum by N. In our previous age scores, the variance (S_X^2) was 4. Then take the square root of the variance to find the standard deviation. In this case,

$$S_X = \sqrt{4.0}$$

so

$$S_X = 2.0$$

The standard deviation of the age scores is 2.0.

Computational Formula for the Sample Standard Deviation

The computational formula for the standard deviation merely adds the square root symbol to the computational formula for the variance.

THE COMPUTATIONAL FORMULA FOR THE SAMPLE STANDARD DEVIATION IS

$$S_X = \sqrt{\frac{\Sigma X^2 - \frac{(\Sigma X)^2}{N}}{N}}$$

For example, again consider those age scores. From back in Table 8.3, ΣX is 35, ΣX^2 is 203, and N is 7. Putting these values in the formula gives

$$S_X = \sqrt{\frac{203 - \frac{(35)^2}{7}}{7}}$$

Completing the computations inside the square root symbol, we have the variance, which is 4.0. Thus

$$S_X = \sqrt{4.0}$$

Taking the square root of 4.0, we once again find that the standard deviation is

$$S_X = 2.0$$

Interpreting the Standard Deviation

Computing the standard deviation is as close as we come to computing the "average of the deviations." Thus, in our age scores we interpret a S_X of 2.0 as indicating that the scores differ, or deviate, from the mean by an "average" of about 2. Some scores will deviate by more and some by less, but overall the scores deviate from the mean by something like an average of 2. Further, the standard deviation measures in the same units as the raw scores, so here the scores differ from the mean age by an "average" of 2 *years*.

The standard deviation allows us to gauge the extent to which the scores are consistently close to each other, and correspondingly, the degree to which they are accurately summarized by the mean score. Because the average deviation is influenced by the size of the individual deviations, if S_X is relatively large, we know that a relatively large proportion of scores are rather far away from the mean and that few scores are close to it. If, however, S_X is relatively small, then most scores are close to the mean, and relatively few are far from it.

Finally, the standard deviation indicates how much the scores below the mean deviate from it and how much the scores above the mean deviate from it. Therefore, we can further summarize a distribution by describing the scores that lie at "plus one standard deviation from the mean" ($+1S_X$) and "minus one standard deviation from the mean" ($-1S_X$). For example, the age scores of 2, 3, 4, 5, 6, 7, 8 produced a $\overline{X}$ of 5.0 and a S_X of 2.0. The score that is $+1S_X$ from the mean is the score at 5 + 2, or 7. The score that is $-1S_X$ from the mean is the score at 5 − 2, or 3. As you can see, a good way to summarize these scores is to say that the mean score is 5 and the majority of the scores are between 3 and 7.

Applying the Standard Deviation to the Normal Curve

There is a precise mathematical relationship between the standard deviation and the normal curve, so that describing a distribution in terms of the scores that are between $-1S_X$ and $+1S_X$ is especially useful. For example, say that in a statistics class with a mean score of 80, the S_X is 5. The score at $80 - 5$ is the score of 75, and the score at $80 + 5$ is the score of 85.

Figure 8.2 shows about where these scores are located on a normal distribution. First notice that there is an easy way to locate where the scores at $-1S_X$ and $+1S_X$ are located on any normal curve. Over the scores that are close to the mean, the curve forms a downward convex shape (∩). As you travel away from the mean, at a certain point the curve changes its pattern to an upward convex shape (∪). The points at which the curve changes its shape are called "inflection points." Because of the mathematical relationship between a normal curve and the standard deviation, the scores at the inflection points are always the scores that are one standard deviation away from the mean.

Now Figure 8.2 shows how we summarize the distribution. First, saying that the mean is 80 implies that most scores are *around* 80. Then, finding the scores at $-1S_X$ and $+1S_X$ defines "around": Most of the scores are between 75 and 85 (from the parking lot perspective we used in Chapter 6, this is where most people are standing). In fact, because of the relationship between the standard deviation and the normal distribution, approximately 34% of the scores in a perfect normal distribution are *always* between the mean and the score that is one standard deviation from the mean. Thus, as in Figure 8.2, 34% of the students have scores between 75 and 80, and 34% have scores between 80 and 85. Altogether, approximately 68% of the scores are between the scores at $+1S_X$ and $-1S_X$ from the mean, so approximately 68% of the students have scores between 75 and 85. Conversely, only about 32% of the scores are outside this range, with about 16% below 75 and 16% above 85. Thus, saying that most scores are between 75 and 85 is an accurate summary, because the majority of scores (68%) are here.

Of course it is very unlikely that the scores from a small statistics class would produce an ideal normal distribution. However, here's how we apply the *normal curve model* to real data. If these scores are at least approximately normally distributed, we

FIGURE 8.2 Normal Distribution Showing Scores at Plus or Minus One Standard Deviation

With $S_X = 5.0$, the score of 75 is at $-1S_X$ and the score of 85 is at $+1S_X$. The percentages are the approximate percentages of the scores falling in each portion of the distribution.

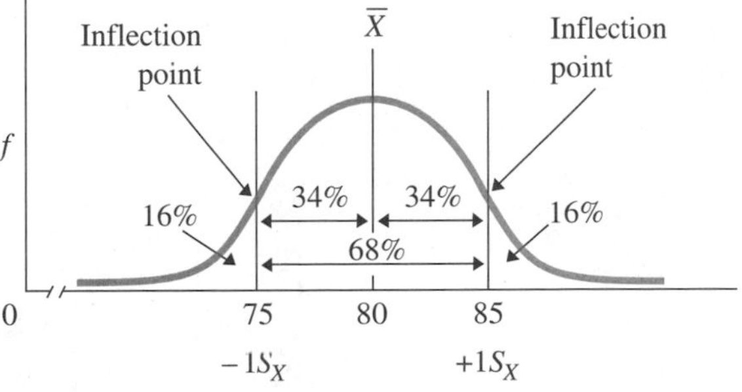

can operate as if the class formed the ideal normal distribution. Then we *expect* about 68% of the scores in the class to fall between 75 and 85. The more that the distribution conforms to a perfect normal curve, the more that precisely 68% of the scores will be between 75 and 85.

Recall that different samples of data will produce variations in the shape of a normal distribution. By finding the scores at $-1S_X$ and $+1S_X$ from the mean, we can envision and communicate these differences. Figure 8.3 shows three variations of the normal curve for scores having a mean of 50. The size of the standard deviation (and variance) indicates how spread out a distribution is. For example, in Distribution A, the S_X is 4.0 so the most frequent scores are between 46 and 54. Essentially, the most frequent scores are bunched close to the mean and so most frequently there are small deviations and thus a relatively small "average" or standard deviation. But look at Distribution B. Here the relatively frequent scores are spread out over a wider range between about 43 and 57, and with the majority of the deviations up to +7 and −7, the "average" deviation is the S_X of 7. Finally, in Distribution C, the frequent scores are between about 38 and 62, and with so many large deviations between +12 and −12, we have the large "average" deviation of S_X equal to 12.

> *REMEMBER* The larger the value of S_X (or S_X^2), the more the scores are spread out around the mean, and the wider the distribution.

Despite differences in shape, any normal distribution will still have approximately 68% of the scores between the scores at $+1S_X$ and $-1S_X$ from the mean. This is

FIGURE 8.3 Three Variations of the Normal Curve

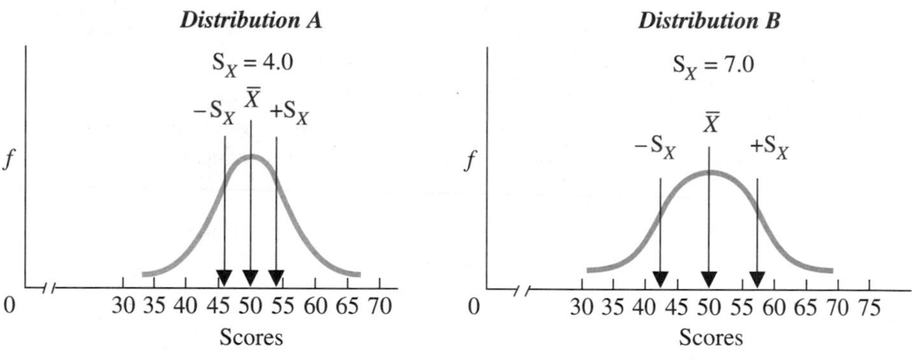

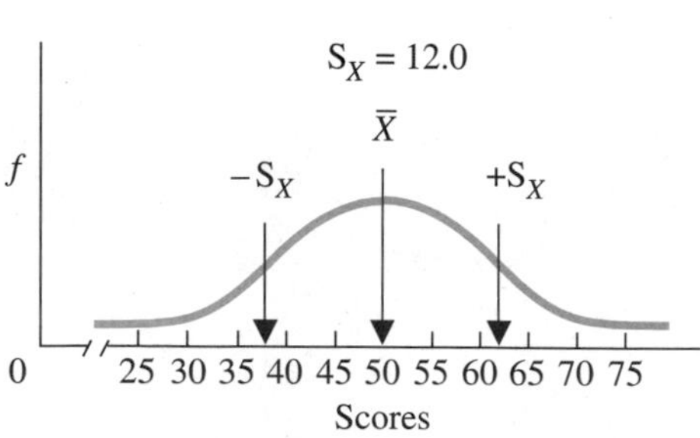

because, mathematically, 68% of the *area* under the normal curve falls between $+1S_X$ and $-1S_X$ (between the inflection points). Thus, 68% of the area under the curve in Distribution A lies between 46 and 54, while because Distribution C is much wider, 68% of its area lies between 38 and 62.

In summary, then, here is how the standard deviation (and variance) add to our description of a distribution. For the distributions in Figure 8.3, for example, if you know that they form normal distributions, you can envision their general shape. If you know that the mean is 50 in each, you know where the center of the distribution is and what the typical score is. And if you know, for example, that $S_X = 4$, then you know that those participants who did not score 50 missed it by an "average" of 4 points, the distribution is relatively narrow, and most scores (68% of them) are within 4 points of 50, or between 46 and 54. Conversely, if you know that the $S_X = 12$, you know that participants who did not score 50 missed it by an "average" of 12 points, the distribution is relatively wide, and 68% of the scores are within 12 points of 50, or between 38 and 62.

Avoiding Some Common Errors in Computing S_X

Because the computations for S_X (and S_X^2) are somewhat involved, always examine your answers to be sure that they are correct. First, variability can never be a negative number. Claiming that S_X is -5, for example, indicates that scores are a negative distance from the mean! Also, the formulas involve squared numbers, and squared numbers cannot be negative.

Second, watch for answers that simply do not make sense. If the raw scores range between 35 and 65, what should you conclude if you find that S_X is 20? *My* first guess is that you made a mistake! If these scores form anything like a normal distribution, the mean will be about 50 and the largest deviations for the scores of 35 and 65 will be only 15 points. Therefore, the "average deviation" for all scores cannot be 20. Similarly, it is unlikely that the S_X for these scores would be something like .25. If there are only two extreme deviations of 15, imagine how many small deviations it would take for the average to be only .25. (You can evaluate any variance by first finding its square root, the standard deviation.)

Strange numbers for S_X and S_X^2 may be correct for strange distributions, but always be alert to whether statistical answers make sense. Checking your calculations is the best way to ensure that you have the correct answer. Further, a general rule of thumb for any normal distribution is that the standard deviation equals about one-sixth of the overall range of scores in the data.

THE POPULATION STANDARD DEVIATION AND THE POPULATION VARIANCE

Recall that the ultimate goal is to describe the population of scores represented by a sample, and that Greek letters symbolize numbers that describe population parameters. The symbol for the true population standard deviation is σ_X. (The σ is the lowercase Greek letter s, or sigma.) Because the squared standard deviation is the variance, the symbol for the true population variance is σ_X^2. The definitional formulas for σ_X and σ_X^2 are similar to those we saw previously for describing a sample:

POPULATION STANDARD DEVIATION

$$\sigma_X = \sqrt{\frac{\Sigma(X - \mu)^2}{N}}$$

POPULATION VARIANCE

$$\sigma_X^2 = \frac{\Sigma(X - \mu)^2}{N}$$

The only novelty here is that we determine how far each score deviates from the population mean, μ. Otherwise, the population standard deviation and variance indicate exactly the same things about the population that the sample standard deviation and variance indicate about the sample. Both are ways of measuring how much the scores differ from μ, indicating how much the scores are spread out in the population.

REMEMBER The symbols σ_X and σ_X^2 are used when describing the true population variability.

Of course, we usually think of a population as being infinitely large, so we usually cannot use the above formulas and compute the true values of σ_X and σ_X^2. Instead, we make estimates about the population based on a random sample of scores. First, as in the previous chapter, we estimate that if we could measure the entire population, the population mean (μ) would equal the value of our sample mean ($\overline{X}$). Now we ask the question "What is our estimate of the variability of the scores around μ?"

Estimating the Population Variance and Population Standard Deviation

You might think that we would compute the sample variance as described previously and then use it to estimate the population variance (and do likewise for the standard deviation). If, for example, the sample variance is 4, should we then guess that the population variance is also 4? Nope! The sample variance and sample standard deviation are used *only* to describe the variability of the scores in a sample. They are *not* the best way to estimate the corresponding population parameters.

To understand why this is true, say that we measure an entire population of scores and compute its true variance, σ_X^2. We then draw a series of random samples of scores from the population and compute the variance of each sample, S_X^2. Sometimes the sample variance will equal the actual population variance, but other times the sample will not be perfectly representative of the population. Then either the sample variance will be smaller than the population variance and will underestimate it, or the sample variance will be larger than the population variance and will overestimate it. Over many random samples, however, more often than not the sample variance will *underestimate* the population variance. The same thing happens if we use the standard deviation.

In statistical terminology, the formulas for S_X^2 and S_X are called the *biased estimators* of the population variance and population standard deviation: They are biased toward underestimating the true population parameters. Using the biased estimators is a problem because, as you saw in Chapter 7, if our estimates cannot be perfectly accurate, we at least want the underestimates and overestimates to cancel out over the long run. (Remember the two statisticians shooting targets?) With the biased estimators, the underestimates and overestimates will not cancel out to equal the true population parameter. Instead, although the sample variance and sample standard deviation accurately describe the variability of a sample of scores, they are too often too small compared to the true population variance and standard deviation.

Why do S_X^2 and S_X produce biased estimates of the population variance and standard deviation? Because their formulas are for describing the variability in the sample, not for estimating the population. Remember that to accurately estimate a population, we should have a random sample. Here, we want the variability, or deviation, of each score to be random so that it accurately represents the variability of the scores in the population. Yet, when we measure the variability of a sample, we use the mean as our reference point. In doing so, we encounter the mathematical restriction that the sum of the deviations, $\Sigma(X - \overline{X})$, always equals zero. Because of this, not all of the variability in the sample reflects *random* variability.

For example, say that the mean of five scores is 6, and that four of the scores are 1, 5, 7, and 9. Their deviations are -5, -1, $+1$, and $+3$, so the sum of their deviations is -2. Without even looking at the final score, we know that it must be 8, because it has a deviation of $+2$ and so then the sum of all deviations will be zero. Thus, given the sample mean and the deviations of the other scores, this final deviation is not random. Rather, it is determined by those of the other scores. Therefore, only the deviations produced by the scores of 1, 5, 7, and 9 reflect the random variability found in the population. The same would be true for any four of the five scores. Thus, when N is 5, only four of the scores actually reflect the random variability of scores in the population. In general, out of N scores in a sample, only $N - 1$ of them actually reflect the variability in the population.

In calculating the biased estimators (S_X and S_X^2), the formulas divide by an N of 5. Because we divide by too large a number, the answer tends to be too small, underestimating the actual variability in the population. However, to estimate the variability based on four scores, we should divide by 4, or $N - 1$. By doing so, we compute the unbiased estimators of the population variance and standard deviation.

THE DEFINITIONAL FORMULAS FOR THE UNBIASED ESTIMATORS OF THE POPULATION VARIANCE AND STANDARD DEVIATION ARE

Estimated Population Variance

$$s_X^2 = \frac{\Sigma(X - \overline{X})^2}{N - 1}$$

Estimated Population Standard Deviation

$$s_X = \sqrt{\frac{\Sigma(X - \overline{X})^2}{N - 1}}$$

As you can see, we are still computing a number that is analogous to the average of the deviations in the sample. We include all of the scores when computing the sum of the squared deviations in the numerator, *but* (and this is the big but), then we divide by $N - 1$, the number of scores in the sample minus one.

Notice that the symbol for the unbiased estimator of the population standard deviation is s_X and the symbol for the unbiased estimator of the population variance is s_X^2. To keep all of your symbols straight, remember that the symbols for the sample variance and standard deviation involve the capital or "big" S, and in those formulas you divide by the *big* value of N. The estimated population variance and standard deviation involve the lowercase or *small s*, and you divide by the *smaller* number, $N - 1$. Further, the *small s* is used to estimate the true population value, symbolized by the *small* Greek s, σ. Finally, think of s_X^2 and s_X as the inferential versions of the variance and standard deviation, because the *only* time you use them is to estimate, or infer, the variance or standard deviation of the population based on a sample. You can think of S_X^2 and S_X as the descriptive variance and standard deviation, because they are used to describe the sample.

REMEMBER S_X^2 and S_X describe the variability in a sample, and s_X^2 and s_X estimate the variability in the population.

For future reference, the quantity $N - 1$ is called the degrees of freedom. The **degrees of freedom** is the number of scores in a sample that reflect the random variability in the population. The symbol for degrees of freedom is *df*, so here $df = N - 1$.

In the final analysis, you can think of $N - 1$ as simply a correction factor. Because $N - 1$ is a smaller number than N, dividing by $N - 1$ produces a slightly larger answer than does dividing by N. This larger answer will tend not to underestimate, so that over the long run we have a more accurate estimate of the population variability.

Computational Formula for the Estimated Population Variance

The only difference between the computational formula for the estimated population variance and the previous computational formula for the sample variance is that here the final division is by $N - 1$.

THE COMPUTATIONAL FORMULA FOR ESTIMATING THE POPULATION VARIANCE IS

$$s_X^2 = \frac{\Sigma X^2 - \frac{(\Sigma X)^2}{N}}{N - 1}$$

In our previous examples using age scores, $N = 7$, $\Sigma X^2 = 203$, and $\Sigma X = 35$. Putting these quantities into the above formula gives

$$s_X^2 = \frac{\Sigma X^2 - \frac{(\Sigma X)^2}{N}}{N - 1} = \frac{203 - \frac{(35)^2}{7}}{6}$$

Work through this formula in exactly the same way you worked through the formula for the sample variance, except that here the final division involves $N - 1$, or 6. Since 35^2 is 1225, and 1225 divided by 7 equals 175,

$$s_X^2 = \frac{203 - 175}{6}$$

Now 203 minus 175 equals 28, so

$$s_X^2 = \frac{28}{6}$$

and the final answer is

$$s_X^2 = 4.67$$

This answer is slightly larger than the one we obtained when we computed the sample variance for these age scores. There, S_X^2 was 4.0. Although 4.0 accurately describes the sample variance, it is likely to underestimate the actual variance of the population: 4.67 is more likely to be the population variance. In other words, if we could measure all scores in the population from which this sample was drawn and then compute the true population variance, we would expect σ_X^2 to be 4.67.

Computational Formula for the Estimated Population Standard Deviation

A standard deviation is always the square root of the corresponding variance, so the formula for the estimated population standard deviation merely adds the square root sign to the above formula for the estimated population variance.

THE COMPUTATIONAL FORMULA FOR ESTIMATING THE POPULATION STANDARD DEVIATION IS

$$s_X = \sqrt{\frac{\Sigma X^2 - \frac{(\Sigma X)^2}{N}}{N - 1}}$$

In the previous section, the estimated population variance from our sample of age scores was $s_X^2 = 4.67$. Using the above formula, s_X is $\sqrt{4.67}$, or 2.16. Thus, if we could compute the standard deviation using the entire population of scores, we would expect σ_X to be 2.16.

Interpreting the Estimated Population Variance and Standard Deviation

We interpret the estimated variance and standard deviation in the same way as S_X^2 and S_X, except that now we are describing how much we *expect* the scores to be spread out in the population, how consistent or inconsistent we *expect* the scores to be, and how accurately we *expect* the population to be summarized by μ.

Notice that, assuming a sample is representative of the population, we can pretty much reach our ultimate goal of describing an unknown population of scores. If we can assume that the distribution is normal, we have described its overall shape. The sample mean, $\overline{X}$, provides a good estimate of the population mean, μ. The size of s_X (or s_X^2) is our estimate of how spread out the population is—an estimate of the "average amount" that the scores in the population deviate from μ. Further, we expect approximately 68% of the scores in the population to lie between the scores at $+1s_X$ and $-1s_X$ from μ. Then, because scores reflect behaviors, we have a good idea of how most individuals in the population behave in a given situation (which is why we conduct research in the first place).

Recall that in addition to being able to describe an overall distribution of scores, another goal of research is to be able to predict any individual's score. Variability plays an important role in meeting this goal as well.

VARIANCE AS THE ERROR IN PREDICTIONS

In Chapter 7, we saw that the sample mean is the best single score to use to predict unknown scores. However, everyone will not obtain the predicted mean score. Measures of variability are used to describe how well we can predict scores.

In statistics, we estimate the amount of error in predicting unknown scores based on how well we can predict the scores in a known sample: Pretending that we don't know the scores, we predict them, and then see how close we came to the actual scores. For example, if that statistics class has a mean score of 80, then our best guess is that any student in the class has a grade of 80. Of course, not every student will actually obtain a grade of 80. When we use the mean as the predicted score, the amount we are wrong in a single prediction is the quantity $(X - \overline{X})$, the amount that the actual score *deviates* from the predicted mean score. Because some predictions in a sample will contain more error than others, it makes sense to summarize the error by finding the average amount that the actual scores deviate from the mean. As we've seen, the way to find the "average" amount that scores deviate from the mean is to compute the variance and standard deviation. Thus, here is a slightly novel way of thinking about measures of variability. Because these statistics measure the difference between each score and the mean, they also measure the error in our predictions when we predict the mean for everyone in a sample. The larger the variability, the larger the differences between the mean and the scores, so the larger the error when we use the mean to predict scores.

Thus, if the standard deviation in the statistics class is 5, then the actual scores differ from the mean of 80 by an "average" of 5 points. Therefore, if we predict a score of 80 for every member of the class, the actual scores will differ from the predicted score by an "average" of 5 points. Sometimes we will be wrong by more, sometimes by less, but overall we will be wrong by an amount equal to the sample standard deviation (S_X).

Similarly, the variance (S_X^2) indicates the average of the squared deviations from the mean, so the variance is the average of the "squared errors" we have when predicting the mean score for everyone in a sample. Unfortunately, the concept of squared errors is rather strange. This is too bad, because in statistics the proper way to describe the error in predictions is to compute the variance. In fact, the variance is sometimes called *error variance*: it is our way of measuring the average error between the predicted mean score and the actual raw scores. Thus, when we use the mean to predict scores, the larger the variance, the larger the error, and the smaller the variance, the smaller the error. To keep this concept in focus, think of the extreme case in which all the scores in the statistics class are the same score of 80. Then the mean is 80, and the variance is zero. In this case, predicting the mean of 80 for each student produces zero error: There is no difference between what we predict for students (the $\overline{X}$) and what they got (the *Xs*) and that is exactly what $S_X^2 = 0$ indicates.

REMEMBER The sample variance, S_X^2, is error variance, the "average" error when using the sample mean as the predicted score for everyone in the sample.

Estimating the Error in Predictions in the Population

As you know, we can also predict scores in the population. Our best estimate of any score in the population is the population mean, μ, which we assume is equal to our sample mean. Thus, based on the statistics class mean of 80, we estimate that the μ is 80 and we predict that any student in the population who takes this class will receive a grade of 80.

To determine the error in these predictions, we use the same logic used above for the sample. The error in predictions equals the population variance, because it describes the differences between the population mean we predict for each individual and the actual scores in the population. However, we usually cannot compute the true population variance, so instead we compute the estimated population variance, s_X^2. Thus, say that for the statistics class the s_X^2 is 5.75: We expect the scores to differ from μ by about 5.75. Therefore, when we predict that other students taking this class will receive a score equal to the μ of 80, the amount we expect to be wrong is also 5.75.

REMEMBER The estimated population variance, s_X^2, is the error we expect when, based on the sample mean, we predict μ as the score for anyone in the population.

Researchers can always measure a sample of scores, compute the mean, and use it to predict scores. Therefore, the value of S_X^2 is the maximum error we are forced to accept when predicting scores in a sample (and s_X^2 is the maximum error when predicting scores in a population). As you will see in later chapters, because the variance is the worst that we can do, it is our reference point. Anything that improves the accuracy of predictions is measured relative to the variance.

SUMMARIZING RESEARCH USING THE MEAN AND STANDARD DEVIATIONS

The standard deviation is the most commonly reported measure of variability in psychological research because it is easiest to interpret. Essentially, the standard deviation is interpreted as indicating how consistently close individual scores are to the mean. Thus, in an observational study, the mean may describe the number of times participants exhibited a particular behavior, but the standard deviation indicates how consistently all participants exhibited the behavior that often. Or, in summarizing a survey, the mean rating from Likert-type questions may describe the typical opinion held by participants, but the standard deviation indicates the extent of disagreement among them.

The same approach is taken when interpreting the results of an experiment. For example, in the previous chapter we conducted an experiment on the effect of recalling either a 5-, 10-, or 15-item list. We saw that the mean summarized the dependent variable of recall scores (the mistakes participants make when recalling a list), indicating *around* where participants in each condition scored. For the complete picture, we should also include the standard deviation in each condition, as shown in Table 8.4. By merely knowing the $\overline{X}$ and S_X in each condition, we can easily summarize the results. For example, typically a score of 3 was observed in the 5-item condition, with, on average, scores varying above or below this by about .82. Or, in terms of error variance, by squaring the S_X of .82 we have $S_X^2 = .67$. Thus, when we predict that people will score a 3 when recalling a 5-item list and we are wrong, we expect to be off by about .67. Likewise we expect scores in the 10-item list to be around 6, but we may be off by about the S_X of 1.41, or by the S_X^2 of 1.99, and so on.

REMEMBER Summarize an experiment by calculating the mean and standard deviation (or variance) in each condition.

The pattern of differences between the means of the conditions shows the relationship between recall scores and list length. The next question to ask, however, is *how well* do the data fit this relationship? Recall that not all relationships are created equal, differing in the "strength of relationship" or "degree of association." The more consistently scores on the dependent variable change as conditions of the independent variable change, the stronger the relationship. Because the standard deviation and variance describe the inconsistency in scores in each condition, they communicate the strength of a relationship.

For example, say that our experiment produced either the strong or weak relationship shown in Table 8.5. Part A shows a perfectly consistent relationship because there is one dependent score associated with each condition. In other words, there is no difference or variability among the scores *within* each condition, and each S_X equal to zero indicates this fact. In part B, however, the values of S_X are relatively large. This indicates that the individual scores within the conditions are rather inconsistent and variable, so that overall this is a relatively weak and inconsistent relationship.

REMEMBER The smaller the values of S_X for the conditions in an experiment, the stronger the relationship.

TABLE 8.4 Mean and Standard Deviation in Each Condition of Recalling 5-, 10-, or 15-Item Lists

5-item list	*10-item list*	*15-item list*
3	5	9
4	5	11
2	8	7
$\overline{X} = 3$	$\overline{X} = 6$	$\overline{X} = 9$
$S_X = .82$	$S_X = 1.41$	$S_X = 1.63$

TABLE 8.5 Means and Standard Deviations Recall Experiment with a Perfectly Strong or a Very Weak Relationship

(A) Perfectly Strong Relationship

5-item list	*10-item list*	*15-item list*
3	6	9
3	6	9
3	6	9
$\overline{X} = 3$	$\overline{X} = 6$	$\overline{X} = 9$
$S_X = 0$	$S_X = 0$	$S_X = 0$

(B) Very Weak Relationship

5-item list	*10-item list*	*15-item list*
3	6	6
6	2	8
0	10	13
$\overline{X} = 3$	$\overline{X} = 6$	$\overline{X} = 9$
$S_X = 2.45$	$S_X = 3.27$	$S_X = 2.94$

Evaluating S_X as above involves a rather subjective approach. In later chapters, we'll discuss more objective ways to evaluate the strength of a relationship.

Considering the strength of the relationship in the sample data is an important component of interpreting an experiment. Above, the degree of consistency in the relationship suggests the extent to which recall scores are "caused" by list length. The perfectly consistent relationship would suggest that list length has a major, controlling effect on recall: It seems to be *the* variable that determines a score. Conversely, the weak relationship would suggest that there are other factors in addition to list length that cause or influence recall (perhaps intelligence or motivation is important). Following this line of reasoning, we interpret the results "psychologically," considering what the consistency and inconsistency in the relationship indicate about the variables of the amount of information in a list and its recall, and how this relates to theoretical explanations, models, and constructs involving memory storage and retrieval.

Variability and the Power of a Design

Recall from the previous chapter that a problem for researchers is that we may fail to find convincing evidence of a relationship in the sample data, even when there is a relationship in nature. Then, because the sample data form a relationship so poorly, we may erroneously decide there really is no relationship between the variables. To avoid this problem, recall that we seek a *powerful design.* With greater power, we are more likely to obtain clear and convincing evidence of a relationship, so that we don't miss relationships in nature.

You already know that one aspect of power is to aim for large differences in scores *between* the conditions. The next step is to aim for the strongest, most consistent relationship possible: This is because the clearest, most convincing indication of a relationship in nature comes from the strongest sample relationship. Above, the perfectly strong relationship in Table 8.5 will be very difficult for us—and our statistics—to misinterpret as anything other than a relationship between list length and recall. However, the weak relationship is not so convincing: We see a lot of overlap in scores between the conditions, and there is not a consistent pairing of one score with one list length. Therefore, we might erroneously conclude from these data that there is not a relationship here.

Whether you observe a weak or strong relationship depends on the amount of variability among scores *within* each condition. Although this variability can be described by the standard deviation, it is generally referred to as *error variance*. Because we assume that nature is lawful, we assume that something causes error variance. Therefore, a key issue in creating a powerful design is for you to eliminate anything that might cause error variance—differences in scores within conditions. Then you will have the strongest relationship possible.

REMEMBER For power, we seek the strongest relationship possible. This occurs with large differences in scores *between* conditions, but little difference in scores *within* conditions.

To minimize error variance, we again focus on our old friends, the issues of reliability, validity, and control. For example, are the measurements unreliable, reflecting random error? If the above list items can sometimes be guessed, then even when participants actually remember the list in a condition to the same degree, we will see different scores. Or do the measurements lack content validity? If the memory test partially tests language skills, then people within the same condition will have different recall scores. Likewise, do we lack internal validity because of fluctuating extraneous variables? If, for example, the physical environment changes unsystematically, this could differentially influence participants so that they have different scores within the same condition. In addition, individual differences produce error variance. Thus, differences between participants in terms of their memory ability, their motivation, or their reactivity to being observed will produce different recall scores in the same condition.

As these examples illustrate, anything that influences participants can produce variability—*differences*—in scores within each condition, resulting in a weak relationship. Therefore, we have yet another reason for designing a study that accurately and exclusively measures the variables of interest, while controlling all other factors. In earlier chapters, we saw how design flaws can mislead us when we interpret results conceptually. Now we see that these same flaws also influence statistical interpretations. And, clearly, our statistical interpretations influence our conceptual interpretations, because we must find a statistical relationship before we can draw any inferences about the underlying behavior. Therefore, by dealing with the issues of reliability, validity, and control, we: (1) create a powerful design so we draw the correct statistical interpretations, and (2) eliminate design flaws so we draw the correct psychological interpretations as well.

FIGURE 8.4 Organizational Chart of Descriptive and Inferential Measures of Variability

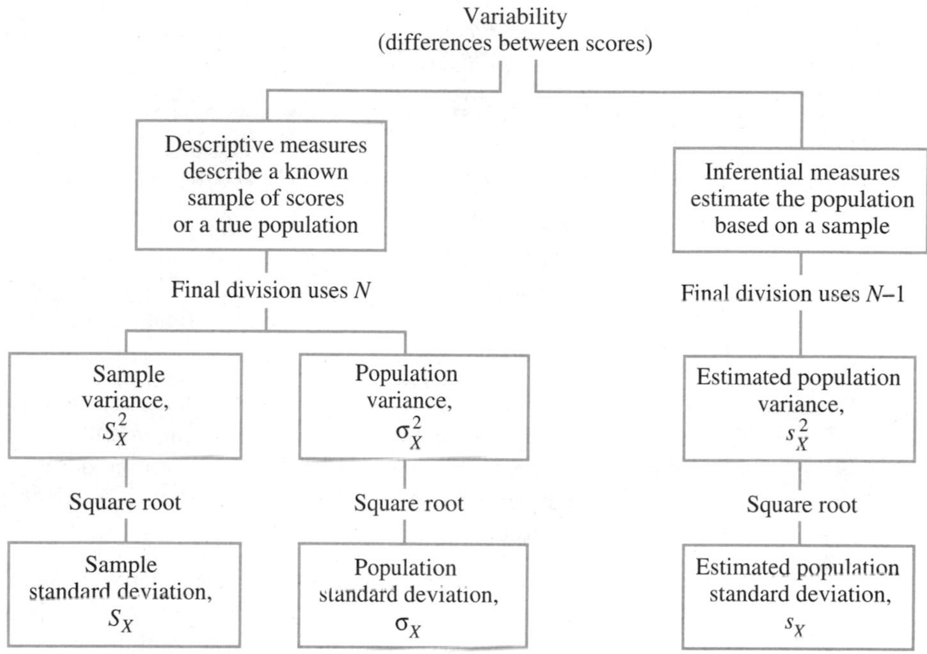

PUTTING IT ALL TOGETHER

As you might suspect, the concept of variability plays a major role throughout later discussions. You can organize your thinking about the different measures of variability using the diagram in Figure 8.4 above. Remember that variability refers to the differences between scores. The variance and standard deviation are simply two methods for describing variability, or differences. Constructing any formula for the standard deviation merely requires adding the square root sign to the corresponding formula for variance. For either the variance or the standard deviation, from the population we can compute the true population parameter, or from the sample data compute the descriptive sample version or the estimated population version. The difference when using inferential formulas is that they require a final division by $N - 1$ instead of by N.

CHAPTER SUMMARY

1. Unless otherwise indicated in a formula, the order of mathematical operations is to square or find the square root first, then multiply or divide, and then add or subtract.

2. *Measures of variability* describe how much the scores differ from each other, or how much the distribution is spread out.

3. The *range* is a measure of variability based on the difference between the highest score and the lowest score.

4. The *variance* is used to describe symmetrical or normal distributions. It is the average of the squared deviations of scores around the mean.

5. The *standard deviation* is also used to describe symmetrical or normal distributions of interval ratio scores. It is the square root of the variance. It can be thought of as the "average" amount that scores deviate from the mean.

6. There are three versions of the formula for variance: S_X^2 describes how far sample scores are spread out around the sample mean, σ_X^2 describes how far the population of scores is spread out around μ, and s_X^2 is computed using sample data, but is the inferential unbiased estimate of how far the population is spread out around μ.

7. There are three versions of the formula for the standard deviation: S_X describes how far the sample scores are spread out around the sample mean, σ_X describes how far the population is spread out around μ, and s_X is computed using sample data, but is the inferential, unbiased estimate of how far the population is spread out around μ.

8. Formulas for descriptive measures of variability (for S_X^2 and S_X) use N as the final denominator but the inferential formulas (for s_X^2 and s_X) use $N - 1$. The quantity $N - 1$ is the *degrees of freedom* in the sample.

9. The sample variance, S_X^2, is the amount of error when the value of $\overline{X}$ is the predicted score for everyone in the sample. The estimated population variance, s_X^2, is the expected error when the value of μ is the predicted score for everyone in the population.

10. The smaller the value of S_X (or S_X^2) in each condition of an experiment, the stronger the relationship.

11. *Error variance* is the variance in each condition, reflecting the error when using the mean of a condition to predict scores in that condition.

12. Power requires the strongest relationship possible. This occurs with large differences in scores *between* conditions, but small differences in scores *within* conditions.

13. Threats to reliability and validity and individual differences produce error variance, and the greater the error variance, the weaker the relationship. Therefore, a powerful design minimizes error variance by controlling such threats and minimizing individual differences.

PRACTICE PROBLEMS

(Answers for odd-numbered problems are provided in Appendix D.)

1. If given no other information when completing a formula, what is the order in which to perform mathematical operations?

2. In any research, what three characteristics of a distribution must the researcher describe?
3. What do measures of variability communicate about: (a) The size of the differences between the scores in a distribution? (b) How consistently participants behaved?
4. (a) What is the range? (b) Why is it not the most accurate measure of variability? (c) When is it used as the sole measure of variability?
5. (a) What do both the variance and standard deviation tell you about a distribution? (b) Which measure will you usually want to compute? Why?
6. (a) What is the mathematical definition of the variance? (b) How is a sample's variance related to its standard deviation and vice versa?
7. (a) What do S_X, s_X, and σ_X have in common in terms of what they communicate? (b) How do they differ in terms of their use?
8. Why are your estimates of the population variance and standard deviation always larger than the corresponding values for describing a sample from that population?
9. You correctly compute the variance of a distribution to be $S_X^2 = 0$. What should you conclude about this distribution?
10. In a condition of an experiment, a researcher obtains the following creativity scores.

3	2	1	0	7	4	8	6	9	1
6	8	6	9	4	5	0	8	7	6

In terms of creativity, interpret the variability of these data using: (a) The range (b) The variance (c) The standard deviation.
11. If you could test the entire population in problem 10, what would you expect each of the following to be? (a) The typical creativity score (b) The variance (c) The standard deviation (d) The two scores between which about 68% of all creativity scores fall in this situation.
12. Say the sample in problem 10 had an N of 1000. How many people would you expect to score below 2.10? Why?
13. (a) In an experiment, what is error variance?
14. As part of studying the relationship between mental and physical health, you obtain the following heart rates.

73	72	67	74	78	84	79	71	76
78	76	79	81	75	80	78	76	

In terms of differences in heart rates, interpret these data using: (a) The range (b) The variance (c) The standard deviation.
15. If you could test the population in problem 14, what would you expect each of the following to be? (a) The shape of the distribution (b) The typical heart rate (c) The variance (d) The standard deviation (e) The two scores between which about 68% of all heart rates fall.
16. Foofy has a normal distribution of scores ranging from 2 to 9. (a) She has computed the variance here to be −.06. What should you conclude from this answer and why? (b) She recomputes the standard deviation to be 18. What should you conclude and why? (c) She recomputes the variance to be 1.36. What should you conclude and why?
17. For his test grades, Guchi has a $\overline{X}$ of 60 and $S_X = 20$. Pluto has a $\overline{X}$ of 60 and $S_X = 5$. (a) Who is the more consistent student and why? (b) Who is more accurately described as a 60 student and why?
18. (a) For which student in problem 17 can you more accurately predict the next test score and why? (b) Who is more likely to do either extremely well or extremely poorly on the next exam, and why?
19. On a final exam, the $\overline{X} = 65$ and $S_X = 6$. What score would you predict for each student, and if you're wrong, what do you expect will be the error in your prediction?
20. (a) How does error variance decrease the power of a study? (b) What causes error variance?

21. Consider the results of the following experiments.

Experiment 1

Condition 1	*Condition 2*	*Condition 3*
12	30	45
11	33	48
14	36	49
10	35	44

Experiment 2

Condition 1	*Condition 2*	*Condition 3*
18	8	3
13	11	9
9	6	5

(a) What should you do to summarize these experiments? (b) Using the means, summarize the relationship in each study. (c) Which experiment produced the stronger relationship? (d) Confirm your answer to (c) mathematically. (e) Which experiment contains the greater amount of error variance? How do you know? (f) What does the amount of error variance suggest about the impact of the independent variable in each experiment?

22. A researcher obtains a large degree of error variance in an experiment on children's ability to remember a passage they read as a function of the number of hours they have previously watched television. (a) What does this statement indicate about the scores in each condition? (b) What might cause such variability? (c) What impact will such variability have on the statistical results? (d) What impact will it have on the conceptual interpretation of the variable of television watching?

SUMMARY OF FORMULAS

1. The formula for the range is

$$\text{Range} = \text{highest score} - \text{lowest score}$$

2. The computational formula for the sample variance is

$$S_X^2 = \frac{\Sigma X^2 - \frac{(\Sigma X)^2}{N}}{N}$$

3. *The computational formula for the sample standard deviation is*

$$S_X = \sqrt{\frac{\Sigma X^2 - \frac{(\Sigma X)^2}{N}}{N}}$$

4. *The computational formula for estimating the population variance is*

$$s_X^2 = \frac{\Sigma X^2 - \frac{(\Sigma X)^2}{N}}{N - 1}$$

5. *The computational formula for estimating the population standard deviation is*

$$s_X = \sqrt{\frac{\Sigma X^2 - \frac{(\Sigma X)^2}{N}}{N - 1}}$$

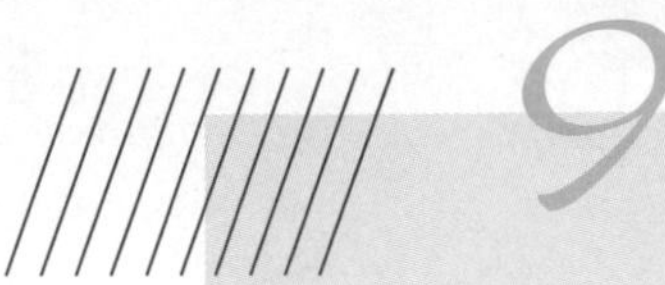

9 Summarizing Data with *z*-Score Transformations and the Normal Curve Model

To understand this chapter:

- From Chapter 6, recall that relative frequency is the proportion of the time certain scores occur, that it corresponds to the proportion of the total area under the normal curve, and that a score's percentile equals the percentage of the total curve to the left of the score.
- From Chapter 7, recall that the larger a score's deviation from the mean, the lower the score's frequency.
- From Chapter 8, recall that S_X and σ_X indicate the "average" deviation of scores around $\overline{X}$ and μ, respectively.

Then your goals in this chapter are to learn:

- What a *z*-score is and what it tells you about a score's relative location.
- How the standard normal curve is used with *z*-scores to determine expected relative frequency, simple frequency, and percentile.
- The characteristics of a sampling distribution of means and what the standard error of the mean is.
- How a sampling distribution of means is used with *z*-scores to determine the expected relative frequency of sample means.

The techniques discussed in the preceding chapters for graphing, measuring central tendency, and measuring variability comprise the descriptive procedures used in most psychological research. In this chapter, we combine these procedures to answer another question about data: How does any one particular score compare to the other scores in a sample or population? We answer this question by transforming raw scores into what are known as z-scores. Recall that we transform scores to compare scores on different variables and to make scores within the same distribution easier to interpret. The z-transformation is the Rolls-Royce of transformations, because it allows us to interpret and compare scores from virtually *any* normal distribution of interval or ratio scores.

The following sections first examine the logic of z-scores and discuss how to compute them. Then we'll look at their uses, both in describing individual scores and in describing sample means.

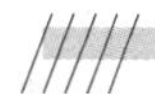

MORE STATISTICAL NOTATION

The answers here will involve negative or positive numbers. However, sometimes we ignore a number's sign. The *absolute value* of a number is the size of a number, regardless of its sign.

Also, you'll encounter the symbol $\pm$, which means "plus or minus." It provides a shorthand method for describing two numbers or the range of numbers between them. Saying ± 1, means $+1$ and -1. Saying the scores "between ± 1," means all possible scores from -1, through 0, up to and including $+1$.

UNDERSTANDING z-SCORES

As an example, let's say that we conduct a study at Prunepit University in which we measure the attractiveness of a sample of males. We train several judges to evaluate participants on the variable of attractiveness, and each male's score is the total number of points assigned by the judges. We want to analyze these attractiveness scores, especially those of three men: Slug, who scored 35; Binky, who scored 65; and Biff, who scored 90.

How do we interpret such scores? We might develop an absolute definition of attractiveness: If a male scores above X, he is attractive. The problem with this is that we cannot easily justify such a definition. Remember that psychologists usually do not know what a particular raw score specifically indicates about a variable in nature. Thus, it is difficult to know whether, in the grand scheme of things, a particular attractiveness score is high or low, good or bad, or what. Instead, like most variables in psychology, we interpret each score in *relative* terms: Whether a score is good, bad, or indifferent depends on the other scores in the distribution. Essentially, we determine what a particular score indicates about a particular situation found in nature by comparing it to all other possible scores found in that situation. To illustrate, let's say that our sample of attractiveness scores forms the normal curve shown in Figure 9.1. By looking at the distribution and using the statistics we've discussed previously, we can make several statements about each man's score. Let's review.

FIGURE 9.1 Frequency Distribution of Attractiveness Scores at Prunepit U

Scores for three individual subjects are identified on the X axis.

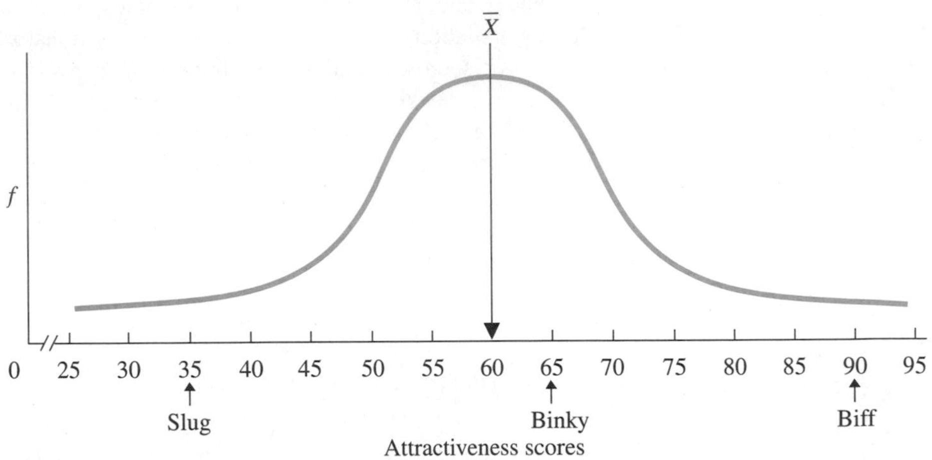

What would we say to Slug? "Bad news, Slug. Your score is to the left of the mean, so you're below average in attractiveness in this sample. What's worse, you're *way* below the mean. Down in the tail of the distribution, the height of the curve above your score is not large, indicating a low *frequency*: Not many men received this low score. Also, the proportion of the total area under the curve at your score is also small, so the *relative frequency*—the proportion of men receiving your score—is low. (In terms of our parking lot approach to the normal curve described in Chapter 6, the proportion of all men at Slug's score is small.) Finally, Slug, your *percentile* is low: A small percentage scored below your attractiveness score, while a large percentage scored above it. So, Slug, scores like yours are relatively infrequent, and few scores are lower than yours. Relative to the others, you're ugly!"

What would we tell Binky? "Binky, there's good news and bad news. The good news is that your score of 65 is above the mean of 60, which is also the median, or 50th percentile: You are better looking than more than 50% of the sample. The bad news is that your score is not *that* far above the mean. The area under the curve at your score is relatively large, and thus the relative frequency of your score is large (the proportion of equally attractive men is large). What's worse, there is a relatively large part of the distribution with higher scores."

And then there is Biff. "Yes, Biff, as you expected, you are above average in attractiveness. In fact, as you have repeatedly told everyone, you are one of the most attractive men around. The area under the curve at your score, and thus the relative frequency of your score, is quite small, meaning that only a small proportion of the men are equally attractive. Also, the area under the curve to the left of your score is relatively large. This means that if we cared to figure it out, we'd find that you scored at a very high percentile: A large percentage of scores are below your score, while a small percentage are above your score."

Recognize that although Biff is better looking than most men in our study, he may in fact have a face so ugly that it could stop a clock! Or in some other sample, Slug might be considered highly attractive. Because we have defined attractive and ugly relative to this distribution, we have described each score's relative standing. The **relative standing** of a score reflects a systematic evaluation of the score relative to the characteristics of the sample or population in which the score occurs.

In describing each man's relative standing above, we first determined his score's location relative to the mean. Based on this information, we then determined other measures of relative standing, such as the score's relative frequency and percentile. Thus, the essence of determining a score's relative standing is to determine how far the score is above or below the mean.

Describing a Score's Relative Location as a *z*-Score

To quantify a score's relative standing, we begin by computing the score's *deviation*. In previous chapters, we saw that the distance between a raw score and the mean is the amount the score deviates from the mean, which equals $(X - \overline{X})$. Thus, for example, Biff's raw score of 90 deviates from the mean of 60 by +30 (90 − 60 = +30). The + sign indicates that he is above the mean. A deviation of +30 *sounds* as if it might be a large deviation, but is it? As with a raw score, we do not necessarily know whether a particular deviation score should be considered large or small. Therefore we need a frame of reference. When we examine the entire distribution, we see that only a few scores deviate by as much as Biff's score, and that is what makes his an impressively high score. Similarly, Slug's score of 35 deviates from the mean by −25 (35 − 60 = −25). This too is impressive, because only a few scores deviate below the mean by such an amount. Thus, a score is impressive if it is far from the mean, and "far" is determined by how frequently other scores deviate from the mean by that same amount.

To interpret a score's location, then, we need a way to compare its deviation to all deviations. For Biff, we need to quantify whether his deviation of +30 is impressive relative to all deviations in the sample. To do this, we need a *standard* to compare to Biff's *deviation*: We need a standard deviation! As we saw in Chapter 8, calculating the standard deviation is our way of computing the "average deviation" of the scores around the mean. By comparing a score's deviation to the standard deviation, we can describe the location of an individual score in terms of the average deviation.

For example, say that for the data from Prunepit U., the sample standard deviation is 10 (10 attractiveness points). Biff's deviation of +30 attractiveness points is equivalent to 3 standard deviations. Thus, another way to describe Biff's raw score is to say that it is located 3 standard deviations above the mean.

We have simply described Biff's score in terms of its distance from the mean, measured in standard deviation units. We do the same type of thing when we convert inches to feet. In that case, the unit of measurement called a foot is defined as 12 inches, so a distance of 36 inches is equal to 3 of those units, or 3 feet. In our study, the unit of measurement called a standard deviation is defined as 10 attractiveness points. Biff's deviation from the mean is 30 attractiveness points, so his raw score is a distance of 3 of those units, or 3 standard deviations, from the mean.

By transforming Biff's deviation into standard deviation units, we have performed the *z*-score transformation and computed Biff's *z*-score. A ***z*-score** is the distance a raw score deviates from the mean when measured in standard deviations. It is a single number that summarizes a raw score's relative standing: Biff's raw score deviates from the mean by an amount that is three times the "average" amount that all scores in the sample deviate from the mean. Because his *z*-score is positive, he is 3 standard deviations above the mean.

REMEMBER A z-score describes a raw score's location in terms of the number of standard deviations the score is from the mean.

Computing z-Scores

The symbol for a z-score is z. Above, we performed two mathematical steps in computing Biff's z. First, we found his score's deviation by subtracting the mean from the raw score. Then we divided the score's deviation by the standard deviation.

THE FORMULA FOR TRANSFORMING A RAW SCORE IN A SAMPLE INTO A z-SCORE IS

$$z = \frac{X - \overline{X}}{S_X}$$

This is both the definitional and computational formula for z. When starting from scratch with a sample of raw scores, first compute $\overline{X}$ and S_X and then substitute their values into the formula. Notice that we are computing a z-score from a *sample* of scores, so we use the descriptive sample standard deviation, S_X (the formula in Chapter 8 that involves dividing by N, not $N - 1$).

To find Biff's z-score, we substitute his raw score of 90, the $\overline{X}$ of 60, and the S_X of 10 into the above formula:

$$z = \frac{X - \overline{X}}{S_X} = \frac{90 - 60}{10}$$

Find the deviation in the numerator first, so subtract 60 from 90 (always subtract $\overline{X}$ from X). Rewriting the formula gives

$$z = \frac{+30}{10}$$

Then divide, and

$$z = +3.00$$

Likewise, Binky's raw score of 65 produces a z-score of

$$z = \frac{X - \overline{X}}{S_X} = \frac{65 - 60}{10} = \frac{+5}{10} = +.50$$

Binky's raw score is literally one-half of 1 standard deviation above the mean.

And finally, Slug's raw score is 35, so his z is

$$z = \frac{X - \overline{X}}{S_X} = \frac{35 - 60}{10} = \frac{-25}{10} = -2.50$$

Here 35 minus 60 results in a deviation of *minus* 25, which, when divided by 10, results in a z-score of -2.50. Slug's raw score is 2.5 standard deviations *below* the mean. In working with z-scores, always pay close attention to the positive or negative sign: It is part of the answer.

Usually a z-score is for describing the relative standing of a score within a sample, so we use the previous formula. However, we can also compute a z-score for a score in a population, if we know the population mean, μ, and the true standard deviation of the population, σ_X.

THE FORMULA FOR TRANSFORMING A RAW SCORE IN A POPULATION INTO A z-SCORE IS

$$z = \frac{X - \mu}{\sigma_X}$$

This formula is identical to the previous formula except that now the answer indicates how far the raw score lies from the population mean, measured in units of the true population standard deviation. (Note: z-scores are a descriptive statistic, so we *never* compute z-scores using the *estimated* population standard deviation, s_X.)

Computing a Raw Score When z Is Known

Sometimes we know a z-score and want to find the corresponding raw score. For example, in the Prunepit U. study, say that another student, Bucky, scored $z = +1.00$. What is his raw score? With $\overline{X} = 60$ and $S_X = 10$, his z-score indicates that he is 1 standard deviation above the mean, or, in other words, 10 points above 60. Therefore, his raw score is 70. What did we just do? We multiplied his z-score times the S_X and then added the mean.

THE FORMULA FOR TRANSFORMING A z-SCORE IN A SAMPLE INTO A RAW SCORE IS

$$X = (z)(S_X) + \overline{X}$$

Using the formula to transform Bucky's z-score of $+1.0$, we have

$$X = (+1.0)(10) + 60$$

so

$$X = +10 + 60$$

so

$$X = 70$$

To check this answer, compute the z-score for the raw score of 70. You should end up with the z-score you started with: $+1.0$.

Say that Fuzzy has a negative z-score of -1.3 (with $\overline{X} = 60$ and $S_X = 10$). Then

$$X = (-1.3)(10) + 60$$

so

$$X = -13 + 60$$

Adding a negative number is the same as subtracting its positive value, so

$$X = 47$$

Fuzzy has a raw score of 47.

As with any statistic, after transforming a raw score or z-score, always determine whether your answer makes sense. At the very least, negative z-scores must correspond to raw scores smaller than the mean, and positive z-scores must correspond to raw scores larger than the mean. Further, as you'll see, we seldom obtain z-scores greater than ± 3 (plus or minus 3). Although they are possible, be very skeptical if you compute a z greater than ± 3, and double-check your work.

How Variability Influences z-Scores

The size of a particular z-score depends on both the amount that the raw score deviates from the mean *and* on the variability (the standard deviation) in the distribution. For example, Biff's deviation of +30 produced an impressive z-score of +3, because the standard deviation was 10: On "average" scores deviated from the mean by only 10, so a deviation of +30 is unusually large. If, however, the data had produced a standard deviation of 30, then Biff's would have a $z = (90 - 60)/30 = +1.00$. In this case Biff's score would not be as impressive because his deviation would equal the "average" deviation, indicating that his raw score is among the more common scores.

Thus, bear in mind that two factors produce a z-score having a large absolute value: (1) A large deviation from the mean, and (2) A small standard deviation.

INTERPRETING z-SCORES: THE z-DISTRIBUTION

The easiest way to interpret scores is to create a z-distribution. A ***z*-distribution** is the distribution produced by transforming an entire raw score distribution into z-scores. By transforming all attractiveness scores in the previous study into z-scores, we get the z-distribution shown in Figure 9.2.

Notice the three ways the X-axis is labeled. This shows that by creating a z-distribution, we have only transformed the way in which we identify each score. Saying that Biff has a z of +3.00 is merely another way to say that he has a deviation of +30, or a raw score of 90. Recognize this very important fact: Because we are still looking at the same point on the distribution, Biff's z-score of +3.00 and his deviation of +30 have the same frequency, relative frequency, and percentile as his raw score of 90.

The advantage of looking at z-scores, however, is that they directly communicate a score's relative location in the distribution. The z-score of 0 here corresponds to the mean raw score of 60: A person having the mean score is zero distance from the mean. For any other score, think of a z-score as having two parts: The number and the sign in front of it. The number indicates the distance the score is from the mean, measured in standard deviations. The sign indicates the *direction* the score lies relative to the mean.

FIGURE 9.2 Distribution of Attractiveness Scores at Prunepit U

The labels on the X axis show first the raw scores, then the deviations, and then the z-scores.

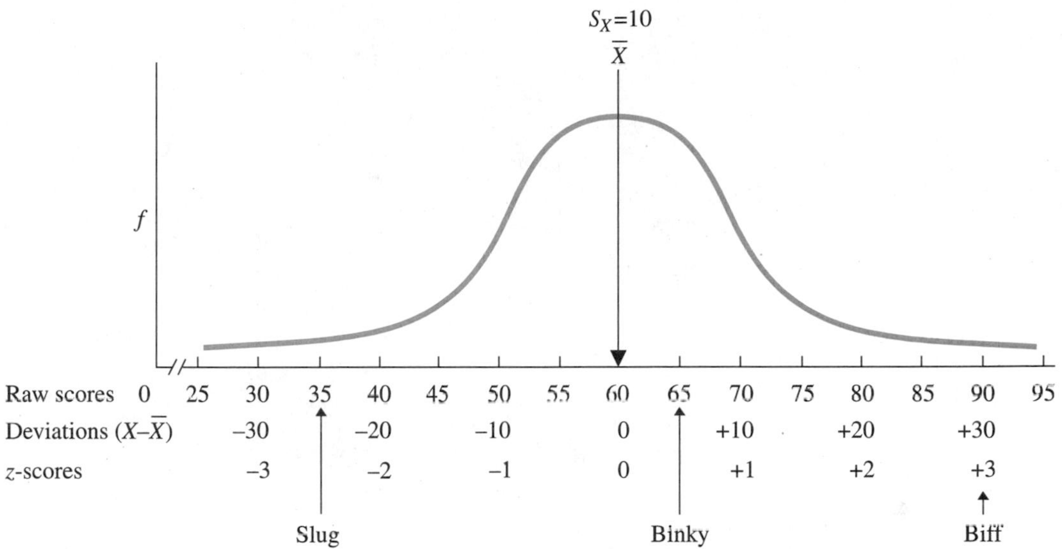

A "+" indicates that the score is above and graphed to the right of the mean. A "−" indicates that the score is below and graphed to the left of the mean. From this perspective, z-scores become increasingly larger numbers with a positive sign as you proceed to raw scores farther above and to the right of the mean. Conversely, z-scores become increasingly larger numbers with a negative sign as you proceed to raw scores farther below and to the left of the mean. However, do not be misled by negative z-scores: A raw score that is farther below the mean is a *smaller* raw score, but it has a *larger* negative z-score. Thus, for example, a z-score of −2 corresponds to a lower raw score than does a z-score of −1.

> **REMEMBER** The farther a raw score is from the mean, the larger its corresponding z-score. On a normal distribution, the larger the z-score, whether positive or negative, the less frequently that z-score and the corresponding raw score occur.

It is important to recognize that a negative z-score is not automatically a bad score. It depends on the nature of the variable being measured. For some variables, the goal is to have a low raw score (errors on a test, number of parking tickets, amount owed on a bill). With these variables, negative z-scores are best. For example, say we measure the number of times that people get depressed during a one-month period, and we find that the $\overline{X}$ is 5 and the S_X is 1. Figure 9.3 shows the z-distribution for this sample. Ideally, we'd like never to be depressed, so the goal is to have the score that is the greatest distance *below* the mean. Thus, a z-score of −3 would be good because it corresponds to only 2 episodes of depression. Conversely, a z-score of +3 would not be good, because it corresponds to 8 episodes of depression.

FIGURE 9.3 Distribution of Number of Episodes of Depression and the Corresponding z-Scores

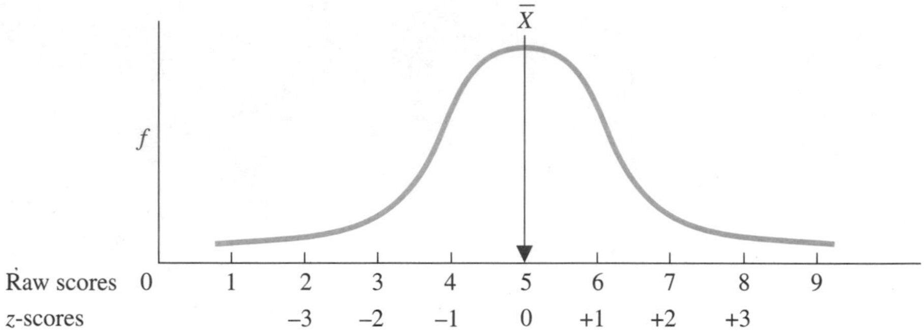

Characteristics of the z-Distribution

The previous examples illustrate three important characteristics of any z-distribution.

1. *A z-distribution always has the same shape as the raw score distribution.* The preceding distributions form normal distributions only because the raw scores formed normal distributions. Transforming a nonnormal distribution into z-scores will not make it form a normal curve.
2. *The mean of any z-distribution always equals 0.* The mean of the raw scores transforms into a z-score of 0. Likewise, if we compute the mean of all z-scores, the result would be 0: The sum of all z-scores is the sum of the positive and negative deviations around the mean, and this sum always equals 0, so the mean is 0.
3. *The standard deviation of any z-distribution always equals 1.* One standard deviation for raw scores—regardless of its value—transforms into 1 z-score unit. Whether the standard deviation in the raw scores is 10 (as in Figure 9.2) or 1 (as in Figure 9.3), it is still one standard deviation, and one standard deviation is one z-score unit.

Because of these characteristics, all normal z-distributions are similar. In particular, it's very important that you recognize that any particular z-score will be at the same relative location on *any* normal distribution. For example, look again at Figures 9.2 and 9.3. Recall from Chapter 8 that the raw scores that are $\pm 1S_X$ from the mean are always directly underneath the "inflection points" of a normal curve. Regardless of the value of S_X, because z-scores of ± 1 are $\pm 1\ S_X$ from the mean, these z-scores are also always underneath the inflection points on any normal curve. Likewise, a z of ± 2 (2 S_Xs from the mean) will always be twice as far from the $\overline{X}$ as zs of ± 1, and this is about halfway to the tail of the distribution. And a z of ± 3 will always be located in the extreme tail, because regardless of the value of S_X, it is always a distance of three times the average deviation, and when a score is that far from $\overline{X}$, it's in the extreme tail.

Now here's the nifty part: Any *raw score* that produces a particular z-score will always have the same relative location in the raw score distribution. For example, because a z of +3 is always in the extreme upper tail, any time a raw score produces a z of +3, that raw score will, like Biff's, be in the upper tail of the raw score distribution. Or, because a z of .50 will, like Binky's, be halfway between the mean and a z of +1 (or halfway to the inflection point), any raw score on any distribution that produces a z of +.50 will also be located halfway to the inflection point.

And here's the *really* nifty part: We don't have to create the entire z-distribution to see this. In previous chapters, we've seen how statistical procedures allow us to envision a distribution and the location of particular scores within the distribution. By transforming a raw score to a z-score and envisioning the z-distribution, we can identify the location of *any* raw score in *any* distribution.

REMEMBER A z-score allows us to locate the corresponding raw score on any normal distribution.

As we see in the following sections, the z-distribution provides us with a way to compare distributions, describe individual scores, and even describe sample means.

USING THE z-DISTRIBUTION TO COMPARE DIFFERENT DISTRIBUTIONS

In research, comparing a score on one variable to a score on another variable is a problem if the two variables are very different. For example, say that Althea received a grade of 38 on a statistics quiz and a grade of 45 on an English paper. These grades reflect scores on different kinds of tasks, assigned by different instructors using different criteria, so it's like comparing apples to oranges. To avoid this problem, we transform the raw scores from each class into z-scores. This produces two z-distributions, each with a mean of 0, a standard deviation of 1, and a range of between about −3 and +3. Each z-score indicates an individual's relative standing in his or her respective class. Therefore, we can compare Althea's relative standing in English to her relative standing in statistics, and we are no longer comparing apples and oranges. (The z-transformation equates, or standardizes, the distributions, and so z-scores are often referred to as **standard scores**.)

Say that for the statistics quiz, the $\overline{X}$ was 30 and the S_X was 5. We transform all grades to z-scores, including Althea's grade of 38, which becomes $z = +1.60$. For the English paper, the $\overline{X}$ was 40 and the S_X was 10, so Althea's grade of 45 becomes $z = +.50$. Figure 9.4 shows the locations of Althea's z-scores on the respective z-distributions. The larger English distribution reflects a larger class size than in statistics. Regardless, a z-score of +1.60 is farther above the mean than a z of +.50. Thus, in terms of her relative standing in each class, Althea did better in statistics, because she is farther above the statistics mean than she is above the English mean.

Another student, Millie, obtained raw scores that produced zs of −2.00 in statistics and −1.00 in English. In which class did she do better? Her z-score of −1.00 in English is better, because it is less distance below the mean.

Of course, it would be easier to compare these two distributions if we plotted them on the same set of axes: z-scores enable us to do just that.

FIGURE 9.4 Comparison of Two Distributions for Statistics and English Grades, Showing Raw Scores and z-Scores

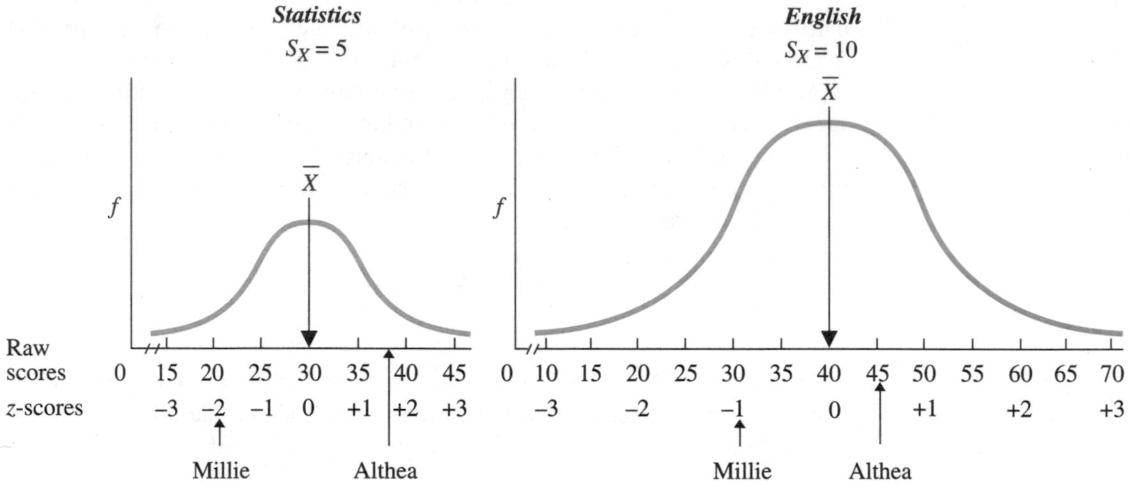

Plotting Different z-Distributions on the Same Graph

By transforming both the statistics and English scores into z-scores, we establish a common variable. Then, to see each student's relative location in each class, we can graph both of the previous distributions on one set of axes, as shown in Figure 9.5.

FIGURE 9.5 Comparison of Two Distributions for Statistics and English Grades, Plotted on the Same Set of Axes

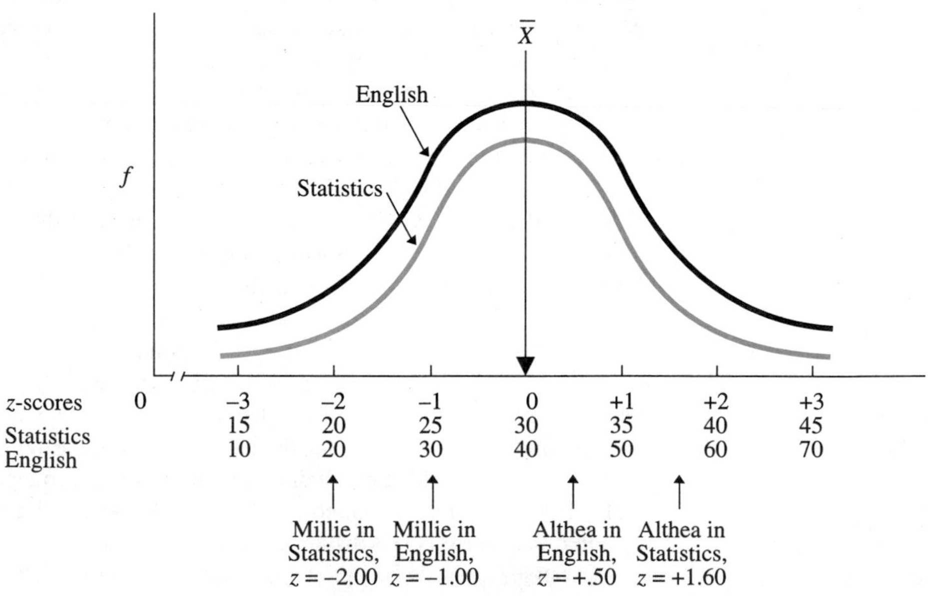

As we've noted, all normal *z*-distributions are similar, so there are only two minor differences between the curves. First, the classes produced different standard deviations, so the raw scores for each class are spaced differently along the *X*-axis. For example, going from the *z*-score of +1 to +2 corresponds to going from the raw scores of 35 to 40 in statistics, but from 50 to 60 in English. Second, the greater height of the English distribution reflects a larger *f* for each score, and overall, the English class simply had a larger *N*. If the two raw score distributions had contained the same *N*, the curves would be identical.

By plotting the two distributions on the same set of axes, we can easily compare anyone's scores. Clearly, Althea scored better in statistics than in English, and Millie scored better in English than in statistics.

The fact that different *z*-distributions can be plotted on the same set of axes leads to our second use of the *z*-distribution: Determining the relative frequency of raw scores in any normal distribution.

USING THE *z*-DISTRIBUTION TO DESCRIBE SCORES

An important statistical use of *z*-scores is that they allow us to compute the relative frequency of scores. The two *z*-distributions in Figure 9.5 illustrate a critical point: **The relative frequency of a particular z-score is the same on any and all normal z-distributions.** Recall that relative frequency is the proportion of the time that a score occurs, and is computed as the proportion of the total area under the curve. Thus, for example, you know from previous chapters that 50% of all scores on a normal curve are to the left of the mean score. Now you also know that scores to the left of the mean produce negative *z*-scores, so in other words negative *z*-scores make up 50% of a distribution. Thus, regardless of their raw scores, the students in each class in Figure 9.5 with the negative *z*-scores constitute 50% of their respective distributions. Further, 50% of a distribution corresponds to a relative frequency of .50. Thus, on *any* normal *z*-distribution, the total relative frequency of the negative *z*-scores is .50.

Having determined the relative frequency of the *z*-scores, we can work backwards to find the relative frequency of the corresponding raw scores. In the statistics distribution in Figure 9.5, those students having negative *z*-scores have raw scores ranging between 15 and 30, so the relative frequency of scores from 15 to 30 is .50. Likewise, in the English distribution, those students having negative *z*-scores have raw scores ranging between 10 and 40, so the relative frequency of 10 to 40 is .50.

Similarly, recall from Chapter 8 that approximately 68% of all scores on a normal distribution fall between the score that is $1S_X$ below the mean and the score that is $1\ S_X$ above the mean. Now you also know that raw scores that are $\pm 1\ S_X$ from the mean produce *z*-scores of ± 1, respectively. Thus, back in Figure 9.5, students with *z*-scores between ± 1 constitute approximately 68% of their distributions. On *any* normal *z*-distribution, the relative frequency of *z*-scores between ± 1 is approximately .68. Having determined this, we can again work backwards to the raw scores. We see that statistics grades between 25 and 35 constitute approximately 68% of the statistics distribution, and English grades between 30 and 50 constitute approximately 68% of the English distribution.

In the same way, we can determine the relative frequencies within any distribution of scores once we envision it as a z-distribution. For example, in a normal distribution of IQ scores (whatever the $\overline{X}$ and S_X may be), we know that those IQ scores producing negative z-scores have a relative frequency of .50, and about 68% of all IQ scores fall between the two IQ scores corresponding to the z-scores of ± 1. The same will hold true for a distribution of running speeds, or a distribution of personality test scores, or for *any* normal distribution.

We can also determine the relative frequency of scores in any other portion of a distribution. To do so, we employ the standard normal curve.

The Standard Normal Curve

Because the relative frequency of a particular z-score is always the same on any normal z-distribution, we conceptualize all normal z-distributions as conforming to one standard curve. In fact, this curve is called the standard normal curve. The **standard normal curve** is a theoretical perfect normal curve, which serves as a model of the perfect normal z-distribution. (Because it is a z-distribution, the mean of the standard normal curve is 0, and the standard deviation is 1.)

We use the standard normal curve to first determine the relative frequency of particular z-scores on a perfect normal curve. Then, as we saw above, once we know the relative frequency of certain z-scores, we work backwards to determine the relative frequency of the corresponding raw scores. Thus, the first step is to find the relative frequency of the z-scores. To do this, we look at the area under the standard normal curve. Statisticians have determined the proportion of the area under various parts of the normal curve. Look at Figure 9.6.

FIGURE 9.6 Proportions of Total Area Under the Standard Normal Curve

The curve is symmetrical: 50% of the scores fall below the mean, and 50% fall above the mean.

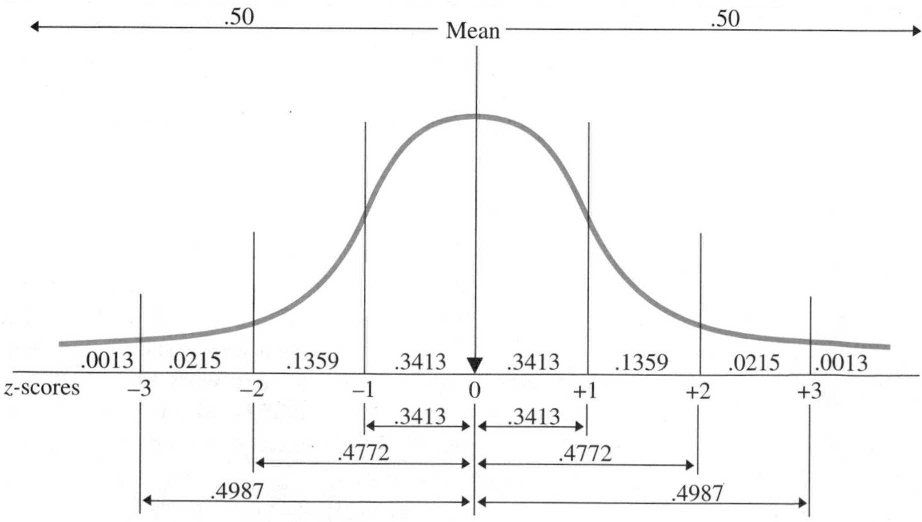

Above the X axis, the numbers between the vertical lines indicate the proportions of the total area between z-scores. Below the X axis, the number on each arrow indicates the proportion of the total area between the mean and the indicated z-score. (Don't worry: You won't need to memorize these proportions.)

The proportion of the total area under the curve is the same as relative frequency, so each proportion above is the relative frequency of the z-scores located in that section of the ideal normal curve. For example, .3413 of the z-scores are located between the z of 0 and the z of $+1$ on a perfect normal distribution. Because we can express a proportion as a percent by multiplying the proportion times 100, we can also say that 34.13% of all z-scores fall between the zs of 0 and $+1$. Similarly, z-scores between $+1$ and $+2$ occur 13.59% of the time, and z-scores between $+2$ and $+3$ occur 2.15% of the time. Because the distribution is symmetrical, the same proportions occur between the mean and the corresponding negative z-scores.

To determine the relative frequency for larger areas, we add the above proportions. For example, .3413 of the distribution is located between $z = -1$ and the mean, and .3413 of the distribution is between the mean and $z = +1$. Thus, a total of .6826, or 68.26%, of the distribution is located between the zs of -1 and $+1$. (See, about 68% of the distribution really is between $\pm 1S_X$ from the mean.) Likewise, we can add together nonadjacent portions of the curve. For example, .0228, or 2.28%, of the distribution is in the tail of the distribution beyond $z = -2$, and 2.28% is beyond $z = +2$. Thus, a total of 4.56% of all scores falls in the tails beyond $z = \pm 2$.

Figure 9.6 shows why we seldom obtain z-scores greater than ± 3. This is a theoretical distribution representing an infinitely large distribution (the tails of the polygon never touch the X axis). Even so, only .0013 of the scores are above $z = +3$ and only .0013 are below $z = -3$. In total, only .0026, (.26 of *1* percent) of the scores are beyond ± 3.00. Between $z = \pm 3$, however, is 99.74% of all scores (100% − .26% = 99.74%). Thus, for all practical purposes, the range of z is between ± 3. Also, now you can see why, in Chapter 8, I said that for normally distributed scores, the value of S_X should be about one-sixth of the range of the raw scores. The range of the raw scores is about between $z = -3$ and $z = +3$, a distance of six times the standard deviation. If the range is six times the standard deviation, then the standard deviation is one-sixth of the range.

Applying the Standard Normal Curve Model

The standard normal curve allows us to quickly determine the relative frequency of scores in any normally distributed sample or population. This is true even though real data will not form a perfect normal curve. As we've seen in previous chapters, with the normal curve model we assume that a population of raw scores "more or less" forms a normal distribution and that a sample represents such a distribution. However, to conceptualize the distribution, we do not draw a "more or less" normal curve. We draw the ideal perfect normal curve and use it as a model of our distribution, operating as if the raw scores form the perfect curve. If we operate as if the raw scores form a perfect normal curve, then after transforming the raw scores to z-scores, the z-distribution will also form a perfect normal curve. As we've seen, the perfect normal z-distribution is the standard normal curve.

> ***REMEMBER*** The standard normal curve is our model for any roughly normal distribution when transformed to z-scores.

We then use this model to determine the relative frequency of scores between any two raw scores. First transform the raw scores into z-scores. Then, from the standard normal curve, determine the proportion of the total area between these z-scores. This proportion is the same as the relative frequency of the z-scores between these scores on a perfect normal curve. And then finally, the relative frequency of these z-scores is the same as the relative frequency of their corresponding raw scores in a perfect normal distribution. Thus, the relative frequency obtained from the standard normal curve is the *expected* relative frequency of the raw scores in the data, if the data formed a perfect normal distribution.

For example, the original sample of attractiveness scores from Prunepit U. forms an approximately normal distribution, so we can apply the standard normal curve model to these data. Say that Cubby has a raw score of 80, which, with $\overline{X} = 60$ and $S_X = 10$, is a z of $+2.00$. We can show Cubby's location on the distribution as illustrated in Figure 9.7. We might first ask what proportion of scores are expected to fall between the mean and Cubby's score. From the standard normal curve we see that .4772 of the total area falls between the mean (a $z = 0$) and $z = +2$. Because .4772 of all z-scores fall between the mean and a z of $+2$, we expect .4772, or 47.72%, of all attractiveness scores at Prunepit U. to fall between the mean score of 60 and Cubby's score of 80.

We might also ask how *many* people scored between the mean and Cubby's score. Then we would convert the above relative frequency to simple frequency by multiplying the N of the sample times the relative frequency. Say that the N at Prunepit was 1000. If we expect .4772 of all scores to fall between the mean and $z = +2$, then $(.4772)(1000) = 477.2$: We expect about 477 scores to fall between the mean and Cubby's raw score of 80.

How accurately the expected relative frequency from the model describes the actual scores depends on three aspects of the data listed opposite:

FIGURE 9.7 Location of Cubby's Score on the z-Distribution of Attractiveness Scores

Cubby is at approximately the 98th percentile.

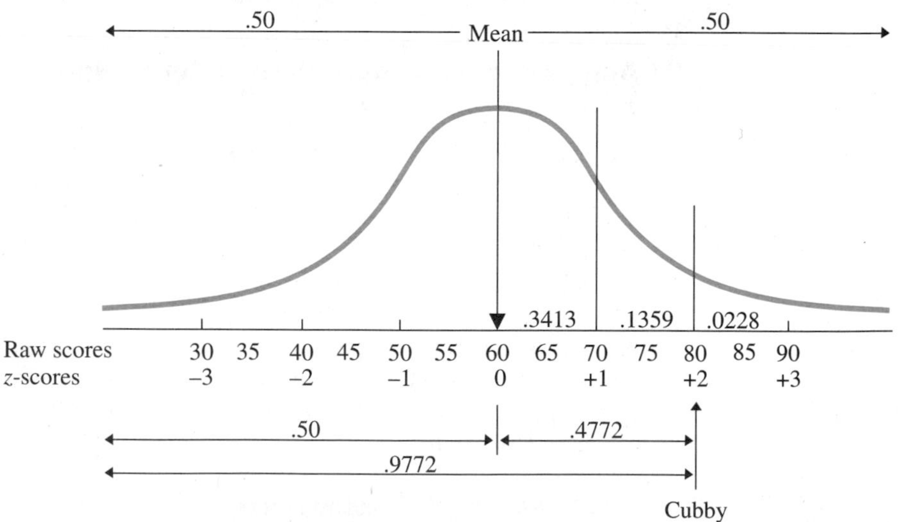

1. The closer the raw scores are to forming a normal distribution, the more accurately the model describes the data. Therefore, the standard normal curve model is appropriate *only* if you can assume that the data are at least approximately normally distributed.
2. The larger the sample N, the more closely the sample tends to fit the normal curve and therefore the more accurate the model will be. The model is most accurate when applied to very large samples or to populations.
3. The model is most appropriate when raw scores are theoretically continuous scores (which can include decimals) measured using ratio or interval scales (which have equal intervals between scores, and measure actual amounts).

The Prunepit data meet these requirements, so the expected results for Cubby should be fairly accurate. And that's important: Don't lose sight of the fact that z-scores allow us to describe Cubby's relative standing accurately, *without* sifting through all 1000 scores.

Finding percentile for a given raw score Look again at Figure 9.7 and you'll see that we can also determine Cubby's expected percentile from the standard normal curve model. Recall that a percentile is the percent of all scores *below* (graphed to the left of) a score. On a normal distribution, the mean is the median (the 50th percentile). Any positive z-score is above the mean, so Cubby's z-score of $+2$ is above the 50th percentile. In addition, as Figure 9.7 shows, Cubby's score is above the 47.72% of the scores that fall between the mean and his z-score. Thus, we add the 50% of the scores below the mean to the 47.72% of the scores between the mean and his score. In total, 97.72% of all z-scores are below Cubby's z-score. We usually round off a percentile to a whole number, so we conclude that Cubby's z-score is at the 98th percentile. Likewise, Cubby's raw score of 80 is at the 98th percentile (so he's relatively very attractive). Conversely, if 97.72% of the curve is below a z of $+2$, then, as shown in Figure 9.7, only 2.28% of the curve is above a z of $+2$ ($100\% - 97.72\% = 2.28\%$). Thus, anyone scoring above $z = +2.00$, or the raw score of 80 would be in about the top 2% of all scores.

On the other hand, say that Elvis obtained an attractiveness score of 40, producing a z-score of -2.00. You can find Elvis's percentile using Figure 9.8. Because .0215 of the distribution is between $z = -2$ and $z = -3$, and .0013 of the distribution is below $z = -3$, there is a total of .0228, or 2.28%, of the distribution below (to the left of) Elvis's score. With rounding, Elvis ranks at the 2nd percentile.

Finding a raw score at a given percentile We can also work in the opposite direction, finding a raw score at a particular relative frequency or percentile. Say that we want to find the raw score at the 16th percentile. Because the 16th percentile is below the 50th percentile, we are looking for a negative z-score. Consider Figure 9.9. If 16% of all scores are below the unknown score, then 34% of all scores are between it and the mean ($50\% - 16\% = 34\%$). We have already seen that about 34% of a normal z-distribution is between a z of -1 and the mean, so that leaves about 16% of the distribution to the left of a z of -1 (although technically it is $50\% - 34.13\% = 15.87\%$). Thus, with rounding, a z of -1 is at approximately the 16th percentile.

FIGURE 9.8 Location of Elvis's Score on the z-Distribution of Attractiveness Scores

Elvis is at approximately the 2nd percentile.

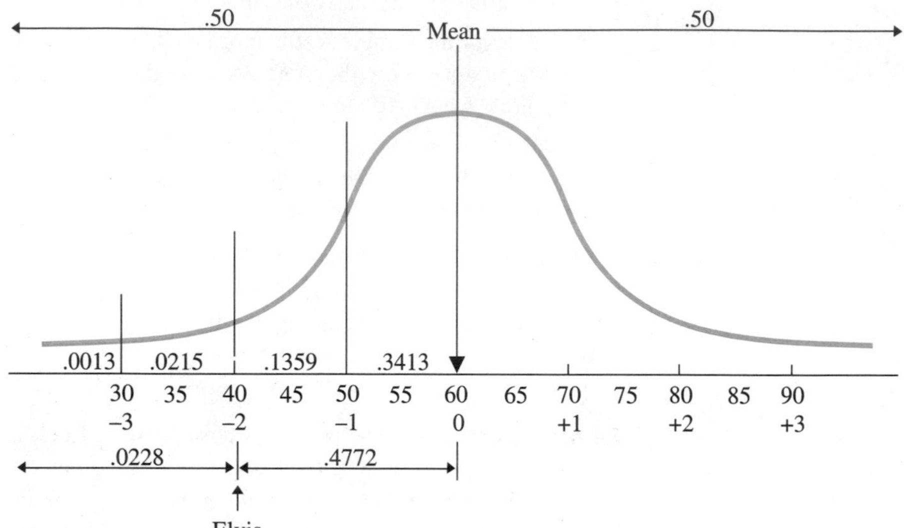

We then use the formula $X = (z)(S_X) + \overline{X}$ to find the raw score at $z = -1$. For this sample, $\overline{X} = 60$ and $S_X = 10$, so $X = (-1)(10) + 60 = 50$. Thus, the raw score of 50 is expected to be at approximately the 16th percentile.

Using the z-Table

In the above examples, we rounded off the z-scores to keep things simple. With real data, however, we do not round z-scores containing decimals to whole numbers. Fur-

FIGURE 9.9 Proportions of the Standard Normal Curve at Approximately the 16th Percentile

The 16th percentile corresponds to a z-score of about −1.0.

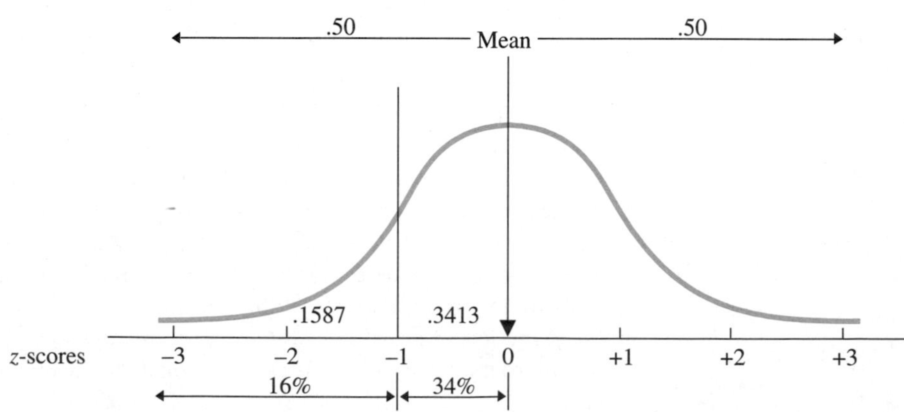

TABLE 9.1 Sample Portion of the z-Table

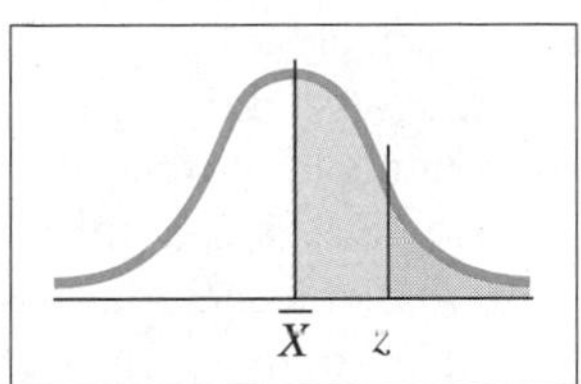

z	*Area between the mean and z*	*Area beyond z*
1.60	.4452	.0548
1.61	.4463	.0537
1.62	.4474	.0526
1.63	.4484	.0516
1.64	.4495	.0505
1.65	.4505	.0495

ther, fractions of z-scores do *not* result in proportional divisions of the previous areas. For example, even though $z = +.50$ is one-half of $z = +1$, the area between the mean and $z = +.50$ is *not* one-half of the area between the mean and $z = +1$. Instead, we find the proportion of the total area under the standard normal curve for any two-decimal z-score by looking in Table 1 of Appendix A. This table is called the *z-table*. A portion of the z-table is reproduced in Table 9.1 above.

Say that you seek the proportions corresponding to $z = +1.63$. First locate $z = 1.63$ in column A, labeled "z," and then move to the right. Column B, labeled "Area between the mean and z," contains the proportion of the total area under the curve between the mean and the z identified in column A. Thus, .4484 of the curve, or 44.84% of all z-scores, is between the mean and $z = +1.63$. This is shown in Figure 9.10. Column C of the table is labeled "Area beyond z" and contains the proportion of the total area under the curve that is in the tail beyond the z-score in column A. Thus, .0516 of the curve, or 5.16% of all z-scores, is in the tail of the distribution beyond $z = +1.63$ (this

FIGURE 9.10 Distribution Showing the Area Under the Curve Above $z = +1.63$ and Between $z = +1.63$ and the Mean

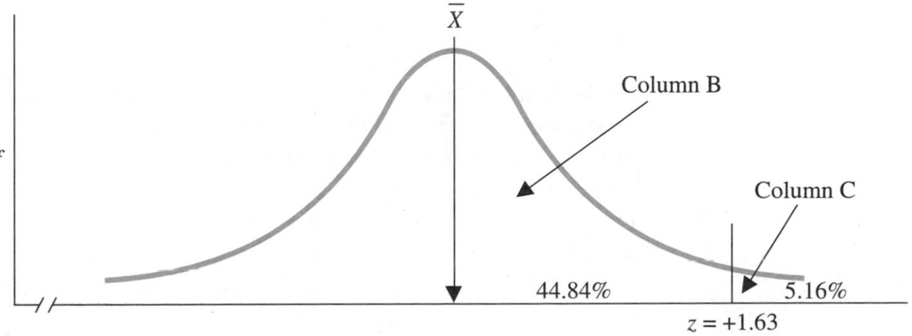

is also shown in Figure 9.10). If you get confused when using the z-table, look at the normal distribution at the top of the table. The shaded portion and arrows indicate the part of the curve described in each column.

To work in the opposite direction and find the z-score that corresponds to a particular proportion, read the columns in the opposite order. First find the proportion in column B or C, depending on the area you seek, and then identify the corresponding z-score in column A. For example, say that you seek the z-score that demarcates 44.84% of the curve between the mean and z. In column B of the table you'll find .4484, which corresponds to the z-score of 1.63.

Notice that the z-table contains no positive or negative signs. Because the normal distribution is symmetrical, only the proportions for one-half of the standard normal curve are given. *You* must decide whether z is positive or negative, based on the problem you're working on.

Sometimes you will need a proportion not given in the z-table, or you will need the proportion for a z-score that has three decimal places. In such cases, you should use the mathematical procedure called *linear interpolation*. Instructions for interpolating z-scores or their corresponding proportions are given in Part 2 of Appendix B.

In summary, then, with the z-table you can answer virtually any question about the relative standing of scores in any normal distribution. If you want to know the relative frequency or percentile at a particular raw score, compute its z-score to then get the corresponding proportion of the area under the curve from the z-table. If you want to know the raw score demarcating a particular relative frequency or percentile, look up the corresponding area under the normal curve in the z-table to get the appropriate z-score, and then calculate the corresponding raw score.

You'll be working with z-scores or their equivalent in most of the remainder of this book, so it is *very* important that you become comfortable with interpreting their location on a normal distribution. To do so, always sketch a normal curve. Then, as we did in the previous examples, label the location of the mean score, identify the relevant portions under the curve, and indicate the corresponding z-scores and raw scores. By giving yourself a curve to look at, you'll greatly simplify the problem.

Using z-Scores to Define Psychological Attributes

Recall that we often have difficulty interpreting what, in the grand scheme of things, a particular raw score indicates. Because of this, we also have difficulty deciding on the cutoff scores to use when classifying people based on their scores. What must someone do to be considered a genius? How do we define an abnormal personality? To answer such questions, psychologists often use a "statistical definition" based on relative standing. Essentially, this involves applying the normal curve model and defining an attribute in terms of a particular z-score. For example, we might statistically define the old-fashioned term "genius" as a person with a z-score of more than $+2$ on an intelligence test. Because a z greater than $+2$ falls in about the highest 2% of the distribution, we have defined a genius as anyone who scores in the top 2% of all IQ scores. Similarly, we might define as "abnormal" any person with a z-score beyond ± 2 on a personality test. Such scores are "abnormal" in a statistical sense, because they are very infrequent in the population and are among the most extreme raw scores.

Instructors who "curve" grades generally do so using the normal curve and *z*-scores. They assume that grades are normally distributed, so they assign letter grades based on proportions of the area under the normal curve. If the instructor defines an A student as one who is in the top 2%, then students with *z*-scores greater than +2 receive As. If the instructor defines B students as those in the next 13%, then students having *z*-scores between +1 and +2 receive Bs, and so on.

Transforming *z*-Scores

The *z*-score transformation is especially common in applied, educational, or clinical diagnostic settings, where we want to describe an individual. Sometimes in such settings we must communicate this information to people who are statistically unsophisticated. Therefore, we may transform *z*-scores into numbers that are easier to understand. Typically, such transformations eliminate negative scores and reduce the number of decimal places. At the same time, a transformed score is still located at the underlying *z*-score, with the corresponding area under the curve.

One such transformation occurs with college entrance exams such as the Scholastic Aptitude Test (SAT). Essentially, a grade on the SAT is a *z*-score that is then transformed using the formula: SAT score = $z(100) + 500$. Because the mean of *z*-scores is 0, the mean in SAT scores is 500. Similarly, the standard deviation of *z* is 1, which becomes 100. You may have heard that the highest possible SAT score is 800. This is because a score of 800 corresponds to a *z* of +3. Because *z*-scores beyond +3 are so infrequent, for simplicity these scores are rounded to +3, or 800. Likewise, the lowest score is 200.

USING *z*-SCORES TO DESCRIBE SAMPLES

So far we've discussed using the standard normal curve model to describe the relative standing of any single raw score. Now, using the same logic, we'll determine the relative standing of an entire sample. This procedure is important, not only because it allows you to evaluate a sample, but also because it is the basis for inferential statistics (and you will definitely be computing some of those in the very near future).

To see how the procedure works, say that we obtained the SAT scores of a random sample of 25 students at Prunepit U. Their mean score is 520. Nationally the mean of *individual* SAT scores is 500 (and $\sigma_X = 100$), so it appears that at least some Prunepit students scored relatively high, pulling the overall mean to 520. But how do we interpret the performance of the sample as a whole? Is a sample mean of 520 impressively above average, or more mundane? By considering only this sample mean, we cannot answer this question, because we have the same problem we had when examining an individual raw score: Without a frame of reference, we don't know whether, in the grand scheme of things, a particular sample mean is high, low, good, bad, or indifferent.

The solution is to evaluate a sample mean in terms of its relative standing. Previously, we compared a particular raw score to all other possible scores in the distribution that might have occurred. (Biff's score was impressive compared to any other score he might have obtained.) Now, we will compare a sample mean to all other possible sample means that might have occurred. We envision all possible sample means that might occur as a distribution of sample means. Then by locating our sample mean on this distribution, we can determine whether the Prunepit sample is relatively impressive.

The first task is to envision the distribution of all possible sample means that might occur, called the sampling distribution of means.

The Sampling Distribution of Means

By saying that nationally the average SAT score is 500, we are saying that in the population of SAT scores, the population mean (μ) is 500. Because we randomly selected a sample of 25 students and obtained their SAT scores, we essentially drew a random sample of 25 scores from this population of scores. To evaluate our sample mean, we will first create a distribution of all other possible means we might have obtained when randomly selecting a sample of 25 scores from this population.

One way to do this would be to record every raw score in the population of SAT scores on a slip of paper and deposit the slips into a very large hat. We could then hire a statistician to sample this population an infinite number of times (she would get very bored, so the pay would have to be good). The statistician would randomly select from the hat a sample with the same size N as in our sample (25), compute the sample mean, replace the scores in the hat, draw another 25 scores, compute the mean, and so on. Because the scores selected in each sample would not be identical, all sample means would not be identical. By constructing a frequency distribution of the different values of sample means she obtained, the statistician would create a sampling distribution of means. The **sampling distribution of means** is the frequency distribution of all possible values of sample means that occur when an infinite number of samples of the same size are randomly selected from one raw score population. Thus, our SAT sampling distribution of means is the population of all possible sample means that can occur when the SAT raw score population is exhaustively sampled using our N of 25. Because this contains all possible SAT sample means, it contains the sample means we might have obtained from our Prunepit sample instead of the mean of 520. Therefore, by looking at this sampling distribution, we can determine the standing of our sample mean relative to all others and thus have a better idea of how to interpret it.

Notice that, as discussed in Chapter 5, the bored statistician used *simple* random sampling. We assume that this sampling technique generally produces the means that we might have obtained, even if in our actual study, we use some other sampling technique.

Subsequent chapters will discuss the sampling distributions of many types of statistics. But in general, a sampling distribution for any statistic has four important characteristics:

1. All samples contain raw scores from the same population.

2. All samples are randomly selected.

3. All samples have the same size N, equal to that in our sample.

4. The sampling distribution is a frequency distribution that reflects the population of all possible values of the sample statistic we might have obtained with our sample.

Of course, the bored statistician could not actually sit down and sample the population an infinite number of times. However, she could create a *theoretical* sampling distribution by applying the central limit theorem. The **central limit theorem** is a statistical principle that defines the mean, the standard deviation, and the shape of a theoretical sampling distribution. From the central limit theorem, we know that the sampling distribution of means will: (1) Form an approximately normal distribution, (2) Have a mean that equals the mean of the raw score population from which the sampling distribution was created, and (3) Have a standard deviation that is mathematically related to the standard deviation of the raw scores.

Thus, the central limit theorem tells us that if the bored statistician did sample the population of SAT scores as described above, she would create a sampling distribution of means that would look like the curve shown in Figure 9.11. You should conceptualize a sampling distribution of means in the same way that you conceptualize a distribution of raw scores. The only difference is that here each "score" along the *X* axis is a sample mean. (You can still think of the distribution as a parking lot full of people, except that now each person is the captain of a sample, having the sample's mean score and thus representing the sample.)

Understand that the different values of the sample means in the sampling distribution occur simply because of the luck of the draw of which scores are selected for each sample. The bored statistician would sometimes obtain a sample mean that is higher than 500 because, *by chance*, she randomly selected a sample of predominantly high scores. At other times she might select predominantly low scores, producing a mean below 500. Or she might, by luck, obtain a sample mean equal to 500. Thus the sampling distribution provides a picture of all of the different sample means—and their frequency—that occur due to chance when randomly sampling the SAT population. Because our Prunepit sample was a random sample obtained by chance from this population, this picture allows us to evaluate our sample mean relative to all other sample means that might have occurred.

FIGURE 9.11 Sampling Distribution of Random Sample Means of SAT Scores

The X axis is labeled to show the different values of $\overline{X}$ *we obtain when we sample a population where the mean for SAT scores is 500.*

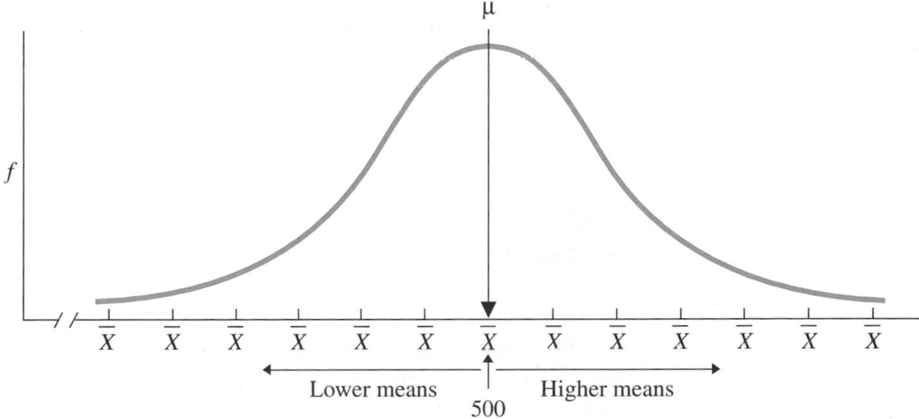

We know from the central limit theorem that the bored statistician would obtain such a sampling distribution, because, first, *regardless of the shape of the raw score distribution*, a sampling distribution is always an approximately normal distribution. This is because, over an infinite number of random samples, most often the sample mean will equal the population mean, μ. Sometimes, however, a sample will contain a few more high scores or low scores relative to the population, so the sample mean will be close to, but slightly greater than or less than μ. Less frequently, random sampling will produce rather strange samples, with sample means farther above or below μ. Once in a great while, a sample of very infrequent and unusual scores will be drawn, resulting in a sample mean that deviates greatly from μ. The larger the N in the samples, the more closely the distribution conforms to the perfect normal curve.

Second, the mean of the sampling distribution *always* equals the mean of the underlying raw score population. Thus, the μ of the SAT sampling distribution is 500, because the μ of individual SAT scores is 500. This is because the raw scores in a population are balanced around μ, so the sample means created from those scores are also balanced around μ. After all, the mean of the sampling distribution is simply the average sample mean, and therefore, those means above and below μ will average out, so the overall mean of the sampling distribution will equal the mean of the raw score population.

Finally, although it's not shown in Figure 9.11, the sample means may be very different from one another and deviate greatly from the average sample mean, μ, or they may be very similar and deviate little from μ. As we'll see, the size of the "standard deviation" of the sampling distribution is related to the size of the standard deviation in the population of raw scores. If the raw scores are highly variable, then each time the statistician samples the population she is likely to get a very different set of scores, and so the various sample means will differ greatly from one another. Conversely, if the raw scores are not variable, then every sample will tend to contain the same scores, and so the sample means will be very similar to one another.

Notice that once we have computed its standard deviation, we will have a rather complete description of the sampling distribution of SAT means: We'll know its μ and its standard deviation, and we'll know that it's a normal distribution. The importance of the central limit theorem is that we obtain this information *without* actually infinitely sampling the population of SAT scores. All we need to know is that in the raw score population, μ is 500 and σ_X is 100, and that we sampled using an N of 25. Further, we produce the *same* sampling distribution with any population of raw scores having these characteristics, *regardless of what they measure*. Thus, if a personality or creativity test has a $\mu = 500$ and σ_X is 100, samples of 25 would produce the identical theoretical sampling distribution as above. Likewise, by knowing the characteristics of any *other* raw score population, we can create a sampling distribution of means for it.

> *REMEMBER* The central limit theorem allows us to envision the sampling distribution of means that would be created by exhaustive random sampling of any raw score distribution.

Once we envision the SAT sampling distribution, as in Figure 9.11, we have a model of all the different sample means—and the frequency with which they occur—when sampling the SAT population. Now to evaluate the original Prunepit U. sample mean, we simply determine *where* a mean of 520 falls on the X-axis of the sampling distribution, and then interpret this accordingly. If 520 lies close to 500, then it is a rather frequent, common mean when sampling SAT scores (the bored statistician would frequently obtain this result). But if 520 lies toward the tail of the distribution, far from 500, then it is a more infrequent and unusual sample mean (even over an *infinite* number of samples, she would seldom encounter such a mean).

The sampling distribution is a normal distribution, and you already know how to determine where any score lies on a normal distribution: We use—you guessed it—z-scores. We simply determine how far the sample mean deviates from the mean of the sampling distribution, measured using the "average" deviation in the distribution. With one number then, the z-score will tell us the sample mean's relative location within the sampling distribution, and thus indicate its relative standing among all possible means.

To calculate the z-score for a sample mean, we need one more piece of information: The standard deviation of the sampling distribution.

The Standard Error of the Mean

The standard deviation of the sampling distribution of means is called the **standard error of the mean**. (The term "standard deviation" was already taken.) Recall that the terms *error* and *deviation* are synonymous. Therefore, like a standard deviation, the standard error of the mean can be thought of as the "average" amount that the sample means deviate from the mean, μ, of the sampling distribution.

For the moment, we'll discuss the *true* standard error of the mean, as if we had actually computed it using the entire sampling distribution. The symbol for the true standard error of the mean is $\sigma_{\overline{X}}$. The σ indicates that we are describing the population, but the subscript $\overline{X}$ indicates that we are describing the population of sample means—what we call the sampling distribution of means. The central limit theorem tells us that $\sigma_{\overline{X}}$ can be found using the following formula:

THE FORMULA FOR THE TRUE STANDARD ERROR OF THE MEAN IS

$$\sigma_{\overline{X}} = \frac{\sigma_X}{\sqrt{N}}$$

This formula involves σ_X because, as we've seen, the variability of the raw score population influences the variability of the sampling distribution. If we know the true standard deviation of the raw scores, we can determine the true "standard deviation" of the sample means.

> *REMEMBER* The true standard error of the mean, $\sigma_{\overline{X}}$, is computed using the true standard deviation of the population of raw scores, σ_X, that is used to create the sampling distribution.

The formula also involves N, the number of scores used to compute our sample mean. We factor in the size of N, because it influences how representative each sample is and thus determines how close each sample mean is to μ. With a very small N (say 2), each sample can easily contain scores that are very unrepresentative of the population, so that each sample mean differs greatly from μ, producing a large "average deviation" in the sampling distribution. With a larger N, however, we should obtain more representative samples. At the furthest extreme, say that each sample contains virtually the entire population. Now each sample mean should equal μ, or be very close to it, so that the average deviation in the sampling distribution will be very small.

We can compute $\sigma_{\overline{X}}$ for the sampling distribution of SAT scores because we *know* that the true standard deviation of the population of SAT scores is 100. Because N was 25, the above formula says that the standard error of the mean for the sampling distribution is

$$\sigma_{\overline{X}} = \frac{100}{\sqrt{25}}$$

The square root of 25 is 5, so

$$\sigma_{\overline{X}} = \frac{100}{5}$$

and thus

$$\sigma_{\overline{X}} = 20$$

A value of $\sigma_{\overline{X}} = 20$ indicates that in the sampling distribution of means created from SAT scores, the individual sample means differ from the μ of 500 by an "average" of 20 SAT points when the N of each sample is 25.

Now, at last, we are ready to calculate a z-score for our sample mean.

Calculating a *z*-Score for a Sample Mean

Previously we saw that when we know the population mean, μ, and we know the true population standard deviation, σ_X, the formula for transforming a raw score into a z-score is

$$z = \frac{X - \mu}{\sigma_X}$$

The sampling distribution of means is a population (of sample means) and we know its μ and true "standard deviation," $\sigma_{\overline{X}}$. Therefore, we transform a sample mean into a z-score using a similar formula.

THE FORMULA FOR TRANSFORMING A SAMPLE MEAN INTO A z-SCORE IS

$$z = \frac{\overline{X} - \mu}{\sigma_{\overline{X}}}$$

Don't be confused by the minor difference in symbols between the preceding formulas. Conceptually we do the same thing in both: We simply find how far a score on a distribution falls from the mean of the distribution, measured in standard deviations of that distribution. For a sample mean, we find how far the sample mean is from the mean of the sampling distribution, μ, measured in standard error units, or $\sigma_{\overline{X}}$.

> *REMEMBER* Use the above formula only when you know the true population standard deviation, σ_X.

Using the above formula, we calculate the *z*-score for the sample from Prunepit U. With $\overline{X} = 520$, $\mu = 500$ and $\sigma_{\overline{X}} = 20$, we have

$$z = \frac{\overline{X} - \mu}{\sigma_{\overline{X}}} = \frac{520 - 500}{20} = \frac{+20}{20} = +1.0$$

Thus, a sample mean of 520 has a *z*-score of +1.0 on the sampling distribution of means that occurs when *N* is 25 and the sampling distribution is created from the SAT raw score population where $\mu = 500$ and $\sigma_X = 100$.

Using the Sampling Distribution to Determine Relative Frequency of Sample Means

Everything we said previously about a *z*-score for an individual raw score applies to a *z*-score for a sample mean. It makes no difference that the *z*-score now refers to the location of a sample mean: A *z*-score is a *z*-score! Thus, because our sample mean has a *z*-score of +1, we know that it's above the μ of the sampling distribution (the average sample mean) by an amount equal to the "average" amount that sample means deviate from μ. Therefore, we know that, although they were not stellar, the Prunepit students did outperform a substantial proportion of comparable samples. Likewise, if another sample of 25 SAT scores (say from Podunk U.) produced a mean of 440, we'd know how poorly these students performed: Here $z = (440 - 500)/20 = -3.0$, so this sample would have produced one of the lowest SAT means we'd ever expect to obtain.

Notice that, as with an individual's *z*-score, the size of a sample mean's *z*-score depends both on the size of the deviation from μ *and* on the variability in the distribution. If, for example, $\sigma_{\overline{X}}$ above had been 5, then our $\overline{X}$ of 520 would have produced a *z* of $(520 - 500)/5 = +4$! Here, with all sample means deviating from 500 by an "average" of only 5, a mean that deviates by 20 would be *extremely* unusual (which is what a *z* of +4 indicates).

To obtain a more precise description of a sample mean, we can go one step further (and here's another nifty part): Because the sampling distribution of means always forms at least an approximately normal distribution, if we transformed *all* of the sample means in the sampling distribution into *z*-scores, we would have a roughly normal *z*-distribution. Recall that the standard normal curve is our model of *any* normal *z*-distribution. This is true even if it is a *z*-distribution of sample means! Therefore, we can use the standard normal curve model to determine the relative frequency of sample means in any portion of a sampling distribution, just as we did previously for raw scores.

Figure 9.12 shows the standard normal curve applied to our SAT sampling distribution. This curve is the same curve, with the same proportions, that we saw when describing individual raw scores. Once again, the farther a score (here a $\overline{X}$) is from the mean of the distribution (here μ), the larger the absolute value of the *z*-score. The larger the *z*-score, the smaller the relative frequency and simple frequency of the *z*-score and

FIGURE 9.12 Proportions of the Standard Normal Curve Applied to the Sampling Distribution of SAT Means

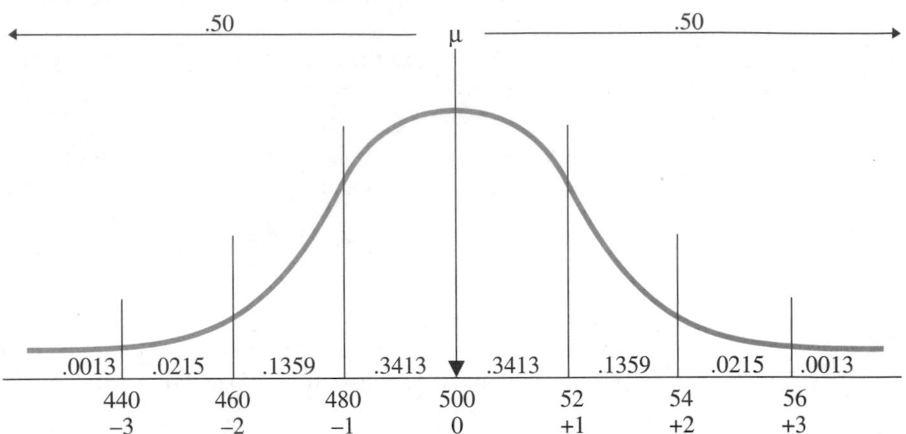

of the corresponding sample mean. Therefore, we again first use the standard normal curve (and the *z*-table) to determine the proportion of the area under any part of the curve. Then this proportion is also the expected relative frequency of the corresponding sample means in that part of the sampling distribution.

For example, the sample from Prunepit U. has a z of $+1$. As you know, .3413, or 34.13%, of all scores fall between the population mean and $z = +1$ on *any* normal distribution. Thus, 34.13% of all SAT sample means will fall between μ and the sample mean at a z of $+1$. Because the μ is 500 and a z of $+1$ is at the sample mean of 520, we can also say that 34.13% of all SAT sample means are expected to be between 500 and 520 (when N is 25). Further, by adding in the 50% of the distribution below μ, we see that about 84% of all means fall below—are to the left of—a z of $+1$. Therefore, our sample mean of 520 ranks at about the 84th percentile among all such sample means. Similarly, a different sample mean of 540 would have a z of $+2$ on this sampling distribution. Because we know that about 2% of all *z*-scores are above a z of $+2$, we expect that only about 2% of all SAT sample means will be larger than 540.

We can use this same procedure to determine the relative frequency of any random sample mean: To determine the relative frequency above or below a particular mean, compute its *z*-score and then look up the corresponding proportion of the area under the curve in the *z*-table. This proportion is also the expected relative frequency of the corresponding sample means in the sampling distribution of all possible sample means that might occur in this situation.

PUTTING IT ALL TOGETHER

The most important concept for you to understand is that any normal distribution of scores can be described using the standard normal curve model and *z*-scores. To paraphrase a famous saying, a normal distribution is a normal distribution is a normal distribution. Any normal distribution contains the same proportions of the total area under the curve between *z*-scores. Therefore, picture a normal distribution and repeat after me: The larger the *z*-score, whether positive or negative, the farther the *z*-score and the corresponding raw score are from the mean of the distribution. The farther they are from the mean of the distribution, the lower the relative frequency of the z-score and of the corresponding raw score. This is true whether the raw score is an individual's score or a sample mean.

By the way, what was Biff's percentile?

CHAPTER SUMMARY

1. The *relative standing* of a score reflects a systematic evaluation of the score relative to the sample or population. A *z-score* indicates a score's relative standing by indicating the distance the score is above or below the mean, measured in standard deviations.

2. The larger a positive z-score, the farther the raw score is above the mean. The larger the absolute value of a negative z-score, the farther the raw score is below the mean. On a normal distribution, the larger the z-score, whether positive or negative, the less frequently it occurs, and thus the less frequently the corresponding raw score occurs.

3. Transforming a raw score distribution into z-scores produces a *z-distribution*. The z-distribution has the same shape as the raw score distribution, but the mean of a z-distribution is always 0 and the standard deviation is always 1.

4. The *standard normal curve* is a mathematically perfect normal curve that can be used as a model of any z-distribution when (a) The distribution is at least roughly normally distributed, and (b) The corresponding raw scores reflect an interval or ratio scale.

5. The proportion of the area under a part of the standard normal curve equals the relative frequency of z-scores falling in that part. This relative frequency leads to the *expected* relative frequency, simple frequency, and percentile of the corresponding raw scores.

6. A *sampling distribution* is a distribution of the values of a sample statistic that are obtained when samples of a particular N are randomly selected from a particular raw score population an infinite number of times.

7. The *sampling distribution of means* is the distribution of sample means obtained when samples of a particular N are randomly selected from a raw score population. Different sample means occur solely because of chance—the luck of the draw when each sample is randomly selected.

8. The *central limit theorem* shows that in a sampling distribution of means, (a) The distribution will be approximately normal, (b) The mean of the sampling distribution will equal the mean of the raw score population, and (c) The variability of the sample means is related to the variability of the raw scores.

9. The true *standard error of the mean*, symbolized by $\sigma_{\overline{X}}$, is the standard deviation of the sampling distribution of means. Computed only when σ_X is known, $\sigma_{\overline{X}}$ indicates the average amount that the sample means deviate from the μ of the sampling distribution.

10. The location of a sample mean on the sampling distribution of means can be described by calculating a *z*-score: The distance the sample mean score, $\overline{X}$, is from the mean of the distribution, μ, measured in standard error units, $\sigma_{\overline{X}}$.

11. The proportion of the area under a part of the standard normal curve is the expected relative frequency of sample means in that part of the sampling distribution of means.

12. Biff's percentile was 99.87.

PRACTICE PROBLEMS

(Answers for odd-numbered problems are provided in Appendix D.)

1. What does a *z*-score indicate?
2. On what factors does the size of a *z*-score depend?
3. What is a *z*-distribution?
4. What are the three uses of *z*-distributions when describing individual scores?
5. Why are *z*-scores referred to as standard scores?
6. In freshman English last semester, Foofy earned a 76 ($\overline{X} = 85$, $S_X = 10$) and her friend Bubbles, in a different class, earned a 60 ($\overline{X} = 50$, $S_X = 4$). Should Foofy be bragging about how much better she did in the course? Why?
7. Poindexter received a grade of 55 on a biology test ($\overline{X} = 50$) and a grade of 45 on a philosophy test ($\overline{X} = 50$). He is considering whether to ask his professors to curve the grades using *z*-scores. (a) What other information should he consider before making his request? (b) Does he want the S_X to be large or small in biology? Why? (c) Does he want the S_X to be large or small in philosophy? Why?
8. Foofy computes *z*-scores for a set of normally distributed exam scores. She obtains a *z*-score of -3.96 for 8 (out of 20) of the students. What does this mean?
9. (a) What is the standard normal curve model? (b) What is it used for? (c) What criteria should be met for the model to give an accurate description of a sample?
10. Why are *z*-scores beyond ± 3 seldom obtained?
11. For the data:

9 5 10 7 9 10 11 8 12 7 6 9

(a) Compute the *z*-score for the raw score of 10. (b) Compute the *z*-score for the raw score of 6.
12. For the data in question 11, find the raw scores that correspond to the following: (a) $z = +1.22$ (b) $z = -0.48$
13. Which *z*-score in each of the following pairs corresponds to the smaller raw score? (a) $z = +1.0$, $z = +2.3$ (b) $z = -2.8$, $z = -1.7$ (c) $z = -.70$, $z = +.20$ (d) $z = 0$, $z = -2.0$
14. For each pair in question 13, which *z*-score has the higher frequency?
15. In a normal distribution of scores, what proportion of all scores would you expect to fall in each of the following areas? (a) Between the mean and $z = +1.89$ (b) Below $z = -2.30$ (c) Between $z = -1.25$ and $z = +2.75$ (d) Above $z = +1.96$ and below -1.96
16. For a distribution in which $\overline{X} = 100$, $S_X = 16$, and $N = 500$, answer the following. (a) What is the relative frequency of scores between 76 and the mean? (b) How many participants are expected to score between 76 and the mean? (c) What is the percentile of someone scoring 76? (d) How many participants are expected to score above 76?

17. Poindexter may be classified as having a math dysfunction—and thus not have to study statistics—if he scores below the 25th percentile on a national diagnostic test. The μ of the test is 75 and $\sigma_X = 10$. Approximately what raw score is the cutoff score for him to avoid statistics?
18. For an IQ test, we know that $\mu = 100$ and $\sigma_X = 16$. We are interested in creating the sampling distribution when $N = 64$. (a) What does that sampling distribution of means reflect? (b) What is the shape of the distribution of IQ means and the mean of the distribution? (c) Calculate $\sigma_{\overline{X}}$ for this distribution. (d) What is your answer in (c) above called, and what does it indicate?
19. A recent graduate has two job offers and must decide which to accept. The job in City A pays \$27,000. The average cost of living there is \$50,000, with a standard deviation of \$15,000. The job in City B pays \$12,000. The average cost of living there is \$14,000, with a standard deviation of \$1,000. Assuming the data are normally distributed, which is the better job offer? Why?
20. A researcher obtained a sample mean of 68.4 with $N = 49$. In the population, the mean is 65 and $\sigma_X = 10$. The researcher believes his sample mean is rather unusual. (a) Is he correct? (b) How often can he expect to obtain a sample mean that is higher than 68.4? (c) Why might he obtain such an unusual mean?
21. If you took 1,000 random samples of 50 participants each from a population where $\mu = 19.4$ and $\sigma_X = 6.0$, how many samples would you expect would produce a mean below 18?
22. Suppose you own shares of a company's stock, with its mean selling price at \$14.89 based on the past 10 trading days. Over the years, the mean price of the stock has been \$10.43 ($\sigma_X = \$5.60$). You wonder if the mean selling price over the next ten days can be expected to go higher. Should you wait to sell, or should you sell now?

SUMMARY OF FORMULAS

1. *The formula for transforming a raw score in a sample into a z-score is*

$$z = \frac{X - \overline{X}}{S_X}$$

where X is the raw score, $\overline{X}$ is the sample mean, and S_X is the sample standard deviation.

2. *The formula for transforming a z-score in a sample into a raw score is*

$$X = (z)(S_X) + \overline{X}$$

3. *The formula for transforming a raw score in a population into a z-score is*

$$z = \frac{X - \mu}{\sigma_X}$$

4. *The formula for the true standard error of the mean is*

$$\sigma_{\bar{X}} = \frac{\sigma_X}{\sqrt{N}}$$

where σ_X is the true standard deviation of the raw score population.

5. *The formula for transforming a sample mean into a z-score on the sampling distribution of means is*

$$z = \frac{\bar{X} - \mu}{\sigma_{\bar{X}}}$$

where $\bar{X}$ is the sample mean, μ is the mean of the sampling distribution (which is also equal to the μ of the raw score distribution), and $\sigma_{\bar{X}}$ is the standard error of the mean of the sampling distribution of means.

PART 3

CORRELATIONAL RESEARCH AND CORRELATIONAL STATISTICS

As you know, most psychological research examines a relationship between variables. Therefore, the final question to answer with descriptive statistics is "What is the nature of the relationship we have found?" There are two major approaches to answering this question, which we'll cover in the next two chapters. Chapter 10 discusses the procedure known as "correlation," and Chapter 11 discusses "linear regression."

If it seems that the topics in the upcoming chapters are different from previous topics, it's because they *are* different. Although correlation and regression are major descriptive statistical procedures, their perspective is different from that of previous techniques. Therefore, think of the upcoming chapters as somewhat of a detour. After we complete the detour, we'll return to describing a sample using the mean and standard deviation, and describing the location of the sample on a sampling distribution of sample means. Don't forget those procedures, because they form the basis for inferential statistics.

10

Correlational Research and the Correlation Coefficient

To understand this chapter:

- From Chapter 2, recall what a data point is, and how to identify a strong relationship.
- From Chapter 3, recall what validity and reliability are and why they are important.
- From Chapter 4, recall that a restriction of range occurs whenever the range of scores on a variable is limited, and understand multiple raters and inter-rater reliability.
- From Chapter 8, understand that greater variability indicates that scores are not typically close to each other.

Then your goals in this chapter are to learn:

- The difference between analyzing correlational data and experimental data.
- How to read and interpret a scatterplot.
- How to interpret a correlation coefficient.
- How correlation is used to demonstrate different types of reliability and validity.
- When to use the Pearson r, the Spearman r_s, and the point-biserial r_{pb}.
- The logic of inferring a population correlation based on a sample correlation.

Recall that in a relationship, as the scores on one variable change, there is a consistent pattern of change in the scores on the other variable. Merely demonstrating a relationship is not enough, however. We also want to know its characteristics—What is the nature of the relationship? How consistently do the scores change together? In what direction do the scores change?—and so on. This chapter discusses the descriptive statistical procedure used to answer such questions, called *correlation*. In the following sections we will consider when such procedures are used and then examine what they tell us. Finally, we'll see how to calculate different types of correlation statistics. First, though, here are some more symbols.

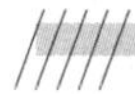

MORE STATISTICAL NOTATION

Correlational analysis requires scores from two variables, some X scores and some Y scores. Usually we obtain an X and a Y score from the same participant and then each participant's X is paired with the corresponding Y. If the X and Y scores are not from the same participant, there must be some other rational system for pairing the scores (for example, pairing the scores of roommates). Also, with pairs of scores, there must be the same number of X and Y scores.

The same conventions are used for Y that were previously used for X. Thus, ΣY is the sum of Y scores, ΣY^2 is the sum of squared Y scores, and $(\Sigma Y)^2$ is the squared sum of Y scores. The mean of the Y scores is $\overline{Y}$ and equals $\Sigma Y/N$. Similarly, the variance of a sample of Y scores is S_Y^2 and the standard deviation of Y is S_Y. To find S_Y^2 or S_Y, use the same formulas used to find S_X^2 or S_X, except now plug in Y scores instead of X scores.

You will also encounter three new notations. First, $(\Sigma X)(\Sigma Y)$ says to find the sum of the Xs and the sum of the Ys, and then multiply the two sums together. However, ΣXY says to first multiply each X score in a pair times its corresponding Y score and then sum all of the resulting products. Finally, D indicates to find the *difference* between the X and Y scores in a pair, which we find by subtracting one from the other.

Now, to correlation!

UNDERSTANDING CORRELATIONAL RESEARCH

Recall that a common descriptive research approach is the correlational design. The term correlation is synonymous with relationship, so in a correlational design we examine the relationship—examine the correlation—between variables. The relationship can involve scores from virtually any variables, regardless of how we obtain them. Often we measure the variables using a questionnaire or other observational technique, but we may also measure responses using any of the methods used in experiments. But recall that experimental and correlational designs differ in terms of how we demonstrate a relationship. For example, it is commonly believed that as people drink more coffee, they become more nervous. To demonstrate this in an experiment, we would first actively manipulate the amount of coffee people consume: We might randomly assign some participants to a condition where they are given 1 cup, others to a condition where they receive 2 cups, and others to where they drink 3 cups. Then we would measure participants' nervousness (perhaps using physiological methods or a questionnaire). Our research hypothesis implies the question, "For a given amount of coffee, what is a person's nervousness?" Recall that the "given" variable is always the *X* variable. The important thing to recognize about an experiment here is that, by randomly assigning participants to a condition, we, the researchers, determine each participant's *X* score—that is, we decide whether their "score" will be 1, 2, or 3 cups on the coffee variable. Then, we show that, for an *assigned X*, participants tend to produce a certain *Y* score.

In correlational research, however, we do not manipulate any variables. Instead, we simply measure two (or more) variables. Thus, we might simply ask each participant to report the amount of coffee consumed that day, and then measure how nervous each participant was. The distinguishing aspect of the correlational design is that, by not assigning participants to an amount of the *X* variable to experience, we do not determine their *X* scores. Rather, the scores on both variables reflect an amount or category of a variable that a participant has *already* experienced. Coffee is still the *X* variable, but here we ask "For a *reported X* score of coffee consumed, what *Y* score of nervousness do people tend to produce?"

Note that, as the name implies, correlational research typically employs correlational statistics to summarize the relationship. However, computing a correlation does *not* automatically create a correlational design. As we will see, correlational *statistics* can be applied to any design, including experiments. A correlational *design* occurs whenever we do not randomly assign participants to the levels of either variable, regardless of how the data are analyzed.

Drawing Conclusions from Correlational Research

Our confidence in the conclusions drawn from correlational designs is influenced by the same concerns we've raised previously about experiments. Thus, the measurement procedures should have construct and content validity, they should be reliable, and they should be sensitive to subtle differences among participants. Extraneous influences such as demand characteristics or order effects should be eliminated. And, as always, we address such concerns by considering the four components of any study—the participants, environment, researcher, and measurement task—looking for any aspect that may threaten reliability and validity.

In addition, recall that compared to experiments, correlational designs have two flaws that severely reduce internal validity for concluding that differences in the *X* variable *cause* differences in *Y* scores: First, often we cannot be certain that *X* occurred before *Y*. For example, above we simply measured scores reflecting amount of coffee and nervousness. But perhaps participants were *first* more nervous and *then* drank more coffee. If so, then greater nervousness may actually cause greater coffee consumption. In any correlation, it is possible that *Y* causes *X*.

The second flaw is that in correlational designs, we do not usually control extraneous variables, so there may be all sorts of confounding variables present. In particular, because we don't randomly assign participants to a score on the *X* variable, differences between them on the *X* variable are likely to be confounded by differences in subject variables. For example, some participants in the coffee study may have had less sleep than others the night before testing. Perhaps the lack of sleep caused those people to be more nervous *and* to drink more coffee. In any correlational study, some other variable may cause both X and Y to change. (Recall, this is called "The Third Variable Problem.")

Thus, the relationship found in a correlational study may mean that changes in X cause changes in Y as we think. But, changes in Y may cause changes in X, or some third variable may produce differences in both the X and Y scores. Therefore, a single correlational study should *not* be interpreted as showing that changes in X cause changes in Y.

REMEMBER Causality is not inferred from a correlational design, because the order of occurrence of the variables is unknown, and potential confoundings are likely.

Reasons for Using the Correlational Approach

Though we must accept the limited evidence for causality found in a correlational design, it is still a legitimate research method. In fact, a correlational design may be preferable for several reasons.

First, because of ethical and practical considerations, some relationships can be studied only through a correlational design. For example, ethically we cannot study such variables as sexual abuse, accidents, or drug use by randomly assigning people to experience them in a true experiment. Likewise, it is impossible to manipulate such variables as a person's career, race, personality traits, or mental and physical illness. Yet such variables and behaviors do occur in the real world, and we can study them through correlational methods. Thus, for example, we can measure the amount of abuse people have experienced and relate it to other behaviors or characteristics. Further, through replication, together with experimental studies of relevant constructs, researchers may eventually develop some degree of confidence in causal statements about such variables.

Correlational procedures are also useful for discovering new relationships. For example, we might measure people on numerous variables that we suspect are related to nervousness (their health history, self-esteem level, physiology, diet, and so on) and then discover which variables are related. Such research is useful not only for describing the behavior but also for identifying possibly causal relationships to then study using experimental designs.

In addition, a laboratory experiment creates a rather artificial situation. An advantage of a correlational study is that often it can be conducted outside of the laboratory where potentially we have greater ecological and external validity. For example, say we are interested in the behavior of workers at a factory. Studying employees while they work in a factory is more valid than studying laboratory subjects in an artificial factory simulation (but see Pritchard, Dunnette & Jorgenson, 1972, for a surprisingly realistic laboratory study).

And finally, recall that part of understanding a behavior is being able to predict when it will occur. Because of better external validity, the relationship found in a correlational study may be better for predicting behaviors than that found in an experiment.

REMEMBER Correlational designs are used to discover relationships, to solve ethical and practical problems, and to provide greater external validity.

Distinguishing Characteristics of Correlational Analysis

There are four major differences between how we handle data in correlational analysis versus in an experiment. First, if we conducted our coffee experiment, we would examine changes in *each* mean nervousness score (the *Y* scores) as a function of changes in the conditions of coffee consumed (the *X* scores). With correlational data, however, we typically have a rather large and unwieldy number of different *X* scores: People would probably report a wide range of coffee consumption beyond only 1, 2 or 3 cups. Comparing the mean nervousness scores from many groups would be very difficult. Therefore, in correlational procedures, we do not compute a mean *Y* score at each *X*. Instead, we summarize the *entire* relationship formed by all pairs of *X*–*Y* scores in the data. This is the major advantage of using correlation, because it simplifies a very complex relationship involving a wide range of scores into an easily interpreted statistic.

A second difference is that because we examine all pairs of *X*–*Y* scores at once, we have *one* sample. Then: ***N* stands for the number of pairs of scores in the data.**

Third, in a correlational study, either variable may be the *X* or *Y* variable. Above we asked, "For a given amount of coffee, what are the nervousness scores?" so amount of coffee was the *X* variable and nervousness was the *Y* variable. Conversely, if we had asked, "For a given nervousness score, what is the amount of coffee consumed?" then nervousness would be the *X* variable and amount of coffee the *Y* variable.

Finally, as we see in the next section, the data are graphed differently in correlational research than in an experiment. The individual pairs of scores are used to create a scatterplot.

Plotting Correlational Data: The Scatterplot

A **scatterplot** is a graph that shows the location of each data point formed by a pair of *X*-*Y* scores. The scatterplot in Figure 10.1 shows data that might occur if we actually

FIGURE 10.1 Scatterplot Showing Nervousness as a Function of Coffee Consumption

Each data point is created using a subject's coffee consumption as the X score and the subject's nervousness as the Y score.

Cups of coffee: X	*Nervousness scores: Y*
1	1
1	1
1	2
2	2
2	3
3	4
3	5
4	5
4	6
5	8
5	9
6	9
6	10

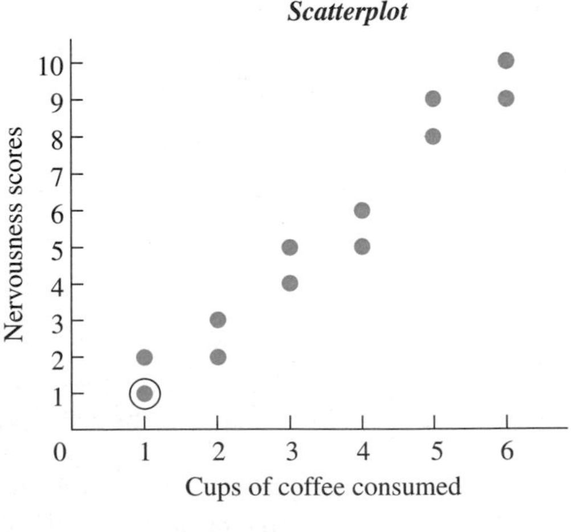

studied how nervousness scores change as a function of coffee consumption. You can see that people drinking 1 cup of coffee tend to have nervousness scores around 1, those drinking 2 cups tend to have nervousness scores around 2, and so on. (Note that in the table on the left, two people scored 1 on coffee consumption and nervousness. As shown in the scatterplot, some researchers circle such a data point to indicate that points are plotted on top of each other.) *Always draw the scatterplot* of a set of correlational data. A scatterplot allows you to see the nature of the relationship that is present and to map out the best way to accurately summarize and describe it.

The shape of the scatterplot is an important indication of both the presence of a relationship and the nature of the relationship. We can summarize a scatterplot visually by drawing a line around its outer edges. A scatterplot may have any one of a number of shapes when a relationship is present. When no relationship is present, however, the scatterplot will be either circular or elliptical, oriented so that the ellipse is parallel to the *X* axis. The scatterplots in Figure 10.2 show what no relationship between coffee consumption and nervousness would look like. There is no relationship here because as the *X* scores increase, the *Y* scores do not consistently change, and no particular value of *Y* tends to be associated with a particular value of *X*. Instead, virtually every value of *Y* is associated with every value of *X*.

The scatterplots in Figure 10.2 are further summarized by a line that seems to best fit through the center of the scatterplot. This line is called the **regression line.** (We discuss the procedures for drawing the regression line in the next chapter.) The orientation of the regression line matches the orientation of the scatterplot. Thus, when no relationship is present, the regression line is a horizontal straight line.

When a relationship is present, the scatterplot will form some shape *other* than a circle or a horizontal ellipse (and the regression line will not be a horizontal straight line).

The Correlation Coefficient

The particular shape and orientation of a scatterplot reflect the characteristics of the relationship formed by the data. To summarize these characteristics, we compute the statistic known as the correlation coefficient. A **correlation coefficient** is a number

FIGURE 10.2 Scatterplots Showing No Relationship Between Coffee Consumption and Nervousness

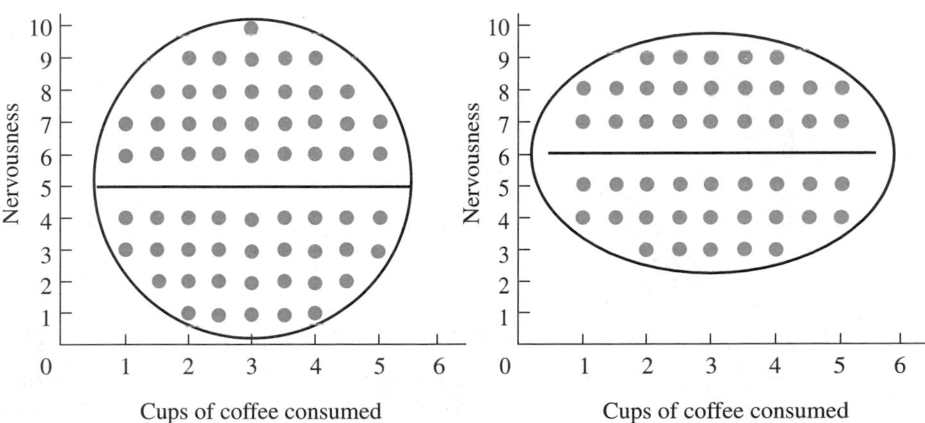

that summarizes and describes the important characteristics of a relationship. There are two important characteristics of a relationship: The type of relationship, and the strength of the relationship. The following sections discuss these characteristics.

TYPES OF RELATIONSHIPS

The **type of relationship** that is present in a set of data is determined by the overall direction in which the *Y* scores change as the *X* scores change. There are two general types of relationships: Linear and nonlinear relationships.

Linear Relationships

The term *linear* means "straight line," and a linear relationship has a regression line that is a straight line. For example, the two scatterplots in Figure 10.3 show the linear relationships between the amount of time that students study and their test performance, and between the number of hours students spend watching television and the amount of time they spend sleeping. Both scatterplots show a linear relationship because: (1) They do not form horizontal ellipses, and (2) They are best summarized by a straight line. We obtain such scatterplots because, in a **linear relationship**, as the *X* scores increase, the *Y* scores tend to change in only one direction. On the left, as students study longer, their grades tend only to increase. On the right, as students watch more television, their sleep tends only to decrease.

The above scatterplots reflect two subtypes of linear relationships, depending on the *direction* in which the *Y* scores change. The study-test relationship is a positive relationship. In a **positive linear relationship**, as the scores on the *X* variable increase, the scores on the *Y* variable also tend to increase. Thus, low *X* scores are paired with low *Y* scores, and high *X* scores are paired with high *Y* scores. Any relationship that fits the general pattern "the more *X*, the more *Y*" is a positive linear relationship, such as the more you eat, the more you weigh. You can remember that such relationships are called

FIGURE 10.3 Scatterplots Showing Positive and Negative Linear Relationships

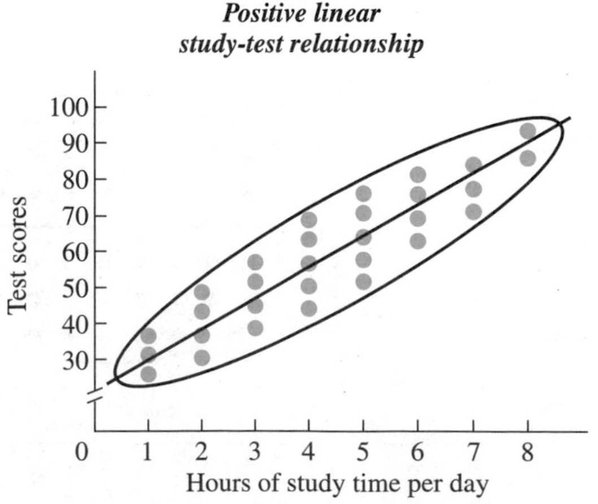

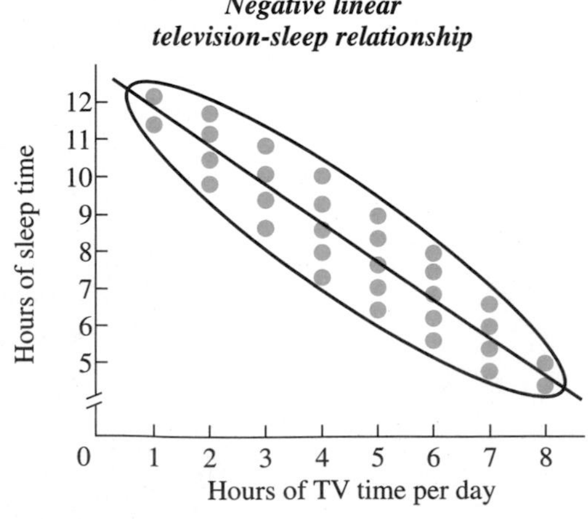

positive by remembering that as the X scores increase, the Y scores change in the direction away from zero, toward higher *positive* scores.

On the other hand, the television-sleep relationship is a negative relationship. In a **negative linear relationship**, as the scores on the X variable increase, the scores on the Y variable tend to decrease. Low X scores are paired with high Y scores, and high X scores are paired with low Y scores. Any relationship that fits the general pattern "the more X, the less Y" is a negative linear relationship, such as the more you study the fewer errors you make on a test. You can remember that such relationships are negative by remembering that as the X scores increase, the Y scores change in the direction toward zero, heading for the *negative* scores.

It is important to understand that the term *negative* does not mean there is something wrong with the relationship. Negative relationships are no different from positive relationships *except* in terms of the direction in which the Y scores change as the X scores increase.

Nonlinear Relationships

If a relationship is not linear, then it is called nonlinear. *Nonlinear* does not mean that the data cannot be summarized by a line; it means that the data cannot be summarized by a *straight* line. Thus, another name for a nonlinear relationship is a curvilinear relationship. In a **nonlinear**, or **curvilinear**, **relationship**, as the X scores change, the Y scores do not tend to *only* increase or *only* decrease: The Y scores change their direction of change.

Nonlinear relationships come in many different shapes. Figure 10.4 shows two common nonlinear relationships. The scatterplot on the left shows the relationship between a person's age and the amount of time required to move from one place to another. Very young children move slowly, but as age increases, movement time decreases. Beyond a certain age, however, the time scores change direction so that as age continues to increase, movement time increases. Because of the shape of the scatterplot, such a

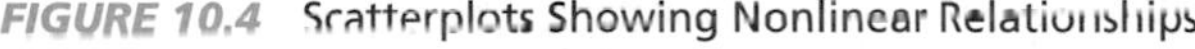

FIGURE 10.4 Scatterplots Showing Nonlinear Relationships

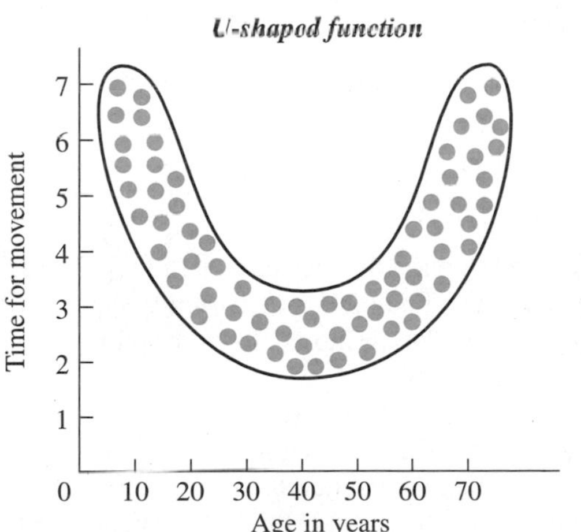

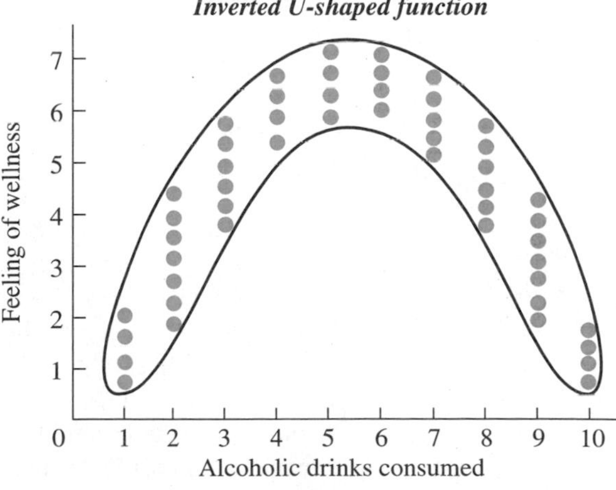

relationship is called *U-shaped*. The scatterplot on the right shows the relationship between the number of alcoholic drinks people consume and their sense of feeling well. At first, people report feeling better as they drink, but beyond a certain point, more drinking makes them feel progressively worse. Such a scatterplot reflects an *inverted U-shaped* relationship. Curvilinear relationships may be even more complex than above, producing a wavy pattern that repeatedly changes direction.

Note that the preceding terminology is also used to describe the type of relationship found in experiments. We have a positive relationship if, as the amount of the independent variable (X) increases, the dependent scores (Y) also increase. We have a negative relationship if the dependent scores decrease. We have a nonlinear relationship if the dependent scores change their direction of change.

How the Correlation Coefficient Describes the Type of Relationship

Correlational research in psychology focuses almost entirely on linear relationships, so we'll discuss only linear correlation. How do you know whether data form a linear relationship? Make a scatterplot! If the scatterplot is best summarized by a straight line, then linear correlation is appropriate. Also, sometimes you may want to describe the extent to which a nonlinear relationship has a linear component and somewhat fits a straight line. Here, too, linear correlation is appropriate. However, do not summarize a nonlinear relationship using a linear correlation coefficient. Describing a nonlinear relationship with a straight line is like putting a round peg into a square hole: The relationship will not fit the straight line very well, and the correlation coefficient will not accurately describe the relationship.

The correlation coefficient communicates two things about a relationship. First, merely by computing a linear correlation coefficient, we communicate that we are describing a linear relationship. Second, the coefficient itself communicates whether the linear relationship is positive or negative. If the coefficient—the number we compute—has a minus sign in front of it, then the relationship is negative. If the coefficient does not have a minus sign, then we put a plus sign in front of it, indicating a positive relationship.

The other characteristic of a relationship communicated by the correlation coefficient is the strength of the relationship.

STRENGTH OF THE RELATIONSHIP

Recall that a relationship can exhibit varying degrees of consistency. The *strength of a relationship* is the extent to which one value of Y is consistently paired with one and only one value of X. The strength of a relationship is also referred to as the *degree of association* between the two variables.

The absolute value of the correlation coefficient (the size of the number we calculate) indicates the strength of the relationship. The largest value you can obtain is 1.0, and the smallest value is 0. (Thus, when we include the positive or negative sign, the correlation coefficient may be any value from -1.0 to $+1.0$.) The *larger* the absolute value of the coefficient, the *stronger* the relationship. In other words, the closer the coefficient is to ± 1, the more consistently one value of Y is paired with one and only one value of X.

> *REMEMBER* The sign of a correlation coefficient indicates either a positive or a negative linear relationship, and the larger the absolute value, the stronger the relationship.

Computing the correlation coefficient is not difficult. The difficulty comes in interpreting the coefficient in terms of the strength of the relationship. Therefore, we'll first discuss how to use the correlation coefficient to envision the scatterplot that the data would form and thus to envision the nature of the relationship that is present. Our starting point is to know how to envision a perfect relationship.

Perfect Association

A correlation coefficient of +1.0 or −1.0 describes a perfectly consistent linear relationship. Figure 10.5 shows an example of each.

FIGURE 10.5 Data and Scatterplots Reflecting Perfect Positive and Negative Correlations

Perfect positive coefficient = +1.0	
X	*Y*
1	2
1	2
1	2
3	5
3	5
5	8
5	8
5	8

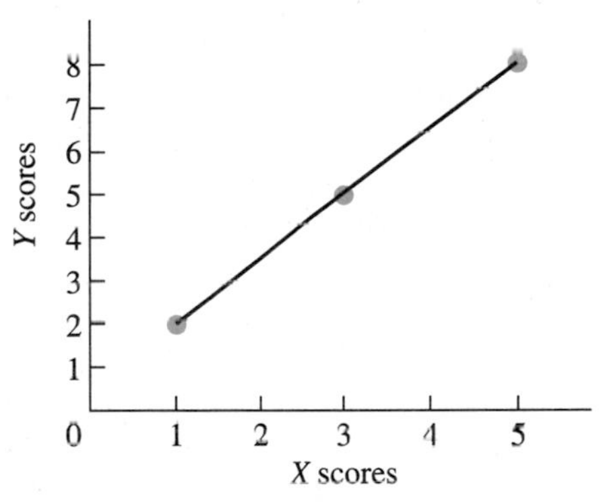

Perfect negative coefficient = −1.0	
X	*Y*
1	8
1	8
1	8
3	5
3	5
3	5
5	2
5	2
5	2

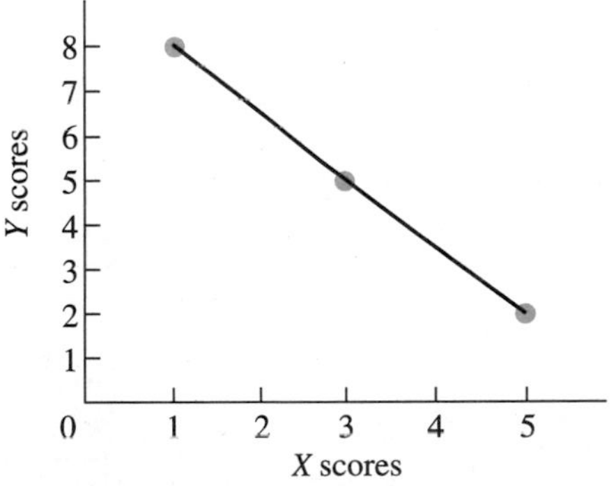

There are four ways to think about what a correlation coefficient of ± 1.0 indicates about a relationship. First, it tells us that *every* participant who obtains a particular X score obtains one and only one value of Y. Every time X changes, the Y scores all change to one new value. Thus, ± 1.0 indicates a perfect one-to-one correspondence between the X and Y scores.

At the same time, a coefficient of ± 1.0 indicates that there are no differences among the Y scores associated with a particular X. Differences between scores is variability, so ± 1.0 indicates that there is no variability in the Y scores at each X.

Third, a coefficient of ± 1.0 ensures perfect predictability of Y scores. Pretend that we do not know someone's Y score. Because each Y score is associated with only one X score, if we know the person's X score, then we know his or her Y score. (You'll see how to predict Y scores in the next chapter.)

Fourth, because it indicates that there is no variability or differences among the Y scores at each X, a coefficient of ± 1.0 indicates that the data points at an X are all on top of one another. And, because it is a perfect straight-line relationship, when we summarize the scatterplot with the regression line, all of the data points fall *on* the line.

Intermediate Association

The way to interpret any other value of the correlation coefficient is in terms of how close it comes to ± 1.0, indicating the extent to which the data come close to forming a perfect linear relationship. For example, Figure 10.6 shows data and the resulting scatterplot that produce a correlation coefficient of $+.98$. Again, we interpret the correlation coefficient in four ways. First, with an absolute value less than 1.0, it indicates that there is not perfectly consistent association: Every participant obtaining a particular X did not obtain the exact same Y. A coefficient of $+.98$ is close to ± 1.0, however, so it indicates that here there is "close" to perfect consistency between the X and Y scores. That is, even though there are different values of Y at the same X, the Y scores are relatively close to each other.

Second, now there are *different* Y scores associated with a single X score, so there is variability between the Y scores at each X. Thus, participants with an X of 1 obtained a Y of 1 or 2, but those with an X of 3 scored a Y of 4 or 5. Once again, however, $+.98$ is close to $+1.0$, indicating that the variability in Y scores at each X is *relatively* small

FIGURE 10.6 Data and Scatterplot Reflecting a Correlation Coefficient of +.98

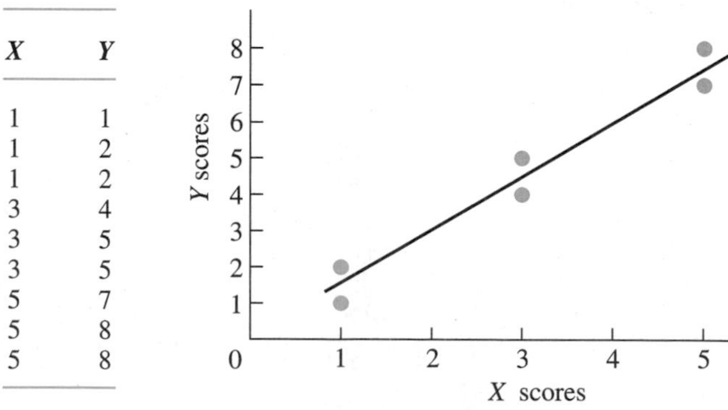

X	*Y*
1	1
1	2
1	2
3	4
3	5
3	5
5	7
5	8
5	8

when compared to the overall variability of all Y scores in the sample. Think of it this way: Over the entire sample, the above Y scores are between 1 and 8, so the overall range is 8. At each X score, however, the Ys span a range of only 1. It is this small variability in Y at each X relative to the overall variability in all Y scores that produces a correlation coefficient close to 1.

Third, when the correlation coefficient is not ± 1.0, knowing each participants' X score only allows us to predict *around* what their Y scores will be: For an X of 1, for example, we'd predict a Y of around 1 or 2. As we've seen in previous chapters, error in predictions because of variability is called "error variance." Here, the error variance is the spread in the Y scores at each X. A coefficient of $+.98$ is close to $+1.0$, however, indicating that the predicted Y score will be close to the actual Y scores that participants obtained, and so there is relatively little error variance.

Fourth, because there is now variability in the Ys at each X, all data points do not fall *on* the regression line: As in Figure 10.6, variability results in vertical spread in the data points above and below the line at each X. But, a coefficient of $+.98$ is close to $+1.0$, indicating that in the data, the spread between the lowest and highest Y scores at each X is small. Therefore, the Y scores are close to, or hug, the regression line, resulting in a scatterplot that is a narrow, or skinny, ellipse. In fact, the absolute value of the correlation coefficient always tells us how skinny the scatterplot is. When the coefficient is ± 1.0, the scatterplot forms a straight line, which is the skinniest ellipse possible. The closer the coefficient is to ± 1.0, the skinnier the scatterplot, and vice versa.

> **REMEMBER** The absolute value of a correlation coefficient indicates the extent to which the data come close to forming a perfect linear relationship.

As the absolute value of the correlation coefficient becomes smaller, it indicates larger variability in the Ys at each X, and thus a weaker, less consistent relationship. Figure 10.7 shows data that produce a much smaller correlation coefficient of $-.28$. The fact that this is a negative relationship has nothing to do with its strength. Rather, as the scatterplot illustrates, the variability or vertical spread in the Y scores at each X is relatively large. This does two things that are the opposite of a relationship. First, instead of the Y scores changing only when an X score changes, here the Y scores

FIGURE 10.7 Data and Scatterplot Reflecting a Correlation Coefficient of $-.28$

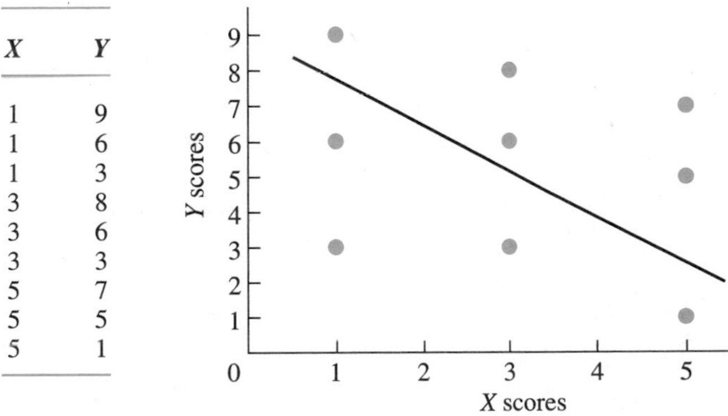

X	Y
1	9
1	6
1	3
3	8
3	6
3	3
5	7
5	5
5	1

change (differ) even though the X score remains the same. Second, there is overlap between the Y scores at the different Xs. Instead of there being one value of Y at one X and a different value of Y at a different X, here one value of Y is paired with many *different* values of X. Thus, in contrast to a relationship, here the Y scores tend to change when X does not, and the Y scores tend to stay the same when X changes.

A coefficient of $-.28$ is not very close to ± 1.0, indicating that this relationship is not very close to forming a perfectly consistent linear relationship. Thus we know that: (1) Only barely does one value of Y or even close to one value of Y tend to be associated with one value of X, (2) The variability in the Ys at each X is almost as large as the variability among all Y scores in the data, (3) Knowing participants' X score will not produce a very accurate prediction of their actual Y score, and (4) The large variability in Y will produce a fat scatterplot that does not hug the regression line.

> *REMEMBER* Greater variability in the Y scores at each X results in a weaker relationship and thus a smaller correlation coefficient.

Although theoretically a correlation coefficient may be as large as ± 1.0, in real research such values do not occur. Remember that the X and Y scores reflect the behaviors of living organisms, and that, because of individual differences, living organisms are never perfectly consistent. Also, there will always be some fluctuating external variables that produce differences in their behavior. Therefore, real psychological research often produces correlation coefficients in the neighborhood of .40. A coefficient around $\pm .50$ is considered to be quite respectable, and coefficients above $\pm .50$ are downright impressive. A correlation of ± 1.0 is so unlikely to occur that, if you ever obtain one, you should assume you've made a computational error. If you obtain a coefficient greater than ± 1.0, you've definitely made an error, because ± 1.0 indicates a perfect relationship and you can't do better than that.

Zero Association

The lowest possible value of the correlation coefficient is 0, indicating that no linear relationship is present. Figure 10.8 shows data that produce a correlation coefficient of 0.

FIGURE 10.8 Data and Scatterplot Reflecting a Correlation Coefficient of 0

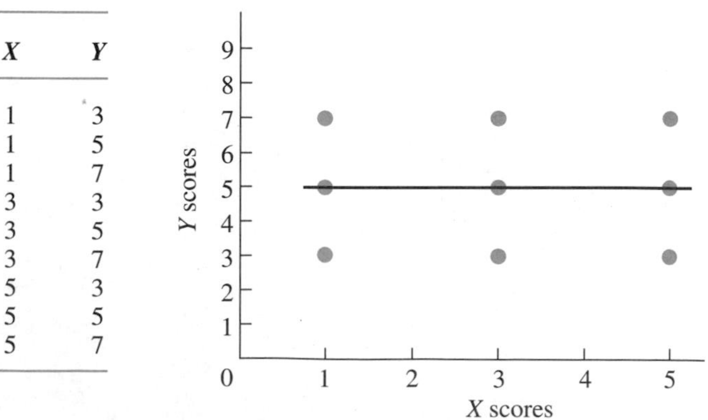

X	Y
1	3
1	5
1	7
3	3
3	5
3	7
5	3
5	5
5	7

A scatterplot having this shape is as far from forming a slanted straight line as possible, and a correlation coefficient of 0 is as far from ± 1.0 as possible. Therefore, we know that there are no values of Y that tend to be consistently associated with only one value of X. Instead the Ys found at one X are the same as the Ys found at any other X. This also means that knowing someone's X score will not in any way help to predict the corresponding Y score. Finally, this indicates that the spread in Y at any X equals the overall spread of Y in the data, producing a scatterplot that is a circle or horizontal ellipse that in no way hugs the regression line.

> ***REMEMBER*** The larger the correlation coefficient, the stronger the relationship because the less the Ys are spread out at each X and the closer the data come to forming a straight line.

Now that you understand the correlation coefficient, let's see how it's used in research.

USING THE CORRELATION COEFFICIENT IN RESEARCH

Say that after collecting our coffee and nervousness data, we compute a correlation coefficient of $+.55$. As usual, the next step is to perform the appropriate inferential statistical procedures. If the correlation coefficient passes the inferential test, it is the basis for interpreting the study. For example, originally our hypothesis led us to predict a positive linear relationship in the coffee study, so the positive coefficient of $+.55$ provides evidence that confirms our hypothesis. A negative coefficient would have been contrary to the relationship predicted by the hypothesis, thus disconfirming it. (In a different study, the hypothesis might lead us to predict only some kind of linear relationship, so then either a positive or negative coefficient would confirm the hypothesis.)

Further, we also know that a coefficient with an absolute value of .55 is, in everyday research, a relatively strong, consistent relationship. The strength of the relationship is also important to our interpretation, because, as with experiments, it suggests whether there are other variables operating that we have not taken into account. If, for example, we had obtained a coefficient of $+.90$, this would suggest that few rather unimportant variables are related to nervousness in addition to coffee consumed. Conversely, a coefficient of $+.09$ would suggest there must be other important variables related to nervousness that we have not considered. As usual, following this line of reasoning, we interpret the results "psychologically," considering what the particular relationship indicates about nervousness and coffee consumption, and how they relate to our theoretical explanations, models, and constructs involving emotions and physiology. This was a correlational design, however, so we do not infer that more coffee causes more nervousness!

Recognize that although a coefficient is interpreted as reflecting the degree of consistency in a relationship, it does not directly measure units of "consistency." Thus, for example, a coefficient of $+.60$ indicates a more consistent relationship than, say, a coefficient of $+.30$, but the relationship with a coefficient of .60 is not twice as consistent as that with .30. We'll discuss how to directly compare correlation coefficients in the next chapter. For now, stick to interpreting a coefficient in relative terms, based on how close it is to ± 1.0.

In addition to describing the overall results of a study, computing a correlation coefficient is the statistical technique for demonstrating the reliability and validity of a measurement procedure in any experimental or correlational design.

Ascertaining Reliability

Recall that if a measurement procedure is reliable, then whenever a participant consistently exhibits a particular behavior, he or she should receive the same score. There are three ways a correlation coefficient is used to show reliability.

First, recall that often a measurement involves judging or rating a participant's behavior, and for reliability, we employ multiple raters. To demonstrate that the behavior is judged reliably, we need to show high *inter-rater reliability*, the consistency of ratings by any two raters. For example, say that we are studying creativity and have two judges rate how creative participants are when "doodling." If the judges are reliable, then we should see a large positive correlation between their ratings: We should find that a low creativity score assigned to a participant by one rater was consistently matched by the other, while a high score given by one was also given by the other.

A similar approach is used to show the reliability of a testing procedure that does not involve raters. **Test-retest reliability** indicates that participants tend to obtain the same score when tested at different times. For example, if a college exam has test-retest reliability, a student who produces a low score now should also produce a low score later, and a student scoring high now should also score high later. In other words, test-retest reliability is evident when there is a high, positive correlation between the scores obtained from the two testings. Likewise, a physiological measurement has test-retest reliability if scores from the same participants tested twice are positively correlated.

Test-retest reliability reflects the reliability of a subject's score over multiple testing *sessions*. Another approach is to determine whether the *trials* within one testing session are reliable. **Split-half reliability** indicates that participants' scores on some trials consistently match their scores on other trials. Typically we first split a test in half, comparing the odd-numbered trials to the even-numbered trials. (This balances out order effects due to fatigue or practice.) Then we compute a summary score—such as a mean or the total number correct—for each subject on each half of the test. Then we correlate the summary scores. For example, if the questions on a college examination have split-half reliability, then students scoring low on the odd questions should also score low on the even questions, and so on. Likewise, we might determine the split-half reliability of a series of reaction-time trials by correlating the times from even- and odd-numbered trials.

> *REMEMBER* Test-retest reliability is the correlation between repeated testings, and split-half reliability is the correlation between different trials within a test.

For any of the above we are expected to exercise substantial control, so that a coefficient of +.80 or better is usually required for the procedure to be considered reliable.

Ascertaining Validity

Recall that the issue of validity is whether a procedure actually measures what it is intended to measure. Believe it or not, one approach to demonstrating validity is simply a researcher's judgement that the procedure is valid. **Face validity** is the extent to which a measurement procedure appears to measure what it's intended to measure. Thus, if we judge that on its "face" a procedure looks reasonable, we have one very limited way of arguing the procedure is valid. For example, a reaction-time task has face validity if it appears to measure the time taken by people to perform a mental operation. An intelligence test has face validity if it appears to measure intelligence.

> ***REMEMBER*** Face validity means that a procedure is valid because it looks valid.

To produce more objective evidence of validity, however, we employ correlational procedures. One approach is to determine the convergent validity of a procedure. **Convergent validity** is the extent to which the scores obtained from one procedure are positively correlated with scores obtained from another procedure that is already accepted as valid. For example, say after developing a new test of creativity, we give the same people both our test and another accepted test. If our test is valid, then there should be a strong, positive correlation between the scores from the two tests. If so, we can argue that both procedures "converge" on and measure creativity.

On the other hand, **discriminant validity** is the extent to which the scores obtained from one procedure are *not* correlated with the scores from another procedure that measures other variables or constructs. Thus, our creativity test is valid if it does not correlate with accepted measures of intelligence or personality. In this case, we would argue that the procedure "discriminates" between what it is and is not intended to measure.

> ***REMEMBER*** With convergent validity, a procedure correlates with other procedures that are valid. With discriminant validity, a procedure does not correlate with other, unintended measures.

Even though two *procedures* correlate with each other, this does not necessarily mean that they reflect the intended *behaviors*. After all, it might be that neither our new creativity test nor the old accepted test actually reflect creativity. Therefore, another approach for demonstrating validity is to correlate the scores from a procedure with an observable behavior. **Criterion validity** is the extent to which a procedure can distinguish between participants on the basis of some behavior. There are two subtypes of criterion validity.

Concurrent validity is the extent to which a procedure correlates with the *present* behavior of subjects. For example, say that our definition of creativity is such that it should be negatively correlated with ability to follow directions. Concurrent validity would be demonstrated if people who scored high on the creativity test were poor at following directions, but those with low creativity scores followed directions well.

On the other hand, **predictive validity** is the extent to which a procedure correlates with a subject's *future* behavior. For example, say we used our creativity test to predict participants' future success at jobs requiring creative skills. The test has predictive validity if, when we later examine their job success, we determine that subjects' actual success scores are positively correlated with their predicted scores.

> ***REMEMBER*** Criterion validity is the extent to which a procedure relates to a specific behavior, either distinguishing the behavior concurrently or predicting future behavior.

TABLE 10.1 Summary of Methods for Ascertaining Reliability and Validity

Reliability	
Inter-rater reliability:	Ratings from two raters are positively correlated.
Test-retest:	Each subject's test and retest scores are correlated positively.
Split-half:	Subjects' scores from half of the trials positively correlate with their scores from the other half of the trials.
Validity	
Face:	Procedure appears valid.
Convergent:	Procedure correlates with other accepted measures.
Discriminant:	Procedure does not correlate with other unintended measures.
Criterion	
Concurrent:	Procedure correlates with a present behavior.
Predictive:	Procedure correlates with a future behavior.

It's important to know the names of the various approaches to reliability and validity, so consult Table 10.1 above.

COMPUTING THE CORRELATION COEFFICIENT

The following sections discuss the three most common linear correlation coefficients: The *Pearson correlation coefficient*, the *Spearman rank-order correlation coefficient*, and the *point-biserial correlation coefficient*. Each coefficient can be a value between 0 and ±1.0, and everything you've seen previously about interpreting a coefficient applies to each one. The major difference among them is that they are calculated differently, because each is designed for different types of variables: As when selecting any other statistical procedure, the specific coefficient to compute in a particular situation depends on the scale of measurement used to measure the variables.

The Pearson Correlation Coefficient

By far the most common correlation coefficient in psychological research is the Pearson correlation coefficient. The **Pearson correlation coefficient** is used to describe the linear relationship between two interval or ratio variables. (Technically, this statistic is the Pearson Product Moment Correlation Coefficient, but it's usually called the Pearson correlation coefficient. It was invented by Karl Pearson.) The symbol for the Pearson correlation coefficient is r. When you see r, think "relationship." All of the example coefficients in previous sections involved r.

The statistical basis for r is that it compares how consistently each value of Y pairs with each value of X in a linear fashion. Previously, we've seen that to compare scores from different variables, we transform the scores into z-scores. Essentially, the calculation of r transforms each Y score into a z-score (call it z_Y), transforms each X score into a z-score (call it z_X), and then determines the "average" amount of correspondence between the z_Ys and the z_Xs. The Pearson correlation coefficient is defined as:

$$r = \frac{\Sigma(z_X z_Y)}{N}$$

Mathematically, multiplying each z_X times the corresponding z_Y of the pair, summing the products, and then dividing by N produces the average correspondence between the pairs, which is the correlation coefficient.

Luckily there's an easier way to compute r. To derive a computational formula from the above definitional formula, we replace the symbols z_X and z_Y with their formulas, and then, for each z, replace the symbols for the mean and standard deviation with their formulas. As you can imagine, this produces a monster of a formula. After reducing the formula, we have the smaller monster below.

THE COMPUTATIONAL FORMULA FOR THE PEARSON CORRELATION COEFFICIENT IS

$$r = \frac{N(\Sigma XY) - (\Sigma X)(\Sigma Y)}{\sqrt{[N(\Sigma X^2) - (\Sigma X)^2][N(\Sigma Y^2) - (\Sigma Y)^2]}}$$

As an example, say that as part of a health-psychology study, we collected scores from ten people on the variables of the number of times they visited a doctor in the last year and the number of glasses of orange juice they drink daily. To describe the linear relationship between the ratio scores of juice-drinking and doctor visits, we will compute r. Table 10.2 shows a good way to set up the data. For the computational formula

TABLE 10.2 Sample Data for Computing the r Between Orange Juice Consumed (the X Variable) and Doctor Visits (the Y Variable)

	Glasses of juice per day		*Doctor visits per year*		
Subject	X	X^2	Y	Y^2	XY
1	0	0	8	64	0
2	0	0	7	49	0
3	1	1	7	49	7
4	1	1	6	36	6
5	1	1	5	25	5
6	2	4	4	16	8
7	2	4	4	16	8
8	3	9	4	16	12
9	3	9	2	4	6
10	4	16	0	0	0
$N = 10$	$\Sigma X = 17$ $(\Sigma X)^2 = 289$	$\Sigma X^2 = 45$	$\Sigma Y = 47$ $(\Sigma Y)^2 = 2209$	$\Sigma Y^2 = 275$	$\Sigma XY = 52$

for r we need the values of ΣX, ΣX^2, $(\Sigma X)^2$, ΣY, ΣY^2, $(\Sigma Y)^2$, ΣXY, and N. First find each XY by multiplying each X times its corresponding Y, as shown in the far right column in Table 10.2. Then, summing the columns gives ΣX, ΣX^2, ΣY, ΣY^2, and ΣXY. Squaring ΣX and ΣY gives $(\Sigma X)^2$ and $(\Sigma Y)^2$.

Then we put these quantities in the formula for r. Thus,

$$r = \frac{N(\Sigma XY) - (\Sigma X)(\Sigma Y)}{\sqrt{[N(\Sigma X^2) - (\Sigma X)^2][N(\Sigma Y^2) - (\Sigma Y)^2]}}$$

becomes

$$r = \frac{10(52) - (17)(47)}{\sqrt{[10(45) - 289][10(275) - 2209]}}$$

Let's compute the numerator first. Multiplying 10 times 52 gives 520, and 17 times 47 is 799. Rewriting the formula, we have

$$r = \frac{520 - 799}{\sqrt{[10(45) - 289][10(275) - 2209]}}$$

Complete the numerator by subtracting 799 *from* 520, which is -279. (Note the negative sign.)

Now compute the denominator. First we perform the operations within each bracket. In the left bracket, 10 times 45 is 450, and from that we subtract 289, obtaining 161. In the right bracket, 10 times 275 is 2750, and from that we subtract 2209, obtaining 541. Rewriting one more time, we have

$$r = \frac{-279}{\sqrt{[161][541]}}$$

Now multiply the quantities in the brackets together: 161 times 541 equals 87,101. After taking the square root of 87,101, we have

$$r = \frac{-279}{295.129}$$

We divide, and there you have it: $r = -.95$.

This r is not greater than ± 1, so our calculations *may* be correct. Also, we have computed a negative r, and in the raw scores we see that there is a negative relationship: As orange juice scores increase, number of doctor visits decreases. (If you have any doubt, make a scatterplot.) Had this been a positive relationship, the numerator of the formula would not contain a negative number and r would not be negative.

Thus, we conclude that there is a negative linear relationship between juice-drinking and doctor visits. On a scale of 0 to ± 1, where 0 is no relationship and ± 1 is a perfect linear relationship, this relationship is a $-.95$. Relatively speaking, this is a very strong linear relationship: Each amount of orange juice is associated with one relatively small range of doctor visits, and as juice scores increase, doctor visits consistently decrease. Now, as usual, we proceed to interpret this result psychologically. (If the correlation were this large in real life, we'd all be drinking a lot more orange juice, incorrectly thinking that this would prevent doctor visits.)

The Spearman Rank-Order Correlation Coefficient

Sometimes data involve ordinal or rank-order scores (first, second, third, etc.). The **Spearman rank-order correlation coefficient** describes the linear relationship between two variables measured using ranked scores. The symbol for the Spearman correlation coefficient is r_S (the s stands for Charles Spearman, who invented this one).

Recall that in psychological research, ranked scores often arise because a variable is difficult to measure quantitatively. Therefore, we evaluate each participant by making qualitative judgments, and then use these judgments to rank-order the subjects. We use r_S to correlate the ranks on two such variables. If we want to correlate one ranked variable with one interval or ratio variable instead, we transform the interval or ratio scores into ranked scores (we might rank the participant with the highest interval score as 1, those with the second highest score as 2, and so on). Either way that we obtain the ranks, r_S tells us the extent to which participants' ranks on one variable consistently match the ranks on the other variable to form a linear relationship. If every subject has the same rank on both variables, r_S will equal $+1.0$. If everyone's rank on one variable is the opposite of their rank on the other variable, r_S will equal -1.0. If there is only some degree of consistent pairing of the ranks, r_S will be between 0 and ± 1.0, and if there is no consistent pairing, r_S will equal 0.

Because r_S describes the consistency with which ranks match, or agree, one use of r_S is to determine the extent to which two observers agree when they rank participants: That is, to determine the observers' inter-rater reliability. For example, say we employ two observers to determine the aggressiveness of children after they have watched a confederate "beat up" a Bobo-the-clown punching bag doll (as in Bandura, Ross, and Ross, 1961). After observing how each child plays with Bobo, each observer assigns the rank of 1 to his or her choice for most aggressive child, 2 to the second-most-aggressive child, and so on. Figure 10.9 shows data the two observers might produce for 9 subjects. In creating the scatterplot and computing r_S, we treat each observer as a variable: The scores on one variable are the rankings assigned by one observer to the children, and the scores on the other variable are the rankings assigned by the other observer. Judging from the scatterplot, it appears that there is a positive relationship here. To describe this relationship, we compute r_S.

FIGURE 10.9 Sample Data for Computing r_S Between Rankings of Observer A and Rankings of Observer B

Subject	*Observer A:* *X*	*Observer B:* *Y*
1	4	3
2	1	2
3	9	8
4	8	6
5	3	5
6	5	4
7	6	7
8	2	1
9	7	9

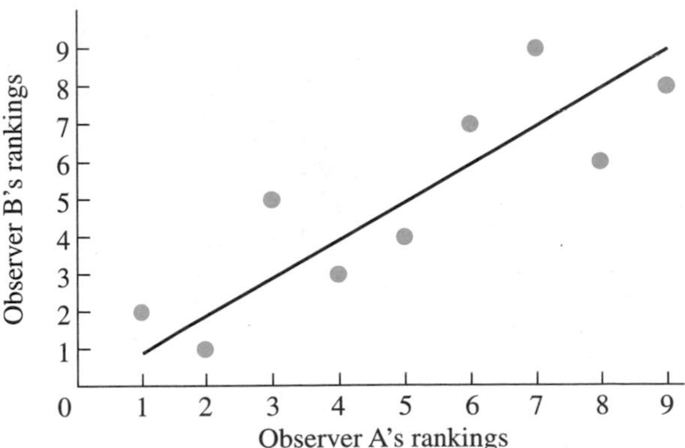

THE COMPUTATIONAL FORMULA FOR THE SPEARMAN RANK-ORDER CORRELATION COEFFICIENT IS

$$r_s = 1 - \frac{6(\Sigma D^2)}{N(N^2 - 1)}$$

N is the number of pairs of ranks, and D is the difference between the two ranks in each pair. (Because of the mathematical properties of ranks, the formula *always* contains the 6 in the numerator.)

When using this formula, first arrange the data as shown in Table 10.3. The column labeled D contains the difference between the rankings in each pair. You can either subtract each X from the corresponding Y, or, as shown here, subtract each Y from the corresponding X. In the rightmost column, after finding the Ds, compute D^2 by squaring the difference in each pair. Finally, determine the sum of the squared differences, ΣD^2 (here ΣD^2 is 18). To compute r_s, you also need N, the number of X–Y pairs (here $N = 9$), and N^2 ($9^2 = 81$). Placing these quantities in the formula, we have

$$r_s = 1 - \frac{6(\Sigma D^2)}{N(N^2 - 1)} = 1 - \frac{6(18)}{9(81 - 1)}$$

In the numerator, 6 times 18 equals 108. In the denominator, 81 − 1 is 80, and 9 times 80 is 720. Now

$$r_s = 1 - \frac{108}{720}$$

After dividing,

$$r_s = 1 - .15$$

Subtracting yields

$$r_s = +.85$$

TABLE 10.3 Data Arrangement for Computing r_s

Subject	***Observer A:*** X	***Observer B:*** Y	D	D^2
1	4	3	1	1
2	1	2	−1	1
3	9	8	1	1
4	8	6	2	4
5	3	5	−2	4
6	5	4	1	1
7	6	7	−1	1
8	2	1	1	1
9	7	9	−2	4
				$\Sigma D^2 = 18$

Thus, on a scale of 0 to ± 1.0, the rankings form a linear relationship to the extent that $r_S = +.85$. This indicates that a child receiving a particular ranking from one observer tended to receive close to the same ranking from the other observer, so the observers did demonstrate high inter-rater reliability. Notice, however, that a high negative r_S (such as $-.85$) would not indicate high inter-rater reliability, because then low ranks by one observer would be paired with high ranks by the other observer, and vice versa.

Recognize that you cannot calculate r_S until after you have dealt with any tied ranks that occur in the data. A **tied rank** occurs when two participants receive the same rank-order score on the *same* variable. The problem with tied ranks is that they result in an incorrect value of r_S. Therefore you must first resolve—correct—these tied ranks. As an example, say that we wish to correlate the finishing positions of the runners in two races. Table 10.4 shows such data with runners A and B tied for first place in race *Y*.

Resolve tied ranks using the following logic: If runners A and B had not tied for first place, then one of them would have been first and one would have been second. Therefore, *assign to each participant at a tied rank the mean of the ranks that would have been used had there not been a tie*. The mean of 1 and 2 is 1.5, so, as in Table 10.4, Runners A and B are each assigned a new *Y* score of 1.5. Now, in a sense, you have used up first and second place (1 and 2), so runner C is assigned a new *Y* of 3. (After all, he was the third person to cross the finish line.) Likewise, assign runner D the new rank of 4. If there had been additional runners, we would assign them new ranks based on what we did above. For example, say there were four additional runners. If runners E, F, and G were originally tied for fourth place in race *Y*, they would now be tied for fifth. We'd resolve this tie by assigning them the mean of 5, 6, and 7, and runner H would be ranked 8.

Once you have resolved all ties in the *X* and *Y* variables, compute r_S using the new ranks and the above formula.

The Point-Biserial Correlation Coefficient

Sometimes we want to correlate the scores from a continuous interval or ratio variable with the scores from a dichotomous variable (recall, this is a variable having only two categories). The **point-biserial correlation coefficient** describes the linear relationship between the scores from one continuous variable and one dichotomous variable. The symbol for the point-biserial correlation coefficient is r_{pb} (the pb stands for point-biserial, and no, Mr. Point and Mr. Biserial didn't invent this one).

TABLE 10.4 Sample Data Containing Tied Ranks

Runner	*Race X*	*Race Y*		*To resolve ties*		*New Y*
A	4	1}	. . . →	Tie uses up ranks 1 and 2,	. . . →	{1.5
B	3	1}		becomes 1.5		{1.5
C	2	2}	. . . →	Becomes 3rd	. . . →	{.3
D	1	3}	. . . →	Becomes 4th	. . . →	{.4
.	.	.				.
.	.	.				.
.	.	.				

Say that we wish to correlate the dichotomous variable of gender (male/female) with the interval scores from a personality test. We cannot quantify "male" and "female," so first we assign numbers to represent these categories. We can assign any numbers, but an easy system is to use 1 to indicate male and 2 to indicate female. Think of each number as indicating whether a person scored "male" or "female." Then r_{pb} will describe how consistently certain personality test scores are paired with each gender score.

THE COMPUTATIONAL FORMULA FOR THE POINT-BISERIAL CORRELATION COEFFICIENT IS

$$r_{pb} = \left(\frac{\overline{Y}_2 - \overline{Y}_1}{S_Y}\right)(\sqrt{pq})$$

Always call the dichotomous variable the X variable and the interval or ratio variable the Y variable. Then $\overline{Y}_1$ stands for the mean of the Y scores for one of the two groups of the dichotomous variable (in our example, $\overline{Y}_1$ will be the mean personality score for males). The symbol $\overline{Y}_2$ stands for the mean of the Y scores for the other group ($\overline{Y}_2$ will be the mean personality score for females). The S_Y is the standard deviation of *all* Y scores in the data. The p stands for the proportion of the sample that is in one of the groups of the dichotomous variable. The q stands for the proportion of the subjects in the other group. Each proportion is equal to the number of individuals in the group divided by the total N of the study.

Say that we tested 10 people and obtained the data shown in Figure 10.10. First compute S_Y, the standard deviation of Y. Placing into the formula the data in Figure 10.10,

$$S_Y = \sqrt{\frac{\Sigma Y^2 - \frac{(\Sigma Y)^2}{N}}{N}} = \sqrt{\frac{26019 - \frac{(503)^2}{10}}{10}} = 8.474$$

Next, the first four people scored "male," and their mean test score, $\overline{Y}_1$, is 45.50. The remaining six people scored "female," and their mean test score, $\overline{Y}_2$, is 53.50. The S_Y for all Y scores is 8.474. Let's call p the proportion of people who scored "male," so p is 4/10, or .40. Then q is the proportion of participants scoring "female," which is 6/10, or .60.

Now putting these values into the formula for r_{pb} gives:

$$r_{pb} = \left(\frac{\overline{Y}_2 - \overline{Y}_1}{S_Y}\right)(\sqrt{pq}) = \left(\frac{53.50 - 45.50}{8.474}\right)(\sqrt{(.40)(.60)})$$

Subtracting 45.50 from 53.50 gives 8.00, so

$$r_{pb} = \left(\frac{8.00}{8.474}\right)(\sqrt{(.40)(.60)})$$

Dividing 8.00 by 8.474 gives .944. Also, .40 times .60 is .24, and the square root of .24 is .489. Thus, we have

$$r_{pb} = .944(.489)$$

FIGURE 10.10 Example Data for Computing r_{pb}

Subject	*Gender:* *X*	*Test:* *Y*	
	Males		
1	1	50	
2	1	38	$\overline{Y}_1 = 45.50$
3	1	41	
4	1	53	
	Females		
5	2	60	
6	2	50	
7	2	44	
8	2	68	$\overline{Y}_2 = 53.50$
9	2	53	
10	2	46	
$N = 10$	2	$\Sigma Y = 503$	

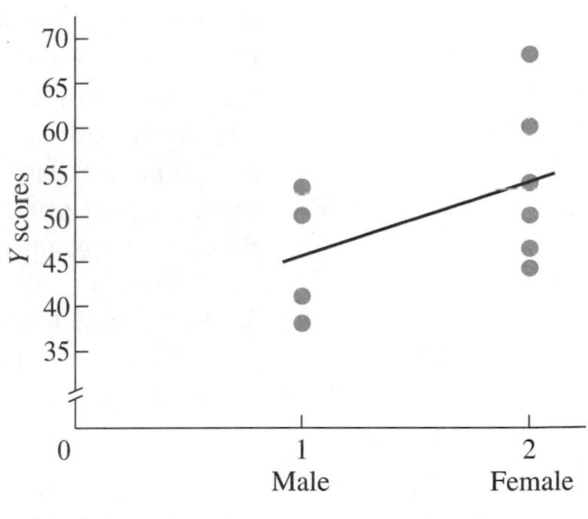

Multiplying, we find that

$$r_{pb} = +.462$$

Thus, with rounding, r_{pb} is $+.46$. We interpret this as indicating that, on a scale of 0 to ± 1.0, we have a reasonably strong relationship: Close to one value of test scores tend to be associated with one gender, and test scores close to a different value tend to be associated with the other gender.

Note that in this example the dichotomous variable is a qualitative variable, so the nominal scores of 1 and 2 do not actually reflect more or less of the gender variable. Therefore, the r_{pb} is positive only because we arbitrarily assigned a 1 to males and a 2 to females. Had we assigned females a 1 and males a 2, their locations on the X axis of the scatterplot would be reversed and we would have a negative relationship. Likewise, the formula is $\overline{Y}_2 - \overline{Y}_1$, so above we had $53.50 - 45.50$. Had we chosen to call the female mean $\overline{Y}_1$ and the male mean $\overline{Y}_2$, we would have had $45.50 - 53.50$, which would have resulted in a negative r_{pb} of $-.46$. Thus, for any qualitative X variable, the absolute value of r_{pb} will accurately describe the strength of the relationship, but whether it is positive or negative depends on how you have arranged the data.

Creating a Powerful Correlational Design

Recall that we always try to create a powerful design. This boils down to obtaining the strongest, most convincing relationship in our sample, so we won't miss a relationship that occurs in nature. This logic also applies to correlational research. Here, obtaining the most convincing relationship translates into obtaining the largest correlation coefficient possible.

As we've seen, the strength of a relationship and the size of the correlation coefficient depends on having the minimum variability or differences in Y scores at each X.

There are two things that produce such error variance. First, individual differences will operate: Way back in our coffee study, for example, some people will exhibit different nervousness scores for the same amount of coffee, because of individual differences in physiology, arousal levels, and so on. Second, any measurement will lack perfect reliability and validity, and there will be momentary fluctuations in variables that influence participants differently. Thus, our participants may incorrectly report their coffee consumption, the environment may differentially influence them, the researcher may behave inconsistently, or there may be inconsistency in measuring nervousness. Any of these factors can produce a less consistent relationship and a lower correlation coefficient. Thus, as with experiments, the key to obtaining a powerful correlational design is to obtain reliable and valid measurements, with a minimum of uncontrolled extraneous variables that may produce variability.

In addition, we increase power by avoiding the restriction of range problem.

The Restriction of Range Problem

Recall from Chapter 4 that we always seek a sensitive measurement procedure, and that part of obtaining a sensitive measure is to avoid restricting the range. Restricted range occurs when the range between the lowest and highest scores on one or both variables is small, or restricted. In addition, restricted range reduces the accuracy of the correlation coefficient. To understand this, look at Figure 10.11. Recall that the correlation coefficient reflects the spread in *Y* at each *X* *relative* to the overall spread in all *Y*s. When we consider the full range of *X* scores in the figure, the spread in the *Y* scores at each *X* is small relative to the overall variability in *Y*, and the data form a narrow ellipse that hugs the regression line. Therefore, *r* will be relatively large, and we will correctly conclude that there is a strong relationship between these variables.

If, however, we restrict the range of the *X* scores by collecting scores only between score A and score B in Figure 10.11, we'll have just the data in the shaded part of the scatterplot. Now the spread in *Y*s at each *X* is large relative to the overall spread of all

FIGURE 10.11 Scatterplot Showing Restriction of Range in *X* Scores

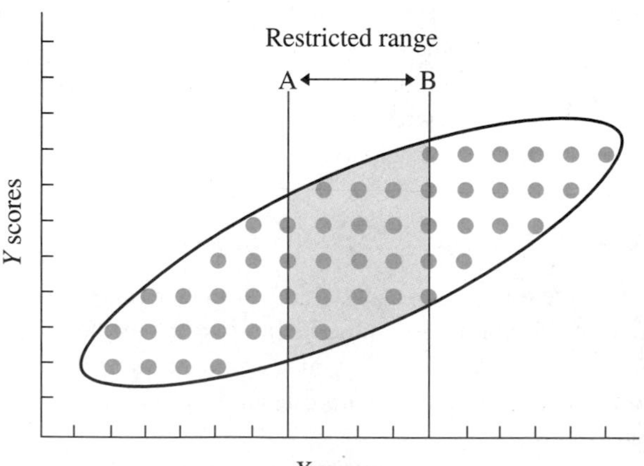

*Y*s in the shaded area. This makes the overall shape of the scatterplot now more circular. Therefore, a correlation coefficient using only the data in the shaded portion will be relatively small and we will conclude that there is a weak relationship here. This conclusion will be erroneous, however, because we would have found a much stronger relationship if we had not restricted the range.

Thus, restriction of range leads to an erroneous *underestimate* of the degree of association between two variables. (Because either variable can be called the *X* or *Y* variable, restricting the range of *Y* has the same effect as above.) This is important, not only because we always want accuracy, but also because restricted range may result in such a small correlation coefficient that we erroneously conclude the variables are not related in nature, when in fact they are.

How do you avoid restricting the range? Previously we saw that the measurement task should produce a wide range of scores and, especially, avoid ceiling and floor effects (where scores are all very high or all very low). In addition, you must not be too selective when obtaining participants. Thus, if you're interested in the relationship between a participant's high school grade average and subsequent salary, do not restrict the range of grades by including only honor students: Measure all students to get the entire range of grades. If you're correlating personality types with degree of emotional problems, don't restrict the sample to college students only. People with severe emotional problems tend not to be admitted to or remain in college, so using a college sample eliminates their scores and thus restricts the range. Instead, include the full range of people from the general population. In all cases, the goal is to allow a wide range of scores to occur on both variables, so that you have a complete description of the relationship.

CORRELATIONS IN THE POPULATION

As you know, ultimately we use a sample to describe the population we would find if we could measure it. After we perform the appropriate inferential statistics, we use a sample correlation coefficient to estimate the correlation coefficient we would obtain if we could correlate the *X* and *Y* scores of everyone in the population.

The symbol for a population correlation coefficient is ρ. This is the Greek letter rho (the Greek r). Technically, ρ is the symbol for the population correlation coefficient when the Pearson r is used. Thus, we compute r for a random sample, which gives us an estimate of the value of ρ. If the data involve ranked scores, we compute r_s and have an estimate of the population coefficient symbolized by ρ_s. If the data involve one continuous variable and one dichotomous variable, we compute r_{pb} to estimate ρ_{pb}.

A population correlation coefficient is interpreted in the same way as a sample correlation coefficient. Thus, ρ stands for a number between 0 and ± 1.0, indicating either a positive or a negative linear relationship in the population. The larger the absolute value of ρ, the stronger the relationship: The more that one and only one value of *Y* tends to be associated with each *X* and the more closely the scatterplot for the population hugs the regression line. We interpret ρ_s and ρ_{pb} in the same manner.

PUTTING IT ALL TOGETHER

By far, the importance of a correlation coefficient is that it is one number that allows you to envision and summarize the important information in a scatterplot. If, for example, I tell you that a particular study produced an r equal to $+.70$, *without even seeing the data*, you know the essential aspects of the relationship in terms of its strength and direction. No other type of statistic so directly summarizes a relationship. Therefore, as we'll see in later chapters, even when you conduct an experiment, always think "correlation coefficient" to describe the strength and type of relationship you have observed.

CHAPTER SUMMARY

1. In a *correlational design*, the researcher measures but does not manipulate the variables. The demonstrated relationship is not necessarily a causal relationship.
2. A *scatterplot* is a graph showing the location of each data point formed by a pair of X–Y scores.
3. A scatterplot is summarized by the *regression line*.
4. In a *positive linear relationship*, as the X scores increase, the Y scores tend to increase. In a *negative linear relationship*, as the X scores increase, the Y scores tend to decrease. In a *nonlinear relationship*, as the X scores increase, the Y scores do not only increase or only decrease.
5. A horizontal scatterplot, with a horizontal regression line, indicates no relationship. Sloping scatterplots with regression lines oriented so that Y increases as X increases indicate a positive linear relationship. Sloping scatterplots with regression lines oriented so that Y decreases as X increases indicate a negative linear relationship. Scatterplots producing curved regression lines indicate nonlinear relationships.
6. A *linear correlation coefficient* communicates the *type* of relationship (either positive or negative) and the *strength* of the relationship (the extent to which one value of Y is consistently paired with one and only one value of X).
7. The smaller the absolute value of the coefficient, the greater the variability in the Ys at each X, the greater the vertical width of the scatterplot, and the less accurately Y scores can be predicted from X scores.
8. *Test-retest reliability* indicates that participants tend to obtain the same score when repeatedly tested at different times. *Split-half reliability* indicates that subjects' scores on some trials tend to match their scores on other trials.
9. With *face validity*, a measurement procedure appears valid. With *convergent validity*, a procedure is correlated with other procedures that are already accepted as valid. With *discriminant validity*, a procedure is not correlated with procedures that measure other things.

10. With *criterion validity*, the scores from a measurement procedure correlate with an observable behavior in one of two ways: With *concurrent validity*, a procedure correlates with a current behavior, and with *predictive validity*, a procedure correlates with a future behavior.

11. The *Pearson correlation coefficient*, r, describes the linear relationship between two interval or ratio variables.

12. The *Spearman rank-order correlation coefficient*, r_s, describes the linear relationship between two variables that have been measured using ranked scores.

13. The *point-biserial correlation coefficient*, r_{pb}, describes the linear relationship between scores from one continuous interval or ratio variable and one dichotomous variable.

14. The power of a correlational design is increased by minimizing error variance and avoiding restricting the range of scores, so that the largest possible coefficient is obtained.

15. When the range of scores on one or both variables is restricted, the correlation coefficient underestimates the strength of the relationship that would be found if the range were not restricted.

16. If it passes the appropriate inferential procedure, a sample correlation coefficient is used to estimate the corresponding population correlation coefficient: r is used to estimate ρ, r_s is used to estimate ρ_s, and r_{pb} is used to estimate ρ_{pb}.

PRACTICE PROBLEMS

(Answers for odd-numbered problems are provided in Appendix D.)

1. What is the difference between an experiment and a correlational study in terms of how the researcher: (a) Collects the data? (b) Examines the relationship?
2. What is the advantage of computing a correlation coefficient?
3. What are the two reasons why we can't conclude that we have demonstrated a causal relationship based on correlational research?
4. A researcher has just completed a correlational study, measuring the number of boxes of tissue purchased per week and the number of vitamin tablets consumed per week by each participant. (a) Which is the independent and the dependent variable? (b) Which variable is X and which is Y?
5. (a) What is a scatterplot? (b) What is a regression line?
6. What two characteristics of a linear relationship are described by a correlation coefficient?
7. (a) Define a positive linear relationship. (b) Define a negative linear relationship. (c) Define a curvilinear relationship.
8. As the value of r approaches ± 1.0, what does it indicate about the following? (a) The shape of the scatterplot. (b) The variability of the Y scores at each X. (c) The closeness of Y scores to the regression line. (d) The accuracy with which we can predict Y if X is known.
9. What does a correlation coefficient equal to 0 indicate about the four characteristics in problem 8?

10. For each of the following, indicate whether it is a positive linear, negative linear, or nonlinear relationship: (a) Quality of performance (*Y*) increases with increased arousal (*X*) up to an optimal level, then performance decreases with increased arousal. (b) Heavier jockeys (*X*) tend to win fewer horse races (*Y*). (c) As number of minutes of exercise per week (*X*) increases, dieting individuals lose more pounds (*Y*). (d) The number of bears in an area (*Y*) decreases as the area becomes increasingly populated by humans (*X*).

11. Poindexter sees the data in problem 10, part d, and concludes, "To preserve the bear population we should stop people from moving into bear country." What is the problem with his conclusion?

12. (a) What is the difference between test-retest reliability and split-half reliability? (b) Why would a professor prefer to show the split-half reliability of a college exam instead of the test-retest reliability?

13. (a) What is the difference between convergent and discriminant validity? (b) How does criterion validity differ from the types of validity in (a)? (c) What are the two types of criterion validity, and how is each determined?

14. (a) When do you compute a Pearson correlation coefficient? (b) When do you compute a Spearman coefficient? (c) When do you compute a point-biserial coefficient?

15. For each of the following pairs of variables, give the symbol for the correlation coefficient you should compute. (a) SAT scores and IQ scores. (b) The taste rankings of different types of tea by an expert and a novice. (c) Presence or absence of a head injury and scores on a vocabulary test. (d) Finishing position in a race and number of glasses of liquid consumed during the race.

16. In the correlation between orange juice consumed and number of doctor visits discussed earlier in this chapter, does drinking more orange juice cause people to be healthier so that they don't have to go to the doctor?

17. (a) What does ρ stand for? (b) How is the value of ρ determined? (c) What does ρ tell you?

18. (a) How do you maximize the power of a correlational design? (b) What produces a restricted range? (c) Why should it be avoided? (d) How is it avoided?

19. Foofy and Poindexter investigated the relationship between IQ score and high school grade average, studying students in a government program for exceptionally smart students. They found an $r = +.03$, and concluded that there is virtually no relationship between IQ and grade average. Should you agree or disagree with this conclusion? Is there a problem with their study?

20. A researcher measures the following scores for a group of students where *X* is their number of errors on a math test, and *Y* is their satisfaction with their grades.

Participant	*Errors X*	*Satisfaction Y*
1	9	3
2	8	2
3	4	8
4	6	5
5	7	4
6	10	2
7	5	7

(a) With such ratio scores, what is the relationship here?
(b) How well will he be able to predict satisfaction scores using this relationship?

21. The data below reflect whether a participant is a college graduate (Y or N) and the score he or she obtained on a self-esteem test. To what extent is there a positive or negative linear relationship here?

Participant	College Graduate X	Self-Esteem Y
1	Y	8
2	Y	7
3	Y	12
4	Y	6
5	Y	10
6	N	2
7	N	8
8	N	6
9	N	1
10	N	9

22. In the data below, the X scores reflect students' academic rankings in their freshman class, and the Y scores reflect their rankings in their sophomore class. To what extent do these data form a linear relationship? (Caution: Think before you calculate.)

Participant	Fresh. X	Soph. Y
1	2	3
2	9	7
3	1	2
4	5	7
5	3	1
6	7	8
7	4	4
8	6	5
9	8	6

23. You want to know if a nurse's absences from work in one month (Y) can be predicted by knowing his or her score on a test of psychological "burnout" (X). What do you conclude from the following ratio data?

Participant	Burnout X	Absences Y
1	2	4
2	1	7
3	2	6
4	3	9
5	4	6
6	4	8
7	7	7
8	7	10
9	8	11

24. You hypothesize that students who sit toward the front of a classroom (those with a 0 on the *X* variable) perform better than those who sit toward the back of the classroom (1 on *X*) when given a brief quiz (the *Y* scores). Do these data support your hypothesis? (Call the group with 0s group 2.)

Participant	*Location* *X*	*Quiz* *Y*
1	0	4
2	0	6
3	0	11
4	0	5
5	1	8
6	1	5
7	1	8
8	1	11
9	1	7
10	1	4

25. A researcher observes the behavior of a group of monkeys in the jungle. She determines each monkey's relative position in the dominance hierarchy of the group (with an *X* of 1 being most dominant), and also notes each monkey's relative weight (with a *Y* of 1 being the lightest). What is the relationship between dominance and weight?

Participant	*Dominance* *X*	*Weight* *Y*
1	1	10
2	2	8
3	5	6
4	4	7
5	9	5
6	7	3
7	3	9
8	6	4
9	8	1
10	10	2

SUMMARY OF FORMULAS

1. The computational formula for the Pearson correlation coefficient is

$$r = \frac{N(\Sigma XY) - (\Sigma X)(\Sigma Y)}{\sqrt{[N(\Sigma X^2) - (\Sigma X)^2][N(\Sigma Y^2) - (\Sigma Y)^2]}}$$

where *X* and *Y* stand for the scores on the *X* and *Y* variables and *N* is the number of pairs in the sample.

2. *The computational formula for the Spearman rank-order correlation coefficient is*

$$r_s = 1 - \frac{6(\Sigma D^2)}{N(N^2 - 1)}$$

where N is the number of pairs of ranks and D is the difference between the two ranks in each pair.

3. *The computational formula for the point-biserial correlation coefficient is*

$$r_{pb} = \left(\frac{\overline{Y}_2 - \overline{Y}_1}{S_Y}\right)(\sqrt{pq})$$

where
$\overline{Y}_1$ is the mean of the scores on the continuous variable for one group of the dichotomous variable,
$\overline{Y}_2$ is the mean of the scores on the continuous variable for the other group of the dichotomous variable,
S_Y is the standard deviation of all Y scores,
ρ is the proportion of all subjects in the sample in one dichotomous group and q is the proportion of all subjects in the sample in the other dichotomous group. Each is found by dividing the number of subjects in the group by N, the total number of X–Y pairs in the study.

11

Using Linear Regression to Predict Scores

To understand this chapter:

- From Chapter 7, recall that without additional information, the $\overline{X}$ is used to predict all scores in a sample.
- From Chapter 8, understand that when using the mean to predict scores, variance is the "average error" in predictions.
- From Chapter 10, recall how to interpret r and how it describes how the scores hug the regression line.

Then your goals in this chapter are to learn:

- How a regression line summarizes a scatterplot.
- How the regression equation is used to predict the Y scores at a given X.
- Which statistics measure errors in prediction when using regression.
- How the strength of the relationship determines how accurately Y scores can be predicted.
- What the term "proportion of variance accounted for" means.

When the laws of nature produce a relationship, certain Y scores are naturally paired with certain X scores. Therefore, if we know an individual's X score and the relationship between X and Y, we can predict the corresponding Y score. Statistically, we make such predictions based on the linear regression line. In the following sections we examine the logic behind the regression line and see how to use it to predict scores. Then we'll look at ways of measuring the errors in prediction.

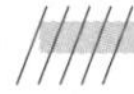

MORE STATISTICAL NOTATION

Be sure that you understand that the variance in Y scores, S_Y^2, and the standard deviation of Y scores, S_Y, are computed in the same ways that we compute these values for X scores. Also recognize that, when graphed, the variability between any two Y scores is reflected by their different vertical locations along the Y-axis. And, that the larger the values of S_Y^2 or S_Y, the more the Y scores are vertically spread out in the scatterplot.

UNDERSTANDING LINEAR REGRESSION

Linear regression is the procedure for describing the best-fitting straight line that summarizes a linear relationship. Therefore, we use this procedure in conjunction with the Pearson correlation. While r is the *statistic* that summarizes the linear relationship, regression produces the *line* on the scatterplot that summarizes the relationship. Always compute r first, however, to determine whether a relationship exists. If $r = 0$ then the regression line is unnecessary: All it would tell you is "Yup, there really is no relationship here." If r is not 0, then linear regression helps to further describe and understand the relationship.

Summarizing the Scatterplot with the Regression Line

An easy way to understand the regression line is to compare it to a line graph. In Chapter 7, we created a line graph by plotting the mean of the Y scores at each X, and then connecting adjacent data points with straight lines. The scatterplot on the left in Figure 11.1 shows the line graph of an experiment with four conditions of X.

FIGURE 11.1 Comparison of a Line Graph and a Regression Line

Each data point is formed by a subject's X–Y pair. Each asterisk (*) indicates the mean Y score at an X.

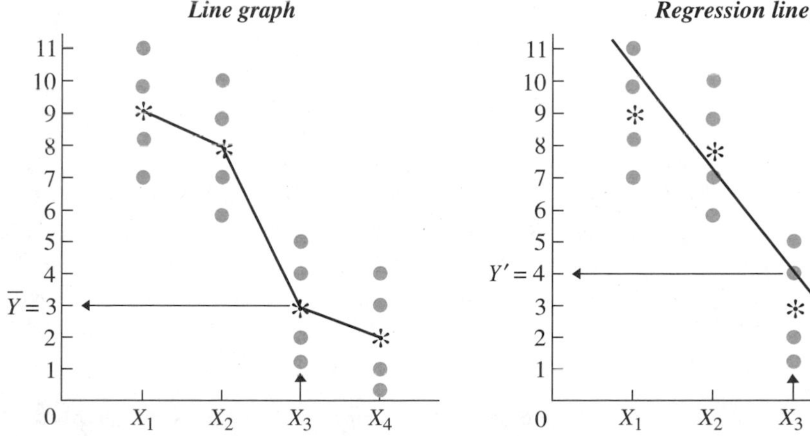

Then for example, as the arrows indicate, the mean of Y at X_3 is 3. Because the mean is the central score, we assume that those participants scoring at X_3 also scored *around* a Y of 3, so: (1) 3 is our best description of their scores, and (2) 3 is our best prediction of Y for anyone else scoring at that X.

It is difficult, however, to see the linear relationship in the above data, because the means do not fall in a straight line. Therefore, as in the graph on the right in Figure 11.1, we can summarize the linear relationship by drawing a regression line. Think of the regression line as a straightened-out version of the line graph: It is drawn so that it comes as close as possible to connecting the mean of Y at each X while still producing a straight line. Although all means are not on the line, the distance that some means are above the line averages out with the distance that others are below the line. Thus, the regression line is the *best-fitting* line, because "on average" it passes through the center of the various Y means. Because each Y mean is located in the center of the corresponding Y scores, by passing through the center of the means, the regression line passes through the center of the Y scores. Thus, the **linear regression line** may be defined as the straight line that summarizes the linear relationship in a scatterplot by, on average, passing through the center of the Y scores at each X.

As usual, this is another descriptive procedure that allows us to summarize and envision data. Here, think of the regression line as reflecting the linear relationship hidden in the data. Because the actual Y scores fall above and below the line, the data only more or less fit this line, but we have no system for drawing a more or less linear relationship. Therefore, the regression line is how we envision what a perfect version of the linear relationship in our data would look like.

You read the regression line in the same way you read a line graph: Travel vertically from any X until you intercept the regression line, then travel horizontally until you intercept the Y axis. For example, as indicated in the graph on the right in Figure 11.1 now the value of Y at X_3 is 4. The symbol for this value is Y', pronounced "Y prime." Like the means of a line graph, each Y' is a summary of the Y scores for that X. The advantage of the regression line, however, is that each Y' is a summary score based on the *entire* linear relationship formed across *all* X–Y pairs in the data. Because the Y scores are evenly spread out around (above and below) the regression line, all Y scores are evenly spread out around (more or less than) each value of Y'. Thus, considering the entire linear relationship in Figure 11.1, we find that participants at X_3 scored around 4, so 4 is our best description of their scores, and 4 is our best prediction of Y for anyone else scoring at that X.

Thus, the symbol Y' stands for a **predicted Y score**: Each Y' is our best prediction of the Y scores at a corresponding X, based on the linear relationship that is summarized by the regression line. (You may also encounter another common symbol for the predicted Y score—$\hat{Y}$—because, believe it or not, there is no one, agreed-upon symbol.)

It is important to recognize that the Y' at a value of X is the value of Y falling *on* the regression line. A line is made up of an infinite series of adjacent dots, so in essence, the regression line is created by plotting the data points formed by pairing all possible values of Y' with the corresponding possible values of X. If we think of the line as reflecting a perfect version of the linear relationship hidden in the data, then each value of Y' is the Y score everyone would have at each X if a perfect version of this relationship were present.

> *REMEMBER* The linear regression line summarizes the linear relationship in a sample, and is used to predict a subject's Y score—Y'—at any X.

Predicting Scores Using Regression

Why are we interested in predicting Y scores? After all, we have the Y score of everyone in the sample right in front of us. The answer is that regression techniques are the primary device for predicting *unknown* scores based on a linear relationship. To do this, first we establish the relationship by measuring a sample of X-Y scores and computing the correlation. Then, we perform inferential procedures for r (described in Chapter 14) to be sure we can draw inferences from the sample. If we can, we use the regression line to determine the Y' for each X. Then we measure the X scores of other subjects who were not in our sample, and the corresponding Y' is our best prediction of their Y scores. Thus, the importance of linear regression is that it allows us to predict unknown Y scores if we know the corresponding X score on a correlated variable.

Regression is usually applied in correlational studies because the observed relationship may be better for predicting behaviors than that found in an experiment: They usually provide greater external validity and a more complete description of the relationship. For this reason, regression techniques are common in applied research, especially when creating "selection tests." A selection test is for selecting people with certain qualifications or attributes. For example, students are taking a selection test when they take the Scholastic Aptitude Test (SAT) to be admitted to college. We know from previous students that SAT scores are somewhat positively correlated with college grades. Therefore, through regression techniques, the SAT scores of those students applying for college are used to predict their future college performance: A student's X score on the SAT leads to a predicted Y' of the student's college grade average. If the predicted grades are too low, the student is not admitted to the college. Similar selection tests are involved when people take a test when applying for a job so that the employer can predict who will be better workers, or when clinical patients take diagnostic tests for identifying those at risk of developing emotional problems. Not surprisingly, predictive validity (discussed in the previous chapter) is of paramount importance when creating selection tests.

The emphasis on prediction in correlation and regression leads to two important terms. In our discussions, we'll use the X variable to predict scores on the Y variable. (There are procedures out there for predicting X scores from Y.) In statistical lingo, when the X variable is used to predict Y scores, X is called the **predictor variable** (it does the predicting). When the unknown scores being predicted are on the Y variable, Y is called the **criterion variable**. Thus, above, SAT scores are the predictor variable and college grade average is the criterion variable. (To remember "criterion," remember that your predicted grades must meet a certain criterion for you to be admitted to the college.)

After determining that there is a correlation in the sample, we create the regression line.

THE LINEAR REGRESSION EQUATION

To draw a regression line, we don't simply eyeball the scatterplot and sketch in something that looks good. Instead, we use the linear regression equation. The **linear regression equation** is the equation that produces the value of Y' at each X and thus defines the straight line that summarizes a relationship. When we plot the data points formed by the X–Y' pairs, they all fall in a perfectly straight line. Then we draw a line connecting the data points and we have the regression line. The regression equation describes two characteristics of the regression line: Its slope and its Y-intercept.

The **slope of a line** is a number that indicates how slanted the line is and the direction in which it slants. Figure 11.2 shows examples of regression lines that have different slopes. When there is no relationship, the regression line is horizontal—such as line A—and then the slope equals 0. A positive linear relationship—such as lines B and C—yields a slope equal to a positive number. Because line C is steeper, its slope is a larger positive number. A negative linear relationship—such as line D—yields a slope that is a negative number.

The ***Y*-intercept of a line** is the value of Y at the point where the regression line intercepts, or crosses, the Y axis. In other words, the Y-intercept is the value of Y' when X equals 0. In Figure 11.2, line B intercepts the Y axis at 2, so its Y-intercept is 2. If we extended line C, it would intercept the Y axis at a point below the X axis, so its Y-intercept is a negative Y score. Because line D reflects a negative relationship, its Y-intercept is the relatively high Y score of 9. Finally, line A exhibits no relationship, and its Y-intercept equals 8. Notice that here the value of Y' for any value of X is always 8. **When there is no relationship, every value of *Y′* equals the *Y*-intercept.**

The regression equation works like this: The slope summarizes the *direction* in which Y scores change as X increases, and also the *rate* at which they change. As in line C in Figure 11.2, the steeply sloped line reflects a relatively large change in Y scores for each increase in X, as compared to, say, line B, which reflects a smaller change in Y for each increase in X. The Y-intercept then indicates the starting point from which the Y scores begin to change as the X scores increase. Thus, together, the slope and intercept describe how, starting at a particular value, the Y scores change with changes in X. Then, the summary of the Y scores at each X is Y'.

The symbol for the slope of the regression line is b. The symbol for the Y-intercept is a.

FIGURE 11.2 Regression Lines That Have Different Slopes and *Y*-Intercepts

Line A indicates no relationship, lines B and C indicate positive relationships having different slopes and Y-intercepts, and line D indicates a negative relationship.

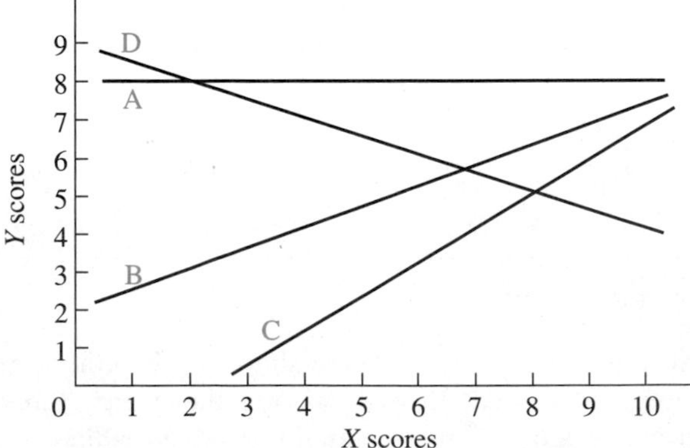

> *THE LINEAR REGRESSION EQUATION IS*
>
> $$Y' = bX + a$$

This formula says that to find the value of Y' for a given X, multiply the slope, b, times X and then add the Y-intercept, a.

As an example, say that a researcher has developed a paper-and-pencil selection test to identify people who will be productive "widget-makers." To demonstrate that the test has predictive validity, the researcher examines whether test scores are correlated with widget-making ability. The researcher gives the test to an unrealistically small N of 11 people and then, in a simulation of a factory setting, measures the number of widgets each participant makes in an hour. Figure 11.3 shows the resulting scatterplot, with test scores as the predictor (X) variable and number of widgets produced per hour as the criterion (Y) variable. The raw scores are also listed, arranged for computing r.

FIGURE 11.3 Scatterplot and Data for Widget-Making Study

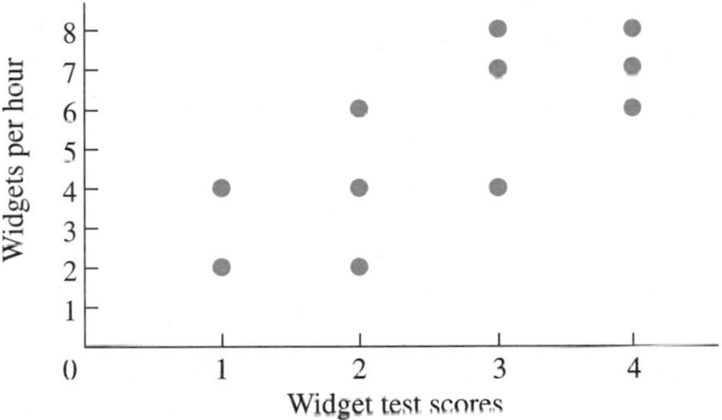

Participant	*Widget test score:* X	*Widgets per hour:* Y	XY
1	1	2	2
2	1	4	4
3	2	4	8
4	2	6	12
5	2	2	4
6	3	4	12
7	3	7	21
8	3	8	24
9	4	6	24
10	4	8	32
11	4	7	28
$N = 11$	$\Sigma X = 29$	$\Sigma Y = 58$	$\Sigma XY = 171$
	$\Sigma X^2 = 89$	$\Sigma Y^2 = 354$	
	$(\Sigma X)^2 = 841$	$(\Sigma Y)^2 = 3364$	
	$\bar{X} = 29/11 = 2.64$	$\bar{Y} = 58/11 = 5.27$	

The first step is to find r:

$$r = \frac{N(\Sigma XY) - (\Sigma X)(\Sigma Y)}{\sqrt{[N(\Sigma X^2) - (\Sigma X)^2][N(\Sigma Y^2) - (\Sigma Y)^2]}}$$

$$r = \frac{11(171) - (29)(58)}{\sqrt{[11(89) - 841][11(354) - 3364]}}$$

The result is $r = +.736$, which rounds to $r = +.74$. Thus, this is a quite strong positive linear relationship.

To predict widget-making scores, we compute the linear regression equation. To do that, we compute the slope and the Y-intercept. Compute the slope first.

Computing the Slope

THE FORMULA FOR THE SLOPE OF THE LINEAR REGRESSION LINE IS

$$b = \frac{N(\Sigma XY) - (\Sigma X)(\Sigma Y)}{N(\Sigma X^2) - (\Sigma X)^2}$$

N is the number of pairs of scores in the sample, and X and Y are the scores in the sample. This is not a difficult formula, because we typically compute the Pearson r first. The numerator in the formula for b is the same as the numerator in the formula for r, and the denominator in the formula for b is the left-hand quantity in the denominator of the formula for r. (An alternative formula for the slope is $b = (r)\,(S_Y / S_X)$.)

Substituting the appropriate values from our computations of r in the widget study into the formula for b, we have

$$b = \frac{N(\Sigma XY) - (\Sigma X)(\Sigma Y)}{N(\Sigma X^2) - (\Sigma X)^2} = \frac{11(171) - (29)(58)}{11(89) - 841}$$

After multiplying and subtracting in the numerator

$$b = \frac{199}{11(89) - 841}$$

After completing the denominator

$$b = \frac{199}{138} = +1.44$$

Thus, the slope of the regression line for the widget-making data is $b = +1.44$. This positive slope indicates a positive relationship, which fits with the positive r of $+.74$. Had the relationship been negative, the formula would have produced a negative number for the slope.

We're not finished yet. Now compute the Y-intercept.

Computing the *Y*-Intercept

THE FORMULA FOR THE Y-INTERCEPT OF THE LINEAR REGRESSION LINE IS

$$a = \overline{Y} - (b)(\overline{X})$$

First, multiply the mean of all X scores, $\overline{X}$, times the slope of the regression line, b. Then subtract that quantity from the mean of all Y scores, $\overline{Y}$. For the widget-making data, b is +1.44 and from Figure 11.3, $\overline{Y}$ is 5.27 and $\overline{X}$ is 2.64. Filling in the formula for a, we have

$$a = 5.27 - (+1.44)(2.64)$$

After multiplying

$$a = 5.27 - (+3.80) = +1.47$$

Thus, the Y-intercept of the regression line for the widget-making study is $a = +1.47$.

We're still not finished!

Describing the Linear Regression Equation

Once you have computed the Y-intercept and the slope, rewrite the regression equation, substituting the computed values for a and b. Thus, for the widget-making study,

$$Y' = +1.44X + 1.47$$

This is the finished regression equation that describes the regression line for the relationship between widget test scores and widgets-per-hour scores.

Putting all of this together, the preceding computations are summarized in Table 11.1.

We're still not finished! The final step is to graph the regression line.

Plotting the Regression Line

To plot the regression line, you need some pairs of X and Y' scores to use as data points. Therefore, choose some values of X, insert each in the finished regression equation, and calculate the value of Y' for that X. Actually, you need only two data points to draw a straight line: One where X is low and one where X is high. (An easy low X to use is $X = 0$ because then Y' equals the Y-intercept, a.)

To see how the calculations work, we'll compute Y' for all X scores from the widget-making study. We begin with the finished regression equation:

$$Y' = +1.44X + 1.47$$

First we find Y' for $X = 1$, so

$$Y' = +1.44(1) + 1.47$$

TABLE 11.1 Summary of Computations for the Linear Regression Equation

1. Compute r.
2. Compute the slope, where $b = \dfrac{N(\Sigma XY) - (\Sigma X)(\Sigma Y)}{N(\Sigma X^2) - (\Sigma X)^2}$.
3. Compute the Y-intercept, where $a = \overline{Y} - (b)(\overline{X})$.
4. Substitute the values of a and b into the formula for the regression equation $Y' = (b)(X) + a$.

Multiplying 1 times +1.44 and adding 1.47 yields a Y' of 2.91. Thus, we predict that anyone scoring 1 on the widget-making test will make 2.91 widgets per hour. Using the same procedure, we obtain the values of Y' for the remaining X scores of 2, 3, and 4. These are shown in Figure 11.4.

To graph the regression line, simply plot the data points for the X-Y' pairs and draw the line. (Note that as shown, we typically do not include the scatterplot, nor do we draw the regression line through the Y-intercept.)

Now we're finished. (Really.)

Using the Regression Equation to Predict *Y* Scores

The Y' for a particular X is the predicted Y score for all participants having that X score. Above, for example, people scoring an X of 1 had a Y' of 2.91. Therefore, we predict that anyone else not in our sample who scores an X of 1 will also have a Y' of around 2.91. Further, we can compute Y' for any value of X that falls within the range of Xs observed in the sample, even if it's a score that was not obtained by our original subjects. No participant obtained a test score of 1.5. Yet, inserting an X of 1.5 into the regression equation yields a Y' of 3.63, so we predict that anyone scoring an X of 1.5 will have a Y of 3.63. (Notice that we'd obtain the same result from Figure 11.4 if we could see decimal scores on the graph: If we traveled vertically from an X of 1.5 to the regression line and then traveled horizontally to the Y axis, we would find Y' is 3.63.)

That's all there is to computing the regression line and predicting scores. All, that is, except for the fact that any prediction we make may be wrong. Therefore, in the remainder of the chapter we'll deal with describing the amount of error in the predictions. This topic is important for two reasons. First, a complete description of a relationship includes the descriptive statistics that summarize the error we have when using linear regression to predict Y scores. Second, understanding how to describe the error in predictions when using the regression equation sets the stage for understanding how to describe the errors in predictions when using other statistical procedures. (In other words, the following topics will not go away.)

FIGURE 11.4 Regression Line for Widget-Making Study

Widget test scores: X	*Predicted widgets per hour: Y*
1	2.91
2	4.25
3	5.79
4	7.23

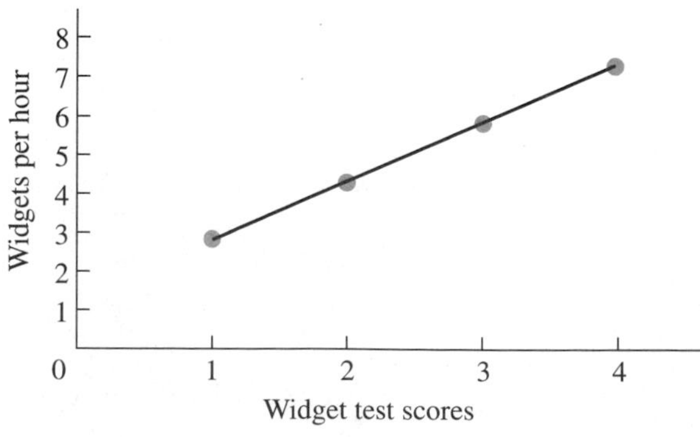

DESCRIBING THE ERRORS IN PREDICTION

Recall that to describe the amount of error we expect in our predictions, we first describe the errors when predicting the actual Y scores in a sample: We'll pretend we don't know the scores, predict them, and then compare the predicted Y' scores to the actual Y scores. The predictions for some participants will be close to their actual Y scores, while predictions for others may contain considerably more error. Therefore, to summarize the error across the entire relationship, we compute something like the average error in the predictions.

The error in any single prediction is the amount that a participant's actual score, Y, differs, or *deviates*, from the corresponding predicted Y' score: In symbols, this is $(Y - Y')$. To find the average error, we find something like the average of the deviations (yes, here we go again). The first step is to find the deviations for all participants. We compute Y' for each individual in the data and then subtract each Y' from its corresponding Y score. Then we would like to sum these differences, getting $\Sigma(Y - Y')$, and then find the average. But we can't. Like a mean, the Y' scores are in the center of the Y scores, so the actual Ys are equally spread out above and below the Y' scores. Therefore, like a mean, the positive and negative deviations cancel out, and so the sum of the deviations of the Y scores from their Y' scores is always zero. Thus, on the one hand, we like using the regression equation to predict scores because, over the long run, the overestimates and underestimates cancel each other out. On the other hand, the problem is that if $\Sigma(Y - Y')$ is always zero, then the average error is always zero.

To solve this problem, we *square* each deviation. The sum of the squared deviations of $Y - Y'$ is not necessarily zero, so neither is the average squared deviation. (Does this sound familiar?) When we find the average of the squared deviations between the Y and corresponding Y' scores, we are computing a type of variance that describes the "average" spread of the actual Y scores around—above and below—their corresponding Y' scores.

Computing the Variance of the Y Scores Around Y'

The symbol for the variance of a sample of Y scores around Y' is $S^2_{Y'}$. The S^2 indicates sample variance or error, and the subscript Y' indicates it is error when using Y' to predict Y scores.

THE DEFINITIONAL FORMULA FOR THE VARIANCE OF THE Y SCORES AROUND THEIR CORRESPONDING Y′ SCORES IS

$$S^2_{Y'} = \frac{\Sigma(Y - Y')^2}{N}$$

This formula says to subtract each Y' predicted for a participant from his or her corresponding Y score, square each deviation, sum the squared deviations, and then divide by N. The answer is the average squared difference between the Y and corresponding Y' scores.

Remember the widget-making study? Table 11.2 shows the actual X and Y scores obtained, as well as the Y' scores for each X produced by the regression equation. In the column labeled $Y - Y'$, we subtract each Y' from the corresponding Y. In the column labeled $(Y - Y')^2$, we square each difference. Then we sum the squared differences to find $\Sigma(Y - Y')^2$.

Filling in the formula for $S^2_{Y'}$ we have

$$S^2_{Y'} = \frac{\Sigma(Y - Y')^2}{N} = \frac{22.07}{11}$$

After dividing,

$$S^2_{Y'} = 2.006$$

With rounding,

$$S^2_{Y'} = 2.01$$

Thus, in these data the average squared difference between the actual Y scores and their corresponding values of Y' is 2.01. In other words, when using variance to measure error, we are "off" by something like an "average" of 2.01 when we predict participants' widget-making score (Y) based on their test score (X).

Recall that differences between a predicted score and the corresponding actual scores goes by the generic name of *error variance*. Therefore, $S^2_{Y'}$ describes the error variance when using regression to predict Y scores: It is literally a formula for variance that measures error in prediction. (The non-generic name for error variance in regression is "the residual variance" or just "the residual.")

REMEMBER $S^2_{Y'}$ is a way to describe the "average error" when predicting Y scores using the corresponding Y' scores.

TABLE 11.2 Widget-Making Data with Computed Y' Scores

Participant	*Widget test score:* X	*Widgets per hour:* Y	*Predicted widgets:* Y'	$Y - Y'$	$(Y - Y')^2$
1	1	2	2.91	−.91	.83
2	1	4	2.91	1.09	1.19
3	2	4	4.35	−.35	.12
4	2	6	4.35	1.65	2.72
5	2	2	4.35	−2.35	5.52
6	3	4	5.79	−1.79	3.20
7	3	7	5.79	1.21	1.46
8	3	8	5.79	2.21	4.88
9	4	6	7.23	−1.23	1.51
10	4	8	7.23	.77	.59
11	4	7	7.23	−.23	.05
$N = 11$		$\Sigma Y = 58$ $\Sigma Y^2 = 354$ $(\Sigma Y)^2 = 3364$			$\Sigma(Y - Y')^2 = 22.07$

In statistical language, the formulas for computing the regression line are called the *least-squares regression method.* This is because with them, the sum of the squared deviations $[\Sigma(Y - Y')^2]$ is zero, the least that it can be. The term "sum of squared deviations" is shortened to "squares." The least-squares method produces *squares* between Y and Y' that is the *least* that it can be. Any other method leads to greater error in predictions, resulting in larger differences between Y and Y' and a larger value of $S^2_{Y'}$.

Using the definitional formula for $S^2_{Y'}$ is very time-consuming. If in place of Y' in the definitional formula for $S^2_{Y'}$ we put in all of the formulas for finding Y' (for finding a, b, and so on) and then simplified that equation, we would find all of the components for the following.

THE COMPUTATIONAL FORMULA FOR THE VARIANCE OF Y SCORES AROUND Y′ IS

$$S^2_{Y'} = S^2_Y(1 - r^2)$$

This formula says to find the variance of all Y scores in the data (S^2_Y) and to square r. Subtract r^2 from 1 and then multiply the result times S^2_Y. The answer is $S^2_{Y'}$.

In the widget study, r was $+.736$. Using the data from Table 11.2, the S^2_Y is 4.38. Placing these numbers in the above formula gives

$$S^2_{Y'} = 4.38(1 - .736^2)$$

After squaring $+.736$ and subtracting the result from 1, we have

$$S^2_{Y'} = 4.38(.458)$$

so

$$S^2_{Y'} = 2.01$$

Again, we expect to be off by an "average" of about 2.01 when predicting participants' widget-making (Y) using their widget test score (X).

There are, however, the usual problems when interpreting a variance such as $S^2_{Y'}$. Squaring the difference between each Y and Y' produces an unrealistically large number. Also, the error is measured in squared units, so technically, above we were off by 2.01 squared widgets. (Sound familiar?) As usual, the solution is to find the square root of the variance, producing a type of standard deviation. To distinguish the standard deviation found in regression from other standard deviations, we call this one the *standard error of the estimate.*

Computing the Standard Error of the Estimate

The **standard error of the estimate** indicates the amount that the Y scores in a sample differ from their corresponding Y' scores. It is the clearest way to describe the "average" error when using Y' to predict Y scores. The symbol for the standard error of the estimate is $S_{Y'}$.

THE DEFINITIONAL FORMULA FOR THE STANDARD ERROR OF THE ESTIMATE IS

$$S_{Y'} = \sqrt{\frac{\Sigma(Y - Y')^2}{N}}$$

This is the same formula we used previously for the variance of Y scores around Y', except with the added square root sign. Thus, to compute $S_{Y'}$, compute $S^2_{Y'}$ and then find its square root. For example, in the widget-making study, we computed that $S^2_{Y'} = 2.01$. Taking the square root produces an $S_{Y'} = 1.42$.

To create a computational formula, recall that the shortcut to computing the variance of Y around Y' was

$$S^2_{Y'} = S^2_Y(1 - r^2)$$

By taking the square root of each component, we construct the computational formula.

THE COMPUTATIONAL FORMULA FOR THE STANDARD ERROR OF THE ESTIMATE IS

$$S_{Y'} = S_Y\sqrt{1 - r^2}$$

This formula says to find the square root of the quantity $1 - r^2$ and multiply it times S_Y, the standard deviation of the Y scores. The answer is $S_{Y'}$.

For the widget study, S^2_Y was 4.38 so S_Y is 2.093. r was $+.736$. Placing these numbers in the above formula gives

$$S_{Y'} = 2.093\sqrt{1 - .736^2}$$

Squaring $+.736$ yields .542, which when subtracted from 1 gives .458. The square root of .458 is .677. Thus, we have

$$S_{Y'} = 2.093(.677)$$

so

$$S_{Y'} = 1.42$$

Again the standard error of the estimate is 1.42. Because the Y scores measure the variable of widgets per hour, the standard error of the estimate is 1.42 *widgets per hour*. Therefore, we conclude that when using the regression equation to predict the number of widgets produced per hour based on a person's widget test score, we will be wrong by an "average" of about 1.42 widgets per hour.

> *REMEMBER* The standard error of the estimate most clearly describes the amount of error when using the regression equation (and Y') to predict Y scores.

It is appropriate to compute the standard error of the estimate anytime you compute a correlation coefficient, even if you do not perform regression—it is still important to know the average prediction error that regression would produce.

Assumptions of Linear Regression

Understand that because $S_{Y'}$ measures the differences between all Y and Y' scores, it indicates the amount the Y scores are spread out around—above and below—the Y' scores. In order for $S_{Y'}$ to accurately describe this spread, however, we must be able to make two assumptions about how the Y scores are distributed.

First, we assume that the Y scores are equally spread out around the regression line throughout the relationship. This assumption goes by the funny little name of homoscedasticity. **Homoscedasticity** occurs when the Y scores are spread out to the same degree at every X. The scatterplot on the left in Figure 11.5 shows what homoscedastic data from the widget study would look like. Because the vertical distance separating the Y scores is the same at each X, the spread of the Y scores around the regression line—and around each Y'—is the same at each X. Therefore, $S_{Y'}$ will accurately describe this spread, which is the error in predicting Y scores at any X score.

Conversely, the scatterplot on the right in Figure 11.5 shows an example of heteroscedastic data. **Heteroscedasticity** occurs when the spread in Y is not equal throughout the relationship. In such cases, the value of $S_{Y'}$ will not accurately describe the "average" spread or error throughout the entire relationship. In the right in Figure 11.5, for example, $S_{Y'}$ will be much larger than the actual average error in predicting widget-making scores associated with low test scores, and much less than the average error in predicting widget-making scores associated with high test scores.

The second assumption is that the sample of Y scores at each X represents an approximately normal distribution. That is, if we constructed a frequency polygon of the Y scores at each X, we would expect to have a normal distribution centered around Y'. Figure 11.6 illustrates this assumption for the widget-making study. Recall that if a distribution is normal, we expect approximately 68% of all scores to fall between ± 1 standard deviation from the mean. Because $S_{Y'}$ is like a standard deviation, if the Y scores are normally distributed around each Y', we expect approximately 68% of all Y scores to be between $\pm 1S_{Y'}$ from the regression line. Thus, in the widget-making study, the

FIGURE 11.5 Illustrations of Homoscedastic and Heteroscedastic Data

The vertical width of the scatterplot above an X indicates how spread out the corresponding Y scores are. On the left, the Ys have the same spread at each X.

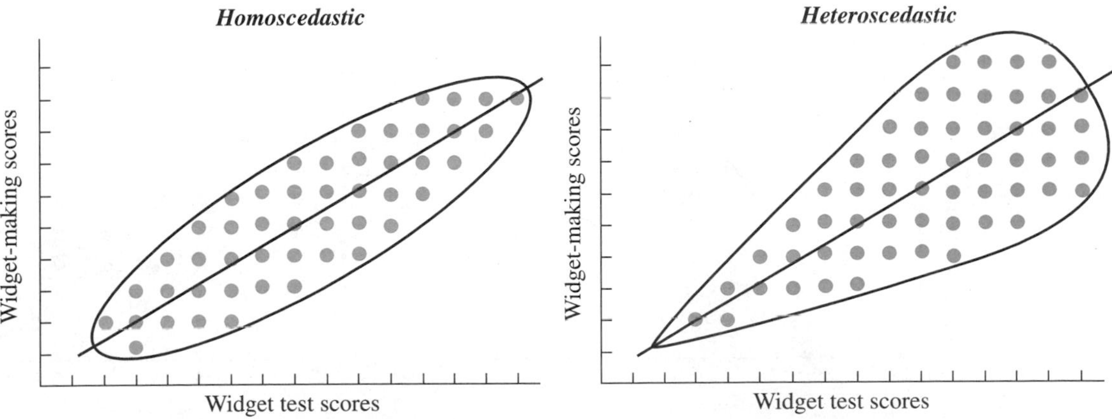

FIGURE 11.6 Scatterplot Showing Normal Distribution of Y Scores at each X

At each X, there is a normal distribution of different Y scores centered around Y′.

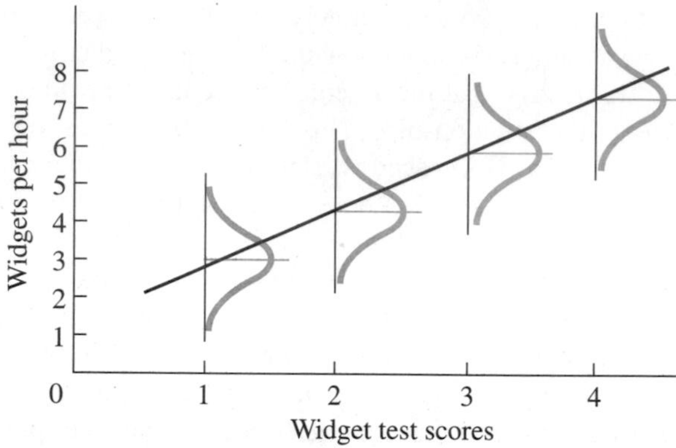

standard error of the estimate is 1.42, so we expect approximately 68% of the actual Y scores to fall between ± 1.42 from their respective Y' scores.

In summary, $S_{Y'}$ and $S^2_{Y'}$ indicate how much the Y scores are spread out around the Y' scores and thus describe the error when predicting Y scores. Now we can discuss how the strength of the relationship (and the size of r) influences the amount of error in predictions.

Strength of the Relationship and Prediction Error

The stronger the relationship, the smaller the error when we use Y' to predict Y scores. The minimum error occurs when r is ± 1.0, because there is no spread or differences in the Y scores at each X and the scatterplot *is* the regression line. The graph on the left in Figure 11.7 shows such a perfect relationship.

FIGURE 11.7 Scatterplots and Regression Lines When $r = +1$ and when $r = 0$

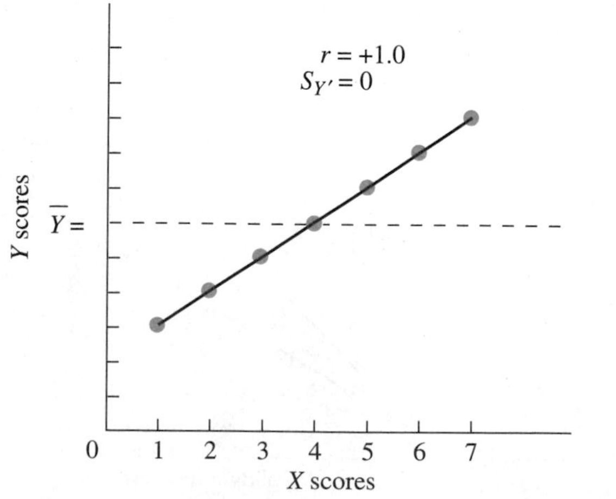

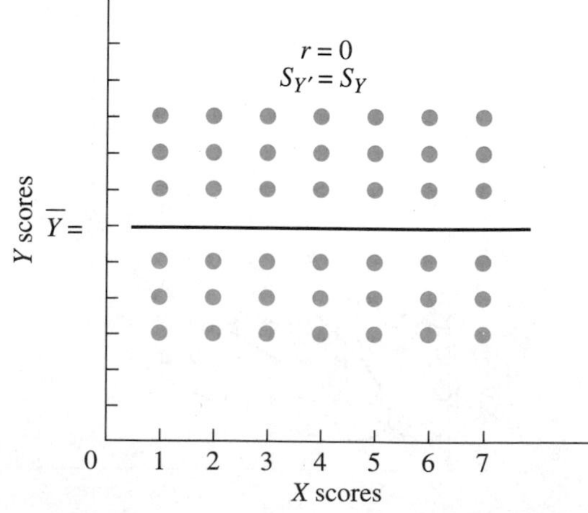

The Y' for each X equals the one Y score every subject actually obtained for that X, so the error in predictions, $S_{Y'}$, equals zero (as does $S^2_{Y'}$).

At the other extreme, r equals 0 in the graph on the right in Figure 11.7. Here the regression line is horizontal, and all values of Y' equal the Y-intercept. Recognize that the Y-intercept is the overall mean of all Y scores: The regression line always passes through the center of the Y scores, and when there is no relationship, that center is always at the sample's mean score. *Thus, when $r = 0$, the Y-intercept is equal to the mean of all Y scores, and this mean of Y is the predicted Y' for all participants.*

Note that this state of affairs simply reflects what we've seen all along: We always predict the sample mean for each participant whenever we have no additional information. When there is no relationship, knowing someone's X score does not provide additional information about the corresponding Y score. So, even if we use the regression equation to compute Y', we still end up predicting the value of $\overline{Y}$ for each participant.

When $r = 0$, the standard error of the estimate is at its maximum, and that is equal to the standard deviation of the Y scores in the sample (to the value of S_Y). Why? Because here we predict the $\overline{Y}$ for everyone in the sample. We can call each prediction Y' or we can call it $\overline{Y}$, but it is the same score. Therefore, we can replace the symbol Y' with the symbol $\overline{Y}$ in the formula for the standard error of the estimate, as shown here:

$$S_{Y'} = \sqrt{\frac{\Sigma(Y - Y')^2}{N}} = \sqrt{\frac{\Sigma(Y - \overline{Y})^2}{N}} = S_Y$$

The formula on the right is the formula for the standard deviation of all Y scores, S_Y. Thus, when $r = 0$, the standard error of the estimate, $S_{Y'}$, is equal to the total variability in the Y scores, S_Y. (Likewise, $S^2_{Y'}$ is equal to S^2_Y).

Thus, the size of $S_{Y'}$ is *inversely* related to the absolute size of r: When r is 0, we have the maximum amount of prediction errors, with $S_{Y'}$ equal to S_Y. When r is ± 1.0, we have the minimum prediction errors, with $S_{Y'}$ equal to 0. When r is between 0 and ± 1.0, $S_{Y'}$ is less than S_Y but greater than zero. Figure 11.8 shows two such intermediate relationships.

FIGURE 11.8 Scatterplots of Strong and Weak Relationships

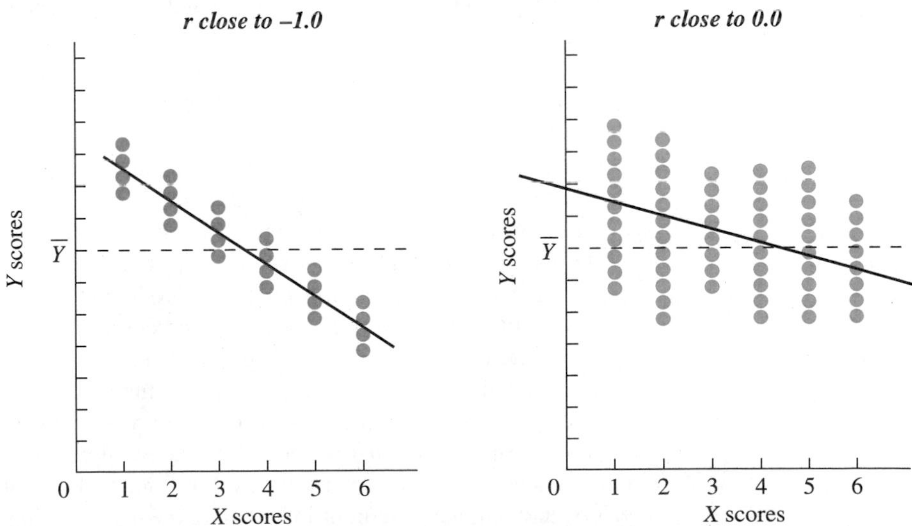

In the scatterplot on the left, r is close to -1.0. The Y scores at each X are not spread out much around the regression line, so the actual Y scores are relatively close to their corresponding Y'. Therefore, our error ($S_{Y'}$ and $S^2_{Y'}$) will be relatively small. Conversely, in the scatterplot on the right, r is much closer to 0. The Y scores at each X are spread out around the regression line, meaning that many of the actual Y scores are relatively far from their corresponding Y'. Therefore, there will be considerable error, and $S_{Y'}$ (and $S^2_{Y'}$) will be relatively large.

> *REMEMBER* The larger the absolute value of r, the smaller the error in prediction if we use regression, $S_{Y'}$.

Thus, any relationship will produce some value of $S_{Y'}$ between 0 (when $r = \pm 1.0$,) and S_Y (when $r = 0$). We use this fact as the basis for evaluating a relationship. This evaluation process goes by the strange (and not quite grammatical) name of computing "the proportion of variance accounted for."

PREDICTING VARIABILITY: THE PROPORTION OF VARIANCE ACCOUNTED FOR

In the widget-making study, r was a relatively large $+.74$. This means that if we know someone's X score, we have a good idea of his or her Y score. The problem is, we don't know in any quantitative way how much information this relationship provides, and so we don't know how useful the relationship is. Although from $S_{Y'}$ we expect our predictions to be off by an "average" of 1.42 (and using $S^2_{Y'}$, by an "average" of 2.01), we don't know if, in the grand scheme of things, such an error rate is a lot or a little: Although 1.42 may seem like a small number, maybe in nature, with these variables, this is actually a huge error.

From this chapter, however, we do *know* one thing: This relationship is better than nothing. If we were unaware of the relationship, we would end up predicting $\overline{Y}$ for everyone and then our error would equal the S_Y of 2.093 (or $S^2_Y = 4.38$). (Because we can always compute $\overline{Y}$ and use it to predict scores, S_Y and S^2_Y reflect the greatest error we can have.) Using the relationship, however, we predict Y' and then have less error: The $S_{Y'}$ of 1.42 is less than the S_Y of 2.93 (and $S^2_{Y'}$ is less than S^2_Y.) Further, the *stronger* the relationship, the better off we are. You can see this by returning to Figure 11.8 and noting that when r is larger, participants' actual Y scores are closer to the values of Y' we predict for them than they are to the overall mean of Y. Therefore, basing our predictions on a stronger relationship reduces the error ($S_{Y'}$ and $S^2_{Y'}$) even more so relative to the error we'd have (S_Y and S^2_Y) if we used $\overline{Y}$ to predict scores.

We use the above logic to evaluate any relationship: We compare our prediction errors when we *use* the relationship to our prediction errors if there was *no* relationship. Recall that the proper way to measure prediction errors is with *variance*. So, first we pretend that there is no relationship and determine the size of the prediction errors: With no relationship, we'd use $\overline{Y}$ as the predicted score and then our error equals S^2_Y. Next we determine the prediction errors when we use the relationship with X to predict Y' scores: The prediction error here is $S^2_{Y'}$ and will be less than S^2_Y. By comparing these two variances, we can determine by how much the relationship reduces prediction errors, and thus how useful or informative the relationship is.

By comparing how well we predict scores when using a relationship to how well we predict scores without the relationship, we will, in statistical language, be calculating the proportion of variance accounted for. The **proportion of variance accounted for** is the proportion of error in predictions when we use the mean to predict scores that is eliminated when we use the relationship with another variable to predict scores. In other words, the proportion of variance accounted for is the proportional improvement in predictions achieved when we use a relationship to predict scores, compared to if we did not use the relationship.

To understand the term "proportion of variance accounted for," understand that it is a shortened version of "the proportion of total variance in Y scores that is accounted for by the relationship with X." The total variance in Y is S_Y^2. Recall that S_Y^2 is simply a way to measure the *differences* among the scores. When we use $\overline{Y}$ as everyone's predicted score, we will not predict any of the different Y scores that participants actually obtain, so we will not predict any of the differences in Y scores. When we use the relationship, however, we are at least closer to knowing when participants will have one value of Y and when they will have a *different* value of Y. Therefore, using the relationship helps us to predict, or "account for," some of the differences in Y scores, so we say that we account for some proportion of the variance in Y.

REMEMBER Accounting for the variance in Y means that we can predict differences in Y scores.

To understand computing the proportion of variance accounted for, we'll first calculate it for only one individual. Then we'll see how to compute it for an entire sample.

The Logic of Calculating the Variance Accounted For

Say that in a study where we have discovered a relationship, we want to predict the Y score for "Dorcas." Dorcas actually obtained a score of 4. The overall mean of the Y scores in the sample is 9 and so, by predicting the $\overline{Y}$ score of 9 for her, we are wrong by 5 points ($9 - 4 = 5$). Because the error when predicting the mean of Y is the worst that we can do, think of this amount as the *total error.*

However, say that when using the relationship and the regression equation, we predict a Y' of 6 for Dorcas. We are now 2 points away from her score of 4. Thus, of the 5 points of error we had without using the relationship, we are still off by 2 points when we use the relationship. Think of this amount as the *error remaining*.

By reducing the error from 5 points to 2 points, however, the relationship gets us 3 points closer to Dorcas's actual Y score than we were when we predicted the mean as her score. Think of this improvement as the *error eliminated.*

As usual, it's difficult to judge whether a quantity like 3 points of error should be considered large or not. Therefore, we need a frame of reference: We describe the above components in terms of the proportion each represents of the total error we have when not using the relationship. For Dorcas, the error remaining when using the relationship, 2, divided by the total error without using the relationship, 5, is 2/5, or .40. Thus, when we use the relationship, we have .40, or 40%, of the error remaining that we have when we do not use the relationship. However, if .40 of the total error remains when we use the relationship, then we have eliminated $1 - .40$, or .60 of the error. Thus, .60, or 60%, of the total error we have when we do not use the relationship is eliminated when we do use the relationship.

Altogether, to find the proportion of the total error eliminated, we formed a ratio where the error remaining when using the relationship is divided by the total error when not using the relationship, and then that quantity is subtracted from 1. Thus:

$$\text{Proportion of total error eliminated} = 1 - \left(\frac{\text{error remaining}}{\text{total error}}\right)$$

so for Dorcas

$$1 - \left(\frac{2}{5}\right) = 1 - .40 = .60$$

This proportion means that we are 60% closer to Dorcas's actual Y score if we use the regression equation to predict her score than if we use the overall $\overline{Y}$ as her predicted score.

Of course, we never examine only one participant. Instead, we compute the proportion of variance accounted for in the entire sample.

Computing the Proportion of Variance Accounted For

To describe the entire sample, we use the above logic, except that we compute the appropriate measures of variance.

When we ignore the relationship and predict the overall mean Y score for every participant, we do not accurately predict any of the different scores. In that case, based on $\Sigma(Y - \overline{Y})^2$, the error is S_Y^2, the "average" error we have when we use the mean score to predict Y scores. Thus, S_Y^2 is analogous to the total error we have without using the relationship.

When we use the relationship and predict Y' for each participant, we may not perfectly predict each Y score. Based on $\Sigma(Y - Y')^2$, the error is $S_{Y'}^2$, the "average" error when we use Y' to predict Y scores. Thus, $S_{Y'}^2$ is analogous to the error remaining when we use the relationship.

From here, we follow the same procedure as we did with Dorcas. There we saw that

$$\text{proportion of total error eliminated} = 1 - \left(\frac{\text{error remaining}}{\text{total error}}\right)$$

Substituting the symbols $S_{Y'}^2$ and S_Y^2 in the above formula gives us

$$\text{proportion of total error eliminated} = 1 - \left(\frac{S_{Y'}^2}{S_Y^2}\right)$$

Because the total error is measured using the variance, we can change the left part of the equation from the proportion of total error eliminated to the proportion of variance accounted for. Thus,

THE DEFINITIONAL FORMULA FOR THE PROPORTION OF VARIANCE IN Y THAT IS ACCOUNTED FOR BY A LINEAR RELATIONSHIP WITH X IS

$$\text{Proportion of variance accounted for} = 1 - \left(\frac{S_{Y'}^2}{S_Y^2}\right)$$

Using this formula, we can return to the widget study and determine the proportion of variance in widget-making scores that is accounted for by the relationship with widget test scores. From the data way back in Table 11.2, we computed that S_Y^2 was 4.38, and $S_{Y'}^2$ was 2.01. Then, the proportion of variance accounted for is

$$1 - \frac{2.01}{4.38} = 1 - .46 = .54$$

By dividing 2.01 (the error remaining when using the relationship) by 4.38 (the total error when not using the relationship), we find that .46 of the total error remains, even when we use the relationship. But, if only .46 of the total error remains, then using the relationship eliminates .54 of the error we have when not using the relationship. Thus, we say that this relationship accounts for .54 of the variance in Y scores: If we know participants' scores on the widget test, we are, "on average," .54, or 54%, more accurate in predicting their individual widget-making scores than we would be if we did not use this relationship. In other words, when we use the relationship to compute Y', our predictions are, on average, 54% closer to the actual Y scores than when we do not use the relationship and instead use $\overline{Y}$ as the predicted Y score for each subject.

> *REMEMBER* The proportion of variance accounted for is the proportional improvement in accuracy when predicting Y scores by using the relationship with X, compared to when we use $\overline{Y}$ to predict Y scores.

There is one other way to conceptualize the variance accounted for: It is the proportion of all variance in a sample that is *not* error variance. Error variance reflects those differences in Y scores that occur when X does not change, so essentially, it reflects those differences in Y that do not match up with changes in X. The proportion of variance accounted for reflects those differences in Y that *do* match—are correlated—with changes in X.

Using r to Compute the Proportion of Variance Accounted For

Computing the proportion of variance accounted for with the above definitional formula is rather time-consuming. We've seen, however, that the size of r is related to the amount of error in our predictions, $S_{Y'}^2$, by the formula:

$$S_{Y'}^2 = S_Y^2(1 - r^2)$$

In fact, this formula contains all the components of the previous definitional formula, so solving for the proportion of variance accounted for, we have

$$r^2 = 1 - \frac{S_{Y'}^2}{S_Y^2}$$

On the right of the equation, 1 minus the ratio $S_{Y'}^2/S_Y^2$ is the definitional formula for the proportion of variance accounted for. Thus,

THE COMPUTATIONAL FORMULA FOR THE PROPORTION OF VARIANCE ACCOUNTED FOR IS

$$\text{Proportion of variance accounted for} = r^2$$

Not too tough! All you do is compute r (which you would do anyway) and square it. Yes, it took a long time to get to such a simple method, but to understand r^2 you must understand $1 - (S^2_{Y'}/S^2_Y)$.

Previously in the widget-making study, we computed that the relationship accounted for .54 of the variance in Y scores. Because r for this study was +.736, we can compute the proportion of variance accounted for as $(.736)^2$ which also equals .54.

In statistical language, r^2 is called the *coefficient of determination*, which is merely another name for the proportion of variance accounted for. The proportion of variance *not* accounted for is called the *coefficient of alienation*, and equals $1 - r^2$. This value equals the ratio $S^2_{Y'}/S^2_Y$, which is the proportion of total error remaining (the proportion of all differences in Y that *is* error variance). In the widget-making study, r^2 was .54, so we still cannot account for 1 − .54, or .46 of the variance in the Y scores.

Note that r^2 describes the proportion of *sample* variance that is accounted for by the relationship. If r passes the inferential statistical test, then r^2 is a *rough* estimate of the proportion of variance in Y that is accounted for by the relationship in the population. Thus, we expect to be roughly 54% more accurate if we use the relationship and our widget test scores to predict any other, unknown widget-making scores in the population.

Uses of the Variance Accounted For

The reason we make such a big deal out of the proportion of variance accounted for is that it is *the* statistical measure of how "important" a particular relationship is. After all, it is variance and variability that lead to scientific inquiry in the first place. Thus, when researchers ask, "Why does a person do this instead of that?" they are trying to predict and explain differences in scores. In other words, they are trying to account for variance. The greater the proportion of variance accounted for by a relationship, the more accurately we can identify and predict differences in behavior, and thus the more scientifically useful the relationship is.

For example, at the beginning of this discussion we wondered how to evaluate the widget-making study with an r of +.74 and a $S_{Y'}$ of 1.42. With r^2 equal to .54, we now know. We are .54 or 54% better off using this relationship than if we did not, and our error of 1.42 is 54% less than we'd have without this relationship. All in all then, this is a rather useful and thus important relationship, and the widget test should prove to be a valuable selection test.

This is especially so because, although theoretically a relationship may account for any proportion of the variance, in real research we get very excited when a relationship accounts for around 25% of the variance. Remember, this is an r of ±.50, which is pretty good. Given the complexity of nature and the behaviors of living organisms, we are unlikely to find an r that is very close to ±1.0, so we are also unlikely to find a relationship that accounts for close to 100% of the variance.

We use r^2 when comparing different relationships to see which is more informative. Say that we find one relationship between the length of a person's hair and his or her creativity, but r is only +.02. Yes, this r indicates a relationship, but such a weak relationship is virtually useless. The fact that $r^2 = .0004$ indicates that knowing someone's hair length improves predictions about creativity by only four-hundredths of *1* percent! However, say that we also find another relationship between a person's age and his or her creativity, and here r is −.40. This relationship is more important, at least in a statistical sense, because $r^2 = .16$. Age is the more important variable for understanding differences in creativity, because knowing participants' ages gets us an average of 16% closer to accurately predicting their creativity. Knowing their hair length gets us only .04% closer to predicting their creativity.

The logic of r^2 is applied to any relationship. For example, in the previous chapter we discussed r_s and r_{pb}. Squaring these coefficients also indicates the proportion of variance accounted for. (It is as if we performed the appropriate regression analysis, computed $S^2_{Y'}$ and S^2_Y, and so on.) Likewise, as you'll see in later chapters, we also determine the proportion of variance accounted for in experiments: We describe the proportion of variance in the dependent variable (the Y scores) that is accounted for by using the relationship with the independent variable (the X scores).

> *REMEMBER* The proportion of variance accounted for is the basis for evaluating the scientific importance or usefulness of any relationship.

A WORD ABOUT MULTIPLE CORRELATION AND REGRESSION

Often researchers examine relationships involving more than two variables, and then there are a number of advanced correlation and regression procedures to use. Although these procedures are appropriate regardless of how the variables are measured, they are frequently found in questionnaire and interview research. Usually questions measuring several variables will be included in one test, from which we relate respondents' scores on each variable.

The most common procedure is to correlate several X variables with one Y variable. For example, there is a positive correlation between a person's height and his or her ability to play basketball: Taller people tend to make more baskets. There is also a positive correlation such that the more people practice basketball, the more baskets they tend to make. Obviously, for maximum accuracy in predicting how well people shoot baskets, we should consider both how tall they are and how much they practice. In this example there are two predictor variables (height and practice) that predict the criterion variable (basket shooting). When we simultaneously use multiple predictor variables for one criterion variable, we use the statistical procedures known as **multiple correlation** and **multiple regression**.

The logic here is the same as the logic of the procedures we've discussed. The multiple correlation coefficient, called R, indicates the strength of the relationship between the multiple predictors taken together and the criterion variable. The multiple regression equation allows us to predict someone's Y score by simultaneously considering his or her scores on all X variables. The squared multiple R is the proportion of variance in Y that is accounted for by using the relationship with the X variables to predict Y scores.

PUTTING IT ALL TOGETHER

This chapter and the previous one have introduced many new symbols and concepts. However, they boil down to three major topics:

1. *Correlation*. When there is a relationship between the X and Y variables, a particular value of Y tends to be paired with one value of X. The stronger the relationship, the more consistently one value of Y is paired with one value of X.
2. *Regression*. With a relationship, knowing each X score helps us predict the corresponding Y score. We predict Y scores by calculating Y' scores using the linear regression equation, and graph these predictions as the linear regression line.
3. *Error in prediction*. The proportion of variance in Y that is accounted for by X is the amount we reduce errors in predicting Y scores when we use the relationship, compared to what they would be if we did not use the relationship. This proportion equals r^2.

CHAPTER SUMMARY

1. *Linear regression* is the procedure for predicting unknown Y scores based on a correlated X score. It produces the *linear regression line*, which is the best-fitting straight line that summarizes a linear relationship.
2. The *linear regression equation* includes the *slope*, which indicates how much and in what direction the regression line slants, and the *Y-intercept*, which indicates the value of Y at the point where the line crosses the Y axis.
3. For each X, the regression equation produces Y', which is the predicted Y score for that X. The regression line connects all X–Y' data points.
4. The *standard error of the estimate* ($S_{Y'}$), describes the spread of the actual Y scores around the Y' scores, and thus the error in predictions across the entire relationship. The differences (and error) between Y and Y' also may be summarized by the *variance of the Y scores around Y'* ($S^2_{Y'}$).
5. Regression requires the assumptions that: (1) The Y scores are *homoscedastic*, meaning that the spread in the Y scores around all Y' scores is the same, and (2) The Y scores at each X are normally distributed around the corresponding value of Y'.
6. When r equals ± 1.0, there is zero error in predictions and both $S_{Y'}$ and $S^2_{Y'}$ equal zero. When r equals 0, $S_{Y'}$ equals S_Y and $S^2_{Y'}$ equals S^2_Y. When r is between 0 and ± 1.0, the value of $S_{Y'}$ is between the value of S_Y and 0.
7. The *proportion of variance accounted for* is the proportional improvement in accuracy achieved by using a relationship to predict Y scores, rather than using $\overline{Y}$ to predict Y scores. This coefficient of determination equals the squared correlation coefficient.
8. The proportion of variance not accounted for—the coefficient of alienation—equals $1 - r^2$. This is the proportion of the prediction error that is not eliminated when Y' is the predicted score instead of $\overline{Y}$.
9. The proportion of variance accounted for indicates the statistical importance of a relationship.

10. *Multiple regression* and *multiple correlation* are used to predict scores on a Y variable by using the scores from multiple X variables.

PRACTICE PROBLEMS

(Answers for odd-numbered problems are provided in Appendix D.)

1. What is the linear regression line?
2. What is the linear regression procedure used for?
3. (a) What is Y'? (b) How do you obtain it?
4. What is the general form of the linear regression equation? Identify its component symbols.
5. (a) What does the Y-intercept of the regression line indicate? (b) What does the slope of the regression line indicate?
6. (a) How do you decide which variable to call X in a correlation? (b) What other names are given to the X and Y variables?
7. (a) What is the name for $S_{Y'}$? (b) What does $S_{Y'}$ tell you about the spread in the Y scores? (c) What does $S_{Y'}$ tell you about your errors in prediction?
8. (a) What two assumptions must you make about the data in order for the standard error of the estimate to be accurate, and what does each mean? (b) How does heteroscedasticity lead to an inaccurate description of the data?
9. (a) How is the value of $S_{Y'}$ related to r? (b) When is $S_{Y'}$ at its maximum value? Why? (c) When is $S_{Y'}$ at its minimum value? Why?
10. (a) Conceptually, why is the proportion of variance accounted for equal to 1 with a perfect correlation? (b) Why is it 0 when there is no relationship?
11. (a) What are the two statistical names for r^2? (b) How do you interpret r^2?
12. A researcher determined that the correlation between statistics grades and scores on an admissions test to graduate school in psychology is $r = +.41$, $S_{Y'} = 3.90$. (a) Compute the standard error of the estimate for these data (b) If the researcher predicts the overall mean score on the admissions test for each student, using variance, on average how much error can she expect? (c) If she predicts admissions test scores using the regression equation and statistics grades, using variance, on average how much error can she expect? (d) What proportion of the error in (b) remains even after using the regression equation? (e) What proportion of the error in (b) is eliminated by using the regression equation? (f) What is your answer to part (e) called?
13. Bubbles has a statistics grade of 70, and Foofy has a grade of 98. (a) Based on the data in problem 12, who is predicted to have a higher grade on the admissions test? Why? (b) Subsequently, Bubbles received the higher test score. Explain how this can occur.
14. Poindexter conducted a correlational study measuring subjects' ability to concentrate and their ability to remember, finding $r = +.30$. He also correlated their ability to visualize information and their memory ability, obtaining an $r = +.60$. He concludes that there is twice as consistent, and therefore twice as informative, a relationship between visualization ability and memory ability as there is between concentration ability and memory ability. Why do you agree or disagree?
15. (a) In problem 14, what advanced statistical procedures can Poindexter employ to improve his predictions about memory ability even more? (b) Say that the resulting correlation coefficient is .67. Using the proportion of variance accounted for, explain what this means.
16. A researcher finds that variable A accounts for 25% of the variance in variable B. Another researcher finds that variable C accounts for 50% of the variance in variable B. Why does variable C produce a relationship that is scientifically more important?
17. A student complains that it is unfair to use scores from the Scholastic Aptitude Test (SAT) to determine college admission because she might do much better in college than predicted

(a) What statistic(s) will indicate whether her complaint is correct? (b) What concern about the test's validity is she actually addressing?

18. What do you know about a research project when you see that multiple correlation and regression procedures were performed?

19. In a study, you measure how much participants are initially attracted to a person of the opposite sex (X) and how anxious they become during their first meeting with him or her (Y). For the following ratio data:

Participant	*X*	*Y*
1	2	8
2	6	14
3	1	5
4	3	8
5	6	10
6	9	15
7	6	8
8	6	8
9	4	7
10	2	6

(a) Compute the statistic that describes the nature of the relationship formed by the data. (b) Compute the linear regression equation. (c) What anxiety score do you predict for any subject who produces an attraction score of 9? (d) When using this data, what is the "average" amount of error you should expect in your predictions?

20. (a) For the relationship in problem 19, what is the proportion of variance in Y that is accounted for by X? (b) What is the proportion of variance not accounted for? (c) Why is or is not this a valuable relationship? (d) Is how much people are attracted to others a major cause of how nervous they become during their initial meeting?

21. A researcher computes a Spearman r_s of $+.20$ when correlating the rankings of students in their research class (X) with their rankings in terms of how studious they are (Y). Another researcher computes a point-biserial correlation of $-.20$ when correlating a participant's gender (X) with his or her studiousness. Using the proportion of variance accounted for, interpret each result.

22. Using two questionnaires, a researcher measures how positive a participant's mood is (X) and how creative he or she is (Y), obtaining the following interval scores.

Participant	*X*	*Y*
1	10	7
2	8	6
3	9	11
4	6	4
5	5	5
6	3	7
7	7	4
8	2	5
9	4	6
10	1	4

(a) Compute the statistic that summarizes this relationship. (b) What is the predicted creativity score for anyone scoring 3 on the mood questionnaire? (c) Assuming that your prediction is in error, what is the amount of error you expect to have? (d) How much smaller will your error be if you use the regression equation than if you merely use the overall mean creativity score as the predicted score for everyone?

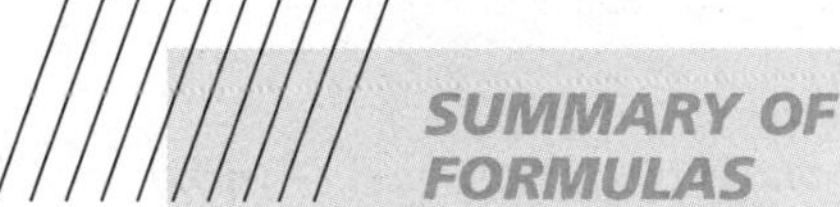

SUMMARY OF FORMULAS

1. *The formula for the linear regression equation is*

$$Y' = bX + a$$

where b stands for the slope of the line, X stands for an X score, and a stands for the Y-intercept.

2. *The formula for the slope of the linear regression line is*

$$b = \frac{N(\Sigma XY) - (\Sigma X)(\Sigma Y)}{N(\Sigma X^2) - (\Sigma X)^2}$$

where N is the number of pairs of scores in the sample and X and Y are the scores in the sample.

3. *The formula for the Y-intercept of the linear regression line is*

$$a = \overline{Y} - (b)(\overline{X})$$

where $\overline{Y}$ is the mean of all Y scores, b is the slope of the regression line, and $\overline{X}$ is the mean of all X scores.

4. *The definitional formula for the standard error of the estimate is*

$$S_{Y'} = \sqrt{\frac{\Sigma(Y - Y')^2}{N}}$$

where Y is each score in the sample, Y' is the corresponding predicted Y score, and N is the number of pairs in the sample.

5. *The computational formula for the variance of Y scores around Y′ is*

$$S^2_{Y'} = S^2_Y(1 - r^2)$$

6. *The computational formula for the standard error of the estimate is*

$$S_{Y'} = S_Y\sqrt{1 - r^2}$$

where S_Y is the standard deviation of the Y scores in the sample.

7. *The definitional formula for the proportion of variance in Y that is accounted for by a linear relationship with X is*

$$\text{Proportion of variance accounted for} = 1 - \left(\frac{S_{Y'}^2}{S_Y^2}\right)$$

where $S_{Y'}^2$ is the squared standard error of the estimate and S_Y^2 is the variance of all Y scores in the sample.

8. *The computational formula for the proportion of variance in Y that is accounted for by a linear relationship with X is*

$$\text{Proportion of variance accounted for} = r^2$$

9. *The computational formula for the proportion of variance not accounted for is*

$$\text{Proportion of variance not accounted for} = 1 - r^2$$

PART 4

INTRODUCTION TO INFERENTIAL STATISTICS

Believe it or not, you now know the common descriptive methods used in psychological research. We are finally ready to discuss inferential statistical procedures. But descriptive statistics actually form the basis for *all* statistical interpretations, so we'll use these techniques with inferential procedures too. We use inferential statistics to answer these questions: Based on our sample, what would we expect to find if we could perform this study on the entire population? Would we find that the population has approximately the same mean as our sample? If we observe different samples that produce different means, would we find approximately the same difference between the population means? Would we find that the correlation coefficient in the population is about the same as the correlation coefficient in our sample?

The problem is that we cannot *know* what the population contains, so the best we can do is to place an intelligent bet. In essence, inferential procedures are ways to make decisions about the population that have a high probability of being correct. The first step in understanding these procedures is to understand probability. In Chapter 12, we discuss the concept of probability and see how it is used to make statistical decisions. Subsequent chapters then deal with the various inferential procedures used with different designs. As you read each chapter and become familiar with the different procedures, remember to keep one eye on the big picture: All inferential procedures involve making decisions about the scores and relationship we would find in the population, if we could study it.

12

Probability and Making Decisions About Chance Events

To understand this chapter:

- From Chapter 8, understand computing a z-score for a raw score in a distribution of scores, or for a sample mean in a sampling distribution of means. Also recall that the proportion of the area under a part of the normal curve equals the relative frequency of the scores or sample means in that part of the curve.

Then your goals are to learn:

- How probability is computed based on an event's relative frequency in the population.
- How the probability of raw scores is computed using z-scores and the standard normal curve.
- How the probability of sample means is computed using z-scores and the standard normal curve.
- How to set up and use a sampling distribution of means to determine whether a sample mean is likely to represent a particular population.

This chapter introduces you to the wonderful world of probability. As we'll see, psychologists use probability in conjunction with the standard normal curve model to make decisions about their data. We'll keep the discussion simple because you do not need to be an expert in probability. However, you do need to understand the basic logic of chance.

MORE STATISTICAL NOTATION

In daily conversation, we use the words *chances*, *odds*, and *probability* interchangeably. In statistics, however, there are differences among these terms. Odds are expressed as fractions or ratios ("The odds of winning are 1 in 2"). Chance is expressed as a percentage ("There is a 50% chance of winning"). Probability is expressed as a decimal ("The probability of winning is .50"). You should express the answers you compute as probabilities.

The symbol used to represent probability is the lowercase letter p. When describing the probability of a particular event—such as event A—we write it as $p(\text{A})$, which is pronounced "p of A" or "the probability of A."

THE LOGIC OF PROBABILITY

Probability is used to describe random, chance events. Such events occur when nature is being fair—when there is no bias toward one event over another (no rigged roulette wheels or loaded dice). Thus, a chance event occurs or does not occur merely because of the luck of the draw. In statistical work, chance is a very important concept, and probability is our way of mathematically describing how chance operates to produce an event.

How can we describe an event that happens only by chance? By paying attention to how often the event occurs when chance is operating. The probability of any event is based on how often the event occurs *over the long run*. Intuitively, we use this logic all the time: If event A happens frequently over the long run, then we tend to think that it is likely to happen at any moment, and we say that A has a high probability. If event B happens infrequently, then we tend to think that it is unlikely to happen, and we say that B has a low probability.

When we decide that event A happens frequently, we are making a relative judgment. Compared to anything else that might happen in this situation, event A happens frequently. Intuitively, we determine the *relative frequency* of event A: The proportion of time that A occurs out of all possible events that might occur in this situation. In statistical terminology, all possible events that can occur in a given situation form the *population* of events. Thus, the **probability of an event** is equal to the event's relative frequency in the population of all possible events that can occur.

REMEMBER The probability of an event equals the relative frequency of the event in the population.

If a population contains all possible events that might occur, then the events that do occur make up a sample from that population. Thus, when we ask if a particular event will occur, we are actually asking what is the probability that a sample will contain the event when we randomly sample from a particular population. You already know that a score's relative frequency in the population is also the score's expected relative frequency in any sample. We use probability to express a particular event's expected relative frequency in any single random sample. For example, I am a rotten typist, and while typing the manuscript for this book, say I randomly made typos 80% of the time. This means that in the population of my typing, typos occur with a relative frequency of .80. We expect the relative frequency of typos to continue at a rate of .80 in any random sample of my typing. We express this expected relative frequency as a probability, so the probability is .80 that I will make a typo when I type the next woid.

As the above example illustrates, a **probability** is a mathematical statement indicating the likelihood of an event when a particular population is randomly sampled. It is how we express what we expect to occur when only chance is operating, and it indicates our confidence that a particular event will occur. For example, if event A has a relative frequency of zero in a particular situation, then the probability of event A is zero. This means that we do not expect A to occur in this situation, because it never does. But if event A has a relative frequency of .10 in this situation, then A has a probability of .10. This means that we have some—but not much—confidence that the event will occur in any particular sample. Because it occurs only 10% of the time in the population, we expect it to occur in only 10% of our samples. On the other hand, if event A has a probability of .95, we are confident that A will occur. Because it occurs 95% of the time in the population, we expect it to occur in 95% of our samples. At its most extreme, event A may occur 100% of the time so it is 100% of the population and its probability is 1. Then we are positive it will occur in this situation because it always does.

Notice that an event cannot happen less than 0% of the time nor more than 100% of the time, so a probability can *never* be less than 0 or greater than 1. Notice also that all events in a population together constitute 100% of the time, so the relative frequencies of all events must add up to 1, and the probabilities of all events in the population must also add up to 1. Thus, if the probability of my making a typo at any moment is .80, then because $1 - .80 = .20$, the probability is .20 that any word I type will be error free.

It is important to remember that except when p equals either 0 or 1, we are never certain that an event will or will not occur in a particular situation. The probability of an event is its relative frequency *over the long run*, or in the infinite population. It is up to chance whether a particular sample contains the event. For example, even though I make typos at a rate of .80 over the long run, I may go for quite a while without making a typo. That 20% of the time I make no typos has to occur sometime. Thus, although the probability is .80 that I will make a typo in each word, it is only over the long run that we truly expect to see precisely 80% typos.

COMPUTING PROBABILITY

Computing the probability of an event is simple: We need only determine its relative frequency in the population. When we know the relative frequency of every event in a population, we have a probability distribution. A **probability distribution** tells us the probability of every possible event in a population.

Creating Probability Distributions

There are two ways to create a probability distribution. One way is to measure the relative frequency of every event in the population. If we "empirically" determine the relative frequencies of all events in a population, we create an *empirical probability distribution*. Typically, however, we cannot measure all of the events in a population, so we create an empirical probability distribution from samples of those events. We assume that the relative frequency of an event in a random sample represents the relative frequency of the event in the population.

For example, say that Dr. Fraud is sometimes a very cranky individual, and as near as we can tell, his crankiness is random. We observe him on 18 days and determine that he is cranky on 6 of these days. Relative frequency equals the frequency of an event divided by N, so the relative frequency of Dr. Fraud's crankiness is 6/18, or .33. Assuming that this sample represents the population of how Dr. Fraud always behaves, we expect that he will continue to be cranky on 33% of all days. Thus, the probability that he will be cranky today is $p = .33$. Conversely, he is not cranky on 12 of the 18 days we observe him, so the relative frequency of his not being cranky is 12/18, or .67. Thus, the probability that he will not be cranky today is $p = .67$. Because his cranky days plus his noncranky days constitute all possible events, we have an empirical probability distribution for his crankiness.

The other way to create a probability distribution is to create a theoretical probability distribution. A **theoretical probability distribution** is a theoretical model of the relative frequencies of events in a population. We devise theoretical probability distributions based on how we assume nature distributes events in the population. Then we can look at the model to determine the relative frequency of each event in the population. This relative frequency is then the probability of the event in any random sample.

Let's look first at some common random events such as tossing a coin and drawing playing cards. (As you'll see, determining probability in such situations is analogous to determining probability in research.) With coin tossing, we assume that we are discussing random flips of a fair coin. After tossing a coin in the air, there are two possible outcomes that can occur: A head or a tail. We assume that nature has no bias toward heads or tails, so that over the long run we will see 50% heads and 50% tails. In other words, we expect the relative frequency of heads to be .50 and the relative frequency of tails to be .50. Because relative frequency in the population *is* probability, we have the theoretical probability distribution: The probability of a head is $p = .50$, and the probability of a tail is $p = .50$.

Similarly, when we draw a playing card, we assume that there is no bias favoring any 1 of the 52 cards in the deck, so each has the same relative frequency in the population. Over the long run, we expect each card to occur at a rate of once out of every 52 draws, so each card has a relative frequency of 1/52, or .0192. Therefore, the probability of drawing any specific card on a single random draw is $p = .0192$.

And that, ladies and gentlemen, is the logic of probability. We either theoretically or empirically devise a model of how events are assumed to be distributed in the population. This model gives us the expected relative frequency of each event in the population. From this, we derive the expected relative frequencies of events in any sample, expressed as p. By computing the probability for all possible events in a particular situation, we create the probability distribution for that situation.

General Formula for Computing Probability

Hidden in the above examples is a method for computing probability. When we assume nature has no bias towards one or another event, we are assuming that all events in the population are equally likely. Then:

THE FORMULA FOR COMPUTING PROBABILITY WHEN EVENTS ARE EQUALLY LIKELY IS

$$p(\text{event}) = \frac{\text{number of outcomes that satisfy event}}{\text{total number of possible outcomes}}$$

In the numerator is the number of possible outcomes that satisfy the requirements of the event we are describing. In the case of flipping a coin, for example, there is one outcome that satisfies the condition of showing a head. In the denominator is the total number of possible outcomes that can occur. There are two possible outcomes that can occur when flipping a coin: either head or tail. Therefore, the probability of a head is 1/2, which equals .5.

Likewise, we might define the event as randomly drawing a king from a deck of cards. There are four kings in a deck, and any one would satisfy us. With a total of 52 possible outcomes, the probability of randomly drawing a king on one draw is 4/52, or .0769.

You can even apply this formula to real life! For example, raffle tickets are sometimes sold one for a dollar and sometimes sold three for a dollar. Many people think they have a better chance of winning when everyone gets three tickets for a dollar. They're wrong. Say that 100 people each spend a dollar and get one ticket. Then each person's probability of winning is equal to 1/100, or .01. If 100 people each buy 3 tickets for a dollar, then each person's probability of winning equals 3/300. But 3/300 equals 1/100, so each person still has a probability of winning equal to .01.

The above formula can be used to find the probability of equally likely events. However, the formula becomes tedious when the event being described is a complex sequence of alternatives. Therefore, shortcut formulas for three of the most common complex events are provided in Appendix B.

Factors Affecting the Probability of an Event

All random events are not the same. First, events may be either independent or dependent. Two events are **independent events** when the probability of one event is *not* influenced by the occurrence of the other event. For example, contrary to popular belief, washing your car does *not* make it rain. These are independent events, so the probability of rain does not increase when you wash your car. On the other hand, two events are **dependent events** when the probability of one event is influenced by the occurrence of the other event. For example, whether you pass an exam usually depends on whether you study: The probability of passing increases or decreases depending on whether studying occurs.

The probability of an event is also affected by the type of sampling performed. When **sampling with replacement**, any previously selected samples are replaced back into the population before drawing additional samples. For example, say we are selecting two cards from a deck of playing cards. Sampling with replacement occurs if, after drawing the first card, we return it to the deck before drawing the second card. The probabilities on the first and second draw are each based on 52 possible outcomes, and the probability of any particular card's being selected either time is constant. On the other hand, when **sampling without replacement** any previously selected samples are *not* replaced in the population before selecting again. Thus, sampling without replacement occurs if, after a card is drawn, it is then discarded. Now, although the probability of drawing a particular card on the first draw is based on 52 possible outcomes, the probability of any remaining card being selected on the second draw is based on only 51 outcomes. With fewer possible outcomes, the probability is slightly larger on the second draw than on the first draw.

In statistics we usually assume that events are independent and sampled with replacement.

The reason we discuss probability is not that we have an uncontrollable urge to flip coins and draw cards. Researchers use probability to make decisions about scores and samples of scores. To do that, they use the standard normal curve.

OBTAINING PROBABILITY FROM THE STANDARD NORMAL CURVE

As you've seen, once we know the distribution of events in the population, we can determine their relative frequencies and thus their probabilities. In statistics, the events we describe are scores, and we usually assume that they form a normal distribution. Then our theoretical probability distribution is based on the standard normal curve. Here's how it works.

Determining the Probability of Individual Scores

In Chapter 8, we used *z*-scores to find the proportion of the total area under the normal curve in any part of a distribution. This proportion corresponds to the relative frequency of the scores in that part of the population. But we've just seen that the relative frequency of scores in the population equals the probability of those scores. Thus,

> **The proportion of the area under the standard normal curve for scores in any part of the distribution equals the probability of those scores!**

For example, look at Figure 12.1.

FIGURE 12.1 *z*-Distribution Showing the Area for Scores Below the Mean and Between 0 and 1

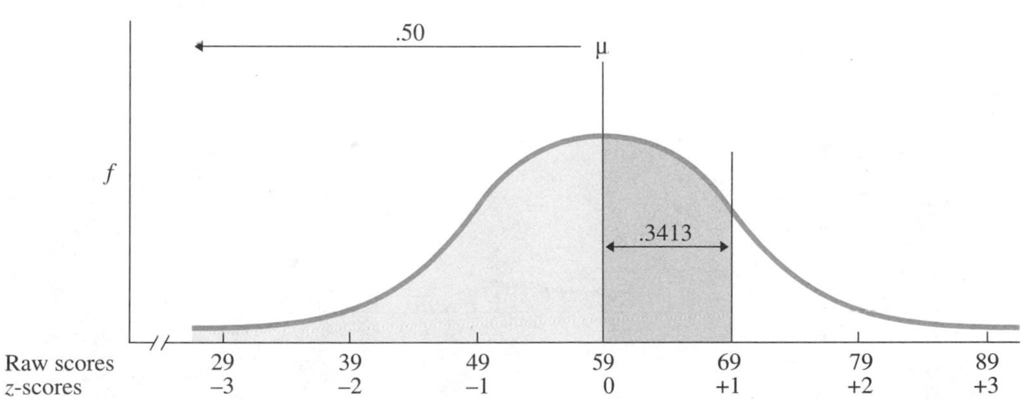

We know that 50% of the area and thus 50% of all scores fall below the mean, producing negative z-scores. Therefore, because negative z-scores occur .50 of the time in the population, the probability of randomly selecting a negative z-score—or a participant who produces a negative z-score—is .50. In other words, because we expect to run into negative z-scores 50% of the time, the probability is .50 that the next score we draw will be a negative z-score.

Having determined the probability of selecting certain z-scores, we automatically know the probability of selecting the corresponding raw scores. In Figure 12.1, the raw scores have a mean of 59, so raw scores below 59 produce negative z-scores. Thus, the probability is also .50 that any individual we select will have a raw score below 59.

We can obtain the probability for any part of a normal distribution by using the z-tables in Appendix C. For example, the z-tables indicate that z-scores between the mean and a z of $+1.0$ occur .3413 of the time in the population. Thus, the probability of randomly selecting any one of these z-scores is .3413. Further, looking back at the shaded part of Figure 12.1, we see that a score of 69 produces a z-score of $+1.0$ (and the mean is 59). Therefore, the probability of randomly selecting a raw score between 59 and 69 in this distribution is also .3413.

Likewise, we can determine the probability of randomly selecting a score greater than a certain z-score. For example, what is the probability of selecting a z-score larger than $+2.0$? In Figure 12.2, z-scores above $+2.0$ are in the shaded area on the right. The z-tables indicate that z-scores above $z = +2.0$ constitute .0228 of the population and thus occur .0228 of the time. Therefore, the probability of randomly selecting a z-score above $+2.0$ is .0228 (which in this case is also the probability of selecting a raw score above 79).

What is the probability of selecting a z-score that is beyond a z of ± 2.0? The phrase "beyond ± 2.0" means that we will be satisfied by any score above $+2.0$ or below -2.0, so we must determine the probability of drawing a score from either of the shaded areas shown in Figure 12.2. All scores above $z = +2.0$ constitute .0228 of the curve, and all scores below $z = -2.0$ constitute an additional .0228 of the curve. In total, .0228 + .0228, or .0456, of the curve contains scores that will satisfy us. Because z-scores beyond ± 2.0 occur a total of .0456 of the time, the probability of randomly selecting a z beyond ± 2.0 is .0456. In Figure 12.2, a raw score of 39 corresponds to a z of -2.0, and a raw score of 79 corresponds to a z of $+2.0$. Therefore, the probability is .0456 that we will randomly select a raw score below 39 or above 79.

FIGURE 12.2 Area Under the Curve Beyond $z = +2.0$

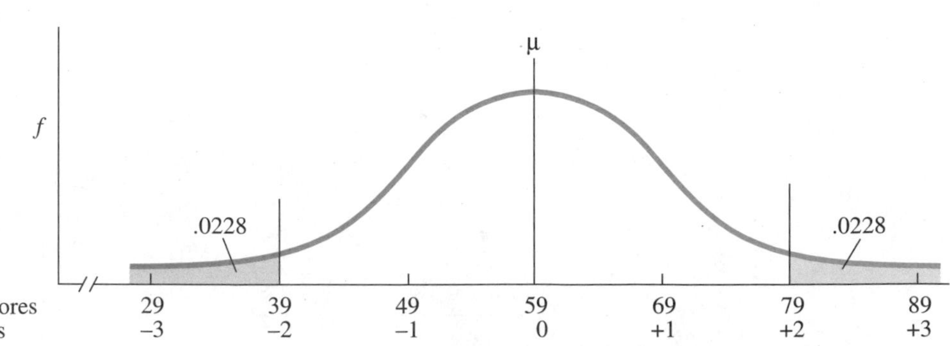

REMEMBER When you will accept scores from two parts of a distribution—when you say "or"—*add* the individual areas or probabilities together.

Using the above procedures, we can determine the probability of randomly selecting any scores from any portion of any normal distribution. In fact, we can even find the probability of randomly selecting "scores" that are sample means.

Determining the Probability of Sample Means

In Chapter 8, we conceptualized the sampling distribution of means as the frequency distribution of all possible sample means that would result if a statistician randomly sampled a raw score population an infinite number of times using a particular sample *N*. Thus, think of a sampling distribution as a "picture of chance," showing how often we can expect random chance to produce samples having certain sample means when we randomly sample a particular raw score population. Then we found a mean's location on a sampling distribution by first computing the standard error of the mean (the standard deviation of the sampling distribution) using the formula $\sigma_{\overline{X}} = \sigma_X / \sqrt{N}$. Then we computed the mean's *z*-score using the formula $z = \overline{X} - \mu / \sigma_{\overline{X}}$. And finally, by applying the standard normal curve model, we determined the relative frequency of sample means falling above or below that *z*-score. But, because a sampling distribution shows the relative frequency of sample means in the population, it also forms a theoretical probability distribution. Therefore, in the same way that we determined the probability of randomly selecting certain raw scores, we can determine the probability of randomly selecting certain sample means.

For example, let's look again at the sampling distribution of SAT means discussed in Chapter 8, which is shown in Figure 12.3. The population mean (μ) of this distribution is 500; and when *N* is 25, the standard error of the mean is 20. Say we're interested in the portion of the sampling distribution that contains means with *z*-scores between 0 and +1.0. The relative frequency of such *z*-scores is .3413, so the relative frequency of sample means that produce these *z*-scores is also .3413. Because an event's relative frequency in the population equals its probability, the probability of randomly selecting a sample mean with a *z*-score between 0 and +1.0 from this population is $p = .3413$.

FIGURE 12.3 Sampling Distribution of SAT Means When $N = 25$

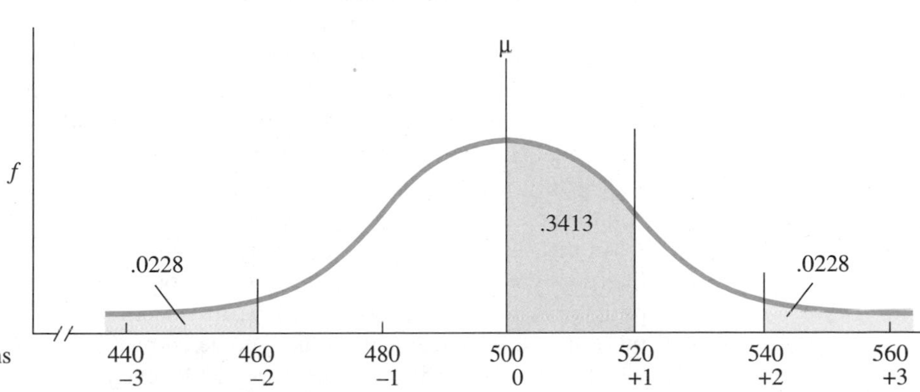

Further, on this SAT sampling distribution, sample means between 500 and 520 produce z-scores between 0 and +1.0. Thus, when N is 25, the probability is .3413 that we will obtain a sample mean between 500 and 520 from this SAT population.

Think about this: Randomly selecting a sample mean is the same as randomly selecting a sample of raw scores that produce that mean. Likewise, randomly selecting a sample of raw scores is the same as randomly selecting a sample of participants and then measuring their raw scores. Therefore, because the area under the normal curve provides the probability of selecting certain sample means, it also provides the probability of selecting the corresponding samples that produce those means. Thus, we can rephrase our finding above: When we randomly select 25 students from this SAT population, the probability of selecting a sample that produces a mean between 500 and 520 is .3413.

Here's another example. Previously we determined that z-scores beyond ±2.0 have a probability of .0456. Thus, the probability of selecting a random sample that produces a sample mean having a z-score beyond ±2.0 is also .0456. On our SAT sampling distribution, z-scores of ±2.0 correspond to means of 540 and 460, respectively. Therefore, when we sample this SAT population, the probability is .0456 that we will randomly select a sample of 25 scores that produces a mean below 460 or above 540. In the same way, we can determine the probability of obtaining particular sample means from any normally distributed raw score population.

Recall that the reason for learning about probability is so that you can make decisions about data. Now that you understand how to compute probability, the next step is to understand how to make decisions using it.

MAKING DECISIONS BASED ON PROBABILITY

The probability that I will produce a typo at any moment is .80. Should you bet that I'll make a typo on the next word? Yes, because it's a good bet. Over the long run, I make typos 80% of the time, so on any word you are likely to win the bet. What is the probability that you made the correct decision (that you'll win the bet?) The probability of a correct decision equals the probability that chance will produce the event you are predicting. To figure this probability, assume that you'll make the same decision 100% of the time you're in this situation. Then because I make typos 80% of the time, you'll be correct 80% of the time. Thus, the relative frequency of your winning the bet is .80, so the probability of winning any single bet is $p = .80$. Remember, though, that I do not make typos 20% of the time, so you'll lose the bet 20% of the time. Because the relative frequency of your losing is .20, the probability of losing any single bet is $p = .20$. Conversely, if you bet that I will not make a typo, the probability is only .20 that you'll win the bet, but .80 that you'll lose the bet.

As this illustrates, we bet in favor of high-probability events because then we are likely to win the bet. We bet against low-probability events, because otherwise the probability of winning would be low. In statistics, we also make bets and again, we make the decision that is most likely to be correct. The bet we make involves deciding whether random chance produced a particular event.

Deciding Whether Chance Produced a Particular Event

It is essential to remember that probability always implies "over the long run." For example, say we flip a coin 7 times and obtain 7 heads in a row. This will not concern us, assuming that we have a fair coin. If we see 7 heads (or 70 heads) in a row, we still assume that over the long run, the relative frequencies of heads and tails will each be .50. We understand that any single sample of coin tosses may not be *representative* of the population of coin tosses, because a head's relative frequency in the sample may not equal its relative frequency in the population. We have merely shown up at a time when luck produced an unrepresentative sample.

People who fail to understand this principle fall victim to the "gambler's fallacy." If we have obtained 7 heads in a row, the fallacy is that a head is now less likely to occur, because it has already occurred too frequently (as if the coin said, "Hold it. That's enough heads for a while!"). If it is a fair coin, the probability of a head on the next toss is still .50. The mistake of the gambler's fallacy is failing to recognize that each coin toss is an *independent* event and we have merely observed an unrepresentative sample of coin tosses.

But what if you are not certain whether the coin is a fair coin? If you are betting money on the flips of a coin and another player obtains 7 heads in a row, you should stop playing: It's a good bet that the coin is rigged. How do we reach this decision? By computing the probability of obtaining such an outcome if the coin is *not* rigged.

To compute a probability, we need to know the total number of possible outcomes, so here we first determine all possible series of heads and tails that could occur when flipping a coin 7 times. There is one way to obtain all 7 heads, but there are many ways to obtain 6 heads and 1 tail, such as

	tail,	head,	head,	head,	head,	head,	head
or:	head,	tail,	head,	head,	head,	head,	head
or:	head,	head,	tail,	head,	head,	head,	head

and so on. Next we determine all the ways to obtain 2 tails in 7 tosses, 3 tails in 7 tosses, and so on. We will find a total of 128 different arrangements of heads and tails that can be obtained when a coin is tossed 7 times in a row. With only one way to obtain 7 heads, the probability of obtaining 7 heads in 7 coin tosses is 1/128, or about .008. (A shortcut formula for computing such probabilities, called the binomial expansion, is presented in Appendix B, part B4.)

Now we can decide about the fairness of the coin. If this coin is fair, then 7 heads in a row will hardly ever happen by chance. We expect its relative frequency to be 8 out of every 1000 times, or .8% of the time. Because it is difficult to believe a player would be that lucky, we reject the idea that random chance produced such an unlikely outcome, and we leave the game.

Notice that we made a definitive, all-or-nothing decision: The probability suggests that the coin is crooked, so we decide that it is definitely crooked. In statistics, whenever you use probability to make a decision, your decision is a definite yes or no. Period!

However, we could be wrong! Even though 7 heads in a row occurs infrequently, we cannot be *certain* that the coin is rigged. Unlikely events are just that: They are unlikely, not impossible. Random chance does produce unusual samples every once in a while. In fact, we computed that chance would produce 7 heads in a row on .8% of the throws. Perhaps this was one of them. Thus, we can never *know* whether random chance was operating or not. The best we can do is to base a decision on the most likely outcome.

We determine the probability that we incorrectly rejected the chance explanation in the same way that we determine the probability of losing any bet. Because 7 heads in a row is so unlikely to occur with a fair coin, anytime we see this we'll bet that chance did not produce the result. Because chance *will* produce 7 heads on .8% of those times, then .8% of the time we'll lose the bet, so $p = .008$ of losing any single bet. The probability of winning the bet, however, is $1 - .008$, or .992.

Deciding Whether Chance Produced a Representative Sample

Recall that in research, we use a random sample to draw conclusions about a population. In essence, we want to say that the way in which the sample behaves indicates the way in which the population would behave if we could observe it. However, we can never be certain how the population would behave, because there is no guarantee that the sample accurately reflects the population. In other words, we are never certain that a sample of scores is *representative* of a particular population of scores.

Representativeness is not all or nothing, however. A sample can be more or less representative, having more or less the same characteristics as the population. This is because how representative a sample is depends on random *chance*—the luck of the draw of which scores or participants were selected. By chance the sample may be somewhat different from the population from which it is selected, and thereby represent that population somewhat poorly.

This, then, is the central problem for researchers: When a sample is different from the population it actually represents, it has the characteristics of some *other* population and appears to represent that other population. Thus, although a sample always represents some population, we are never sure *which* population it represents: The sample may represent one population poorly, or it may represent another population altogether.

This was actually the problem we faced with the crooked coin: We were uncertain whether the sample of coin tosses poorly represented the population of tosses with a fair coin or represented the population of tosses with a crooked coin. We decided whether the sample represented the population of fair coin tosses by determining the probability of obtaining the sample from that population. Implicitly, we used this logic: If we flip a coin several times and obtain 51% heads and 49% tails, this sample still appears to represent the population of fair coin tosses containing 50% heads and 50% tails. Even 60% heads and 40% tails is reasonably representative of fair coin tosses. In these situations, the fact that we do not have a 50-50 split between heads and tails is written off as being due to chance: Chance produced a less than perfectly representative sample.

Beyond a certain point, however, we begin to doubt that only random chance is at work. For example, with 70% heads and 30% tails, we grow suspicious that mere luck produced such an unrepresentative sample of the population of fair coin tosses. And with 100% heads—7 in a row—we seriously doubted the honesty of the other player: It was *too unlikely* for us to get such a sample by chance if the sample actually represented the population of fair coin tosses. Therefore, we rejected the idea that luck merely produced a less than perfectly representative sample from that population. The only alternative was that the sample represented the population of tosses produced by a crooked coin.

Here is another example. You obtain a sample paragraph of someone's typing, but you do not know if it's mine. Does this sample represent the population of my typing? Say that you find zero typos in the paragraph. You cannot know for certain whether I typed this paragraph, but the chances are not good: I type errorless words only 20% of the time, so the probability that I could type an entire errorless paragraph is very small. Thus, because chance is *too unlikely* to produce such a sample from the population of my typing, you should decide against this low-probability event. You would reject the idea that chance produced this highly unrepresentative sample from the population of my typing and instead conclude that the sample represents some other population where such a sample is more likely, such as that of a competent typist.

On the other hand, say that there are typos in 79% of the words in the paragraph. This paragraph is reasonably representative of my typing. Although you expect 80% typos from me over the long run, you do not expect precisely 80% typos in every sample. Rather, a sample with 79% errors seems likely to occur simply by chance when the population of my typing is sampled. Thus, you can accept that this paragraph is more or less representative of my typing, but because of random chance, there are slightly fewer typos in the sample than in the population.

We use this same logic in research to decide whether a sample of scores is representative of a particular population of scores. This is the essence of all inferential statistical procedures: Based on the probability of obtaining a particular sample from a particular population, we decide whether the sample represents that population. If the sample is likely to occur when that population is sampled, then we decide that it may represent that population. If the sample is too unlikely to occur when that population is sampled, then we decide that the sample does not represent that population and instead represents some other population.

> REMEMBER The essence of all inferential statistical procedures is to decide whether a sample of scores is likely or unlikely to occur in a particular population of scores.

The next chapter puts all of this into a research context. In the final sections of this chapter, we'll examine the mechanics of deciding whether a sample of scores represents a particular population of scores.

MAKING DECISIONS ABOUT A SAMPLE MEAN

Say that we return to Prunepit University and obtain a random sample of SAT scores that produces a mean of 550. This is surprising because we think that students at Prunepit U. are terminally average. Because the ordinary national population of SAT scores has a μ of 500, we should have obtained a sample mean of 500 if our sample was perfectly representative of this population. How do we explain having a sample mean of 550? On the one hand, the simplest explanation is that we obtained a sample of relatively high SAT scores merely because of random chance—the luck of the draw that determined who was selected to be in the sample. Thus, it is possible that chance produced a less than perfectly representative sample of the population where μ is 500. On the other hand, perhaps the sample does not come from or represent the ordinary, national population of SAT scores: After all, these *are* Prunepit students, so they may belong to a very different population of students, having some other μ.

To decide whether the sample represents the population of SAT scores where μ is 500, we'll determine the probability of obtaining a sample mean of 550 from this population. As we saw previously, to determine the probability of obtaining a particular sample mean from a particular population, we first create the sampling distribution of means for that population. For our Prunepit U. problem, we envision the sampling distribution as showing the frequencies of all the different means that the bored statistician from Chapter 9 would obtain if, using our *N*, she randomly sampled the ordinary SAT population an infinite number of times. Any differences between the sample means are due to the luck of the draw—which scores she happened to select for each sample. Thus, the sampling distribution is a model of how often random chance will produce a particular sample mean when the samples *do* represent a population in which μ is 500.

Then, using *z*-scores, we will determine the location of our sample mean on the sampling distribution and thereby determine the probability of obtaining such a mean. If it is likely that random chance would produce a sample mean of 550 from the ordinary SAT population in which μ is 500, we will accept that the Prunepit sample may represent this population. Conversely, if it is too unlikely that chance would produce a sample mean of 550 from this population, we will reject the idea that it was merely the luck of the draw that produced such a mean. Instead, we conclude that the sample represents some other population of scores, having some other μ.

Be sure you understand the above logic before proceeding. If you do, you're ready to perform the mechanics of the procedure. Notice that the procedure involves two tasks: We must determine the probability of obtaining our sample from the original population, and we must decide whether the sample is too unlikely for it to represent this population. We perform both tasks simultaneously, once we have set up the sampling distribution.

Identifying the Region of Rejection

Earlier we decided that we did not have a fair coin because 7 heads in a row was too unlikely an event for us to believe that chance produced it. Likewise, we must decide what it will take to convince us that a sample mean is too unlikely to be representing a particular population. To do so, we define the *criterion* for making our decision. (In Chapter 11, the *Y* variable was called the criterion variable, but here the term "criterion" refers to a probability.) The **criterion** is the probability that is used to decide whether a sample is too unlikely to occur by chance for it to represent a particular population. If the probability of obtaining a sample from a particular population is *less* than the criterion, then we decide that it is too unlikely for the sample to represent that population. If it is too unlikely that the sample represents that population, then the sample must represent some other population.

Psychologists usually use the probability of .05 as their criterion. Because we must make a yes or no decision, we reject the idea that chance produced *any* of the sample means that, together, have only a .05 probability of occurring in a particular population. In other words, if the Prunepit sample mean is among those means that occur only 5% of the time, then it is too unlikely to be considered representative of the ordinary SAT population where $\mu = 500$.

REMEMBER The criterion probability that defines samples as too unlikely is usually $p = .05$.

As you know, on a normal distribution, the scores having a low frequency and thus a low probability lie in the tails of the distribution. Therefore, those sample means having a probability of .05 make up the extreme 5% of the curve. Thus, all samples falling in the extreme 5% of the sampling distribution are treated the same and considered too unlikely to accept as representing the particular population.

Because there are two tails, we can divide this 5% in half and demarcate the extreme 2.5% in each tail of the sampling distribution. Figure 12.4 shows how to do this for the SAT example. Quite literally, I've drawn a line in each tail of the distribution that defines "too unlikely." In statistical terms, the shaded areas beyond the lines make up the region of rejection. The **region of rejection** is the portion of a sampling distribution containing values that are considered too unlikely to occur to be representing the underlying raw score population. The size of the area under the curve that comprises the region of rejection is always equal to our criterion. Usually, our criterion is .05, so the total region of rejection usually comprises .05 of the curve.

Be sure that you understand what the region of rejection represents. Because the μ of the population is 500, sample means of 500 are perfectly representative. The more the means in Figure 12.4 differ from 500, the more they are both unrepresentative of μ and unlikely to occur. With our criterion, we have simply defined the most unrepresentative means as the most unlikely, extreme 5%. Thus, the region of rejection contains the worst, most unrepresentative sample means we could obtain when sampling from the ordinary SAT population. Very infrequently are samples *so* poor at representing this population that they have means lying in the region of rejection. In fact, they are so unrepresentative, it is a better bet that they represent some *other* population.

But remember, we already have a sample mean. The question is, what population does it represent? The sampling distribution shows the means the statistician obtained when she definitely *was* representing the SAT population. Therefore, if the Prunepit U. sample mean lies within the region of rejection, then such an unrepresentative sample is very unlikely to occur, if we—like the statistician—were also representing this population. Essentially, we "shouldn't" get such an unlikely sample mean when representing

FIGURE 12.4 Setup of Sampling Distribution of Means Showing the Region of Rejection

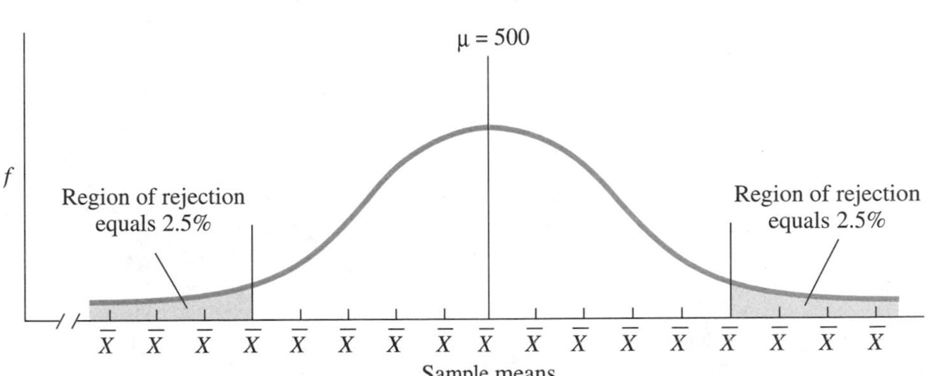

this population. If we do get such a mean, then we probably aren't representing this population. Therefore, in statistical terms, when a sample mean falls in the region of rejection, we *reject* the idea that the sample is merely somewhat unrepresentative of the original population. Instead, we decide that the sample mean represents some other population.

Conversely, if our sample mean is not in the region of rejection, then it is not too unlikely to be representing the ordinary SAT population. In fact, by our criterion, sample means not in the region of rejection *are* likely to occur when this population is sampled and thus *are* likely to represent it. In statistical terms, in this case we *retain* the idea that chance may have produced any differences between our sample mean and μ, and that the sample may represent, although poorly, this population of SAT scores.

Identifying the Critical Value

We locate our sample mean on the sampling distribution by computing a z-score for it. Because the absolute values of z-scores get larger as we go farther into the tail of the distribution, if the absolute value of our z-score is large enough, then our sample mean falls in the region of rejection. How large must the z-score be? With a criterion of .05, we set up the region of rejection so that .025 of the total area under the curve was in each tail. From the z-table, we see that the extreme .025 of the curve lies beyond the z-score of ± 1.96. Thus, the region of rejection starts at ± 1.96. This is shown in Figure 12.5. For a sample mean to lie in the region of rejection, its z-score must lie *beyond* ± 1.96. In this example, ± 1.96 is the critical value of z. A **critical value** marks the edge of the region of rejection and thus defines the value required for a sample to fall in the region of rejection.

Now, deciding whether our sample represents the SAT population in which μ is 500 boils down to comparing the sample's z-score with ± 1.96. Any sample mean with a z-score that is beyond the critical value of $\pm$ 1.96 lies in the region of rejection. Then we will *reject* the idea that chance produced such an unrepresentative sample mean from this population, and thus reject the idea that the sample represents this SAT population. Instead, it is more likely that the sample represents some other population.

FIGURE 12.5 Setup of Sampling Distribution of SAT Means Showing Region of Rejection and Critical Values

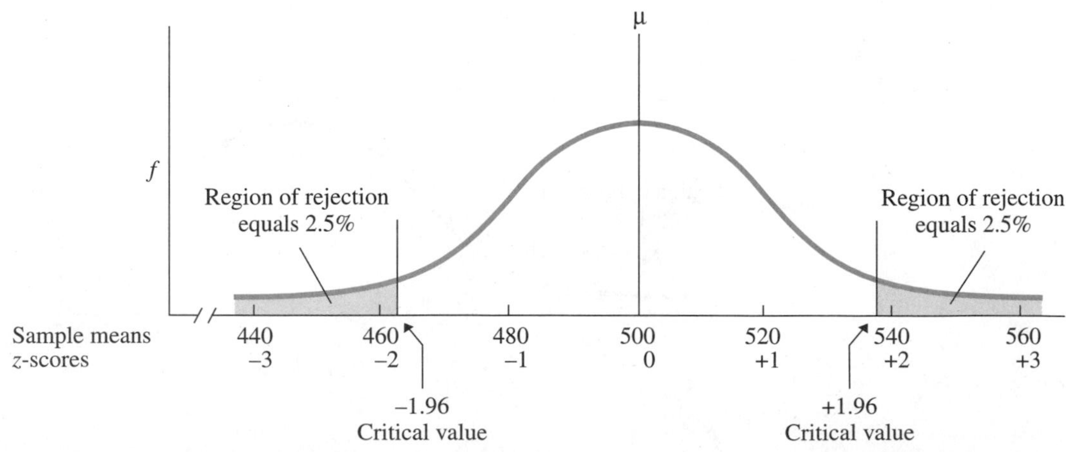

Conversely, any sample mean with a z-score that is smaller than or equal to the critical value is *not* in the region of rejection. Then we will *retain* the idea that chance may have produced an unrepresentative sample mean from this population and retain the idea that our sample may represent this SAT population.

> *REMEMBER* When the z-score for a sample lies beyond the critical value, we *reject* the idea that the sample represents the raw score population reflected by the sampling distribution. When the z-score does not lie beyond the critical value, we *retain* the idea that the sample may represent that raw score population.

Deciding If a Sample Is Representative

Now, at long last, we can evaluate the sample mean of 550 from Prunepit U. On the sampling distribution created from the population of SAT scores for which μ is 500, σ_X is 100, and N is 25, the standard error of the mean is $100/\sqrt{25}$, which is 20. Then the sample mean of 550 has a z-score of $(550 - 500)/20$, which is $+2.5$. Think about this z-score. If the sample represents the ordinary population of SAT scores, it seems to be doing a very poor job of it. With a population mean of 500, a perfectly representative sample would also have a mean of 500 and thus have a z-score of 0. Good old Prunepit University produced a z-score of $+2.5$!

To confirm our suspicions, we examine the sampling distribution shown in Figure 12.6. The sample's z-score of $+2.5$ is beyond the critical value of ± 1.96, so it's in the region of rejection. Because the z-score is in the region of rejection, the corresponding sample mean of 550 is also in the region of rejection. Thus, it is not a good bet that, through random chance, we obtained a somewhat unrepresentative sample from the SAT population: samples that are this unrepresentative hardly ever happen. After all, our bored statistician hardly ever got such a mean, and she sampled SAT scores an *infinite* number of times. So, if we conclude that our sample represents this population, then we must also conclude that we had a tremendous amount of luck in getting a

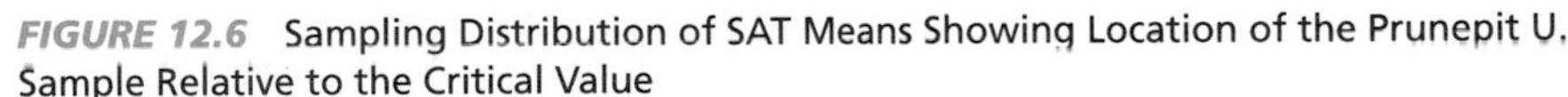

FIGURE 12.6 Sampling Distribution of SAT Means Showing Location of the Prunepit U. Sample Relative to the Critical Value

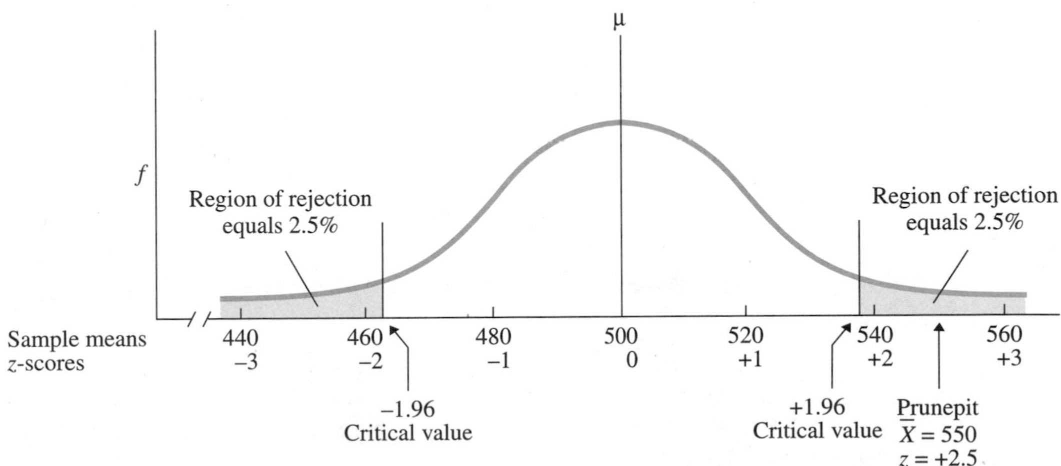

sample mean of 550. But we don't believe we're that lucky! Therefore, we reject that chance produced such a weird sample from this population, rejecting that our sample represents the ordinary population of SAT raw scores having a μ of 500.

Notice that, as with the crooked coin, we make a yes or no decision. Because the sample would be unlikely to occur if it represented the SAT raw score population where μ is 500, we decide that no, it definitely does not represent that population.

Once we reject the idea that a sample represents the raw score population reflected by the sampling distribution, it is only logical that the sample must represent some other population of scores where such a sample mean is more likely. For example, perhaps our Prunepit students obtained the high mean of 550 because they lied about their scores, so they may represent the population of students who lie about the SAT. Perhaps they are more intelligent or motivated than the typical student, and thus represent the population of such over-achievers. Or, because Prunepit U. is located on a toxic waste dump, perhaps our sample represents the population of SAT scores of students living on toxic waste dumps.

Once we reject that our sample represents the population where μ is 500, we use the sample mean to estimate the μ of the population that the sample does represent. A sample having a mean of 550 is most likely to come from a population having a μ of 550. Therefore, our best guess is that, for whatever reason, the SAT population represented by the Prunepit sample has a μ of 550.

Thus, in sum, as we did with previous bets, we decide against the low-probability event that the sample represents the SAT population where μ is 500, and we decide in favor of the high-probability event that the sample represents a population where μ is 550.

On the other hand, say that instead, our sample mean had been 474, resulting in a z-score of $(474 - 500)/20 = -1.30$. Because -1.30 does not lie beyond the critical value of ± 1.96, the sample mean is not in the region of rejection. Looking back at Figure 12.6, we see that when the bored statistician sampled the population, such a sample mean was relatively frequent and thus likely. Because of this, we can accept that random chance produced a less than perfectly representative sample for us but that the sample probably represents the ordinary SAT population.

Notice that, in essence, this is a system for evaluating the *difference* between an obtained sample mean and what we would expect if the sample represents a particular population. We expect a representative sample mean to be "close" to the population mean. We do not intuitively know, however, which values of $\overline{X}$ should be considered close to a particular μ. Because a z-score describes a sample mean's relative standing, it allows us to determine whether the sample mean is relatively close to the population mean. From this perspective, the critical value of ± 1.96 defines "close." If a sample mean has a z-score lying beyond the critical value, the sample mean is not close to the population mean. Then, the difference between the sample mean and the population mean is so large that we do not consider the sample to represent that population.

Other Ways to Set Up the Sampling Distribution

In the above example, the region of rejection was in both tails of the distribution because we wanted to make a decision about any sample mean having an extreme positive or negative z-score. However, we can also set up the distribution to examine only negative z-scores or only positive z-scores. (The next chapter discusses why you would do this.)

Say that we are interested only in sample means *less than* 500. Our criterion is still .05, but means below μ have negative z-scores, so we place the entire region of rejection constituting 5% of the curve in the lower left-hand tail of the sampling distribution, as shown in Figure 12.7. Notice that we have a different critical value. From the z-table (and using the interpolation procedures described in Part 3.A of Appendix B), the extreme 5% of a distribution lies beyond a z-score of 1.645. Therefore, our sample's z-score must lie beyond *minus* 1.645 for the sample mean to be in the region of rejection. If it does, we will again conclude that such a sample is too unlikely to occur when sampling the SAT raw score population where $\mu = 500$, so we'll reject the idea that the sample represents this population.

On the other hand, say that we're interested only in sample means *greater* than 500, having positive z-scores. Here we place the entire region of rejection (the entire 5%) in the upper, right-hand tail of the sampling distribution, as shown in Figure 12.8. Now the critical value is *plus* 1.645. If our sample's z-score is beyond $+1.645$, then the sample mean lies in the region of rejection. Then we again reject the idea that the sample represents the underlying raw score population.

On Being Wrong When We Decide About a Sample

When we rejected the idea that we were playing with a fair coin, we might have been wrong: Maybe the player obtained 7 heads in a row with a fair coin by chance. In the same way, we might be wrong when we reject that the sample from Prunepit U. represents the ordinary SAT population where μ is 500. The sampling distribution clearly shows that sample means out there in the region of rejection *do* occur. Therefore, it is possible to obtain a sample mean this unusual when we *are* representing the ordinary SAT population where μ is 500. Maybe our Prunepit U. mean was one of those means. *Anytime* we reject the idea that a sample represents a particular population, we may be wrong.

We can also be wrong when we retain the idea that a sample represents a particular population. Consider the most extreme case, in which we obtain a sample mean of 500!

FIGURE 12.7 Setup of SAT Sampling Distribution to Test Negative z-Scores

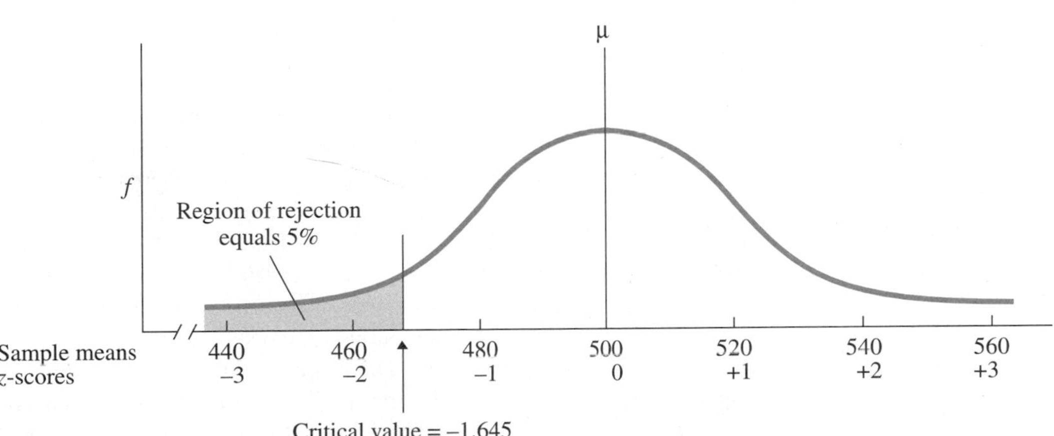

FIGURE 12.8 Setup of SAT Sampling Distribution to Test Positive *z*-Scores

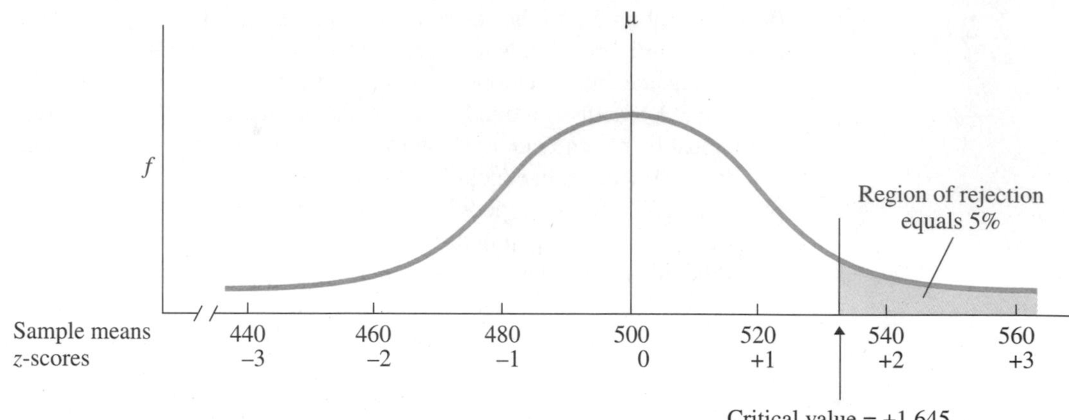

This sample certainly appears to represent the ordinary population of SAT scores where μ is 500. Using the above procedures, we would compute a *z*-score of 0, so we would retain this idea. But it is still possible that this sample actually represents some other population. Perhaps, for example, the sample is actually a very unrepresentative sample from the population where μ is 550! That is, maybe, simply by the luck of the draw, the sample contains too many low scores, so the mean is 500 instead of 550, and so it appears to represent the population where μ is 500. In the same way, anytime we retain the idea that a sample represents a particular population, we may be wrong.

Thus, regardless of which population we decide a sample represents, we have not proven anything, and there is always the possibility that we have made an incorrect decision. In the next chapter, we'll determine the probabilities of such errors. For now, recognize that even though such errors are possible, they are not likely. That's why we perform inferential statistical procedures. By incorporating probability into our decision making, we have greater confidence that we have correctly identified the population that a sample represents. Then, we have greater confidence that we are correctly interpreting and generalizing the results of our research.

PUTTING IT ALL TOGETHER

The decision-making process discussed in this chapter is used in all inferential statistics. The basic question always addressed is "Does our sample data represent a particular raw score population?" To answer this question, we create a sampling distribution from that particular population. (This will be *much* easier if you sketch the distribution and label the region of rejection, as I did in this chapter.) Then we compute a statistic, such as a *z*-score, to describe the results of our study. If the *z*-score lies beyond the critical value, then the corresponding sample data fall in the region of rejection. Any sample in the region of rejection is too unlikely to occur for us to accept as representing the underlying raw score population. Therefore, we reject the idea that chance produced such an unrepresentative sample from this raw score population, and we conclude that the sample represents some other population that is more likely to produce such data.

CHAPTER SUMMARY

1. *Probability*, or p, indicates the likelihood of an event when random chance is operating. The probability of an event equals its relative frequency in the population.

2. The probability of equally likely events equals the number of outcomes that can satisfy the event out of the total number of possible outcomes that can occur.

3. The probability of correctly predicting a chance event equals the probability that the event will occur. The probability of incorrectly predicting a chance event equals the probability that the event will not occur.

4. Events are *independent events* if the probability of one event is not influenced by the occurrence of the other event(s). Events are *dependent events* if the probability of one event is influenced by the occurrence of the other event(s).

5. *Sampling with replacement* is replacing a sample in the population before another sample is selected. *Sampling without replacement* is *not* replacing one sample in the population before another sample is selected.

6. A *theoretical probability distribution* is a theoretical model of the relative frequencies of all possible events in a population when random chance is operating.

7. The standard normal curve model is a theoretical probability distribution that can be applied to any normal raw score distribution. Raw scores are transformed to z-scores, and the proportion of the area under the curve is the probability of randomly selecting those z-scores. This is also the probability of selecting the corresponding raw scores from the underlying raw score distribution.

8. A sampling distribution of means is also a theoretical probability distribution. Sample means are transformed to z-scores, and the proportion of the area under the curve is the probability of randomly selecting those z-scores. This is also the probability of selecting the corresponding sample means from the underlying raw score population.

9. The probability of randomly selecting a particular sample mean is the same as the probability of randomly selecting a sample of participants whose scores produce that sample mean.

10. The *region of rejection* is located in the extreme tail or tails of a sampling distribution. Any z-score (and corresponding sample mean) in the region of rejection is too unlikely to accept as representing the raw score population reflected by the sampling distribution.

11. The edge of the region of rejection closest to the mean of the sampling distribution is at the *critical value*. A sample falls in the region of rejection if the sample's z-score is beyond the critical value.

12. The size of the region of rejection is determined by the *criterion*, which is the probability that defines a sample as too unlikely. Usually, the criterion is .05. This produces a region of rejection that constitutes the extreme .05 of the sampling distribution.

PRACTICE PROBLEMS

(Answers for odd-numbered problems are provided in Appendix D.)

1. (a) What does a probability convey about a random event in a sample? (b) What is the probability of a random event based on?
2. (a) What is the difference between an empirical probability distribution and a theoretical probability distribution? (b) Why is the proportion of the area under the normal curve equal to probability?
3. What is the probability of: (a) Getting a six when rolling a die? (b) Selecting a diamond when cutting a deck of cards? (c) Randomly guessing the correct answer to a multiple choice question with four choices? (d) Selecting the ace of diamonds twice in a row when sampling a deck of cards without replacement?
4. (a) When are events independent? (b) When are they dependent? (c) How does sampling without replacement affect the probability of events, compared to sampling with replacement?
5. A couple with eight children, all girls, decides to have one more baby, because the next one is bound to be a boy! Is this reasoning accurate?
6. Foofy read in the newspaper that there is a .05% chance of swallowing a spider while you sleep. She subsequently developed insomnia. (a) What is the probability of swallowing a spider? (b) Why isn't her insomnia justified on the basis of this probability? (c) Why is her insomnia justified on the basis of this probability?
7. Poindexter's uncle is planning to build a house on land that has been devastated by hurricanes 160 times in the past 200 years. Because there hasn't been a major storm there in 13 years, his uncle is certain this is a safe investment. His nephew argues that there definitely will be a hurricane in the next year or so. What are the fallacies in the reasoning of both men?
8. Four airplanes from different airlines have crashed in the past two weeks. Bubbles must travel on a plane, but is terrified that it will crash. Her travel agent claims that the probability of a plane crash is minuscule. Who is correctly interpreting the situation? Why?
9. For each of the following, indicate whether the first event is dependent on or is independent of the second event: (a) Playing golf; the weather. (b) Buying new shoes; buying a new car. (c) Losing weight; eating fewer calories. (d) Winning the lottery; playing the same numbers each time.
10. What is the probability of randomly selecting a subject who scores the following? (a) $z = +2.03$ or above (b) $z = -2.8$ or above (c) z between -1.5 and $+1.5$ (d) z beyond ± 1.72
11. For a distribution in which $\overline{X} = 43$ and $S_X = 8$, what is the probability of randomly selecting the following? (a) A score of 27 or below (b) A score of 51 or above (c) A score between 42 and 44 (d) A score below 33 or above 49
12. You are shopping for a used car. Over the life of the car you are thinking of buying, the probability of engine trouble is .65. (a) If you conclude that the engine will malfunction, what is the probability that you are correct? What is the probability that you are incorrect? (b) If you conclude that the engine will not malfunction, what is the probability that you are correct? What is the probability that you are incorrect? (c) Should you purchase this car? Why?
13. The mean of a population of raw scores is 18 ($\sigma_X = 12$). What is the probability of randomly selecting a sample of 30 scores having a mean above 24?
14. The mean of a population of raw scores is 50 ($\sigma_X = 18$). What is the probability of randomly selecting a sample of 40 scores having a mean below 46?

15. A sample produces a mean that is different from the μ of the population that we think the sample represents. What are the two possible reasons for this difference?

16. When testing the representativeness of a sample mean, (a) What is the criterion? (b) What is the region of rejection? (c) What is the critical value of z and what is it used for?

17. Suppose that for the data in problem 13, you obtained a sample mean of 24. Using the .05 criterion with the region of rejection in both tails of the sampling distribution, should you consider the sample to be representative of the population in which $\mu = 18$? Why?

18. Suppose that for the data in problem 14 you obtained a sample mean of 46. Using the .05 criterion with the region of rejection in both tails of the distribution, should you consider the sample to be representative of the population in which $\mu = 50$? Why?

19. In a study, you use a questionnaire and obtain the following data representing the aggressive tendencies of some football players:

40 30 39 40 41 39 31 28 33

(a) Researchers have found that in the population of nonfootball players, μ is 30 ($\sigma_X = 5$). Using both tails of the sampling distribution, determine whether your football players represent a different population. (b) What do you conclude about the population of football players and its μ?

20. On a standard test of motor coordination, a sports psychologist found that the population of average bowlers had a mean score of 24, with a standard deviation of 6. A random sample of 30 bowlers at Fred's Bowling Alley had a mean of 26. A sample of 30 bowlers at Ethel's Bowling Alley had a mean of 18. Using the criterion of $p = .05$ and both tails of the sampling distribution, what should she conclude about each sample's representativeness of the population of average bowlers?

21. (a) In problem 20, if each sample did not represent the population of average bowlers, what would be your best estimate of the μ of the population it does represent? (b) Explain the logic behind this conclusion.

22. Foofy computes the $\overline{X}$ from data that her professor says is a random sample drawn from population Q. She determines that this sample mean has a z-score of +41 on the sampling distribution for population Q (and she computed it correctly!). Foofy claims she has proven that this could not be a random sample from population Q. Do you agree or disagree? Why?

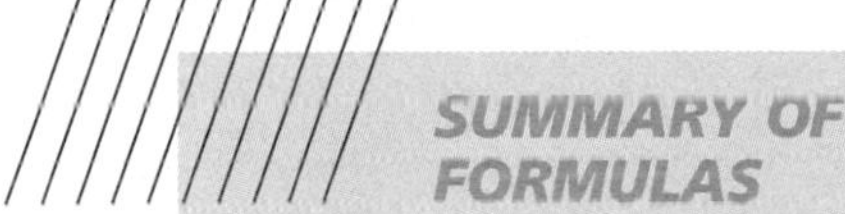

SUMMARY OF FORMULAS

1. The formula for computing probability for equally likely events is

$$p(\text{event}) = \frac{\text{number of outcomes that satisfy event}}{\text{total number of possible outcomes}}$$

2. The formula for the true standard error of the mean is

$$\sigma_{\overline{X}} = \frac{\sigma_X}{\sqrt{N}}$$

where σ_X is the true standard deviation of the raw score population.

3. *The formula for transforming a sample mean into a z-score on the sampling distribution of means is*

$$z = \frac{\overline{X} - \mu}{\sigma_{\overline{X}}}$$

where $\overline{X}$ is the sample mean, μ is the mean of the sampling distribution (which is also equal to the μ of the raw score distribution), and $\sigma_{\overline{X}}$ is the standard error of the mean of the sampling distribution of means.

13

Overview of Statistical Hypothesis Testing: The *z*-Test

To understand this chapter:

- From Chapter 7, recall that a relationship in the population occurs when different means from the conditions of an independent variable represent different μs and thus different distributions of dependent scores.
- From Chapter 12, recall that when a sample's *z*-score falls in the region of rejection, the sample is too unlikely to accept as representing the underlying raw score population.

Then your goals in this chapter are to learn:

- Why the possibility of sampling error leads a researcher to perform inferential statistical procedures.
- When your experimental hypotheses lead to either a one-tailed or two-tailed statistical test.
- How to set up a sampling distribution for one- and two-tailed tests.
- How to interpret significant and nonsignificant results.
- What impact Type I errors, Type II errors, and power have on the interpretations a researcher makes.

In Chapter 12, we discussed the basic process behind inferential statistics. In this chapter, we'll put these procedures into a research context and present the statistical language and symbols used to describe them. Until further notice, we'll be talking about experiments.

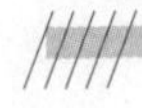

MORE STATISTICAL NOTATION

Five new symbols will be used in stating mathematical relationships:

1. The symbol for *greater than* is $>$. We read from left to right, so, for example, $A > B$ means that A is greater than B. (The large opening in the symbol $>$ is always on the side of the larger quantity, and the symbol "points toward" the smaller quantity.)
2. The symbol for *less than* is $<$. Thus, $B < A$ means that B is less than A.
3. The symbol for *greater than or equal to* is $\geq$. Thus, $B \geq A$ indicates that B is greater than or equal to A.
4. The symbol for *less than or equal to* is $\leq$. Thus $B \leq A$ indicates that B is less than or equal to A.
5. The symbol for *not equal to* is $\neq$. Thus, $A \neq B$ indicates that A is different from B.

THE ROLE OF INFERENTIAL STATISTICS IN RESEARCH

As you saw in the previous chapter, a random sample may be more or less representative of a population because, just by the luck of the draw, the sample contains too many high scores or too many low scores relative to the population. Because the sample is not perfectly representative of the population from which it is selected, the sample mean does not equal the population mean.

The shorthand term for communicating that chance produced an unrepresentative sample is to say that the sample reflects sampling error. **Sampling error** results when chance produces a sample statistic (such as $\overline{X}$) that is not equal to the population parameter it represents (such as μ). Because of the luck of the draw, the *sample* is in *error* to some degree in representing the population.

> *REMEMBER* Sampling error results when, by chance, the scores that are selected result in a sample statistic that is different from the population parameter it represents.

Sampling error is the reason researchers perform inferential statistics. There is always the possibility of sampling error in any sample. Whenever a sample mean is different from a particular population mean, it may be that: (1) The sample poorly represents that population of scores due to sampling error, or (2) The difference is not due to sampling error, and instead the sample represents some other population. This creates a dilemma for a researcher who is trying to infer that a relationship exists in nature. Recall that in an experiment, we change the conditions of the independent variable in hopes of seeing scores on the dependent variable change in a consistent fashion. We want to infer that this relationship is found in the population: If we measured the entire population, we would find a different population of scores located around a different value of μ under each condition. But here is where the possibility of sampling error arises. Even though we obtain a different sample mean for each condition, perhaps we are being misled by sampling error. Maybe the samples are actually unrepresentative of the *same* population, so that if we tested everyone in the population under each condition, we would find the same population of scores, having the same μ, in each condition (and then there would not be a relationship).

Thus, before we can generalize about what a relationship represents in terms of variables and constructs, we must decide whether the sample relationship represents anything at all. That is, even though we may find a mathematical relationship in our sample, it may not represent a real relationship in nature. By a "real" relationship, I mean there is some underlying aspect of nature that actually associates a particular *Y* score with a particular *X* score. Here we are not considering whether the data are reliable or valid. Regardless of what the scores may measure, our most basic inference is that there is some unseen process that actually ties the *X* and *Y* scores together, so that the relationship in the sample represents a relationship that actually exists in nature. However, when the differences between the means of our conditions are actually due to sampling error, then the sample relationship does not reflect a real relationship. Instead, the data *coincidentally* form the pattern of a particular relationship by chance, in the same way that if you drop a handful of coins they may land in a particular pattern by chance. Thus, even though sample data show a relationship, it is possible that the *X* and *Y* scores are paired up simply by luck, and there is no real relationship in the population. It is also possible that there may be a real relationship in the population, but because of sampling error, it is different from the relationship in the sample data.

To deal with the possibility of sampling error, we apply inferential statistics. **Inferential statistics** are procedures for deciding whether sample data represent a relationship in the population. Using the process discussed in the previous chapter, we decide whether our samples are likely to represent populations that form a particular relationship or whether they are likely to represent populations that do not form the relationship.The specific inferential procedure we employ in a given research situation depends on the research design and on the scale of measurement used when measuring the *dependent variable*. There are two general categories of inferential statistics: Parametric and nonparametric.

Parametric statistics are procedures that require certain assumptions about the parameters of the raw score populations represented by the sample data. Recall that parameters describe the characteristics of a population, so parametric procedures are used when we can assume the population has certain characteristics. In essence, the assumptions of a procedure are the rules for using it, so think of them as a checklist for selecting the procedure to use in a particular study. There are specific assumptions for each parametric procedure, but two assumptions are common to them all: (1) The population of dependent scores forms a normal distribution, and (2) The dependent scores are interval or ratio scores. Thus, parametric procedures are used when it's appropriate to summarize the scores by calculating their mean. (In this and upcoming chapters, we'll focus on parametric procedures.)

Nonparametric statistics are inferential procedures that do not require assumptions about the populations represented by the sample data. Usually, these procedures are employed with nominal or ordinal dependent scores or with a skewed distribution of interval or ratio scores (when the data are most appropriately described by the median or mode).

As you'll see, we prefer to use parametric procedures, and we can use them even if data only come close to meeting their assumptions. This is because parametric procedures are robust. With a **robust procedure**, if we do not meet the assumptions of the procedure perfectly, we will have only a negligible amount of error in the inferences we draw. So, for example, if data represent a population that is approximately normally distributed, we can still use a parametric procedure. In essence, we perform nonparametric procedures only when the data clearly violate the assumptions of parametric procedures.

In subsequent chapters, you'll become immersed in the details of inferential statistics. Don't lose sight of the fact that every procedure ultimately is used to decide whether the data represent a "real" relationship in the population. A study is *never* finished until you have performed the appropriate inferential procedure and made that decision. In this chapter, we'll discuss one such procedure to let you see the general format and terminology used in all inferential procedures.

SETTING UP INFERENTIAL STATISTICAL PROCEDURES

It's not appropriate to think about statistical analysis only after you have collected data. There are several decisions that you should make beforehand to be sure the data will be "analyzable." It is possible to collect data and then find out that there are no appropriate statistical procedures to apply.

As you know, in setting up a study, we first create a hypothesis (a statement about how nature operates) that leads us to predict a relationship that the experiment should demonstrate. Of course, there is an implicit second and opposite hypothesis we might create in which we state that this is not how nature operates, and therefore, that the experiment will not demonstrate the predicted relationship. Let's call these two hypotheses the experimental hypotheses. **Experimental hypotheses** describe the predicted outcome we may or may not find in an experiment. One hypothesis implies that manipulating the independent variable will work as expected, and the other implies that the manipulation will not work as expected. These two hypotheses are important because they form the basis for setting up any statistical analysis.

We may predict a relationship in one of two ways. The simplest prediction is that there is some kind of relationship, but we are not sure whether scores will increase or decrease as we change the independent variable (essentially we're unsure whether we'll see a positive or negative linear relationship). The other, more complicated prediction not only predicts a relationship, but also predicts the *direction* in which the scores will change: We may predict that as we change the independent variable, the dependent scores will increase (producing a positive relationship), or we may predict that they will decrease (producing a negative relationship).

First, let's examine a study in which we merely predict some kind of a relationship. Say that we're interested in brain physiology, and have discovered a neural transmitter substance that is related to learning ability and intelligence. After several successful replications with animal subjects, we are ready to test this substance with humans in an "IQ pill." The amount of the pill is our independent variable, and the person's IQ is our dependent variable. For the moment, say that we believe the pill will affect IQ, but we're not sure whether it will make people smarter or dumber. Therefore, we predict a relationship where the more of the pill a person consumes, the more his or her IQ will change. In a nutshell, the possible outcomes are:

1. We will demonstrate that the pill works by either increasing or decreasing IQ scores.

2. We will not demonstrate that the pill works, because IQ scores will not change.

Once we know the hypothesis and prediction, we design the study.

Designing a Single-Sample Experiment

Although there are many ways to design the IQ pill study, the simplest approach is a single-sample experiment. We will randomly select one sample of participants and say we give each one pill. After waiting for the pill to work, we'll test the participants using a standard IQ test. Then the sample will represent the population of IQ scores of all people when they have taken one pill, and the sample's $\overline{X}$ will represent that population's μ.

To demonstrate a relationship, however, we must demonstrate that *different* amounts of the pill produce *different* populations of IQ scores, having different μ's. Therefore, we must compare the population represented by our sample to some other population receiving some other amount of the pill. *To perform any single-sample experiment, we must already know the population mean under some other condition of the independent variable.* One amount of the pill is zero pills. The IQ test we are using has been given to many people over the years who have not taken the pill, and say this population of IQ scores has a μ of 100. We will compare this population without the pill to the population with the pill represented by our sample. (Essentially, the population that has taken the IQ test is equivalent to a control group). If the population represented by the sample has a different μ than the population without the pill, then we will have demonstrated a relationship in the population.

Once you have designed the study, the next step is to translate the experimental hypotheses into their corresponding statistical hypotheses.

Creating the Statistical Hypotheses

Statistical hypotheses are statements that describe the population parameters that sample data represent if the predicted relationship does or does not exist. There are always two statistical hypotheses, the alternative hypothesis and the null hypothesis.

The alternative hypothesis Although you can create the hypotheses in either order, it is often easier to create the alternative hypothesis first, because it corresponds to the experimental hypothesis that the experiment *does* work as predicted. The **alternative hypothesis** describes the population parameters that the sample data represent if the predicted relationship exists. The alternative hypothesis is always the hypothesis of a difference; it says that changing the independent variable produces the predicted *difference* in the population of scores. Therefore, it is the hypothesis that the sample data represent a "real" relationship in nature.

If we could test the entire population and if the IQ pill works as predicted, then we would find one of two outcomes. Figure 13.1 shows the change in the population if the pill *increases* IQ scores. This shows a relationship because, by changing the conditions, everyone's IQ is increased so that the distribution moves to the right, over to the higher IQ scores. We don't know how much IQ scores will increase, so we don't know the

FIGURE 13.1 Relationship in the Population If the IQ Pill Increases IQ Scores

As the amount of the pill changes from 0 pills to 1 pill, the IQ scores in the population tend to increase in a consistent fashion.

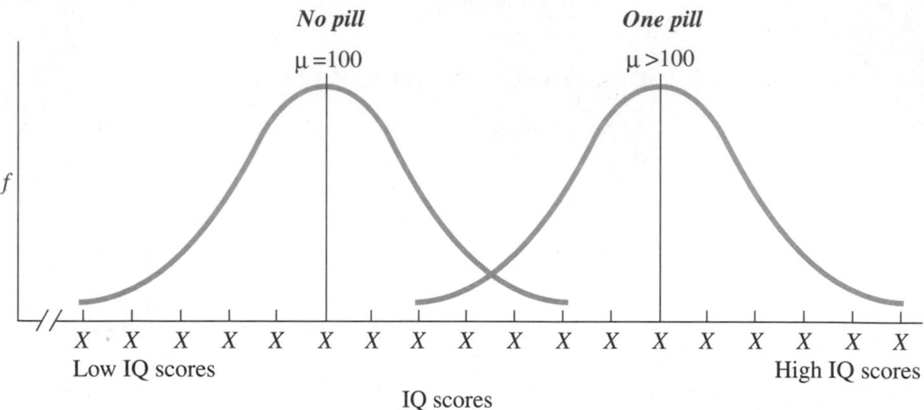

value of μ with the pill. But we do know that if the pill increases IQ, then the μ of the population with the pill will be *greater* than 100, because 100 is the μ of the population without the pill.

On the other hand, Figure 13.2 shows the change in the population if the pill *decreases* IQ scores. Here there is a relationship because everyone's IQ decreases so the distribution is moved to the left, over to the lower IQ scores. Again, we do not know the value of μ with the pill, but if the pill decreases IQ, then the μ of the population with the pill will be *less than* 100 (because 100 is the μ of the population without the pill).

FIGURE 13.2 Relationship in the Population If the IQ Pill Decreases IQ Scores

As the amount of the pill changes from 0 pills to 1 pill, the IQ scores in the population tend to decrease in a consistent fashion.

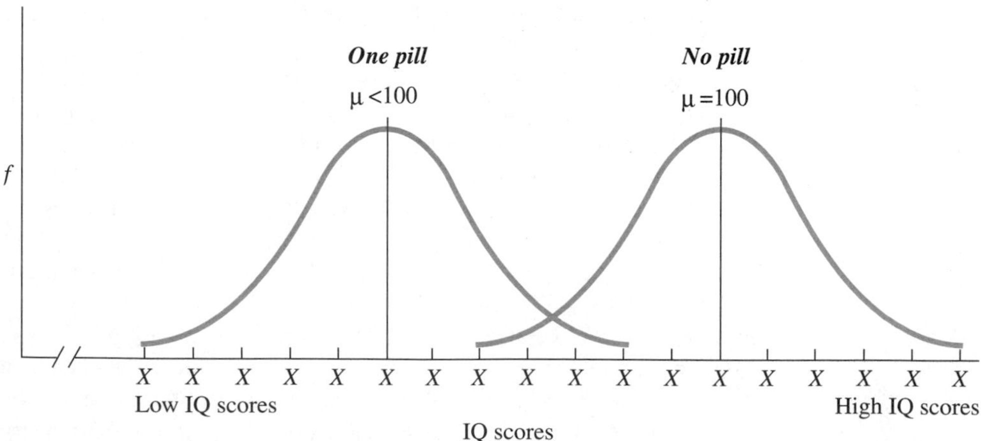

The alternative hypothesis provides a shorthand way of communicating all of the above. If the pill works as predicted, then the population with the pill will have a μ that is either greater than or less than 100. In other words, the population mean with the pill will *not equal* 100. The symbol for the alternative hypothesis is H_a. (The H stands for hypothesis, and the a stands for alternative.) For the IQ pill experiment, the alternative hypothesis is

$$H_a: \mu \neq 100$$

H_a implies that the sample mean represents a population mean not equal to 100. If with the pill μ is not 100, then there is a real relationship in the population. Thus, we can interpret H_a as implying that our independent variable really does work as predicted.

The null hypothesis The statistical hypothesis that corresponds to the experimental hypothesis that the independent variable does *not* work as predicted is called the null hypothesis. The **null hypothesis** describes the population parameters that the sample data represent if the predicted relationship does *not* exist. The null hypothesis is the hypothesis of "no difference," saying that changing the independent variable does *not* produce the predicted difference in the population of scores. Instead, it implies that the data poorly represent the situation where no relationship exists, so although the sample data may form a relationship, this is due to sampling error and the data do not represent a "real" relationship in nature.

If the IQ pill does nothing, then it would be as if the pill were not present. We already know that the population of IQ scores without the pill has a μ of 100. Therefore, if the pill does not work, then after everyone has taken the pill, IQ scores will be unchanged and μ will still be 100. Thus, if we measured the population with and without the pill, we would have one population of scores, located at the μ of 100, as shown in Figure 13.3.

FIGURE 13.3 Population of Scores If the IQ Pill Does Not Affect IQ Scores

Here there is no relationship.

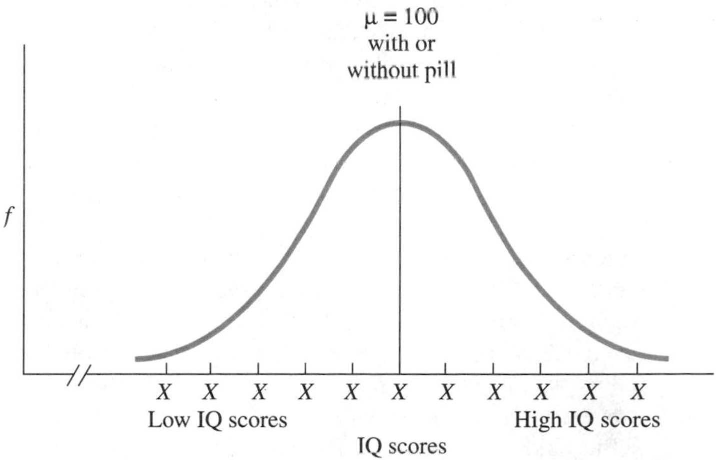

The null hypothesis is a shorthand way of communicating the above. The symbol for the null hypothesis is H_0. (The subscript is 0 because *null* means zero, as in zero relationship.) The null hypothesis for the IQ pill study is

$$H_0: \mu = 100$$

H_0 implies that the sample mean represents a population mean equal to 100. If μ is still 100 with the pill, then we have not demonstrated the predicted relationship. Thus, H_0 implies that there really is no relationship here and so the independent variable does not work as predicted.

> *REMEMBER* The alternative hypothesis implies that the sample data represent populations that form the predicted relationship. The null hypothesis implies that the sample data poorly represent populations which do not form the predicted relationship.

After you've created the statistical hypotheses, the final step prior to collecting data is to select and set up the appropriate statistical procedure. We'll violate the order of things, however, and go directly to some data so that you can understand what it is we're setting up.

The Logic of Statistical Hypothesis Testing

The statistical hypotheses for the IQ pill study are

$$H_0: \mu = 100$$

$$H_a: \mu \neq 100$$

Notice that, together, H_0 and H_a always include all possibilities, so one or the other of them must be true for a given sample: Our sample mean represents either a μ equal to 100 or a μ not equal to 100. Each is a hypothesis—a guess—that may or may not be correct. We use inferential procedures to test (choose between) the hypotheses. To see why we need "statistical hypothesis testing," say that we randomly selected a sample of 36 people, gave them the IQ pill, measured their IQ, and found that the mean IQ score was 105. Can we conclude that the IQ pill works?

We would *like* to say this: People who have not taken the IQ pill have a mean IQ of 100, so if the pill did not work, then the sample mean should have been 100. However, our sample mean was 105. This suggests that the pill does work, raising IQ scores about 5 points. If the pill does this for the sample, we assume that it would do this for the population. Therefore, we expect that the population that received the pill would have a μ of 105. Our results appear to be consistent with our alternative hypothesis, which states that the sample represents a population mean not equal to 100. Thus, it seems that if we were to measure everyone in the population with and without the pill, we would have the distributions shown previously in Figure 13.1, with the population that received the pill located at a μ of 105. Conclusion: We have demonstrated that the pill works. In nature there is a relationship such that increasing the amount of the pill from 0 to 1 is associated with increasing IQ scores.

But hold on! Not so fast! Remember sampling error? We just assumed that the sample is *perfectly* representative of the population it represents. We said that if the pill did not work the sample mean *should* have been 100. What we should have said is "If the pill did not work *and there was no sampling error*, the sample mean should have been 100." But what if there *was* sampling error? Maybe we obtained a mean of 105 not because the pill works, but because we inaccurately represented the situation where the pill does *not* work. Maybe the pill does nothing, but we happened to select smarter-than-average participants who already had a high IQ. Maybe the null hypothesis is correct: Even though it does not look like it, maybe the sample represents the population where μ is still 100. Maybe we have not demonstrated that the pill works.

In fact, we can never *know* whether the IQ pill works based on the results of one study. Whether the sample mean is 105, 1050, or 105,000, it is still possible that the null hypothesis is true: The pill doesn't work, the sample actually represents the population where μ is 100, and the sample mean is different from 100 because of sampling error.

> ***REMEMBER*** The null hypothesis always maintains that the sample data reflect sampling error and thus do not really represent the predicted relationship.

To know for certain whether the pill actually works, we would have to give it to the entire population and see whether μ was 100 or 105. We cannot do that. Instead, the best we can do is to use our study to increase our *confidence* that the pill works. The null hypothesis says that sampling error gave us the misleading impression that the pill works. If we can reject the null hypothesis, then we will be more confident that the pill works. From this perspective, statistical hypothesis testing is merely a tool to increase our confidence that a study demonstrates the predicted relationship.

So first we must make a decision about that pesky sampling error. In statistical hypothesis testing, the hypothesis that we test is always the null hypothesis: Here, we will test the idea that there is sampling error in representing the population having a μ of 100, and the pill does not really work. If sampling error can reasonably explain our results, then we will *not* accept that the pill works. After all, it makes no sense to believe that the pill works if the results can be easily explained as sampling error from the population of IQ scores *without* the pill. As scientists trying to describe nature, we want to be very careful not to conclude that we have demonstrated a relationship when actually the results are due to chance factors. Therefore, we won't buy that the pill works unless we are convinced that the results are *not* due to sampling error. Therefore, only if we can disconfirm the null hypothesis do we accept the alternative hypothesis (that the pill works).

Although we cannot determine whether the null hypothesis is true, we can determine how *likely* it is that a sample mean of 105 will occur through sampling error when the sample actually represents the population where μ is 100. If such a mean is too unlikely, then we reject H_0, rejecting the idea that the sample poorly represents the population where μ is 100. If this sounds familiar, it's because this is the procedure discussed in the previous chapter. In fact, all parametric and nonparametric inferential statistics involve the same logic: We test the null hypothesis by determining the probability of obtaining the sample data from the population described by the null hypothesis.

Whether you should use a particular inferential procedure is determined by whether the study meets the assumptions of the procedure. The IQ pill study meets the assumptions of the parametric inferential procedure known as the z-test.

TESTING A MEAN WHEN σ_X IS KNOWN: THE *z*-TEST

You already know how to perform the *z*-test. The ***z*-test** is the procedure for computing a *z*-score for a sample mean on the sampling distribution of means that we've discussed in previous chapters. The formula for the *z*-test is the formula used in Chapter 9 and again in Chapter 12 (and we'll see it again in a moment). The assumptions of the *z*-test are:

1. We have randomly selected one sample.
2. The dependent variable is at least approximately normally distributed in the population and it involves an interval or ratio scale (the mean is the appropriate measure of central tendency.
3. We know the mean of the population of raw scores under some other condition of the independent variable.
4. We know the true standard deviation of the population, σ_X, described by the null hypothesis. (It is not estimated using the sample.)

We can use the *z*-test for the IQ pill study because we have one sample of IQ scores, such scores are from an interval variable, the population of IQ scores is normally distributed, and we know σ_X. Say that in the research literature, in the population where μ is 100, the standard deviation is 15.

> ***REMEMBER*** The *z*-test is used only if the raw score population's σ_X is known.

Now that we have chosen the statistical procedure, we can set up the sampling distribution.

Setting Up the Sampling Distribution for a Two-Tailed Test

In the IQ pill study, H_0 is that the sample represents the population where μ is 100 (and σ_X is 15). To test it, we examine the sampling distribution of means created from the raw score population where μ is 100 and σ_X is 15. To create the sampling distribution, assume we have again hired the (very) bored statistician. Using the N of 36 that we used, she infinitely samples the raw score population of IQ scores where μ is 100. The central limit theorem tells us that her sampling distribution will be normally distributed and that the μ of the distribution will equal the μ of the IQ raw score population, 100. Recall that this is also the value of μ in our null hypothesis (H_0: $\mu = 100$). H_0 says that because we are sampling from the raw score population where μ is 100, the average sample mean will also be 100.

> ***REMEMBER*** The mean of the sampling distribution always equals the value described by H_0.

You can call this sampling distribution the null, or H_0, sampling distribution, because it describes the situation *when null is true*: Here it describes random samples from a population where μ *is* 100 and any sample mean not equal to 100 occurs *solely* because of sampling error—the luck of the draw that determined who was selected for that particular sample. (Always add the phrase "when null is true" to any information you obtain from a sampling distribution.)

Recall that those means that are considered to be "too unlikely" are located in the region of rejection. Once you envision the sampling distribution, set up the statistical test by identifying the size and location of the region of rejection.

Choosing alpha: The size of the region of rejection Recall from Chapter 12 that the *criterion* is the probability that defines sample means that are "too unlikely" to represent the underlying raw score population. The criterion is also the theoretical size of the region of rejection. (Previously, when our criterion was .05, the region of rejection comprised the extreme 5% of the curve.) Here is a new symbol: The symbol for the criterion and the theoretical size of the rejection region is α, the Greek letter **alpha**. Psychologists usually set their criterion at .05, so in code they say $\alpha = .05$.

Locating the region of rejection in the two-tailed test The region of rejection can be located either in both tails of the distribution or in only one tail. Which arrangement to use depends on your experimental hypotheses and thus on your statistical hypotheses—in particular, the alternative hypothesis. We predicted that the pill would make people either smarter or dumber, resulting in the alternative hypothesis that the sample mean represents either a μ larger than or smaller than 100. We will be correct if the sample mean is either larger than 100 or smaller than 100 and, in either case, too unlikely to represent the population for which $\mu = 100$. Sample means that are either larger or smaller than 100 and also too unlikely to represent that population are those in either the upper or lower tails of the distribution. Therefore, as shown in Figure 13.4, we place part of the region of rejection in the tail above $\mu = 100$ and part of it in the tail below $\mu = 100$.

Placing the region of rejection in the two tails of the distribution creates a two-tailed test. A **two-tailed test** is used to test statistical hypotheses in which we predict that

FIGURE 13.4 H_0 Sampling Distribution of IQ Means for a Two-Tailed Test

There is a region of rejection in each tail of the distribution, marked by the critical values of ± 1.96.

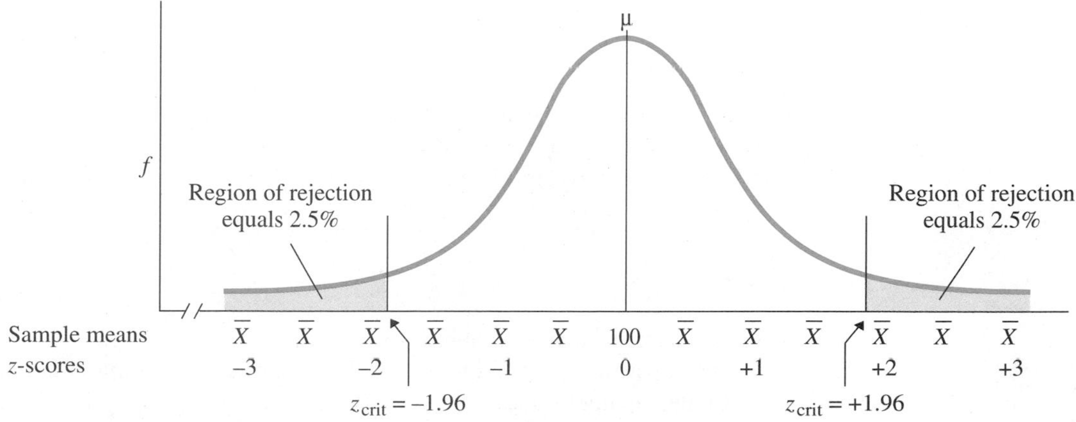

there is some kind of relationship, but we do not specifically predict that scores will only increase or only decrease. You have a two-tailed test whenever your null hypothesis simply states that the population parameter equals some value and your alternative hypothesis states that the population parameter does not equal that value.

> *REMEMBER* A two-tailed test is used when we do not predict the direction in which scores will change.

Determining the critical value Once you have selected the two-tailed test, you identify its critical value. We'll abbreviate the critical value of z as z_{crit}. Therefore, as shown back in Figure 13.4, z_{crit} demarcates the region of rejection. With $\alpha = .05$, the region of rejection in each tail contains the extreme 2.5% of the distribution. From the z-tables we see that a z-score of 1.96 demarcates the extreme 2.5% of the curve. Because either $+1.96$ or -1.96 demarcates a part of our region of rejection, z_{crit} is ± 1.96.

Now the test of H_0 boils down to this: If the z-score for our sample mean of 105 lies beyond ± 1.96, then the mean lies in the region of rejection. Thus, the final step is to compute the z-score for the sample.

Computing the z-Test

Here is some more code. The z-score we compute is "obtained" from the data, so it is called *z obtained*, which we abbreviate as z_{obt}.

As in previous chapters,

THE COMPUTATIONAL FORMULA FOR THE z-TEST IS

$$z_{obt} = \frac{\overline{X} - \mu}{\sigma_{\overline{X}}}$$

$\overline{X}$ is the value of our sample mean. μ is the mean of the sampling distribution when H_0 is true (the μ of the raw score population that H_0 says the sample represents). $\sigma_{\overline{X}}$ is the standard error of the mean, which is computed as

$$\sigma_{\overline{X}} = \frac{\sigma_X}{\sqrt{N}}$$

where N is the N of our sample and σ_X is the true population standard deviation.

First find the standard error of the mean, $\sigma_{\overline{X}}$. For the IQ pill study, the population standard deviation, σ_X, is 15, and the sample N is 36. Putting these values into the formula for $\sigma_{\overline{X}}$ gives

$$\sigma_{\overline{X}} = \frac{\sigma_X}{\sqrt{N}} = \frac{15}{\sqrt{36}} = \frac{15}{6} = 2.5$$

Thus, the sampling distribution has a $\sigma_{\overline{X}}$ of 2.5.

Now compute the z-score for our sample mean of 105 by putting the appropriate values into the formula for z_{obt}:

$$z_{obt} = \frac{\overline{X} - \mu}{\sigma_{\overline{X}}} = \frac{105 - 100}{2.5} = \frac{+5}{2.5} = +2.0$$

Thus, on the sampling distribution from the population where $\mu = 100$ (and $\sigma_{\overline{X}} = 2.5$), our sample mean of 105 has a z-score of +2.

If H_0 is correct, then ideally we'd expect a sample mean of 100 and thus a z_{obt} of 0. Even allowing for some sampling error, we'd expect a mean close to 100 and thus a z_{obt} close to 0. However, our sample produced a z_{obt} of +2! To confirm our suspicions, we compare z_{obt} to z_{crit}.

INTERPRETING z_{obt}

If we are to believe H_0, then a mean of 105 should be likely to occur when sampling from the raw score population where μ is 100. If so, then the sampling distribution should show that a mean of 105 occurs relatively frequently and is thus likely in this situation. However, the sampling distribution in Figure 13.5 shows just the opposite. The location of a z_{obt} of +2.0 tells us that the bored statistician would hardly ever obtain a sample mean of 105 when she was drawing samples that represent the population where μ is 100. This makes it difficult to believe that we obtained our sample mean by drawing from the population where μ is 100. In fact, because our z_{obt} of +2.0 is beyond the z_{crit} of ± 1.96, the sample mean is in the region of rejection. Therefore, we conclude that our mean of 105 is "too unlikely" to accept as representing the population where $\mu = 100$. That is, the idea that our sample is merely a poor representation of the population where $\mu = 100$ is not reasonable: Samples are seldom *that* poor at representing this population. Therefore, because it is too unlikely that we would obtain such

FIGURE 13.5 Sampling Distribution of IQ Means

The sample mean of 105 is located at $z_{obt} = +2.0$.

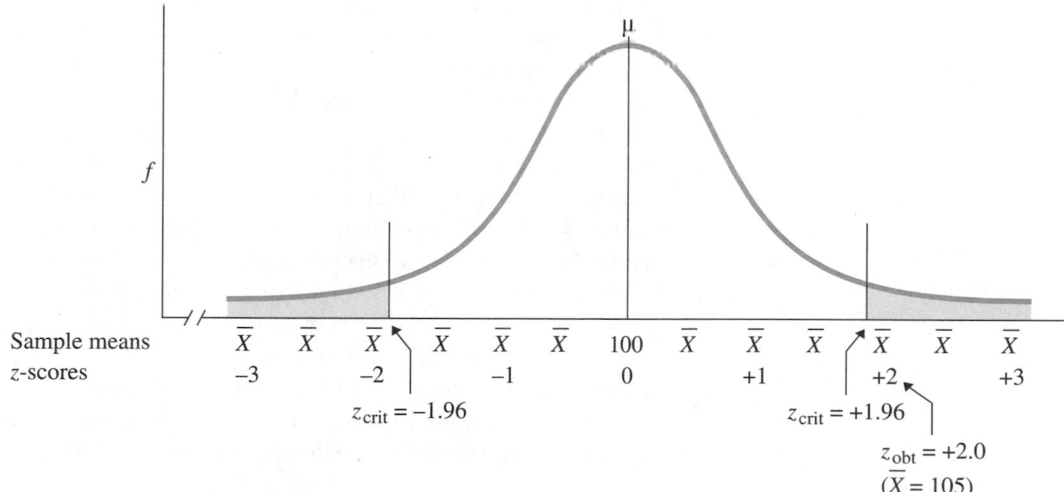

a sample when representing the population where $\mu = 100$, we reject the idea that our sample represents this population.

We have just rejected the null hypothesis. H_0 states that our sample represents the population where μ is 100, and we have found that this is not a reasonable hypothesis. Whenever our sample mean falls in the region of rejection, we say that we "reject H_0." If we reject H_0, then we are left with H_a, and we also say that we "accept H_a." Here H_a is $\mu \neq 100$, so we accept that the sample represents a population where μ is not 100. Thus, in sum, we have determined that our sample is unlikely to represent the population where μ is 100, so we conclude that it is likely to represent a population where μ is not 100.

> *REMEMBER* When a sample statistic falls beyond the critical value, the statistic lies in the region of rejection on the H_0 sampling distribution. This indicates that the statistic is too unlikely to accept as representing the parameter described by H_0, so reject H_0 and accept H_a.

Reporting Significant Results

The shorthand way of communicating that we have rejected H_0 and accepted H_a is to use the term *significant*. (Statistical hypothesis testing is sometimes called significance testing.) *Significant* does *not* mean important or impressive. **Significant** indicates that the results are too unlikely to occur if the predicted relationship does not exist in the population. Therefore, it implies that the relationship found in a study is "believable," representing a "real" relationship found in nature and not a chance pattern resulting from sampling error.

> *REMEMBER* The term *significant* indicates that we have rejected the null hypothesis and believe that our data reflect a real relationship.

The term *significant* can be used in several ways. In the IQ study, we might say that the pill produced a "significant difference" in IQ scores. This indicates that the difference between our sample mean and the μ found without the pill is too large to accept as occurring by chance if the data represent that μ. Or we might say that we obtained a "significant z": Our sample mean has too large a z-score to accept as occurring by chance if the sample represents the μ described by H_0. Or we can say that there is a "significant effect of the pill": We have decided that the change in IQ scores reflected by the sample mean is not caused by chance sampling error, so presumably it is an effect caused by changing the conditions of the independent variable.

It is very important to remember that your decision is simply either yes, to reject H_0, or no, to not reject H_0. All z-scores in the region of rejection are treated the same, so one z_{obt} cannot be "more significant" than another. Likewise, there is no such thing as "very significant" or "highly significant." (That's like saying "very yes" or "highly yes.") If z_{obt} is beyond z_{crit}, regardless of how far it is beyond, you completely and fully reject H_0 and the results are simply significant, period!

But also recognize that whether a result is significant depends on how you define "too unlikely." *Significant* implicitly means that *given your* α and therefore the size of your region of rejection, you have decided that the data are too unlikely to represent the situation described by the null hypothesis. Therefore, anytime you report the results of a statistical test, indicate the statistic computed, the obtained value, and the α used. Thus, to report our significant z_{obt} of +2.0, we would write

$$z_{obt} = +2.0, \ p < .05$$

Notice that instead of indicating that α equals .05, we indicate that the probability, *p*, is less than .05, or $p < .05$. We'll discuss the reason for this shortly.

Interpreting Significant Results

In accepting H_a, we also accept the corresponding experimental hypothesis that the independent variable works as predicted: Apparently, the IQ pill study demonstrated that the pill works. However, there are three very important restrictions on how far we can go in claiming that the pill works.

First, we did not *prove* that H_0 is false. All we have "proven" is that a sample of 36 scores is unlikely to produce a mean of 105 if the scores represent the population where $\mu = 100$ and σ_X is 15. But unlikely does not mean impossible. As the sampling distribution shows, unrepresentative means of 105 do occur once in a while when we *are* representing this population. Maybe the pill did not work, and our sample was simply very unrepresentative. There is *always* that possibility.

Second, the term "significant" indicates only that the *numbers* in the data are unlikely to occur if the sample represents a population of *numbers* in which the relationship does not exist. It does not mean that these numbers accurately reflect our variables, or that the behaviors and constructs are related in the way that we've hypothesized. Thus, by saying we have demonstrated a "real" relationship, we're saying that the sample data reflect *some* variable in nature that is associated with the higher scores we call IQ scores. But that variable may *not* be the pill. In other words, we might have had a confounding: The higher IQ scores in our sample might actually have occurred because participants cheated on the IQ test, or because there was something in the air that made them smarter or because there were sunspots, or who-knows-what! Only a well-controlled design can eliminate such factors, so as usual, we still must *argue* that it is the pill that produced higher IQ scores.

Finally, assuming that the pill increased IQ scores and produced the mean of 105, then it is logical to assume that if we gave the pill to everyone in the population, the resulting μ would be 105. Even if the pill works, however, the μ is probably not *exactly* 105. Our sample may reflect (you guessed it) sampling error! That is, the sample may accurately indicate that the pill influences IQ, but it may not perfectly represent how much the pill influences scores. Therefore, if we gave the pill to the population, we might find a μ of 104, or 106, or *any* other value. However, a sample mean of 105 is most likely when the population μ is around 105, so we would conclude that the μ resulting from our pill is probably *around* 105.

Bearing these qualifications in mind, we can return to our sample mean of 105 and interpret it the way we wanted to several pages back: It looks as if the pill increases IQ scores by about 5 points. But now, because we know that the sample mean is *significantly* different from a μ of 100, we are confident that we are not being misled by sampling error. Therefore, we are more confident that a relationship exists in the population and that we have discovered something about how nature operates. (But stay tuned, because we could be wrong.) Nonetheless, we now proceed to interpret and explain the relationship "psychologically," generalizing the results back to the hypothetical constructs, theories, or models which generated the study.

Retaining H_0

For the sake of illustration, let's say that the IQ pill had instead produced a sample mean of 99. Should we conclude that the pill decreases IQ scores, or should we conclude that these data reflect sampling error from the population that occurs with no pill where μ is 100? Using the z-test, we compute the z-score for the sample as

$$z_{obt} = \frac{\overline{X} - \mu}{\sigma_{\overline{X}}} = \frac{99 - 100}{2.5} = \frac{-1.0}{2.5} = -.40$$

Thus, this sample mean has a z_{obt} of $-.40$. We again examine the sampling distribution, shown in Figure 13.6. The z_{obt} of $-.40$ is *not* beyond the z_{crit} of ± 1.96, so the sample mean does not lie in the region of rejection on the H_0 sampling distribution where μ is 100. As the figure shows, when the bored statistician sampled this population, she frequently obtained a mean of 99 because of sampling error. Thus, *our* sample mean of 99 was likely to have occurred because of sampling error when representing this population. Therefore, the null hypothesis—that our sample is merely a poor representation of the population where μ is 100—is a reasonable hypothesis. At the very least, we certainly have no reason to think that H_0 is *not* true. So, in the language of statistics, we say that we have "failed to reject H_0" or we "retain H_0." Sampling error from the population where μ is 100 can explain these results just fine, thank you, so we will not reject this explanation.

FIGURE 13.6 Sampling Distribution of IQ Means

The sample mean of 99 has a z_{obt} $-.40$.

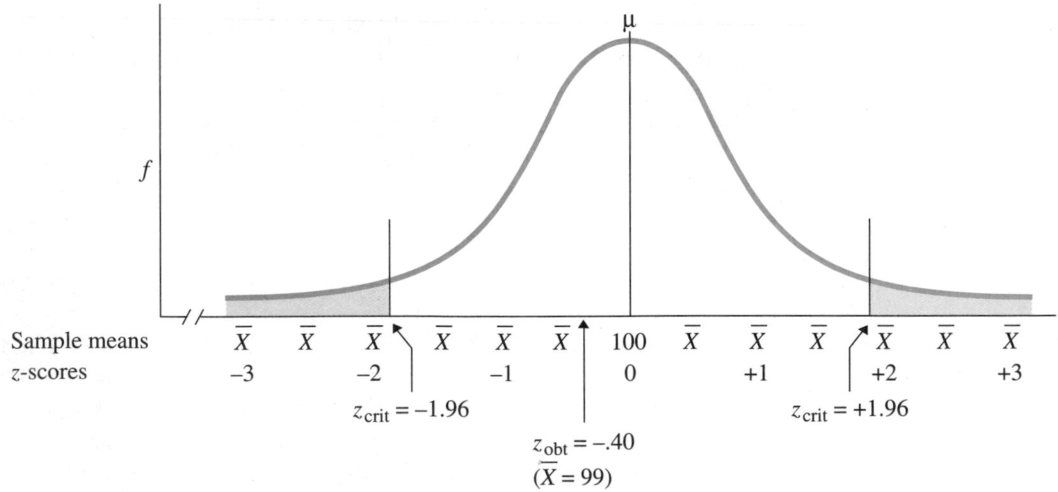

REMEMBER Retain H_0 when z_{obt} does not lie beyond z_{crit} and thus does not lie in the region of rejection.

The shorthand way to communicate all of this is to say that we have nonsignificant results (we don't say insignificant). **Nonsignificant** indicates that results are not too unlikely to accept as being due to sampling error. In other words, the differences reflected by the results were likely to have occurred even if there is no real relationship in the population.

When a result is not significant, we again report the α level used in making the decision. Thus, to report our nonsignificant z_{obt} when $\alpha = .05$, we would write

$$z_{obt} = -0.40, \ p > .05$$

Notice that with nonsignificant results, we indicate that p is *greater* than .05.

Interpreting Nonsignificant Results

When we retain H_0, we also retain the corresponding experimental hypothesis that the independent variable did not work as expected (we cannot yet become rich and famous). However, we have not proven that the predicted relationship does not exist in the population or that the pill does *not* work. We have simply failed to find convincing evidence that the pill *does* work. The only thing we are sure of is that sampling error *could* have reasonably produced data such as ours when the relationship does not exist. Because we are likely to obtain a sample with a mean of 99 without the pill, why should we think the pill works?

By failing to reject H_0, we *still* have two contradictory hypotheses that are both viable interpretations of our results: (1) H_0, that the data only reflect sampling error and show no real relationship, and (2) H_a, that the data do not reflect sampling error and do show a relationship. Thus, maybe in fact the pill did not work. Or maybe the pill did work, but it changed scores by so little that we did not see that it worked. Or maybe the pill changed IQ scores greatly, but we did not see the change because we have so much sampling error in representing the different population μ that would be created. We simply don't know if the pill works or not.

Thus, when we do not reject H_0, we cannot decide whether the relationship exists in the population or not. Therefore, we do not say anything about whether the independent variable actually influences behavior or not, and we do not even begin to interpret the results psychologically. All that we can say is that we have failed to find a significant difference, and thus we have failed to demonstrate the predicted relationship in the population.

REMEMBER Nonsignificant results provide no convincing evidence—one way or the other—as to whether a "real" relationship exists.

You cannot design a study that is intended to show that no relationship exists. For example, you could not run a study to demonstrate that the pill does not work, because, at best, you'll end up retaining both H_0 and H_a.

SUMMARY OF STATISTICAL HYPOTHESIS TESTING

The steps and logic we've discussed are used in all inferential procedures, so it's worthwhile to review them. In any research project, after creating the experimental hypotheses and designing the study to demonstrate the predicted relationship, you must:

1. Create the statistical hypotheses: The null hypothesis (H_0) describes the population parameter that the sample statistic represents if the predicted relationship does *not* exist. The alternative hypothesis (H_a) describes the population parameter that the statistic represents if the predicted relationship *does* exist.
2. Select the appropriate parametric or nonparametric procedure by matching the assumptions of the procedure to the study.
3. Select the value of α, which determines the size of the region of rejection.
4. Collect the data and compute the obtained value of the inferential statistic. This is analogous to finding a z-score for the sample data on the sampling distribution.
5. Set up the sampling distribution and, based on α and the way the test is set up, determine the critical value.
6. Compare the obtained value to the critical value.
7. If the obtained value lies beyond the critical value in the region of rejection, reject H_0, accept H_a, and describe the results as significant. Then describe and interpret the relationship in the population based on the sample data.
8. If the obtained value does not lie beyond the critical value in the region of rejection, do not reject H_0, and describe the results as nonsignificant. Because the data may reflect sampling error, do *not* draw any conclusions about the relationship.

THE ONE-TAILED TEST

In some experiments, we predict that scores will only increase or only decrease, and then we perform a one-tailed test. A **one-tailed test** is used when we predict the *direction* in which scores will change. For a one-tailed test, you must set up the statistical hypotheses and the sampling distribution differently than for a two-tailed test.

The One-Tailed Test for Increasing Scores

Say that we had developed a "smart" pill. Then the experimental hypotheses would be: (1) We will demonstrate that the pill makes people smarter by increasing IQ scores or (2) We will not demonstrate that the pill makes people smarter.

The alternative hypothesis again follows the experimental hypothesis that the independent variable works as predicted. Here, because the population without the pill has a μ of 100, if the smart pill worked and we gave it to everyone, it would *increase* IQ scores, so the population μ would be greater than 100. In symbols, this alternative hypothesis is:

H_a: $\mu > 100$

H_a implies that our sample mean represents the *larger* population μ that would occur if the pill worked as predicted.

The null hypothesis again implies that the independent variable does not work as predicted. A "smart pill" does not work if it either leaves IQ scores unchanged or *decreases* IQ scores (making people dumber). Therefore, if the smart pill did not work as predicted, the population would have a μ either equal to 100 or less than 100. This null hypothesis is

H_0: $\mu \leq 100$

H_0 implies that our sample mean represents one of these populations.

We again test H_0 by examining the sampling distribution that describes the sample means that occur when H_0 is true. The specific population of raw scores we use to create the sampling distribution is the one in which $\mu = 100$. This is because, if the population of IQ scores with the pill is above 100, then it is automatically above any value less than 100.

> **REMEMBER** In a one-tailed test, the null hypothesis always includes a population μ equal to some value. Test H_0 by testing whether the sample data represent that population.

We again set $\alpha = .05$. Because we have a one-tailed test, however, we place the entire region of rejection in only *one tail* of the sampling distribution. You can identify which tail to put it in by identifying what data you must see in order to claim that your independent variable works as predicted. To say that the smart pill works, we must conclude that the sample mean represents a μ larger than 100. To do that, first the sample mean itself must be larger than 100 (if not, then not even the people in the sample are smarter than people not given the pill). Second, the sample mean must be *significantly* larger than 100. The means that are significant and larger than 100 are in the region of rejection that is in the upper tail of the sampling distribution. Therefore, we place the entire region of rejection in the upper tail of the distribution, to form the sampling distribution shown in Figure 13.7. We don't place anything in the lower tail, because for this study, sample means down there are not different from means close to 100: All show that the smart pill doesn't work, and we are not interested in distinguishing between whether it doesn't work because it has no influence on IQ or because it makes people dumber. As in the previous chapter, the region of rejection in the upper tail of the distribution that constitutes 5% of the curve is marked by a z_{crit} of +1.645.

Say that after testing the smart pill on a sample of 36 participants, we find $\overline{X} = 106.58$. The sampling distribution is still based on the population of IQ scores where $\mu = 100$ and $\sigma_X = 15$, so using the previous data and formula for the z-test, this sample mean has a z-score of $(106.58 - 100)/2.5$, which is +2.63. As shown in Figure 13.7, this z_{obt} is beyond the z_{crit}, so our sample mean is among those sample means that occur only 5% of the time when samples are representing the population where μ is 100. Therefore, we conclude that the sample mean is too unlikely to accept as representing the population where $\mu = 100$. Further, if the sample is too unlikely to represent the population where μ is 100, then the sample is way too unlikely to represent a population where μ is *less than* 100. Therefore, we reject the null hypothesis that

FIGURE 13.7 Sampling Distribution of IQ Means for a One-Tailed Test of Whether Scores Increase

The region of rejection is entirely in the upper tail.

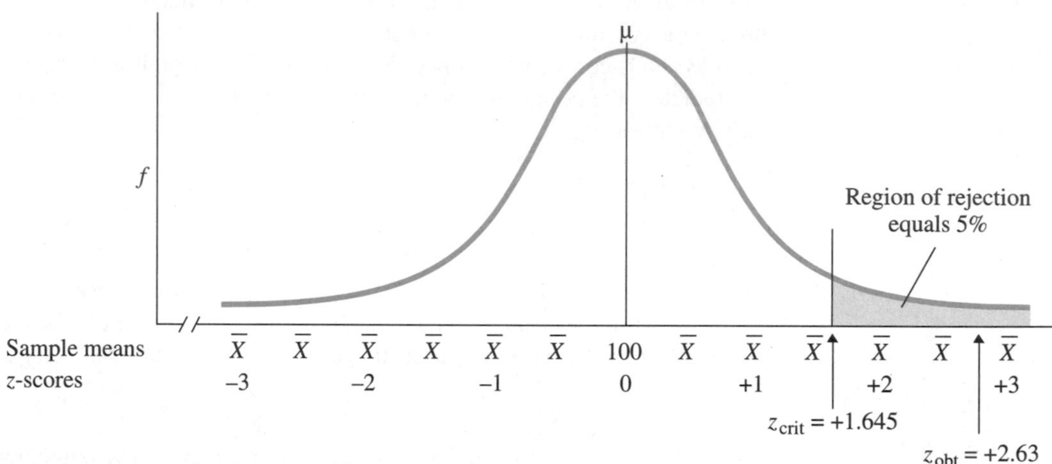

$\mu \leq 100$, and accept the alternative hypothesis that $\mu > 100$. We are now confident that these results do not reflect sampling error, but rather represent the predicted relationship where the smart pill works. We conclude that the pill produces a significant increase in IQ scores, and we estimate that with the pill, μ would equal about 106.58 (keeping in mind all of the cautions and qualifications for interpreting significant results that we discussed previously).

If the z_{obt} did not lie in the region of rejection, we would retain H_0, and we would have no evidence as to whether the smart pill works.

The One-Tailed Test for Decreasing Scores

We can also arrange a one-tailed test using the lower tail. Say that we expect the pill to lower IQ scores. Now if we gave the pill to the entire population and it works, the population μ would be *less than* 100. On the other hand, if the pill does not work, it would produce the same population as no pill (with $\mu = 100$) or it would make people smarter (with $\mu > 100$). Thus the hypotheses tested here are

Pill does not work: H_0: $\mu \geq 100$
Pill works: H_a: $\mu < 100$

Again we test the null hypothesis, using the sampling distribution from the raw score population where μ is 100. But the only way to conclude that the pill lowers IQ is if our sample mean is significantly *less* than 100. Therefore, the entire region of rejection is in the lower tail of the distribution, to form the sampling distribution in Figure 13.8

With $\alpha = .05$, the z_{crit} is now *minus* 1.645. If the sample mean produces a *negative* z_{obt} beyond -1.645 (for example, $z_{obt} = -1.69$), then we reject the H_0 that the sample mean represents a μ equal to or greater than 100, and we accept the H_a that the sample represents a μ less than 100. This significant result would increase our confidence that we have demonstrated a relationship in nature where the pill lowers IQ scores. If,

FIGURE 13.8 Sampling Distribution of IQ Means for a One-Tailed Test of Whether Scores Decrease

The region of rejection is entirely in the lower tail.

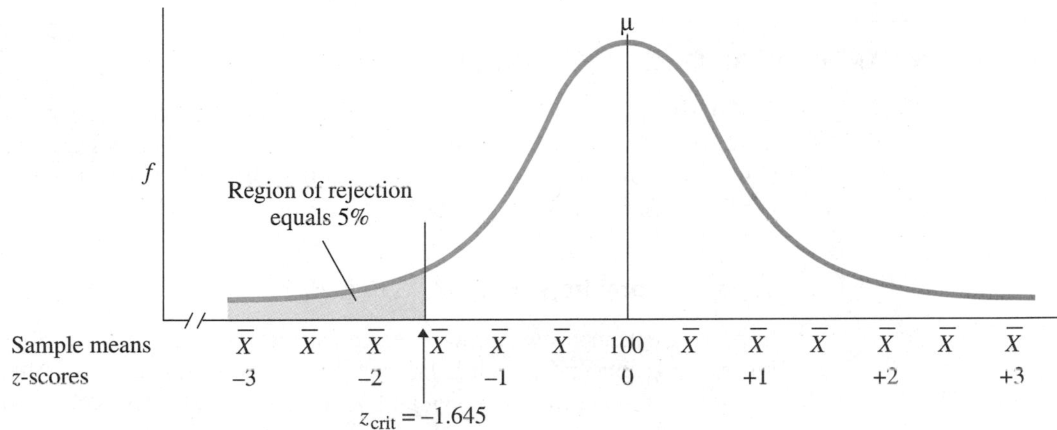

however, the z_{obt} did not fall in the region of rejection (for example, $z_{obt} = -1.25$), we would not reject H_0 and would have no evidence as to whether the pill works or not.

Choosing One-Tailed Versus Two-Tailed Tests

Look again at the previous sampling distributions for the one-tailed tests. There is a danger here, because there is no region of rejection in one tail of each distribution. This means that if the pills work opposite to the way we predict we *cannot* reject H_0. If the smart pill produces a mean that is below 100, then the pill does not *increase* scores as predicted. If the pill to lower IQ produces a mean above 100, then that pill does not *decrease* scores as predicted. In either case, regardless of the size of z_{obt}, we would retain H_0 and the corresponding experimental hypothesis that we have not demonstrated that the pill works *as predicted.*

> **REMEMBER** In a one-tailed test, z_{obt} is significant only if it lies beyond z_{crit} *and* has the same sign as z_{crit}.

You cannot switch the region of rejection and the sign of z_{crit} after the results are in to make them significant. If the smart pill produced lowered scores, you could not switch and say, "Whoops, I meant to say it decreased scores." Remember, the tail you use is determined by your hypothesis, and, after years of developing the theoretical and biochemical basis for a "smart pill," it would make no sense to suddenly say that the same basis leads you to predict it's a "dumb pill." Likewise, it makes no sense to switch between a one-tailed and a two-tailed test after the fact. Therefore, a one-tailed test should be used *only* if you have a *convincing* reason for predicting the direction in which the independent variable will change scores. Otherwise, use a two-tailed test. This is the safer approach because it allows you to conclude that an independent variable produced a change, even if you could not accurately predict whether it would increase or decrease scores.

REMEMBER Use a two-tailed test unless you have a definite prediction about the direction in which scores will change.

ERRORS IN STATISTICAL DECISION MAKING

One other major issue to consider when performing hypothesis testing involves potential errors. These are not errors in our calculations, but rather they are errors in our decisions: Regardless of whether we conclude that sample data do or do not represent the predicted relationship, we may be wrong.

Type I Errors: Rejecting H_0 When H_0 Is True

In previous examples where we rejected H_0 and claimed that the various pills worked, it is still possible that the pill did not work and that we obtained an unlikely and unrepresentative sample from the population where μ is 100. That is, it is possible that the sample so poorly represented the situation where the pill did not work that we mistakenly thought the pill did work. If so, then we made a Type I error. A **Type I error** is defined as rejecting H_0 when H_0 is true. In a Type I error, we have so much sampling error that we—and our statistical procedures—are fooled into concluding that the predicted relationship exists when it really does not. In practical terms, we conclude that the independent variable works when it doesn't.

Whenever we discuss Type I errors, it is a given that H_0 is true. Therefore, a Type I error can occur *only* when we reject H_0. Then, the probability of making a Type I error is the probability that we have rejected H_0. This probability is determined by the size of the region of rejection, so, to use the lingo, it is α that determines the probability of a Type I error. For example, Figure 13.9 shows the region of rejection in a one-tailed test when α is .05.

FIGURE 13.9 Sampling Distribution of Sample Means Showing That 5% of All Sample Means Fall in the Region of Rejection When H_0 Is True

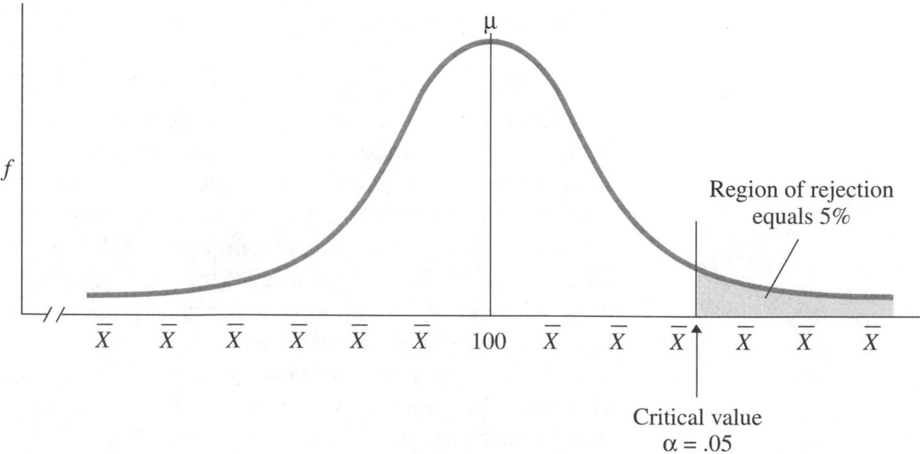

Making a Type I error is losing the "bet" that sampling error did not produce our sample mean. Therefore, the logic here is the same as in the previous chapter when we determined the probability of losing any bet. As the sampling distribution shows, sample means in the region of rejection occur 5% of the time when H_0 is true. But we reject H_0 100% of the time when a sample mean falls in the region of rejection. Therefore, sample means that cause us to reject H_0 occur 5% of the time when H_0 is true. If we reject H_0 when it is true 5% of the time, then, in other words, 5% of the time we will make a Type I error. Thus, over the long run, the relative frequency of Type I errors is .05. Because relative frequency equals probability, the theoretical probability of making a Type I error is .05 whenever we reject H_0. (For a two-tailed test, when α is .05, the total region of rejection is still .05, so again the theoretical probability of making a Type I error is .05.)

Now you can understand why a study must meet the assumptions of a statistical procedure: If we violate the assumptions, then the true theoretical probability of a Type I error will be *greater* than we think it is. But recall that parametric tests are robust. This means that if we violate their assumptions somewhat, the probability of a Type I error will still be close to our α. For example, if we set α at .05 and violate the assumptions of a parametric test somewhat, then the theoretical probability of a Type I error will be close to .05, such as .051. If we greatly violate the assumptions, however, we may think that α is .05 when in fact it is much larger, such as .20!

Here is an important distinction: Although the *theoretical* probability of a Type I error equals α, the *actual* probability of a Type I error is slightly less than α. This is because in figuring the size of the region of rejection, we include the critical value: In Figure 10.9, 5% of the curve is *at* or above +1.645. Yet to reject H_0, a z_{obt} must be *larger* than the critical value. We cannot determine the precise area under the curve for the infinitely small point located at 1.64500, so we can't remove it from the 5%. All that we can say is that when α is .05, any z_{obt} that falls in the region of rejection is in slightly less than 5% of the curve. Because the actual area of the region of rejection is less than α, the actual probability of a Type I error is also slightly less than α.

Thus, in any of the previous examples when we rejected H_0, the probability that we made a Type I error was slightly less than .05. That is why we reported our significant result as $z_{obt} = +2.0$, $p < .05$. Think of $p < .05$ as a shortened form of "p(Type I error) $< .05$". This indicates that the probability is slightly less than .05 that we made a Type I error by calling this result significant when we shouldn't have. It is very important to always report the probability that you have made a Type I error.

On the other hand, we reported our nonsignificant result as $z_{obt} = -.40$, $p > .05$. Here, $p > .05$ communicates that the reason we did not call this result significant is because in order to do so, our α and the region of rejection would have to be larger than .05. Then the probability of a Type I error would be greater than .05, and that's unacceptable.

We typically set α at .05 because this is an acceptably low probability of making a Type I error. If α is larger than .05, then we are too likely to conclude that a relationship exists when in fact it does not. This may not sound like a big deal, but for example, the next time you fly in an airplane, consider that the designer's belief that the wings will stay on may actually be a Type I error: Sampling error misled the designer into *erroneously* thinking the wings will stay on. A 5% chance of this being the case is scary enough—we certainly don't want more than a 5% chance. Recall that in science, we are skeptical and make decisions like a jury: We want to be convinced "beyond a reasonable doubt" that sampling error did not produce our results, and having only a 5% chance that it did is convincing.

Sometimes, however, making a Type I error can be especially dangerous, so we want to reduce even further the probability of making one. Then we usually set α at .01. For example, we might have set α at .01 if the smart pill had some dangerous side effects. We would be especially concerned about needlessly subjecting the public to these side effects if the pill doesn't work, so we would want to make it even less likely that we will conclude that the pill works when it really does not. When α is .01, the region of rejection is in the extreme 1% of the sampling distribution, so we have a larger absolute critical value than when α is .05. This means that the z_{obt} must be larger in order for us to reject H_0. Intuitively, it takes even more to convince us that the pill works, and thus there is a lower probability that we will make an error. Statistically, now we will reject H_0 only 1% of the time when H_0 is true, so the probability of making a Type I error is now $p < .01$.

Remember however, that we use the term *significant* in an all-or-nothing fashion: a result is *not* "more" significant when $\alpha = .01$ than when $\alpha = .05$. If z_{obt} falls in the region of rejection that was used to define significant, then the result is significant, period! The *only* difference is that when $\alpha = .01$, there is a smaller probability that we have made a Type I error.

Many computer programs compute the exact probability of a Type I error for a particular result. For example, we might see "$p = .02$." This indicates that the sample mean falls in the extreme 2% of the sampling distribution, and thus there is a .02 probability of a Type I error here. If our α is .05, then this result is even more convincing than we need, so it is significant. Conversely, we might see "$p = .07$," To call this result significant, we'd need a region of rejection that is the extreme 7% of the sampling distribution. This implies an α of .07, however, which is greater than an α of .05 and thus too big. Therefore this result is not significant.

We can make a Type I error only when we *reject* H_0 when H_0 is true. If we *retain* H_0 when H_0 is true, then we avoid making a Type I error by making the correct decision. Because α is the theoretical probability of making a Type I error, $1 - \alpha$ is the probability of avoiding a Type I error. In other words, if 5% of the time we obtain samples in the region of rejection when H_0 is true, then 95% of the time we will obtain samples that are not in the region of rejection when H_0 is true. Thus, 95% of the time when H_0 is true, we will not reject it and so we will not make a Type I error.

> *REMEMBER* A Type I error is rejecting H_0 when H_0 is true, and its theoretical probability equals α. Avoiding a Type I error is retaining H_0 when H_0 is true, and the theoretical probability here equals $1 - \alpha$.

Type II Errors: Retaining H_0 When H_0 Is False

In addition to Type I errors, it is possible to make a totally different kind of error. This error can occur only when H_0 is false, and it is called a Type II error. A **Type II error** is retaining H_0 when H_0 is false (and H_a is true). With a Type II error, we conclude that we have no evidence for the predicted relationship in the population, when, in fact, the relationship exists. Here, in practical terms, we fail to identify an independent variable that really does work.

Anytime we discuss Type II errors, it is a given that H_0 is false and H_a is true. We make a Type II error because our sample mean is so close to the mean of the population described by H_0 that we conclude that the sample reflects sampling error from that population. We might have made a Type II error previously when we obtained a sample mean of 99 and did not reject H_0. Perhaps the μ of the population with the pill would be 99, and our sample represents this μ. It's also possible that the pill would increase IQ scores greatly, say to a μ of 115, but we obtained a very unrepresentative sample with a mean of 99. In either case, the sample mean of 99 was so close to 100 (the μ without the pill), that we retained H_0 and made a Type II error.

The computation of the probability of Type II errors is beyond the scope of this discussion, but you should know that the symbol for the theoretical probability of a Type II error is β, the Greek letter **beta**. Whenever we retain H_0, β is the probability that we have made a Type II error.

We can make a Type II error only when we retain H_0 and H_0 is false. If we reject H_0 (say it is false) when H_0 is false, we have avoided a Type II error and made a correct decision. This would be the case, for example, if we concluded that the IQ pill works, and it really does work. If β is the probability of making a Type II error, then $1 - \beta$ is the probability of avoiding a Type II error. Thus, anytime we reject H_0, the probability is $1 - \beta$ that we have made a correct decision and rejected a false H_0.

REMEMBER A Type II error is retaining H_0 when H_0 is false, and its theoretical probability equals β. Avoiding a Type II error is rejecting H_0 when H_0 is false, and the theoretical probability here equals $1 - \beta$.

Comparing Type I and Type II Errors

There's no doubt that Type I and Type II errors are two of the most confusing inventions ever devised. So, first recognize that Type I and Type II errors are mutually exclusive: If you could have made one type of error, then you cannot have made the other. Sometimes (although we never know when) the null hypothesis really is true. In that case, rejecting null is a Type I error and retaining it is the correct decision. But sometimes (we also never know when) the null hypothesis is really false. In this case, retaining null is a Type II error and rejecting it is the correct decision.

Second, recognize that when we consider the possible true state of affairs regarding H_0, there are actually four possible outcomes in any study. Look at Table 13.1.

TABLE 13.1 Possible Results of Rejecting or Retaining H_0

		Our decision	
		We reject H_0 (claim H_a is true)	*We retain H_0 (claim H_0 may be true)*
The truth about H_0	*H_0 is true (H_a is false: no relationship exists)*	We make a Type I error ($p = \alpha$)	We are correct, avoiding a Type I error ($p = 1 - \alpha$)
	H_0 is false (H_a is true: a relationship exists)	We are correct, avoiding a Type II error ($p = 1 - \beta$)	We make a Type II error ($p = \beta$)

The first row describes the situation when H_0 is really true. If we reject H_0 when H_0 is true, we make a Type I error, and the probability of this is α. If we retain H_0 when H_0 is true, we avoid a Type I error by making a correct decision, and the probability of this is $1 - \alpha$. The second row describes the situation when H_0 is really false. If we retain H_0 when H_0 is false, we make a Type II error, and the probability of this is β. If we reject H_0 when H_0 is false, we avoid a Type II error by making a correct decision, and the probability of this is $1 - \beta$.

In any experiment, the results of our hypothesis testing procedures will place us in one of the two columns. If we reject H_0, either we've made a Type I error or we've made the correct decision and avoided a Type II error. If we retain H_0, either we've made a Type II error or we've made the correct decision and avoided a Type I error.

Statistical procedures are designed to minimize the probability of Type I errors because it is most serious to conclude that an independent variable works when really it does not. Basing scientific "facts" on what are actually Type I errors can cause untold damage. On the other hand, Type II errors are also important. In order for us to learn about nature, we must avoid Type II errors and conclude that an independent variable works when it really does.

The Power of a Statistical Test

The goal of scientific research is to reject H_0 when H_0 is false: We conclude that the pill works, and the truth is that the pill does work. Not only have we avoided making an error, but we have also learned something about nature. This ability is so important that it has a special name: **Power** is the probability that we will reject H_0 when it is actually false, correctly concluding that the sample data reflect a real relationship. When null is false, the only other thing we might do is retain it, making a Type II error. Therefore, in other words, power is the probability that we will avoid a Type II error, so power equals $1 - \beta$.

The concept of power has shown up in previous chapters when we discussed techniques for producing a powerful design: This is a design that is likely to produce a strong, convincing sample relationship. We seek such a relationship because then, using our present terminology, our inferential procedures are likely to reject H_0 when it is false. Power is important because after all, why bother to conduct a study if we are unlikely to reject the null hypothesis even when the predicted relationship really *does* exist? Therefore, whether we have sufficient power is crucial whenever we decide to retain H_0. For example, previously, when we failed to conclude that the pill produced significant differences when the sample mean was 99, maybe the problem was that we lacked power: maybe the probability was not high that we would reject H_0, *even if the pill really worked.* Therefore, we always want to maximize our power so that we have confidence in our decision if we end up retaining H_0.

The logic behind maximizing power is this: When H_0 is false, we "should" reject it. In other words, when H_0 is false, our results "should" be significant. Thus, we maximize power by maximizing the probability that results will be significant. If it sounds like we're rigging the decision to reject H_0, remember this: With α at .05 or less, we minimize the probability of making the *wrong* decision when H_0 is true (not making a Type I error). At the same time, by maximizing power, we maximize the probability of making the *correct* decision when H_0 is false (not making a Type II error).

Results are significant if z_{obt} is larger than z_{crit}. Therefore, we maximize power by maximizing the size of the obtained value relative to the critical value. This increases the probability that the results will be significant and thus increases the probability of rejecting H_0 when it is false.

> *REMEMBER* The larger the obtained value, the more likely it is to be significant and thus the greater the power.

There are several things researchers do to maximize power, but they boil down to two approaches. First, as we've seen in previous chapters, we create a powerful design that is likely to result in powerful data: Data that show a strong, convincing sample relationship. Inferential statistics essentially consider the strength of the relationship, with a stronger relationship producing a larger obtained value.

Second, we employ powerful statistical procedures. For example, previously I said we use nonparametric procedures only if we cannot use parametric procedures. This is because parametric tests are more powerful than nonparametric tests: If we analyze data using a parametric test, we are more likely to reject H_0 than if we analyze the same data using a nonparametric test. Therefore, we try to design research that incorporates parametric procedures.

In addition, a one-tailed test is more powerful than a two-tailed test. This is because the z_{crit} for a one-tailed test is smaller (closer to the mean of the sampling distribution) than the z_{crit} for a two-tailed test. Previously, we used a z_{crit} of 1.645 for a one-tailed test and 1.96 for a two-tailed test. All other things being equal, a z_{obt} is more likely to be beyond 1.645 than beyond 1.96. Thus, we are more likely to conclude that results are significant with a one-tailed test than with a two-tailed test. (With $\alpha = .05$ for both, the probability that we will incorrectly call the result significant is the same.) Of course, remember that we can reject H_0 only if z_{obt} has the same sign as z_{crit}.

In later chapters we'll see additional ways to maximize power.

PUTTING IT ALL TOGETHER

Essentially, the purpose of inferential statistics is to minimize the probability of making Type I errors. If we had not performed the z-test for the studies in this chapter, we might have concluded that the IQ pills worked, even though we were actually being misled by sampling error. We would have no idea if this occurred, nor even the chances that it occurred. After finding a significant result, however, we can be confident that we have not made a Type I error, because we know the probability of doing so is less than .05. (Similarly, if results are not significant, through the concept of power we can estimate the probability that we've made a Type II error, so we can be confident that we have not missed the fact that the pill works.)

All statistical hypothesis testing procedures follow the logic described here: H_0 is the hypothesis that says your data represent the populations you would find if the predicted relationship does not exist; H_a says that your data represent the predicted relationship. You then compute something like a z-score for the data on the sampling distribution when H_0 is true. If the z-score is larger than the critical value, it is unlikely that the results represent the populations described by H_0, so reject H_0 and accept H_a. That's it! That's inferential statistics (well, not quite).

CHAPTER SUMMARY

1. *Sampling error* occurs when random chance produces a sample statistic that is not equal to the population parameter it represents.

2. *Inferential statistics* are procedures for deciding whether sample data represent a particular relationship in the population.

3. *Parametric statistics* are inferential procedures that require assumptions about the parameters of the underlying raw score populations the data represent. They are performed with normally distributed interval or ratio scores.

4. *Nonparametric statistics* are inferential procedures that do not require assumptions about the population parameters represented by a sample. They are performed when interval or ratio scores are not normally distributed, or when nominal or ordinal scores are being measured.

5. The *alternative hypothesis*, H_a, is the statistical hypothesis that describes the population μs being represented if the predicted relationship exists in the population. H_a implies that the sample mean represents one of these μs.

6. The *null hypothesis*, H_0, is the statistical hypothesis that describes the population μs being represented if the predicted relationship does not exist. H_0 implies that the sample mean represents one of these μs.

7. A *two-tailed test* is used when the direction in which the dependent scores will change is not predicted. A *one-tailed test* is used when the direction of the relationship is predicted.

8. *Alpha*, α, is the theoretical size of the region of rejection. Typically, α equals .05.

9. The *z-test* is the parametric procedure used in a single-sample experiment if (a) The population of interval or ratio scores is normally distributed and (b) The standard deviation of the raw score population is known.

10. If z_{obt} lies beyond z_{crit}, then the corresponding sample mean lies in the region of rejection. This indicates that the mean is unlikely to occur when randomly sampling the population described by H_0. Therefore, *reject* H_0 and *accept* H_a. This is called a *significant* result and is taken as evidence that the predicted relationship exists in the population.

11. If z_{obt} does not lie beyond z_{crit}, then the corresponding sample mean is *not* located in the region of rejection. This indicates that the mean is likely to occur when randomly sampling the population described by H_0. Therefore, *retain* H_0. This is called a *nonsignificant* result and is taken as a failure to obtain evidence that the predicted relationship exists in the population.

12. A *Type I error* occurs when a true H_0 is rejected. The theoretical probability of a Type I error equals α. If a result is significant, the probability of a Type I error is $p < \alpha$. The theoretical probability of avoiding a Type I error by retaining a true H_0 is $1 - \alpha$.

13. A *Type II error* occurs when a false H_0 is retained. The theoretical probability of making a Type II error is β. The theoretical probability of avoiding a Type II error by rejecting a false H_0 is $1 - \beta$.

14. The *power* of a statistical test is the probability of rejecting a false H_0, and it equals $1 - \beta$. When used appropriately, parametric procedures are more powerful than nonparametric procedures, and one-tailed tests are more powerful than two-tailed tests. The manner in which a study is designed and conducted also influences power.

PRACTICE PROBLEMS

(Answers for odd-numbered problems are provided in Appendix D.)

1. (a) What is sampling error? (b) Why does the possibility of sampling error present a problem to researchers when inferring a population μ from a sample $\overline{X}$?
2. What are inferential statistics used for?
3. (a) What is the difference between a real relationship and one in a sample that results from sampling error? (b) By a real relationship, do we mean that the independent variable is necessarily related to the dependent variable?
4. For each of the following, decide whether the researcher should perform parametric or nonparametric procedures. (a) When ranking the intelligence of a group of participants given a smart pill. (b) When comparing the median income for a sample to that of the national population. (c) When comparing the mean reading speed for a sample of deaf children to the average reading speed of the population of hearing children. (d) When measuring interval scores from a personality test given to a group of emotionally troubled people, and comparing them to the normal distribution of such scores found in the population of emotionally healthy people.
5. (a) Why do researchers prefer parametric procedures? (b) Why can we use parametric procedures even if we cannot perfectly meet their assumptions?
6. (a) What does H_0 communicate? (b) What does H_a communicate?
7. (a) When do you use a one-tailed test? (b) When do you use a two-tailed test?
8. (a) What are the advantage and the disadvantage of two tailed tests? (b) What are the advantage and the disadvantage of one-tailed tests?
9. For each of the following experiments, describe the experimental hypotheses (identifying the independent and dependent variables): (a) Studying whether the amount of pizza consumed by college students during finals week increases relative to the rest of the semester. (b) Studying whether performing breathing exercises alters blood pressure. (c) Studying whether sensitivity to pain is affected by increased levels of hormones. (d) Studying whether frequency of dreaming decreases as a function of more light in the room while sleeping.
10. For each study in problem 9, indicate whether a one-tailed or a two-tailed test should be used, and state the H_0 and H_a. Assume that $\mu = 50$ when the amount of the independent variable is zero.
11. (a) What does α stand for, and what two things does it determine? (b) How does the size of α affect whether a result is significant or nonsignificant?
12. What does it mean to have significant results in terms of (a) The obtained and critical value? (b) The region of rejection of the sampling distribution? (c) The likelihood of obtaining our sample mean when H_0 is true? (d) α?

13. A researcher predicts that listening to music while taking a test is beneficial. He obtains a sample mean of 54.63 when 49 subjects take a test while listening to music. The mean of the population of students who have taken this test without music is 50 ($\sigma_X = 12$). (a) Should he use a one-tailed or two-tailed test? Why? (b) What are H_0 and H_a for this study? (c) Compute z_{obt}. (d) With $\alpha = .05$, what is z_{crit}? (e) Does the researcher have evidence of a relationship in the population? If so, describe the relationship.

14. A researcher wonders whether attending a private high school leads to higher or lower performance on a test of social skills. A random sample of 100 students from a private school produces a mean score of 71.30 on the test, and the national mean score for students from public schools is 75.62 ($\sigma_X = 28$). (a) Should she use a one-tailed or a two-tailed test? Why? (b) What are H_a and H_a for this study? (c) Compute z_{obt}. (d) With $\alpha = .05$, what is z_{crit}? (e) What should the researcher conclude about this relationship in the population?

15. (a) What is the probability that the researcher in problem 13 made a Type I error, and what would the error be in terms of the independent and dependent variables? (b) What is the probability that the researcher in problem 13 made a Type II error, and what would the error be in terms of the independent and dependent variables?

16. (a) What is the probability that the researcher in problem 14 made a Type I error? What would the error be in terms of the independent and dependent variables? (b) What is the probability that the researcher in problem 14 made a Type II error? What would the error be in terms of the independent and dependent variables?

17. A researcher hypothesizes that males and females are the same when it comes to intelligence. Why is this hypothesis impossible to test?

18. Researcher A finds a significant negative relationship between increasing stress level and ability to concentrate. Researcher B replicates this study but finds a nonsignificant relationship. Identify the statistical error that each researcher may have made.

19. (a) What is power? (b) Why do researchers want to maximize power? (c) Why is a one-tailed test more powerful than a two-tailed test?

20. A report indicates that Brand X toothpaste significantly reduced tooth decay relative to other brands, with $p < .44$. (a) What does this indicate about the researcher's decision about Brand X? (b) What makes you suspicious of the claim that Brand X works better than other brands?

21. Foofy claims that using a one-tailed test is cheating, because a one-tailed test produces a smaller absolute value of $z_{crit,}$ and therefore it is easier to reject H_0 than it is with a two-tailed test. If the independent variable doesn't work, she claims, we are more likely to make a Type I error. Why is she correct or incorrect?

22. Poindexter claims that the real cheating occurs when we increase power by increasing the likelihood that results will be significant. He reasons that we are more likely to reject $H_{0,}$ when H_0 is true, and therefore we are more likely to make a Type I error. Why is he correct or incorrect?

23. Bubbles reads a report of Study A, in which, using a two-tailed test, the results are significant: $z_{obt} = +1.97, p < .05$. She also reads about Study B, in which $z_{obt} = +14.21$, $p < .0001$. (a) She concludes that the results of Study B are way beyond the critical value used in Study A, falling in a region of rejection containing only .0001 of the sampling distribution. Why is she correct or incorrect? (b) She concludes that the results of Study B are more significant than those of Study A. Why is she correct or incorrect? (c) In terms of their conclusions, what is the difference between the two studies?

24. A researcher measures the self-esteem scores of a sample of statistics students, reasoning that their frustration with this topic may lower their self-esteem relative to that of the typical college student (where $\mu = 55$ and $\sigma_X = 11.35$). He obtains the following scores.

44 55 39 17 27 38 36 24 36

(a) Should he use a one-tailed or two-tailed test? Why? (b) What are H_0 and H_a for this study? (c) Compute z_{obt}. (d) With $\alpha = .05$, what is z_{crit}? (e) What should the researcher conclude about the relationship between the self-esteem of statistics students and that of other students?

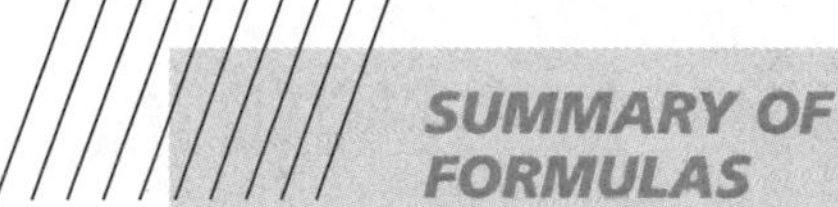

SUMMARY OF FORMULAS

1. *The computational formula for the z-test is*

$$z_{obt} = \frac{\overline{X} - \mu}{\sigma_{\overline{X}}}$$

where $\overline{X}$ is the value of our sample mean and μ is the mean of the sampling distribution when H_0 is true (the μ of the raw score population described by H_0).

2. *The computational formula for the standard error of the mean, $\sigma_{\overline{X}}$, is*

$$\sigma_{\overline{X}} = \frac{\sigma_X}{\sqrt{N}}$$

where N is the N of our sample and σ_X is the true population standard deviation.

14

The Single Sample Study: Testing a Sample Mean or Correlation Coefficient

To understand this chapter:

- From Chapter 2, recall the difference between a true and quasi-independent variable.
- From Chapter 8, recall that s_X is the *estimated* population standard deviation, that s_X^2 is the *estimated* population variance, and that both involve the degrees of freedom, *df*, which equals $N - 1$.
- From Chapter 10, recall the uses and interpretation of r, r_s, and r_{pb}.
- From Chapter 13, recall the basics of significance testing, including one-tailed and two-tailed tests, H_0 and H_a, Type I and Type II errors, and power

Then your goals in this chapter are to learn:

- The difference between the z-test and the t-test.
- How to perform hypothesis testing using the t-test.
- What the confidence interval for μ is.
- How to perform significance testing of r, r_s, and r_{pb}.
- How to increase the power of a t-test or correlation coefficient.

Statistical hypothesis testing is second nature to behavioral researchers, whether they're conducting descriptive or experimental research, in the laboratory or in the field. Therefore, the major issue for you is to learn the different procedures that are used in different research designs. This chapter introduces the procedure known as the t-test. Like the z-test, the t-test is used for significance testing in a single-sample experiment. You'll see that the t-test also forms the basis for significance testing of a correlation coefficient. Finally, this chapter introduces the confidence interval, a new procedure for describing a population μ.

MORE STATISTICAL NOTATION

The t-test's official name is Student's t-test (although it was developed by a statistician named W. S. Gosset). The answer obtained from the t-test is symbolized by t_{obt}. The critical value of t is symbolized by t_{crit}.

UNDERSTANDING THE t-TEST FOR A SINGLE SAMPLE MEAN

As Chapter 13 described, an assumption of the z-test is that we know the true standard deviation of the raw score population, σ_X, so that we can compute the true standard error of the mean of the sampling distribution, $\sigma_{\bar{X}}$. Then z_{obt} indicates the location of a sample mean on the sampling distribution of means. In most actual research, however, we do *not* know the standard deviation of the raw score population. Instead, we estimate σ_X by computing s_X. Then we use this estimated population standard deviation to compute an *estimate* of the standard error of the mean. With this estimated standard error of the mean, we again compute something *like* a z-score to locate the sample mean on the sampling distribution. Because this location is based on an estimate of the standard error, however, we have not computed a z-score and performed the z-test. Instead we have computed t_{obt} and performed the single-sample t-test. The **single-sample t-test** is the parametric procedure used to test the null hypothesis for a single-sample experiment when the standard deviation of the raw score population must be estimated.

We are still asking the same question here that we asked in Chapter 13, but the characteristics of our design simply dictate that we use a slightly different procedure to answer it. Here is an example. Say that in one of those "home-and-gardening/good-housekeeper" magazines, there is a test of one's housekeeping abilities. The magazine is targeted at women, and it reports that nationally, the average test score for women is 75 (so their μ is 75), but it does not bother to report the standard deviation. Our question is "How do men score on this test?" To answer it, we'll randomly select a sample of adult males (perhaps even using stratified sampling for representativeness) and give them the test. The resulting sample of test scores represents the population of test scores for men. Therefore, we will compare the μ of the population of men represented by the sample to the μ of 75 for the population of women. If men score differently from women, then we'll have a relationship, where as the (quasi-) independent variable of gender changes, the population of test scores changes.

We're open-minded, so we won't predict the direction in which test scores will change. Therefore, we have the following two-tailed experimental hypotheses: (1) The experiment will demonstrate that men score either higher or lower than women, and (2) The experiment will not demonstrate differences in test scores as a function of gender.

The null hypothesis always indicates that we have not demonstrated the predicted relationship, so here it indicates that gender has no effect on scores. If so, the μ for men will equal the μ for women, which is 75. Thus, the null hypothesis is

$$H_0: \mu = 75$$

H_0 implies that our sample represents the population of *people* where $\mu = 75$. You can call the population men or call it women, it makes no difference because it's the same batch of scores.

The alternative hypothesis always indicates that we have demonstrated the predicted relationship, so here it indicates that gender is related to test scores. If so, then the μ for men should not equal the μ for women of 75. Therefore, the alternative hypothesis is

$$H_a: \mu \neq 75$$

H_a implies that the sample represents a different population for men where $\mu \neq 75$.

The next step is to select and set up the appropriate statistical test. First, we choose alpha: $\alpha = .05$ sounds good. Second, we check that the study meets the assumptions of the single-sample *t*-test.

The Assumptions of the Single-Sample *t*-Test

To perform the single sample t-test, you should be able to assume the following about your dependent variable:

1. There is one random sample of interval or ratio scores.

2. The raw score population forms a normal distribution.

3. The standard deviation of the raw score population is estimated by the s_X computed from the sample.

You're OK if the data are only somewhat normally distributed, because like all parametric tests, the *t*-test is robust, meaning that it produces minimal error even if you violate its assumptions somewhat. This is especially true if N is at least 30.

Based on similar types of research, let's assume that the housekeeping study will meet these assumptions, so we collect the data.

The Logic of the Single-Sample *t*-Test

Say that we test 9 men. (As you'll see, for maximum power, we never collect so few scores in an actual experiment.) The sample mean turns out to be $\overline{X} = 65.67$, so we'd expect the population of men to score around a μ of 65.67. Because the population of females scores around a μ of 75, maybe we have demonstrated a relationship in which changing gender from female to male results in lower housekeeping scores.

But hold on! What about sampling error? Maybe gender has no influence on test scores, and we are being misled by sampling error: Maybe our sample represents the same population of scores as found for women, but by chance we selected some exceptionally sloppy men who just happened to have low test scores. If so, then the sample poorly represents the population where $\mu = 75$. Maybe the null hypothesis is true, and the study did not demonstrate a relationship.

To be confident in our conclusions about this relationship, we first test this null hypothesis by performing the *t*-test. We use exactly the same logic as in the *z*-test: H_0 says that the mean of 65.67 is different from the μ of 75 because of sampling error. By computing t_{obt}, we determine the location of the sample mean on the sampling distribution of means that occurs when a sample *does* represent the population described by H_0. This allows us to determine the likelihood of obtaining our sample mean if the sample represents a population where μ is 75. If t_{obt} is beyond t_{crit}, our sample mean lies in the region of rejection, so we will reject the idea that the sample represents the population where μ equals 75.

The only differences between the *z*-test and the *t*-test are that t_{obt} is not calculated in the same way as z_{obt} and the value of t_{crit} is obtained from the *t*-distribution instead of from the *z*-distribution.

CALCULATING THE SINGLE-SAMPLE *t*-TEST

The computation of t_{obt} consists of three steps that parallel the three steps performed in the *z*-test. There, we first determined the true standard deviation (σ_X) of the raw score population. For the *t*-test, we compute an estimated standard deviation of the raw score population (s_X) based on the sample data and the formula

$$s_X = \sqrt{\frac{\Sigma X^2 - \frac{(\Sigma X)^2}{N}}{N - 1}}$$

The second step of the *z*-test was to compute the true standard error of the mean by dividing the true standard deviation (σ_X) by the square root of *N*:

$$\sigma_{\overline{X}} = \frac{\sigma_X}{\sqrt{N}}$$

For the *t*-test, we compute the estimated standard error of the mean. The symbol for the **estimated standard error of the mean** is $s_{\overline{X}}$ (The *s* stands for an estimate of the standard deviation, and the subscript $\overline{X}$ indicates that it is for a distribution of means.) We compute the estimated standard error of the mean by dividing the estimated standard deviation by the square root of *N*.

THE DEFINITIONAL FORMULA FOR THE ESTIMATED STANDARD ERROR OF THE MEAN IS

$$s_{\overline{X}} = \frac{s_X}{\sqrt{N}}$$

Notice the similarity between the preceding formulas for $\sigma_{\overline{X}}$ and $s_{\overline{X}}$. The only difference is whether we divide into σ_X or s_X.

The third step in the z-test was to compute z_{obt} using the formula

$$z_{obt} = \frac{\overline{X} - \mu}{\sigma_{\overline{X}}}$$

Similarly, the final step in the t-test is to compute t_{obt}.

THE DEFINITIONAL FORMULA FOR THE t*-TEST FOR A SINGLE SAMPLE MEAN IS*

$$t_{obt} = \frac{\overline{X} - \mu}{s_{\overline{X}}}$$

$\overline{X}$ is the sample mean, μ is the mean of the H_0 sampling distribution (which also equals the value of μ described in the null hypothesis), and $s_{\overline{X}}$ is the estimated standard error of the mean.

Notice the similarity between the formulas for z_{obt} and t_{obt}. The z_{obt} indicates the distance that the sample mean is from the μ of the sampling distribution, measured in units called the standard error of the mean. The t_{obt} measures this distance in estimated standard error units.

Computational Formulas for the Single-Sample *t*-Test

You can compute t_{obt} using the three steps given above, or you can use one of the following computational formulas. First, replacing the symbol $s_{\overline{X}}$ with the formula for computing it gives

$$t_{obt} = \frac{\overline{X} - \mu}{\frac{s_X}{\sqrt{N}}}$$

This simply shows the computation of $s_{\overline{X}}$ in the denominator of the t-test by dividing s_X by the square root of N.

To shorten the computations, you can avoid taking the square root when computing s_X. Instead, replace the standard deviation with the estimated variance, s_X^2, and take the square root of the entire denominator. Thus,

THE COMPUTATIONAL FORMULAS FOR THE SINGLE SAMPLE t*-TEST ARE*

$$t_{obt} = \frac{\overline{X} - \mu}{\sqrt{\frac{s_X^2}{N}}} \quad \text{and} \quad t_{obt} = \frac{\overline{X} - \mu}{\sqrt{(s_X^2)\left(\frac{1}{N}\right)}}$$

You can use either of these formulas. The formula on the left computes $s_{\bar{X}}$ by dividing s_X^2 by N, and the formula on the right computes $s_{\bar{X}}$ by multiplying s_X^2 times the quantity $1/N$. Remember, after finding the square root, the final number in the denominator of any of these formulas is still the estimated standard error, $s_{\bar{X}}$. Then, dividing $s_{\bar{X}}$ into the difference found in the numerator gives t_{obt}.

For our housekeeping study, say we obtained the data in Table 14.1. First compute s_X^2. Substituting the data from Table 14.1 into the formula gives

$$s_X^2 = \frac{\Sigma X^2 - \frac{(\Sigma X)^2}{N}}{N - 1} = \frac{39289 - \frac{349281}{9}}{9 - 1} = 60.00$$

Thus, we estimate that the variance of the population of test scores represented by the sample is 60.

Now compute t_{obt} for the mean of 65.67 when the population μ is 75, s_X^2 is 60, and N is 9. Filling in a computational formula, we have

$$t_{obt} = \frac{\bar{X} - \mu}{\sqrt{(s_X^2)\left(\frac{1}{N}\right)}} = \frac{65.67 - 75}{\sqrt{(60)\left(\frac{1}{9}\right)}}$$

In the denominator, 1/9 is .11, which multiplied times 60 is 6.667. So

$$t_{obt} = \frac{65.67 - 75}{\sqrt{6.667}} = \frac{-9.33}{2.582} = -3.61$$

The square root of 6.667 is 2.582, so the estimated standard error of the mean, $s_{\bar{X}}$ is 2.582. Dividing -9.33 by 2.582, we arrive at a t_{obt} of -3.61.

Thus, our sample mean produced a t_{obt} of -3.61 on the sampling distribution of means where $\mu = 75$. This is very similar to having a *z*-score of -3.61. Now the question is, "Is a t_{obt} of -3.61 (and the underlying difference between the sample mean and μ) significant?" To answer this question, we compare the t_{obt} to the appropriate t_{crit}, and for that we must examine the *t*-distribution.

TABLE 14.1 Test Scores of Nine Participants

Participant	*Scores (X)*	X^2
1	50	2500
2	75	5625
3	65	4225
4	72	5184
5	68	4624
6	65	4225
7	73	5329
8	59	3481
9	64	4096
$N = 9$	$\Sigma X = 591$	$\Sigma X^2 = 39289$
	$(\Sigma X)^2 = 349281$	
	$\bar{X} = 65.67$	

The t-Distribution

In previous chapters, we described the sampling distribution of means using z-scores, because the *z*-distribution is the appropriate model of the sampling distribution when σ_X is known. The *t*-distribution is the appropriate model when σ_X is estimated. Think of the *t*-distribution in the following way. One last time, we hire our *very* bored statistician. She draws an infinite number of samples of the same size *N* from the raw score population described by H_0. For each sample, she computes $\overline{X}$, s_X, $s_{\overline{X}}$ and t_{obt}. She then plots the frequency distribution of the different means, labeling the *X* axis with the corresponding values of t_{obt} as well. Thus, the ***t*-distribution** is the distribution of all possible values of *t* computed for random sample means selected from the raw score population described by H_0.

You can envision a *t*-distribution as shown in Figure 14.1. As with *z*-scores, a sample mean equal to μ has a *t* equal to zero. Means greater than μ have positive values of *t*, and means less than μ have negative values of *t*. The larger the absolute value of *t*, the farther it and the corresponding sample mean are from the μ of the distribution. Therefore, the larger the *t*, the lower a mean's relative frequency and thus the lower its probability.

The t_{obt} locates our sample mean on this model, telling us the probability of obtaining the mean when H_0 is true. To complete the *t*-test, we find t_{crit} and create the region of rejection. If t_{obt} is beyond t_{crit}, then our sample mean is too unlikely a mean to accept as representing the population described by H_0.

There is one important novelty here: There are actually *many* versions of the *t*-distribution, each having a slightly different shape. The shape of a particular *t*-distribution depends on the size of the *N* that the statistician uses when creating the *t*-distribution. If she selects samples with a small *N*, the *t*-distribution will be a very rough approximation to the standard normal curve. This is because her samples will often contain large sampling error, so she will often obtain a value of s_X that is different from σ_X, and this inconsistency produces a *t*-distribution that is only roughly normal. Conversely, however, very large samples will tend to represent the population accurately, so s_X will be close to the one true value of σ_X. As with the *z*-test, using the true value of σ_X produces a sampling distribution that closely conforms to the true normal *z*-distribution. In between, as sample size increases, each *t*-distribution will be a successively closer approximation to the true normal curve.

FIGURE 14.1 Example of a *t*-Distribution of Random Sample Means

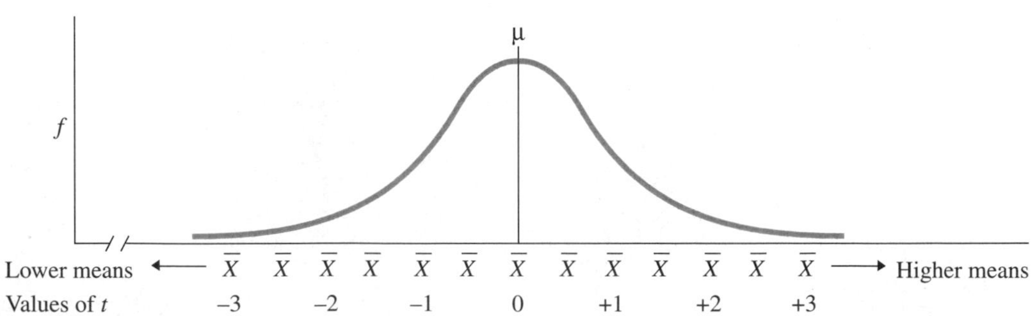

The fact that there are differently shaped *t*-distributions is important for one reason: When we set up the region of rejection, we want it to contain precisely that portion of the curve defined by our α. If $\alpha = .05$, then we want to mark off precisely the extreme 5% of the curve. On distributions that are shaped differently, we mark off that 5% at different locations. Because the size of the region of rejection is marked off by the critical value, *with differently shaped t-distributions we will have different critical values.* For example, Figure 14.2 shows a one-tailed region of rejection in two *t*-distributions. Say that the extreme 5% of Distribution A is beyond a t_{crit} of +2.5. If we also use +2.5 as t_{crit} on Distribution B, however, the region of rejection will contain *more* than 5% of the distribution. Conversely, the t_{crit} marking off 5% of Distribution B will mark off *less* than 5% of Distribution A. (The same problem also exists for a two-tailed test.)

This issue is important because not only is α the size of the region of rejection, it also determines the probability of a Type I error. Unless we use the t_{crit} that marks off the correct region of rejection, the actual probability of a Type I error will not equal our α (and that's not supposed to happen!). Thus, there is only one version of the *t*-distribution that we should use in a particular situation: The one that the bored statistician would have created when using the *same N* as in our sample.

The Degrees of Freedom

To be precise, it's not actually *N* that determines the appropriate *t*-distribution for a study. Instead, the shape of a particular *t*-distribution is determined by the size of $N - 1$, what we call the degrees of freedom, or *df*. Because we compute s_X using $N - 1$, it is *df* that determines how closely s_X estimates the true σ_X. Therefore, the larger the *df*, the closer the *t*-distribution is to forming a normal curve.

It does not take a tremendously large *df*, however, to produce a truly normal *t*-distribution. When *df* is greater than 120, the *t*-distribution is virtually identical to the standard normal curve. But for each value of *df* between 1 and 120 there is a differently shaped *t*-distribution. Thus, if a sample has a *df* between 1 and 120, we must determine

FIGURE 14.2 Comparison of Two *t*-Distributions Based on Different Sample *N*s

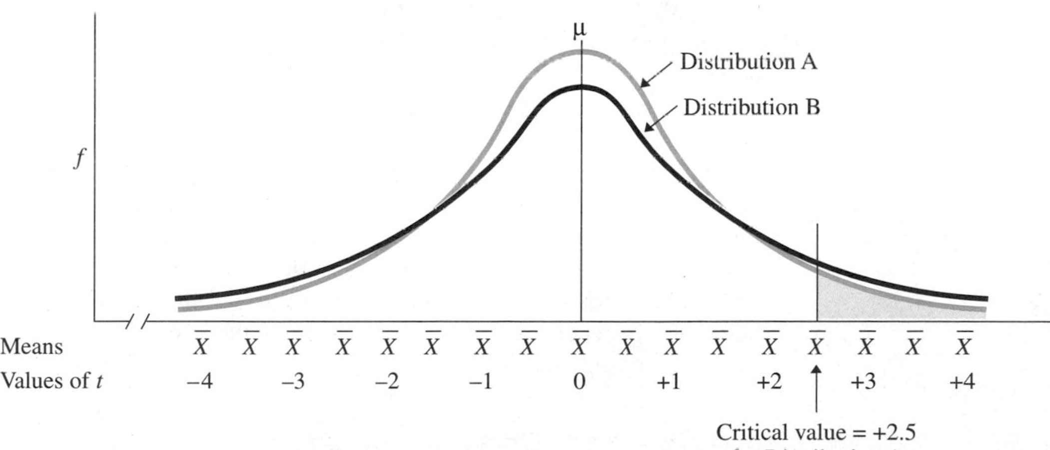

the appropriate critical value by first identifying the appropriate sampling distribution, and for that we use the *df* in the sample. Only then will t_{crit} accurately mark off the region of rejection so that the probability of a Type I error truly equals α.

> *REMEMBER* The appropriate t_{crit} for the single-sample *t*-test comes from the *t*-distribution that has *df* equal to $N - 1$, where *N* is the number of scores in our sample.

The *t*-Tables

You obtain the appropriate values of t_{crit} from Table 2 in Appendix C, entitled "Critical Values of *t*." Take a look at these "*t*-tables." You'll find separate tables for two-tailed and one-tailed tests. In each, identify the appropriate column for your value of α. Then find the value of t_{crit} in the row opposite the *df* in your sample. For example, in the housekeeping study, *N* is 9, so *df* is $N - 1 = 8$. For a two-tailed test with $\alpha = .05$ and $df = 8$, t_{crit} is ± 2.306.

As usual, the table contains no positive or negative signs. In a one-tailed test, you must decide whether t_{crit} is positive or negative. Also, notice that the table uses the symbol for infinity (∞) for *df* greater than 120. With a sample of this size, using the sample to estimate the population's standard deviation is virtually as good as knowing the population's true standard deviation. Then the *t*-distribution matches the standard normal curve, and the critical values are those we saw with the *z*-test.

Using the *t*-Tables

If you peruse the *t*-tables (a little light reading for the terminally bored), you will *not* find a critical value for every *df* between 1 and 120. When the *df* of your sample does not appear in the table, there are two approaches you can take.

First, remember that all you need to know is whether t_{obt} lies beyond t_{crit} (and thus lies in the region of rejection). Often you can determine this by examining the critical values for the *df* above and below the *df* of your sample. For example, say that we perform a one-tailed *t*-test at $\alpha = .05$ with 49 *df*. The *t*-tables give t_{crit} for 40 *df* (+1.684) and for 60 *df* (+1.671). Because 49 *df* lies between 40 *df* and 60 *df*, the critical value we seek lies between +1.671 and +1.684. It's a good idea to draw a picture of this, as shown in Figure 14.3.

Notice that the actual region of rejection starts at a point in between these two critical values. Therefore, if t_{obt} lies beyond the t_{crit} of +1.684, then it is already in the region of rejection for 49 *df*, and so it is significant. On the other hand, if t_{obt} is *not* beyond the t_{crit} of +1.671, then it is way short of the region of rejection for 49 *df*, and so it is not significant. In the same way, you can evaluate any obtained value that falls *outside* of the bracketing critical values given in the tables.

If t_{obt} falls *between* the bracketing values of t_{crit} in the table, however, then you should use the second approach, which is to use the interpolation procedure described in part B.3 of Appendix B.

Interpreting the Single-Sample *t*-Test

Remember the housekeeping study? Our purpose is to decide whether men produce a different population of test scores than women. Once we have the values of t_{obt} and t_{crit},

FIGURE 14.3 One *t*-Distribution Showing the Location of Three Values of t_{crit}

The t_{crit} of +1.684 is for 40 df (dashed line), the t_{crit} of +1.671 is for 60 df (dotted line), and the t_{crit} for 49 (solid line) is between them.

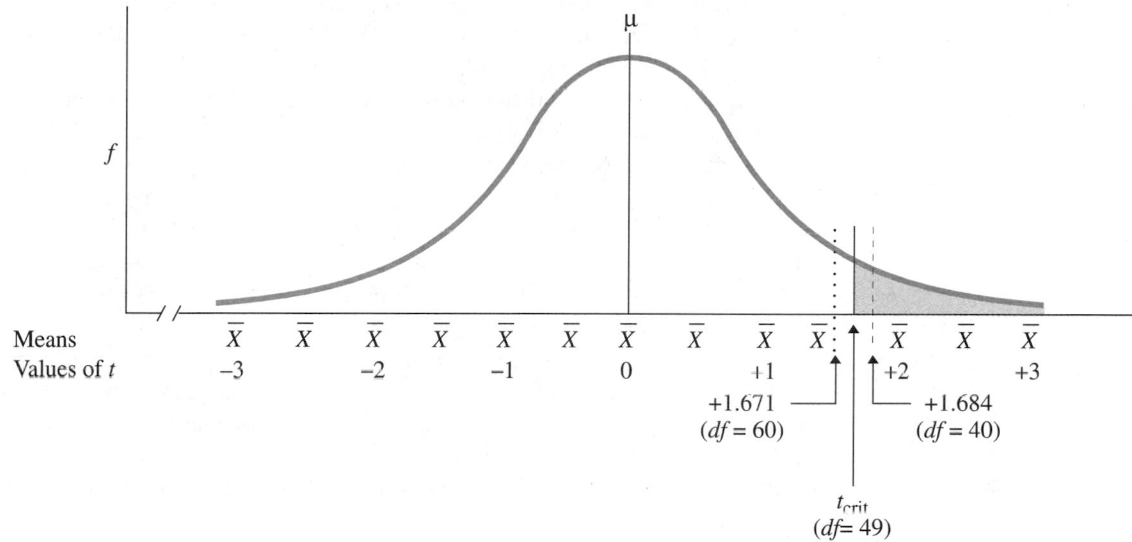

the single-sample *t*-test is identical to the *z*-test. Our t_{obt} is -3.61 and the two-tailed t_{crit} is ± 2.306. With this information, you can envision the sampling distribution shown in Figure 14.4. Remember, this is the sampling distribution when H_0 is true and the samples *do* represent the population where μ is 75 (and $s_{\overline{X}}^2 = 60$ and $N = 9$). H_0 implies that our mean of 65.67 poorly represents this μ. But, as shown, seldom is a mean *that* poor at representing this μ. In fact, the t_{obt} lies beyond the t_{crit}, so the mean is in the region of rejection. Thus, our $\overline{X}$ of 65.67 is too unlikely to occur when H_0 is true, so we reject H_0, rejecting the idea that the sample represents the population where μ is 75.

FIGURE 14.4 Two-Tailed *t*-Distribution for $df = 8$ When H_0 Is True and $\mu = 75$

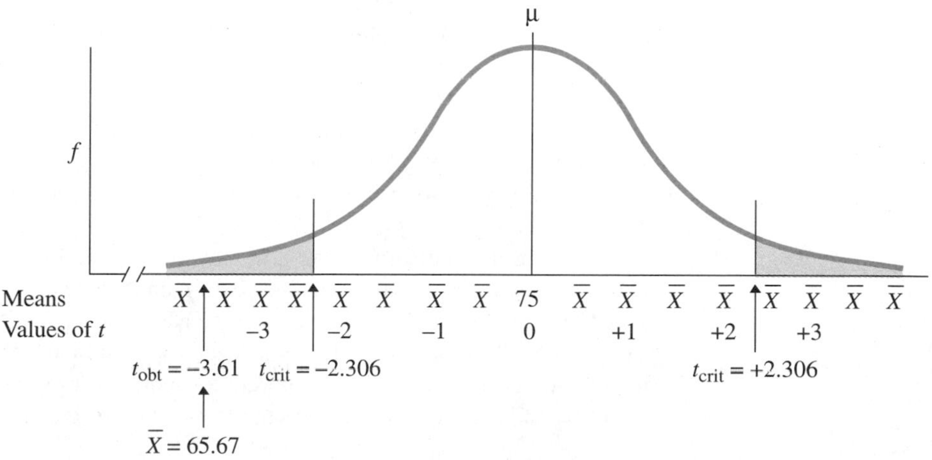

We report these results as

$$t(8) = -3.61, p < .05$$

This statement communicates that: We performed the *t*-test, we had 8 degrees of freedom, the t_{obt} is -3.61, and we judged the t_{obt} and the underlying $\overline{X}$ to be significant, with the probability less than .05 that we made a Type I error (rejected a true H_0).

Thus, we conclude that the sample mean of 65.67 is significantly different from the population mean of 75. In other words, we accept the alternative hypothesis that the sample mean represents a μ not equal to 75. Because a sample mean of 65.67 is most likely to occur when the sample represents the population where μ is 65.67, our best estimate is that the sample represents a population of scores for men located at around 65.67. Because we expect one population of scores for men located at around 65.67 and a different population of scores for women located at a μ of 75, we conclude that we have demonstrated a relationship in the population between the independent variable (gender) and the dependent variable (test scores).

Then, as always, we proceed to interpret the relationship psychologically, generalizing to relevant behaviors and constructs. Are men more ignorant about housekeeping, and if so, why? Or are they pretending to be ignorant so they can avoid housework, and if so, why? And so on. (Regardless of the interpretation, be especially cautious about concluding that gender *causes* differences in test scores. This quasi-independent variable is probably confounded by all sorts of subject variables—like differences in peer-pressure, stereotyping, or social desirability—that may cause men to have lower test scores.

Of course, if t_{obt} had not fallen beyond t_{crit} (for example, $t_{obt} = +1.32$), then it would not lie in the region of rejection and we would not reject H_0. We would conclude that the sample was likely to represent the population where μ is 75, so we would have no convincing evidence for a relationship between gender and test scores. We report such a nonsignificant result as $t(8) = +1.32, p > .05$.

Testing One-Tailed Hypotheses in the Single-Sample *t*-Test

We would perform a one-tailed test if, for example, we had specifically predicted that men score *higher* than women. Then H_a would be that the sample represents a population μ greater than 75 (or H_a: $\mu > 75$). The H_0 would be that the sample represents a population μ less than or equal to 75 (or H_0: $\mu \leq 75$). We would calculate t_{obt} as above and compare it to the one-tailed t_{crit} from the *t*-tables for our *df* and α. For the sample to represent the predicted population of higher scores, the sample mean must be *significant* and *larger* than 75. Therefore, the region of rejection is in the upper tail of the sampling distribution, as shown in the graph on the left in Figure 14.5. For t_{obt} to be significant, it must be positive and beyond t_{crit}. If it is, then our mean is too unlikely to represent a μ equal to or less than 75. Therefore, we reject H_0 and accept H_a—that the sample mean represents a μ greater than 75.

We would also perform a one-tailed test if we had predicted that men score *lower* than women, using the sampling distribution on the right in Figure 14.5. Now H_a is that μ is less than 75 and H_0 is that μ is greater than or equal to 75. Because we predict a sample mean less than 75, to be significant t_{obt} must be negative and beyond $-t_{crit}$. If it is, the mean is too unlikely to represent a μ equal to or greater than 75. Therefore, we reject H_0 and accept H_a—that the sample mean represents a μ less than 75.

FIGURE 14.5 H_0 Sampling Distributions of *t* for a One-Tailed Test

On the left, we predict an increase in scores. On the right, we predict a decrease in scores.

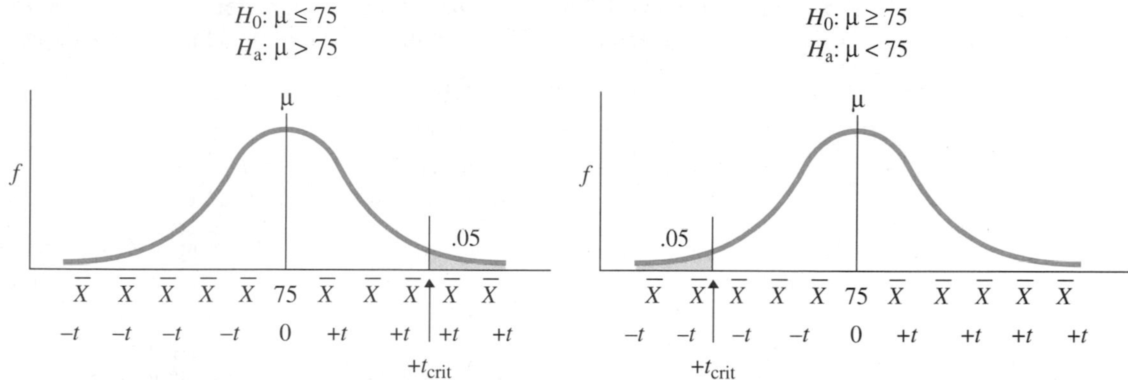

As you've seen, anytime we reject H_0, we conclude that the sample mean represents a μ that is different from the one described by H_0. The next important step is to get a better estimate of the value of that μ by computing a confidence interval.

ESTIMATING THE POPULATION μ BY COMPUTING A CONFIDENCE INTERVAL

There are two ways to estimate a population μ. One way is to say that μ is equal to the sample mean, performing **point estimation**. For example, earlier we estimated that the μ for men is located on the variable of test scores at the *point* identified as 65.67. However, no one really believes that if we actually tested the entire population, μ would be *exactly* 65.67. The problem with point estimation is that it is extremely vulnerable to sampling error. Our sample of men probably does not *perfectly* represent the population of men. Realistically, therefore, we can say only that the population μ for men is probably *around* 65.67.

The other, better way to estimate μ is to include the possibility of sampling error and perform **interval estimation**. With interval estimation, we specify an interval—a range of values—within which we expect the population parameter to fall. You often encounter such intervals in real life: They are usually phrased in terms of "plus or minus" some amount (called the **margin of error**). For example, when the evening news reports that a sample survey showed that 45% of the voters support the President, you may hear that the margin of error is plus or minus 3%. This means that the pollsters created an interval around 45%. They expect that if they asked the entire population, the μ would be within ±3% of 45%. In other words, they believe that between 42% and 48% of all voters in the population actually support the President.

We perform interval estimation by creating a confidence interval. Confidence intervals can be used to describe various population parameters, but the most common is the confidence interval for a single population μ. The **confidence interval for a single**

population μ describes an interval that contains values of μ, any one of which our sample mean is likely to represent. Thus, instead of merely saying the sample mean represents a μ *around* 65.67, a confidence interval is our way of statistically defining "around." As shown in the following diagram, a confidence interval identifies those values of μ above and below 65.67 that the sample mean is likely to represent.

$$\underbrace{\mu_{low} \quad \ldots \quad \mu \quad \mu \quad \mu \quad \mu \quad 65.67 \quad \mu \quad \mu \quad \mu \quad \mu \quad \ldots \quad \mu_{high}}$$

values of μ, one of which is likely to be represented by our sample mean

The symbol μ_{low} stands for the lowest value of μ that the sample mean is likely to represent, and μ_{high} stands for the highest value of μ that the mean is likely to represent. When we compute these two values of μ, we have the confidence interval.

How do we know if a sample mean is likely to represent a particular value of μ? It depends on sampling error. For example, we intuitively know that sampling error is too unlikely to produce a sample mean of 65.67 if μ is, say, 500. In other words, 65.67 is significantly different from 500. But sampling error *is* likely to produce a sample mean of 65.67 if, for example, μ is 65. In other words, a mean of 65.67 is not significantly different from this μ. Thus, a sample mean is likely to represent a particular value of μ if the sample mean is *not* significantly different from that μ. The logic behind computing a confidence interval is simply to compute the highest and lowest values of μ that are not significantly different from our sample mean. All values of μ between these two values are also not significantly different from the sample mean, so the mean is likely to represent one of them.

> REMEMBER A confidence interval contains the values between the highest and lowest values of μ that are not significantly different from a sample mean.

Computing the Confidence Interval for a Single μ

Because the *t*-test was appropriate for testing the significance of our sample mean, it also forms the basis for computing the confidence interval. We are looking for the highest and lowest possible values of μ that are not significantly different from the sample mean. For a sample mean to differ significantly from μ, its t_{obt} must be *beyond* t_{crit}. Therefore, the most a sample mean can differ from μ and still not differ significantly is when its t_{obt} *equals* t_{crit}. We can state this using the formula for the *t*-test:

$$t_{obt} = \frac{\overline{X} - \mu}{s_{\overline{X}}} = t_{crit}$$

To find the largest and smallest values of μ that do not differ significantly from our sample mean, we simply determine the values of μ that we can put into this formula with our sample mean and our $s_{\overline{X}}$ so that t_{obt} equals t_{crit}. Because we want to describe values above and below the sample mean, we use the two-tailed value of t_{crit}. Thus, we first want to find the value of μ that produces a $-t_{obt}$ equal to $-t_{crit}$. Rearranging the above formula, we have the formula for finding this value of μ:

$$\mu = (s_{\overline{X}})(+t_{crit}) + \overline{X}$$

We also want to find the value of μ that produces a $+t_{obt}$ equal to $+t_{crit}$. The formula for finding this value of μ is

$$\mu = (s_{\overline{X}})(-t_{crit}) + \overline{X}$$

Our sample mean represents a μ *between* these two values of μ, so we put the above formulas together into one formula.

THE COMPUTATIONAL FORMULA FOR THE CONFIDENCE INTERVAL FOR A SINGLE μ IS

$$(s_{\overline{X}})(-t_{crit}) + \overline{X} \leq \mu \leq (s_{\overline{X}})(+t_{crit}) + \overline{X}$$

The symbol μ stands for the unknown value represented by our sample mean. Replace the other symbols with the values of $\overline{X}$ and $s_{\overline{X}}$ that were computed from the sample data. Find the two-tailed value of t_{crit} in the *t*-tables at your α for $df = N - 1$, where N is the sample N.

> *REMEMBER* In computing a confidence interval, use the two-tailed critical value, even if you performed one-tailed hypothesis testing.

We can use the above formula to compute the confidence interval for our sample of housekeeping scores. There, $\overline{X} = 65.67$ and $s_{\overline{X}} = 2.582$. The two-tailed t_{crit} for $df = 8$ and $\alpha = .05$ is ± 2.306. Filling in the formula for the confidence interval, we have

$$(2.582)(-2.306) + 65.67 \leq \mu \leq (2.582)(+2.306) + 65.67$$

After multiplying 2.582 times −2.306 and +2.306 we have

$$-5.954 + 65.67 \leq \mu \leq +5.954 + 65.67$$

Adding −5.954 is the same as subtracting 5.954, so the formula at this point indicates that the sample mean represents a μ that is greater than or equal to the quantity 65.67 − 5.954, but less than or equal to the quantity 65.67 + 5.954. In other words, our mean represents a μ of 65.67, plus or minus 5.954.

After adding ±5.954 to 65.67, we have

$$59.72 \leq \mu \leq 71.62$$

This is the finished confidence interval for our sample of men's scores. We can now return to the previous diagram of the confidence interval and replace the symbols μ_{low} and μ_{high} with the numbers 59.72 and 71.62, respectively.

59.72 . . . μ μ μ μ 65.67 μ μ μ μ . . . 71.62

values of μ, one of which is likely to be represented by our sample mean

As shown, we've determined that the sample mean probably represents a μ around 65.67, such that μ is greater than or equal to 59.72, but less than or equal to 71.62.

Confidence Intervals and the Size of Alpha

Why do we call this a "confidence" interval? We defined this interval using $\alpha = .05$, so .05 is the theoretical probability of making a Type I error. Thus, 5% of the time the interval will be in error and will not contain the μ represented by our $\overline{X}$. Recall, however, that the quantity $1 - \alpha$ is the probability of avoiding a Type I error. Thus, $1 - .05$, or 95%, of the time the interval will contain the μ represented by our $\overline{X}$. Therefore, the probability is .95 that the interval contains the μ. Recall that probability is our way of expressing confidence in an event. Thus, above we are 95% confident that the interval between 59.72 and 71.62 contains the μ represented by our sample of men.

> *REMEMBER* The amount of confidence we have that a confidence interval contains the μ represented by our sample mean equals the quantity $1 - \alpha$ multiplied by 100.

The smaller the α, the smaller the probability of an error, so the greater our confidence. Had we set α equal to .01 above, we would have a 99% confidence interval. With $\alpha = .01$ and $df = 8$, t_{crit} would be ± 3.355, and the 99% confidence interval would be

$$57.01 \leq \mu \leq 74.33$$

Note that this 99% confidence interval spans a wider range of values of μ than did the 95% confidence interval. Logically, the larger the range within the interval, the greater our confidence that the interval contains the μ represented by the sample mean. (Think of a confidence interval as a fishing net. The larger the net, the greater our confidence that we'll catch the μ represented by the sample mean.) There is, however, an inevitable tradeoff: A wider range of values of μ less precisely identifies the specific value of μ represented by our sample. Usually we compromise between sufficient confidence and sufficient precision by using $\alpha = .05$ and creating the 95% confidence interval.

Thus, we conclude our single-sample *t*-test by saying, with 95% confidence, that the sample of men represents a μ between 59.72 and 71.62. Because the center of the interval is at 65.67, we still communicate that μ is *around* 65.67, but we have much more information than if we merely said that μ is somewhere around 65.67. Therefore, anytime you are describing the μ represented by a sample mean, you should compute a confidence interval.[1]

SIGNIFICANCE TESTS FOR CORRELATION COEFFICIENTS

Another type of single-sample study arises when we compute a correlation coefficient. For example, say we studied the relationship between a man's age and his housekeeping test score in a correlational design where we measured the test scores and the age of a sample of 25 men. Using the formula from Chapter 10, say we compute a Pearson correlation coefficient of $r = -.45$. This indicates that the older a man, the lower his housekeeping score.

[1] To compute a confidence interval when performing the *z*-test, use the formula given above, except use the critical values from the *z*-tables. If $\alpha = .05$, then $z_{\text{crit}} = \pm 1.96$. If $\alpha = .01$, then $z_{\text{crit}} = \pm 2.575$.

Remember, though, that a correlation coefficient is a statistic that describes the relationship found in the sample. Ultimately, we want to describe the relationship in the population. Therefore, we use the sample coefficient to estimate the population parameter we would expect to find if we computed the correlation for the entire population. Recall that the population correlation coefficient is called rho and its symbol is ρ. Thus, in our study we might estimate that ρ would equal $-.45$ if we measured the entire population of men.

But hold on, there's a problem here: That's right, sampling error. The problem of sampling error applies to *all* statistics. Here, even though the r suggests a relationship, it is possible that, because of sampling error, in the population there is either no relationship or a different relationship than in the sample. Therefore, regardless of the value of any sample correlation coefficient we compute, before we can be confident that the relationship exists in the population, we must answer the question "Is the correlation coefficient significant?"

REMEMBER Never accept that a sample correlation coefficient demonstrates a real relationship unless it is significant.

Statistical Hypotheses for the Correlation Coefficient

As usual, you should create your experimental and statistical hypotheses before collecting the data, and you can perform either a one-tailed or a two-tailed test. Use a two-tailed test if you do not predict the direction of the relationship. For example, in the housekeeping study, we might predict that the study demonstrates either that older men produce higher test scores (a positive correlation), or that older men produce lower test scores (a negative correlation). Conversely, we might hypothesize that the study will not demonstrate any relationship.

This latter hypothesis translates into the null hypothesis, because H_0 always implies that the predicted relationship does not exist in the population. If there is neither the predicted positive correlation nor the predicted negative correlation, then there is zero correlation. Most of the time, behavioral researchers test a null hypothesis involving zero correlation in the population. Therefore, in this book,

THE TWO-TAILED NULL HYPOTHESIS FOR SIGNIFICANCE TESTING OF A CORRELATION COEFFICIENT IS

$$H_0: \rho = 0$$

H_0 implies that r represents a ρ equal to zero. If r does not equal zero, the difference is written off as sampling error. You can understand how such sampling error can occur by looking at the hypothetical scatterplot in Figure 14.6. Assume this shows the population of X and Y scores when H_0 is true: There is no relationship and so ρ is 0. Recall that a circular scatterplot reflects zero correlation, while a slanting elliptical scatterplot reflects an r not equal to zero. The null hypothesis implies that, by chance, we selected an elliptical sample scatterplot from the circular population plot. Thus, r may not equal 0 but this is because it poorly represents a population where ρ equals 0. (Age and

FIGURE 14.6 Scatterplot of a Population for Which $\rho = 0$, as Described by H_0

Any r *is a result of selecting a sample scatterplot from within this scatterplot.*

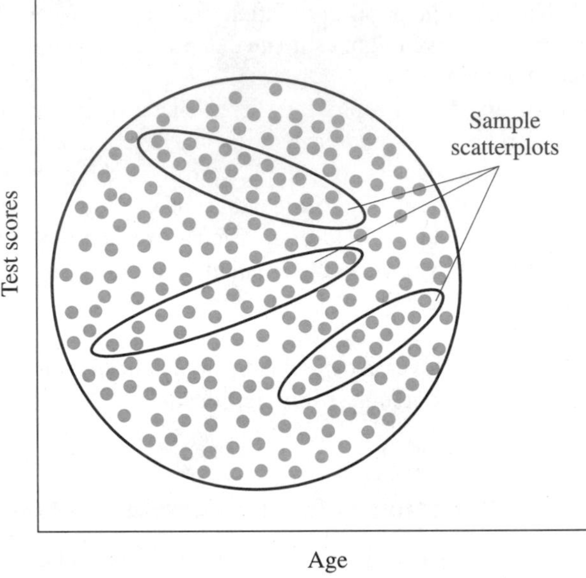

housekeeping scores are not really related, but the scores in the sample by chance happen to pair up so that it looks like they're related.)

On the other hand, the alternative hypothesis always implies that the predicted relationship exists in the population. If we predict that there is either a positive or a negative relationship, we predict that ρ does *not* equal zero.

THE TWO-TAILED ALTERNATIVE HYPOTHESIS FOR SIGNIFICANCE TESTING OF A CORRELATION COEFFICIENT IS

$$H_a: \rho \neq 0$$

H_a implies that r represents a population where ρ is not zero and thus reflects a real relationship in the population. If so, we assume that the population's scatterplot would be similar to the sample's scatterplot.

As usual, we test H_0, so here we test whether the sample correlation represents a ρ of 0. If r is too unlikely to be accepted as representing a ρ of 0, then we reject H_0 and accept H_a (that the sample represents a population where ρ does not equal zero).

Recall from Chapter 10 that there are three types of correlation coefficients: The Pearson r, the Spearman r_s, and the point-biserial r_{pb}. The logic and format of the above statistical hypotheses are used regardless of which correlation coefficient you compute (you merely change the subscripts). The following sections discuss the particulars of hypothesis testing for each type of coefficient.

The Significance Test for the Pearson *r*

As usual, before you collect any data, you should make sure the study meets the assumptions of the statistical procedure. The assumptions for hypothesis testing of the Pearson correlation coefficient are:

1. There is a random sample of *X–Y* pairs, and each variable is an interval or ratio variable.
2. The *Y* scores and the *X* scores in the sample each represent a normal distribution. Further, they represent a *bivariate* normal distribution. This means that the *Y* scores at each value of *X* form a normal distribution and that the *X* scores at each value of *Y* form a normal distribution. (If *N* is larger than 25, however, violating this assumption is of little consequence.)
3. The null hypothesis states that the population correlation is zero. (When H_0 states that ρ is some value other than 0, a different procedure is used.)

Our housekeeping and age data meet these assumptions, so we set α at .05 and then test *r*. To do that, we examine the sampling distribution.

The sampling distribution of *r* The H_0 sampling distribution of *r* shows the different values of *r* that occur when H_0 is true and we are sampling the population where ρ is 0. By now, you could create this sampling distribution yourself. Using the same *N* as in our study, you would select an infinite number of samples of *X–Y* pairs from the population where $\rho = 0$. For each sample, you would compute *r*. If you then plotted the frequency of the various values of *r*, you would have the sampling distribution of *r*. The **sampling distribution of a correlation coefficient** is a frequency distribution that shows all possible values of the coefficient that can occur when samples of size *N* are drawn from a population where ρ is zero. Such a sampling distribution is shown in Figure 14.7.

FIGURE 14.7 Distribution of Random Sample *rs* When $\rho = 0$

It is an approximately normal distribution, with values of r plotted along the X axis.

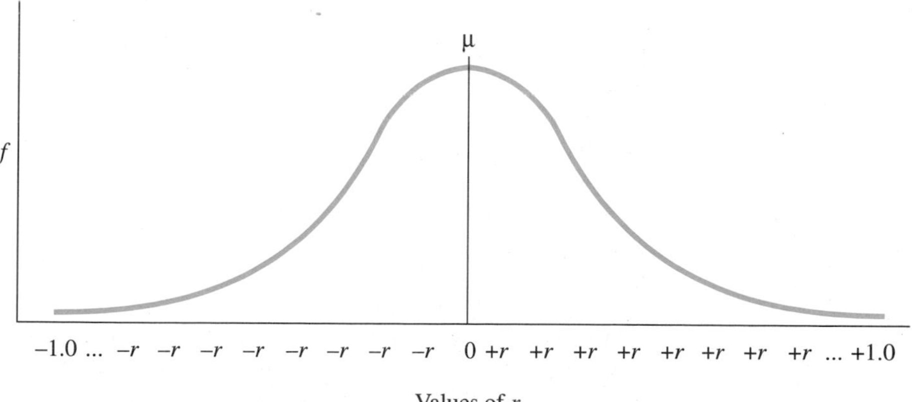

The only novelty here is that instead of showing different sample means along the X axis, the different values of r are plotted. As shown, when H_0 is true and ρ is 0, most frequently r will also equal zero, so the mean of the sampling distribution—the average r—is 0. Because of sampling error, however, sometimes r will not equal zero. The larger the r (whether positive or negative), the less frequently it occurs, and thus the less likely it is to occur when the sample represents a population where ρ is zero.

To test H_0, we simply determine where on this distribution our r lies. To do so, we could perform a variation of the t-test, but luckily that is not necessary. Instead, the value of r is just like a z-score, directly communicating its location on the sampling distribution relative to the mean of the distribution. The mean of the sampling distribution is always an r of 0. So, for example, our r of $-.45$ is a distance of .45 below the mean. Therefore, we test H_0 simply by examining the value of the obtained r. The symbol for an obtained r is r_{obt}. To determine whether r_{obt} lies in the region of rejection, we compare it to the critical value of r, which is symbolized as r_{crit}.

Testing the Pearson *r* As with the t-distribution, the shape of the sampling distribution of r is slightly different for each *df*, so there is a different value of r_{crit} for each *df*. Table 3 in Appendix C, "Critical Values of the Pearson Correlation Coefficient," gives the critical values of r. To use these r-tables, we first need the appropriate degrees of freedom. *But*, here's a new one: With the Pearson correlation coefficient, the degrees of freedom equals $N - 2$, where N is the number of *pairs* of scores in the sample. (Previously, we saw that $N - 1$ was a correction factor when drawing inferences involving one sample's variance. Different statistics, however, require different correction factors, so the computation of their *df* will change.)

> ***REMEMBER*** In significance testing of r, the degrees of freedom equals $N - 2$, where N is the number of pairs of scores.

To find r_{crit}, enter Table 3 for either a one-tailed or two-tailed test at the appropriate α and *df*. For our housekeeping correlation, N was 25, so $df = 23$. We are performing a two-tailed test with $\alpha = .05$ and $df = 23$, so r_{crit} is $\pm.396$. Armed with this information, we set up the sampling distribution in Figure 14.8. An r_{obt} of $-.45$ is beyond the r_{crit} of $\pm.396$, so r_{obt} is in the region of rejection. The sampling distribution shows that when samples *do* represent the population where ρ is 0, seldom is a sample so poor at representing this ρ that an r_{obt} of $-.45$ is produced. Because such an r_{obt} is just too unlikely, we reject the null hypothesis that r_{obt} represents the population ρ of 0, concluding that our r_{obt} is significantly different from 0.

As usual, α is the theoretical probability that we've made a Type I error. With $\alpha =$ .05, over the long run we will obtain values of r_{obt} that cause us to erroneously reject H_0 a total of 5% of the time. Thus, the probability that we've made a Type I error this time is slightly less than .05. We report our significant r_{obt} as

$$r(23) = -.45, p < .05$$

(Note the *df* in parentheses.)

Remember that by rejecting H_0, we have not proven anything. In particular, we conducted a correlational study, so we have not proven that changes in age *cause* test scores to change. In fact, we have not even proven that there is a relationship in nature (we may have made a Type I error). Instead, we are simply more confident that the r_{obt}

FIGURE 14.8 H_0 Sampling Distribution of r When H_0: $\rho = 0$

For the two-tailed test, there is a region of rejection for positive values of r_{obt} *and for negative values of* r_{obt}.

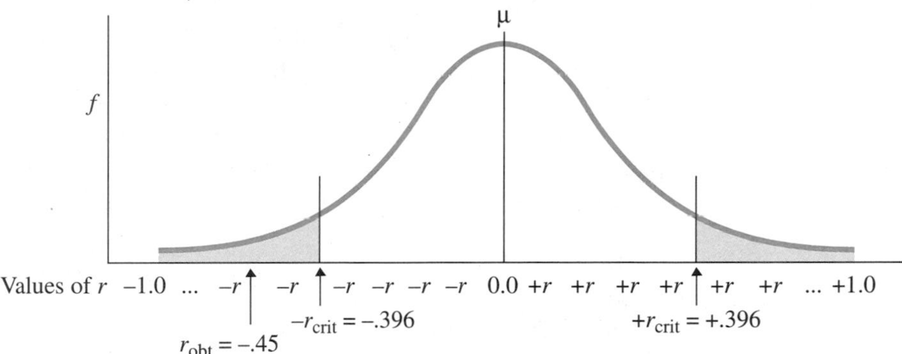

does not merely reflect some quirk of sampling error but rather represents a "real" relationship in the population.

Because the sample r_{obt} is $-.45$, our best estimate is that in the population, ρ equals $-.45$. Recognizing that the sample may contain sampling error, however, we expect that ρ is probably *around* $-.45$. (We could more precisely identify the value of ρ by computing a confidence interval, which describes the values of ρ that r_{obt} is likely to represent. The procedure for computing confidence intervals for ρ, however, is very different from the one discussed previously.)

Thus, we believe there is a relationship here, and so we interpret it psychologically, generalizing to relevant behaviors and constructs. For example, perhaps older men score lower on the housekeeping test because they come from generations where typically wives did the housekeeping, while men were the "breadwinners."

In Chapter 11, we saw that we further describe such a relationship by computing the linear regression equation and r^2. We do this *only* when r_{obt} is significant, however, because only then are we confident that we're describing a "real" relationship. Therefore, we would now compute the linear regression equation for predicting housekeeping test scores if we know a man's age. We also compute r^2, which equals $-.45^2$ or .20. Recall that this is the proportion of variance in the sample's Y scores that is accounted for by the relationship with the X scores. Here it indicates that we are, on average, 20% more accurate when we use the relationship with age to predict test scores than we are when we do not use the relationship.

Remember that the term *significant* does not indicate an *important* relationship. Although a relationship *must* be significant in order to be important, it can be significant and still be unimportant. The importance of a relationship is determined by computing r^2. The greater the proportion of variance accounted for, the more that knowing participants' X scores improves the accuracy of predicting and understanding differences in their Y scores—their behavior—and so the greater the scientific importance of the relationship. For example, we might find an r_{obt} of $+.10$ that is significant. In saying that r_{obt} is significant, we are saying *only* that it is unlikely to occur by chance. A relationship where r_{obt} equals $+.10$, however, is *not* statistically important: Because

$+.10^2$ is only .01, this relationship accounts for only 1% of the variance, and thus it is virtually useless in explaining differences in Y scores. (We're only 1% better off with it than without it.) Therefore, although this is a relationship that is not likely to occur through sampling error, at the same time it is a statistically unimportant and not very useful relationship.

Of course, if r_{obt} had not fallen beyond r_{crit}, we would retain H_0 and conclude that the sample may represent a population where $\rho = 0$. As usual, here we have not proven that *no* relationship exists in the population—we have simply failed to demonstrate convincingly that one *does* exist. Therefore, we make no claims about the relationship that may or may not exist in the population, nor do we describe it using the regression equation or r^2. We report nonsignificant results as above, except that $p > .05$.

One-tailed tests of r Say that in a different study, we specifically predicted only a positive correlation or only a negative correlation between age and housekeeping scores. Then we would perform a one-tailed test.

THE ONE-TAILED HYPOTHESES FOR SIGNIFICANCE TESTING OF A CORRELATION COEFFICIENT ARE

Predicting positive correlation	***Predicting negative correlation***
H_0: $\rho \leq 0$	H_0: $\rho \geq 0$
H_a: $\rho > 0$	H_a: $\rho < 0$

We test each H_0 by again testing whether r_{obt} represents a population where there is zero relationship—so we again examine the sampling distribution for $\rho = 0$. From Table 3 of Appendix C, we find the one-tailed critical value for our *df* and α, and set up one of the sampling distributions shown in Figure 14.9.

When predicting a positive correlation, we use the sampling distribution on the left. The r_{obt} is significant if it is positive and falls beyond the positive r_{crit}. If so, we accept the H_a that the sample represents a ρ greater than 0. Conversely, when predicting a negative correlation, we use the distribution on the right in Figure 14.9. Here, r_{obt} is significant if it is negative and falls beyond the negative r_{crit}. If so, we accept the H_a that the sample represents a ρ less than 0.

If r_{obt} is not beyond the appropriate r_{crit}, then r_{obt} is not significant, and we conclude that the study failed to demonstrate the predicted relationship.

Significance Testing of the Spearman r_s and the Point-Biserial r_{pb}

Perform a similar procedure when testing the Spearman correlation coefficient, r_s, or the point-biserial correlation coefficient, r_{pb}. These correlations describe relationships in a *sample*, but—that's right—perhaps they merely reflect sampling error. Perhaps if we computed the correlation in the population, we would find that our r_s actually represents a population correlation, symbolized by ρ_s, that is 0. Likewise, perhaps any r_{pb} actually represents a population correlation, symbolized by ρ_{pb}, that is 0. Therefore,

FIGURE 14.9 H_0 Sampling Distribution of r Where $\rho = 0$ for One-Tailed Test

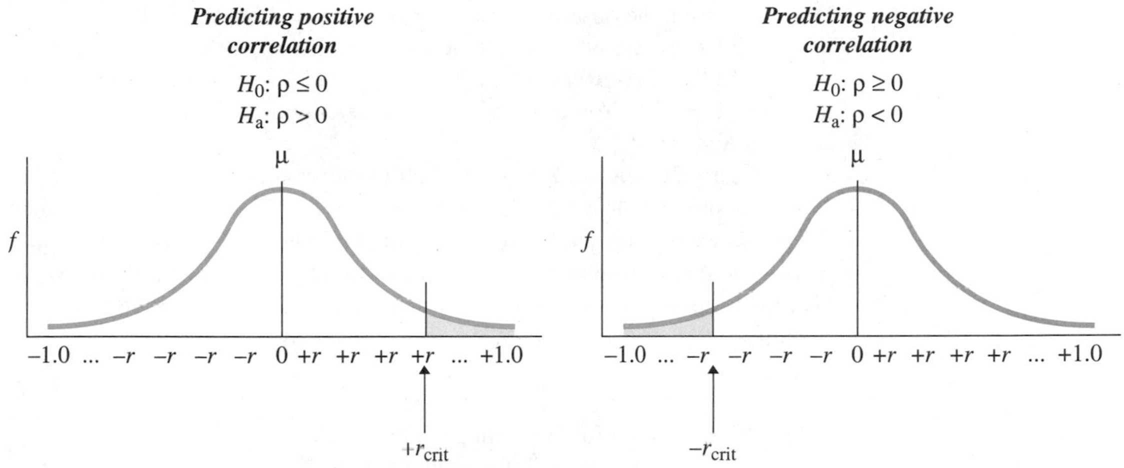

before we can conclude that either of these types of correlations represents a relationship in nature, we must perform the appropriate hypothesis testing.

To test each sample correlation coefficient, perform the following steps:

1. Set alpha: How about .05?
2. Consider the assumptions of the test. The Spearman r_s requires a random sample of pairs of *ranked* (ordinal) scores. The point-biserial r_{pb} requires random scores from one dichotomous variable and one interval or ratio variable. (Because of the type of data involved and the lack of parametric assumptions, r_s and r_{pb} are technically nonparametric procedures.)
3. Create the statistical hypotheses. You can test either the one- or two-tailed hypothesis we had with ρ, except now substitute ρ_s or ρ_{pb}.

The only new aspect in testing r_s or r_{pb} is in their respective sampling distributions.

Significance testing of r_s To test r_s, we use a new family of sampling distributions and a different table of critical values. Table 4 in Appendix C, called "Critical Values of the Spearman Rank-Order Correlation Coefficient," contains the critical values of r_s for one- and two-tailed tests for an α of .05 and .01. Obtain the critical value from this table as in previous tables, except here use N, *not* degrees of freedom.

REMEMBER The critical value of r_s is obtained using N, the number of pairs of scores in the sample.[2]

Let's use the example from Chapter 10, where we determined the inter-rater reliability of two observers by correlating their rankings of the aggressiveness of 9 children. We found $r_s = +.85$. To believe that the observers are really reliable, however, we must believe that this is a real relationship, and not that their rankings happened to pair up by

[2] Table 4 in Appendix C contains critical values for N up to 30. When N is greater than 30, transform r_s to a z-score using the formula $z_{obt} = (r_s)(\sqrt{N-1})$. For $\alpha = .05$, the two-tailed $z_{crit} = \pm 1.96$ and the one-tailed $z_{crit} = 1.645$.

chance (not that in the complete population of their rankings, these observers would produce an $\rho_s = 0$). In other words, this r_s must be significant. We performed this correlation assuming the observers' rankings would agree, so we predicted a positive correlation. Therefore, we have a one-tailed test with the hypotheses H_0: $\rho_s \leq 0$ and H_a: $\rho_s > 0$. From Table 4 in Appendix C, with $\alpha = .05$ and $N = 9$, the critical value for the one-tailed test is $+.600$. You can envision the H_0 sampling distribution of r_s when $N = 9$ as shown in Figure 14.10.

We determine whether r_s is significant in the same way we have determined significance for previous statistics. The obtained r_s of $+.85$ is beyond the critical value of $+.600$, so we reject H_0: an r_s of $+.85$ is too unlikely to accept as representing the population where ρ_s is zero or less than zero, so we accept H_a (that $\rho_s > 0$). We have a significant r_s, and we estimate that the value of the correlation in the population of such rankings, ρ_s, is around $+.85$. We report this result as

$$r_s(9) = +.85, p < .05$$

(The N of the sample follows r_s in parentheses.)

In a different study, with different predictions, we might have performed a one-tailed test using the other tail, or a two-tailed test (with the appropriate two-tailed critical value).

Significance testing of r_{pb} We test r_{pb} using the same logic as above. The H_0 sampling distributions of r_{pb} are identical to the distributions for the Pearson r, so critical values of r_{pb} are obtained from Table 3 in Appendix C, "Critical Values of the Pearson Correlation Coefficient." Again use degrees of freedom, which equals $N - 2$.

> ***REMEMBER*** The critical values for r_{pb} are the same as those for r.

In Chapter 10, we computed the r_{pb} between participants' scores on the dichotomous variable of gender (male or female) and their scores on a personality test. We obtained $r_{pb} = +.46$, with $N = 10$. Say we perform a two-tailed test with the hypotheses H_0: $\rho_{pb} = 0$ and H_a: $\rho_{pb} \neq 0$. From Table 3 in Appendix C, with $\alpha = .05$ and $df = 8$, the critical value for the two-tailed test is $\pm.632$. You can envision the sampling distribution of r_{pb} as shown in Figure 14.11. The obtained r_{pb} of $+.46$ is not beyond the critical value,

FIGURE 14.10 One-Tailed H_0 Sampling Distribution of Values of r_s When H_0 is $\rho_s = 0$

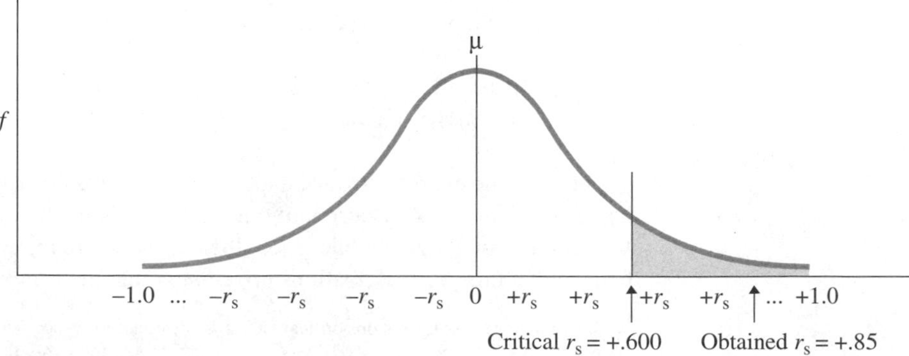

FIGURE 14.11 H_0 Sampling Distribution of r_{pb} When H_0 is $\rho_{pb} = 0$

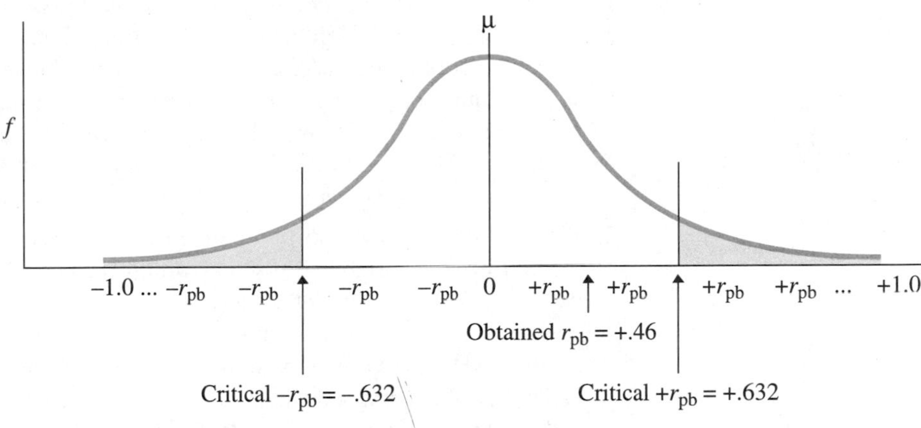

so we do not reject H_0 and we conclude that our sample may represent the population where ρ_{pb} equals zero. Thus, we have no evidence that gender and personality scores are related. We report this result as

$$r_{pb}(8) = +.46, p > .05$$

In a different study, if we had predicted only a positive or only a negative correlation, we would perform a one-tailed test. Remember, though, as discussed in Chapter 10, *we* determine whether r_{pb} is positive or negative by how we arrange the categories of the dichotomous variable. Therefore, one-tailed tests of r_{pb} must be set up so that they are consistent with how the data are arranged.

MAXIMIZING THE POWER OF THE t-TEST AND CORRELATION COEFFICIENT

Recall that *power* is the probability of *not* committing a Type II error: Sometimes H_0 is false and then we avoid the error by rejecting H_0. Any discussion of power assumes that we have one of those times when H_0 *is* false, so our job is to reject it. Therefore, we maximize the power of a statistical procedure by maximizing the probability of obtaining significant results. To do that, we maximize the absolute size of the obtained statistic relative to the critical value. The larger the obtained value, the more likely it is to fall beyond the critical value and thus the more likely it is to be significant, causing us to reject H_0 when it is false.

Previously, we've seen that we try to design a powerful study that provides convincing evidence of a relationship. Statistically, such a relationship translates into maximizing the size of the obtained value relative to the critical value. Thus, looking at the formula for t_{obt}:

$$t_{obt} = \frac{\overline{X} - \mu}{\frac{s_X}{\sqrt{N}}}$$

We see that there are three aspects of a study that increase power:

1. *The greater the differences produced by changing the independent variable, the greater the power.* Recall that we seek a strong manipulation of the independent variable that produces large differences in scores between the conditions. In the single-sample experiment, this translates into a larger difference between $\overline{X}$ and μ. Mathematically, a larger difference between $\overline{X}$ and μ produces a larger numerator. All other things (N and s_X) being equal, dividing into a larger numerator results in a larger t_{obt}.
2. *The smaller the variability of the raw scores, the greater the power.* We also seek to minimize the variability of scores within each condition (minimizing the error variance). This is because the t-test measures the difference between $\overline{X}$ and μ relative to the standard error of the mean, $s_{\overline{X}}$, and the size of the standard error is determined by the variability of the raw scores. Therefore, the smaller the variability of the raw scores, s_X, the smaller the denominator. Dividing by a smaller denominator produces a larger t_{obt}.
3. *For small samples, the greater the N, the greater the power.* Another aspect of increasing the power of a study is to test a large N. Conceptually, the larger the N, the more accurately we will represent the true situation in the population and the less likely we are to make any type of error. Mathematically, the size of N influences the results in two ways. First, in the above formula, dividing s_X by a larger value of $\sqrt{N}$ produces a smaller denominator, which results in a larger t_{obt}. Second, a larger N produces larger df, and the larger the df, the smaller the t_{crit}. The smaller the t_{crit}, the more likely that t_{obt} lies beyond it, so the more likely that t_{obt} is significant. However, the impact of sample size on power is greatest in *small* samples. Generally, an N of 30 is needed for minimal power, and increasing N up to 121 increases power substantially. However, an N such as 500 is not substantially more powerful than an N of, say, 450.

We also attempt to design a powerful correlational study that provides convincing evidence of a relationship. This too translates into maximizing the size of the obtained value relative to the critical value, so it is likely to be significant. The following are ways to maximize the power of a correlation coefficient.

1. *Avoiding a restricted range maximizes power.* Recall that a restricted range occurs when there is a small range of scores on the X or Y variable. This produces a smaller coefficient than would be found without a restricted range. Thus, we try to avoid a restriction of range, producing a larger coefficient.
2. *Minimizing the variability of the Y scores at each X increases power.* Variability in Y scores (error variance) decreases the strength of the relationship and thus decreases the size of the correlation coefficient. Therefore, minimizing variability produces a larger coefficient.
3. *Describing linear relationships with linear correlations increases power.* The coefficients we've used describe the extent to which the data form a linear relationship. If the relationship is nonlinear, then a coefficient may be small and thus not significant. We lose power because we conclude that there is no relationship when, in fact, a nonlinear relationship exists.
4. *Increasing the N of small samples maximizes power.* Because of lessened control, correlational designs frequently produce substantial error variance and a weak

relationship. Therefore, we may compensate by employing a large N, often in the range of several hundred participants. If a given relationship occurs for a larger number of subjects, the coefficient is larger. Also, a larger N produces larger df, so the critical value is smaller and the coefficient is more likely to be significant.

REMEMBER Designing a powerful study results in a larger obtained value relative to the critical value, so we are more likely to reject H_0 at those times when it is false.

PUTTING IT ALL TOGETHER

In one sense I hope that you found this chapter rather boring. Not because it *is* boring, but because, for each statistic, we performed virtually the same operations. In significance testing of *any* statistic, we ultimately do and say the same things. In all cases, if the obtained statistic is out there far enough on the H_0 sampling distribution, it is too unlikely to accept as occurring if H_0 is true, so we reject that H_0 is true. Any H_0 implies that the sample does not represent the predicted relationship, so rejecting H_0 increases our confidence that the data do represent the predicted relationship, with the probability of a Type I error equal to $p < \alpha$.

CHAPTER SUMMARY

1. The t-test is for significance testing of a single sample mean when (a) there is one random sample of interval or ratio data, (b) the raw score population is a normal distribution, and (c) the standard deviation of the raw score population is estimated by computing s_X from the sample data.
2. A t-distribution is a theoretical sampling distribution of all possible values of t when a raw score population is infinitely sampled using a particular N. The appropriate t-distribution for a single-sample t-test is the distribution identified by $N - 1$ degrees of freedom (df).
3. In *point estimation*, the value of the population parameter is assumed to equal the value of the corresponding sample statistic.
4. In *interval estimation*, the value of a population parameter is assumed to lie within a specified interval. The *confidence interval for a single* μ describes a range of μs, any one of which the sample mean is likely to represent. The interval contains the highest and lowest values of μ that are not significantly different from the sample mean. The confidence that the interval actually contains the value of μ is equal to $(1 - \alpha)100$.
5. The *sampling distribution of a correlation coefficient* is a frequency distribution showing all possible values of the coefficient that occur when samples of size N are drawn from a population where the correlation coefficient is zero.
6. Significance testing of the *Pearson r* assumes that (a) a random sample of pairs of scores from two interval or ratio variables and (b) the Y scores are normally distributed at each value of X and the X scores are normally distributed at each value of Y.

7. Significance testing of the *Spearman* r_s assumes a random sample of pairs of ranked-order (ordinal) scores.

8. Significance testing of the *point-biserial* r_{pb} assumes a random sample of pairs of scores where one score is from a dichotomous variable and one score is from an interval or ratio variable.

9. Only when a correlation coefficient is significant is it appropriate to compute the linear regression equation and the proportion of variance accounted for.

10. Maximize the power of the *t*-test by (a) creating large differences between the conditions of the independent variable, (b) minimizing the variability in the sample, and (c) increasing the *N* of small samples.

11. Maximize the power of a correlation coefficient by (a) avoiding a restricted range, (b) minimizing the variability in *Y* at each *X*, (c) using the appropriate coefficient, and (d) increasing the *N* of small samples.

PRACTICE PROBLEMS

(Answers for odd-numbered problems are provided in Appendix D.)

1. A scientist has conducted a single-sample experiment. (a) What two parametric procedures are available to her? (b) What is the deciding factor for selecting between them?
2. What are the other assumptions of the *t*-test?
3. (a) What is the difference between $s_{\bar{X}}$ and $\sigma_{\bar{X}}$? (b) How is their use the same?
4. (a) Why are there different values of t_{crit} when samples have different *N*s? (b) What must you compute prior to finding t_{crit}?
5. You wish to compute the 95% confidence interval for a sample with a *df* of 80. Using interpolation, determine the value of t_{crit} you should use.
6. Say you have a sample mean of 44 in a study. (a) Estimate the corresponding μ using point estimation. (b) What does a confident interval computed for this μ tell you? (c) Why is computing a confidence interval a better approach than using a point estimate?
7. (a) What is power? (b) Why is power especially important when we fail to reject H_0? (c) What do researchers do to avoid this dilemma?
8. Poindexter performed a two-tailed experiment in which $N = 20$. He couldn't find his *t*-tables, but he remembered that the t_{crit} at $df = 10$. He decided to compare his t_{obt} to this t_{crit}. Why is this a correct or incorrect approach? (*Hint*: Consider whether t_{obt} turns out to be significant or nonsignificant at this t_{crit}.)
9. You wish to determine whether this textbook is beneficial or detrimental to students learning research methods and statistics. On a national exam, $\mu = 68.5$ for students who have used other textbooks. A random sample of students who have used this book has the following scores:

   64 69 92 77 71 99 82 74 69 88

   (a) What are H_0 and H_a for this study? (b) Compute t_{obt}. (c) With $\alpha = .05$, what is t_{crit}? (d) What do you conclude about the use of this book? (e) Compute the confidence interval for μ.
10. A researcher predicts that smoking cigarettes decreases a person's sense of smell. On a standard test of olfactory sensitivity, the μ for nonsmokers is 18.4. By giving this test to a random sample of people who smoke a pack a day, the researcher obtains the following scores:

16 14 19 17 16 18 17 15 18 19 12 14

(a) What are H_0 and H_a for this study? (b) Compute t_{obt}. (c) With $\alpha = .05$, what is t_{crit}? (d) What should the researcher conclude about this relationship? (e) Compute the confidence interval for μ.

11. Foofy conducts a study to determine if hearing an argument in favor of an issue alters subjects' attitudes toward the issue one way or the other. She presents a thirty-second speech in favor of an issue to 8 participants. In a national survey, the mean attitude score in favor of the issue was $\mu = 50$. Testing her subjects with this survey, she obtains the following scores.

10 33 86 55 67 60 44 71

(a) What are H_0 and H_a? (b) What is the value of t_{obt}? (c) With $\alpha = .05$, what is the value of t_{crit}? (d) What are the statistical results? (e) If appropriate, compute the confidence interval for μ. (f) Using the preceding statistics, what conclusions should Foofy draw about the relationship between such arguments and their impact on attitudes?

12. For the study in problem 11, (a) What statistical principle should Foofy be concerned with? (b) Identify three problems with her study from a statistical perspective. (c) Why would correcting the problems identified in (b) improve her study?

13. Poindexter examined the relationship between the quality of sneakers worn by volleyball players and their average number of points scored per game. Studying 20 people who owned sneakers of good to excellent quality, he computed $r = +.41$. He immediately claimed to have support for the notion that better-quality sneakers are related to better performance on a somewhat consistent basis. He then computed r^2 and the regression equation. Do you agree or disagree with his approach? Why?

14. Eventually, for the study in problem 13, Poindexter reported that $r(18) = +.41, p > .05$. (a) What should he conclude about this relationship? (b) What other computations should he perform to describe the relationship in these data? (c) What statistical principle should he be concerned with? (d) What aspects of the study can he improve to better deal with this principle? (e) What will correcting these things from (d) do in regard to his finding significant results?

15. A scientist suspects that as his subjects' stress level changes, so does the amount of their impulse buying. He collects data from 72 participants and obtains an r of +.38. (a) What are H_0 and H_a? (b) With $\alpha = .05$, what is r_{crit}? (c) What are the statistical results of this study? (d) What conclusions should be drawn about the relationship in the population? (e) What other calculations should be performed to describe the relationship in these data?

16. Foofy examines the relationship between an individual's physical strength and his or her college grade point average. She computes the correlation for a sample of 2,000 subjects and obtains $r(1998) = +.08, p < .0001$. She claims she has uncovered a useful tool for predicting which college applicants are likely to succeed academically. Do you agree or disagree? Why?

17. A researcher investigates the relationship between handedness and strength of personality. She tests 42 participants, assigning left-handers a score of 1 and right-handers a score of 2. She obtains a correlation coefficient of +.33 between subjects' handedness and their scores on a personality test. (a) Which type of correlation coefficient did she compute? (b) What are H_0 and H_a? (c) With $\alpha = .05$, what is the critical value? (d) What should the researcher conclude about the strength of the relationship in the population? What should she conclude about the direction of the relationship in the population? (e) Are the results of this study relatively useful?

18. A newspaper article claims that for all U.S. colleges, the academic rank of the college is negatively related to the rank of its football team. You examine the accuracy of this claim. From a sample of 28 colleges, you obtain a correlation coefficient of −.32. (a) Which type of correlation coefficient did you compute? (b) What are H_0 and H_a? (c) With $\alpha = .05$, what is the critical value? (d) What are the statistical results? (e) What should you conclude about the accuracy of the newspaper claim for all colleges in the United States? (f) In trying to determine a particular school's academic ranking in your sample, how important is it that you look at the school's football ranking?

19. To predict a person's sense of humor from their mathematical ability, a researcher measures participants' math skills and how funny they find three puns to be. He tests 10 math majors and finds a nonsignificant *r*. (a) What characteristic of his participants may account for this result? (b) What problem with his criterion variable may account for this result? (c) What other obvious improvement in power can he achieve?

20. In question 19, say that previous research has shown that people with very high or very low math skills tend to find puns humorous, but those with intermediate skills do not. How can this finding account for the nonsignificant *r*?

21. Summarize the steps involved in describing and interpreting the results of a correlational study.

22. Summarize the steps involved in describing and interpreting the results of a single sample experiment.

SUMMARY OF FORMULAS

1. *The definitional formula for the single sample* t-*test is*

$$t_{obt} = \frac{\overline{X} - \mu}{s_{\overline{X}}}$$

The value of $s_{\overline{X}}$ is computed as

$$s_{\overline{X}} = \frac{s_X}{\sqrt{N}}$$

and s_X is computed as

$$s_X = \sqrt{\frac{\Sigma X^2 - \frac{(\Sigma X)^2}{N}}{N - 1}}$$

2. *The computational formulas for the single sample* t-*test are*

$$t_{obt} = \frac{\overline{X} - \mu}{\sqrt{\frac{s_X^2}{N}}} \quad \text{and} \quad t_{obt} = \frac{\overline{X} - \mu}{\sqrt{(s_X^2)\left(\frac{1}{N}\right)}}$$

where s_X^2 is the estimated variance computed for the sample. Values of t_{crit} are found in Table 2 of Appendix C, "Critical Values of t," *for* df = N − *1.*

3. *The computational formula for the confidence interval for a single* μ *is*

$$(s_{\overline{X}})(-t_{crit}) + \overline{X} \leq \mu \leq (s_{\overline{X}})(+t_{crit}) + \overline{X}$$

where t_{crit} is the two-tailed value for $df = N - 1$ and $\overline{X}$ and $s_{\overline{X}}$ are computed using the sample data.

4. *To test the significance of a correlation coefficient*, compare the obtained correlation coefficient to the critical value.
   (a) *Critical values of r* are found in Appendix C, Table 3, for $df = N - 2$, where N is the number of pairs of scores in the sample.
   (b) *Critical values of* r_s are found in Appendix C, Table 4, for *N*, the number of pairs in the sample.
   (c) *Critical values of* r_{pb} are found in Appendix C, Table 3, for $df = N - 2$, where *N* is the number of pairs in the sample.

PART 5

DESIGNING AND ANALYZING TWO-SAMPLE EXPERIMENTS

So far, we've discussed statistical procedures for a single-sample experiment, which compares the $\overline{X}$ measured under one condition of the independent variable to a known value of μ from another condition. Because research focuses on the cutting edge of unknown psychological processes, however, we usually do not already know any values of μ. Instead the simplest approach is to conduct a two-sample experiment: We measure dependent scores under two conditions of the independent variable, estimate the population μ for each, and determine if they reflect a relationship. In the next two chapters, we discuss the major design issues as well as the parametric statistics of two-sample experiments.

15

The Two-Sample Between-Subjects Experiment and the Independent Samples *t*-Test

To understand this chapter:

- From Chapter 3, recall what we mean by reliability and internal and external validity.
- From Chapter 10, understand the uses of the point-biserial correlation coefficient, r_{pb}.
- From Chapter 11, recall how to conceptualize the proportion of variance accounted for.
- And, of course, don't forget what you've learned about inferential statistics.

Then your goals in this chapter are to learn:

- What constitutes a between-subjects design and independent samples.
- How to control subject variables in a between-subjects design.
- How to perform the independent-samples *t*-test.
- How to compute a confidence interval for the difference between two μs.
- How r_{pb} describes the effect size in a two-sample experiment.

In this chapter, we discuss how to design a two-sample experiment, including how to deal with the issue of subject variables. Then we discuss how to analyze such research using the two-sample t-test, a test similar to the single-sample t-test you learned in the previous chapter. Also, we'll discuss additional procedures for describing a significant relationship.

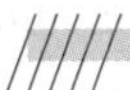

MORE STATISTICAL NOTATION

It's time to pay very close attention to subscripts. We will compute the mean of each of two conditions, identifying one as $\overline{X}_1$ and the other as $\overline{X}_2$. Likewise, we'll compute an estimate of the variance of the raw score population represented by the sample in each condition, identifying the variance from one as s_1^2 and the variance from the other as s_2^2. Finally, recall that N actually indicates the total number of scores in a study. Therefore, we will use the lowercase n with a subscript to indicate the number of scores in each sample: n_1 is the number of scores in condition 1, and n_2 is the number of scores in condition 2.

DESIGNING THE TWO-SAMPLE EXPERIMENT

As an example study, say that we're interested in the notion that people who witness a crime or other event may recall the event better when they are hypnotized. We decide to create two samples of participants who will each watch a videotape of a supposed robbery. Later, one group will be hypnotized to the same extent and then answer 30 questions about the details of the event. The other group—the control condition—will answer the questions without benefit of hypnosis. Thus, the conditions of the independent variable are the presence or absence of hypnosis, and the dependent variable is the number of questions answered correctly. You can envision this design as shown in Table 15.1. The recall scores are normally distributed ratio scores, so we compute the mean of each condition, summing vertically in each column. If the means differ, we'll

TABLE 15.1 Diagram of Hypnosis Study

The independent variable is degree of hypnosis and the dependent variable is recall

	No Hypnosis	*Hypnosis*
Recall scores →	X	X
	X	X
	X	X
	"	"
	"	"
	$\overline{X}$	$\overline{X}$

have evidence of a relationship where, as the amount of hypnosis changes, recall accuracy also changes.

As usual, to allow a clear interpretation of the variables and behaviors under study, we apply all of the controls discussed previously. Thus, we seek a reliable manipulation of hypnosis so that when under hypnosis, all participants are deeply and consistently hypnotized. We seek reliable and valid questions so that we really measure each person's memory for the robbery. We seek internal validity so that differences in recall scores between the conditions are really due to differences in hypnosis. And we seek external validity so that the way the samples operate is really the way the general population operates. Essentially, we accomplish these goals by controlling any extraneous variable that may fluctuate *within* a condition or *between* the conditions.

At the same time, we cannot claim to have demonstrated anything about a behavior unless we find a statistically significant relationship in the sample data. But our statistical decisions are only as good as the data we collect (garbage in, garbage out!). Therefore, we also seek a powerful design that produces powerful and likely-to-be-significant data. For example, compare the powerful and un-powerful examples in Table 15.2. The powerful data provide a much more convincing demonstration that hypnosis improves memory: Hypnosis makes a relatively big difference in recall scores between conditions, and there is perfect consistency—no error variance—within each condition because no differences between scores occur. To achieve such data, the hypnotized group should be deeply hypnotized so that the strong manipulation produces large differences in recall scores between groups. We also seek to eliminate fluctuating extraneous variables so that we minimize differences in recall scores within each condition.

After collecting our data, the parametric statistical procedure for determining whether the results of a two-sample experiment are significant is the two-sample t-test. However, there are two distinctly different ways that we can create samples, so there are two very different versions of the t-test. One version, called the *independent samples t-test*, is discussed in this chapter. (The other version, called the *dependent samples t-test*, is discussed in the next chapter.)

> ***REMEMBER*** There are two versions of the two-sample t-test, depending on how the researcher creates the samples.

We employ the independent samples t-test when we have two independent samples, which is another name for a between-subjects design with two conditions. In a

TABLE 15.2 Possible Data from Hypnosis Study Showing Little Power and Maximum Power

Un-Powerful		*Powerful*	
No Hypnosis	*Hypnosis*	*No Hypnosis*	*Hypnosis*
30	20	10	30
10	8	10	30
23	29	10	30
15	23	10	30
17	20	10	30
$\overline{X} = 19$	$\overline{X} = 20$	$\overline{X} = 10$	$\overline{X} = 30$

between-subjects design, a group of participants is selected for each condition, without regard for those selected in the other condition, and each participant serves in only one condition. You can recognize this design by the *absence* of anything fancy used to create the samples: We do not match up participants between the different conditions, nor do we test the same participants in different conditions. Instead, in this design we have one random sample of participants in one condition, and another, separate, and "independent" sample in the other condition.

Whether a between-subjects design is appropriate depends first on whether it is appropriate to compare participants in one condition to an entirely different batch of participants in the other condition. For example, in the hypnosis study, we do not want participants to have "practice" with the different conditions (by answering the memory questions more than once), so we choose a separate sample for each condition, producing a between-subjects design.

For a between-subjects design to be appropriate, however, it must also provide acceptable control over extraneous subject variables.

CONTROLLING SUBJECT VARIABLES IN A BETWEEN-SUBJECTS DESIGN

So far, we've taken the participants in a study pretty much for granted. But participants are thinking, feeling and behaving organisms who can modify any measurement procedure. Therefore, an important aspect of designing a study is to consider subject variables—individual differences—that may unduly influence the results. Remember, we want to control any extraneous variable that may fluctuate *within* a condition or *between* the conditions. Subject variables that fluctuate *within conditions* threaten reliability because differences between participants cause them to respond differently and inconsistently, resulting in error variance. Subject variables that fluctuate *between conditions* threaten internal validity, because we cannot know whether differences in scores between the conditions are due to a manipulation or to the subject variables. Ideally, therefore, we seek to control all extraneous subject variables so that all participants are equivalent on any variable that may influence the results.

To identify subject variables that you should control, look for any characteristic that is substantially correlated with the independent and dependent variables. Such a correlation *may* indicate that the extraneous variable has a causal influence. First, look for differences among participants that may influence the impact of your treatments. In the hypnosis study, for example, people differ in how easily they are hypnotized, how long they remain "under," and so on. If the people in our study differ greatly along this variable, then all people in the hypnosis condition will not experience the same hypnosis, and we may not produce consistent, significant differences in recall between the conditions.

Second, look for differences among participants that may influence responses on the dependent variable. For example, when measuring recall of the videotaped robbery, a person's inherent memory ability will influence his or her retention of the details. This will produce differences in recall scores between or within conditions, although they will have nothing to do with the effect of our manipulation.

Generally, when the stimulus is rather concrete and elicits a physical response, look for subject variables that influence physical responses. These may be physiological, such as participants' height or degree of coordination, or psychological, such as their cognitive abilities or motivation. For stimuli and responses that involve social behaviors or attitudes, look for variables that influence social processes, such as personality or cultural differences. Of course, the research literature is helpful in identifying important subject variables to control. Research specifically related to your study will indicate variables that others believed needed controlling. General research investigating individual differences will indicate subject variables that can influence the behavior you are studying.

> *REMEMBER* A subject variable that is correlated with the influence of the independent variable or with performance on the dependent variable is a potential subject variable for you to control.

With a between-subjects design, our first line of defense for controlling subject variables is random assignment.

Random Assignment

In a between-subjects design involving a true independent variable, we randomly assign participants to each condition. This controls subject variables by randomly mixing them, so that differences in a subject variable are balanced out in each condition. For example, by randomly assigning people to our hypnosis conditions, some who have a good memory and some who do not should end up in each condition. Overall, differences in recall scores between the conditions should not be due to differences in the memory ability of participants, so this potential confounding should be eliminated.

Be careful, however, to assign participants in a truly random way, avoiding any hidden variable that determines their assignment. For example, we wouldn't assign students who sit in the front of a class all to one condition and those sitting in the back to the other. Where students sit is not random, so we might confound the conditions with various personal characteristics. Similarly, do not assign to the same condition all people who first volunteer for a study. Those who participate early in a study may be more prompt, compulsive, or ambitious than later subjects. Instead, randomly assign participants to different conditions as they arrive so that such characteristics are spread out between conditions.

Pros and cons of random assignment Given how frequently research findings can be replicated, random assignment—and random selection—are powerful tools for producing balanced, representative samples in each condition. This is especially heartening because we often cannot identify the important subject variables to be controlled. With random assignment, we don't need to know the variables that are being controlled, because whatever they are in the population, we allow them to occur in a balanced way in each condition.

There are, however, three potential problems with random assignment. First, random assignment is not guaranteed to balance subject variables within each condition. Thus, for example, by chance we may still have people in one hypnosis condition who all have a much better memory than those in another, so that the independent variable is confounded. Second, random assignment works less well with small samples, so we are still likely to have groups that differ along important subject variables. Third, when random assignment does balance out a variable effectively, the variable then fluctuates *within* each condition. Fluctuating variables produce increased error variance and reduce the strength of the relationship. For example, differences in memory ability can cause any two people in the *same* hypnosis condition to have *different* recall scores, so that we end up with a weaker, less consistent relationship.

Because of these potential problems, researchers sometimes actively control subject variables. One approach is to balance subject variables.

Balancing Subject Variables

We do not leave the balancing of a critical subject variable to random chance, because the possibility of a serious confounding is too great. Instead, we control the subject variable by systematically balancing or "counterbalancing" its influence within each condition.

To balance a subject variable, we first make the variable part of our selection criteria. For obvious physical or personal characteristics (gender, age), we merely solicit participants who meet the criteria. For less obvious characteristics, we **pretest** participants: Prior to conducting the study, we measure potential subjects on the variable to be controlled. For example, we may measure a physical attribute (strength), a cognitive skill (reading ability), or a personality trait (anxiety level). Recognize that conducting a pretest is no different from measuring participants on a dependent variable, so we need a valid and reliable measurement technique that takes into account such issues as scoring criteria, sensitivity, demand characteristics, order effects, and so on.

Using the pretest information, we create a separate subject "pool" for each aspect of the subject variable we wish to balance. For example, in the hypnosis study, we could control participants' gender by creating a pool of males and a pool of females. To control for memory ability, we could identify those males and females who have good and poor eyewitness memory using an appropriate pretest. Then we assign participants so that each pool is represented in each condition in a balanced way. For example, we could randomly select and randomly assign participants so that 25% of those assigned to each condition are from the male-good memory pool, 25% are from the female-good memory pool, and so on. This design is shown in Table 15.3.

TABLE 15.3 Diagram of the Hypnosis Experiment Showing Balancing of Gender and Memory Ability

Participant Pool	*No Hypnosis*	*Hypnosis*
Male-Good memory	X X X ...	X X X ...
Female-Good memory	X X X ...	X X X ...
Male-Poor memory	X X X ...	X X X ...
Female-Poor memory	X X X ...	X X X ...
	$\overline{X}$	$\overline{X}$

The *X*s in each row represent participants' scores from the corresponding pool. To determine the effect of the conditions of hypnosis, we ignore gender and memory ability and average all scores vertically in a condition. Then the mean score in each condition should be equally influenced by differences in memory ability and gender, so that any differences between conditions cannot be confounded by these variables.

This procedure introduces an important new term: Above, when we ignore the gender and memory ability of participants and obtain an overall mean score in each column, we are "collapsing" across the subject variables. **Collapsing** across a variable means that we combine scores from the different amounts or categories of that variable. Above, we collapsed across gender and memory ability. If we used a male experimenter with half the participants in each condition and a female experimenter with the other half, we would collapse across experimenter gender by combining the scores of people tested by both experimenters, computing one overall mean for each condition. Likewise, when we test participants on multiple trials in a condition, we average them together, so we collapse across trials.

> *REMEMBER* Collapsing across a variable means that we combine the scores from the different levels of the variable.

Pros and cons of balancing If we find a significant relationship, then, because we've balanced the subject variable, we can be sure that it was not a confounding, so we have greater internal validity. We also have greater external validity, because we demonstrate the relationship even with the different levels of the variable present. Above, we are sure gender and memory ability do not confound the results, and we demonstrate a more general relationship because different genders and memory abilities are present.

Of course, that's if we find a significant relationship. As with random assignment, the drawback to counterbalancing is that it involves changing a subject variable *within* conditions, so there may be greater variability in the scores within each condition. Above, by including in each condition the scores of males and females who have good and poor memories, we are likely to see larger variability in recall scores than if we tested only males or only people with a good memory. Thus, counterbalancing may produce a relatively large error variance, so we obtain a less consistent relationship and have less power.

If you think about it long enough, you can identify any number of variables to counterbalance in any study. The drawback is that extensive balancing schemes greatly complicates the design of a study. Further, because different participants must be tested with each level of a balanced variable, this may dramatically increase the number of individuals required in each condition. And finally, a pretest may alert participants to the variables under study or to the purpose and predictions of the research. This knowledge may communicate demand characteristics that participants respond to during the experiment proper. (To avoid such problems, some form of deception in the pretest may be necessary.)

> *REMEMBER* Balancing a subject variable ensures that it cannot confound the results, but it may result in increased error variance and pretesting may communicate demand characteristics.

Limiting the Population

An alternative to counterbalancing a subject variable is to limit the population based on that variable, so that we keep the variable constant. Then the variable cannot influence the results. For example, if we expect males and females to differ greatly in how hypnosis influences their memory, we might limit the population to males only or to females only.

We limit the population through selection criteria. We pretest participants to identify those who meet the criteria and are approximately the same on the subject variable. For example, we might create a pool of males who have a very good eyewitness memory. Then, from this pool, we would randomly select and assign participants to the conditions.

Pros and cons of limiting the population There are two advantages to selecting participants from a more limited population. First, this increases internal validity by eliminating a potential confounding that might occur with random assignment: By testing only males, for example, we need not be concerned about whether gender is balanced in each condition. Second, this increases power by reducing the error variance: The more similar the participants, the less variable the scores are likely to be within each condition. Above, differences in scores within a condition that might occur between males and females will not occur when all participants are males.

There are also two drawbacks to limiting the population. First, if we become too selective, we may overly restrict the range of scores. For example, by limiting the study to just men with a very good memory, we might see little or no difference in recall scores between the hypnosis conditions. Second, because we are more selective in choosing participants, they represent a more limited population and so external validity is reduced. Thus, if we test only males, we will have no evidence for generalizing to females.

Researchers usually opt for increased power and internal validity, even at the expense of external validity. Therefore, the advantages of limiting the population usually outweigh the disadvantages.

> *REMEMBER* Limiting the population eliminates potential confounding by a subject variable and reduces error variance, but at the possible cost of restricted range and reduced external validity.

Selecting the Approach for Dealing with Subject Variables

Random assignment, counterbalancing subject variables, and limiting the population are not mutually exclusive procedures. We could, for example, limit the population to only one gender and then balance memory ability. And regardless of the extent to which we counterbalance or limit the population, we still rely on random selection and random assignment to balance any other subject variables within and between conditions.

In selecting a procedure or combination of procedures, you are faced with two considerations. First, how important is the variable? The more likely it is to influence the results, the more you must actively control it. Never leave the control of a highly influential variable to random assignment: Either counterbalance it or limit the population to keep it constant.

Second, weigh the goal of having the statistical power to find a significant relationship with the goal of making internally and externally valid inferences about the relationship. The larger the number of variables that are counterbalanced, the more that variables that can influence scores are changing, so the greater the error variance may be. Conversely, keeping a variable constant by limiting the population reduces potential error variance, but at the cost of reduced external validity because a more unique type of participant is being tested.

The same problems arise when counterbalancing *any* environmental, researcher, or measurement variable as well. For example, we could balance experimenter gender, employing a male experimenter with half of the subjects in each hypnosis condition and a female experimenter with the other half. This would produce greater generalizability, because we demonstrate the relationship with both types of experimenter. But we would balance this variable because we expect that whether a male or female experimenter is present *makes a difference* to subjects and to their scores. Therefore, counterbalancing will produce greater variability in scores within the conditions. If only a male or only a female experimenter were present throughout, however, we would see less error variance, but we would also demonstrate the relationship in a more unique situation.

There is no easy solution to this predicament. You should strive for a happy medium, but, if pushed, researchers generally risk producing a unique situation. So, counterbalance only those few variables that are *likely* to confound the independent variable or that *seriously* bias dependent scores. Control other, more minor variables by keeping them constant.

> *REMEMBER* Whether to counterbalance an extraneous variable depends on how much it threatens internal validity.

If any of the above approaches do not solve your problems with subject variables, you may instead employ the within-subjects designs discussed in the next chapter. If you choose a between-subjects design with two conditions, the results are analyzed using the independent samples *t*-test.

THE INDEPENDENT SAMPLES *t*-TEST

Remember that independent samples is simply another name for a between-subjects design: Samples are independent when we randomly select and assign a participant to one condition, without regard to who else has been selected for either condition. The samples are then composed of independent events, which, as we saw in Chapter 12, means that the probability of a particular score occurring in one sample is not influenced by a particular score occurring in the other sample.

Here's how to perform the *t*-test: Say that we conducted the hypnosis study, producing a mean recall in the control condition of 20, and a mean recall in the hypnosis condition of 23. On the one hand, perhaps the sample means accurately represent the population for each condition and we have demonstrated a relationship in nature: As we change the conditions of hypnosis, the recall scores in the population change in a consistent fashion. On the other hand, maybe we are being misled by sampling error. Maybe, even though we have different sample means, if we could test the entire population there would actually be the *same* population of scores under both conditions. Therefore, before we make any conclusions about the experiment, we must test the hypothesis that the difference between our sample means merely reflects sampling error.

By now you know the routine: (1) Check the assumptions of the statistical procedure and create the statistical hypotheses. (2) Set up and perform the statistical test. (3) If the results are significant, describe the relationship.

Assumptions of the Independent Samples *t*-Test

In addition to requiring two independent samples, the *t*-test requires that:

1. The dependent variable measures interval or ratio scores.
2. The population of raw scores represented by each sample forms a normal distribution. (If each sample n is greater than 30, the populations need form only roughly normal distributions.)
3. The variance of any raw score population must be estimated using the sample data.
4. And here's a new one: The populations represented by the samples have homogeneous variance. **Homogeneity of variance** means that if we could compute the true variance of the population represented by each sample, the value of σ_X^2 in each population would be the same. (Although it is not required that each condition have the same n, the more the ns differ from each other, the more important it is to have homogeneity of variance.[1])

From the research literature, say we find that the hypnosis study meets the above assumptions reasonably well. Now for our statistical hypotheses.

Statistical Hypotheses for the Independent Samples *t*-Test

Depending on the experimental hypotheses, we may have a one-tailed or a two-tailed test. For now, say that we don't specifically predict whether hypnosis will produce higher or lower recall scores than no hypnosis. This is a two-tailed test because we do not predict the *direction* of the difference between the two conditions: We merely predict that the sample from each condition represents a different population of recall scores, having a different value of μ.

First, the alternative hypothesis: In general terms, we expect condition 1 to produce $\overline{X}_1$ which represents μ_1, the μ we would find if we tested everyone in the population under condition 1. Condition 2 should produce a different $\overline{X}_2$, representing μ_2, the μ we would find if we tested everyone in the population under condition 2. A possible outcome from such an experiment is shown in Figure 15.1. For statistical purposes, the specific values of μ_1 and μ_2 are not our first concern. What *is* important is that μ_1 and μ_2 are *different* from each other. If the sample means represent a different population

[1] Chapter 17 introduces a test for determining whether you can assume homogeneity of variance.

FIGURE 15.1 Relationship in the Population in a Two-Sample Experiment

As the conditions change, the population tends to change in a consistent fashion.

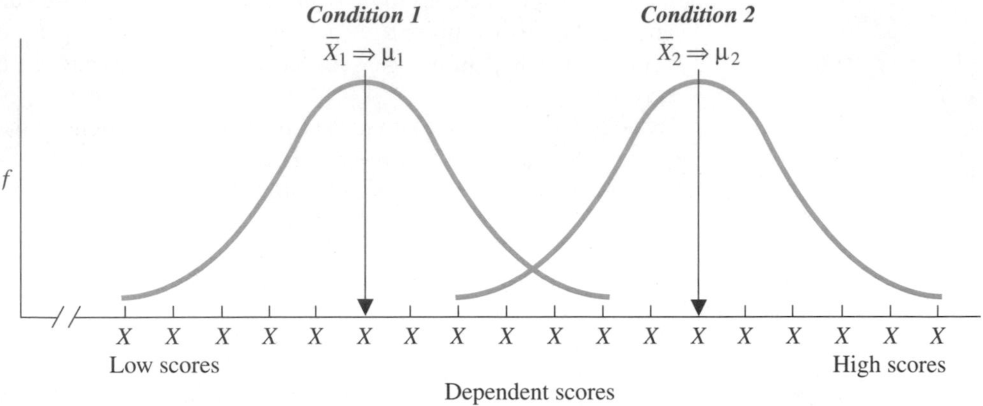

for each condition, then the experiment has demonstrated a relationship in nature: Changing the independent variable produces a change in the population of dependent scores.

Because we do not predict which of the two μs will be larger, the predicted relationship exists if one population mean is larger or smaller than the other. That is, μ_1 should not equal μ_2. We could state the alternative hypothesis as H_a: $\mu_1 \neq \mu_2$, but there is a better way to do it. If the two μs are not equal, then their *difference* does not equal zero. Thus, the two-tailed alternative hypothesis is

$$H_a\text{: } \mu_1 - \mu_2 \neq 0$$

H_a implies that the means from the two conditions each represent a different population of recall scores, having a different μ.

Of course, there's our old nemesis, the null hypothesis. Perhaps there is no relationship, so if we tested the entire population under the two conditions, we would find the same value of μ. In Figure 15.1, for example, you might find only the lower or upper distribution, or you might find one in the middle. Then the conditions of the independent variable literally would not make a difference in the population and there would be only one value of μ: You could call it μ_1 or μ_2, but it wouldn't matter because it would be the *same* μ. In this case, μ_1 *equals* μ_2. We could state the null hypothesis as H_0: $\mu_1 = \mu_2$, but again there is a better way. If the two μs are equal, then their difference is zero. Thus, the two-tailed null hypothesis is

$$H_0\text{: } \mu_1 - \mu_2 = 0$$

H_0 implies that both sample means represent the same population of recall scores, having the same μ.

Notice that we derived these hypotheses without specifying the value of either μ, so we have the same hypotheses regardless of the dependent variable being measured. Therefore, the above hypotheses are the two-tailed hypotheses for *any* independent samples *t*-test when you are testing whether there is no relationship in the population.

As usual, we test the null hypothesis. The conditions of the hypnosis study produced sample means of 20 and 23, respectively. To summarize these results, we look at the *difference* between the means: Changing from no hypnosis to hypnosis results in a difference in mean recall scores of 3 points. We always test H_0 by finding the probability of obtaining our results when H_0 is true. Therefore, to test H_0 we will determine the probability of obtaining a difference of 3 between $\overline{X}$s when both sample means actually represent the same population μ. In other words, we determine whether a difference of 3 is significantly different from zero. If it is, then there is a significant difference between the two sample means: Sampling error is too unlikely to produce such different sample means if they represent the same population μ. Then we can be confident that the sample means represent two different μs.

As usual, to test H_0, we need a sampling distribution.

The Sampling Distribution for the Independent Samples *t*-Test

Any sampling distribution is the distribution of sample statistics when H_0 is true. Here you can think of creating the sampling distribution as follows. Using the same *n*s as in our study, we select *two* random samples at a time from one raw score population. We compute the two sample means and arbitrarily subtract one from the other. The result is the *difference between the means*, symbolized by $\overline{X}_1 - \overline{X}_2$. We do this an infinite number of times and plot a frequency distribution of all the differences. We then have the **sampling distribution of differences between the means**, which is the distribution of all possible differences between two means when they are drawn from the raw score population described by H_0. You can envision the sampling distribution of differences between the means as shown in Figure 15.2. This distribution is just like any other sampling distribution except that along the *X* axis are the *differences* between two sample means, each labeled $\overline{X}_1 - \overline{X}_2$. As usual, the mean of the sampling distribution is the value stated in H_0, and here it is 0. The mean is zero because most often both sample means will equal the μ of the population of raw scores, so the difference between the sample means will be zero. Sometimes by chance, however, both sample means will not equal μ or each other. Depending on whether $\overline{X}_1$ or $\overline{X}_2$ is larger, the difference will be greater than zero (positive) or less than zero (negative). The larger the absolute

FIGURE 15.2 Sampling Distribution of Differences Between Means When H_0: $\mu_1 - \mu_2 = 0$

The mean of this distribution is zero. Larger positive differences are to the right, and larger negative differences are to the left.

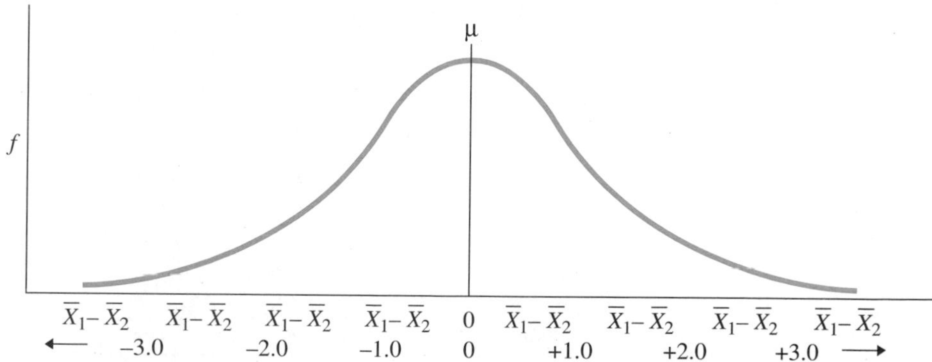

difference between the means, the farther into a tail of the distribution the difference falls, so the less likely such a difference is when H_0 is true.

To test H_0, we simply determine where the difference between our sample means lies on this sampling distribution. Our model of this sampling distribution is the t-distribution, so we locate this difference on the sampling distribution by computing t_{obt}.

Computing the Independent Samples t-Test

In the previous chapter, you computed t_{obt} by performing three steps: You computed the estimated variance of the raw score population, then you computed the estimated standard error of the sampling distribution, and then you computed t_{obt}. In performing the two-sample t-test, you'll complete the same three steps.

Estimating the population variance First compute s_X^2 for each sample, using the formula

$$s_X^2 = \frac{\Sigma X^2 - \frac{(\Sigma X)^2}{n}}{n - 1}$$

Let's label the hypnosis condition as Sample 1, so its mean, variance, and n are $\overline{X}_1$, s_1^2 and n_1, respectively. For the no-hypnosis condition, we have $\overline{X}_2$, s_2^2 and n_2. Say that we obtain the results shown in Table 15.4.

Each s_X^2 estimates the population variance, but each estimate may contain sampling error. (Because of this, if s_1^2 does not equal s_2^2, we have not necessarily violated the assumption of homogeneity of variance in the population.) To obtain the best estimate of the population variance, we'll compute a weighted average of the two values of s_X^2. Each variance is weighted based on the size of df in the sample. The weighted average of the sample variances is called the **pooled variance**, and its symbol is s_{pool}^2.

THE COMPUTATIONAL FORMULA FOR THE POOLED VARIANCE IS

$$s_{pool}^2 = \frac{(n_1 - 1)s_1^2 + (n_2 - 1)s_2^2}{(n_1 - 1) + (n_2 - 1)}$$

This formula says to multiply the s_X^2 from each sample times $n - 1$ for that sample, then add the results together and divide by the sum of $(n_1 - 1) + (n_2 - 1)$.

TABLE 15.4 Data from the Hypnosis Study

	Sample 1: subjects under hypnosis	*Sample 2: subjects under no hypnosis*
Mean details recalled	$\overline{X}_1 = 23$	$\overline{X}_2 = 20$
Number of subjects	$n_1 = 17$	$n_2 = 15$
Sample variance	$s_1^2 = 9.0$	$s_2^2 = 7.5$

Using the data in Table 15.4 to fill in the above formula, we have

$$s^2_{pool} = \frac{(17 - 1)9.0 + (15 - 1)7.5}{(17 - 1) + (15 - 1)}$$

In the numerator, 16 times 9 is 144, and 14 times 7.5 is 105. In the denominator, 16 plus 14 is 30. Now we have

$$s^2_{pool} = \frac{144 + 105}{30} = \frac{249}{30} = 8.30$$

Thus, we estimate that the variance of any of the populations of recall scores represented by our samples is 8.30.

Now use the value of s^2_{pool} to compute the standard error of the sampling distribution.

Computing the standard error of the difference The standard error of the sampling distribution of differences between the means is called the standard error of the difference. The **standard error of the difference** is the estimated standard deviation of the sampling distribution of differences between the means. It indicates how spread out the values of $\overline{X}_1 - \overline{X}_2$ are when the distribution is created using samples having our n and our value of s^2_{pool}. The symbol for the standard error of the difference is $s_{\overline{X}_1 - \overline{X}_2}$. (The subscript indicates we are dealing with differences between pairs of means.)

In the previous chapter, you saw that for the single-sample *t*-test, a formula for the standard error of the mean was

$$s_{\overline{X}} = \sqrt{(s^2_X)\left(\frac{1}{N}\right)}$$

The formula for the standard error of the difference is very similar.

THE DEFINITIONAL FORMULA FOR THE STANDARD ERROR OF THE DIFFERENCE IS

$$s_{\overline{X}_1 - \overline{X}_2} = \sqrt{(s^2_{pool})\left(\frac{1}{n_1} + \frac{1}{n_2}\right)}$$

To compute $s_{\overline{X}_1 - \overline{X}_2}$, it is easiest if you first reduce the fractions $1/n_1$ and $1/n_2$ to decimals and then add them together and multiply the sum times s^2_{pool}. Then find the square root.

For our hypnosis study, s^2_{pool} is 8.3, and n_1 is 17 and n_2 is 15. Filling in the above formula, we have

$$s_{\overline{X}_1 - \overline{X}_2} = \sqrt{8.3\left(\frac{1}{17} + \frac{1}{15}\right)}$$

Since 1/17 is .059 and 1/15 is .067, their sum is .126. Then, we have

$$s_{\overline{X}_1 - \overline{X}_2} = \sqrt{8.3(.126)} = \sqrt{1.046} = 1.02$$

Thus, for our data, the standard error of the difference, $s_{\overline{X}_1 - \overline{X}_2}$, equals 1.02.

We can take the above definitional formula for $s_{\bar{X}_1 - \bar{X}_2}$ and replace the symbol for s^2_{pool} with the formula for s^2_{pool}. Then,

THE COMPUTATIONAL FORMULA FOR THE STANDARD ERROR OF THE DIFFERENCE IS

$$s_{\bar{X}_1 - \bar{X}_2} = \sqrt{\left(\frac{(n_1 - 1)s_1^2 + (n_2 - 1)s_2^2}{(n_1 - 1) + (n_2 - 1)}\right)\left(\frac{1}{n_1} + \frac{1}{n_2}\right)}$$

In the parentheses on the left, we compute s^2_{pool}, and by multiplying it times the value in the parentheses on the right and then taking the square root, we compute $s_{\bar{X}_1 - \bar{X}_2}$.

After computing the standard error of the difference, you can finally compute t_{obt}.

Computing t_{obt} In previous chapters, we found how far the result of our study, $\bar{X}$, was from the mean of the H_0 sampling distribution, μ, measured in standard error units. In general, this formula is

$$t_{\text{obt}} = \frac{(\text{result of the study}) - (\text{mean of } H_0 \text{ sampling distribution})}{\text{standard error}}$$

Now we will perform the same computation, but here the "result of the study" is the *difference* between the two sample means. So in place of "result of the study" we put the quantity $\bar{X}_1 - \bar{X}_2$. Also, now the mean of the H_0 sampling distribution is the *difference* between μ_1 and μ_2 described by H_0. Thus, we replace "mean of H_0 sampling distribution" with the quantity $\mu_1 - \mu_2$. Finally, we replace "standard error" with $s_{\bar{X}_1 - \bar{X}_2}$. Putting this all together, we have the formula.

THE DEFINITIONAL FORMULA FOR THE INDEPENDENT-SAMPLES t-TEST IS

$$t_{\text{obt}} = \frac{(\bar{X}_1 - \bar{X}_2) - (\mu_1 - \mu_2)}{s_{\bar{X}_1 - \bar{X}_2}}$$

$\bar{X}_1$ and $\bar{X}_2$ are the sample means, $s_{\bar{X}_1 - \bar{X}_2}$ is computed as shown above, and the value of $\mu_1 - \mu_2$ is the difference specified by the null hypothesis. The reason for writing H_0 as $\mu_1 - \mu_2 = 0$ is that it directly tells you the mean of the sampling distribution and thus the value of $\mu_1 - \mu_2$ to put in the above formula.

Now we can compute t_{obt} for our hypnosis study. Our sample means were 23 and 20, the difference between μ_1 and μ_2 specified by H_0 is 0, and we computed $s_{\bar{X}_1 - \bar{X}_2}$ to be 1.02. Putting these values into the above formula, we have

$$t_{\text{obt}} = \frac{(23 - 20) - 0}{1.02}$$

Then we have:

$$t_{obt} = \frac{(+3) - 0}{1.02} = \frac{+3}{1.02} = +2.94$$

Our t_{obt} is +2.94. Thus, the difference of +3 between our sample means is located at something like a z-score of +2.94 on the sampling distribution of differences when H_0 is true and both samples represent the same population μ.

You can save a little paper by combining the steps of computing s^2_{pool}, $s_{\bar{X}_1 - \bar{X}_2}$, and t_{obt} into one formula.

THE COMPUTATIONAL FORMULA FOR THE t-TEST FOR INDEPENDENT SAMPLES IS

$$t_{obt} = \frac{(\bar{X}_1 - \bar{X}_2) - (\mu_1 - \mu_2)}{\sqrt{\left(\frac{(n_1 - 1)s_1^2 + (n_2 - 1)s_2^2}{(n_1 - 1) + (n_2 - 1)}\right)\left(\frac{1}{n_1} + \frac{1}{n_2}\right)}}$$

The numerator is the same as in the previous formula for t_{obt}. In the denominator, however, the symbol for the standard error has been replaced by its computational formula.

Thus, for the hypnosis study, substituting the data into the above formula, gives

$$t_{obt} = \frac{(23 - 20) - 0}{\sqrt{\left(\frac{(17 - 1)9.0 + (15 - 1)7.5}{(17 - 1) + (15 - 1)}\right)\left(\frac{1}{17} + \frac{1}{15}\right)}}$$

which becomes

$$t_{obt} = \frac{+3}{\sqrt{(8.3)(.126)}} = \frac{+3}{1.02} = +2.94$$

Thus, again $t_{obt} = +2.94$. Note that the denominator in the t-test is the standard error of the difference, so $s_{\bar{X}_1 - \bar{X}_2}$ is again 1.02.

Interpreting t_{obt}

As usual, to determine if t_{obt} is significant, we compare it to t_{crit}, which is found in Table 2 in Appendix C. We again obtain t_{crit} using degrees of freedom, but we have two samples, so the *df* are computed differently: Now degrees of freedom equals $(n_1 - 1) + (n_2 - 1)$.

> ***REMEMBER*** Critical values of t for the independent samples t-test have $df = (n_1 - 1) + (n_2 - 1)$.

Another way of expressing the *df* is $(n_1 + n_2) - 2$.

For the hypnosis study, $n_1 = 17$ and $n_2 = 15$, so *df* equals $(17 - 1) + (15 - 1)$, which is 30. With alpha at .05, the t-tables show that for a two-tailed test, t_{crit} is ± 2.042. Figure 15.3 locates these values on the sampling distribution of differences.

FIGURE 15.3 H_0 Sampling Distribution of Differences Between Means When $\mu_1 - \mu_2 = 0$

The t_{obt} shows the location of $\overline{X}_1 - \overline{X}_2 = +3.0$.

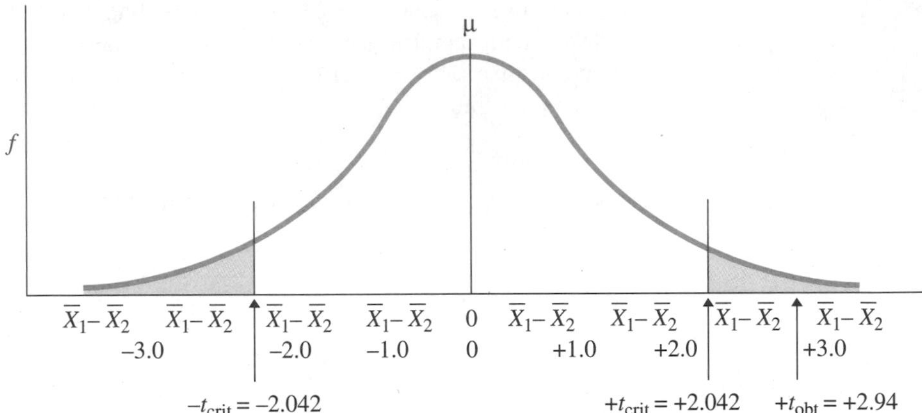

Although H_0 says that the difference between our sample means is merely a poor representation of no difference between μ_1 and μ_2, the location of t_{obt} on the sampling distribution shows that samples such as ours are seldom *that* poor at representing no difference. In fact, t_{obt} lies beyond t_{crit}, so it—and the corresponding difference of +3—is in the region of rejection. Therefore, we reject H_0 and conclude that the difference between our sample means is too unlikely to accept as representing no difference in the population. In other words, the difference between our means is significantly different from zero. We communicate this result as

$$t(30) = +2.94, p < .05$$

As usual, *df* is reported in parentheses, and because $\alpha = .05$, the probability is less than .05 that we have made a Type I error (rejected a true H_0).

We can now accept the alternative hypothesis, which says that $\mu_1 - \mu_2 \neq 0$: The difference between the sample means represents a difference between two μs that is not zero. Because the sample means produced a significant difference, this indicates that the sample means differ significantly from each other. Further, the mean for hypnosis (23) is larger than the mean for no hypnosis (20). So, to be precise, we conclude that hypnosis leads to significantly higher recall scores than no hypnosis. Thus, we have evidence of a relationship in the population where increasing the amount of hypnosis is associated with higher recall scores. Now, as usual, we proceed to interpret this relationship "psychologically": Using the constructs, theories, or models that are relevant, we explain why hypnosis increases recall scores an average of 3 points.

If t_{obt} were not beyond t_{crit}, we would not reject H_0. Then we would not have convincing evidence that the difference between the sample means was anything other than sampling error, so we could not say that there was a relationship between hypnosis and memory.

Because we did find a significant difference, we now describe the relationship. From the previous chapter, you already know that we could compute a confidence interval for the μ that is likely to be represented by each sample mean. However, we can also describe the populations our samples represent in another way. Because we found a difference of 3 between the sample means, we would expect to find a difference of *around* 3 between the corresponding μs. We can create a confidence interval to define this difference more specifically.

Confidence Interval for the Difference Between Two μs

As you know, a confidence interval describes a range of population values, any one of which is likely to be represented by the sample data. The **confidence interval for the difference between two μs** describes a range of *differences* between two μs, any one of which is likely to be represented by the *difference* between our two sample means. To compute the interval, we find the largest and smallest values of the quantity $\mu_1 - \mu_2$ that are not significantly different from the difference between our sample means.

THE COMPUTATIONAL FORMULA FOR THE CONFIDENCE INTERVAL FOR THE DIFFERENCE BETWEEN TWO μs IS

$$(s_{\bar{X}_1-\bar{X}_2})(-t_{\text{crit}}) + (\bar{X}_1 - \bar{X}_2) \leq \mu_1 - \mu_2 \leq (s_{\bar{X}_1-\bar{X}_2})(+t_{\text{crit}}) + (\bar{X}_1 - \bar{X}_2)$$

Here $\mu_1 - \mu_2$ stands for the unknown difference we are estimating, t_{crit} is the two-tailed value found for the appropriate α at $df = (n_1 - 1) + (n_2 - 1)$, and the values of $s_{\bar{X}_1-\bar{X}_2}$ and $(\bar{X}_1 - \bar{X}_2)$ are computed in the *t*-test from the sample data.

In our hypnosis study, the two-tailed t_{crit} for $df = 30$ and $\alpha = .05$ is ± 2.042. We computed that $s_{\bar{X}_1-\bar{X}_2}$ is 1.02 and $\bar{X}_1 - \bar{X}_2$ is +3. Filling in the above formula, we have

$$(1.02)(-2.042) + (+3) \leq \mu_1 - \mu_2 \leq (1.02)(+2.042) + (+3)$$

Multiplying 1.02 times ± 2.042 gives

$$-2.083 + (+3) \leq \mu_1 - \mu_2 \leq +2.083 + (+3)$$

By adding and subtracting 2.083 from +3, we obtain the final confidence interval:

$$.0917 \leq \mu_1 - \mu_2 \leq 5.083$$

Because we used the .05 alpha level, this is the 95% confidence interval. It indicates that, if we performed this experiment on the entire population, we are 95% confident that the interval between .0917 and 5.083 contains the difference we would find between the μs for no hypnosis and hypnosis. In essence, if someone asked us how big a difference hypnosis makes for everyone when answering questions about our videotape, our response would be that we are 95% confident that the difference is, on average, between about .09 and 5.08 correct answers.

Performing One-Tailed Tests on Independent Samples

We could have performed a one-tailed test if we had predicted the direction of the difference between the two conditions. Say that we predicted a positive relationship where hypnosis produces higher recall scores. Everything discussed above applies here, but beware: We *arbitrarily* call one mean $\overline{X}_1$ and one mean $\overline{X}_2$ and then subtract $\overline{X}_1 - \overline{X}_2$. How we assign the subscripts determines whether we have a positive or a negative difference, and thus whether we have a positive or a negative t_{obt}. As you know, the sign of t_{obt} is very important in a one-tailed test.

To prevent confusion, use more meaningful subscripts than 1 and 2. For example, you could use the subscript *h* for the hypnosis condition and *n* for the no-hypnosis condition. Then decide which $\overline{X}$ and corresponding μ you expect to be larger. If you think that the μ for no-hypnosis, μ_n, is *smaller* than the μ for hypnosis, μ_h, and you decide to subtract the smaller μ_n from the larger μ_h, you are predicting a difference that is greater than zero. The alternative hypothesis would then be written as H_a: $\mu_h - \mu_n > 0$ (and H_0 as $\mu_h - \mu_n \leq 0$). Then be sure to subtract the sample means in the *same* way that you subtract the μs, so if you subtract $\mu_h - \mu_n$, then subtract $\overline{X}_h - \overline{X}_n$. Finally, locate the region of rejection based on your prediction and subtraction. If the means represent the predicted relationship, then subtracting the smaller $\overline{X}_n$ from the larger $\overline{X}_h$ should yield a positive difference, producing a positive t_{obt}. To reject H_0, t_{obt} must lie beyond t_{crit}, so t_{crit} must be a *positive* number. Therefore, the region of rejection is in the positive, right-hand tail of the sampling distribution.

One-tailed tests are confusing because, while still predicting a larger μ_h, we could have reversed H_a, expressing it as H_a: $\mu_n - \mu_h < 0$. Now subtracting the larger μ_h from the smaller μ_n produces a difference that is less than zero, and subtracting the sample means this way should produce a negative t_{obt}. Now the region of rejection is in the negative tail of the distribution and t_{crit} is negative.

If you subtract the sample means in the opposite way that you subtract the μs, the sign of t_{obt} will be the opposite of what it should be for your H_a. Then you may erroneously conclude that there is not a significant difference when there is, or vice versa.

DESCRIBING THE RELATIONSHIP IN A TWO-SAMPLE EXPERIMENT

By now you understand how to determine whether results are significant. However, the fact that t_{obt} is significant is not the end of the story. If you stop after hypothesis testing, then you've found a "real" relationship, but you have not described it. Essentially, it's the same as saying, "I've computed a correlation coefficient, but I'm not going to tell you what it is." (Frustrating, isn't it?) Therefore, you're not finished until you fully describe the characteristics of the relationship.

Your focus should always be on the sample means (or other measures of central tendency, when they are used). Then you can begin to understand the behavior by summarizing the typical score—and typical behavior—found in each condition. Confidence intervals then allow you to generalize your findings, describing the range of typical scores—and typical behaviors—you expect in the population (in nature).

In addition, you should also describe the relationship by graphing it. As you know, we graph experimental results by plotting the mean of each condition on the *Y* axis and the conditions of the independent variable on the *X* axis. Thus, we would plot the results of the hypnosis study as shown in Figure 15.4.

In essence, we then interpret the graph and relationship by applying *correlational* statistics. That's right! We performed the *t*-test because it is the more powerful way to

FIGURE 15.4 Line Graph of the Results of the Hypnosis Study

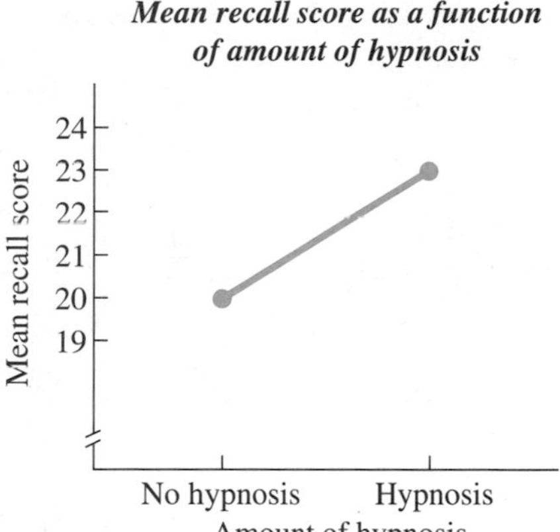

test H_0 in a two-sample experiment. However, recall that the advantage of correlational procedures is that they describe the type and strength of a relationship.

Thus, first, the slanting line graph shows a positive relationship. By knowing the *type* of relationship present, we have more information when we try to understand the behaviors reflected by the variables. In fact, the line graph summarizes the relationship exactly like a regression line (discussed in Chapter 11). With two conditions of the independent variable, we have two values of X. When there are only two values of X, the regression line connects the mean of Y at each X. Further, we can envision the scatterplot that the individual data points would form. Because the mean is the center of dependent scores, we envision data points above and below the mean Y score at each X. Thus, we literally expect participants to score around—above and below— the mean of 20 in the no-hypnosis condition and around the mean of 23 in the hypnosis condition. Further, we would also predict these mean scores for any other people we might test under these conditions.

What's missing, of course, is how much error we have in our predictions or, conversely, what the *strength* of the relationship is. Therefore, whenever the results of an experiment are significant, the next step is to compute a correlation coefficient, correlating the scores on the dependent variable with the conditions of the independent variable. This describes how consistently participants scored at or close to the mean score for each condition. Thus, computing the correlation coefficient in the hypnosis study, for example, will tell us how consistently close recall scores were to the mean of 20 under no hypnosis and to the mean of 23 under hypnosis.

> *REMEMBER* Whenever you obtain significant results by testing means, the first step in describing the strength of the relationship is to compute the appropriate correlation coefficient.

Describing the Strength of the Relationship in a Two-Sample Experiment Using r_{pb}

The point-biserial correlation coefficient (r_{pb}) is usually the appropriate coefficient for describing a two-sample experiment. This is because r_{pb} is used when we have one dichotomous X variable (that is, consisting of two categories) and one continuous interval or ratio Y variable. In a two-sample experiment, the conditions of the independent variable form a dichotomous X variable, and scores on the dependent variable are usually a continuous interval or ratio Y variable.

You could compute r_{pb} using the formula in Chapter 10, but instead you can compute r_{pb} using t_{obt}.

THE FORMULA FOR COMPUTING r_{pb} FROM t_{obt} IS

$$r_{pb} = \sqrt{\frac{(t_{obt})^2}{(t_{obt})^2 + df}}$$

Insert the value of the significant t_{obt} and the df used in the t-test.

In our hypnosis study, we found $t_{obt} = +2.94$ with $df = 30$, so

$$r_{pb} = \sqrt{\frac{(2.94)^2}{(2.94)^2 + 30}} = \sqrt{\frac{8.64}{8.64 + 30}} = \sqrt{\frac{8.64}{38.64}} = \sqrt{.224} = .47$$

Thus, the relationship between amount of hypnosis and recall scores can be summarized as $r_{pb} = .47$. But notice that the final step in the formula for computing r_{pb} is to find the square root, so the answer will always be positive. Therefore, depending on the data, you must determine whether the positive or negative sign is appropriate. The hypnosis study produced a positive relationship, so $r_{pb} = +.47$. This is interpreted as any other correlation coefficient: On a scale of 0 to ± 1, the hypnosis data produced a reasonably strong, consistent relationship.

Recall, however, that a problem with a correlation coefficient is that it can be difficult to interpret: It does not measure in units of consistency, so we can only subjectively evaluate it relative to 0 and ± 1. Instead, the most direct way to evaluate a relationship is to square the correlation coefficient, computing the proportion of variance accounted for.

Describing the Effect Size in a Two-Sample Experiment

The ultimate way to describe the relationship in a two-sample experiment is to compute the proportion of variance accounted for. To do so, square the value of r_{pb} (or don't find the square root in the previous formula). The resulting value is interpreted in the same way that the proportion of variance accounted for was interpreted back in Chapter 11. Essentially, r^2_{pb} indicates how much more accurately we can predict the dependent (Y) scores when we know the relationship and the condition under which participants were tested (their X scores), compared to when we are unaware of the relationship and instead predict the overall mean of all Y scores for everyone in the study. For the hypnosis study, squaring r_{pb} gives $(.47)^2$, or .22. Thus, on average we are 22% closer to knowing participants' actual recall scores when we use the mean score of each condition to predict their scores than we are when we use the overall mean recall score of the experiment.

In an experiment, the proportion of variance accounted for goes by a different name: It is called the *effect size*. This is because when an experiment demonstrates a significant relationship, our explanation is that the independent variable has an *effect* on the dependent variable, "causing" dependent scores to change. (I've used quotation marks because we never *prove* that changing the independent variable causes the scores to change.) Thus, we explain the variability in recall scores, for example, as seeming to be caused by changing the conditions of hypnosis, and how consistently hypnosis does this is its effect size. The **effect size** indicates how consistently differences in the dependent scores are caused by changes in the independent variable. The larger the effect size, the more consistently the raw scores for each condition are located at or close to the mean score for that condition, so the more consistent is the influence of the independent variable.

The greater the effect size, the more that knowing the condition of the independent variable improves our accuracy in describing and predicting participants' dependent scores, and so the greater the scientific importance of the relationship. For example, we found that r^2_{pb} was .22, so only 22% of the differences in recall scores are accounted for by hypnosis. We assume that everything has a cause, so there must be *other* variables that are causing the substantial unaccounted-for differences in recall scores (for example, perhaps attention to detail or concentration ability plays an important role). Therefore, hypnosis is only modestly important in determining recall scores in this study.

Although a large effect size indicates an important relationship in a statistical sense, it does not indicate importance in a practical sense: Statistics will never tell you if you have performed a silly study. In the real world, our conclusion that memory was improved by hypnosis has limited practical importance. (To improve my memory, should I walk around under hypnosis all the time?) Statistical importance addresses a different issue: If you want to understand how nature works when it comes to hypnosis and memory, *then* this relationship is relevant and important. It is relevant because the results were significant, so we are confident that there is a "real" relationship found in nature. It is important to the extent that hypnosis accounts for 22% of the differences in recall scores.

REMEMBER Effect size indicates how big a role the conditions of the independent variable play in determining scores on the dependent variable.

POWER AND THE INDEPENDENT SAMPLES *t*-TEST

Remember that we maximize power by designing a powerful study. This in turn maximizes the size of t_{obt} relative to t_{crit}, maximizing the probability of rejecting H_0 when it is false. To see how powerful data translates into a larger t_{obt}, look again at the formula.

$$t_{obt} = \frac{(\overline{X}_1 - \overline{X}_2) - (\mu_1 - \mu_2)}{\sqrt{\left(\frac{(n_1 - 1)s_1^2 + (n_2 - 1)s_2^2}{(n_1 - 1) + (n_2 - 1)}\right)\left(\frac{1}{n_1} + \frac{1}{n_2}\right)}}$$

Anything that increases the numerator or decreases the denominator produces a larger t_{obt}.

First, recall that we seek a strong manipulation to maximize the difference between the two conditions. Selecting very different conditions of the independent variable is likely to produce a large difference in dependent scores, producing a larger value of $(\overline{X}_1 - \overline{X}_2)$. Then the numerator in the formula will be larger, so t_{obt} will be larger.

Second, we seek to minimize the variability of the raw scores. Minimizing this error variance within conditions minimizes each s_X^2, producing a smaller s_{pool}^2 and thus a smaller $s_{\bar{X}_1 - \bar{X}_2}$. Dividing by a smaller denominator produces a larger t_{obt}. (Also, with less variability, there is a stronger relationship, and so r_{pb} and r_{pb}^2 are larger.)

Finally, we maximize the sample *n*s. The larger the number of participants per condition, the smaller the denominator when calculating t_{obt}. In addition, larger *n*s give a larger *df*, resulting in a smaller value of t_{crit}. Generally, for minimum acceptable power, at least 15 to 30 participants in each condition are needed. Increasing the *n*s beyond this tends to substantially increase power, until we have about 60 scores per condition. Beyond this point, increasing *n* only modestly increases power. Of course, if more participants can be tested easily, they will further increase the external validity and power in a study. Also, if we cannot produce relatively large differences between conditions and/or we expect large error variance, we compensate by testing even larger *n*s. Reading the literature related to your research will show you the *n*s other researchers find acceptable. For help in making decisions regarding power, you can utilize advanced statistical procedures called "power analysis" (e.g., Cohen, 1988).

ELIMINATING PARTICIPANTS FROM THE DATA

Even with all our precautions, research never runs as smoothly as planned (Murphy's Law always applies). Some participants will behave strangely (I've had subjects go to sleep!). Through debriefing, you may find some participants were biased by a demand characteristic or discover others who were told the details of the study by a previous subject. A fire drill may occur while testing someone or your tape-recorder may blow a fuse.

In such situations, participants are not participating in the study that you designed: They are not being exposed to the independent variable reliably, they are not responding as directed, or they are being influenced by extraneous variables. Therefore, you may exclude their data from your analysis. You must be very sure that these participants do not belong in the study, and you cannot exclude them just because their scores do not confirm your prediction. (That's rigging the results and committing fraud: You might as well make up the data.) When in doubt, therefore, include a participant's data. But, if scores are clearly inappropriate, you can exclude them. Then for the proper *n*, test additional participants to fill in for those you've excluded. (Any study that produces many excluded participants, however, is a problem, because it contains a consistent hidden factor that is selecting a biased sample.)

PUTTING IT ALL TOGETHER

Two-sample experiments are fairly common in psychological research, so you are likely to use two-sample *t*-tests in your own research, and you'll need to understand them when reading the research of others. Effect size is another procedure you'll frequently encounter, because it is the only way to determine whether a relationship is important or is much ado about nothing. In fact, as of 1994 the American Psychological

Association now requests that all published research include a measure of effect size.[2] Because effect size has not always been reported in psychological research, elaborate experiments have often been performed to study what are actually very minor variables. (Researchers sometimes forget the old adage "Things that are not worth doing are not worth doing well!") Therefore, compute effect size whenever you have significant results in an experiment. Then, by using r^2_{pb} or other measures you will learn about, you'll know whether your variable is worth studying—and you'll also be on the cutting edge of statistical sophistication.

CHAPTER SUMMARY

1. A *between-subjects design* contains *independent samples*. Two samples are independent when participants are randomly selected for a condition, without regard for who else has been selected for either condition, and each subject participates in only one condition.
2. Researchers control subject variables that are correlated with the influence of the independent variable or with performance on the dependent variable.
3. In a between-subjects design, subject variables are controlled by balancing them through random assignment, by directly counterbalancing them, and by keeping them constant through limiting the population being sampled.
4. Counterbalancing subject variables tends to increase internal and external validity, but it may also increase error variance and reduce power. Limiting the population reduces error variance and increases power, but it may also reduce external validity.
5. A *pretest* is used to identify participants who meet certain selection criteria.
6. *Collapsing across a variable* means to combine scores from the different amounts or categories of that variable.
7. The *independent samples t-test* assumes that (a) For two random, independent samples, the dependent scores measure an interval or ratio variable; (b) The population of raw scores represented by each sample forms a normal distribution; and (c) The populations represented by the samples have homogeneous variance.
8. *Homogeneity of variance* means that the values of σ^2_X in the populations being represented by the sample data are equal.
9. A significant t_{obt} from the independent samples t-test indicates that the difference between $\overline{X}_1 - \overline{X}_2$ is unlikely to represent the difference between μ_1 and μ_2 described by H_0. Therefore, the results are assumed to represent the predicted relationship in the population.
10. The *confidence interval* for $\mu_1 - \mu_2$ contains a range of differences between two μs, any one of which is likely to be represented by the difference between two independent sample means.

[2] See the fourth edition of the *Publication Manual of the American Psychological Association*, 1994, published by the American Psychological Association, Washington, D.C.

11. The strength of a significant relationship between the independent and dependent variables in a two-sample experiment is described by computing the *point-biserial correlation coefficient,* r_{pb}.

12. The squared point-biserial coefficient, r^2_{pb}, measures the proportion of variance in the dependent scores that is accounted for, or explained, by changing the conditions of the independent variable.

13. The proportion of variance accounted for in an experiment is called the *effect size*. The larger the effect size, the more consistently the dependent scores change as the conditions of the independent variable change.

14. The power of the two-sample *t*-test increases with (a) larger differences in scores between the conditions, (b) smaller variability of scores within each condition, and (c) larger *n*.

PRACTICE PROBLEMS

(Answers for odd-numbered problems are provided in Appendix D.)

1. A scientist has conducted a two-sample experiment. (a) What two versions of a parametric procedure are available to him? (b) What is the deciding factor for selecting between them?

2. (a) How do you create independent samples? (b) How do you create a between-subjects design?

3. How do you identify subject variables that may confound a study?

4. (a) How does random assignment to conditions control subject variables? (b) When is counterbalancing subject variables more appropriate than random assignment? Why?

5. (a) What positive impact does counterbalancing subject variables have on internal and external validity? (b) What negative impact does it have on the data?

6. (a) What positive impact does limiting the population have on the results of a study? (b) What negative impact does it have?

7. What problems arise from pretesting participants?

8. (a) What are the assumptions for the independent samples *t*-test? (b) What is homogeneity of variance?

9. What does it mean to collapse across a variable?

10. (a) When is it acceptable to exclude a participant's data? (b) How might you reduce external validity by doing so?

11. Foofy has obtained a statistically significant two-sample t_{obt}. Now, what three things should she do to complete her analysis?

12. (a) What does a measure of effect size indicate? (b) How is it computed in a two-sample experiment?

13. In an experiment, a researcher seeks to demonstrate a relationship between hot or cold baths (the independent variable) and the amount of relaxation they produce (the dependent variable). He obtains the following relaxation scores:

Sample 1 (hot): $\overline{X} = 43$, $s^2_X = 22.79$, $n = 15$

Sample 2 (cold): $\overline{X} = 39$, $s^2_X = 24.6$, $n = 15$

(a) What are H_0 and H_a for this study? (b) Compute t_{obt}. (c) With $\alpha = .05$, what is t_{crit}? (d) What should the researcher conclude about this relationship? (e) Compute the confidence interval for the difference between the μs. (f) How big of an effect does bath temperature have on relaxation? (g) Describe how you would graph these results.

14. A researcher investigates whether a period of time feels longer when people are bored than when they are not bored. The researcher obtains the following estimates of the time period (in minutes):

Sample 1 (bored): $\overline{X} = 14.5$, $s_X^2 = 10.22$, $n = 28$

Sample 2 (not bored): $\overline{X} = 9.0$, $s_X^2 = 14.6$, $n = 34$

(a) What are H_0 and H_a for this study? (b) Compute t_{obt}. (c) With $\alpha = .05$, what is t_{crit}? (d) What should the researcher conclude about this relationship? (e) Compute the confidence interval for the difference between the μs. (f) How important is boredom in determining how quickly time seems to pass?

15. Foofy predicts that students who use a computer program that corrects spelling errors will receive higher grades on a term paper. She uses an independent samples design in which Group A uses a spelling checker and Group B does not. She tests H_0: $\mu_A - \mu_B \leq 0$ and H_a: $\mu_A - \mu_B > 0$. She obtains a negative value of t_{obt}. (a) What should she conclude about this outcome? (b) Assuming that her sample means actually support her predictions, what miscalculation is she likely to have made?

16. What does a confidence interval for the difference between two μs indicate?

17. A researcher investigates whether classical background music is more or less soothing to air-traffic controllers than Top-40 background music. He plays classical background music to one group and Top-40 music to another. At the end of the day, he gives each subject an irritability questionnaire and obtains the following data:

Sample 1 (classical): $\overline{X} = 14.69$, $s_X^2 = 8.4$, $n = 6$

Sample 2 (Top-40): $\overline{X} = 17.21$, $s_X^2 = 11.6$, $n = 6$

After computing the independent samples *t*-test, he finds $t_{obt} = +1.38$. (a) With $\alpha = .05$, report the statistical results. (b) What should the researcher conclude about these results? (c) What other statistics should be computed? (d) What statistical flaw is likely in the experiment? (e) What could the researcher do to improve the experiment? (f) What effect might this have?

18. An experimenter investigated the effects of sensitivity training on police effectiveness at resolving domestic disputes. He randomly sampled a group of police officers who had completed a sensitivity course and a group who had not. Participants were tested on their ability to successfully resolve a simulated domestic dispute. The following success scores were obtained.

No Course	*Course*
11	13
14	16
10	14
12	17
8	11
15	14
12	15
13	18
9	12
11	11

(a) Should a one-tailed or a two-tailed test be used? (b) What are the null and alternative hypotheses? (c) Subtracting *course* from *no course*, compute t_{obt} and determine whether it is significant. (d) Compute the confidence interval for the difference between μs. (e) What conclusions can the experimenter draw from these results? (f) Compute the effect size and interpret it.

SUMMARY OF FORMULAS

1. *The formula for computing the independent samples t-test is*

$$t_{\text{obt}} = \frac{(\overline{X}_1 - \overline{X}_2) - (\mu_1 - \mu_2)}{\sqrt{\left(\frac{(n_1 - 1)s_1^2 + (n_2 - 1)s_2^2}{(n_1 - 1) + (n_2 - 1)}\right)\left(\frac{1}{n_1} + \frac{1}{n_2}\right)}}$$

Values of t_{crit} are found in Table 2 in Appendix C for $df = (n_1 - 1) + (n_2 - 1)$.

In the formula, $(\overline{X}_1 - \overline{X}_2)$ is the difference between the sample means, $(\mu_1 - \mu_2)$ is the difference described in H_0, s_1^2 and n_1 are from one sample, and s_2^2 and n_2 are from the other sample. Values of s_1^2 and s_2^2 are found using the formula

$$s_X^2 = \frac{\Sigma X^2 - \frac{(\Sigma X)^2}{n}}{n - 1}$$

2. *The computational formula for the confidence interval for the difference between two μs is*

$$(s_{\overline{X}_1 - \overline{X}_2})(-t_{\text{crit}}) + (\overline{X}_1 - \overline{X}_2) \leq \mu_1 - \mu_2 \leq (s_{\overline{X}_1 - \overline{X}_2})(+t_{\text{crit}}) + (\overline{X}_1 - \overline{X}_2)$$

where t_{crit} is the two-tailed value for $df = (n_1 + n_2) - 2$, the quantity $(\overline{X}_1 - \overline{X}_2)$ is the difference between the sample means, and $s_{\overline{X}_1 - \overline{X}_2}$ is the standard error of the difference found using the formula

$$s_{\overline{X}_1 - \overline{X}_2} = \sqrt{\left(\frac{(n_1 - 1)s_1^2 + (n_2 - 1)s_2^2}{(n_1 - 1) + (n_2 - 1)}\right)\left(\frac{1}{n_1} + \frac{1}{n_2}\right)}$$

3. *The formula for computing r_{pb} from t_{obt} is*

$$r_{\text{pb}} = \sqrt{\frac{(t_{\text{obt}})^2}{(t_{\text{obt}})^2 + df}}$$

With independent samples, $df = (n_1 - 1) + (n_2 - 1)$, where each n is the number of scores in a sample.

4. *The proportion of variance accounted for* in a two-sample experiment equals the squared value of r_{pb}, or r_{pb}^2.

16

The Two-Sample Within-Subjects Experiment and the Dependent Samples *t*-Test

To understand this chapter:

- From Chapter 4, recall why we employ multiple trials and what order effects are.
- From Chapter 15, recall the reasons for controlling subject variables and how counterbalancing increases internal validity but also increases error variance. Also recall the logic of two-sample experiments, how to graph the results, and how r^2_{pb} measures effect size.

Then your goals in this chapter are to learn:

- The difference between matched groups and repeated measures designs.
- What constitutes partial and complete counterbalancing.
- How to perform the dependent-samples *t*-test.
- How to compute a confidence interval for the μ of difference scores.
- How to compute effect size in a two-sample, within-subjects experiment.

In the previous chapter, we discussed one version of the two-sample *t*-test—the independent samples *t*-test—that is used with a between-subjects design. In this chapter, we discuss the other version of the two-sample *t*-test—the dependent samples *t*-test—that is used with a within-subjects design. As usual, the crux of the matter is in the design that leads to a particular statistic. Therefore, in the following sections, we'll first see the logic and methods used for creating within-subjects designs and deal with some special problems they create. Then we'll cover the mechanics of analyzing the data using the *t*-test, which you'll be happy to know is very similar to previous procedures.

MORE STATISTICAL NOTATION

The major novelty in this chapter is that the statistics focus on difference scores: D stands for the difference score produced when we subtract one score from another score. We'll also compute the mean of the difference scores, symbolized as $\overline{D}$. And, finally, pay attention to subscripts again, because when we compute the variance of the difference scores we use the symbol s_D^2. When we discuss the population of difference scores, the mean of the population is symbolized by μ_D.

USING DESIGNS THAT DIRECTLY CONTROL SUBJECT VARIABLES

In the between-subjects designs discussed previously, the approach was to control subject variables through random assignment and counterbalancing (for example, 50% of the participants in a condition are male and 50% are female, or participants who score high on an extraneous variable are mixed in with those who score low). Then our *hope* is that the various levels of the variables cancel each other out, so that *overall*, the group tested in one condition is comparable to the group in the other condition. The problem is that this balancing act simply may not work. Then if the participants in one condition are different from those in the other condition, the independent variable is confounded by subject variables. Also, we saw that by intentionally changing a subject variable within a condition, we include different influences that will produce different scores within the conditions, producing greater error variance and a weaker relationship.

Instead, to gain greater control of subject variables and to reduce error variance, we can more directly ensure that the individuals in one condition are comparable to those in the other conditions. We employ this approach when there is one or more crucial extraneous subject variable that must be controlled. As usual, such variables are correlated with the impact of the independent variable, or with the behavior measured by the dependent variable. To control such variables, unlike in the between-subjects design, we *do* use fancy procedures for selecting participants and creating the samples. There are two types of designs we may use: Matched groups designs and repeated measures designs.

Matched-Groups Designs

One way to guarantee we have comparable participants in each condition is to create a matched-groups design. In a **matched-groups design**, each participant in one condition "matches" a participant in the other condition(s) on one or more extraneous subject variables. Here is a simple example.

Say that we want to measure how well people in two samples shoot baskets in basketball, with one sample using a standard ball and the other using a new type of ball (with handles). Think of all the subject variables that we would want participants to be comparable on (physical size, basketball skill level, amount of practice at basketball, motivation and interest). Consider the participants' heights. If by the luck of random assignment, one sample contained taller people than the other, then differences in shooting baskets could be due to differences in height rather than to using a different ball. Therefore, because height is so crucial in this setting, we won't leave its control up to the luck of random assignment. Instead, we create two samples containing participants who are the same—have "matching"—height.

The procedure for creating matched samples is to first identify pairs of participants who have the same matching score on the variable to be controlled (using a pretest if necessary). Then randomly assign each member of the pair to one condition. Thus, we would identify matching pairs of people who are the same height, and assign one member of each pair to a condition. If we select two people who are six feet (72") tall, we randomly assign one person to one condition and the other person to the other condition. Likewise, a five-foot (60") person in one condition is matched with a five-footer in the other condition, and so on.

We can also match participants on more than one variable. For example, we might want to create samples matched for gender as well as height. This would produce the design shown in Table 16.1.

TABLE 16.1 Diagram of the Basketball Experiment

Each row represents two people who are matched on gender and height. Each X *represents a person's basket shooting score.*

	Standard Basketball	*New Basketball*
Pair 1: Male, 72"	X	X
Pair 2: Female, 72"	X	X
Pair 3: Male, 60"	X	X
Pair 4: Female, 60"	X	X
" "	"	"
" "	"	"
	$\overline{X}$	$\overline{X}$

Each row contains the scores of a matched pair of participants: the first pair consists of two males, both 72" tall, the second pair is two females, both 72" tall, and so on. Think of a row as representing a very small experiment (with one participant per condition). Any difference in basket shooting in the row cannot be due to differences in subjects' height or gender, because these variables are constant. Then, because an experiment with one participant per condition is not a reliable approach, we replicate this study with other pairs.

In all other ways, we conduct the experiment as we would any other, applying the controls we've discussed previously. Thus, for example, for reliability, we'd test participants on multiple trials, having them shoot baskets with their particular ball a number of times. Then we assume that random differences in a their motivation, anxiety, or skill on different trials balance out. Here each person's score would actually be a summary score, *collapsing* across trials. When we collapse the data by averaging vertically in a column, the overall mean becomes a reliable estimate of the typical basket-shooting score in each condition. And, because the potential influence of different heights and gender is equally represented in each condition, any differences between conditions cannot be due to these variables.

You can match participants using any relevant extraneous subject variable, such as weight, age, physical ability, or the school they attend. You may also rely on natural pairs to match participants. For example, roommates or husband-and-wife teams are already matched in terms of having the same housing arrangements. Another common approach is to test identical twins assigning one of each pair to each condition. Because genetic influences are equated, any differences in a behavior between groups must be due to environmental causes. Likewise, in animal research, pairs may be created from litter mates to match them on variables related to their experiences.

If it is difficult to find participants who have the same score on the matching variable, you can rank-order participants and create pairs using adjacent ranked scores. Thus, above, the two males whose heights rank as 1st and 2nd would form one pair, the next two tallest men would form the next pair, and so on.

> ***REMEMBER*** In a matched-groups design, each participant in one condition matches a participant in every other condition in terms of one or more extraneous variables.

Pros and cons of matched groups The advantage to matching is that it ensures that in every condition there is a participant with virtually the same score on the subject variable(s) we wish to control. This keeps these variables constant across the conditions, eliminating this potential confounding. By controlling height and gender in the above experiment, for example, we have greater internal validity for inferring that differences between the conditions are due to our treatment.

There are, however, limitations to a matched-groups design. Again, we have the problem that pretesting may communicate a study's hypothesis and thus increase demand characteristics. In addition, to find matching participants, we may have to pretest many individuals and/or settle for a very small N. Finally, with a design that involves several conditions of the independent variable, matching triplets or larger numbers can be very difficult.

The biggest problem, however, is in measuring and matching important subject variables. First, we may not know the important variables on which to match participants.

Second, we may not have a valid and reliable method for measuring the matching variable. And finally, if there are many subject variables to control, it may be almost impossible to find individuals who match on all variables.

When matching is unworkable but the study calls for tightly controlling subject variables, we can instead employ a repeated-measures design.

Repeated-Measures Designs

The more variables we can match participants on, the more potential confoundings that are eliminated. In fact, the ideal would be to have participants in all conditions who are identical in every respect. The way to have identical participants is to test the *same* individual in each condition. This technique is called repeated measures. In a **repeated-measures design**, each participant is tested under all conditions of an independent variable.

Repeated measures are different from the "multiple trials" discussed previously. In multiple trials, the same participant is repeatedly observed in *one* condition. With repeated measures, the same participant is repeatedly observed under all conditions of an independent variable, regardless of the number of trials per condition.

Here's a different example for a repeated measures design, a classic called the "Stroop interference task" (Stroop, 1935). There, participants were presented colored ink patches, and the researcher measured their reaction time to identify the color present. In the control condition, only color patches were presented. In the experimental condition, the word name for a color was superimposed on a patch of a different color. To see how this works, quickly identify the color of the ink used to print the word below.

red

Identification of the ink color is interfered with by processing the word, so that reaction times are slowed.

Say that we wish to perform this type of study, with an experimental condition where color names are presented in different colored inks, compared to a control condition where nonsense words (such as BJB) are presented with the colored inks. Consider all of the experiential, motivational, physical, and cognitive differences between participants that might influence their performance in this situation. We're not sure which of these differences are most important to control, and even if we were, counterbalancing in a between-subjects design would be unworkable, limiting the population would limit generalizability, and finding participants who match on so many variables would be impossible. The solution is to test the same people under every condition, thus providing the "perfect" control of these subject variables. Therefore, we will measure participants' reaction time when presented the nonsense words and also measure the same people when presented the color names. This design is shown in Table 16.2. Each row represents the scores from one participant tested under both conditions, so although we have one sample of participants, we have two samples of scores. Again think of a row as representing a very small experiment with the identical participant in both conditions. Any difference in reaction time cannot be due to differences in subject variables because they are all constant. Then, for greater reliability, we replicate this study with other participants.

TABLE 16.2 Diagram of Repeated Measures Design of Stroop Experiment

Each row represents one person tested under both conditions. Each X represents a person's mean reaction time for the trials in the condition.

	Nonsense word (e.g. BJB)	*Color Name (e.g. RED)*
Participant 1 (male, order1)	X	X
Participant 2 (male, order2)	X	X
Participant 3 (female, order1)	X	X
Participant 4 (female, order2)	X	X
" "	"	"
	$\overline{X}$	$\overline{X}$

All of the usual design requirements apply, so participants' gender is also counterbalanced. For reliability, we test multiple trials (say 50), using different word and color combinations (such as "yellow" printed in green ink, "blue" printed in red ink, and so on.) Each participant's score in a condition is actually the mean reaction time for all trials in that condition. Recall that one problem with multiple trials, however, is order effects, the influence of practice, fatigue, or experience over a sequence of trials. Any one order—with its peculiar effects—could bias scores so they are especially high or low. Therefore, we might control order effects by testing each half of the participants under one of two orders (indicated in Table 16.2 as order1 and order2).

Collapsing vertically in each column, the mean score for each condition is the typical reaction time per condition. Because the same people are being observed, all subject variables should be constant between the two conditions, Therefore, any differences in mean reaction time between the samples cannot be due to differences in participants' abilities, physical quickness, and so on, because such subject variables are represented equally in both conditions.

A special type of repeated-measures design is employed when participants are measured before and after some event or treatment. This is called a **pretest-posttest design**. Thus, for example, to test whether meditation reduces physical stress, we might measure the same person's blood pressure before and after a period of meditation. Or to determine the effectiveness of a new weight-loss diet, we could measure each person's weight before and after a period of dieting.

> *REMEMBER* A repeated-measures design matches participants along all subject variables by testing each participant under all conditions of an independent variable.

Pros and cons of repeated measures The strength of repeated measures is that they should eliminate potential confounding from virtually any subject variable:

Differences between the conditions should not be due to differences in subject variables because each participant, with the same characteristics, is in every condition. Repeated-measures designs may also produce a number of problems however.

Repeated measures "should" keep subject variables constant, but individuals change from moment to moment. First, participants eventually become aware of all our conditions. Therefore, at some point, participants may identify what they think is the purpose and hypothesis of the study, creating demand characteristics that lead to unnatural behavior. In the Stroop study, for example, people may erroneously decide they are supposed to respond slower to the nonsense words.

Second, because we must test the various conditions in a sequence that occurs over time, scores will be influenced by several factors. **Subject history** refers to the fact that participants continue to have a life and experience things that can change them and influence their responses. Similarly, scores are influenced by **subject maturation**: As someone grows older and more mature, he or she changes in ways that influence responses. Such factors may *severely* reduce internal validity: A response to one condition measured now and a response to another condition measured later may be confounded by changes in a participant due to history and maturation.

Further, repeated measures produce greater "subject mortality." This doesn't mean that participants literally die (usually). Rather, **subject mortality** refers to the loss of subjects because their participation dies out before the study is completed. This can occur in a between-subjects design or anytime that participants refuse to continue in a condition. It is most prominent when repeated testing requires a considerable amount of time per participant, so to reduce fatigue and overload we spread out testing over several days. Then people show up for initial sessions but do not return for later ones. The problem is, mortality effects are selective: People who continue to participate may find the study more interesting, perform better at the task, be more committed to helping science, or be more desperate for college credit or money. In any case, the results are biased. For example, in testing a new diet, the people who give up on dieting are likely to disappear. Those who stay may be so motivated that *any* diet would work well. Then, what is apparently the influence of the test diet may actually be due to a characteristic of the participants. Likewise, in any type of design, subject mortality results in only a certain type of participant, and so we lose external validity for generalizing to the broader population.

To counter these influences, you should try to obtain repeated measures (and multiple trials) within a short time span. Also, attempt to make the mechanics of participating in the study easy, with a task that is interesting and brief, so that its completion does not require extremely dedicated volunteers. And, pay attention to the degree of subject mortality, and during debriefing question participants about their reasons for participating, so that you can gauge how biased the sample is.

REMEMBER The results of a repeated-measures design may be biased by subject history, subject maturation, subject mortality, and demand characteristics.

Finally, a major problem with repeated measures is that they produce a new kind of order effects. As we've seen, order effects are the influence of performing a series of trials. These effects include: (1) Practice effects—getting better at responding over trials; (2) Fatigue effects—getting worse at responding over trials; (3) Carry-over effects—the experience of any one trial that influences scores on subsequent trials; and (4) Response sets—from previous trials, developing a habitual response for subsequent trials.

Previously, we discussed order effects as they occur over multiple trials *within* a particular condition. With repeated measures, however, these same effects also occur *between* conditions. After all, from a participant's perspective, changing from one condition to another largely involves an additional sequence of trials—more of the same, as it were. Thus, in the Stroop study above, by the time participants get to the second condition they may be tired and inattentive, or they may be very good at identifying ink color, or they may be very reactive to having the experimenter around. If the second condition were performed first, however, these influences might not be present.

Further, order effects interact with the previous problems of subject maturation, history, and mortality. If a particular condition was performed at a different point in the sequence, perhaps maturation and history would not have changed participants so much. Or, perhaps subject mortality would have selected a different type of participant who performs differently, if other, less boring conditions were performed first.

Thus, we never know whether performance in one condition is higher or lower than in another condition just because of the order effects operating on that condition. We cannot prevent order effects, but we can attempt to balance their influence.

METHODS FOR CONTROLLING ORDER EFFECTS

To control for order effects in a repeated measures design with two conditions, we counterbalance the order in which participants perform the conditions: Half the participants perform condition 1 followed by condition 2, and the remaining participants perform condition 2 followed by condition 1. Thus, in the Stroop study, half the participants will be tested first with the control condition, and half will start with the experimental condition, as shown in Table 16.3. All other design requirements still apply, so the participants' gender is counterbalanced. Because we're testing multiple trials, we also control for order effects within conditions by testing different participants under different orders (indicated as order1 and order2). Collapsing vertically in each column, the mean score for each condition is the typical reaction time per condition. Because the same person is being observed, however, virtually all subject variables should be constant. Further, any differences between the two conditions is not due to a particular order in which the conditions were performed, because both possible orders of conditions are present.

With only two conditions, the above counterbalancing scheme is quite adequate. For more than two conditions, there are several common techniques for controlling order effects. To see them, say that we expand the Stroop study by adding a third condition. We'll also present words that imply a color, printed in a different, interfering color, such as the word "grass" printed in yellow ink (as in Klein, 1964). We're still using repeated measures, but with more than two conditions we may deal with order effects using either complete counterbalancing, partial counterbalancing, or randomization.

TABLE 16.3 Diagram of Stroop Experiment Showing Counterbalancing of Order of Conditions

The top half of the diagram shows participants tested with one order, and the bottom half shows subjects tested in the reverse order.

Participants Tested in Control Condition First	*Nonsense word (e.g. BJB)*	*Color Name (e.g. RED)*
Participant 1 (male, order1)	X	X
Participant 2 (male, order2)	X	X
Participant 3 (female, order1)	X	X
Participant 4 (female, order2)	X	X
" "	"	"
Participants Tested in Experimental Condition First		
Participant 1 (male, order1)	X	X
Participant 2 (male, order2)	X	X
Participant 3 (female, order1)	X	X
Participant 4 (female, order2)	X	X
" "	"	"
	$\overline{X}$	$\overline{X}$

Complete Counterbalancing of Conditions

Complete counterbalancing is the process of balancing order effects by testing different participants with different orders, so that all possible orders are present. For example, let's call the three conditions of the new Stroop study, A, B, and C. Three conditions produce six possible orders:

ABC ACB BCA BAC CAB CBA

Notice two things about these orders. First, each condition appears in every position within the sequence: A appears twice as the first condition, twice as the second condition, and twice as the third condition (likewise for B and C). Second, every possible sequence is included: For the sequence beginning with A, the two possible orders ABC and ACB are included, and so on. Thus, complete counterbalancing balances both a condition's position in the sequence and the order of the conditions coming before and after it. Applying this technique to the Stroop study, we have the diagram shown in Table 16.4.

TABLE 16.4 Diagram of Completely Counterbalanced Expanded Stroop Experiment

Each row represents those subjects tested under a particular sequence of the three conditions.

	Conditions		
Orders	*A* *Nonsense word (BJB)*	*B* *Color Name (RED)*	*C* *Implied Color (GRASS)*
Participants tested using ABC	X X X...	X X X..	X X X...
Participants tested using ACB	X X X...	X X X..	X X X...
Participants tested using BCA	X X X...	X X X..	X X X...
Participants tested using BAC	X X X...	X X X..	X X X...
Participants tested using CAB	X X X...	X X X..	X X X...
Participants tested using CBA	X X X...	X X X..	X X X...
	$\overline{X}$	$\overline{X}$	$\overline{X}$

For one-third of the participants, condition A is first, so performance under this condition may be biased because of its location in the sequence. However, condition B also occurs first for one-third of the participants, as does condition C, so these conditions are equally biased. Likewise, each condition occurs second at times, and third at times. Note that, within each order of conditions, we still balance other variables of concern: We balance gender, so that both males and females perform each order, and we counterbalance the order in which the trials in each condition are performed. Therefore, when we collapse vertically in each column, each mean reflects the typical response to the condition, with no confounding (1) By subject gender and other subject variables; (2) By order effects within the 50 trials per condition; or (3) By history, maturation, mortality, or order effects occurring because of when in the sequence each condition was performed.

The term "complete counterbalancing" is also applied when counterbalancing the order of multiple trials *within* a condition. If above, for example, all possible orders of the individual trials within a condition were represented, then we'd have also completely counterbalanced the order of trials.

You may have noticed that the design of this study is very complex! Complete counterbalancing tends to create a design that rapidly becomes unworkable, especially if we are also counterbalancing other variables. Further, we may need many more participants so that we can test some under each order. Such problems suggest that we may be going overboard by completely counterbalancing order effects. After all, the goal is simply to ensure that the results are not confounded or unduly influenced by one particular order. Therefore, instead we may only partially counterbalance.

Partial Counterbalancing of Conditions

Partial counterbalancing is the process of balancing order effects by testing different participants using only some of the possible orders. For example, with three conditions, it is common to use the following three orders of conditions:

ABC BCA CAB

As here, typically a partial counterbalancing scheme systematically changes the position at which a condition occurs in the sequence, but does not change the conditions coming before or after the condition. (The procedure for creating this type of counterbalancing scheme is sometimes called a "Latin square design.") Applying this scheme to the Stroop study, we have the diagram shown in Table 16.5. Much simpler! But we have balanced only practice effects or other biases that occur because of where in the sequence a condition occurs (first, second, or third). Partial counterbalancing does not control for carry-over effects (the influence on one condition resulting from having performed a previous condition). The scheme shown in Table 16.5 does not balance potential carry-over effects, because B follows A and C follows B in two thirds of the sequences, but C never immediately follows A. Therefore, when carry-over effects are likely to occur between conditions, we include all orders, using complete counterbalancing.

> ***REMEMBER*** Partial counterbalancing presents some of the possible orders of conditions to control practice effects. Complete counterbalancing includes all possible orders to control practice and carry-over effects.

The term partial counterbalancing can also be applied to multiple trials within each condition. If in the above designs half the participants per row performed the individual trials in order 1 through 50 and the remaining participants per row performed 50 through 1, we'd have partially counterbalanced the order of trials within conditions.

TABLE 16.5 **Diagram of Partially Counterbalanced Expanded Stroop Experiment**

Each row represents those participants tested under a particular sequence of the three conditions.

		Conditions	
Orders	*A* Nonsense word (BJB)	*B* Color Name (RED)	*C* Implied Color (GRASS)
Participants tested using ABC	X X X . . .	X X X . .	X X X . . .
Participants tested using BCA	X X X . . .	X X X . .	X X X . . .
Participants tested using CAB	X X X . . .	X X X . .	X X X . . .
	$\overline{X}$	$\overline{X}$	$\overline{X}$

Counterbalancing is used typically when there are only a few possible orders, so that order effects can be handled efficiently and systematically. When there are many possible orders, and the goal is simply to include some different sequences in an unsystematic way, an alternative approach is randomization.

Randomizing the Order of Conditions

Randomization is the process of balancing order effects by randomly creating different orders under which different participants are tested. In the Stroop task, we could randomize the order of conditions by randomly creating a sequence for each participant. Likewise, we could randomize the order of trials within a condition by essentially shuffling the 50 trials per condition. In fact, we may accomplish both tasks simultaneously by randomly mixing the trials from all conditions. Thus, in the Stroop study, rather than have participants complete all 50 trials in a condition at once, we might intermix all 150 trials. For example, in every 15 trials, we can randomly select and mix 5 trials from each condition. When we unscramble responses to each condition and combine the scores, each condition will have been performed at the beginning, middle, and end of the sequence, and each particular trial will be sometimes performed early in the sequence and sometimes performed later in the sequence.

We do not usually intermix conditions when each condition requires us to stop and change the instructions or procedure. Instead, to prevent confusion, participants complete all trials in one condition before going on to the next condition. This method is known as "blocking" trials (trials are performed as a group or "block"). Then we control order by having different participants perform the conditions in different orders.

CHOOSING A DESIGN

From this and the previous chapter, you've seen that there is much to consider when deciding whether to employ a between-subjects, matched-groups, or repeated-measures design. Usually, the choice is between repeated-measures and between-subjects designs. Matched groups are typically used for controlling only one subject variable—and when subject variables are an issue, there are usually many that are critical.

A repeated-measures design is preferred when participants' responses are likely to be strongly influenced by individual differences in cognitive strategies, physical abilities, or experiences. Essentially, this design is used when it makes sense to compare a person or animal in one condition to the same person or animal in the other conditions. Thus, studies involving memory and learning are usually conducted in this way, as are studies that examine a sequence, as when we study the effects of practice or maturation. (As always, reading the relevant research literature will help you to make this decision.) Further, we may select such designs because, as we'll see, repeated-measures (and matched groups) are analyzed in a way that results in reduced error variance, producing a more powerful design than a comparable between-subjects design.

Other design considerations, however, may prevent you from using repeated measures: First, consider whether a particular condition may produce rather permanent changes in behavior so that the condition has unique carry-over effects. **Nonsymmetrical carry-over effects** occur when the carry-over effects from one order of conditions do not balance out those of another order. Such effects occur whenever performing task A and then task B is not the same as performing task B and then task A. For example, we would not want to test the same participants in more than one condition if each entailed some sort of surprise (you can surprise someone only once). Likewise, once you have taught participants something in one condition, you cannot "unteach" them in a subsequent condition. Or, as is often the case in animal research, a condition may involve some surgical technique that cannot be undone. In such situations, counterbalancing will not effectively balance out the bias produced by a particular order of conditions. Instead, use a between-subjects design, because no carry-over effects are possible.

Also, do not underestimate the influence of subject history, maturation, and mortality. These may be more detrimental to a study than the lessened control occurring in a between-subjects design. In addition, a repeated measures design often places greater demands on your ability to create stimuli. Typically you will need many more different yet comparable stimuli when participants are exposed to all conditions than when they are exposed to only one condition.

Finally, consider the advantages of repeated measures versus the disadvantages of counterbalancing. As you saw in the previous chapter, controlling *any* variable through counterbalancing changes the variable within a condition, and thus tends to produce greater variability among scores within the conditions. Repeated measures almost always require extensive counterbalancing (or randomization) and so including the influence of many changing variables tends to increase error variance, reducing the strength of the relationship and power.

The key to selecting your design depends on the number and importance of subject variables that must be controlled. If numerous uncontrolled subject variables could seriously reduce reliability and validity, then a repeated measures design is preferred despite the difficulties it entails. If there are only a few crucial subjects variables, however, a repeated measures design may create more problems than it solves, and so a better choice may be to identify the most serious subject variable to control and to match participants on that variable in a matched-groups design. You can also still obtain valid and reliable results from a between-subjects design, especially if you balance subject variables and/or limit the population.

> **REMEMBER** Between-subjects designs are preferred if carry-over effects are nonsymmetrical, if the task does not allow repeated testing, or if extensive counterbalancing is unwise.

With either a matched groups or repeated measures design, if there are only two conditions, then the results are analyzed using the dependent-samples *t*-test.

THE DEPENDENT SAMPLES *t*-TEST

From a statistical point of view, there are only two major types of designs. On the one hand, in a between-subjects design, we randomly select a different group of participants to serve in each condition, without considering those selected in the other conditions. On the other hand, both a matched-groups design and a repeated-measures design are a within-subjects design. In a **within-subjects design**, for each participant in one condition there is a comparable participant in the other conditions. (In research terminology, a matched-groups design is technically a between-subjects design, because it involves different participants in each condition.)

The matched-groups and the repeated-measures designs are statistically similar, because both involve dependent samples. **Dependent samples** occur when each score in one sample is paired with a particular score in the other sample. In both designs we essentially compare the effect of a manipulation by comparing each pair of scores. What makes matched and repeated measures samples dependent is the fact that, because there is some underlying reason for pairing the scores, the probability that one score in a pair is a particular value is influenced by the value of the paired score. (We saw this earlier in Chapter 12, where two events were dependent when the probability of one was influenced by the occurrence of the other.) For example, if a five-foot-tall male shooting baskets scored close to 0 in one sample, the probability is high that the matching five-footer also scored close to 0. This is not the case with independent samples. There, the fact that someone has a particular score in one condition does not influence the probability of someone else having that score in the other condition.

When you have dependent samples, you should employ the dependent samples *t*-test. The ***t*-test for dependent samples** is the parametric procedure used for significance testing of sample means from two dependent samples. This test is slightly different from the *t*-test for independent samples discussed in the previous chapter. With independent samples, the sampling distribution described the probability of obtaining a particular difference between two means from independent samples. With dependent samples, this probability is different, so we create the sampling distribution differently, and compute t_{obt} differently.

Assumptions of the Dependent Samples *t*-Test

Except for requiring dependent samples, the assumptions of the dependent samples *t*-test are the same as those for the independent samples *t*-test: (1) The dependent variable involves an interval or ratio scale; (2) The populations represented by the samples form normal distributions; (3) The variance of the raw score populations is estimated by s_X^2; and (4) The populations represented by the data have homogeneous variance.

If the data meet the assumptions, then, as usual, you: (1) Create the statistical hypotheses. (2) Set up and perform the statistical test. (3) If the results are significant, describe the relationship you have demonstrated.

The Logic of the Dependent Samples *t*-Test

Enough about basketball and colored ink! Let's say we are interested in phobias (irrational fears of objects or events). We have a new therapy we want to test on spider-phobics. From the local phobia club, we randomly select a decidedly unpowerful *N* of five spider-phobics, and test the therapy using repeated measures of two conditions, before therapy and after therapy. Before therapy we will measure each participant's fear response to a picture of a spider. We'll measure heart rate, breathing rate, perspiration, etc., and then compute a "fear" score between 0 and 20. After providing the therapy, we will again measure the person's fear response to the picture. Say that from reading previous research, we can assume the study meets the assumptions of the dependent samples *t*-test, and we set alpha at .05.

Obviously, we expect to demonstrate that the therapy will decrease a person's fear, so our hypotheses are one-tailed:

1. The experiment will demonstrate a relationship in which the population μ for scores after therapy is lower than the μ for scores before therapy, or

2. The experiment will not work, either showing no difference between the μs for before and after scores or showing that the μ for scores after therapy is larger than the μ for scores before therapy.

So far, it sounds as if we are performing the same procedure we had in the independent samples *t*-test. However, instead of directly comparing the means from the samples to determine the population μs they represent, we must first transform the data. Then we test the hypotheses using the transformed scores.

With dependent samples, we transform the raw scores by finding the *difference score*, *D*, between the two raw scores in each pair. Say that in the phobia study we collected the data shown in Table 16.6. We will find each difference score by subtracting each phobic's after-therapy score from the corresponding before-therapy score.

To summarize the sample of difference scores, we compute the mean difference score, $\overline{D}$. We add the positive and negative differences to find the sum of the differences, ΣD. Then we divide this amount by *N*, the number of difference scores. For the phobia data, $\overline{D}$ equals 18/5, which is +3.6. Thus, the before scores were, on average, 3.6 points higher than the after scores.

Now here's the strange part: Forget about the before and after scores for the moment, and consider *only* the difference scores. From a statistical standpoint, we have *one* sample mean from *one* random sample of scores. As in Chapter 14, when we have one sample, we perform the single-sample *t*-test! The fact that we have difference scores here in no way violates this *t*-test. So we will create the statistical hypotheses and then test them in virtually the same way we did with the single-sample *t*-test.

> *REMEMBER* The *t*-test for two dependent samples is the single-sample *t*-test performed on the sample of difference scores.

Statistical Hypotheses for the Dependent Samples t-Test

Any sample of difference scores represents the population of difference scores that would result if we could measure the population of raw scores under each of our conditions

TABLE 16.6 Scores for the Before-Therapy and After-Therapy Conditions

Each D equals (before − after).

Subject	*Before therapy*	−	*After therapy*	=	*Difference, D*
1 (Dorcas)	11	−	8	=	+3
2 (Biff)	16	−	11	=	+5
3. (Millie)	20	−	15	=	+5
4. (Attila)	17	−	11	=	+6
5. (Slug)	10	−	11	=	−1
					$\Sigma D = 18$

and then subtract the scores in one population from the corresponding scores in the other population. The mean of the population of difference scores is μ_D. To create our statistical hypotheses, we simply determine the predicted values of μ_D in H_0 and H_a.

Start with H_a. In the one-tailed phobia study, we predict that the population of scores after therapy will contain lower fear scores than the population of scores before therapy. If we subtract the lower after scores from the higher before scores (as we did in the sample), then we should have a population of difference scores that contains positive numbers. The resulting μ_D should also be a positive number. An alternative hypothesis always implies that the predicted relationship exists, so here

$$H_a\colon \mu_D > 0$$

H_a implies that the sample $\overline{D}$ represents a population of differences having a μ_D greater than zero, and thus demonstrates that after-therapy fear scores are lower than before-therapy scores in the population.

On the other hand, according to the null hypothesis, there are two ways we may fail to demonstrate the predicted relationship. First, the therapy may do nothing to fear scores, so that the population of before scores contains the same scores as the population of after scores. If we subtract the after scores from the before scores, the population of difference scores will have a μ_D of zero, although every individual difference won't equal zero. Because of chance fluctuations in physiological or psychological factors, all participants will not produce the same fear score on two observations, and thus their difference scores will not be zero. On average, however, the positive and negative differences will cancel out to produce a μ_D of zero.

Second, the therapy fails if it increases fear scores, so that subtracting larger after scores from smaller before scores produces a population of negative difference scores, with a μ_D that is less than zero. Thus, given the way we are subtracting to find the difference scores, our null hypothesis is

$$H_0\colon \mu_D \leq 0$$

H_0 implies that $\overline{D}$ represents such a μ_D and thus the predicted relationship between therapy and fear scores does not exist.

As usual, we test H_0 by testing whether our sample mean is likely to represent the μ described by H_0. Here, H_0 says that the sample mean represents the population of difference scores where μ_D at most equals zero. If the sample perfectly represents this population, then $\overline{D}$ should equal zero. However, because of those chance fluctuations in fear scores, all of the difference scores in the sample may not equal zero, so neither will $\overline{D}$. Thus, H_0 says that $\overline{D}$ represents a population where μ_D is zero, and if $\overline{D}$ is not zero, it's because of sampling error.

We always test H_0 by examining the sampling distribution of means, so here we examine the sampling distribution of $\overline{D}$. We simply locate our $\overline{D}$ on this sampling distribution by computing t_{obt}.

Computing the Dependent Samples *t*-Test

We'll compute t_{obt} using the data from the phobia study, presented in Table 16.7. Note that we need N, ΣD, $\overline{D}$, and ΣD^2.

TABLE 16.7 Summary of Data from Phobia Study

Subject	*Before therapy*	−	*After therapy*	=	*Difference, D*	D^2
1	11		8		+3	9
2	16		11		+5	25
3	20		15		+5	25
4	17		11		+6	36
5	10		11		−1	1
	$\overline{X} = 14.80$		$\overline{X} = 11.20$		$\Sigma D = +18$	$\Sigma D^2 = 96$
$N = 5$					$\overline{D} = +3.6$	

Computing t_{obt} here is identical to computing the single-sample *t*-test discussed in Chapter 14, except that whenever a formula had the symbol *X* it now has the symbol *D*. To compute t_{obt}, perform the following three steps.

First, find s_D^2, which is the estimated variance of the population of difference scores.

THE FORMULA FOR s_D^2 IS

$$s_D^2 = \frac{\Sigma D^2 - \dfrac{(\Sigma D)^2}{N}}{N - 1}$$

Using the phobia data, we fill in the formula:

$$s_D^2 = \frac{\Sigma D^2 - \dfrac{(\Sigma D)^2}{N}}{N - 1} = \frac{96 - \dfrac{(18)^2}{5}}{4} = 7.80$$

The second step is to find $s_{\overline{D}}$. This is the **standard error of the mean difference,** or the "standard deviation" of the sampling distribution of $\overline{D}$.

THE FORMULA FOR THE STANDARD ERROR OF THE MEAN DIFFERENCE, $s_{\overline{D}}$, IS

$$s_{\overline{D}} = \sqrt{(s_D^2)\left(\frac{1}{N}\right)}$$

For the phobia study, with $s_D^2 = 7.8$ and $N = 5$, we have

$$s_{\overline{D}} = \sqrt{(s_D^2)\left(\frac{1}{N}\right)} = \sqrt{(7.8)\left(\frac{1}{5}\right)} = \sqrt{1.56} = 1.25$$

The third step is to find t_{obt}.

THE DEFINITIONAL FORMULA FOR THE t-*TEST FOR RELATED SAMPLES IS*

$$t_{obt} = \frac{\overline{D} - \mu_D}{s_{\overline{D}}}$$

For the phobia study, $\overline{D}$ is +3.6, $s_{\overline{D}}$ is 1.25, and H_0 says that μ_D equals 0. Putting these values into the above formula, we have

$$t_{obt} = \frac{\overline{D} - \mu_D}{s_{\overline{D}}} = \frac{+3.6 - 0}{1.25} = +2.88$$

This tells us that our sample $\overline{D}$ is located at a t_{obt} of +2.88 on the sampling distribution of $\overline{D}$ when $\mu_D = 0$.

Computational Formula for the Related Samples *t*-Test We can combine the above steps of computing $s_{\overline{D}}$ and t_{obt} into one formula.

THE COMPUTATIONAL FORMULA FOR THE t-*TEST FOR RELATED SAMPLES IS*

$$t_{obt} = \frac{\overline{D} - \mu_D}{\sqrt{(s_D^2)\left(\frac{1}{N}\right)}}$$

The numerator is the same as in the definitional formula. The denominator simply contains the formula for the standard error of the mean difference, $s_{\overline{D}}$.

Thus, putting the data for the phobia study into the above formula gives:

$$t_{obt} = \frac{+3.6 - 0}{\sqrt{(7.8)\frac{1}{5}}} = \frac{+3.6}{\sqrt{1.56}} = \frac{+3.6}{1.25} = +2.88$$

so t_{obt} is again +2.88.

Interpreting the Dependent Samples *t*-Test

To interpret t_{obt}, first find t_{crit} in the *t*-tables (Table 2 in Appendix C) for $df = N - 1$, where N is the number of difference scores. For the phobia study, with $\alpha = .05$ and $df = 4$, the one-tailed t_{crit} is +2.132. This t_{crit} is positive, because we predicted that therapy works to decrease fear. When we subtract the smaller after scores from the larger before scores, we should obtain a positive value of $\overline{D}$, producing a positive t_{obt}. Figure 16.1 shows the completed sampling distribution. This is another sampling distribution of means, except that each mean is the mean of a sample of difference scores drawn from the population of difference scores where μ_D equals zero. The average sample mean will be 0, so the μ of the distribution is zero. The values of $\overline{D}$ that are farther from zero (whether positive or negative) are less likely to occur when H_0 is true

FIGURE 16.1 One-Tailed Sampling Distribution of Random $\overline{D}$s When $\mu_D = 0$

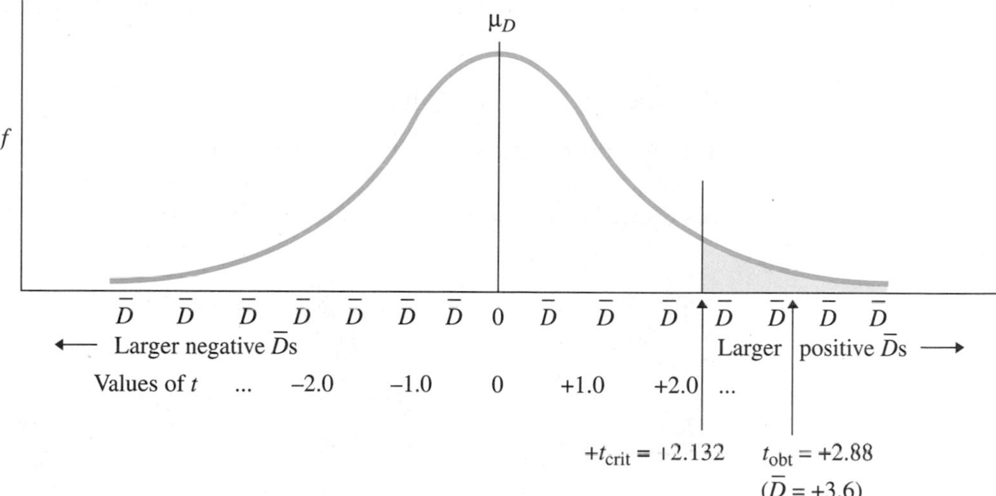

and the sample represents a μ_D of zero. As shown, our t_{obt} of +2.88 is in the region of rejection, so sampling error is too unlikely to have produced a $\overline{D}$ of +3.6 if the sample actually represents a μ_D of zero (and even less likely if the sample represents a μ_D less than zero). Therefore, we reject the null hypothesis that the sample represents $\mu_D \leq 0$. Our results are significant, and we report them as

$$t(4) = +2.88, p < .05$$

By rejecting H_0, we accept H_a that the sample represents a μ_D that is greater than zero. We would see such a population of difference scores only if the population of before-therapy scores contained scores that were larger than those in the population of after-therapy scores. Because the before and after conditions produce significant differences in fear scores, we can also say that the means for each of these conditions differ significantly: The mean of the before scores, 14.80, differs significantly from the mean of the after scores, 11.20.

Thus, we conclude that the before-therapy and after-therapy scores represent a relationship in the population, so we are confident that the therapy works. Therefore, as usual, we now begin the task of interpreting the results psychologically, using relevant hypothetical constructs, models, and theories to explain why the fear scores are as observed in each condition and how the therapy works to decrease a person's spider phobia. (If t_{obt} had not been beyond t_{crit}, then we would retain H_0, and we would not have convincing evidence that the therapy reduces fear scores.)

Based on the sample means, our best estimate is that if we measured the fear scores of everyone in the population, before therapy the μ would be *around* 14.80, and after therapy the μ would be *around* 11.20. It would be nice to define "around" more precisely by computing a confidence interval, but we *cannot* compute a confidence interval for each μ (as in Chapter 14), nor can we compute the confidence interval for the difference between two μs (as in Chapter 15). Our sample means are from *dependent samples*, and this violates the rules for these confidence intervals. Instead, we can deal only with the difference scores, because only they do not violate any assumptions. Because our $\overline{D}$ is +3.6, it probably represents a population of difference scores where μ_D is "around" +3.6. We can compute a confidence interval to describe this μ_D.

Computing the Confidence Interval for μ_D

The **confidence interval for μ_D** describes a range of values of μ_D, one of which our $\overline{D}$ is likely to represent. We compute the interval by computing the highest and lowest values of μ_D that are not significantly different from $\overline{D}$.

THE COMPUTATIONAL FORMULA FOR THE CONFIDENCE INTERVAL FOR μ_D OF THE POPULATION OF DIFFERENCE SCORES IS

$$(s_{\overline{D}})(-t_{crit}) + \overline{D} \leq \mu_D \leq (s_{\overline{D}})(+t_{crit}) + \overline{D}$$

The value of t_{crit} is the two-tailed value for $df = N - 1$, where N is the number of difference scores. $s_{\overline{D}}$ is the standard error of the mean difference as computed above, and $\overline{D}$ is the mean of our difference scores.

To compute the confidence interval for the phobia study, we have $s_{\overline{D}} = 1.25$ and $\overline{D} = +3.6$. Because we seek values above and below +3.6, we switch to the two-tailed critical value here, and with $\alpha = .05$ and $df = 4$, t_{crit} is ± 2.776. Filing the above formula, we have

$$(1.25)(-2.776) + 3.6 \leq \mu_D \leq (1.25)(+2.776) + 3.6$$

which becomes

$$0.13 \leq \mu_D \leq 7.07$$

Thus, we are 95% confident that the sample mean difference of +3.6 represents a population μ_D within this interval. In other words, if we performed this study on the entire population, we would expect the average difference in before and after scores to be between 0.13 and 7.07.

Testing Other Hypotheses with the Dependent Samples *t*-Test

In the preceding study, we could have reversed how we computed the difference scores, subtracting the predicted larger before-scores from the smaller after-scores. Then, if the therapy did not work, we would expect a $\overline{D}$ of zero or greater, so we'd have H_0: $\mu_D \geq 0$. If the therapy worked, we would expect a negative $\overline{D}$ representing a μ_D less than zero, and we'd have H_a: $\mu_D < 0$.

As usual, you should perform a one-tailed test only if you can reasonably predict the direction of the obtained relationship. If not, create two-tailed hypotheses: The sample of difference scores either does or does not reflect differences between the populations for the two conditions. If the populations of raw scores do not differ, then μ_D is zero, so H_0: $\mu_D = 0$. If the populations of raw scores differ, then μ_D is not zero, so H_a: $\mu_D \neq 0$.

DESCRIBING THE RELATIONSHIP IN A WITHIN-SUBJECTS EXPERIMENT

Concluding that t_{obt} is significant is not the end of the story. You then must describe the relationship. As usual, the focus is on the sample means from each condition. Thus, beginning with the mean fear score before therapy of 14.80 and the mean fear score of 11.20 after therapy, you have summarized the typical score—and typical behavior—found in each condition. Recall that the next step is to apply correlational techniques and to describe the relationship by graphing it and computing the effect size.

To graph the results, ignore the *D*s and plot the mean of the original raw scores in each condition on the *Y* axis and the conditions of the independent variable on the *X* axis. The results of the phobia study are shown in Figure 16.2. The slanting line indicates a negative linear relationship. You can also envision the scatterplot that the data would form if individual fear scores in each condition were plotted, envisioning data points above and below the mean *Y* score at each *X*. The best prediction of a participant's *Y* score in a particular condition would again be the mean of that condition.

What's missing, of course, is an indication of the *strength* of the relationship. As in the previous chapter, whenever the results of a two-sample experiment are significant, you should calculate the point-biserial correlation coefficient, r_{pb}. It is the appropriate coefficient when there is one dichotomous *X* variable (consisting of two categories) and one continuous interval or ratio *Y* variable, regardless of whether we have a within-subjects or between-subjects design.

THE FORMULA FOR COMPUTING r_{pb} FROM t_{obt} IS

$$r_{pb} = \sqrt{\frac{(t_{obt})^2}{(t_{obt})^2 + df}}$$

FIGURE 16.2 Line Graph of the Results of the Phobia Study

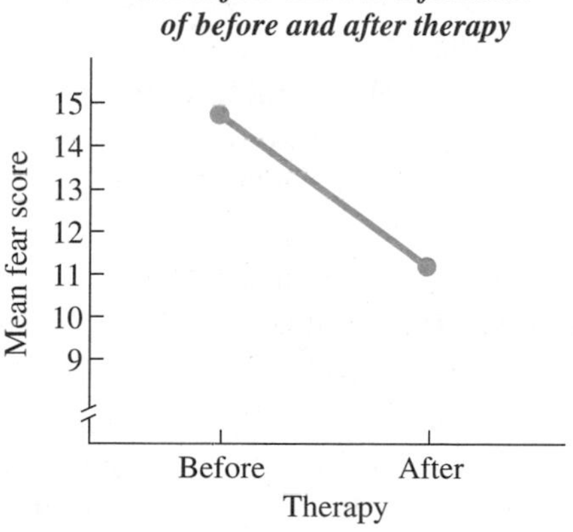

This is the same formula you saw in the previous chapter, except here the *df* equal $N - 1$, where N is the number of difference scores.

More to the point, the more informative statistic is the effect size (the proportion of variance accounted for). Again it is the squared correlation coefficient, or r^2_{pb}. Thus, for the phobia study with $t_{obt} = +2.88$ and $df = 4$, $r_{pb} = -.82$. Then, $r^2_{pb} = (-.82)^2$, or .67. Thus, we can account for 67% of the variance in fear scores by knowing whether participants have undergone the therapy. In other words, on average we are 67% closer to predicting participants' actual fear scores when we predict the mean score of each condition for them rather than when we predict the overall mean fear score of the experiment. Recall that the greater the effect size, the greater the scientific importance of a relationship. Thus, we have stumbled upon a rather important therapy, because so much of the changes in fear scores can be attributed to whether or not participants have experienced it.

REMEMBER Always determine the effect size in the significant results of any between-subjects or within-subjects design.

POWER AND THE DEPENDENT SAMPLES *t*-TEST

Remember power? Power is the probability of *not* making a Type II error, or the probability of rejecting H_0 when it's false. Recall that we maximize power by designing a powerful study, which in turn maximizes the size of t_{obt} relative to t_{crit}. This maximizes the probability of rejecting H_0 when it is false. Look again at the formula for t_{obt}.

$$t_{obt} = \frac{\overline{D} - \mu_D}{\sqrt{(s_D^2)\left(\frac{1}{N}\right)}}$$

As with the independent samples *t*-test, producing a powerful design and powerful data translates into a larger t_{obt} in the following ways:

1. A strong manipulation maximizes the difference between the two conditions, maximizing the difference in dependent scores. This produces a larger value of $\overline{D}$ and a larger difference between $\overline{D}$ and μ_D. Then the numerator in the formula is larger, so t_{obt} will be larger.
2. Controlling extraneous variables minimizes the variability of the raw scores in each condition. This minimizes the differences between the D scores, producing a smaller s^2_D and thus a smaller $s_{\overline{D}}$. Dividing by a smaller denominator produces a larger t_{obt}.
3. Maximizing N produces a smaller denominator and thus a larger answer when calculating t_{obt}. Also, a larger N gives a larger *df*, resulting in a smaller value of t_{crit}.

In addition, a dependent samples *t*-test is intrinsically more powerful than the independent samples *t*-test. For example, say we reanalyzed the phobia study, violating the assumptions and treating the two samples as independent samples of fear scores.

TABLE 16.8 Scores of Subjects 2 and 3 from Phobia Study

Subject	Before therapy	−	After therapy	=	Difference, D
2	16		11		+5
3	20		15		+5

Comparing the two procedures, we would find that transforming the raw scores to *D*s reduces the variability in the difference scores relative to the variability of the original fear scores. For example, Table 16.8 above presents the scores of Subjects 2 and 3 that were part of the original phobia study. Although there is variability (differences) between their before scores and between their after scores, there is no variability in their difference scores. Reduced variability produces a larger t_{obt}, so the t_{obt} for dependent samples will be larger than for independent samples. A larger t_{obt} is more likely to be significant, so a dependent samples design has greater power.

The increased power of a repeated-measures design will help to compensate if, for any reason, you can test only a small number of participants. For comparable power in a between-subjects design, you need a larger *N*. In essence, the strategy is either to observe a few participants many times or to observe many participants on fewer occasions.

PUTTING IT ALL TOGETHER

When you start to counterbalance order and other extraneous variables, the process of designing a study may seem overwhelming. As a friend once remarked, "First you have to consider *everything*!" This is true, but you don't have to control everything. Although you want data from a well-controlled experiment, you must actually get the data! Don't try to control so many variables that you can't conduct the study. Instead, control those variables that seriously confound the results or severely reduce reliability. Keep in mind that when you institute a control to eliminate one problem, you often produce other problems. Therefore, you are never going to produce the perfect study, so produce the best study you can within practical limits.

CHAPTER SUMMARY

1. In a *within-subjects* design, for each participant in one condition there is a comparable participant in the other condition(s). In a *matched-groups design*, each participant in one condition is matched with a participant in every other condition along an extraneous variable. In a *repeated-measures* design, each participant is measured under all conditions of an independent variable.
2. A repeated-measures design that entails measuring participants before and after some event is called a *pretest-posttest design*.

3. Repeated measures are especially prone to confounding by subject history and maturation, by subject mortality, and by order effects.

4. *Subject history* refers to the fact that participants continue to have experiences that can change them and influence their responses.

5. *Subject maturation* refers to the fact that as an individual grows older and more mature, he or she changes in ways that influence responses.

6. *Subject mortality* refers to the loss of participants because their participation dies out before the study is completed.

7. With *complete counterbalancing*, different participants are tested with different orders so that all possible orders of conditions or trials are present. With *partial counterbalancing*, participants are tested using only some of the possible orders of conditions or trials. With *randomization*, different participants are tested using different random orders of conditions or trials.

8. *Nonsymmetrical carry-over effects* occur when the carry-over effects from one order of conditions do not balance out those of another order.

9. The *dependent samples t-test* is the parametric inferential procedure used to test two sample means from either a matched groups or repeated measures design. A significant t_{obt} indicates that the mean of the difference scores between the conditions, $\overline{D}$, is significantly different from the μ_D described by H_0. Then the means of the raw scores in each condition also differ significantly.

10. The *confidence interval* for μ_D contains a range of values of μ_D, any one of which is likely to be represented by the sample mean difference, $\overline{D}$.

11. The strength of a significant relationship between the independent and dependent variables in a two-sample experiment is described by the *point-biserial correlation coefficient*, r_{pb}.

12. The *effect size* in a two-sample experiment is the squared value of r_{pb}. The larger the effect size, the more consistently the dependent scores change as the conditions of the independent variable change.

13. The power of the dependent samples *t*-test increases with (a) larger differences in scores between the conditions, (b) smaller variability of scores within each condition, and (c) larger *N*. All other things being equal, the *t*-test for dependent samples is more powerful than the *t*-test for independent samples.

PRACTICE PROBLEMS

(Answers for odd-numbered problems are provided in Appendix D.)

1. A scientist has conducted a two-sample experiment. (a) What two versions of a parametric procedure are available to him? (b) What is the deciding factor for selecting between them?

2. (a) What is the difference between independent and dependent samples? (b) What are the two ways to create dependent samples? (c) What other assumptions must be met before using the dependent samples t-test?
3. Which is more powerful, a dependent samples or independent samples design? Why?
4. How do you identify subject variables that may confound a study?
5. (a) How can subject variables influence external validity? (b) How can they influence internal validity?
6. What is meant by: (a) subject mortality; (b) subject history; (c) subject maturation?
7. How do each of the influences in question 6 bias your results?
8. (a) How is a matched-groups design created, and how does it control subject variables? (b) How is a repeated-measures design created, and how does it control subject variables?
9. What are the problems produced by repeated-measures designs?
10. (a) What is the difference between partial counterbalancing and complete counterbalancing? (b) What is the advantage of each?
11. (a) What is the advantage of counterbalancing or randomizing order in terms of internal and external validity? (b) What is the disadvantage of both procedures?
12. For each of the following, identify the type of design and the type of t-test that is required. (a) An investigation of the effects of a new memory-enhancing drug on the memory of Alzheimer's patients, testing a group of patients before and after administration of the drug. (b) An investigation of the effects of alcohol on motor coordination, comparing one group of subjects given a moderate dose of alcohol to the population μ for people given no alcohol. (c) An investigation of whether males and females rate differently the persuasiveness of an argument delivered by a female speaker. (d) The study described in (c) but with the added requirement that for each male of a particular age, there is a female of the same age.
13. Foofy has obtained a statistically significant two-sample t_{obt}. What three things should she do to complete her analysis?
14. You conduct a repeated-measures design, comparing a condition in which you train people to improve their memory to a control condition. (a) What problem will counterbalancing order of conditions not solve? (b) What is the technical name for this problem?
15. A rather dim student proposes testing the conditions of "male" and "female" as a repeated measures study. (a) What's wrong with this idea? (b) What control techniques can be applied instead?
16. You conduct a study in which the dependent variable is the degree of a participant's helpfulness in aiding a confederate to study for an exam in psychology. In each condition, participants are tested on five consecutive days. (a) What characteristics of the people who complete the study may bias the results? (b) What is the technical name for this problem?
17. You conduct a repeated measures study of the effects of three different types of motivational messages. Everyday, people listen to one message, and after two weeks, they complete a motivation test. Then they start with another message. (a) Specify your scheme for dealing with order effects. (b) What three problems will otherwise arise with your participants during the study that may confound the results? (c) What problem is likely to be created by any one message that makes repeated measures difficult? (d) What alternative can you choose to eliminate the problems in (b) and (c)?
18. What does the confidence interval for μ_D indicate?
19. A researcher predicts that participants will score higher on a questionnaire measuring well-being when they are exposed to lots of sunshine than when not exposed to much sunshine. A sample of 8 participants is first measured after low levels of sunshine exposure and then again after high levels of exposure. The researcher collects the following well-being scores:

Low:	14	13	17	15	18	17	14	16
High:	18	12	20	19	22	19	19	16

(a) Subtracting low from high, what are H_0 and H_a? (b) Compute the appropriate *t*-test. (c) With $\alpha = .05$, report your results. (d) Compute the appropriate confidence interval. (e) What is the predicted well-being score for a subject tested under low sunshine? Under high sunshine? (f) On average, how much more accurate are these predictions than if we did not know how much sunshine subjects experience? (g) What should the researcher conclude about these results?

20. A researcher investigates whether children exhibit a higher number of aggressive acts after watching a violent television show. The number of aggressive acts for the same ten children before and after watching the show are as follows:

Sample 1 (After)	*Sample 2 (Before)*
5	4
6	6
4	3
4	2
7	4
3	1
2	0
1	0
4	5
3	2

(a) Subtracting before scores from after scores, what are H_0 and H_a for this study? (b) Compute t_{obt}. (c) With $\alpha = .05$, what is t_{crit}? (d) What should the researcher conclude about this relationship? (e) Compute the confidence interval for μ_D. (f) If you want to understand children's aggression, how important is it to consider whether they watch violent television shows?

21. You investigate whether the older or younger male in pairs of brothers tends to be more extroverted. You obtain the following extroversion scores:

Sample 1 (Younger)	*Sample 2 (Older)*
10	18
11	17
18	19
12	16
15	15
13	19
19	13
15	20

(a) What are H_0 and H_a for this study? (b) Compute t_{obt}. (c) With $\alpha = .05$, what is t_{crit}? (d) What should you conclude about this relationship? (e) Is this a scientifically informative relationship?

22. What would you do to increase the power of the phobia study discussed in this chapter?

SUMMARY OF FORMULAS

1. *The computational formula for the* t-*test for related samples is*

$$t_{obt} = \frac{\bar{D} - \mu_D}{\sqrt{(s_D^2)\left(\frac{1}{N}\right)}}$$

Values of t_{crit} are found in Table 2 in Appendix C for $df = N - 1$, where N is the number of difference scores.

In the formula, $\bar{D}$ is the mean of the difference scores, μ_D is the value described by H_0, and s_D^2 is the variance of the difference scores, found using the formula

$$s_D^2 = \frac{\Sigma D^2 - \frac{(\Sigma D)^2}{N}}{N - 1}$$

where D is each difference score and N is the number of difference scores.

2. *The computational formula for the confidence interval for* μ_D *is*

$$(s_{\bar{D}})(-t_{crit}) + \bar{D} \leq \mu_D \leq (s_{\bar{D}})(+t_{crit}) + \bar{D}$$

t_{crit} is the two-tailed value for $df = N - 1$, where N is the number of difference scores, and $s_{\bar{D}}$ is the standard error of the mean difference, found using the formula

$$s_D = \sqrt{(s_D^2)\left(\frac{1}{N}\right)}$$

where s_D^2 is the variance of difference scores and N is the number of difference scores.

3. *The formula for computing* r_{pb} *from* t_{obt} *is*

$$r_{pb} = \sqrt{\frac{(t_{obt})^2}{(t_{obt})^2 + df}}$$

where $df = N - 1$, where N is the number of difference scores.

4. *The proportion of variance accounted for* in a dependent-samples experiment equals the squared value of r_{pb}, *or* r_{pb}^2.

PART 6

DESIGNING AND ANALYZING COMPLEX EXPERIMENTS

You may have noticed that each new type of experiment and its statistical procedure is more complex than the previous one. In Chapters 15 and 16, you saw how to analyze a two-sample experiment involving two conditions of an independent variable. However, researchers often conduct experiments that involve more than two conditions. In fact, they often conduct experiments that involve more than one independent variable, each of which involves more than two conditions. In the following three chapters, we discuss the logic and methods of such designs. We'll also see that they are analyzed using the procedure called the Analysis of Variance. This procedure can be applied to an almost limitless number of different designs, and it is by far the most common statistical procedure found in psychological research.

17

The One-Way Between-Subjects Experiment and the One-Way Analysis of Variance

To understand this chapter:

- From Chapter 2, recall the difference between a linear and nonlinear relationship.
- From Chapter 7, recall that variance measures the differences between scores by measuring their distance from the mean.
- From Chapter 13, understand why we limit the probability of a Type I error to .05.
- From Chapter 15, understand why we compute the effect size for significant results.
- And from our various discussions, understand that all inferential procedures simply involve testing H_0 by calculating an obtained statistic that locates our data on the appropriate sampling distribution.

Then your goals in this chapter are to learn:

- The terminology of analysis of variance.
- Why we compute the ANOVA and then *post hoc* tests.
- What is meant by treatment variance and error variance.
- Why F_{obt} should equal 1 if H_0 is true, and why F_{obt} is greater than 1 if H_0 is false.
- When to compute Fisher's protected *t*-test or Tukey's *HSD*.
- How "eta squared" describes effect size.

In this chapter, we'll discuss experiments that involve more than two conditions. The design issues concerning *any* experiment (with two or more conditions) are largely the same, however, and you already know them. Therefore, our focus will be on the statistical procedure used with such experiments, called *analysis of variance*. Analysis of variance can be used with between-subjects or within-subjects designs and with any number of conditions or independent variables. At first glance, this procedure may appear to be very different from previous procedures. Ultimately, however, the logic of all hypothesis testing is essentially the same.

MORE STATISTICAL NOTATION

The analysis of variance has its own language. First, analysis of variance is abbreviated as **ANOVA**. Second, when only one independent variable is involved in the experiment, a **one-way ANOVA** is performed. Third, an independent variable is called a **factor**. Thus, an experiment with one independent variable has one factor and we perform a one-way ANOVA. Finally, each condition of the independent variable or factor is also called a **level**. It is important to know the number of levels in a factor, and the symbol for the number of levels in a factor is k. The levels of a factor are also called **treatments**. When the different conditions of the independent variable produce significant differences in dependent scores, we have a *treatment effect* of the factor.

As usual, the specific procedure to employ in a particular study depends on the type of data you have and on the study's design. ANOVA is a parametric procedure, so you must have the appropriate dependent scores. Then you may have either a between- or within-subjects design (with independent or dependent samples, respectively). In ANOVA, a factor that is studied using independent samples in all conditions is called a **between-subjects factor**. A factor studied using dependent samples in all conditions is a **within-subjects factor**. In this chapter we discuss the between-subjects, one-way ANOVA. (You'll see other types of ANOVAs in later chapters.)

DESIGNING MULTI-LEVEL EXPERIMENTS

Analysis of variance is the parametric procedure for determining whether significant differences exist in an experiment that contains two or more sample means. There's a bit of confusion here, because when you have only two conditions of the independent variable you can use either a two-sample t-test or ANOVA: You'll reach exactly the same conclusions about the null hypothesis with each procedure, and both have the same probability of making Type I and Type II errors. You *must* use ANOVA, however, when you have more than two conditions of an independent variable.

Sometimes researchers examine more than two conditions of an independent variable because of the logic of their study: For example, we would have three levels of the factor of mood if we attempted to either elevate participants' moods by reading them a positive story, depress their moods by reading them a negative story, or have no influence on mood by presenting them a neutral control condition. As with all experiments, the purpose here is to demonstrate a relationship between the independent variable and the dependent variable. The only novelty is that now we have three samples of scores, because we need to make comparisons between positive and neutral, between negative and neutral, and between positive and negative (adding a control condition often leads to 3 levels of a factor).

In addition, researchers may include more than two conditions so that they can accurately describe the *type* of relationship between the variables. Recall that the type of relationship may be either linear (following a straight line) or nonlinear (following a line that changes direction). In a study with just two conditions, one group's mean can be only higher or lower than the other group's mean, so the data can depict only a linear relationship, even if the relationship in nature is nonlinear. You must have at least three conditions to see a nonlinear relationship. For example, earlier we discussed a study in which we manipulated room temperature in order to influence a participant's aggressiveness. Figure 17.1 shows the graphs we might obtain with two or four temperature conditions. When interpreting graph A, we assume that aggression scores at other, untested temperatures (such as at 80 and 100 degrees) would change in the same way as they did at the observed temperatures of 70 and 90 degrees, so that all means would fall on the straight line. But this is an assumption! By testing the additional temperature conditions, we might confirm this linear relationship, or we might instead find a nonlinear relationship, as in graph B. The 70- and 80-degree conditions show that increasing temperature does not always increase aggression. Increases beyond 80 degrees, however, do produce increases in aggression, with the rate of increase greater for 100 degrees than for 90 degrees.

Thus, to accurately describe the type of relationship formed by the variables, you may want to add more conditions. When practical, research typically involves at least three conditions, in case the relationship is nonlinear. The maximum tends to be six to eight conditions, which is considered more than adequate for describing most relationships.

Finally, there is one other common reason for including multiple levels of a factor: Once you have gone to all the effort of creating the materials and testing two

FIGURE 17.1 Two-Condition and Four-Condition Temperature-Study Results

(A) The inferred linear relationship with two conditions, and (B) the demonstrated nonlinear relationship with four conditions.

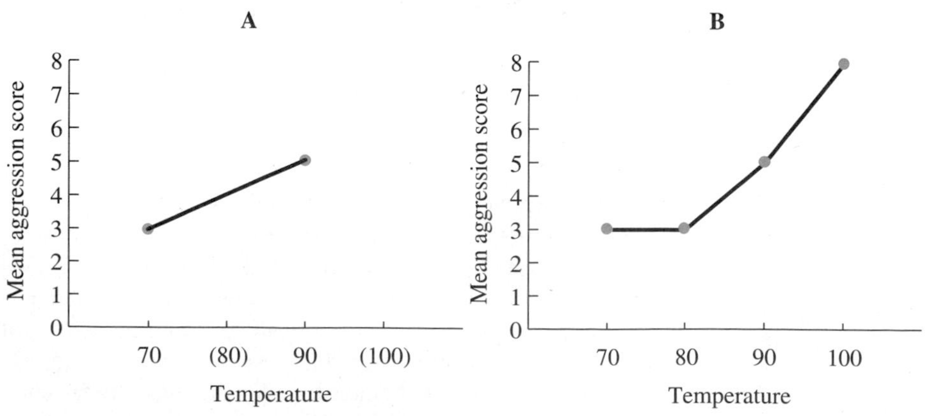

conditions, often little additional effort is needed to test additional conditions (especially if you are running a repeated measures design). You'll often find that it requires little work to alter stimuli, instructions, or the measurement task, so that you can study additional conditions in a very cost-effective way. After all, the reason for conducting research in the first place is to get data that describe behaviors, so the more data you can get, the more you learn.

> *REMEMBER* Researchers study multiple conditions of an independent variable to test their hypothesis adequately, to demonstrate a nonlinear relationship, and to obtain the maximum information from a study.

OVERVIEW OF ANOVA

As an example study, let's study how well people perform a task depending on how difficult they believe the task will be (the perceived difficulty of the task). We select three random samples that contain the unpowerful n of 5 participants each and we provide them with the same 10 easy math problems. To influence their perceptions, however, we tell participants in Level 1 that the problems are easy, in Level 2 that the problems are of medium difficulty, and in Level 3 that the problems are difficult. The dependent measure is the number of problems that participants correctly solve within an allotted time. If participants are tested under only one condition, and we do not match subjects, then this is a one-way, between-subjects design. You can see this design in Table 17.1. Each column is a level of the factor, containing the scores of participants tested under that condition. Averaging vertically in a column gives the mean of each level. We again identify the n and $\overline{X}$ for each level using a subscript. Thus, $\overline{X}_1$ and n_1 are the mean and n from level 1, the easy condition. Because there are three levels in this factor, $k = 3$. The total number of scores in the experiment is symbolized by N, and here $N = 15$. Further, the overall mean of all scores in the experiment will be the mean of all 15 scores. (If we don't have a specific independent variable in mind, the general format for diagramming such a one-way ANOVA is to label the factor as factor A, with levels A_1, A_2, and A_3.)

TABLE 17.1 Diagram of a Study Having Three Levels of One Factor

Factor A: independent variable of perceived difficulty			
Level A_1: easy	*Level A_2: medium*	*Level A_3: difficult*	← *Conditions* *$k = 3$*
X	X	X	
X	X	X	
X	X	X	
X	X	X	
X	X	X	
$\overline{X}_1$ $n_1 = 5$	$\overline{X}_2$ $n_2 = 5$	$\overline{X}_3$ $n_3 = 5$	Overall $\overline{X}$ $N = 15$

All of the usual design considerations apply for creating a strong and reliable manipulation of the independent variable, with valid and reliable measurement of the dependent variable (considering a sensitive measure, demand characteristics, construct validity, etc.). Thus, we would attempt to convince participants that under "easy," the problems are *very* easy, but under "difficult," the problems are *very* difficult. We would also counterbalance the order in which different participants within each condition perform the math problems, and we might limit the population or counterbalance for important subject variables (such as participants' math ability or their "math phobia").

Ideally, we will find a different sample mean for each condition. Then we want to conclude that if the entire population was tested under each level of difficulty, we would find three different populations of scores located at three different μs. But there's the usual problem: Differences between the sample means may reflect sampling error, in which case we would actually find the same population of scores, having the same μ, for each level of perceived difficulty. Therefore, as usual, before we can conclude that a relationship exists, we must eliminate the idea that the differences among the sample means merely reflect sampling error.

How ANOVA Controls for Experiment-Wise Error Rate

You might think that we could use the independent samples *t*-test to determine whether there are significant differences between the three means above. That is, we might perform "multiple *t*-tests," testing whether $\overline{X}_1$ differs from $\overline{X}_2$, then whether $\overline{X}_2$ differs from $\overline{X}_3$, and finally whether $\overline{X}_1$ differs from $\overline{X}_3$. We cannot use this approach, however, because of the resulting probability of making a Type I error (rejecting a true H_0). With $\alpha = .05$, the theoretical probability of a Type I error in a *single* *t*-test is .05. We can make a Type I error when comparing $\overline{X}_1$ to $\overline{X}_2$, when comparing $\overline{X}_2$ to $\overline{X}_3$, or when comparing $\overline{X}_1$ to $\overline{X}_3$. Therefore, the *overall* probability of making a Type I error somewhere in the experiment is considerably greater than .05.

This overall probability of making a Type I error is called the experiment-wise error rate. The **experiment-wise error rate** is the probability of making a Type I error in an experiment. We can use the *t*-test when comparing only two means because with only one comparison, the experiment-wise error rate equals α. When there are more than two levels in a factor, however, performing multiple *t*-tests results in an overall α that is greater than the one we selected: If we think $\alpha = .05$, it will actually be greater than .05. Because of the importance of avoiding Type I errors, we cannot afford to have the actual α greater than .05. Therefore, we *must* perform ANOVA. Only then will the experiment-wise error rate after we have compared all sample means equal the α we have chosen.

As usual, before proceeding with a study, you should check that it meets the assumptions of the statistical procedure.

Assumptions of the One-Way, Between-Subjects ANOVA

In a one-way, between-subjects ANOVA, the experiment has only one independent variable, and all of the conditions contain independent samples. We assume that:

1. Each condition contains a random sample of interval or ratio scores.

2. The population represented by the scores in each condition forms a normal distribution.

3. The variances of all populations represented in the study are homogeneous.

Like other parametric procedures, ANOVA is robust, so you can use it even if the populations are not perfectly normal or homogeneous. Although the number of participants in each condition need not be equal, violations of the assumptions are even less serious when all *ns* are equal. Also, certain procedures are *much* easier to perform with equal *ns*.

If the study meets the assumptions of ANOVA, set α (usually at .05) and create the statistical hypotheses.

Statistical Hypotheses of ANOVA

ANOVA tests only two-tailed hypotheses. The null hypothesis is that there are no differences among the populations represented by the conditions. Thus, for our perceived difficulty study with the three levels of easy, medium, and difficult, we have

$$H_0\text{: } \mu_1 = \mu_2 = \mu_3$$

As usual, the null hypothesis says that all μs are equal, so there is no real relationship here: Changing the perceived difficulty from easy to medium to difficult won't really influence how people perform the math problems.

In general, when we perform ANOVA on a factor with *k* levels, the null hypothesis is

$$H_0\text{: } \mu_1 = \mu_2 = \ldots = \mu_k$$

The "$\ldots = \mu_k$" indicates that there are as many μs as there are levels.

H_0 always implies that the means for all levels represent the same population mean, and therefore the sample means should be equal. If the means are not equal, it is because of sampling error in representing the one value of μ.

You might think that the alternative hypothesis would be that the various μs are not equal, or $\mu_1 \neq \mu_2 \neq \mu_3$. However, a study may demonstrate a relationship in which only *some* but not *all* conditions differ (as previously in Figure 17.1). Thus, it's possible our data will represent a difference between μ_1 and μ_2, but not between μ_1 and μ_3, or perhaps between μ_2 and μ_3 only. To communicate this idea, the alternative hypothesis is

$$H_a\text{: not all } \mu\text{s are equal}$$

H_a implies that there is a relationship in the population involving at least two of the conditions: The population mean represented by one of our levels of difficulty will be different from the population mean represented by at least one other level of difficulty.

We always test H_0, so in ANOVA, we test whether all sample means represent the same population mean.

The Order of Operations in ANOVA: The *F* Statistic and *Post Hoc* Comparisons

The statistic that forms the basis for ANOVA is called *F*. The ***F* statistic** is used to determine whether two or more sample means represent different μs. The *F* we calculate is the F_{obt}. We compare F_{obt} to the critical value of *F*, the F_{crit}.

When F_{obt} is not significant, it indicates that there are no significant differences between any of the sample means and that all means are likely to represent the same μ.

When this occurs, the experiment has failed to demonstrate a relationship, we are finished with the statistical analysis, and it's back to the drawing board.

When F_{obt} is significant, it indicates that *somewhere* among the level means *at least two* means are likely to represent different μs. The problem is that F_{obt} does not indicate *which* specific means differ significantly. Perhaps more than two means are significantly different, or maybe all of them are. Thus, for example, if F_{obt} for the perceived difficulty study is significant, it will indicate only that there is at least one significant difference somewhere among the means of the easy, medium, and difficult levels.

Obviously, to understand the relationship between our variables, we must determine which levels actually differ significantly. Therefore, we perform a second statistical procedure, called *post hoc* comparisons. (*Post hoc* means "after the fact," which here is after F_{obt} is significant.) ***Post hoc* comparisons** are like *t*-tests, in which we compare *pairs* of sample means from a factor, one pair at a time. By comparing all possible pairs, we determine which means differ significantly from each other. Thus, if F_{obt} for the perceived difficulty study is significant, in our *post hoc* comparisons we'll compare the means from easy and medium, from easy and difficult, and from medium and difficult. The results of these comparisons will indicate which means differ significantly from each other.

Note that we perform *post hoc* comparisons *only* when F_{obt} is significant. This two-step procedure ensures that the total, experiment-wise probability of a Type I error will be less than .05 (or whatever alpha we selected).

> *REMEMBER* If F_{obt} is significant, perform *post hoc* comparisons to determine which specific means differ significantly.

There is one exception to this rule. When there are only two levels in the factor, the significant difference indicated by F_{obt} must be between the only two means in the study, so it is unnecessary to perform *post hoc* comparisons.

The first step is to compute F_{obt}. Therefore, the following sections present the statistical basis for ANOVA and the logic of its computation.

COMPONENTS OF THE *F* STATISTIC

Analysis of variance does just that: It analyzes variance. Recall that computing variance is simply a way to measure the differences among scores. (As you read the following, keep saying to yourself: "Variance is differences.") ANOVA involves partitioning the variance. That is, we take the total variability of the scores in an experiment and break it up, or partition it, in terms of its source. There are two potential sources of variance. First, scores may differ from each other even when participants are in the same condition. We call this variability the **variance within groups**. Second, scores may differ from each other because they are from different conditions. We call this variability the **variance between groups**.

Using the sample data, we compute estimates of the value that each of these variances would have in the population. But we do not *call* each an estimated variance. Instead, we call each a **mean square**. This is a shortened name for variance, which is literally the mean squared deviation. The symbol for a mean square is MS. Because we estimate the variance within groups and the variance between groups, we compute two mean squares, the mean square within groups and the mean square between groups.

The Mean Square Within Groups

The **mean square within groups** is an estimate of the variability of scores as measured by differences within the conditions of an experiment. The symbol for the mean square within groups is MS_{wn}. The word *within* says it all: Think of MS_{wn} as the "average variability" of the scores within each condition around the mean of that condition. Table 17.2 illustrates how to conceptualize the computation of MS_{wn}. You can think of MS_{wn} in this way: First we find the variance in level 1 (finding the squared differences between the scores in level 1 and $\overline{X}_1$), then we find the variance of scores in level 2 around $\overline{X}_2$, and then we find the variance of scores in level 3 around $\overline{X}_3$. Then we "pool" the variances together, just like we did in the independent samples *t*-test back in Chapter 15. Thus, the MS_{wn} is the "average" variability of the scores in each condition around the mean of that condition.

We compare the scores in each condition with the mean for that condition, so MS_{wn} reflects the inherent variability between scores that arises from individual differences or from other random factors when participants are all treated the same. We've seen in previous chapters that such variance is called error variance. Thus, the MS_{wn} estimates the **error variance**, which we'll symbolize as σ^2_{error} (MS_{wn} is also known as the *error term*). In symbols,

Sample	*Estimates*	*Population*
MS_{wn}	$\rightarrow$	σ^2_{error}

Because we assume homogeneity of variance, MS_{wn} estimates the variance among the scores in any population that may be represented by a condition. For example, an MS_{wn} equal to 4 indicates that we estimate the variance of the scores in any of the populations represented by our conditions to be 4.

> *REMEMBER* The MS_{wn} is an estimate of the error variance, the inherent variability within any population represented by the samples.

TABLE 17.2 How to Conceptualize the Computation of MS_{wn}

Factor A		
Level A_1:	*Level A_2:*	*Level A_3:*
X	X	X
X	X	X
X	X	X
X	X	X
X	X	X
$\overline{X}_1$	$\overline{X}_2$	$\overline{X}_3$

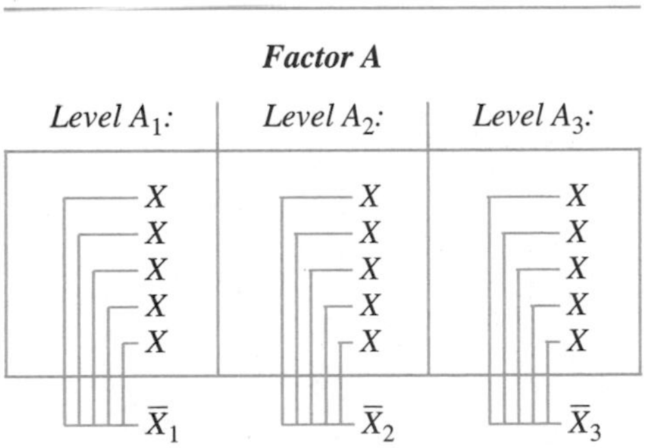

The Mean Square Between Groups

The other variance computed in ANOVA is the mean square between groups. Here, the word *between* says it all. The **mean square between groups** is an estimate of the variability in scores that occurs *between* the levels in a factor. The mean square between groups is symbolized by MS_{bn}. You can conceptualize the computation of MS_{bn} as shown in Table 17.3. We summarize the scores in each level by computing the mean of that level, and then we determine how much it deviates from the overall mean of all scores in the experiment. In the same way that the squared deviations of raw scores around their mean describe how different the scores are from each other, the squared deviations of the level means from the overall mean indicate how different the level means are from each other. *Thus, MS_{bn} is our way of measuring how much the means in a factor differ from each other.*

The key to understanding ANOVA is to understand what MS_{bn} represents when H_0 is true and when it is false. First, when H_0 is true, the scores in the different conditions all come from the same population. When we sample the population, not every score will equal μ or each other. You can see this represented in the distribution in Figure 17.2. First look at the two *X*s. Any two scores may differ because of inherent variability, Here, for example, by chance one X is above μ and the other is below μ. If the population was available, we would measure such differences between all scores using the differences between the scores and μ to calculate σ^2_{error}. Instead, we estimate it by calculating MS_{wn} from the sample data.

Here's the crucial part: If H_0 is correct and changing the independent variable does not work, then the reason we get a difference between any two *means* is *also* due to this inherent variability. For example, as in Figure 17.2, by chance we may get a *batch* of scores above μ in one condition that produce a relatively high $\overline{X}$, but in another condition we may get a batch of scores below μ that produce a relatively low $\overline{X}$. If the population was available, we would measure the differences between all means. Instead, we estimate this by calculating MS_{bn}.

When H_0 is true, the differences between level means—their variability—is determined by how variable the raw scores are. If most scores are rather close to each other, then we will tend to get very similar scores in each condition, so the means will be close to each other. If the scores are highly variable, however, then we'll get a very different batch of scores from one condition to the next, producing very different $\overline{X}$s.

TABLE 17.3 How to Conceptualize the Computation of MS_{bn}

Factor A			
Level A_1	*Level A_2*	*Level A_3*	
X	X	X	
X	X	X	
X	X	X	
X	X	X	
X	X	X	
$\overline{X}_1$	$\overline{X}_2$	$\overline{X}_3$	Overall $\overline{X}$

FIGURE 17.2 The One Population of Scores When H_0 is True, Showing the Logic of MS_{wn} and MS_{bn}

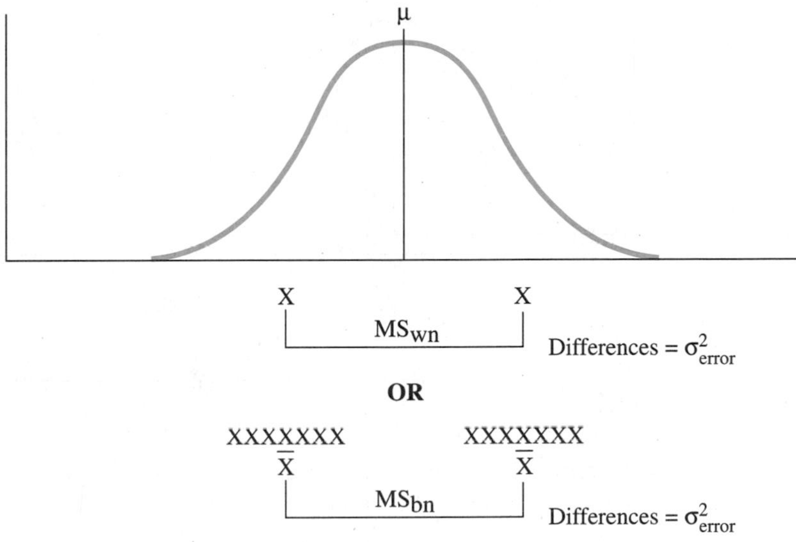

Because the differences between the means are related to the differences between the scores, we could calculate σ^2_{error} in the population *either* by finding how much some randomly selected raw scores differ or by finding how much some random sample means differ. It doesn't matter, because *both* reflect the inherent differences between scores. Likewise, when we use our sample data to calculate MS_{bn} and MS_{wn}, they are (1) both estimating the one value of σ^2_{error}, so (2) they should be *equal.*

REMEMBER If H_0 is true, then ideally MS_{bn} and MS_{wn} should be equal.

When H_0 is false, however, the scores from different conditions do not come from the same population, as shown in Figure 17.3. Now two reasons combine to determine the value of MS_{bn}. First, any two sample means differ because we have a *treatment effect* (what we have called a "real" relationship, in which different conditions produce different populations of scores and each sample mean represents a different population μ). Differences due to treatment are called treatment variance, which we can symbolize by σ^2_{treat}. **Treatment variance** reflects differences between scores that occur because the scores are from different populations. In Figure 17.3, the distance separating the μs of the distributions can be seen as reflecting treatment variance. We summarize these differences as the differences between the means of the conditions, which we determine when calculating MS_{bn}. Therefore, MS_{bn} contains an estimate of the treatment variance.

Second, the difference between any two means also reflects the inherent variability among the scores in each population. As in Figure 17.3, by chance we might obtain a $\overline{X}$ below μ in one population, and a $\overline{X}$ above μ in the other. Thus, the total distance separating these sample means reflects two things: Differences due to treatment, σ^2_{treat} (the solid line) *PLUS* differences due to the inherent variability of the scores in each population, σ^2_{error} (the dotted lines). So here's the crucial part: Because we calculate MS_{bn} as

FIGURE 17.3 Two Populations of Scores When H_0 Is False, Showing the Logic of MS_{bn}

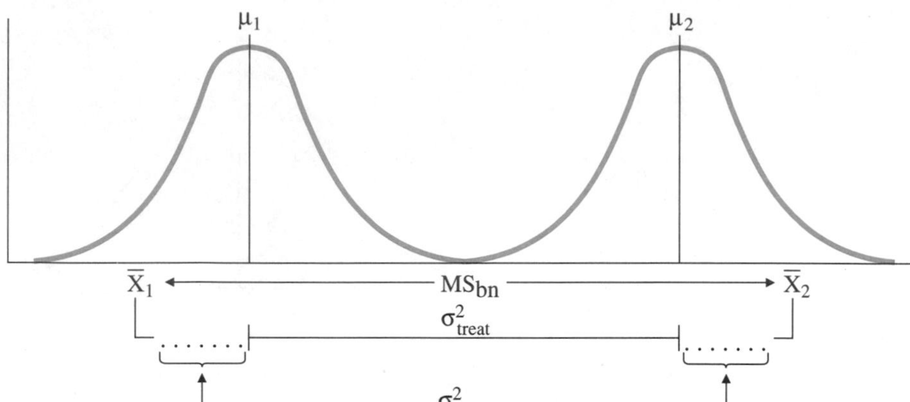

the differences between the sample means, the MS_{bn} contains an estimate of *both* error variance plus treatment variance. In symbols, this is

Sample	***Estimates***	***Population***
MS_{bn}	$\rightarrow$	$\sigma^2_{error} + \sigma^2_{treat}$

REMEMBER If H_0 is false, then MS_{bn} contains estimates of both error variance (which measures differences *within* each population) and treatment variance (which measures differences *between* the populations).

The σ^2_{error} component is again the same value as estimated by MS_{wn}. But with the added σ^2_{treat} component, MS_{bn} will now be *larger* than MS_{wn}. Thus, for example, say that when MS_{wn} equals 4, we compute that MS_{bn} equals 10. We can think of MS_{bn} as indicating that, in addition to an error variance of 4, there is an "average difference" of 6 between the populations. Of course, the problem of sampling error makes the issue a little more complicated.

Comparing the Mean Squares: The Logic of the *F*-Ratio

We are not interested in the individual values of MS_{bn} and MS_{wn}. Instead, we examine their ratio, called the *F*-ratio and compute F_{obt}. The ***F*-ratio** equals the mean square between groups divided by the mean square within groups. Thus,

THE COMPUTATIONAL FORMULA FOR THE F-RATIO IS

$$F_{obt} = \frac{MS_{bn}}{MS_{wn}}$$

MS_{bn} is always on top!

We can conceptualize the *F*-ratio as representing this:

$$\begin{array}{cccc} & \textbf{\textit{Sample}} & \textbf{\textit{Estimates}} & \textbf{\textit{Population}} \\ F_{obt} = & \dfrac{MS_{bn}}{MS_{wn}} & \begin{array}{c}\rightarrow \\ \rightarrow\end{array} & \dfrac{\sigma^2_{error} + \sigma^2_{treat}}{\sigma^2_{error}} \end{array}$$

The MS_{bn} represents the inherent differences among scores in any population, σ^2_{error}, *plus* whatever differences there are between the populations represented by the conditions, σ^2_{treat}. This value is divided by the MS_{wn}, which is only an estimate of the error variance in the populations, σ^2_{error}.

Now you can understand what the *F*-ratio indicates about the null hypothesis. If H_0 is true and all conditions represent the same population, then there are zero differences due to treatment. Instead, any differences between the sample means totally reflect the inherent variability of scores that occurs naturally in the one population. Thus, when H_0 is true, MS_{bn} contains solely σ^2_{error}, and the σ^2_{treat} component of MS_{bn} equals zero. In symbols, when H_0 is true, we have

$$\begin{array}{cccc} & \textbf{\textit{Sample}} & \textbf{\textit{Estimates}} & \textbf{\textit{Population}} \\ F_{obt} = & \dfrac{MS_{bn}}{MS_{wn}} & \begin{array}{c}\rightarrow \\ \rightarrow\end{array} & \dfrac{\sigma^2_{error} + 0}{\sigma^2_{error}} = \dfrac{\sigma^2_{error}}{\sigma^2_{error}} = 1 \end{array}$$

Both mean squares are merely estimates of the one value of σ^2_{error}. If our data represent the situation where H_0 is true, then the mean square between groups should *equal* the mean square within groups. When two numbers are equal, their ratio equals 1. *Therefore, when H_0 is true and all conditions represent one population, the F-ratio should equal 1.*

On the other hand, H_a says that not all μs are equal. If H_a is true, then at least two conditions represent different populations, and there are differences between at least two means that are due to treatment. Therefore, when H_a is true, the σ^2_{treat} component of MS_{bn} does not equal zero, so we have

$$\begin{array}{cccc} & \textbf{\textit{Sample}} & \textbf{\textit{Estimates}} & \textbf{\textit{Population}} \\ F_{obt} = & \dfrac{MS_{bn}}{MS_{wn}} & \begin{array}{c}\rightarrow \\ \rightarrow\end{array} & \dfrac{\sigma^2_{error} + \text{some amount of } \sigma^2_{treat}}{\sigma^2_{error}} = F > 1 \end{array}$$

Here MS_{bn} contains error variance *plus* some amount of treatment variance, so MS_{bn} will be *larger* than MS_{wn}, which contains only error variance. Placing a larger number in the numerator of the *F*-ratio produces an F_{obt} greater than 1. *Thus, when H_a is true, MS_{bn} is larger than MS_{wn} and F_{obt} is greater than 1.* The larger the differences between the populations represented by the level means, the larger the σ^2_{treat} component, and thus the larger MS_{bn} will be. Regardless of the effect of the independent variable, however, the size of MS_{wn} remains constant. Therefore, the larger the differences between means, the larger F_{obt} will be.

Regardless of whether we have a positive or negative linear relationship or a nonlinear relationship, MS_{bn} will be larger than MS_{wn} so that F_{obt} is greater than 1. (This is why we have only two-tailed hypotheses in ANOVA.) An F_{obt} between 0 and 1 is possible, but it occurs when the denominator of the *F*-ratio, MS_{wn}, is larger than the numerator, MS_{bn}. Here we assume that MS_{bn} and/or MS_{wn} are merely poor estimates of error variance and therefore not equal. F_{obt} cannot be less than zero, because the mean squares are variances which cannot be negative numbers.

Thus, on the one hand, F_{obt} should equal 1 if the level means represent the same μ. On the other hand, F_{obt} should be greater than 1 if the level means represent two or more different μs. But hold on! There's one other reason F_{obt} might be greater than 1, and that is (here we go again) sampling error! When H_0 is true, F_{obt} "should" equal 1 *if* the mean squares are perfectly representative. *But*, through sampling error with *one* population, we might obtain differences between level means that are larger than the differences between the scores in the population. This will produce an MS_{bn} that is larger than MS_{wn}, so F_{obt} will be larger than 1 simply because of sampling error.

This all boils down to the same old problem of significance testing. An F_{obt} greater than 1 may accurately reflect the situation where two or more conditions represent different populations of raw scores (and there is a "real" treatment effect). Or, because of sampling error, an F_{obt} greater than 1 may inaccurately reflect the situation where all conditions represent the same population (and there only *appears* to be a treatment effect). Therefore, whenever F_{obt} is greater than 1, we must test H_0: We determine the probability of obtaining such an F_{obt} when H_0 is true. To do this, we examine the *F*-distribution.

The *F*-Distribution

The ***F*-distribution** is the sampling distribution that shows the various values of *F* that occur when H_0 is true and all conditions represent one population μ. We could create a sampling distribution in the following way: Using the same number of levels, *k*, and the same *ns* as in our study, we randomly sample one raw score population repeatedly. Each time, we compute MS_{bn}, MS_{wn}, and F_{obt}. After doing this an infinite number of times, we plot the various values of F_{obt}. The resulting distribution can be envisioned as shown in Figure 17.4. This is more or less the same old H_0 sampling distribution, except that now the *X* axis shows the values of *F* that occur when all level means *do* represent the same μ. The *F*-distribution is skewed, because there is no limit to how large F_{obt} can be, but it cannot be less than zero. The mean of the distribution is 1 because, most often when H_0 is true, MS_{bn} will equal MS_{wn} and *F* will equal 1. We are concerned with the upper tail, which shows that sometimes the means are unrepresentative, producing an MS_{bn} that is larger than MS_{wn} and thus an *F* greater than 1. As shown, however, the larger the *F*, the less frequent and thus the less likely it is when H_0 is true.

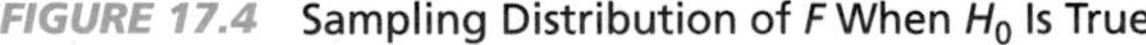

FIGURE 17.4 Sampling Distribution of *F* When H_0 Is True

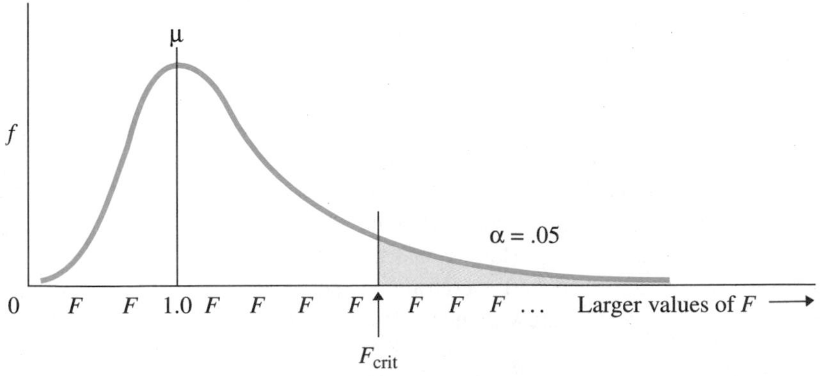

The F_{obt} can reflect a relationship in the population only when it is greater than 1, so the entire region of rejection is in the upper tail of the *F*-distribution. (With ANOVA, we always have two-tailed hypotheses, but we test them using a one-tailed test.) If F_{obt} is larger than F_{crit}, then F_{obt}—and the differences between the conditions that produced it—is too unlikely to accept as having occurred by chance when H_0 is true and all levels of the factor represent the same population. Therefore, we reject H_0 and have a significant F_{obt}.

Like the *t*-distribution, the *F*-distribution consists of a family of curves. Each *F*-distribution has a slightly different shape depending on the degrees of freedom in the data, and thus there is a different value of F_{crit} for each *df*. However, *two* values of *df* determine the shape of each *F*-distribution: The *df* for the mean square between groups and the *df* for the mean square within groups. The symbol for the *df* between groups is df_{bn}, and the symbol for the *df* within groups is df_{wn}. We use both df_{bn} and df_{wn} when finding F_{crit}.

To obtain F_{crit}, turn to Table 5 in Appendix C, entitled "Critical Values of *F*." Across the top of these *F*-tables, the columns are labeled "*df* between groups," and along the left side, the rows are labeled "*df* within groups." Locate the appropriate column and row using the *df*s from your study. The critical values of *F* in dark type are for $\alpha = .05$, and the values in light type are for $\alpha = .01$. For example, say we eventually determine that $df_{bn} = 2$ and $df_{wn} = 12$. Then for $\alpha = .05$, F_{crit} is 3.88.

Be careful to keep your "withins" and "betweens" straight: $df_{bn} = 2$ and $df_{wn} = 12$ is very different from $df_{bn} = 12$ and $df_{wn} = 2$.

Don't be overwhelmed by the details of ANOVA. Buried in here is the simple idea that the larger the differences between the means for the conditions of the independent variable, the larger the MS_{bn} and thus the larger the F_{obt}. If the F_{obt} is larger than F_{crit}, then the F_{obt} (and the corresponding differences between the means) is too unlikely to occur if the means from different conditions represent the same population μ. In such cases, we reject H_0 and confidently conclude that the data represent a "real" relationship.

Now all we need to do is to get through the computations, so hold on, here we go.

COMPUTING THE *F*-RATIO

When we computed the estimated variance in Chapter 8, the quantity $\Sigma(X - \overline{X})^2$ was called the sum of the squared deviations. In ANOVA, this is shortened to the **sum of squares**. The symbol for the sum of squares is *SS*, so in the numerator of the formula for variance, we can replace the sum of the squared deviations with *SS*:

$$s_X^2 = \frac{\Sigma(X - \overline{X})^2}{N - 1} = \frac{SS}{df} = MS$$

In the denominator, $N - 1$ is the degrees of freedom, so we replace $N - 1$ with *df*. Because variance is called a mean square in ANOVA, the fraction formed by the sum of squares, *SS*, divided by the degrees of freedom, *df*, is the general formula for a mean square.

Adding subscripts, we will compute the mean square between groups, MS_{bn}, by computing the sum of squares between groups, SS_{bn}, and then dividing by the degrees of freedom between groups, df_{bn}. Likewise, we will compute the mean square within groups, MS_{wn}, by computing the sum of squares within groups, SS_{wn}, and then dividing by the degrees of freedom within groups, df_{wn}. Once we have MS_{bn} and MS_{wn}, we compute F_{obt}.

If all this strikes you as the most confusing thing ever devised by humans, you'll find creating an ANOVA summary table very helpful. (Most computer programs print the results of ANOVA in a summary table, so you'll need to understand this or it'll all be gibberish.) The general format of the summary table for a one-way ANOVA is:

Summary Table of One-Way ANOVA

Source	*Sum of squares*	*df*	*Mean square*	*F*
Between	SS_{bn}	df_{bn}	MS_{bn}	F_{obt}
Within	SS_{wn}	df_{wn}	MS_{wn}	
Total	SS_{tot}	df_{tot}		

The source column identifies each component. Eventually we'll fill in the values in the other columns. Notice that as we do, the computations are built in. Dividing SS_{bn} by df_{bn} produces MS_{bn}. Dividing SS_{wn} by df_{wn} produces MS_{wn}. Finally, the fraction formed by putting MS_{bn} "over" MS_{wn} produces F_{obt}. (Along the way, we also compute the total sum of squares, SS_{tot}, and the total *df*, df_{tot}). The F_{obt} is always placed in the row labeled "Between." In place of the word "Between," you can use the name of the independent variable. Also, in place of the word "Within," you will sometimes see "Error."

Computational Formulas for the One-Way, Between-Subjects ANOVA

To provide some example data, say that we actually performed the perceived difficulty study that we discussed earlier: We told three conditions of five participants each that some math problems were easy, of medium difficulty, or difficult, and we measured the number of problems they solved correctly. The data are presented in Table 17.4.

TABLE 17.4 Data from Perceived Difficulty Experiment

Factor A: perceived difficulty			
Level A_1: easy	*Level A_2: medium*	*Level A_3: difficult*	
9	4	1	
12	6	3	
4	8	4	
8	2	5	
7	10	2	*Totals*
$\Sigma X = 40$	$\Sigma X = 30$	$\Sigma X = 15$	$\Sigma X = 85$
$\Sigma X^2 = 354$	$\Sigma X^2 = 220$	$\Sigma X^2 = 55$	$\Sigma X^2 = 629$
$n_1 = 5$	$n_2 = 5$	$n_3 = 5$	$N = 15$
$\bar{X}_1 = 8$	$\bar{X}_2 = 6$	$\bar{X}_3 = 3$	$k = 3$

As shown in the table, the first step in performing ANOVA is to compute ΣX, ΣX^2, and $\overline{X}$ for each level. Adding the ΣX from each level gives the total ΣX, and adding the ΣX^2 from each level gives the total ΣX^2. Then, as shown in the following sections, we compute the sum of squares, the degrees of freedom, the mean squares, and then F_{obt}. So that you do not get lost, as you complete each step, fill in the results in the ANOVA summary table created above. (There *will* be a test later.)

Computing the sums of squares The first task is to compute the sum of squares. We do this in three steps.

Step 1 is to compute the total sum of squares, SS_{tot}.

THE COMPUTATIONAL FORMULA FOR SS_{tot} *IS*

$$SS_{tot} = \Sigma X^2_{tot} - \left(\frac{(\Sigma X_{tot})^2}{N}\right)$$

Here we treat the entire experiment as if it were one big sample. Thus, ΣX_{tot} is the sum of all *X*s, and ΣX^2_{tot} is the sum of all squared *X*s. *N* is the total *N* in the study.

Using the data from Table 13.4, we have $\Sigma X^2_{tot} = 629$, $\Sigma X_{tot} = 85$, and $N = 15$, so

$$SS_{tot} = 629 - \frac{(85)^2}{15}$$

$$SS_{tot} = 629 - \frac{7225}{15}$$

$$SS_{tot} = 629 - 481.67$$

Thus, $SS_{tot} = 147.33$.

Step 2 is to compute the sum of squares between groups, SS_{bn}.

THE COMPUTATIONAL FORMULA FOR SS_{bn} *IS*

$$SS_{bn} = \Sigma\left(\frac{(\text{sum of scores in the column})^2}{n \text{ of scores in the column}}\right) - \left(\frac{(\Sigma X_{tot})^2}{N}\right)$$

When we diagram the study, each column represents a level of the factor. Thus, the formula says to find the ΣX for each level, to square ΣX, and then to divide by the *n* in that level. After doing this for all levels, add the results together and subtract the quantity $(\Sigma X_{tot})^2/N$. From Table 17.4, we have

$$SS_{bn} = \left(\frac{(40)^2}{5} + \frac{(30)^2}{5} + \frac{(15)^2}{5}\right) - \left(\frac{(85)^2}{15}\right)$$

so

$$SS_{bn} = (320 + 180 + 45) - 481.67$$

and

$$SS_{\text{bn}} = 545 - 481.67$$

Thus, $SS_{\text{bn}} = 63.33$.

Step 3 is to compute the sum of squares within groups, SS_{wn}. Mathematically, SS_{tot} equals SS_{bn} plus SS_{wn}. Therefore, the total minus the between leaves the within.

THE COMPUTATIONAL FORMULA FOR SS_{wn} IS

$$SS_{\text{wn}} = SS_{\text{tot}} - SS_{\text{bn}}$$

In our example, SS_{tot} is 147.33 and SS_{bn} is 63.33, so

$$SS_{\text{wn}} = 147.33 - 63.33 = 84.00$$

so, $SS_{\text{wn}} = 84.00$.

Filling in the first column of the ANOVA summary table, we have

Summary Table of One-Way ANOVA

Source	***Sum of squares***	***df***	***Mean square***	***F***
Between	63.33	df_{bn}	MS_{bn}	F_{obt}
Within	84.00	df_{wn}	MS_{wn}	
Total	147.33	df_{tot}		

As a double check, make sure that the total equals the sum of the between plus the within. Here, $63.33 + 84.00 = 147.33$.

Now compute the degrees of freedom.

Computing the degrees of freedom We compute df_{bn}, df_{wn}, and df_{tot}. Again, there are three steps.

1. *The degrees of freedom between groups equals $k - 1$*, where k is the number of levels in the factor. In the example, there are three levels in the factor of perceived difficulty (easy, medium, and difficult), so $k = 3$. Thus $df_{\text{bn}} = 2$.
2. *The degrees of freedom within groups equals $N - k$*, where N is the total N of the study and k is the number of levels in the factor. In the example, the total N is 15 and k equals 3, so df_{wn} equals $15 - 3$, or 12.
3. *The degrees of freedom total equals $N - 1$*, where N is the total N in the experiment. In the example, we have a total of 15 scores, so $df_{\text{tot}} = 15 - 1 = 14$.

The df_{tot} is useful because it equals the sum of the df_{bn} plus the df_{wn}. Thus, to check our work in the example, $df_{\text{bn}} + df_{\text{wn}} = 2 + 12$, which equals 14, the df_{tot}.

After *SS* and *df* are recorded, the summary table looks like this:

Summary Table of One-Way ANOVA

Source	*Sum of squares*	*df*	*Mean square*	*F*
Between	63.33	2	MS_{bn}	F_{obt}
Within	84.00	12	MS_{wn}	
Total	147.33	14		

Now find each mean square:

Computing the mean squares Work directly from the summary table to compute the mean squares. Any mean square equals the appropriate sum of squares divided by the corresponding *df*.

THE COMPUTATIONAL FORMULA FOR MS_{bn} IS

$$MS_{\text{bn}} = \frac{SS_{\text{bn}}}{df_{\text{bn}}}$$

From the summary table for our example,

$$MS_{\text{bn}} = \frac{63.33}{2} = 31.67$$

so MS_{bn} is 31.67.

THE COMPUTATIONAL FORMULA FOR MS_{wn} IS

$$MS_{\text{wn}} = \frac{SS_{\text{wn}}}{df_{\text{wn}}}$$

For the example,

$$MS_{\text{wn}} = \frac{84}{12} = 7.00$$

so, MS_{wn} is 7.00.

Do *not* compute the mean square for SS_{tot}.

Placing these values in the summary table, we have

Summary Table of One-Way ANOVA

Source	*Sum of squares*	*df*	*Mean square*	*F*
Between	63.33	2	31.67	F_{obt}
Within	84.00	12	7.00	
Total	147.33	14		

Computing the F_{obt} Last but not least, compute F_{obt}.

THE COMPUTATIONAL FORMULA FOR F IS

$$F_{obt} = \frac{MS_{bn}}{MS_{wn}}$$

In the example, MS_{bn} is 31.67 and MS_{wn} is 7.0, so

$$F_{obt} = \frac{MS_{bn}}{MS_{wn}} = \frac{31.67}{7.00} = 4.52$$

Thus F_{obt} is 4.52.

Now we have the completed ANOVA summary table:

Summary Table of One-Way ANOVA

Source	*Sum of squares*	*df*	*Mean square*	*F*
Between	63.33	2	31.67	4.52
Within	84.00	12	7.00	
Total	147.33	14		

Interpreting F_{obt} in a One-Way ANOVA

To interpret F_{obt}, we must have F_{crit}, so we turn to the F-tables in Appendix C. In the example, df_{bn} is 2 and df_{wn} is 12. With $\alpha = .05$, F_{crit} is 3.88.

Thus, F_{obt} is 4.52 and F_{crit} is 3.88. Lo and behold, as shown in Figure 17.5, we have a significant F_{obt}. The null hypothesis says that the differences between the means of our levels are due to sampling error and that all means poorly represent one μ. However, F_{obt} is out there in the region of rejection, telling us that such differences between $\overline{X}$s

FIGURE 17.5 Sampling Distribution of F When H_0 Is True for $df_{bn} = 2$ and $df_{wn} = 12$

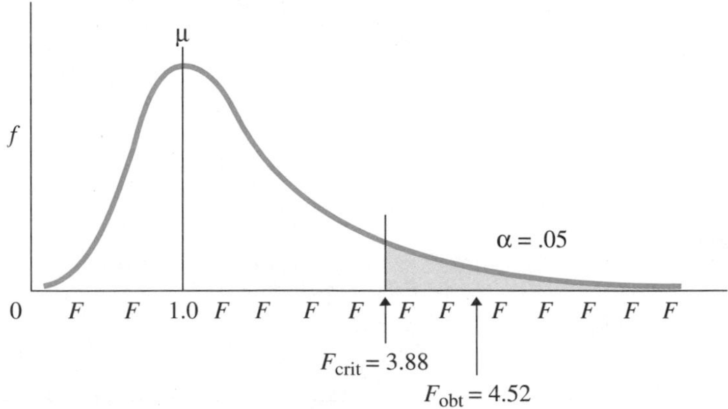

hardly ever happen when H_0 is true. Because F_{obt} is larger than F_{crit}, we reject H_0: We conclude that the differences between these level means are too unlikely for us to accept that the means actually represent one population μ. That is, we conclude that the F_{obt} is significant and that the factor of perceived difficulty produced a significant difference in mean performance scores. We report these results as

$$F(2, 12) = 4.52, p < .05$$

In the parentheses, we report df_{bn} and then df_{wn}. (Get in the habit of saying the df_{bn} first and the df_{wn} second.) As usual, because $\alpha = .05$, the probability that we made a Type I error is $p < .05$.

Of course, had F_{obt} been less than F_{crit}, then the corresponding differences between the level means would *not* be too unlikely to accept as representing no difference in the population, so we would not reject H_0.

Because we rejected H_0 and accepted H_a, we return to the means of the levels of the factor:

Perceived difficulty		
Easy	*Medium*	*Difficult*
$\overline{X}_1 = 8$	$\overline{X}_2 = 6$	$\overline{X}_3 = 3$

We are confident that these means represent a relationship in the population: Changing the level of perceived difficulty would produce different populations of performance scores, having different μs. But we do not know whether *every* sample mean represents a different μ. A significant F_{obt} merely indicates that there is *at least* one significant difference somewhere between the means of a factor. Now we must determine which specific means differ significantly, and to do that we perform *post hoc* comparisons.

PERFORMING *POST HOC* COMPARISONS

There are several versions of *post hoc* tests (each named after its developer). Different procedures differ in how likely they are to produce Type I or Type II errors. For example, one procedure—the Scheffe Test—is very conservative, meaning it is biased toward avoiding Type I errors, even at the risk of making Type II errors. Other common procedures—the Newman-Keuls and Duncan tests—are rather liberal, meaning they are biased toward avoiding Type II errors, even at the risk of making Type I errors. Two in-between procedures we'll discuss that give good protection from both types of errors are Fisher's protected *t*-test and Tukey's *HSD* test.[1] Which test you should use depends on whether or not your *n*s are equal.

Fisher's Protected *t*-Test

Perform Fisher's protected *t*-test when the *n*s in all levels of the factor are *not* equal.

[1] Carmer, S. G., and Swanson, M. R. (1973). An evaluation of ten multiple comparison procedures by Monte Carlo methods. *Journal of the American Statistical Association*, 68, 66–74.

THE COMPUTATIONAL FORMULA FOR THE PROTECTED t-TEST IS

$$t_{obt} = \frac{\overline{X}_1 - \overline{X}_2}{\sqrt{MS_{wn}\left(\frac{1}{n_1} + \frac{1}{n_2}\right)}}$$

This is basically the formula for the independent samples t-test, except that MS_{wn} has replaced the pooled variance, s^2_{pool}, used in the t-test. We are testing H_0: $\mu_1 - \mu_2 = 0$, where $\overline{X}_1$ and $\overline{X}_2$ are the means for any two levels of the factor, and n_1 and n_2 are the corresponding ns in those levels. The t_{crit} is the two-tailed value found in Appendix C, Table 2, for $df = df_{wn} = N - k$.

It is not incorrect to perform the protected t-test even when all ns are equal (it's just time-consuming). Thus, for example, we can compare the mean from our easy level ($\overline{X} = 8.0$) to the mean from the difficult level ($\overline{X} = 3.0$). Each n is 5, and from the ANOVA, MS_{wn} is 7.0. Filling in the formula gives

$$t_{obt} = \frac{8.0 - 3.0}{\sqrt{7.0\left(\frac{1}{5} + \frac{1}{5}\right)}}$$

Then

$$t_{obt} = \frac{+5.0}{\sqrt{7.0(.4)}}$$

which becomes

$$t_{obt} = \frac{+5.0}{\sqrt{2.8}} = \frac{+5.0}{1.67} = +2.99$$

Now compare t_{obt} to the two-tailed value of t_{crit} found in the t-tables. For the example, with $\alpha = .05$ and $df_{wn} = 12$, t_{crit} is ± 2.179. Because the t_{obt} of $+2.99$ is beyond the t_{crit} of ± 2.179, the means from the easy and difficult levels differ significantly (they do not represent the same μ).

To complete the *post hoc* comparisons, perform the protected t-test on all possible pairs of means in the factor. Thus, after comparing the means from easy and difficult, perform the protected t-test comparing the means from easy and medium, and the means from medium and difficult. When you're finished, the experiment-wise error rate will be *protected*, so that the probability of a Type I error for all of these comparisons together is $p < .05$.

If a factor contains many levels, then the protected t-test becomes very tedious. If you're thinking there *must* be an easier way, you're right.

Tukey's *HSD* Multiple Comparisons Test

Perform the Tukey *HSD* multiple comparisons procedure *only* when the ns in all levels of the factor are equal. The *HSD* is a convoluted variation of the t-test in which we compute the minimum difference between any two means that is required for the means to differ significantly (*HSD* stands for the Honestly Significant Difference). There are four steps in performing the *HSD* test.

Step 1: Find q_k. Using the appropriate q_k in the computations is what protects the experiment-wise error rate for the number of means being compared. The value of q_k is found in Table 6 in Appendix C, entitled "Values of the Studentized Range Statistic, q_k." In the table, locate the column labeled with the k that corresponds to the number of means in the factor. Next, find the row labeled with the df_{wn} used when computing F_{obt}. Then find the value of q_k for the appropriate α. For our study above, $k = 3$, $df_{wn} = 12$, and α. = .05, so $q_k = 3.77$.

Step 2: Compute the *HSD*.

THE COMPUTATIONAL FORMULA FOR THE HSD *IS*

$$HSD = (q_k)\left(\sqrt{\frac{MS_{wn}}{n}}\right)$$

MS_{wn} is the denominator from your significant F-ratio, and n is the number of scores in each level of the factor.

In our example, MS_{wn} was 7.0 and n was 5, so

$$HSD = (q_k)\left(\sqrt{\frac{MS_{wn}}{n}}\right) = (3.77)\left(\sqrt{\frac{7.0}{5}}\right) = 4.46$$

Thus, *HSD* is 4.46.

Step 3: Determine the differences between all means. Simply subtract each mean from every other mean. Ignore whether differences are positive or negative (this is a two-tailed test of the H_0 that $\mu_1 - \mu_2 = 0$). The differences for the perceived difficulty study can be diagrammed as shown below:

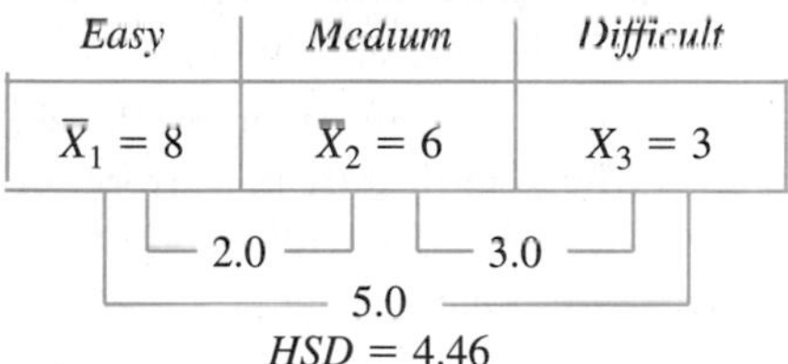

On the line connecting any two levels is the absolute difference between their means.

Step 4: Compare each difference between two means to the *HSD*. If the absolute difference between the two means is *greater than* the *HSD*, then these means differ significantly. (It is as if you performed the protected t-test on these two means and t_{obt} was significant.) If the absolute difference between the two means is less than or equal to the *HSD*, then it is *not* a significant difference (and would not produce a significant t_{obt}).

Above, the *HSD* was 4.46. The means from the easy level ($\overline{X}_1 = 8$) and the difficult level ($\overline{X}_3 = 3$) differ by more than 4.46, so they differ significantly. The mean from the medium level ($\overline{X}_2 = 6$) differs from the other means by less than 4.46, however, so it does not differ significantly from them.

Thus, our final conclusion about this study is that we demonstrated a relationship between participants' scores and perceived difficulty, but only for the easy and difficult conditions. If these two conditions were given to the population, we would expect to find two different populations of scores, having two different μs. We cannot say anything about whether the medium level would produce a different population and μ, because we failed to find that it produced a significant difference. Nonetheless, as usual, we begin to interpret the results in psychological terms, explaining why this manipulation worked as it did, and what underlying constructs, models or behaviors are involved.

> *REMEMBER* You cannot draw any conclusions about the specific influence of your independent variable until you have performed *post hoc* comparisons.

SUMMARY OF THE STEPS IN PERFORMING A ONE-WAY ANOVA

It has been a long haul, but here is everything involved when performing a one-way ANOVA:

1. The null hypothesis is H_0: $\mu_1 = \mu_2 = \ldots \mu_k$, and the alternative hypothesis is H_a: not all μs are equal. Choose α, check the assumptions, and collect the data.
2. First compute the sum of squares between groups, SS_{bn}, and the sum of squares within groups, SS_{wn}. Then compute the degrees of freedom between groups, df_{bn}, and the degrees of freedom within groups, df_{wn}. Dividing SS_{bn} by df_{bn} gives the mean square between groups, MS_{bn}, and dividing SS_{wn} by df_{wn} gives the mean square within groups, MS_{wn}. Finally, dividing MS_{bn} by MS_{wn} gives F_{obt}.
3. Find F_{crit} in Appendix C, Table 5, using the df_{bn} and the df_{wn}. If the null hypothesis is true, F_{obt} "should" equal 1. The larger the value of F_{obt}, the less likely it is that H_0 is true. If F_{obt} is larger than F_{crit}, then F_{obt} is significant, indicating that the means in at least two conditions differ significantly, representing at least two different population μs.
4. If F_{obt} is significant and there are more than two levels of the factor, determine which specific levels differ significantly by performing *post hoc* comparisons. Perform the protected *t*-test if the *n*s in all levels of the factor are not equal or the *HSD* procedure if all *n*s are equal.

If you followed all of that, then congratulations, you're getting *good* at this stuff. Of course, all of this merely determines whether you have a relationship. Now you must describe that relationship.

DESCRIBING THE RELATIONSHIP IN A ONE-WAY ANOVA

As in previous chapters, you are not finished once you have demonstrated a significant relationship. Ultimately, you must understand the relationship, and to help, you should describe the relationship by: (1) Computing a confidence interval for each μ; (2) By graphing the relationship; and (3) By computing the effect size.

The Confidence Interval for Each Population μ

In our example, the mean from the easy condition ($\overline{X} = 8.0$) differs significantly from the mean from the difficult condition ($\overline{X} = 3.0$). Therefore, we expect that the population means represented by these conditions would be "around" 8 and 3, respectively. As usual, to more clearly define "around," we can compute a confidence interval for the μ represented by each sample mean. This confidence interval is the same as the one discussed in Chapter 14, except that here it is computed using the components of ANOVA.

THE COMPUTATIONAL FORMULA FOR COMPUTING THE CONFIDENCE INTERVAL FOR A SINGLE μ, USING THE RESULTS OF A BETWEEN-SUBJECTS ANOVA, IS

$$\left(\sqrt{\frac{MS_{\text{wn}}}{n}}\right)(-t_{\text{crit}}) + \overline{X} \leq \mu \leq \left(\sqrt{\frac{MS_{\text{wn}}}{n}}\right)(+t_{\text{crit}}) + \overline{X}$$

The value of t_{crit} is the two-tailed value found in the t-tables, using the appropriate α and using our df_{wn} as the df. We find MS_{wn} in our ANOVA, and $\overline{X}$ and n are from the level we are describing.

For example, in our easy condition, $\overline{X} = 8.0$, we have $MS_{\text{wn}} = 7.0$, $df_{\text{wn}} = 12$, and $n = 5$. The two-tailed t_{crit} (at $df = 12$ and $\alpha = .05$) is ± 2.179. Placing these values in the above formula, we have

$$\left(\sqrt{\frac{7.0}{5}}\right)(-2.179) + 8.0 \leq \mu \leq \left(\sqrt{\frac{7.0}{5}}\right)(+2.179) + 8.0$$

This becomes

$$(-2.578) + 8.0 \leq \mu \leq (+2.578) + 8.0$$

Finally we have

$$5.42 \leq \mu \leq 10.58$$

Because $\alpha = .05$, this is the 95% confidence interval: If we were to test the entire population of subjects under our easy condition, we are 95% confident that the population mean would fall between 5.42 and 10.58.

Follow the same procedure to describe the μ from any other significant level of the factor.

Graphing the Results in ANOVA

As usual, we graph results by placing the mean dependent score for each condition on the Y axis and the conditions of the independent variable (the levels of the factor) on the X axis. Figure 17.6 shows the line graph for the perceived difficulty study. Notice that

FIGURE 17.6 Mean Number of Problems Correctly Solved as a Function of Perceived Difficulty

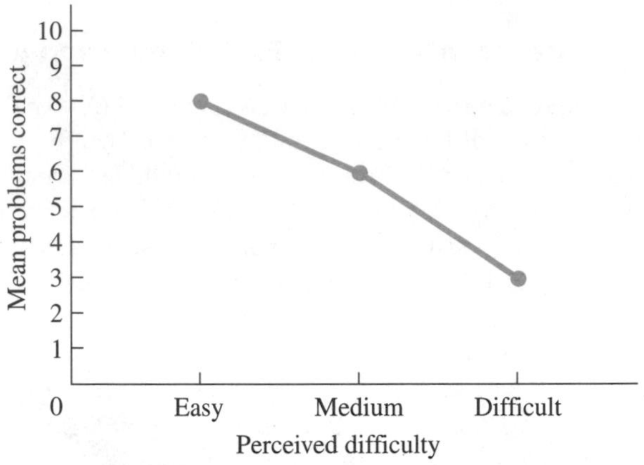

even though the medium level of difficulty did not produce significant differences, we still include it in the graph.

Eta Squared: The Effect Size in the Sample Data

Remember that so far, by only saying that the relationship is significant, it's as if you're saying "I've found a correlation, but I'm not telling what it is." Therefore, first think "correlation coefficient" to describe the strength of the relationship between your independent and dependent variables. However, here you compute a new correlation coefficient called eta (pronounced "ay-tah"). **Eta** is analogous to r_{pb}, except that eta can be used to describe any linear or nonlinear relationship containing two or more levels of a factor.

But remember that we get to the heart of describing a relationship by thinking "squared correlation coefficient." This tells us the *effect size* of the independent variable—the proportion of variance in dependent scores that is associated with changing the conditions. With ANOVA, effect size is computed by squaring the value of eta: **eta squared** indicates the proportion of variance in the dependent variable that is accounted for by changing the levels of a factor. The symbol for eta squared is η^2, and it is calculated as follows:

THE COMPUTATIONAL FORMULA FOR η^2 IS

$$\eta^2 = \frac{SS_{bn}}{SS_{tot}}$$

The SS_{bn} reflects the differences between the means from the various levels of a factor. The SS_{tot} reflects the total differences between all scores in the experiment. Thus, η^2 reflects the proportion of the total differences in the scores that is associated with differences between the sample means. This is interpreted in the same way that we previously interpreted r^2_{pb}. Thus, η^2 indicates how accurately the differences in scores can be predicted using the means of the levels of the factor. The larger the value of η^2, the more consistently the factor "caused" participants to have a particular dependent score in a particular condition, and thus the more scientifically important the factor is for explaining and predicting differences in participants' underlying behavior.

For example, in the perceived difficulty study, SS_{bn} was 63.33 and SS_{tot} was 147.33, so

$$\eta^2 = \frac{SS_{bn}}{SS_{tot}} = \frac{63.33}{147.33} = .43$$

Thus, we are .43, or 43%, more accurate at predicting participants' scores when we predict for them the mean for the particular difficulty level they were tested under, rather than using the overall mean of the study. In other words, 43% of all the variance in these scores is accounted for, or explained, as being the result of changing the levels of perceived difficulty. Because 43% is a substantial amount, the variable of perceived difficulty is relatively important in determining participants' subsequent performance.

η^2 can be used with either equal or unequal *ns*. η^2 is a *descriptive* statistic, however, and only describes the effect size in sample data. Usually this is adequate, but other procedures are needed to accurately estimate the effect size in the population.

Power and the ANOVA

We always want to maximize *power*, the probability of rejecting H_0 when H_0 is false. Here we do so by trying to maximize the size of F_{obt}. Looking at the *F*-ratio,

$$F_{obt} = \frac{MS_{bn}}{MS_{wn}}$$

you'll see that anything that increases the size of the numerator or decreases the size of the denominator will produce a larger F_{obt}. Therefore, as in previous chapters, you should seek to create a powerful design so that: (1) Through a strong manipulation, you maximize the size of the differences between the means of the conditions, thus increasing the size of MS_{bn}; (2) Through control, you minimize the variability—error variance—of scores within conditions, thus reducing the size of MS_{wn}; and (3) you maximize the *ns*, thus increasing df_{wn} and in turn minimizing MS_{wn}. (A larger df_{wn} also results in a smaller value of F_{crit}.) Any of these will increase the probability that F_{obt} is significant, producing greater power to reject H_0 when it is false. The same considerations also increase the power of *post hoc* comparisons to detect a difference between each pair of sample means.

In addition, all of the other issues of power from previous chapters apply to experiments involving ANOVA, except that now you simply have more levels. Thus, on the one hand, you must be careful when randomizing or counterbalancing variables because you may increase error variance and thus decrease power. On the other hand, such techniques can increase internal and external validity. Likewise, as with *t*-tests, the between-subjects ANOVA is less powerful than a comparable within-subjects ANOVA.

PUTTING IT ALL TOGETHER

When all is said and done, the *F*-ratio is a convoluted way of measuring the differences between the means of our conditions and then fitting those differences to a sampling distribution. The larger the F_{obt}, the less likely that the differences between the means are the result of sampling error. A significant F_{obt} indicates that the means are unlikely to represent one population mean. Once you determine that F_{obt} is significant, then determine which of the level means actually differ significantly and describe the relationship. That's all there is to it.

CHAPTER SUMMARY

1. Analysis of variance, or *ANOVA*, has its own vocabulary. Here are the general terms we've used previously, along with the corresponding ANOVA terms:

General term	=	*ANOVA term*
independent variable	=	factor
condition	=	level
sum of squared deviations	=	sum of squares (*SS*)
variance (s^2_X)	=	mean square (*MS*)
effect of independent variable	=	treatment effect

2. Researchers study three or more conditions of an independent variable to test their hypothesis adequately, to demonstrate a nonlinear relationship, and to obtain the maximum information from a study.

3. The *one-way* analysis of variance tests for significant differences between the means from two or more levels of a factor. In a *between-subjects factor*, each independent sample is tested under only one level. In a *within-subjects* factor, dependent samples are present in all conditions (created either by matching or by repeated measures).

4. The *experiment-wise error rate* is the probability that a Type I error will occur in an experiment. ANOVA is used instead of multiple *t*-tests because with ANOVA the experiment-wise error rate will equal α.

5. The *assumptions of the one-way, between-subjects ANOVA* are that (a) the scores in each condition are independent random samples, (b) each sample represents a normally distributed population of interval or ratio scores, and (c) all populations represented in the study have homogeneous (equal) variance.

6. ANOVA tests two-tailed hypotheses. H_0 indicates that the mean of each condition represents the same population mean. H_a indicates that *not* all μs are equal.

7. The *mean square within groups*, MS_{wn}, is an estimate of *error variance*, the inherent variability among scores *within* each population. The *mean square between groups*, MS_{bn}, is an estimate of the error variance plus the treatment variance. *Treatment variance* reflects differences in scores between the populations represented by the levels of the factor.

8. F_{obt} is computed from the *F*-ratio, which equals the mean square between groups divided by the mean square within groups.

9. F_{obt} may be greater than 1 because either: (a) There is no treatment effect, but the sample means are not perfectly representative of this; or (b) Two or more sample means represent different population means.

10. The *F*-distribution is the sampling distribution of all possible values of F_{obt} when H_0 is true.

11. The larger the value of F_{obt}, the less likely it is that the means from all levels represent one population mean. If the F_{obt} is significant, the differences between level means are too unlikely to accept as representing no difference in the population.

12. All of the levels in a factor may not produce significant differences. Therefore, if (a) F_{obt} is significant, and (b) there are more than two levels of the factor, then *post hoc comparisons* determine which specific means differ significantly. When the *ns* of all conditions are *not* equal, perform Fisher's protected *t*-test on each pair of means. If all *ns* are equal, perform *Tukey's HSD test*.

13. When a significant relationship exists between the factor and the dependent scores, *eta squared*, η^2, describes the *effect size*—the proportion of variance accounted for by the factor in the sample data.

14. The *power* of ANOVA increases with increased differences between the conditions of the independent variable, decreased variability of scores within each condition, or increased *n*.

PRACTICE PROBLEMS

(Answers for odd-numbered problems are provided in Appendix D.)

1. What does each of the following terms mean? (a) ANOVA (b) One-way design (c) Factor (d) Level (e) Between-subjects factor (f) Within-subjects factor

2. A researcher conducts an experiment in which scores are measured under two conditions of an independent variable. (a) How will the researcher know whether to perform a parametric or nonparametric statistical procedure? (b) Which parametric procedures are available to her? (c) If the researcher conducts an experiment with three levels of the independent variable, which two versions of a parametric procedure are available to her? (d) How can she select between the choices in (c)?

3. What are the reasons for conducting a study with three or more levels of a factor?

4. (a) What are error variance and treatment variance? (b) What are the two types of mean squares, and what does each one estimate?

5. (a) What is the experiment-wise error rate? (b) Why does performing ANOVA solve the problem of experiment-wise error rates created by performing multiple *t*-tests?

6. Summarize the steps involved in analyzing a multi-level experiment.

7. (a) In a study comparing the effects of four conditions of the independent variable, what is H_0? (b) What is H_a in the same study? (c) Describe in words what H_0 and H_a say for the study?

8. (a) Why should F_{obt} equal 1 if the data represent the H_0 situation? (b) Why is F_{obt} greater than 1 when the data represent the H_a situation? (c) What does a significant F_{obt} indicate about differences between the levels of a factor?

9. (a) Poindexter computes an F_{obt} of .63. How should this be interpreted? (b) He computes another F_{obt} of -1.7. How should this be interpreted?

10. Foofy obtained a significant F_{obt} from an experiment with five levels. She automatically concludes that she has demonstrated a relationship in which changing each condition of the independent variable results in a significant change in the dependent variable. (a) Is she correct? Why or why not? (b) What must she do now?

11. (a) When is it necessary to perform *post hoc* comparisons? Why? (b) When is it unnecessary to perform *post hoc* comparisons? Why?

12. When do you use each of the two types of *post hoc* tests discussed in this chapter?

13. What does η^2 tell you?

14. (a) In the perceived difficulty study discussed in this chapter, how can you maximize the power of the ANOVA? (b) How do these strategies affect the size of F_{obt} and thus increase power? (c) How do these strategies affect *post hoc* tests?

15. Here are data from an experiment studying the effect of age on creativity scores:

Age 4	*Age 6*	*Age 8*	*Age 10*
3	9	9	7
5	11	12	7
7	14	9	6
4	10	8	4
3	10	9	5

(a) Compute F_{obt} and create an ANOVA summary table. (b) With $\alpha = .05$, what do you conclude about F_{obt}? (c) Perform the appropriate *post hoc* comparisons. (d) What should you conclude about this relationship? (e) Statistically, how important is the relationship in this study? (f) Describe how you would graph these results.

16. In a study where $k = 3$, $n = 16$, $\overline{X}_1 = 45.3$, $\overline{X}_2 = 16.9$, and $\overline{X}_3 = 8.2$, you compute the following sums of squares.

Source	*Sum of squares*	*df*	*Mean square*	*F*
Between	147.32	___	___	___
Within	862.99	___	___	
Total	1010.31	___		

(a) Complete the ANOVA summary table. (b) With $\alpha = .05$, what do you conclude about F_{obt}? (c) Perform the appropriate *post hoc* comparisons. What do you conclude about this relationship? (d) What is the effect size in this study, and what does this tell you about the influence of the independent variable?

17. A researcher investigated the number of viral infections people contract as a function of the amount of stress they experienced during a six-month period. She obtained the following data:

Amount of Stress

Negligible stress	*Minimal stress*	*Moderate stress*	*Severe stress*
2	4	6	5
1	3	5	7
4	2	7	8
1	3	5	4

(a) What are H_0 and H_a? (b) Compute F_{obt} and complete the ANOVA summary table. (c) At $\alpha = .05$, what is F_{crit}? (d) Report your statistical results (e) Perform the appropriate post-hoc comparisons. (f) What can you conclude about this study? (g) Describe the effect size and interpret it. (i) Estimate the values of μ that are likely to be found in the severe stress condition.

18. A researcher investigated the effect of volume of background noise on participants' error rates while performing a boring task. He tested three groups of randomly selected students and obtained the following error data and sums of squares:

	low volume	*moderate volume*	*high volume*
$\bar{X}$	61.5	65.5	48.25
n	4	5	7

Summary Table of One-Way ANOVA

Source	*Sum of squares*	*df*	*Mean square*	*F*
Between	652.16			
Within	612.75			
Total	1264.92			

(a) Complete the ANOVA. (b) At $\alpha = .05$, what is F_{crit}? (c) Report the statistical results in the proper format. (d) Perform the appropriate *post hoc* tests. (e) What can you conclude about this study? (f) What other procedures should you perform?

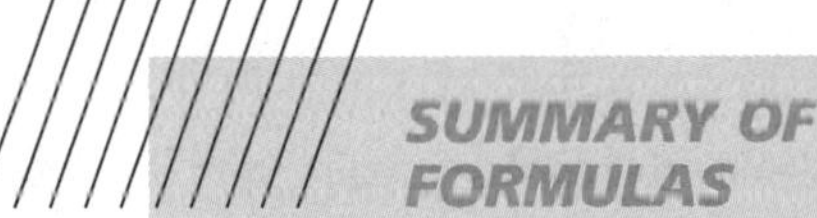

SUMMARY OF FORMULAS

1. *The format for the summary table for a one-way ANOVA is as follows:*

Summary Table of One-Way ANOVA

Source	*Sum of squares*	*df*	*Mean square*	*F*
Between	SS_{bn}	df_{bn}	MS_{bn}	F_{obt}
Within	SS_{wn}	df_{wn}	MS_{wn}	
Total	SS_{tot}	df_{tot}		

2. Computing the sum of squares

***(a)** The computational formula for SS_{tot} is*

$$SS_{tot} = \Sigma X^2_{tot} - \left(\frac{(\Sigma X_{tot})^2}{N}\right)$$

All scores in the experiment are included, and N is the total number of scores.

***(b)** The computational formula for SS_{bn} is*

$$SS_{bn} = \Sigma\left(\frac{(\text{sum of scores in the column})^2}{n \text{ of scores in the column}}\right) - \left(\frac{(\Sigma X_{tot})^2}{N}\right)$$

where each column contains the scores from one level of the factor.

***(c)** The computational formula for SS_{wn} is*

$$SS_{wn} = SS_{tot} - SS_{bn}$$

3. Computing the mean squares

***(a)** The computational formula for MS_{bn} is*

$$MS_{bn} = \frac{SS_{bn}}{df_{bn}}$$

with $df_{bn} = k - 1$, where k is the number of levels in the factor.

***(b)** The computational formula for MS_{wn} is*

$$MS_{wn} = \frac{SS_{wn}}{df_{wn}}$$

with $df_{wn} = N - k$, where N is the total N of the study and k is the number of levels in the factor.

4. *The computational formula for the* F*-ratio is*

$$F_{obt} = \frac{MS_{bn}}{MS_{wn}}$$

Critical values of F are found in Table 5 in Appendix C for df_{bn} and df_{wn}.

5. When F_{obt} is significant and k is greater than 2, *post hoc* comparisons must be performed.

***(a)** The computational formula for the protected* t*-test is*

$$t_{obt} = \frac{\overline{X}_1 - \overline{X}_2}{\sqrt{MS_{wn}\left(\frac{1}{n_1} + \frac{1}{n_2}\right)}}$$

Values of t_{crit} are the two-tailed values found in the t-tables for $df = df_{wn} = N - k$.

***(b)** When the ns is all levels are equal, the computational formula for the HSD is*

$$HSD = (q_k)\left(\sqrt{\frac{MS_{wn}}{n}}\right)$$

Values of q_k are found in Table 6 of Appendix C for df_{wn} and k, where k equals the number of levels of the factor.

6. *The computational formula for the confidence interval for a single μ using the results of a between-subjects ANOVA is*

$$\left(\sqrt{\frac{MS_{wn}}{n}}\right)(-t_{crit}) + \overline{X} \leq \mu \leq \left(\sqrt{\frac{MS_{wn}}{n}}\right)(+t_{crit}) + \overline{X}$$

$\overline{X}$ and n are from the level being described, and t_{crit} is the two-tailed value of t_{crit} at the appropriate α for df_{wn}.

7. *The computational formula for eta squared is*

$$\eta^2 = \frac{SS_{bn}}{SS_{tot}}$$

18

The Two-Way Between-Subjects Experiment and the Two-Way Analysis of Variance

To understand this chapter:

- From Chapter 17, understand the terms *factor* and *level*, what a significant F indicates, when to perform *post hoc* tests, and what η^2 indicates.

Then your goals in this chapter are to learn:

- What a two-way factorial design is and why researchers conduct them.
- How to collapse across a factor and what a significant main effect indicates.
- How to calculate the cell means and what a significant interaction indicates.
- How to compute the Fs in a two-way ANOVA, perform *post hoc* tests, compute η^2, and draw the graph for main and interaction effects.
- How to interpret the results of a two-way experiment.

In the previous chapter, we discussed experiments involving one factor. Researchers often create even larger studies, however, testing the influence of multiple factors. An experiment that contains more than one factor requires a multi-factor ANOVA. To introduce you to multi-factor experiments and multi-factor ANOVA, this chapter discusses experiments with two factors, which are analyzed using a two-way, between-subjects ANOVA. This is like the ANOVA of the previous chapter, except that we compute several values of F. Therefore, be forewarned that the procedure is rather involved (although it is more tedious than it is difficult).

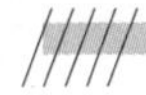

MORE STATISTICAL NOTATION

As with all experiments, the purpose of a two-factor experiment is to determine whether there is a relationship between the independent variable and the dependent variable. The only difference from previous experiments is that now we have two independent variables or factors. Each factor may contain any number of levels, so we have a code for describing each specific design. The generic format is to call one independent variable factor A and the other independent variable factor B. To describe a particular design, we use the number of levels in each factor. If, for example, factor A has two levels and factor B has two levels, we have a two-by-two design or a "two-by-two ANOVA," which is written as 2×2. If one factor has four levels and the other factor has three levels, we have a 4×3 ANOVA, and so on.

To understand the layout of a two-way design, say that for the moment we are again interested in the effects of a "smart pill" on a person's IQ. We'll call the number of smart pills given to participants factor A, and test two levels, 1 or 2 pills, after which we'll measure participants' IQ scores. The basic design of this factor is shown in Table 18.1. Each column in the diagram represents a level of factor A and, within a column, the *X*s represent IQ scores. Averaging the scores in each column vertically yields the mean IQ for each pill level, showing the effect of factor A: How IQ changes as a function of increasing the dosage.

Say that from a totally different perspective, we're also interested in the relationship between age and IQ. We'll call age factor B, and test two levels, 10- and 20-year-olds. We can envision this design as illustrated in Table 18.2. The only novelty here is that the two conditions are arranged horizontally, so that each *row* represents a different level of the factor. Here, averaging the scores in each row horizontally yields the mean IQ for each age level, showing the effect of factor B: How IQ changes as a function of increasing age.

To create a two-way design, we would simultaneously manipulate both the participants' age and the number of pills they receive. A good way to visualize a two-way design is shown in Table 18.3. Each column is still a level of factor A, number of pills. Each row is still a level of factor B, age. But now we have a new term: Each small

TABLE 18.1 Diagram of Factor of Number of Smart Pills

Each X represents a participant's IQ score, and each $\overline{X}$ is the mean for a level of factor A.

Factor A: Number of Pills	
Level A_1: One Pill	*Level A_2: Two Pills*
X	X
X	X
X	X
X	X
X	X
X	X
$\overline{X}$	$\overline{X}$

TABLE 18.2 Diagram of Factor of Age

Each row represents a level of age. Each X represents a participant's IQ score, and each $\overline{X}$ is the mean for a level of factor B.

Factor B: Age	*Level B_1: 10-year-olds*	X X X X X X	$\overline{X}$
	Level B_2: 20-year-olds	X X X X X X	$\overline{X}$

square produced by a particular combination of a level of factor A with a level of factor B is called a **cell**. In this design there are four cells, each containing a sample of participants who are one age and are given one amount of the pill. For example, the highlighted cell contains the scores of a sample of 20-year-olds who receive one pill.

In many computer programs, you must identify each cell using the levels of the two factors. Above, the levels of factor A are A_1 and A_2 and the levels of factor B are B_1 and B_2. Then, for example, the cell formed by combining level 1 of factor A and level 1 of factor B is cell A_1B_1. Likewise, you identify the mean and *n* from each cell in the same way, so, for example, in cell A_1B_1 is $\overline{X}_{A_1B_1}$.

One final consideration: Combining all levels of one factor with all levels of the other factor produces a **complete factorial design**. The design in Table 18.3 is a complete factorial, because all of our levels of drug dose are combined with all of our age levels. On the other hand, in an **incomplete factorial design**, all levels of the two factors are not combined. For example, if for some reason we did not collect data for 20-year-olds given one smart pill, we would have an incomplete factorial design. Incomplete factorial designs require elaborate procedures not discussed here.

TABLE 18.3 Two-Way Design for Studying the Factors of Number of Smart Pills and Subjects' Age

		Factor A: number of pills		
		Level A_1: 1 pill	*Level A_2: 2 pills*	
Factor B: Age	*Level B_1: 10-year-olds*	X X X $\overline{X}_{A_1B_1}$	X X X $\overline{X}_{A_2B_1}$	← scores
	Level B_2: 20-year-olds	X X X $\overline{X}_{A_1B_2}$ (one of the four cells)	X X X $\overline{X}_{A_2B_2}$	← scores

THE REASON FOR MULTIFACTOR STUDIES

Why study two factors in one experiment? After all, we could perform two separate studies, one testing the effect of factor A and one testing the effect of factor B. The answer is that a multifactor design has three major advantages over a single-factor design.

First, in a multi-factor design we can learn everything about the influence of each factor that we would learn if it were the only independent variable. But we can also study something with a multi-factor design that would otherwise be missed—the *interaction effect*. Later we'll discuss interaction effects in detail, but for now, an interaction is the influence of combining the levels from the two factors: It shows how the effect of one independent variable on the dependent variable depends on the level of the other independent variable that is present. Thus, in the previous example, the interaction would indicate the influence of a particular dose of smart pills depending on the age of participants. Interactions are important because, after all, in most natural settings there are varying amounts of many variables present that combine to influence a behavior. By manipulating more than one independent variable in an experiment, we can examine the influence of such combined variables on a behavior. Thus, the primary reason for conducting multi-factor studies is to observe the interaction effect.

A second reason for multi-factor experiments is that often, once we have created a design for studying one independent variable, only a minimum of additional effort is required to study additional factors. Thus, multifactor studies can be an efficient and cost-effective way of determining the effects of—and interactions among—several independent variables.

Finally, multi-factor experiments often are produced because a researcher sets out to study only one variable, but additional factors are created through counterbalancing extraneous variables. For example, let's say that we had set out to only study the influence of one versus two smart pills. But, to prevent confounding, we counterbalanced subject's age, with half 10-year olds and half 20-year-olds in each condition. Instead of collapsing across and ignoring age as we've done in previous chapters, we could treat age as a second factor to create the above 2×2 design, and then examine the influence of age, as well as its interaction with the pill factor. Thus, by *analyzing* the study as a two-way design, you obtain much more information from an experiment that, regardless, you are going to conduct in the same way. Therefore, always consider whether you should analyze any single-factor study as a multi-factor design involving counterbalanced variables.

OVERVIEW OF THE TWO-WAY ANOVA

As usual, regardless of whether we're talking about each factor separately or their interaction, we want to conclude that, if we tested the entire population under the various conditions, the sample means represent the different populations of scores we'd find, located at different μs. But there is the usual problem: Differences between the sample means may simply reflect sampling error, so we might actually find the same population, having the same μ for all conditions. Therefore, once again we must eliminate the idea that the differences between the sample means merely reflect sampling error. To do this, we perform ANOVA. As usual, we first set alpha, usually $\alpha = .05$, and then check the assumptions.

Assumptions of the Two-Way Between-Subjects ANOVA

You perform the two-way, between-subjects ANOVA when you have a complete factorial design and you can assume that:

1. All cells contain independent samples of participants.
2. The dependent variable measures interval or ratio scores.
3. The populations represented by the data are approximately normally distributed.
4. The represented populations all have homogeneous variance.

Logic of the Two-Way ANOVA

Here's a semi-interesting idea for a study. Have you ever noticed that television commercials are much louder than the programs themselves? Advertisers seem to believe that increased volume creates increased viewer attention and so makes the commercial more persuasive. To test whether louder messages are more persuasive, we'll play a recording of an advertising message to participants at each of three volumes. (Volume is measured in decibels, but to simplify things we'll refer to the three levels of volume as soft, medium, and loud.)

Say that we're also interested in the differences between how males and females are persuaded, so we have another factor: The gender of the listener. If, in one study, we examine both the volume of the message and the gender of participants hearing the message, we have a two-factor experiment involving three levels of volume and two levels of gender. We'll test all conditions with independent samples, so we have a 3×2 between-subjects, factorial ANOVA. The dependent variable indicates how persuasive a person believes the message to be on a scale of 0 (not at all) to 25 (totally convincing). We collect the scores and organize them as in Table 18.4. For simplicity, we have a distinctly unpowerful N: nine men and nine women were randomly selected, and

TABLE 18.4 A 3×2 ANOVA for the Factors of Volume of Message and Gender of Subject

		Factor A: volume		
		Level A_1: soft	*Level A_2: medium*	*Level A_3: loud*
Factor B: gender	*Level B_1: male*	9 4 11	8 12 13	18 17 15
	Level B_2: female	2 6 4	9 10 17	6 8 4

$N = 18$

then three men and three women were randomly assigned to hear the message at each volume, so there are three persuasiveness scores per cell.

But now what? How do we make sense out of it all? To answer this question, stop and think about what the study is designed to investigate. We want to show the effects of (1) changing the levels of the factor of volume, (2) changing the levels of the factor of gender, and (3) changing the combination, or interaction, of the factors of volume and gender. As usual, we want to determine whether each of these has a significant effect. Because we want to view each of these effects one at a time, *the way to understand a two-way ANOVA is to treat it as if it consisted of three one-way ANOVAs.* You already understand a one-way ANOVA, so the rest of this chapter simply provides a guide for computing the various Fs.

Any two-way ANOVA breaks down into finding the main effects and the interaction effect.

Main Effects

The **main effect** of a factor is the effect that changing the levels of the factor has on the dependent scores, while ignoring all other factors in the study. In the persuasiveness study, to find the main effect of factor A, volume, we simply ignore the levels of factor B, gender. Literally erase the horizontal line that separates the males and females in Table 18.4, and treat the experiment as if it were as follows:

Factor A: volume

Level A_1: soft	*Level A_2: medium*	*Level A_3: loud*
9	8	18
4	12	17
11	13	15
2	9	6
6	10	8
4	17	4
$\overline{X}_{A_1} = 6$	$\overline{X}_{A_2} = 11.5$	$\overline{X}_{A_3} = 11.33$
$n_{A_1} = 6$	$n_{A_2} = 6$	$n_{A_3} = 6$

By ignoring the fact that there are males and females in each level, we have collapsed across gender, so for our purposes, we simply have six people tested under each volume. In this diagram there is one factor, with three means from the three levels of volume, so $k = 3$, with $n = 6$ in each level.

Whenever we collapse across one factor, we have the *main effect means* for the remaining factor. Thus, collapsing across gender above produces the main effect means for the three levels of volume, $\overline{X}_{A_1} = 6$, $\overline{X}_{A_2} = 11.5$, and $\overline{X}_{A_3} = 11.33$.

Once we have collapsed, we essentially perform a one-way ANOVA on the preceding diagram. To find the main effect of factor A we ask "Do these main effect means represent different μs that would be found if the entire population was tested under each of the three volumes?" To answer this question, we first create the statistical hypotheses for factor A. The null hypothesis is

$$H_0: \mu_{A_1} = \mu_{A_2} = \mu_{A_3}$$

For our study, this says that changing volume has no effect, so the main effect means from the levels of volume represent the same population of persuasiveness scores. If we can reject H_0, then we will accept the alternative hypothesis, which is

H_a: not all μ_As are equal

For our study, this says that at least two main effect means from the levels of volume represent different populations of persuasiveness scores, having different μs.

We now follow the procedures of a one-way ANOVA. To test H_0, we compute an F_{obt} called F_A. If F_A is significant, it indicates that at least two of the main effect means differ significantly. Then we will describe this relationship by graphing the main effect means, performing *post hoc* comparisons to determine which of the specific means of factor A differ significantly, and determining the effect size of changing the levels of factor A.

Once we have examined the main effect of factor A, we move on to the main effect of factor B. To do this, collapse across factor A, volume. That is, erase the vertical lines separating the levels of volume shown earlier in Table 18.4, producing this diagram of the main effect of factor B:

Factor B: gender	*Level B_1: male*	9 4 11	8 12 13	18 17 15	$\overline{X}_{B_1} = 11.89$ $n_{B_1} = 9$
	Level B_2: female	2 6 4	9 10 17	6 8 4	$\overline{X}_{B_2} = 7.33$ $n_{B_2} = 9$

Now we simply have the persuasiveness scores of males and females, ignoring the fact that some of each heard the message at different volumes. In this diagram there is a one-factor, two-sample design. We treat this as another one-way ANOVA to see if there are significant differences between the main effect means of persuasiveness scores for males ($\overline{X}_{B_1} = 11.89$) and females ($\overline{X}_{B_2} = 7.33$). Notice that here there are two levels of gender, so k is now 2 and the n of each level is 9, but for factor A, k was 3 and n was 6.

REMEMBER In a two-way ANOVA, the values of n and k may be different for each factor.

For the main effect of factor B, the null hypothesis is

H_0: $\mu_{B_1} = \mu_{B_2}$

For our study, this says that changing gender has no effect, so the mean for males represents the same μ as the mean for females. If we can reject H_0, then we will accept the alternative hypothesis, which is

H_a: not all μ_Bs are equal

For our study, this says that our samples of males and females represent different populations of persuasiveness scores, having different μs.

To test H_0 for factor B, we compute a separate F_{obt}, called F_B. If F_B is significant, it indicates that the main effect means for factor B differ significantly. Then we graph the main effect means for factor B, perform the *post hoc* comparisons, and compute the effect size of factor B.

After examining the main effects of factors A and B, we turn to the interaction.

Interaction Effects

The interaction of two factors is called a two-way interaction. The **two-way interaction** is the influence on scores when combining each level of factor A with each level of factor B. In our example, the interaction is the effect of each volume when combined with each gender. An interaction is identified as A × B. Our factor A has 3 levels and factor B has 2 levels, so we have a 3 × 2 interaction.

Because we examine the combination of both factors in an interaction, we do not collapse across, or ignore, either factor. Instead, we treat each *cell* in the study as a level of the interaction and compare the cell means. We can diagram the interaction effect this way:

A × B Interaction Effect

Male soft	*Male medium*	*Male loud*	*Female soft*	*Female medium*	*Female loud*
9	8	18	2	9	6
4	12	17	6	10	8
11	13	15	4	17	4
$\overline{X} = 8$	$\overline{X} = 11$	$\overline{X} - 16.67$	$\overline{X} = 4$	$\overline{X} = 12$	$\overline{X} = 6$

$k = 6$
$n = 3$

These are the original six cells that contain three scores per cell from Table 18.4. You can *think* of this as a one-way ANOVA for the six levels of the interaction, so k is now 6 and n is now 3. However, examining an interaction is not as simple as saying that the cell means are significantly different. Here we are testing the extent to which the cell means differ, *after* removing those differences attributable to the main effects of factor A and factor B. Thus, consistent differences between scores that are not due to changing the levels of factor A alone or factor B alone are due to the interaction.

Understanding and interpreting an interaction is difficult, because both independent variables are changing, as well as the dependent scores. To simplify the process, look at the influence of changing the levels of factor A under *one* level of factor B. Then see if this effect of factor A is *different* from when you look at the other level of factor B. For example, the means are grouped in the above diagram so that on the left, you can see how the three means for males change as volume changes. On the right, you can see how the means for females change as volume changes. For males, each increase in volume apparently results in an increase in the mean score. This is not the case for females, however: Increasing volume from soft to medium apparently increases the mean, but increasing volume from medium to loud apparently decreases the mean.

Thus, an interaction exists *when the influence that changing the levels of one factor has on scores depends on which level of the other factor you examine*. (Above, whether increasing volume increases persuasiveness depends on whether males or females are hearing the message.) This dependency leads to two other, equivalent ways of defining an interaction. First, an interaction occurs *when the effect of changing the levels of one factor is not consistent for each level of the other factor*. (Above, increasing volume does not have the same effect on males as on females.) Second, an interaction occurs *when the relationship between one factor and the dependent variable changes as the levels of the other factor change*. (For males, there is a positive linear relationship between volume and scores, but for females there is a curvilinear relationship.)

All of these statements describe the overall pattern produced by the influence of combining the levels of the two factors. Conversely, of course, this influence may not occur. Above, for example, if the scores for females had increased in the same way that the scores for males increased, then there would be no interaction. When there is no interaction, then (1) the influence that changing the levels of one factor has does not depend on which level of the other variable we're talking about, (2) the effect of changing one factor is the same for all levels of the other factor, and (3) we have the same relationship between the scores and one factor at each level of the other factor.

> ***REMEMBER*** A two-way interaction indicates that the influence of one factor depends on the level of the other factor that is present.

As with other effects, the data may appear to represent an interaction, but this may be an illusion created by sampling error. Therefore, we determine whether there is a significant interaction by performing a procedure similar to performing a one-way ANOVA on the cell means. First, we create the statistical hypotheses. As usual, the null hypothesis implies there is no effect. In symbols, these hypotheses are rather complex, but in words the null hypothesis is that any differences between the cell means do not represent an interaction between the population μs.[1] The alternative hypothesis is that at least some differences between the cell means do represent an interaction between the population μs.

To test the null hypothesis for the interaction, we compute yet another separate F_{obt}, called $F_{A \times B}$. If $F_{A \times B}$ is significant, it indicates that at least two of the cell means differ significantly in a way that produces an interaction. Then we graph the interaction effect, perform *post hoc* comparisons to determine which cell means differ significantly, and compute the effect size of the interaction.

Overview of the Computations of the Two-Way ANOVA

As you've seen, a two-way ANOVA involves three Fs: one for the main effect of factor A, one for the main effect of factor B, and one for the interaction effect of A $\times$ B. The logic and calculations for each of these are basically the same as in the one-way ANOVA, because any F_{obt} is the ratio formed by dividing the mean square between groups, MS_{bn}, by the mean square within groups, MS_{wn}.

[1] Technically, H_0 says that differences between scores due to A at one level of B equal the differences between scores due to A at the other level of B. Thus, we have H_0: $\mu_{A_1B_1} - \mu_{A_2B_1} = \mu_{A_1B_2} - \mu_{A_2B_2} = \mu_{A_2B_1} - \mu_{A_3B_1} = \mu_{A_2B_2} - \mu_{A_3B_2}$. H_a is that not all differences are equal.

As usual, MS_{wn} is the variance within groups, and in a two-way ANOVA is computed by computing the "average" variability of the scores in each *cell*. All participants in a cell are treated identically, so any differences among the scores are due to the inherent variability of scores. Thus, the MS_{wn} again estimates the error variance in any of the raw score populations represented by the samples. This is the *one* estimate of the error variance used as the denominator in computing all three *F* ratios.

We measure the variance between groups by computing the MS_{bn}. This is our way of measuring the differences between the means, as an estimate of the treatment effect in the population. Because we have two factors and the interaction, however, there are three sources of between-groups variance. Therefore, we partition the between-groups variance into (1) variance between groups due to factor A, (2) variance between groups due to factor B, and (3) variance between groups due to the interaction. We compute a separate mean square between groups for each of these as an estimate of the treatment variance each produces in the population. Then we compute the appropriate *F* ratio.

Table 18.5 shows all the means from the persuasiveness study that we will compare. For each effect, we compute a mean square between groups by dividing the appropriate sum of squares, *SS*, by the corresponding degrees of freedom, *df*. Thus, first we will collapse across factor B and examine the main effect means of factor A, volume ($\overline{X}_{A_1} = 6$, $\overline{X}_{A_2} = 11.5$, and $\overline{X}_{A_3} = 11.33$). We describe the differences between these means by computing the sum of squares between groups for factor A, called SS_A, and then, after dividing by the degrees of freedom for factor A, called df_A, we have the mean square between groups for factor A, called MS_A.

Likewise, we will collapse across factor A and examine the difference between the main effect means for factor B, gender ($\overline{X}_{B_1} = 11.89$ and $\overline{X}_{B_2} = 7.33$). We compute the sum of squares between groups for factor B, SS_B, and then, dividing by the degrees of freedom between groups for factor B, df_B, we have the mean square between groups for factor B, MS_B.

For the interaction, we do not collapse across either factor: We compare the differences among the six cell means that are not attributable to factor A or factor B alone. We compute the sum of squares between groups for A × B, $SS_{A \times B}$, and after dividing by the degrees of freedom for A × B, $df_{A \times B}$, we have the mean square between groups for the interaction, $MS_{A \times B}$. We also compute MS_{wn} by computing SS_{wn} and then dividing by df_{wn}.

You can see all these components of the two-way ANOVA in the summary table in Table 18.6. To complete the summary table, for factor A we divide MS_A by MS_{wn} to

TABLE 18.5 Summary of Means in Persuasiveness Study

		Factor A: volume			
		A_1: *soft*	A_2: *medium*	A_3: *loud*	
Factor B: gender	B_1: *male*	$\overline{X} = 8$	$\overline{X} = 11$	$\overline{X} = 16.67$	$\overline{X}_{male} = 11.89$
	B_2: *female*	$\overline{X} = 4$	$\overline{X} = 12$	$\overline{X} = 6$	$\overline{X}_{fem} = 7.33$
		$\overline{X}_{soft} = 6$	$\overline{X}_{med} = 11.5$	$\overline{X}_{loud} = 11.33$	

TABLE 18.6 Summary Table of Two-Way ANOVA

Source	*Sum of squares*	/	*df*	=	*Mean square*	*F*
Between						
Factor A (volume)	SS_A		df_A		MS_A	F_A
Factor B (gender)	SS_B		df_B		MS_B	F_B
Interaction (vol × gen)	$SS_{A \times B}$		$df_{A \times B}$		$MS_{A \times B}$	$F_{A \times B}$
Within	SS_{wn}		df_{wn}		MS_{wn}	
Total	SS_{tot}		df_{tot}			

produce F_A. For factor B, dividing MS_B by MS_{wn} produces F_B. For the interaction, dividing $MS_{A \times B}$ by MS_{wn} produces $F_{A \times B}$.

Each F_{obt} in a two-way ANOVA is tested in the same way F_{obt} was tested in the previous chapter. The larger the value of F_{obt}, the less likely it is that H_0 is true. If any F_{obt} is larger than F_{crit}, then F_{obt} is significant and we reject the corresponding H_0.

COMPUTING THE TWO-WAY ANOVA

Having a computer perform the calculations is the best way to perform this ANOVA. Whether or not you use a computer program, however, you should first organize the data in each cell. Table 18.7 shows our persuasiveness scores for the factors of volume and gender, as well as the various components to compute. First compute ΣX and ΣX^2 for each cell and note the n of each cell. Thus, for the male-soft cell, $\Sigma X = 4 + 9 + 11 = 24$, $\Sigma X^2 = 4^2 + 9^2 + 11^2 = 218$, and $n = 3$. Also compute the mean for each cell (e.g., for male-soft, $\overline{X} = 24/3 = 8$). These are the means that are tested in the interaction.

Now collapse across factor B (gender) and look only at the three volumes. Compute ΣX vertically for each column: The ΣX in a column is the sum of the ΣXs from the cells in that column (for soft, $\Sigma X = 24 + 12$). Note the n in each column (here $n = 6$) and compute the sample mean for each column (e.g., $\overline{X}_{soft} = 6$). These are the means that are tested in the main effect of factor A.

Now collapse across factor A (volume) and look only at males versus females. Compute ΣX horizontally for each row: The ΣX in a row equals the sum of the ΣXs from the cells in that row (for males, $\Sigma X = 24 + 33 + 50 = 107$). Note the n in each row (here $n = 9$) and compute the sample mean for each row (e.g., $\overline{X}_{male} = 11.89$). These are the means that are tested in the main effect of factor B.

Finally, compute the total ΣX, called ΣX_{tot}, by adding the ΣX from the three levels of factor A (the three column sums), so $\Sigma X_{tot} = 36 + 69 + 68 = 173$. Alternatively, you can add the ΣX from the two levels of factor B. Also find the total ΣX^2, called ΣX^2_{tot}, by adding the ΣX^2 from each cell, so $\Sigma X^2_{tot} = 218 + 377 + 838 + 56 + 470 + 116 = 2075$. Note the total N is 18.

TABLE 18.7 Summary Data for 3 × 2 ANOVA

		Factor A: volume			
		A_1: *soft*	A_2: *medium*	A_3: *loud*	
Factor B: gender	B_1: *male*	4 9 11 $\overline{X} = 8$ $\Sigma X = 24$ $\Sigma X^2 = 218$ $n = 3$	8 12 13 $\overline{X} = 11$ $\Sigma X = 33$ $\Sigma X^2 = 377$ $n = 3$	18 17 15 $\overline{X} = 16.67$ $\Sigma X = 50$ $\Sigma X^2 = 838$ $n = 3$	$\overline{X}_{male} = 11.89$ $\Sigma X = 107$ $n = 9$
	B_2: *female*	2 6 4 $\overline{X} = 4$ $\Sigma X = 12$ $\Sigma X^2 = 56$ $n = 3$	9 10 17 $\overline{X} = 12$ $\Sigma X = 36$ $\Sigma X^2 = 470$ $n = 3$	6 8 4 $\overline{X} = 6$ $\Sigma X = 18$ $\Sigma X^2 = 116$ $n = 3$	$\overline{X}_{fem} = 7.33$ $\Sigma X = 66$ $n = 9$
		$\overline{X}_{soft} = 6$ $\Sigma X = 36$ $n = 6$	$\overline{X}_{med} = 11.5$ $\Sigma X = 69$ $n = 6$	$\overline{X}_{loud} = 11.33$ $\Sigma X = 68$ $n = 6$	$\Sigma X_{tot} = 173$ $\Sigma X^2_{tot} = 2075$ $N = 18$

As you'll see, these components are used to compute all sums of squares and degrees of freedom. Then you compute the mean squares and, finally, each F_{obt}. To keep track of your computations and prevent brain strain, fill in the ANOVA summary table as you go along.

Computing the Sums of Squares

First compute the various sums of squares. Do this in five steps.

Step 1 is to compute the total sum of squares, SS_{tot}.

THE COMPUTATIONAL FORMULA FOR SS_{tot} IS

$$SS_{tot} = \Sigma X^2_{tot} - \left(\frac{(\Sigma X_{tot})^2}{N}\right)$$

This equation says to divide $(\Sigma X_{tot})^2$ by N and then subtract the answer from ΣX^2_{tot}.

For our persuasiveness study, in Table 18.7 $\Sigma X_{tot} = 173$, $\Sigma X^2_{tot} = 2075$, and $N = 18$. Filling in the formula, we have

$$SS_{tot} = 2075 - \left(\frac{(173)^2}{18}\right)$$

Which becomes

$$SS_{tot} = 2075 - 1662.72$$

so $SS_{tot} = 412.28$.

Note that the quantity $(\Sigma X_{tot})^2/N$ above is also used in the computation of most of the other sums of squares. It is called the *correction* (here the correction equals 1662.72).

Step 2 is to compute the sum of squares for factor A. In our diagram of the two-way ANOVA, the levels of factor A form the columns.

THE COMPUTATIONAL FORMULA FOR THE SUM OF SQUARES BETWEEN GROUPS FOR COLUMN FACTOR A IS

$$SS_A = \Sigma\left(\frac{(\text{sum of scores in the column})^2}{n \text{ of scores in the column}}\right) - \left(\frac{(\Sigma X_{tot})^2}{N}\right)$$

This equation says to square ΣX for each column of factor A and divide by the n in the column. After doing this for all levels, add the answers together and then subtract the correction.

In the example, from Table 18.7 the three columns produced sums of 36, 69, and 68, and n was 6. Filling in the above formula we have

$$SS_A = \left(\frac{(36)^2}{6} + \frac{(69)^2}{6} + \frac{(68)^2}{6}\right) - \left(\frac{(173)^2}{18}\right)$$

$$SS_A = (216 + 793.5 + 770.67) - 1662.72$$

$$SS_A = 1780.17 - 1662.72$$

so $SS_A = 117.45$.

Step 3 is to compute the sum of squares between groups of factor B. In our diagram of the two-way ANOVA, the levels of factor B form the rows.

THE COMPUTATIONAL FORMULA FOR THE SUM OF SQUARES BETWEEN GROUPS FOR ROW FACTOR B IS

$$SS_B = \Sigma\left(\frac{(\text{sum of scores in the row})^2}{n \text{ of scores in the row}}\right) - \left(\frac{(\Sigma X_{tot})^2}{N}\right)$$

This equation says to square ΣX for each level of factor B and divide by the n in the level. After doing this for all levels, add the answers and then subtract the correction.

In our example, the two rows produced sums of 107 and 66 and n was 9. Filling in the above formula gives

$$SS_B = \left(\frac{(107)^2}{9} + \frac{(66)^2}{9}\right) - 1662.72$$

$$SS_B = 1756.11 - 1662.72$$

so $SS_B = 93.39$.

Step 4 is to compute the sum of squares between groups for the interaction, $SS_{A \times B}$. To do this, first compute something called the overall sum of squares between groups, identified as SS_{bn}.

THE COMPUTATIONAL FORMULA FOR SS_{bn} IS

$$SS_{bn} = \Sigma\left(\frac{(\text{sum of scores in the cell})^2}{n \text{ of scores in the cell}}\right) - \left(\frac{(\Sigma X_{tot})^2}{N}\right)$$

Here find $(\Sigma X)^2$ for each cell, divide by the n of the cell, add the answers from all cells together, and subtract the correction.

In our example, filling in the formula gives

$$SS_{bn} = \left(\frac{(24)^2}{3} + \frac{(33)^2}{3} + \frac{(50)^2}{3} + \frac{(12)^2}{3} + \frac{(36)^2}{3} + \frac{(18)^2}{3}\right) - 1662.72$$

$$SS_{bn} = 1976.33 - 1662.72$$

so $SS_{bn} = 313.61$.

The reason we compute SS_{bn} is that it is equal to the sum of squares for factor A plus the sum of squares for factor B plus the sum of squares for the interaction. To find $SS_{A \times B}$, subtract the sum of squares for both main effects (in steps 2 and 3) from the overall SS_{bn}. Thus,

THE COMPUTATIONAL FORMULA FOR THE SUM OF SQUARES BETWEEN GROUPS FOR THE INTERACTION, $SS_{A \times B}$, IS

$$SS_{A \times B} = SS_{bn} - SS_A - SS_B$$

In the example, $SS_{bn} = 313.61$, $SS_A = 117.45$, and $SS_B = 93.39$, so

$$SS_{A \times B} = 313.61 - 117.45 - 93.39$$

so, $SS_{A \times B} = 102.77$.

Step 5 is to compute the sum of squares within groups, SS_{wn}. The sum of squares within groups plus the overall sum of squares between groups equals the total sum of squares. Therefore, subtracting the overall SS_{bn} in step 4 from SS_{tot} in step 1, gives the SS_{wn}.

THE COMPUTATIONAL FORMULA FOR THE SUM OF SQUARES WITHIN GROUPS, SS_{wn}, IS

$$SS_{wn} = SS_{tot} - SS_{bn}$$

In our example, $SS_{tot} = 412.28$ and $SS_{bn} = 313.61$, so

$$SS_{wn} = 412.28 - 313.61$$

Thus, $SS_{wn} = 98.67$.

Placing the various sums of squares in the ANOVA summary table gives Table 18.8. Notice that we do not include the overall SS_{bn}.

Computing the Degrees of Freedom

Now you must determine the various values of *df*.

1. *The degrees of freedom between groups for factor A is* $k_A - 1$, where k_A is the number of levels in factor A. (In our example, k_A is the three levels of volume, so $df_A = 2$.)
2. *The degrees of freedom between groups for factor B is* $k_B - 1$, where k_B is the number of levels in factor B. (In the example, k_B is the two levels of gender, so $df_B = 1$.)
3. *The degrees of freedom between groups for the interaction is the df for factor A multiplied times the df for factor B.* (Above, $df_A = 2$ and $df_B = 1$, so $df_{A \times B} = 2$.)
4. *The degrees of freedom within groups equals* $N - k_{A \times B}$, where N is the total N of the study and $k_{A \times B}$ is the total number of cells in the study. (In our example, N is 18 and we have six cells, so $df_{wn} = 18 - 6 = 12$.)
5. *The degrees of freedom total equals* $N - 1$. Use this to check your previous calculations, because the sum of the above *df*s should equal df_{tot}. (In the example, $df_{tot} = 17$.)

Computing the Mean Squares

It is easiest to perform the remainder of the computations by working directly from the summary table. So far, with the sums of squares and degrees of freedom we have Table 18.9.

Now compute the mean squares. Any mean square equals the appropriate sum of squares divided by the appropriate *df*.

TABLE 18.8 Summary Table of Two-Way ANOVA

Source	*Sum of squares*	*df*	*Mean square*	*F*
Between				
Factor A (volume)	117.45	df_A	MS_A	F_A
Factor B (gender)	93.39	df_B	MS_B	F_B
Interaction (vol × gen)	102.77	$df_{A \times B}$	$MS_{A \times B}$	$F_{A \times B}$
Within	98.67	df_{wn}	MS_{wn}	
Total	412.28	df_{tot}		

TABLE 18.9 Summary Table of Two-Way ANOVA

Source	*Sum of squares*	*df*	*Mean square*	*F*
Between				
Factor A (volume)	117.45	2	MS_{A}	F_{A}
Factor B (gender)	93.39	1	MS_{B}	F_{B}
Interaction (vol × gen)	102.77	2	$MS_{A \times B}$	$F_{A \times B}$
Within	98.67	12	MS_{wn}	
Total	412.28	17		

THE COMPUTATIONAL FORMULA FOR THE MEAN SQUARE FOR FACTOR A IS

$$MS_{A} = \frac{SS_{A}}{df_{A}}$$

In the example,

$$MS_{A} = \frac{117.45}{2} = 58.73$$

THE COMPUTATIONAL FORMULA FOR THE MEAN SQUARE FOR FACTOR B IS

$$MS_{B} = \frac{SS_{B}}{df_{B}}$$

In the example,

$$MS_{B} = \frac{93.39}{1} = 93.39$$

THE COMPUTATIONAL FORMULA FOR THE MEAN SQUARE FOR THE INTERACTION IS

$$MS_{A \times B} = \frac{SS_{A \times B}}{df_{A \times B}}$$

Thus, we have

$$MS_{A \times B} = \frac{102.77}{2} = 51.39$$

THE COMPUTATIONAL FORMULA FOR THE MEAN SQUARE WITHIN GROUPS IS

$$MS_{wn} = \frac{SS_{wn}}{df_{wn}}$$

Therefore,

$$MS_{wn} = \frac{98.67}{12} = 8.22$$

Putting these values in the summary table produces Table 18.10.

Now, finally, compute the Fs.

Computing F_{obt}

Recall that to compute any F, divide the MS_{bn} by the MS_{wn}. Therefore,

THE COMPUTATIONAL FORMULA FOR F_A FOR THE MAIN EFFECT OF FACTOR A IS

$$F_A = \frac{MS_A}{MS_{wn}}$$

In our example

$$F_A = \frac{58.73}{8.22} = 7.14$$

THE COMPUTATIONAL FORMULA FOR F_B FOR THE MAIN EFFECT OF FACTOR B IS

$$F_B = \frac{MS_B}{MS_{wn}}$$

TABLE 18.10 Summary Table of Two-Way ANOVA

Source	*Sum of squares*	*df*	*Mean square*	*F*
Between				
Factor A (volume)	117.45	2	58.73	F_A
Factor B (gender)	93.39	1	93.39	F_B
Interaction (vol × gen)	102.77	2	51.39	$F_{A \times B}$
Within	98.67	12	8.22	
Total	412.28	17		

Thus,

$$F_B = \frac{93.39}{8.22} = 11.36$$

THE COMPUTATIONAL FORMULA FOR $F_{A \times B}$ FOR THE INTERACTION EFFECT IS

$$F_{A \times B} = \frac{MS_{A \times B}}{MS_{wn}}$$

Thus,

$$F_{A \times B} = \frac{51.39}{8.22} = 6.25$$

And now, the finished summary table is shown in Table 18.11.

Interpreting Each F_{obt}

Now determine whether each F_{obt} is significant in the same way you did in Chapter 17: Compare each F_{obt} to the appropriate value of F_{crit}. To find each F_{crit} in the F-tables (Table 5 in Appendix C), you need the df_{bn} and the df_{wn} used in computing the corresponding F_{obt}. (In the F-tables, the columns labeled df_{bn} are also labeled "degrees of freedom in numerator of F ratio." To find each F_{crit}, use the df you used in calculating the numerator of the F as your df_{bn}.)

1. To find F_{crit} for testing F_A, use df_A as the df between groups and df_{wn}. In the example, $df_A = 2$ and $df_{wn} = 12$. For $\alpha = .05$, the F_{crit} for 2 and 12 df is 3.88.
2. To find F_{crit} for testing F_B, use df_B as the df between groups and df_{wn}. In the example, $df_B = 1$ and $df_{wn} = 12$. At $\alpha = .05$, the F_{crit} for 1 and 12 df is 4.75.
3. To find F_{crit} for the interaction, use $df_{A \times B}$ as the df between groups and df_{wn}. In the example, $df_{A \times B} = 2$ and $df_{wn} = 12$. At $\alpha = .05$, the F_{crit} for 2 and 12 df is 3.88.

TABLE 18.11 Summary Table of Two-Way ANOVA

Source	*Sum of squares*	*df*	*Mean square*	*F*
Between				
Factor A (volume)	117.45	2	58.73	7.14
Factor B (gender)	93.39	1	93.39	11.36
Interaction (vol × gen)	102.77	2	51.39	6.25
Within	98.67	12	8.22	
Total	412.28	17		

The df_{bn} for factor B is different from the df_{bn} for factor A. Therefore, the F_{crit} for factor B is different from the F_{crit} for factor A.

REMEMBER Each F_{crit} in the two-way ANOVA will be different if the degrees of freedom between groups are different.

Thus, we end up comparing the values of F_{obt} from the ANOVA summary table with the values of F_{crit}, as follows:

	F_{obt}	F_{crit}
Main effect of volume (A)	7.14	3.88
Main effect of gender (B)	11.36	4.75
Interaction (A × B)	6.25	3.88

By now you can do this in your sleep. Imagine a sampling distribution with a region of rejection and F_{crit} in the positive tail. (If you can't imagine this, look back in Chapter 17 at Figure 17.4.) First, the obtained F_A of 7.14 is larger than the F_{crit}, so F_A falls in the region of rejection. Therefore, conclude that differences between the means for the levels of factor A are significant: Changing the volume of a message produced significant differences in persuasiveness scores. However, a significant F_{obt} only indicates that, *somewhere* in the factor, at least two of the means differ significantly. Therefore, we are confident that changing volume results in at least two different population means. With $\alpha = .05$, this result is reported as

$$F(2, 12) = 7.14, p < .05$$

Likewise, F_B is significant, so we conclude that the males and females in this study represent different populations of scores. Report this result as

$$F(1, 12) = 11.36, p < .05$$

Finally, the $F_{A \times B}$ of 6.25 is significant, so we conclude that the effect of changing the volume on the population *depends on* whether it is a population of males or a population of females. Or, we can conclude that the difference between the male and female populations *depends on* whether a message is played at the soft, medium, or loud level. Report this result as

$$F(2, 12) = 6.25, p < .05$$

Note: It is just a coincidence of the particular data in this example that all three values of F_{obt} were significant. Whether any one F_{obt} is significant does not influence whether any other F_{obt} is significant. With different data, any combination of the main effects and/or the interaction may or may not be significant.

At this point, you have completed the ANOVA. However, you are a long way from being finished with the analysis. Because each significant F_{obt} indicates only that a difference exists somewhere among the means, you now must examine those means.

INTERPRETING THE TWO-WAY EXPERIMENT

To understand and interpret the results of a two-way ANOVA, you should first examine each significant main effect and interaction by graphing their respective means and performing the *post hoc* comparisons on those means.

First let's graph each effect. Look again at the various means, shown in Table 18.12.

Graphing the Main Effects

As usual, you should plot the dependent variable along the *Y* axis and the levels of a factor along the *X* axis. Graph the main effect of factor A by plotting the main effect means from each level of factor A (the column means across the bottom of Table 18.12). Graph the main effect of factor B separately by plotting the main effect means from each level of factor B (the row means at the right of Table 18.12). Figure 18.1 shows the resulting graphs of these main effects. Note that, because volume is measured in decibels, the *X* axis of the volume factor should be labeled in decibels. Also note that the main effect of the gender variable is plotted as a bar graph because, as discussed in Chapter 6, this is how we graph means from a nominal independent variable.

The graphs reflect the same differences we saw when comparing the means in Table 18.12. The graph on the right shows that males scored higher than females. In the graph on the left, the slanting line between soft and medium volume suggests a large (possibly significant) difference between those means. The line between medium and loud volume is close to horizontal, however, and the means are close to equal, so there may not be a significant difference here. To determine which means differ significantly, you must perform *post hoc* comparisons. But first let's graph the interaction.

Graphing the Interaction Effect

An interaction can be a beast to interpret, so always graph it! Graph the interaction by plotting all cell means on a *single* graph. As usual, place the dependent variable along the *Y* axis. However, there are two independent variables (here volume and gender), but only one *X* axis. To solve this problem, place the levels of one factor along the *X* axis. Show the second factor by drawing a separate line on the graph that connects the means for each level of that factor. (Because there's a line for each level of the second factor,

TABLE 18.12 Summary of Means for Persuasiveness Study

		Factor A: volume			
		A_1: *soft*	A_2: *medium*	A_3: *loud*	
Factor B: gender	B_1: *male*	$\overline{X} = 8$	$\overline{X} = 11$	$\overline{X} = 16.67$	$\overline{X}_{male} = 11.89$
	B_2: *female*	$\overline{X} = 4$	$\overline{X} = 12$	$\overline{X} = 6$	$\overline{X}_{fem} = 7.33$
		$\overline{X}_{soft} = 6$	$\overline{X}_{med} = 11.5$	$\overline{X}_{loud} = 11.33$	

FIGURE 18.1 Graphs Showing Main Effects of Volume and Gender

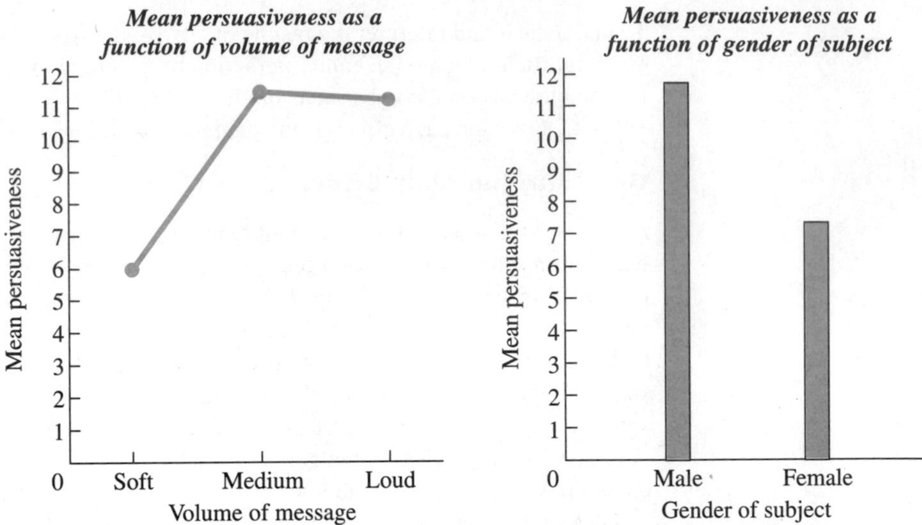

we typically place the factor with more levels on the *X* axis so that there are as few lines as possible.) For the persuasiveness study, the finished graph of the interaction between volume and gender is shown in Figure 18.2.

FIGURE 18.2 Graph of Cell Means, Showing Mean Persuasiveness as a Function of Volume of Message and Subject Gender

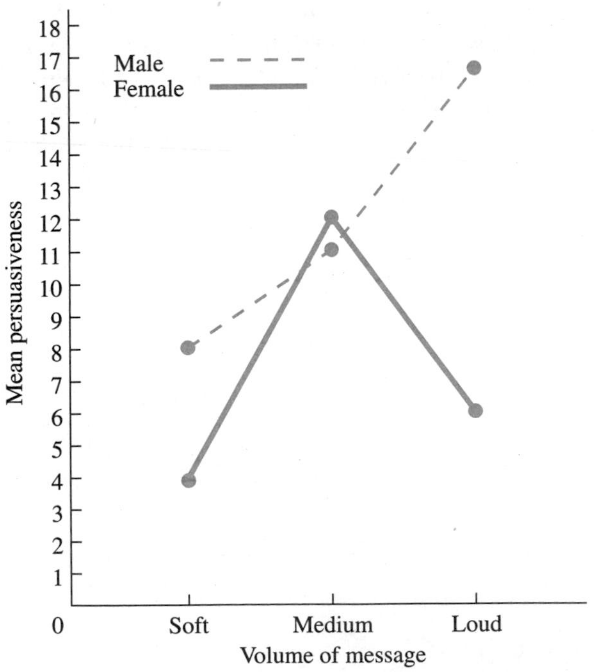

The X axis is labeled with the three volume levels. Then, plotting the cell means from Table 18.12, we connect the three means for males with one line and the three means for females with a different line. Always provide a key to identify each line.

The way to read the graph is to look at one line at a time. Thus, for males (the dashed line), as volume increased, mean persuasiveness scores increased. For females (the solid line), as volume increased, persuasiveness scores first increased but then decreased. Thus, the effect of increasing volume was not the same for males as for females. Instead, the effect of increasing volume on persuasiveness scores *depends* on whether the participants are male or female. There is a positive linear relationship between scores and increasing volume for males and a nonlinear relationship between scores and increasing volume for females.

Of course, if the interaction were not significant, then we would have no reason to believe that changing the volume had different effects on males and females, regardless of what the graph suggested. Further, we do not know which of these cell means actually differ significantly, because we haven't performed the *post hoc* comparisons yet.

Before we do, however, note one final way of recognizing an interaction. An interaction can produce an infinite variety of different graphs, but the key is that *an interaction will produce a graph on which the lines are not parallel*. Remember that each line summarizes the relationship between the X and Y scores and that a line that is shaped or oriented differently from another line indicates a different relationship. Therefore, when the lines produced by graphing the cell means are not parallel, each line depicts a *different* relationship. This indicates that the relationship between X and Y changes depending on the level of the second factor, so an interaction is present. Conversely, when an interaction is not present, the lines will be essentially parallel, with each line depicting essentially the same relationship. To see this distinction, say that our data had produced one of the two graphs in Figure 18.3. The graph on the left is the ultimate in nonparallel lines. Here, as the levels of A change, the mean scores either increase or decrease, *depending* on the level of B we're talking about. Therefore, this graph depicts an interaction. In the graph on the right, however, the lines are parallel. Here, as the levels of A change, the scores increase, regardless of which level of factor B we're talking about. Therefore, this graph does not depict an interaction. (The fact that, *overall*, the scores are higher in B_1 than in B_2 is the main effect of factor B.)

Think of significance testing for the interaction as testing whether the lines are significantly different from parallel. When $F_{A \times B}$ is *not* significant, the lines on the graph

FIGURE 18.3 Two Graphs Showing When an Interaction Is and Is Not Present

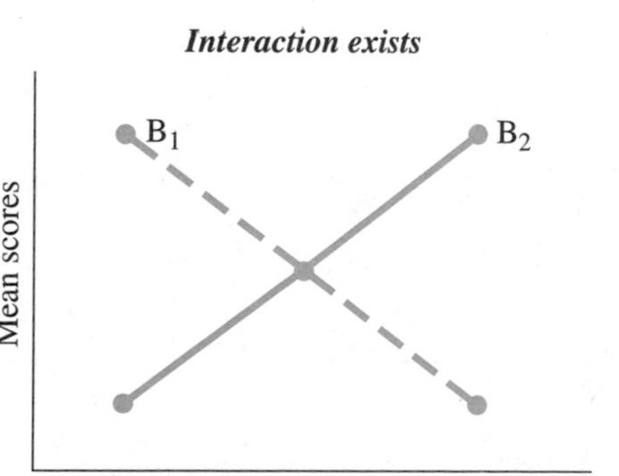

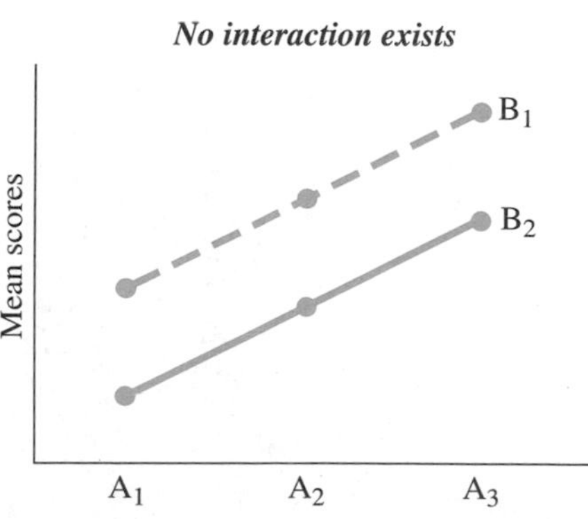

are not significantly different from parallel, so they may represent parallel lines that would be found if we graphed the means of the populations. Conversely, when the interaction is significant, somewhere in the graph the lines *do* differ significantly from parallel. Therefore, if we could graph the means of the populations, the lines probably would not be parallel, and there would be an interaction in the population.

REMEMBER When graphed, interaction effects produce non-parallel line.

Seeing Main Effects in a Graph of the Interaction

In published research articles, often only a graph of the interaction is provided, and you are expected to visualize the main effects from it. Because cell means are averaged together to obtain main effect means, you should envision the data points for the main effect of a factor as the average of the appropriate data points in the interaction graph. To see how this is done, assume we had obtained the data in Figure 18.4. On the graph at the left, the asterisks show where the overall mean for each level of volume is located after we collapse vertically across gender (averaging the two data points in each circle). Plotting the line formed by these points on a separate graph (with the *X* axis labeled for volume level) would show the main effect of volume. On the graph at the right, the asterisks show where the overall mean for each gender level is located after we collapse across volume (averaging the three data points in each circle). Plotting the line formed by these points on a separate graph (with the *X* axis labeled male and female) would show the main effect of gender.

Performing the *Post Hoc* Comparisons

As usual, you should perform *post hoc* comparisons on the means from any *significant* F_{obt}. If there are unequal *n*s between the levels of a factor, perform Fisher's protected *t*-test (from Chapter 17). If the *n*s in all levels of a factor are equal, perform Tukey's *HSD* procedure (also from Chapter 17). The *post hoc* comparisons for an interaction effect, however, are computed differently from those for a main effect.

FIGURE 18.4 Main Effects Seen in the Graph of an Interaction

Here, the dots are cell means and the asterisks are visualized main effect means.

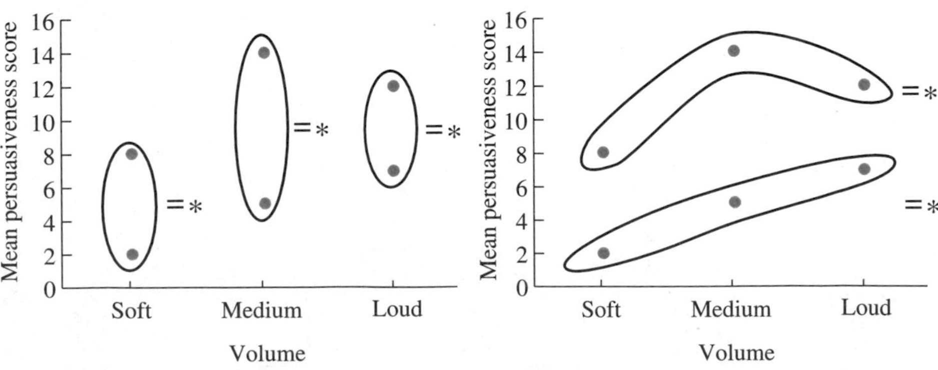

Performing Tukey's *HSD* for main effects Perform the post hoc comparisons on *each* significant main effect, as if it were the only factor in a one-way ANOVA. Recall that the computational formula for the *HSD* is

$$HSD = (q_k)\left(\sqrt{\frac{MS_{wn}}{n}}\right)$$

where MS_{wn} is the denominator of the F_{obt}, q_k is found in Table 6 of Appendix C for df_{wn} and k (where k is the number of levels in the factor), and n is the number of scores in a level. *BUT BEWARE!* For each factor, there may be a different value of n and k. In the persuasiveness study, six scores went into each mean for a level of volume, but nine scores went into each mean for a level of subject gender. *The n is always the number of scores used to compute each mean you are comparing*. Also, because q_k depends on k, different factors having a different number of levels will have different values of q_k. The end result is that you can have a different value of *HSD* for each factor.

> ***REMEMBER*** You must compute a different *HSD* for each main effect when their *k*s or *n*s are different.

In the persuasiveness study, for the volume factor we have three means for the main effect, so $k = 3$, and the n of each mean is 6. In the ANOVA, $MS_{wn} = 8.22$ and $df_{wn} = 12$. From Table 6 in Appendix C, for $\alpha = .05$, $q_k = 3.77$. Placing these values in the above formula gives

$$HSD = (q_k)\left(\sqrt{\frac{MS_{wn}}{n}}\right) = (3.77)\left(\sqrt{\frac{8.22}{6}}\right) = 4.41$$

Thus, the *HSD* for factor A is 4.41.

Finding the differences between all pairs of factor A means, we have

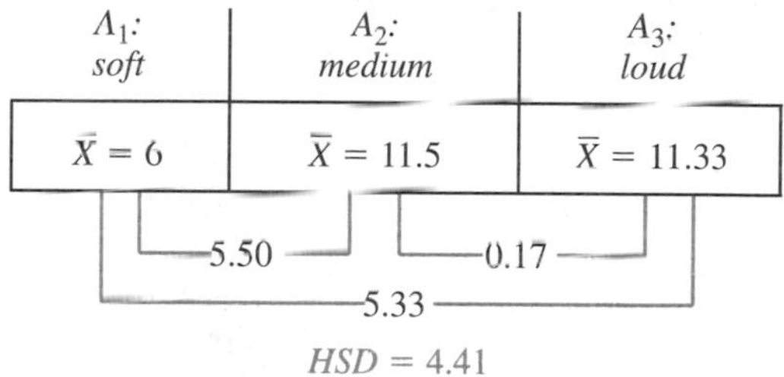

In the middle of each line connecting two means is the absolute difference between them. The differences between the means of soft and medium, 5.50, and between the means of soft and loud, 5.33, are both more than the *HSD* of 4.41, so soft volume produces a significant difference from the other volumes. But, because the means for medium and loud differ by less than 4.41, these conditions do *not* differ significantly.

Because factor B, gender, contains only two levels, you do not need to perform *post hoc* comparisons (it must be that the males differ significantly from the females). If, however, there were more than two levels in a significant factor B, you would then compute the appropriate *HSD* for the n and k in that factor and compare all of these main effect means.

TABLE 18.13 Summary of Interaction Means for Persuasiveness Study

Solid lines connecting two cells show examples of unconfounded comparisons; dashed lines connecting two cells show examples of confounded comparisons.

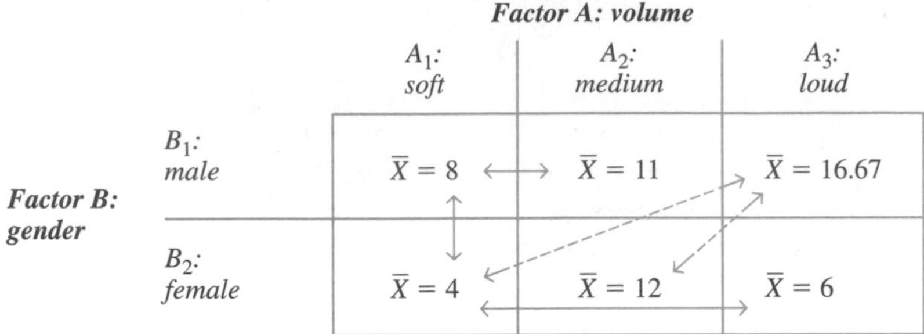

Performing Tukey's *HSD* for the interaction In performing *post hoc* comparisons on a significant interaction, you do *not* compare every cell mean to every other cell mean. Look at the diagram of the interaction means in Table 18.13 above. You would not, for example, compare the mean for males at the loud volume to the mean for females at the soft volume. This is because even if the means do differ significantly, you would not know what caused the difference. You would be comparing apples to oranges, because the two cells differ in terms of both participants' gender *and* volume. In other words, the influence of gender is *confounded* by volume and vice versa, producing a confounded comparison. A **confounded comparison** occurs when two cells differ along more than one factor. When performing *post hoc* comparisons on an interaction, perform only **unconfounded comparisons**, in which two cells differ along only one factor. Therefore, compare only cell means within the same column (comparing means vertically), because you can explain the differences as resulting from factor B. Also compare means within the same row (comparing means horizontally), because you can explain these differences as resulting from factor A. Do not, however, make any diagonal comparisons, because these are confounded comparisons.

> REMEMBER Make only unconfounded comparisons when performing *post hoc* comparisons on the means of an interaction.

When there are equal *ns* in all cells, you can compare the means in the interaction using a slight variation of the Tukey *HSD*.[2] Previously when computing the *HSD*, we found q_k in Table 6 of Appendix C using *k*, the number of means being compared. To compute the *HSD* for an interaction, you must first determine the *adjusted k*. This *k* adjusts for the actual number of unconfounded comparisons being made out of all the cell means in the interaction. Obtain the adjusted *k* from Table 18.14 shown here (and at the beginning of Table 6 of Appendix C). In the left column, locate the design of the study you are examining. Don't be concerned about the order of the numbers. For example, our persuasiveness study was a 3 × 2 design, but we look at the row labeled "2 × 3." Reading across that row, as a double-check confirm that the middle column contains the total number of cell means in the interaction (yes, there are 6). In the right-hand column is the adjusted value of *k* (for this example, it's 5).

[2]Adapted from Cicchetti, 1972.

TABLE 18.14 Values of Adjusted *k*

Design of study	*Number of cell means in study*	*Adjusted value of k*
2 × 2	4	3
2 × 2	6	5
2 × 4	8	6
3 × 3	9	7
3 × 4	12	8
4 × 4	16	10
4 × 5	20	12

The adjusted value of k is the value of k to use in obtaining q_k from Table 6 of Appendix C. Thus, for the persuasiveness study, with $\alpha = .05$, $df_{wn} = 12$, and adjusted $k = 5$, our $q_k = 4.51$. Now compute the *HSD* using the same old formula used previously. Our MS_{wn} is 8.22, but now n is 3, the number of scores in each *cell*. We have

$$HSD = (q_k)\left(\sqrt{\frac{MS_{wn}}{n}}\right) = (4.51)\left(\sqrt{\frac{8.22}{3}}\right) = 7.47$$

Thus, the *HSD* for the interaction is 7.47.

Now determine the differences between all cell means within each column and between all cell means within each row. To see these differences, you can arrange the interaction means as shown in Table 18.15. In the middle of each line connecting two cells is the absolute difference between the two means. Any difference between two means that is larger than the *HSD* of 7.47 is a significant difference. There are only three significant differences here: (1) Between the mean for females at the soft volume and the mean for females at the medium volume; (2) Between the mean for males at the soft volume and the mean for males at the loud volume; and (3) Between the mean for males at the loud volume and the mean for females at the loud volume.

TABLE 18.15 Table of Interaction Means

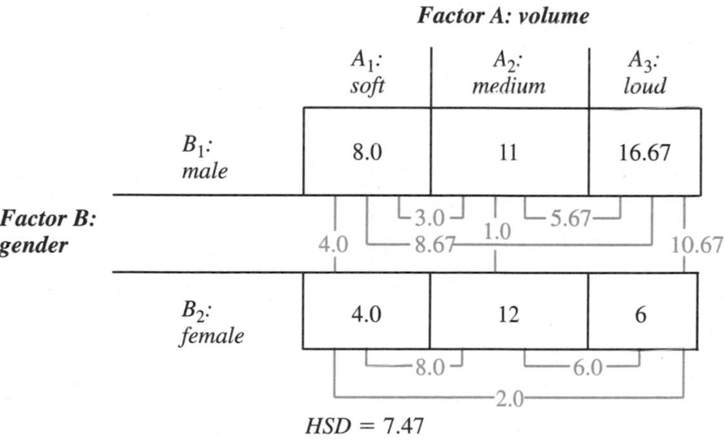

Interpreting the Overall Results of the Experiment

There is no *one* way to interpret all experiments, because the data in each experiment indicate something different. The goal is to come up with a complete, honest, and simplified description of the results of the study. We do that by looking at the significant *post hoc* comparisons within all significant main effects and interaction effects.

All of the differences found in the persuasiveness study can be summarized using the diagram in Table 18.16. Outside of the diagram are the main effect means. Each line connecting two means indicates that they differ significantly. Inside the diagram, each line connecting two cell means indicates a significant difference within the interaction.

Both of the main effects and the interaction produced significant values of F_{obt}. However, any conclusions about significant main effects often must be qualified (or are downright untrue) because of the significant interaction. For example, there is a significant difference between the main effect means of males and females. If there were not a significant interaction, we could conclude that, overall, males score higher than females. When we look at the cell means of the interaction, however, we see that gender differences *depend* on the volume: Only in the loud condition is there a significant difference between males and females. (This difference is so large that it produced an overall mean for males that is larger than the overall mean for females.) Therefore, because the interaction contradicts the overall pattern suggested by the main effect, we *cannot* make an overall, general conclusion about differences between males and females.

Likewise, we cannot make an overall conclusion based on the main effect of volume, which showed that soft volume was significantly different from both the medium and loud volumes. The interaction indicates that increasing the volume from soft to medium produced a significant difference only in females, and increasing the volume from soft to loud produced a significant difference only in males.

Thus, as the above example illustrates, usually you cannot draw any conclusions about significant main effects when the interaction is significant. After all, the interaction indicates that the influence of one factor *depends* on the levels of the other factor and vice versa, so you should not turn around and act as though either factor has a consistent overall effect. Therefore, in such situations, the interpretation of a study is usually limited to the interaction. When the interaction is not significant, then focus on any

TABLE 18.16 Summary of Significant Differences in Our Persuasiveness Study

Each line connects two means that differ significantly.

		Factor A: volume			
		Level A_1: soft	*Level A_2: medium*	*Level A_3: loud*	
Factor B: gender	*Level B_1: male*	8.0	11	16.67	$\overline{X} = 11.89$
	Level B_2: female	4.0	12	6	$\overline{X} = 7.33$
		$\overline{X}_{soft} = 6$	$\overline{X}_{med} = 11.5$	$\overline{X}_{loud} = 11.33$	

significant main effects. (For completeness, however, always perform the entire ANOVA, and report the results for all main effects and interaction effects in the study.)

> *REMEMBER* The primary interpretation of a two-way ANOVA rests on the interpretation of the significant interaction.

Thus, we conclude that increasing the volume of a message beyond soft does tend to increase persuasiveness scores in the population, but this increase occurs for females with medium volume and for males with loud volume. Further, we conclude that differences in persuasiveness scores between males and females do occur in the population, but only if the volume of the message is loud.

Remember experiment-wise error, the probability of a Type I error somewhere in our conclusions? Well, after all of the above shenanigans, the experiment-wise error is protected so that for all of these conclusions together, the probability of a Type I error is still $p < .05$. Also, remember power—the probability of not making a Type II error? All that we said in previous chapters about power applies to the two way ANOVA as well. Thus, for any differences that are not significant, we must be concerned about whether we have maximized power by maximizing the difference between the means, minimizing the variability within each cell, and having a large enough n.

With significant results, we begin to interpret the above factors psychologically. In this and previous examples, however, we've seen both true independent variables like volume (where random assignment to conditions is possible) and quasi-independent variables like gender (where random assignment is not possible). Statistical procedures don't care what type of factors you test. It's up to you to draw the correct conclusions, identifying that while volume may "cause" differences in persuasiveness scores, you cannot conclude that gender is another cause of such differences (some other subject variable that is confounded with gender may be the cause).

As usual, to further describe and understand the relationship, you can compute the effect size and confidence intervals for any significant effects.

Describing the Effect Size: Eta Squared

Whenever you demonstrate a significant relationship, you should then think "squared correlation coefficient" to describe the relationship. Therefore, in the two-way ANOVA, you again compute eta squared, η^2, to describe effect size—the proportion of variance in the dependent scores that is accounted for by the manipulation. Compute a separate eta squared for each *significant* main and interaction effect. The formula for eta squared is again

$$\eta^2 = \frac{\text{sum of squares between groups for the effect}}{SS_{\text{tot}}}$$

To compute each eta squared, divide the sum of squares for the factor, either SS_{A}, SS_{B}, or $SS_{\text{A}\times\text{B}}$, by SS_{tot}. For example, in our persuasiveness study, for factor A (volume), SS_{A} was 117.45 and SS_{tot} was 412.28. Therefore,

$$\eta^2_{\text{A}} = \frac{SS_{\text{A}}}{SS_{\text{tot}}} = \frac{117.45}{412.28} = .28$$

Thus, if we predict participants scores using the main effect mean of the volume condition they were tested under, we can account for .28, or 28%, of the total variance in persuasiveness scores. Following the same procedure for the gender factor, SS_B is 93.39, so η^2_B is .23: Predicting the mean of a gender condition for male and female participants, respectively, will account for an additional 23% of the variance in persuasiveness scores. Finally, for the interaction of volume and gender, $SS_{A\times B}$ is 102.77, so $\eta^2_{A\times B}$ is .25: Using the mean of the cell that subjects were in to predict their scores, we can account for an additional 25% of the variance in persuasiveness scores.

Recall that the greater the proportion of variance accounted for, the more important the factor is in determining participants' scores. Because each of the above has about the same effect size, they are all of equal importance in understanding differences in persuasiveness scores in this experiment. Suppose, however, that one effect accounted for only .01, or 1%, of the total variance. Such a small η^2 indicates that this relationship is very inconsistent. Therefore, it is not a very useful or informative relationship, and we are better served by emphasizing the other, larger, significant effects. In essence, if eta squared indicates that an effect was not a big deal in the experiment, then we should not make a big deal out of it when interpreting the experiment.

The effect size is especially important to consider when dealing with interactions. The one exception to the rule of always focusing on the significant interaction is when it has a very small effect size. Although an interaction contradicts the overall pattern in a main effect, if the interaction has a very small effect size (say, only .02), then it only slightly and inconsistently contradicts the main effect. In such cases, we instead focus the interpretation on any significant main effects that had a more substantial effect size.

> *REMEMBER* Effect size is an important determinant of which factors you should focus on when interpreting a two-way ANOVA.

Confidence Intervals for a Single μ

You can compute the confidence interval for the population μ that is represented by the mean of a level from a main effect or by a cell mean from the interaction. Use the formula presented in the previous chapter, which was

$$\left(\sqrt{\frac{MS_{wn}}{n}}\right)(-t_{crit}) + \overline{X} \leq \mu \leq \left(\sqrt{\frac{MS_{wn}}{n}}\right)(+t_{crit}) + \overline{X}$$

where t_{crit} is the two-tailed value at the appropriate α with $df = df_{wn}$, MS_{wn} is from the ANOVA, $\overline{X}$ is the mean for the level or cell you are describing, and n is the number of scores that the mean is based on.

SUMMARY OF THE STEPS IN PERFORMING A TWO-WAY ANOVA

The following summarizes the steps for performing a two-way ANOVA:

1. Design the experiment, check the assumptions, and collect the data.
2. Compute the sums of squares between groups for each main effect and for the interaction, and compute the sum of squares within groups. Dividing each sum of

squares by the appropriate *df* produces the mean square between groups for each main effect and the interaction, as well as the mean square within groups. Dividing each mean square between groups by the mean square within groups produces each F_{obt}.

3. Find F_{crit} in Table 5 of Appendix C, using the *df* between groups for each factor or interaction and the df_{wn}. If the F_{obt} is larger than F_{crit}, then there is a significant difference between two or more of the means for that factor or interaction.
4. Graph the main effects by plotting the mean of each level of a factor, with the dependent variable on the *Y* axis and the levels of the factor on the *X* axis. Graph the interaction by plotting the cell means. Label the *X* axis with the levels of one factor, and in the body of the graph use a separate line to connect the means from each level of the other factor.
5. Perform *post hoc* comparisons for each significant main effect or interaction.
6. Compute eta squared to describe the proportion of variance in dependent scores accounted for by each significant main effect or interaction.
7. Based on the significant main and/or interaction effects and their values of η^2, develop an overall conclusion regarding the relationships formed by the specific means from cells and levels that differ significantly.
8. Compute the confidence interval for the value of μ represented by the mean in any relevant level or cell.

Congratulations, you're getting *very* good at this stuff.

PUTTING IT ALL TOGETHER

Throughout previous chapters, you may have been misled into thinking that you should perform only the primary analyses that test your predictions directly. Your goal as a researcher, however, is not only to demonstrate the predicted relationship but also to understand it. Therefore, after you have performed your primary analyses, explore the data. You can group scores along any potentially relevant variables. (For example, did the time of day during which participants were tested produce differences?) Or, you may correlate scores with personal information collected from participants. (Does their age relate to their performance?) The more precisely you examine your data and participants, the more precisely you'll be able to explain how a behavior operates and to describe the variables that influence it.

CHAPTER SUMMARY

1. In a two-way, between-subjects ANOVA, there are two independent variables, and all of the conditions of both factors contain independent samples. In a *complete factorial design*, all levels of one factor are combined with all levels of the other factor. Each *cell* is formed by a combination of a level from each factor. The ANOVA examines the *main effects* of manipulating each variable separately, as well as the *interaction effect* of manipulating both variables simultaneously.

2. The assumptions of the two-way, between-subjects ANOVA are that (a) each cell is a random independent sample of interval or ratio scores, (b) the populations represented in the study are normally distributed, and (c) the variances of all populations are homogeneous.

3. A two-way ANOVA is similar to three one-way ANOVAs, with an F_{obt} for each main effect and for the interaction. The means in each main effect result from collapsing across the levels of the other factor. In the interaction, the cell means are examined without collapsing.

4. A significant F_{obt} for a main effect indicates that at least two main effect means from the factor represent significant differences in scores.

5. A significant F_{obt} for an interaction indicates that the effect of changing the levels of one factor *depends on* which level of the other factor is examined. Therefore, the relationship between one factor and the dependent variable changes as the levels of the other factor change. When graphed, an interaction produces *nonparallel lines*.

6. *Post hoc* comparisons are performed on each significant effect that has more than two levels. *Post hoc* comparisons on the interaction are performed for *unconfounded* comparisons only. The means from two cells are unconfounded if the cells differ along only one factor.

7. Conclusions from a two-way ANOVA are based on the significant main and interaction effects and which level or cell means differ significantly. The primary interpretation of a two-way ANOVA usually rests on the significant interaction.

8. *Eta squared* describes the effect size of each significant main effect and interaction. A confidence interval can be computed for the μ represented by any $\overline{X}$ in the study.

PRACTICE PROBLEMS

(Answers for odd-numbered problems are provided in Appendix D.)

1. Identify what the following terms indicate: (a) Two-way design (b) Complete factorial (c) Cell (d) Two-way, between-subjects design
2. What are the reasons for conducting multifactor designs?
3. A student hears that a 2×3 design was conducted and concludes that six factors were examined. Is this conclusion correct? Why or why not?
4. Identify the *F*s that are computed in a two-way ANOVA involving factors A and B.
5. (a) What is the difference between a main effect mean and a cell mean? (b) A significant main effect indicates what about your manipulation? (c) A significant interaction indicates what about your manipulation? (d) Why do we usually base the interpretation of a two-way design on the interaction when it is significant?
6. What does it mean to collapse across a factor?
7. For a 2×2 ANOVA, describe the following in words. (a) The statistical hypotheses for factor A. (b) The statistical hypotheses for factor B. (c) The statistical hypotheses for $A \times B$.

8. Describe how a two-way ANOVA is similar to three one-way ANOVAs.
9. Below are the cell means of three experiments. For each experiment, compute the main effect means and indicate whether there appears to be an effect of A, B, and/or A × B.

Study 1

	A_1	A_2
B_1	2	4
B_2	12	14

Study 2

	A_1	A_2
B_1	10	5
B_2	5	10

Study 3

	A_1	A_2
B_1	8	14
B_2	8	2

10. In problem 9, if you graph the cell means (labeling the *X* axis with factor A), what pattern will you see for each interaction?
11. After performing a 3 × 4 ANOVA with equal *ns*, you find that all *Fs* are significant. What other procedures should you perform?
12. (a) When is it appropriate to compute the effect size in a two-way ANOVA? (b) For each effect, what does the effect size tell us?
13. (a) How can you increase the power of a two-way ANOVA? (b) Doing so will increase the power of F_{obt} and what other procedure?
14. A researcher studies participants' frustration levels when solving problems both as a function of the difficulty of the problem and as a function of whether they are math or logic problems. She finds that logic problems produce significantly more frustration than math problems, that greater difficulty leads to significantly greater frustration, and that more difficult logic problems produce significantly greater frustration than more difficult math problems. In the ANOVA performed for this study, what effects are significant?
15. In question 14, say the researcher instead found no difference between math and logic problems, but that frustration significantly increases with greater difficulty, and that this is true for both math and logic problems. In the ANOVA, what effects are significant?
16. In an experiment, you measure the popularity of two brands of soft drinks (factor A), and for each brand you test males and females (factor B). The following table shows the main effect and cell means from the study:

		Factor A		
		Level A_1: Brand X	*Level A_2: Brand Y*	
Factor B:	*Level B_1: male*	14	23	18.5
	Level B_2: female	25	12	18.5
		19.5	17.5	

(a) Describe the graph of the interaction means when factor A is on the *X* axis. (b) Does there appear to be an interaction? Why? (c) Why will a significant interaction prohibit you from making conclusions based on the main effects?

17. (a) What is a confounded comparison, and when does it occur? (b) What is an unconfounded comparison, and when does it occur? (c) Why do we not perform confounded *post hoc* comparisons?

18. A researcher examines performance on an eye-hand coordination task as a function of three levels of reward and three levels of practice, obtaining the following cell means.

		Reward		
		Low	*Medium*	*High*
Practice	*Low*	4	10	7
	Medium	5	5	14
	High	15	15	15

(a) What are the main effect means for reward, and what do they indicate about this factor? (b) What are the main effect means for practice, and what do they indicate? (c) Is an interaction likely? (d) How would you perform unconfounded *post hoc* comparisons of the cell means?

19. (a) In question 18, why does the interaction contradict your conclusions about the effect of reward? (b) Why does the interaction contradict your conclusions about practice?

20. In question 18, the researcher reports that the effect size of reward is .14, that the effect size of practice is .31, and that the interaction accounts for .01 of the variance. What does each value indicate about the influence of these effects?

21. A study compared the performance scores of males and females tested either early or late in the day. Here are the data:

		Factor A	
		Level A_1: males	*Level A_2: females*
Factor B	*Level B_1: early*	6 11 9 10 9	8 14 17 16 19
	Level B_2: late	8 10 9 7 10	4 6 5 5 7

(a) Using $\alpha = .05$, perform an ANOVA and complete the summary table. (b) Compute the main effect means and interaction means. (c) Perform the appropriate *post hoc* comparisons. (d) What do you conclude about the relationships this study demonstrates? (e) Compute the effect size where appropriate.

22. A researcher investigated the effects of (1) whether participants meditate, and (2) the degree of introversion they exhibit, on the dependent variable of hypnotic suggestibility. She collected the following data:

		Factor A		
		Low Introversion	*Medium Introversion*	*High Introversion*
Factor B	*Meditation*	5 6 2 2 5	7 5 6 9 5	9 8 10 10 10
	No Meditation	10 10 9 10 10	2 5 4 3 2	5 6 5 7 6

(a) Complete the ANOVA and create a summary table. (b) Determine which effects are significant. (c) For each significant effect, compute the means and perform the appropriate *post hoc* comparisons. (d) Compute the effect size where appropriate. (e) What conclusions can be drawn about this study?

23. You conduct an experiment involving two levels of self-confidence (A_1 is low, and A_2 is high) and examine participants' anxiety scores after they speak to one of four groups of differing sizes (B_1 through B_4 represent speaking to a small, medium, large, or extremely large group, respectively). You compute the following sums of squares ($n = 5$ and $N = 40$):

Source	*Sum of squares*	*df*	*Mean square*	*F*
Between				
Factor A	8.42	___	___	___
Factor B	76.79	___	___	___
Interaction	23.71	___	___	___
Within	110.72	___	___	
Total	219.64	___		

(a) Complete the ANOVA summary table. (b) With $\alpha = .05$, what do you conclude about each F_{obt}? (c) Compute the appropriate values of *HSD*. (d) For the levels of factor B, the means are $\overline{X}_1 = 18.36$, $\overline{X}_2 = 20.02$, $\overline{X}_3 = 24.6$, and $\overline{X}_4 = 27.3$. What should you conclude about the main effect of B? (e) How important is the size of the audience in determining a participant's anxiety score? How important is the participant's self-confidence?

24. A researcher investigates the effects of 5 levels of frustration (Factor A) and low and high amounts of sleep deprivation (Factor B) on depression scores ($n = 5$ and $N = 50$). He obtains the following sums of squares.

Source	*Sum of squares*	*df*	*Mean square*	*F*
Between				
Factor A	22.69	___	___	___
Factor B	46.02	___	___	___
Interaction	6.13	___	___	___
Within	109.25	___	___	
Total	184.09	___		

(a) Complete the ANOVA summary table. (b) Which effects are significant? (c) Which *post hoc* comparisons are appropriate? (d) Compute the effect size where appropriate. (e) What conclusions can be drawn about this study?

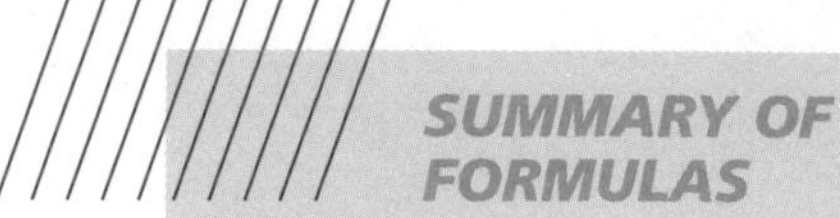

SUMMARY OF FORMULAS

The general format for the summary table for a two-way between-subjects ANOVA is

Summary Table of Two-Way ANOVA

Source	*Sum of squares*	*df*	*Mean square*	*F*
Between				
Factor A	SS_A	df_A	MS_A	F_A
Factor B	SS_B	df_B	MS_B	F_B
Interaction	$SS_{A \times B}$	$df_{A \times B}$	$MS_{A \times B}$	$F_{A \times B}$
Within	SS_{wn}	df_{wn}	MS_{wn}	
Total	SS_{tot}	df_{tot}		

1. Computing the sums of squares

(a) The computational formula for SS_{tot} is

$$SS_{tot} = \Sigma X^2_{tot} - \left(\frac{(\Sigma X_{tot})^2}{N}\right)$$

(b) The computational formula for the sum of squares between groups for the column factor A is

$$SS_A = \Sigma\left(\frac{(\text{sum of scores in the column})^2}{n \text{ of scores in the column}}\right) - \left(\frac{(\Sigma X_{tot})^2}{N}\right)$$

(c) The computational formula for the sum of squares between groups for the row factor B is

$$SS_B = \Sigma\left(\frac{(\text{sum of scores in the row})^2}{n \text{ of scores in the row}}\right) - \left(\frac{(\Sigma X_{tot})^2}{N}\right)$$

(d) The computational formula for the sum of squares between groups for the interaction is

$$SS_{A \times B} = SS_{bn} - SS_A - SS_B$$

where SS_{bn} *is found using the formula*

$$SS_{\text{bn}} = \Sigma\left(\frac{(\text{sum of scores in the cell})^2}{n \text{ of scores in the cell}}\right) - \left(\frac{(\Sigma X_{\text{tot}})^2}{N}\right)$$

(e) *The computational formula for the sum of squares within groups is*

$$SS_{\text{wn}} = SS_{\text{tot}} - SS_{\text{bn}}$$

2. Computing the degrees of freedom
(a) The degrees of freedom between groups for factor A, df_{A}, equals $k_{\text{A}} - 1$, where k_{A} is the number of levels in factor A.
(b) The degrees of freedom between groups for factor B, df_{B}, equals $k_{\text{B}} - 1$, where k_{B} is the number of levels in factor B.
(c) The degrees of freedom between groups for the interaction, $df_{\text{A}\times\text{B}}$, equals df_{A} multiplied times df_{B}.
(d) The degrees of freedom within groups equals $N - k_{\text{A}\times\text{B}}$, where N is the total N of the study and $k_{\text{A}\times\text{B}}$ is the total number of cells in the study.

3. Computing the mean square
(a) *The formula for* MS_{A} *is*

$$MS_{\text{A}} = \frac{SS_{\text{A}}}{df_{\text{A}}}$$

(b) *The formula for* MS_{B} *is*

$$MS_{\text{B}} = \frac{SS_{\text{B}}}{df_{\text{B}}}$$

(c) *The formula for* $MS_{\text{A}\times\text{B}}$ *is*

$$MS_{\text{A}\times\text{B}} = \frac{SS_{\text{A}\times\text{B}}}{df_{\text{A}\times\text{B}}}$$

(d) *The formula for* MS_{wn} *is*

$$MS_{\text{wn}} = \frac{SS_{\text{wn}}}{df_{\text{wn}}}$$

4. Computing F_{obt}
(a) *The formula for* F_{A} *is*

$$F_{\text{A}} = \frac{MS_{\text{A}}}{MS_{\text{wn}}}$$

(b) *The formula for F_B is*

$$F_B = \frac{MS_B}{MS_{wn}}$$

(c) *The formula for $F_{A \times B}$ is*

$$F_{A \times B} = \frac{MS_{A \times B}}{MS_{wn}}$$

5. The critical values of F are found in Table 5 of Appendix C.
(a) To find F_{crit} to test F_A, use df_A and df_{wn}.
(b) To find F_{crit} to test F_B, use df_B and df_{wn}.
(c) To find F_{crit} to test $F_{A \times B}$, use $df_{A \times B}$ and df_{wn}.

6. Performing Tukey's *HSD post hoc* comparisons
(a) *For each significant main effect, the computational formula for the HSD is*

$$HSD = (q_k)\left(\sqrt{\frac{MS_{wn}}{n}}\right)$$

where q_k is found in Table 6 in Appendix C for k equal to the number of levels in the factor, MS_{wn} is the denominator of F_{obt}, and n is the number of scores used to compute each mean in the factor. Any means that differ by an amount that is greater than the value of *HSD* are significantly different.
(b) *For a significant interaction, the HSD is computed as follows.*
(1) Enter the following table for the design (or number of cells), and obtain the adjusted value of k.

Values of Adjusted k

Design of study	*Number of cell means in study*	*Adjusted value of k*
2 × 2	4	3
2 × 3	6	5
2 × 4	8	6
3 × 3	9	7
3 × 4	12	8
4 × 4	16	10
4 × 5	20	12

(2) Enter Table 6 in Appendix C for the value of q_k, using the adjusted k and df_{wn}.
(3) Compute the value of *HSD* as described in step 6(a) above.
(4) Any unconfounded cell means that differ by an amount that is greater than the value of *HSD* are significantly different.

7. *The computational formula for eta squared is*

$$\eta^2 = \frac{\text{sum of squares between groups for the factor}}{SS_{\text{tot}}}$$

When η^2 is computed for factor A, factor B, or the A × B interaction, the sum of squares for the effect is SS_A, SS_B, or $SS_{A\times B}$, respectively.

8. *The computational formula for computing the confidence interval for a single μ, using the results of a between-subjects ANOVA, is*

$$\left(\sqrt{\frac{MS_{\text{wn}}}{n}}\right)(-t_{\text{crit}}) + \overline{X} \leq \mu \leq \left(\sqrt{\frac{MS_{\text{wn}}}{n}}\right)(+t_{\text{crit}}) + \overline{X}$$

where t_{crit} is the two-tailed value at the appropriate α with $df = df_{\text{wn}}$, MS_{wn} is from the ANOVA, and $\overline{X}$ and n are from the level or cell being described.

19

Within-Subjects Experiments and the Analysis of Variance

To understand this chapter:

- From Chapters 15 and 16, understand how to create a between-subjects or a within-subjects factor.
- From Chapter 18, understand how a two-way ANOVA is a series of one-way ANOVAs for the main effects and interaction, and be sure you can interpret an interaction.

Then your goals in this chapter are to learn:

- The logic and analysis of a one-way within-subjects ANOVA.
- The logic and analysis of a two-way within-subjects ANOVA.
- How mixed designs are created and analyzed.
- The logic and analysis of experiments having three or more factors.
- What planned comparisons, simple main effects, and the F_{max} test are.
- What a multivariate analysis and a meta-analysis are.

This chapter explores one-way and two-way designs that involve within-subjects factors. The logic and calculations here are essentially simple variations on designs you've seen in previous chapters. Also, we'll discuss several advanced topics so if you run into them in the literature, you'll understand their basics.

There is no section on more statistical notation here because you already know the symbols. Recall, however, that we control subject variables that are correlated with the influence of the independent variable or with responses on the dependent variable. Although we may limit the population or actively balance such variables, to control subject variables more directly we may either match participants across the conditions if there is one major variable to control, or repeatedly measure the same participants in all levels of a factor if there are numerous variables to control. Usually when subject variables are a concern, many variables are at issue so repeated measures factors are most common.

THE ONE-WAY WITHIN-SUBJECTS ANALYSIS OF VARIANCE

In a one-way, within-subjects ANOVA, we have one independent variable in which we either match different subjects across the conditions or we repeatedly measure the same participants under all the levels of a factor. The other assumptions of this ANOVA are the same as with previous ANOVAs: (1) Participants are randomly selected and the dependent variable is a ratio or interval variable; (2) The populations of scores represented by the conditions are normally distributed; and (3) The population variances are homogeneous.

Here's an example study. Say we're interested in whether participants' manner of dress influences how comfortable they feel in a social setting. There are probably great differences in subject variables that influence how people react in this situation, like how shy they are, their self-perception of their attractiveness, their past social experiences, and so on. Therefore, to isolate only the influence of manner of dress, we'll conduct a repeated measures design. Then we can see how a self-assured and comfortable person reacts when wearing different types of dress, as well as how a nervous and shy person reacts.

Say that on three consecutive days we have each participant act as a "greeter" for people participating in a another experiment. On one day our participants dress very casually, on another day they dress semiformally, and on another they dress very formally. At the end of each day, participants answer a brief questionnaire describing how comfortable they felt greeting people. All of the usual design issues apply here, so we seek a reliable, valid, and strong manipulation of the independent variable, and a reliable and valid measurement of the dependent variable without the presence of confoundings. In particular, therefore, we'd be concerned with counterbalancing the order in which participants perform the three conditions and we might need alternate forms of the questionnaire.

For simplicity, say we test the very unpowerful N of five participants. Labeling the independent variable of type of dress as factor A, you can diagram the design as in Table 19.1. We test the five participants under level A_1, casual, and obtain their comfort scores, we test the same five people under level A_2, semiformal, and obtain their scores, and we also test them under level A_3, formal. We find the level means of factor A by collapsing vertically in each column. As usual, we're testing whether the means from the different levels represent different population μs. Therefore, the hypotheses are the same as in a between-subjects design: H_0: $\mu_1 = \mu_2 = \mu_3$ and H_a: Not all μs are equal. Thus, this is just like a one-way between-subjects design. The only novelty here is in how we calculate F_{obt}.

TABLE 19.1 Diagram of One-Way Repeated Measures Study of the Influence of Type of Dress on Participants' Comfort Levels

		Factor A: Type of Dress		
		Level A_1: Casual	*Level A_2: Semiformal*	*Level A_3: Formal*
Participants Factor	*1*	X	X	X
	2	X	X	X
	3	X	X	X
	4	X	X	X
	5	X	X	X
		$\overline{X}_{A_1}$	$\overline{X}_{A_2}$	$\overline{X}_{A_3}$

> ***REMEMBER*** The one-way within-subjects design is like a between-subjects design, except that usually, participants are tested using repeated measures, and F_{obt} is calculated differently.

Although this is a one-way ANOVA, we can view it as a two-way ANOVA, with factor A as the original factor and the different subjects as a second factor (here with five levels). Then factor A is the type of dress, factor B is the participants factor, and there is an interaction in which each cell is the combination of a level of the subjects factor and a level of the type of dress factor.

It's important to view the design as two-way, because previously when computing F_{obt} we needed a mean square within groups, MS_{wn}. This was the estimate of the error variance, σ^2_{error}, the inherent variability in the scores. We computed MS_{wn} using the differences between the scores in each cell and the mean of the cell. As above, however, in a one-way repeated measures design, each cell contains only one score because there is only one participant. Therefore, the mean of each cell *is* the score in the cell, and differences within a cell are always zero. Obviously, we cannot compute MS_{wn} in the usual way.

In a repeated measures ANOVA, however, the mean square for the *interaction* between factor A and the subjects factor, abbreviated $MS_{A \times S}$, does reflect the inherent variability of scores. Recall that an interaction indicates that the effect of one factor changes depending on which level of the other factor we're talking about. The interaction between subjects and factor A indicates that the influence of factor A changes depending on the subject we're talking about. It is the inherent variability between subjects that produces this change. Therefore, $MS_{A \times S}$ is our estimate of error variance, σ^2_{error}, and so it is the denominator of the F-ratio. (For this reason, instead of being identified as the interaction in the ANOVA summary table, you'll often see this term called the "error" term.)

This is the only novelty in calculating any ANOVA involving within-subjects factors. In the denominator of the F-ratio for the factor is an MS that is based on the interaction between participants and that factor. As usual, the numerator of the F-ratio, MS_A, describes the difference between the means in factor A, and it is our estimate of variability due to error plus variability due to treatment. Thus, the F-ratio for a repeated measures factor is

	Sample	*Estimates*	*Population*
$F_{obt} =$	$\dfrac{MS_A}{MS_{A \times S}}$	$\rightarrow$ $\rightarrow$	$\dfrac{\sigma^2_{error} + \sigma^2_{treat}}{\sigma^2_{error}}$

REMEMBER In a one-way within-subjects ANOVA, the denominator of the F-ratio is the interaction of the factor and subjects.

As usual, if the data perfectly represent the situation where H_0 is true and all μs are equal, then both the numerator and the denominator will estimate only the same error variance, σ^2_{error}, so F_{obt} will equal 1. The larger the F_{obt}, however, the less likely it is that the differences between the means merely reflect sampling error in representing one population μ. If F_{obt} is significant, we conclude that at least two of the means from factor A represent different values of μ.

Computing the One-Way Within-Subjects ANOVA

Say that we obtained the data in Table 19.2. The first step is to compute the ΣX, the $\overline{X}$, and the ΣX^2 for each level of factor A (each column). Then compute ΣX_{tot} and ΣX^2_{tot}. Also compute ΣX_{sub}, which is the ΣX for each participant's scores (each horizontal row). Notice that the ns and N are based on the number of *scores*, not the number of participants.

TABLE 19.2 Data and Initial Computations for Example One-Way Repeated Measures Study

		Factor A: Type of Dress			
		Level A_1: casual	*Level A_2: semiformal*	*Level A_3: formal*	
	1	4	9	1	$\Sigma X_{sub} = 14$
	2	6	12	3	$\Sigma X_{sub} = 21$
Participants	3	8	4	4	$\Sigma X_{sub} = 16$
	4	2	8	5	$\Sigma X_{sub} = 15$
	5	10	7	2	$\Sigma X_{sub} = 19$
		$\Sigma X = 30$ $\Sigma X^2 = 220$ $n_1 = 5$ $\overline{X}_1 = 6$	$\Sigma X = 40$ $\Sigma X^2 = 354$ $n_2 = 5$ $\overline{X}_2 = 8$	$\Sigma X = 15$ $\Sigma X^2 = 55$ $n_3 = 5$ $\overline{X}_3 = 3$	Total: $\Sigma X_{tot} = 30 + 40 + 15 = 85$ $\Sigma X^2_{tot} = 220 + 354 + 55 = 629$ $N = 5$ $k = 3$

Then follow these steps.

Step 1 is to compute the total sum of squares, SS_{tot}.

THE COMPUTATIONAL FORMULA FOR SS_{tot} IS

$$SS_{tot} = \Sigma X^2_{tot} - \left(\frac{(\Sigma X_{tot})^2}{N}\right)$$

Filling in this formula from the example data, we have

$$SS_{tot} = 629 - \left(\frac{85^2}{15}\right)$$

$$SS_{tot} = 629 - 481.67$$

so

$$SS_{tot} = 147.33$$

Note that the quantity $(\Sigma X_{tot})^2/N$ is the *correction* in the following computations. (Here the correction is 481.67.)

Step 2 is to compute the sum of squares for factor A, SS_A.

THE COMPUTATIONAL FORMULA FOR THE SUM OF SQUARES BETWEEN GROUPS FOR FACTOR A IS

$$SS_A = \Sigma\left(\frac{(\text{sum of scores in the column})^2}{n \text{ of scores in the column}}\right) - \left(\frac{(\Sigma X_{tot})^2}{N}\right)$$

This formula says to take the sum of X in each level (column) of factor A, square the sum, and divide by the n of the level. After doing this for all levels, add the results together and subtract the correction.

For the example

$$SS_A = \left(\frac{(30)^2}{5} + \frac{(40)^2}{5} + \frac{(15)^2}{5}\right) - 481.67$$

so

$$SS_A = 545 - 481.67$$

Thus,

$$SS_A = 63.33$$

Step 3 is to find the sum of squares for subjects, SS_{subs}.

THE COMPUTATIONAL FORMULA FOR THE SUM OF SQUARES FOR SUBJECTS, SS_{subs}, IS

$$SS_{\text{subs}} = \frac{(\Sigma X_{\text{sub1}})^2 - (\Sigma X_{\text{sub2}})^2 + \cdots + (\Sigma X_n)^2}{k} - \frac{(\Sigma X_{\text{tot}})^2}{N}$$

This says to take ΣX_{sub}, the sum for each subject, and square it. Then add the squared sums together. Next divide by k, where k is the number of levels of factor A. Finally, subtract the correction.

In the example,

$$SS_{\text{subs}} = \frac{(14)^2 + (21)^2 + (16)^2 + (15)^2 + (19)^2}{3} - 481.67$$

so

$$SS_{\text{subs}} = 493 - 481.67$$

Thus,

$$SS_{\text{subs}} = 11.33$$

Step 4 is to find the sum of squares for the interaction, $SS_{\text{A}\times\text{S}}$. To do this subtract the sums of squares for the other factors from the total.

THE COMPUTATIONAL FORMULA FOR THE INTERACTION OF FACTOR A BY SUBJECTS, $SS_{A\times S}$, IS

$$SS_{\text{A}\times\text{S}} = SS_{\text{tot}} - SS_{\text{A}} - SS_{\text{subs}}$$

Thus,

$$SS_{\text{A}\times\text{S}} = 147.33 - 63.33 - 11.33$$

so

$$SS_{\text{A}\times\text{S}} = 72.67$$

Step 5 is to determine the degrees of freedom.

THE DEGREES OF FREEDOM BETWEEN GROUPS FOR FACTOR A IS

$$df_{\text{A}} = k_{\text{A}} - 1$$

k_{A} is the number of levels of factor A. (In the example, there are three levels of type of dress, so df_{A} is 2.)

THE DEGREES OF FREEDOM FOR THE INTERACTION IS

$$df_{A\times S} = (k_A - 1)(k_{subs} - 1)$$

k_A is the number of levels of factor A, and k_{subs} is the number of subjects. In the example, there are three levels of factor A and 5 subjects, so $df_{A\times S} = (2)(4) = 8$.

Compute df_{subs} and df_{tot} to check the above *df*. The $df_{subs} = k_{subs} - 1$, where k_{subs} is the number of subjects. The $df_{tot} = N - 1$, where N is the total number of *scores* in the experiment. The df_{tot} is equal to the sum of all the other *dfs*.

Step 6 is to place the sum of squares and the *dfs* in the summary table. For the example, we have the following table.

Summary Table of One-Way Repeated Measures ANOVA

Source	*Sum of squares*	*df*	*Mean square*	*F*
Factor A (dress)	63.33	2	MS_A	F_A
Subjects	11.33	4		
Interaction (A × subjects)	72.67	8	$MS_{A\times S}$	
Total	147.33	14		

Because we have only one factor of interest here (type of dress), we find only the F_{obt} for factor A.

Step 7 is to find the mean squares for factor A and the interaction.

THE MEAN SQUARE FOR FACTOR A, MS_A, EQUALS

$$MS_A = \frac{SS_A}{df_A}$$

In the example,

$$MS_A = \frac{SS_A}{df_A} = \frac{63.33}{2} = 31.67$$

THE MEAN SQUARE FOR THE INTERACTION BETWEEN FACTOR A AND SUBJECTS, $MS_{A\times S}$, IS

$$MS_{A\times S} = \frac{SS_{A\times S}}{df_{A\times S}}$$

For our example,

$$MS_{A \times S} = \frac{SS_{A \times S}}{df_{A \times S}} = \frac{72.67}{8} = 9.08$$

Step 8 is to find F_{obt}.

THE COMPUTATIONAL FORMULA FOR THE REPEATED MEASURES F-RATIO IS

$$F_{obt} = \frac{MS_A}{MS_{A \times S}}$$

In our example,

$$F_{obt} = \frac{MS_A}{MS_{A \times S}} = \frac{31.67}{9.08} = 3.49$$

Thus, our finished summary table is as follows:

Source	*Sum of Squares*	*df*	*Mean Square*	*F*
Factor A (Dress)	63.33	2	31.67	3.49
Subjects	11.33	4		
Interaction (A × Subjects)	72.67	8	9.08	
Total	147.33	14		

Step 9 is to find the critical value of F in Table 5 of Appendix C. Use df_A as the degrees of freedom between groups and $df_{A \times S}$ is the degrees of freedom within groups. In the example, for $\alpha = .05$, $df_A = 2$, and $df_{A \times S} = 8$, the F_{crit} is 4.46.

Interpreting the Within-Subjects F_{obt}

At this point, interpret the F_{obt} exactly as you would a between-subjects F_{obt}. Because the above F_{obt} is not larger than F_{crit}, it is not significant. Thus, we do not have evidence that the means from the different levels of type of dress represent different populations of comfort scores.

Had F_{obt} been significant, however, then at least two of the level means differ significantly. To identify which levels differ significantly, you would perform the Tukey *HSD post hoc* test discussed below. (Repeated measures must produce equal *ns* in all levels, so there is no reason to perform Fisher's Protected *t*-test from Chapter 17.) Also, you would graph the results as in previous chapters and compute eta squared (as usual, dividing SS_A by SS_{tot}). You can also compute a confidence interval for the μ estimated by each $\overline{X}$: Use the formula from Chapter 17, but in place of MS_{wn}, use $MS_{A \times S}$. Then, as in every study, interpret the results psychologically, generalizing the relationship in the sample to relevant hypothetical constructs, models, and so on.

In Chapter 16, you saw that the dependent samples t-test is more powerful than the independent samples t-test, because the variability in the scores is less. For the same reason, the repeated-measures ANOVA is more powerful than the between-subjects ANOVA. With repeated measures ANOVA, some variability in the raw scores is deleted by separating out the sum of squares for the subjects factor. This produces an $MS_{A \times S}$ that is smaller than the MS_{wn} would be if a between-subjects ANOVA was computed. Therefore, a larger F_{obt} is produced, which is more likely to be significant, so we have greater power. Thus, perhaps the results in the above example were not significant because it was an un-powerful design: It did involve a small N and the level means did not differ drastically, especially in light of the relatively large variability in the scores. However, this design was still more powerful—more likely to produce significant results—than a comparable between-subjects design.

Performing Tukey *Post Hoc* Tests on a Within-Subjects Factor

Previous formulas for Tukey's *HSD* involved MS_{wn}, which was the denominator when calculating any between-subjects F-ratio. Recall that in ANOVAs involving within-subjects factors, however, the F-ratio does not have MS_{wn} as the denominator. In fact, as you'll see, the denominator of the F-ratio is a different component (with a different symbol) in different types of within-subjects ANOVAs. Therefore, to create one general formula, the term "denominator of F-ratio" has been put into the formula for *HSD*.

THE FORMULA FOR TUKEY'S HSD INVOLVING A WITHIN-SUBJECTS FACTOR IS

$$HSD = (q_k)\left(\sqrt{\frac{\text{denominator of } F\text{-ratio}}{n}}\right)$$

In the previous one-way within-subjects ANOVA, $MS_{A \times S}$ is the denominator of the F-ratio, so the value of $MS_{A \times S}$ would be the numerator in the above formula. Values of q_k are found in Table 6 of Appendix C using k (the number of levels of the factor), and df_{wn}. For the value of df_{wn}, however, use the degrees of freedom you used in computing the denominator MS of the F-ratio (for the above example, this was $df_{A \times S}$). As always, n is the number of scores in each level.

Thus, if F_{obt} had been significant, we'd place the $MS_{A \times S}$ of 9.08 in the numerator of the above formula, and with $k = 3$ and $df_{A \times S} = 8$, q_k is 4.04. The n is 5, so altogether:

$$HSD = (4.04)\left(\sqrt{\frac{9.08}{5}}\right) = (4.04)(1.348) = 5.46$$

Then, as usual, we find the difference between each pair of level means. Here we'd compare the means from the casual and semiformal conditions, the casual and formal conditions, and the semiformal and formal conditions. Any difference that is larger than the *HSD* of 5.46 value would be a significant difference.

> *REMEMBER* Compute the *HSD* in a one-way within-subjects design as you would in a between-subjects design, except replace MS_{wn} with $MS_{A \times S}$ and df_{wn} with $df_{A \times S}$.

THE TWO-WAY WITHIN-SUBJECTS DESIGN

You may also examine more than one factor, each as a within-subjects factor, if it is a sensible way to test hypotheses. As usual, you should create within-subject factors if there are subject variables in serious need of controlling. You have a **two-way within-subjects design** if (1) you repeatedly measure the same group of participants in all conditions of two independent variables, (2) you match participants in all cells, or (3) you have matched groups on one variable and repeated measures on the other. As with any multifactor design, you might examine two independent variables because each is of interest in its own right, or because you wish to examine their interaction. Also, one factor may actually result from counterbalancing some aspect of the situation whose influence you then wish to examine. If the data fit the criteria for parametric statistics, the results are analyzed using the two-way within-subjects ANOVA. The logic and interpretation for this procedure are identical to that of a two-way between-subjects design.

Here's an example from the literature: Flowers, Warner, and Polansky (1979) developed a variation on the Stroop color-word interference task discussed in Chapter 16. To gain a sense of their manipulation, quickly determine the number of numbers present in each row:

2   2   2
THREE   THREE

As in Flowers, et al., we will present to participants one row of numbers at a time. As one factor, we'll either create interference by using a number that is incongruent with the number of numbers present (e.g., 2 2 2), or not create interference because the number presented is congruent with the number of numbers present (e.g., 2 2). Let's call this the Type of Pairing factor, with Congruent-Incongruent levels. As the second factor, we'll compare presenting digits (e.g., 2 2 2) to presenting words (e.g., two two two). Let's call this the Type of Number factor, with Digit-Word levels. The dependent variable is a participant's reaction time to report the number of numbers present in a row. We test each person in all conditions of this 2 × 2 factorial repeated-measures design, which you can envision as shown in Table 19.3. As usual, we should apply the appropriate controls, such as creating comparable stimuli for each condition, testing multiple trials per condition, and counterbalancing order effects due to the order of trials per condition and the order of conditions. Above we've collapsed across these factors, however, computing each person's mean reaction time in each of the four cells.

The computational formulas for performing the two-way within-subjects ANOVA are different from those of the between-subjects design, and are presented in Part 5 of Appendix B. (The example is a 2 × 2, but you can have any number of levels of either factor.) Although the math is somewhat involved, the logic here is simple, because we again view the design as a series of one-way ANOVAs. First, we collapse vertically in each column of Table 19.3, to examine the main effect means for factor A, type of pairing. Apparently, incongruent pairings produced interference, resulting in a slower mean reaction time (.542 seconds) than congruent pairings (.499 seconds). Computing and testing the F_{obt} for Factor A, F_A, will indicate whether these means differ significantly.

Next, we collapse horizontally across type of pairing, examining the main effect means for the type of number presented. Apparently, mean reaction time for digits,

TABLE 19.3 2 × 2 Repeated Measures Design for Digit-Word, and Congruent-Incongruent Pairings

Each X represents a person's mean reaction time. Cell means are taken from Flowers, et al. (1979), Table 1.

		Factor A: Type of Pairing		
		Congruent (2 2)	*Incongruent* (2 2 2)	
Factor B: Type of Number	*Digit* (2)	X X $\bar{X} = .493$ X "	X X $\bar{X} = .543$ X "	$\bar{X} = .518$
	Word (*two*)	X X X = .505 X "	X X $\bar{X} = .542$ X "	$\bar{X} = .524$
		$\bar{X} = .499$	$\bar{X} = .542$	

.518 seconds, was faster than for words, .524 seconds. Computing and testing the F_{obt} for Factor B, F_B, will indicate whether these means differ significantly.

Finally, we examine the interaction, comparing the four cell means. The pattern is difficult to detect by looking at Table 19.3, so take a moment and graph it, placing type of number on the *X* axis. Your graph should look like Figure 19.1. The lower solid line shows that when congruent numbers producing no interference were presented (e.g., 2 2), subjects responded more quickly to digits than to words (.493 seconds versus .505 seconds, respectively). The upper, dashed, line shows that when incongruent numbers producing interference were presented (e.g., 2 2 2), there was virtually no difference in reaction time between digits and words (.543 seconds, versus .542 seconds, respectively). Computing and testing the F_{obt} for the interaction, $F_{A \times B}$, will indicate whether these cell means differ significantly.

FIGURE 19.1 Graph of Mean Reaction Time as a Function of the Interaction Between Type of Pairing and Type of Number

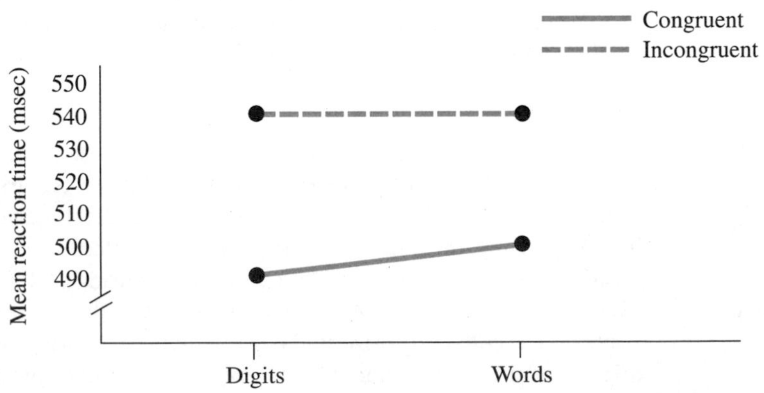

For each effect, we perform the Tukey *HSD post hoc* test if (1) the F_{obt} is significant, and (2) there are more than two levels. As in the one-way within-subjects ANOVA, we again compute the *HSD* using the denominator used in the *F*-ratio for the factor. In this ANOVA, however, the denominator of F_A is $MS_{A \times S}$, the denominator of F_B is $MS_{B \times S}$, and the denominator of $F_{A \times B}$ is $MS_{A \times B \times S}$. Therefore, using the previous formula for *HSD*, for each significant effect we substitute the appropriate *MS* in place of "denominator of *F*-ratio." Also, remember that when finding q_k for *any* interaction, you must first find the adjusted value of *k* using the small table at the beginning of Table 6 in Appendix C.

As always, graph each effect. Also, compute eta squared, η^2, for each significant main effect or interaction, dividing the SS for the factor by the total SS. Then interpret the results psychologically, focusing on the interaction if it's significant: Above, we'll explain why the difference between digits and words disappears when the incongruent, interfering stimuli are presented. If the interaction is not significant, we focus on any significant main effects: Above, we'll explain why, overall, participants responded more quickly to digits than to words, and why any type of incongruent pairing (using words or digits) produced interference.

REMEMBER The two-way within-subjects design is interpreted in the same manner as the two-way between-subjects design.

THE TWO-WAY MIXED DESIGN

In practice, you may find it difficult to match participants when there are several levels of one factor. Also, there can be extensive problems and complexity when trying to counterbalance order effects or other extraneous variables in a repeated measures design with several levels of one factor. These drawbacks can be overwhelming if you try to match participants or have repeated measures in a two-way design. Therefore, it may be better for you to create a within-subjects factor only when it critically requires control of subject variables, and to test any other factor as a between-subjects factor. Whenever a design features a "mix" of within-subjects and between-subjects factors, it is called a **mixed design**. With two factors, it is a two-way mixed design.

REMEMBER A two-way mixed design has one within-subjects factor and one between-subjects factor.

As usual, such a two-way design may arise because we are primarily interested in two variables, one of which is best suited to a within-subjects design, while the other is best suited to a between-subjects design. We may also seek to investigate only one within-subjects factor originally, but then determine the effects due to counterbalancing a second, between-subjects variable. Likewise, one common mixed design arises because we start with one between-subjects variable, but we also include the within-subjects factor of multiple trials so that we can examine practice effects. And finally, a mixed design commonly arises when we begin with one factor set up as a pretest-posttest design—measuring the same participants both before and after a treatment—and create a between-subjects factor by adding a control group.

Here's an example from real research on "subliminal perception." Typically, this research involves presenting a stimulus that is visible to subjects only very briefly—say, for about 5 milliseconds. (It takes about 150 milliseconds to blink.) The hypothesis is that the stimulus can in some way be processed, even though people do not consciously recognize that it was present. There are many misconceptions regarding subliminal perception (there is no accepted evidence that brief messages hidden in advertisements make you buy a product, or that hidden messages in music turn you into a dangerous psychopath). There is, however, well-controlled experimental evidence that a subliminal stimulus can register. For example, in cognitive research, first flashing the word *doctor* allows faster recognition of *nurse* when the latter is subsequently presented at normal speed, than does first flashing the word *bread* (Meyer & Schvanveldt, 1971). In social research, flashing words that describe honesty or meanness produce a corresponding bias in participants' later description of a confederate (Erdley & D'Agostino, 1988). In clinical research, studies have shown an influence of soothing types of subliminal messages (see Silverman & Weinberger, 1985, for a review).

Let's consider the study conducted by Silverman, Ross, Adler, and Lustig (1978). They tested the ability of men at dart-throwing before and after presentation of the subliminal message "Beating Dad is OK." The message was hypothesized to reduce residual guilt developed from childhood feelings of competition with father figures.

A *poor* way to design this study is shown in Table 19.4. After collapsing vertically in each condition, if the mean dart scores are significantly higher after the message, we'd like to conclude that the message improved performance. But! There is something seriously wrong here in terms of potential confoundings! Here's a hint: Remember maturation, history, reactivity, and practice effects? Perhaps the After-message scores improved because of the influence of these variables. That is, maybe the men acclimated to being tested, or their brains matured and developed better eye-hand coordination, or they got better at dart-throwing because of the practice they received during the earlier condition.

To eliminate these competing hypotheses, we need a control group that does everything the experimental group does but, instead of seeing the message "Beating Dad is OK," sees a placebo message that does not alleviate guilt. Adding a control group creates a between-subjects factor, with some individuals tested before and after the experimental message, and others tested before and after the control message. Now we have the much-better, two-way mixed design shown in Table 19.5. If the dependent variable meets the criteria for parametric procedures, we compute a two-way, mixed design ANOVA. The computational formulas for this design are presented in Part 6 of

TABLE 19.4 Diagram of a One-Way Dart-Throwing Experiment

Xs represent each subject's dart score.

Before message	***After message***
X	X
X	X
"	"
$\bar{X}$	$\bar{X}$

TABLE 19.5 Diagram of the Two-Way, Mixed-Design Dart-Throwing Experiment

		Repeated-measures factor	
		Before message	*After message*	
Between-subjects factor	*Participants with "Dad" message*	X X X ⋮	X X X ⋮	$\overline{X}$
	Participants with control message	X X X ⋮	X X X ⋮	$\overline{X}$
		$\overline{X}$	$\overline{X}$	

Appendix B (and again, the design need not be a 2 × 2). The logic here is the same as in previous examples, however, because *all* multifactor ANOVAs are treated as a series of one-way ANOVAs. Therefore, as usual, we'll compute and test an F_A (let's say it compares the Before-After means), an F_B (which indicates whether the "Dad" versus the control message produced significant differences) and an $F_{A \times B}$ (indicating whether the cell means differ significantly).

The main effects in the dart-throwing example, however, are not likely to indicate anything interesting. When we collapse vertically in Table 19.5 and obtain the main effect means for Before-message and After-message, both the experimental and control messages will be included. Therefore, any difference between these two means will show only that scores change between the two testings. Such differences may be due to the experimental message or to maturation and practice effects. We cannot tell.

Likewise, collapsing horizontally across Before-message and After-message, gives the main effect means for the two types of messages. If the "Dad" message produces a higher mean than the control message, we will not know whether this occurred because the Before-message scores were higher and/or because the After-message scores were higher.

The specific test of the hypothesis that the "Dad" message increases performance comes from the interaction where we compare the four cell means. Ideally, we would predict a significant interaction that produces the graph shown in Figure 19.2. These data produce an interaction because the lines are not parallel. Then ideally the *post hoc* comparisons will confirm the following: (1) There is no difference in the Before-message scores for the two groups, suggesting that the study is not contaminated by initial differences in dart-throwing skills between the two groups of participants. (2) There is no change in scores from Before-message to After-message for the control group, suggesting that maturation, acclimation to testing, practice, and so on, are not producing a Before-After difference in the experimental group. (3) The After-message scores of men receiving the "Dad" message are significantly higher than those receiving the control message. This combination of findings would convincingly support the hypothesis that the "Dad" message does improve dart-throwing. (As this example illustrates,

FIGURE 19.2 Ideal Interaction Between Pretest vs. Posttest and Control vs. Experimental Cells

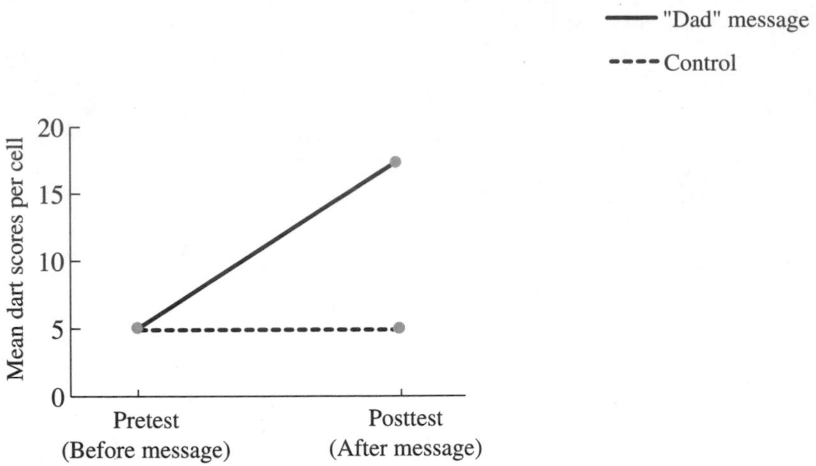

with a little thought you can predict and understand interaction effects, so regardless of the type of design, don't think solely in terms of main effects.)

In a mixed design, you again perform the Tukey *HSD* test for any significant effects with more than two means. Once again, however, computing each of the above *F*s involves a different component as the denominator of the *F* ratio. Therefore, as we did previously, for each *F* substitute the appropriate *MS* into the numerator of the formula for computing the *HSD*.

THE THREE-WAY DESIGN

The beauty of ANOVA is that it can be applied to even more complex experiments with as many factors as you wish, regardless of whether the design is all between-subjects, all within-subjects, or a mixed design. You may add more independent variables, or you may analyze more counterbalanced control factors.

For example, say that we add the variable of participants' gender (male versus female) to the above two-way dart study. With three factors, we have a **three-way design**, which with two levels of each factor is a $2 \times 2 \times 2$ design. Say that we obtained the data for this mixed design shown in Table 19.6. The previous 2×2 design for males is on the left. On the right, that design is replicated, but with females. If the data fit the criteria of a parametric procedure, then a three-way mixed design ANOVA is appropriate. In the following sections, we'll discuss how to interpret such a study (but not how to calculate the *F*s: You really need a computer for that).

Main Effects

Because there are three independent variables, the ANOVA produces a separate F_{obt} for three main effects. As usual, to find a main effect, we collapse across the other factors.

TABLE 19.6 Diagram of a Three-Way Design for the Factors of Before and After Messages, Type of Messages, and Subject Gender

Each mean is the mean dart score of subjects in that cell.

	Males		*Females*	
	Before message	*After message*	*Before message*	*After message*
"Dad" message	$\overline{X} = 10$	$\overline{X} = 20$	$\overline{X} = 8$	$\overline{X} = 12$
Control message	$\overline{X} = 10$	$\overline{X} = 10$	$\overline{X} = 8$	$\overline{X} = 10$

Thus, to find the main effect of gender, we average all of the males' scores together (for the box on the left in Table 19.6, $\overline{X} = 12.5$) and all of the females' scores together (in the box on the right, $\overline{X} = 9.5$). Apparently, overall, males were better at throwing darts than females. To find the main effect of Before-message versus After-message, we average together the columns containing Before scores, regardless of gender ($\overline{X} = 9.0$), and the columns containing After scores ($\overline{X} = 13.0$). Apparently, overall, participants scored higher on the posttest than on the pretest. Finally, to find the main effect of type of message, we average the scores in the rows of the "Dad" message ($\overline{X} = 12.5$) and in the rows of the control message ($\overline{X} = 9.5$). Apparently, overall, people who saw the "Dad" message scored higher than those who saw the control message.

Two-Way Interactions

Three factors produce three two-way interactions, and each has a separate F_{obt}. Collapsing across gender produces the interaction of Before-After and Dad control messages, shown below:

	Before	*After*
"Dad"	$\overline{X} = 9$	$\overline{X} = 14$
Control	$\overline{X} = 9$	$\overline{X} = 10$

The mean in each cell is based on both males' and females' scores. Note that the difference between the Dad and control messages is greater in the After-message condition. (To put it another way, the difference between Before-message and After-message depends on the type of message.) Apparently, therefore, there is an interaction between these two factors.

To produce the other two-way interactions, we collapse across the third factor as shown on the next page.

Gender and Type of Message interaction

	Male	*Female*
"Dad"	$\overline{X} = 15$	$\overline{X} = 10$
Control	$\overline{X} = 10$	$\overline{X} = 9$

Gender and Before-After interaction

	Male	*Female*
Before	$\overline{X} = 10$	$\overline{X} = 8$
After	$\overline{X} = 15$	$\overline{X} = 11$

On the left, collapsing across Before-After produces the interaction between gender and Type of Message. Note that the difference between the Dad and control messages is greater for males than it is for females. (To put it another way, the difference between males and females depends on which message they receive.) Thus, there is an apparent interaction between gender and type of message.

On the right, collapsing across Type of Message produces the interaction between gender and Before-After. Here, males show a greater increase from Before to After than do females. (In other words, the difference between males and females depends on whether we examine the Before scores or the After scores.) Thus, apparently there is also a two-way interaction here.

The Three-Way Interaction

Finally, we do not collapse across any factor, computing an F_{obt} for the three-way interaction: This is the effect of simultaneously changing the levels of all three factors. Previously, we saw that in a two-way interaction, the effect of one variable changes depending on the level of the second factor we examine. In a **three-way interaction**, the two-way interaction between any two variables changes depending on which level of the third factor we examine. (Conversely, if the three-way interaction is not significant, then we have basically the same two-way interaction regardless of the level of the third factor we examine.) By graphing the original cell means from Table 19.6, we have the three-way interaction shown in Figure 19.3. As this shows, the interaction between Before-After and Dad-Control depends on whether participants are male or female. For males, there is a dramatic change from pretest to posttest scores with the "Dad" message, but control subjects show no change. For females, there is a different two-way interaction: There is slight improvement in dart-throwing following the control message, suggesting that the control females benefited from their practice throwing darts. Following the Dad message, experimental females showed a slight, additional improvement beyond the practice effects of the control females. Thus, the Dad message had a minimal positive influence on females, so maybe they aren't intensely guilty about competing with their fathers to begin with. There's dramatic improvement after the Dad message among males, however, so maybe they do feel guilty about this competition, so that the message reduces an otherwise serious restriction on their performance.

> *REMEMBER* A three-way interaction shows that the interaction between any two factors is not consistent across the levels of the third factor.

FIGURE 19.3 Graphs Showing How the Two-Way Interaction Between Pretest and Posttest and "Dad"-Control Message Changes as a Function of Subject's Gender

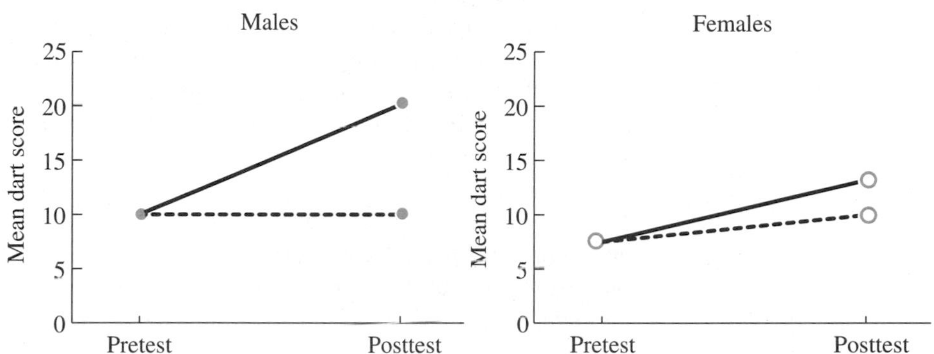

In a published report of this research (using the APA format discussed in Appendix A), this interaction would be graphed on *one* set of X-Y axes, using four different styles of lines. For example, as shown in the legend (or key) in Figure 19.3, we can combine solid lines for experimental groups and dashed lines for controls, and solid dots for males and open dots for females. Thus, •————• connects the means of the male-experimental group, ○————○ connects the means of the female-experimental group, •- - - - - - - - -• connects those of the male-control group, and ○- - - - - - - - -○ connects those of the female-control group.

Of course, we would not believe any of the previous interpretations unless the main effects and interactions were significant, and for each, we'd perform *post hoc* comparisons to determine which specific means differ significantly. Then as usual, we would interpret the results by first focusing on significant interactions, because they contradict main effects. A significant three-way interaction, however, also contradicts any two-way interactions: Above, we saw that the two-way interaction between Dad-Control and Before-After differs depending on whether it involves males or females. Therefore, the interpretation of a three-way design focuses on the significant three-way interaction. Thus, based on Figure 19.3, we would attempt to explain the psychological reasons why the Dad message produced a dramatic improvement in scores for males, but produced a small improvement for females. If the three-way interaction is not significant, we focus on significant two-way interactions. If these interactions are not significant, then we focus on significant main effects.

In sum, the logic and interpretation of all multi-factor ANOVAs is pretty much the same, with the only difference being the details of how the *F*s are computed. Therefore, you're ready for virtually any ANOVA that you may encounter in the literature. Before leaving this topic, however, there are several related procedures to briefly discuss.

THE TEST FOR HOMOGENEITY OF VARIANCE: THE F_{max} TEST

Throughout the discussions of *t*-tests and ANOVA, we have assumed that the populations represented by the data have homogeneous variance: The value of σ_X^2 is the same for each population. This is important, because when we violate this assumption, the actual probability of a Type I error will be greater than our α. Therefore, if we are unsure whether the data meet this assumption, we can perform a homogeneity of variance test. A **homogeneity of variance test** determines whether two estimated variances, s_X^2, are significantly different from each other. A significant difference between two values of s_X^2 indicates that they are likely to represent populations having different values of σ_X^2. Therefore, if the values of s_X^2 differ significantly, we *cannot* assume that the population variances are homogeneous and we should *not* perform the parametric procedure. Instead, we perform the appropriate nonparametric procedure (discussed in Chapter 21).

As an example, say we conduct a study of stress level as a function of listening to various types of music—classical, jazz, rock and roll, and a control condition of silence. After playing five participants each a selection of music, we measure their stress using a questionnaire and then compute the s_X^2 in each condition. We find:

Classical	*Jazz*	*Rock and Roll*	*Silence*
$s_X^2 = 3.8$	$s_X^2 = 8.5$	$s_X^2 = 10$	$s_X^2 = 2.5$

From the literature or from computing these values of s_X^2, we suspect that the data do not meet the assumption of homogeneity of variance.

Although there are several tests of homogeneity, a simple version is Hartley's F_{max} test. This test is used for independent samples when the *n*s in all conditions are equal. To perform the F_{max} test, first compute the estimated population variance, s_X^2, for each condition, using the formula you learned back in Chapter 8. Then select the largest value of s_X^2 and the smallest value of s_X^2. Above, the largest and smallest values of s_X^2 are 10 and 2.5. As usual, the null hypothesis says there is no difference, so here H_0 says that the two sample variances represent the same, homogeneous σ_X^2. Then,

THE COMPUTATIONAL FORMULA FOR THE F_{max} TEST IS

$$F_{max} = \frac{\text{largest } s_X^2}{\text{smallest } s_X^2}$$

where each s_X^2 is computed using the scores in one level of the experiment.

Critical values for F_{max} are found in Table 7 in Appendix C. In the table, columns across the top are labeled "*k*" for the number of levels in the factor. The rows are labeled "$n - 1$," where *n* is the number of scores in each level.

The logic of the F_{max} test is the same as the logic of the *F*-ratio. If both values of s_X^2 perfectly represent the same σ_X^2, then they should be equal, so their ratio, F_{max}, should equal 1. If they are not equal, then F_{max} will be larger than 1. The larger the F_{max}, the greater the difference between the largest and smallest s^2_X. If F_{max} is significant, the difference between the two values of s^2_X is significant.

Placing the largest and smallest values of s^2_X in the above formula gives $10/2.5 = 4$, so the obtained F_{max} is 4.0. From Table 7, the critical value of F_{max} with $k = 4$ and $n - 1 = 4$ is 13.70. Because the obtained F_{max} of 4.0 is less than this critical value, the two variances are *not* significantly different. Therefore, we do not have evidence that the sample variances represent different population variances, so we can assume that they represent equal variances: In other words, we can assume that they represent homogeneous variance, so it is acceptable to perform ANOVA.

OTHER WAYS TO COMPARE THE MEANS IN A FACTORIAL DESIGN

There are two variations to *post hoc* comparisons that you should be aware of, because they will crop up when reading the psychological literature. They are "planned comparisons" and "simple main effects."

Planned Comparisons

Instead of computing an F_{obt} and then performing *post hoc* tests, you may find research that instead only performs planned comparisons. With *post hoc* tests, all means in a factor are compared to each other, but with planned comparisons, the hypotheses lead us to compare the means in certain conditions only: We "plan ahead" which of the conditions should differ, and those are the ones we compare. **Planned comparisons** are procedures for comparing only some levels in a factor. For example, in the music and stress study above, we might use planned comparisons to compare only the control condition to each of the other conditions, without comparing jazz to classical, or jazz to rock-and-roll, and so on. Planned comparisons are also called *a priori* comparisons, and formulas for computing them can be found in advanced statistics texts.

Simple Main Effects

There is also another approach for comparing the cell means in an interaction that you may encounter, called simple main effects. A **simple main effect** is the effect of one independent variable at one level of a second independent variable. For example, say that we add the second factor of the time of day of testing (either AM or PM) to the above music and stress design, producing this experiment.

		Type of Music			
		Classical	*Jazz*	*Rock and Roll*	*Silence*
Time	*AM*	$\overline{X}$	$\overline{X}$	$\overline{X}$	$\overline{X}$
	PM	$\overline{X}$	$\overline{X}$	$\overline{X}$	$\overline{X}$

While *post hoc* comparisons compare all unconfounded pairs of cell means in the diagram, a simple main effect is similar to calculating an F for one row or column. For example, we might examine the simple main effect of changing type of music within the AM only, looking only at the upper row of four cell means in the above diagram. The simple main effect is analyzed as a one-way ANOVA on these cell means, but with somewhat different computations. (See, for example, Hinkle, Wiersma & Jurs, 1994.) If the simple main effect is significant, it indicates that for AM, changing type of music produces a significant relationship, and that somewhere among these four means there are significant differences. This information is helpful if, for example, there is not a significant simple main effect for type of music when testing in the PM. Then we'd know that the interaction reflects a relationship for AM but no relationship for PM.

GOING BEYOND THE ANALYSIS OF VARIANCE

In the literature, you'll also encounter two types of research approaches that expand upon the type of ANOVA we have discussed, including more variables and taking a broader perspective. These approaches are called "multivariate statistics" and "meta-analysis."

Multivariate Analysis

Everything in our discussions so far has involved *one* dependent variable, and the statistics we have performed are called **univariate statistics**. Researchers can, however, measure participants on two or more dependent variables in one experiment. For example, in the music study above, we might have measured the same participants' stress, happiness, and self-esteem, or any other reactions that might be influenced by music. Statistics for multiple dependent variables are called **multivariate statistics**. These include the multivariate t-test and the multivariate analysis of variance (MANOVA). Even though these are very complex procedures, the basic logic still holds: The larger the t_{obt} or F_{obt}, the less likely it is that the samples represent the same population. If the results are significant, the observed relationship between the independent variables and dependent variables is unlikely to be the result of sampling error. Given these significant results, the researcher then examines the influence of the independent variables on each individual dependent variable at a time, using the t-tests or ANOVAs we've discussed.

Meta-Analysis

Recall that ultimately, confidence in the external validity of research is developed through repeated studies that literally and conceptually replicate a finding. Rather than subjectively evaluating the extent to which several studies support a particular hypothesis, however, researchers analyze the studies using meta-analysis. **Meta-analysis** is a statistical procedure for combining, testing, and describing the results from different studies. For example, Carlson, Marcus-Newhall, and Miller (1990) performed a meta-analysis of some 22 published studies that investigated the influence of cues for aggression (e.g., weapons) on participants' aggressive behavior. With meta-analysis, researchers generally take one of two approaches: Either they determine whether the experiments taken together consistently show a significant effect of a particular variable, or they estimate the effect size of a variable based on all of the studies.

On the one hand, a meta-analysis provides objective methods for generalizing a variable's effect, and because the results are based on many participants tested under varying procedures, we have a high degree of confidence in the conclusions. On the other hand, a meta-analysis glosses over many differences in operational definitions, controls, and measurement procedures, glossing over the quality of the studies as well. Therefore, although a meta-analysis adds to our understanding of a behavior, we must necessarily speak in *very* general terms.

PUTTING IT ALL TOGETHER

There is no limit to the number of factors you can have in a study or test in an ANOVA. You can study four independent variables in the same study (using a between subjects, within-subjects or mixed design) and perform a four-way ANOVA, or you can create a five-way design, and so on. There are, however, practical limits to such designs. The number of participants needed becomes quite large, and the counterbalancing scheme or stimulus requirements may be impossibly complex. Although with effort these problems can be solved, researchers are also limited by their ability to interpret such studies. If you conduct a "simple" four-way study—a $(2 \times 2 \times 2 \times 2)$ design—the ANOVA provides separate *F*s for four main effects, six two-way interactions, four three-way interactions, and one monster four-way interaction (there would be eight lines on its graph!). If this sounds very complicated, it's because it *is* very complicated. Three-way interactions are difficult to interpret, and interactions that contain more than three factors are practically impossible to interpret.

Remember that a major concern of science is to simplify the complexity found in nature. Duplicating this complexity in a study is counterproductive. Therefore, unless you have a very good reason for including many factors in one experiment, it is best to examine only two or at most three factors at a time. Then conduct additional experiments to investigate the influence of other variables. In each, you can perform a literal replication of portions of your previous studies, thus greatly increasing their internal and external validity. And, although you will not learn of the simultaneous interactions of many variables, you will understand what you do learn.

CHAPTER SUMMARY

1. A *one-way, within-subjects ANOVA* is applied to a study having one factor in which the same participants are tested under all conditions, or there are matched subjects in each condition.
2. In a *two-way within-subjects design*, matched groups or the same repeatedly measured participants are tested in all conditions of two independent variables.
3. In a *two-way mixed design*, one within-subjects and one between-subjects factor is examined.

4. Any within-subjects factor is more powerful than if it were tested as a between-subjects factor.
5. A *three-way design* produces three main effects, three two-way interactions, and one three-way interaction. In a three-way interaction, the two-way interaction between two factors changes as the levels of the third factor change.
6. All multifactor designs can be viewed as a series of one-way ANOVAs. Differences between particular designs are in how F_{obt} is computed, particularly in the "error term" that is the denominator of the *F*-ratio. When performing Tukey's *post hoc* test on a factor, use the denominator of the *F*-ratio for that factor.
7. The F_{max} *test* tests whether the variances from the levels in an experiment differ significantly. If they do not, then the data meet the assumption of homogeneity of variance.
8. *Post hoc* comparisons compare all pairs of conditions in a factor. *Planned comparisons* compare only some specified pairs of conditions.
9. A *simple main effect* is the effect of one factor within the interaction at one level of a second factor.
10. *Univariate statistics* are performed when a study investigates one dependent variable. *Multivariate statistics* are the inferential statistical procedures performed when participants in one study are measured on two or more dependent variables.
11. *Meta-analysis* involves statistical procedures for combining, testing, and describing the results from different studies.

PRACTICE PROBLEMS

(Answers for odd-number problems are provided in Appendix D.)

1. (a) A researcher conducts a study involving one independent variable. (a) What are the two general types of parametric procedures available to her? (b) She next conducts a study involving two independent variables. What are the three versions of a parametric procedure available to her? (c) In (b), what aspect of her design determines which version she should perform?
2. What is the difference between a two-way within-subjects design and: (a) A two-way between-subjects design? (b) A two-way mixed design?
3. You read in a research report that the repeated measures factor of participants' weight gain led to a decrease in their mood. (a) What does this tell you about the design? (b) What does it tell you about the results?
4. Which of these relationships suggests using a repeated measures design? (a) Examining the improvement in language ability as children grow older. (b) Measuring participants' reaction when the experimenter surprises them by unexpectedly shouting, under three levels of volume of shouting. (c) Comparing the dating strategies of males and females. (d) Comparing memory ability as a function of amount of alcoholic beverages consumed.
5. In a study on the influence of practice on performing a task requiring eye-hand coordination, participants are tested after having no practice, after 1 hour of practice, and again after 2 hours of practice. You obtain the following data, with higher scores indicating better performance.

	Amount of Practice		
Participants	*Zero*	*One hour*	*Two hours*
S1	4	3	6
S2	3	5	5
S3	1	4	3
S4	3	4	6
S5	1	5	6
S6	2	6	7
S7	2	4	5
S8	1	3	8

(a) What are H_0 and H_A? (b) Complete the ANOVA summary table. (c) With $\alpha = .05$, what do you conclude about F_{obt}? (d) Perform the appropriate *post hoc* comparisons. (e) What is the effect size in this study? (f) What should you conclude about this relationship?

6. You measure 21 students' degree of positive attitude toward statistics at four equally spaced intervals during the semester. The mean score for each level is: Time 1, 62.50; Time 2, 64.68; Time 3, 69.32; Time 4, 72.00. You obtain the following sums of squares:

Source	*Sum of Squares*	*df*	*Mean Square*	*F*
Factor A	189.30			
Subjects	402.79			
A × Subjects	688.32			
Total	1280.41			

(a) What are H_0 and H_A? (b) Complete the ANOVA summary table. (c) With $\alpha = .05$, what do you conclude about F_{obt}? (d) Perform the appropriate *post hoc* comparisons. (e) What is the effect size in this study? (f) What do you conclude about this relationship?

7. (a) What is the F_{max} test used for? (b) In a one-way, between-subjects study with three levels, ($n = 16$), $s_X^2 = 43.68$ in level 1, $s_X^2 = 23.72$ in level 2, and $s_X^2 = 9.50$ in level 3. With $\alpha = .05$, does this study meet the assumption of homogeneity of variance? (c) What does this indicate about the inferential procedure that should be used here?
8. One research article reports a meta-analysis and another a multivariate analysis. What do these terms communicate about the procedure used in each study?
9. You measure the dependent variable of participants' hypnotic suggestibility as a function of whether or not they meditate before being tested, and whether they were shown a film containing a low, medium, or high amount of fantasy. The fantasy-level factor is repeated measures, the meditation factor is between-subjects. (a) What is the name for this type of design? (b) Identify the Fs you would obtain (without actually calculating them) and

indicate what each will tell you. (c) For the following data, compute the means for each main effect and interaction. Which effects appear likely to be significant? (d) What will you conclude about this study?

	Amount of fantasy		
	Low	*Medium*	*High*
Meditation	5 6 2 2 5	7 5 6 9 5	9 8 10 10 10
No meditation	10 10 9 10 10	2 5 4 3 2	5 6 5 7 6

(To perform the complete ANOVA, see Practice Problem 1 in Part 6 of Appendix B.)

10. In problem 9 above, say that instead, both meditation and fantasy-level are repeated measures factors. (a) What is the name for this design? (b) Identify the *F*s you will obtain (without actually calculating them) and indicate what each will tell you. (c) Will the power of this design be larger or smaller than in problem 9? Why? (To perform the complete ANOVA, see Practice Problem 1 in Part 5 of Appendix B.)

11. (a) What is a simple main effect? (b) In question 9, what does the simple main effect appear to indicate for the effect of amount of fantasy when participants meditate?

12. Chapter 16 described a study which tested a new therapy on spider-phobics by measuring their fear of a spider in a Before-therapy condition and again in an After-therapy condition. (a) What major design flaw was present in this study? (b) How would you fix it? (c) Specifically identify the type of analysis you would apply to the improved study. (d) Let Before/After be factor A and Therapy/No-therapy be factor B. If significant, what will F_A, F_B, and $F_{A \times B}$ each indicate?

13. Below are the cell means from a study comparing the performance of young boys and girls who are given candy or money as a reward, and who are tested either in the early morning or early afternoon.

	Boys		*Girls*	
	Candy	*Money*	*Candy*	*Money*
Morning	$\bar{X} = 10$	$\bar{X} = 20$	$\bar{X} = 8$	$\bar{X} = 12$
Afternoon	$\bar{X} = 20$	$\bar{X} = 10$	$\bar{X} = 8$	$\bar{X} = 12$

(a) Compute the means for each main effect and interaction. (b) Assuming that any difference between means of at least 10 is a significant difference, which main effects and interactions are significant? (c) What is your overall conclusion about the effect of the variables in this study?

14. You study whether alcohol affects performance on a simple eye-hand coordination task and whether the time of year of testing affects performance. Each participant performed the task immediately after drinking 0 or 3 drinks and each did so once during the summer and once during the winter. With $n = 3$ in each cell the following cell means were obtained.

		Drinks Prior to Task Performance	
		A_1: *0 drinks*	A_2: *3 drinks*
Time of Year	*B1: summer*	16	6
	B2: winter	11	12

(a) What is the name for this type of design? (b) Identify the *F*s you would obtain (without actually calculating them) and indicate what each will tell you. (c) Compute the means for each main effect. (d) Which effects appear likely to be significant? (e) What will you conclude about this study? (To complete the ANOVA summary table of this study, see Practice Problem 2 in Part 5 of Appendix B.)

15. A researcher studies the influence of four doses of a new drug to reduce depression in adult women who either do or do not have the AIDs virus (are HIV+ or HIV−). Dosage is a repeated measures factor and HIV status is a between-subjects factor. With $n = 3$ in each cell, the following overall mean mood improvement scores were obtained.

		Factor B: Dose of Antidepressant			
		B_1: *Control*	B_2: *Low*	B_3: *Med.*	B_4: *High*
Factor A: HIV status	A_1: *HIV−*	4	5	13	17
	A_2: *HIV+*	3	6	12	19

(a) What is the name for this type of design? (b) Identify the *F*s you would obtain (without actually calculating them) and indicate what each will tell you. (c) Compute the means for each main effect. (d) Which effects appear likely to be significant? (e) What will you conclude about this study? (To complete the ANOVA summary table of this study, see Practice Problem 2 in Part 6 of Appendix B.)

SUMMARY OF FORMULAS

1. *The formulas for the one-way within-subjects ANOVA are:*

   Step 1: Determine k, the number of levels of the factor, and the sum of scores, ΣX, the sum of squared scores, ΣX^2, the n, and the $\overline{X}$ in each level. Also calculate the sum of the scores obtained by each subject, ΣX_{sub} (the sum of each row).

   Step 2: Determine the total sum of X, ΣX_{total}, and the total sum of squared X, $\Sigma X^2_{\text{total}}$.

Step 3: Compute the Correction term:

$$\text{Correction Term} = \left(\frac{(\Sigma X_{\text{total}})^2}{N}\right)$$

Step 4: Compute the total sum of squares:

$$SS_{\text{tot}} = \Sigma X^2_{\text{total}} - \text{the Correction Term in Step 3}$$

Step 5: Compute the sum of squares between groups for factor A:

$$SS_{\text{A}} = \Sigma\left(\frac{(\text{sum of scores in each column})^2}{n \text{ of scores in the column}}\right) - \text{Step 3}$$

Step 6: Compute the sum of squares for Subjects:

$$SS_{\text{subs}} = \frac{(\Sigma X_{\text{sub1}})^2 + (\Sigma X_{\text{sub2}})^2 + \ldots (\Sigma X_{\text{subn}})^2}{k} - \text{Step 3}$$

Step 7: Compute the sum of squares for the interaction:

$$SS_{\text{A}\times\text{S}} = \text{Step 4} - \text{Step 5} - \text{Step 6}$$

Step 8: Compute the Degrees of Freedom
(a) Factor A: $df_{\text{A}} = k_{\text{A}} - 1$
(b) For A × subjects: $df_{\text{A}\times\text{S}} = (k_{\text{A}} - 1)(\text{number of subjects} - 1)$
(c) Total: $df_{\text{tot}} = N - 1$

Step 9: Compute the mean square for factor A:

$$MS_{\text{A}} = \frac{SS_{\text{A}}}{df_{\text{A}}} = \frac{\text{Step 5}}{\text{Step 8.a}}$$

Step 10: Compute the mean square for A × subjects:

$$MS_{\text{A}\times\text{S}} = \frac{SS_{\text{A}\times\text{S}}}{df_{\text{A}\times\text{S}}} = \frac{\text{Step 7}}{\text{Step 8.b}}$$

Step 11: Compute F_{obt}

$$F_{\text{obt}} = \frac{MS_{\text{A}}}{MS_{\text{A}\times\text{S}}} = \frac{\text{Step 9}}{\text{Step 10}}$$

Step 12: Critical values of F are found in Table 5 in Appendix C, for df_{A} as the degrees of freedom between groups, and $df_{\text{A}\times\text{S}}$ as the degrees of freedom within groups.
Step 13: The format of the Summary Table of One-Way Within-Subjects ANOVA is:

Source	*Sum of Squares*	*df*	*Mean Square*	*F*
Factor A (Between Groups)	SS_{A}	df_{A}	MS_{A}	F_{obt}
Subjects	SS_{subs}			
A × Subjects	$SS_{\text{A}\times\text{S}}$	$df_{\text{A}\times\text{S}}$	$MS_{\text{A}\times\text{S}}$	
Total	SS_{tot}	df_{tot}		

2. *The formula for Tukey's* HSD *involving a within-subjects factor is*

$$\text{HSD} = (q_k)\left(\sqrt{\frac{\text{denominator of } F\text{-ratio}}{n}}\right)$$

Values of q_k are found in Table 6 in Appendix C using k, the number of levels of the factor, and df_{wn}, the degrees of freedom used when computing the denominator of the F-ratio being tested. n is the number of scores in each mean being tested.

3. *Formulas for the two-way, within-subjects ANOVA are in Part 5 of Appendix B.*

4. *Formulas for the two-way, mixed design ANOVA are in Part 6 of Appendix B.*

5. *To compute eta squared* (η^2) *in any of these designs:*

   Divide the SS between groups for the factor by the SS_{tot}.

6. *The computational formula for the homogeneity of variance test is*

$$F_{max} = \frac{\text{largest } s_X^2}{\text{smallest } s_X^2}$$

Critical values of F_{max} are found in Table 7 in Appendix C using k, the number of levels in the factor, and $n - 1$, where n is the number of scores in each level.

PART 7

ALTERNATIVE APPROACHES TO DESIGN AND ANALYSIS

Believe it or not, you now understand the vast majority of the designs and statistical analyses used in psychological research. Most often, studies involve parametric data in a two- or three-way factorial experiment using ANOVA or a correlational design using the Pearson correlation coefficient. Sometimes, however, researchers do not conduct the typical factorial experiment, and sometimes they don't obtain data that are appropriate for parametric statistics. In the next chapter, we'll discuss some alternative types of research designs. Then, in the following chapter, we will discuss statistical procedures that are used with nonparametric data.

20

Quasi-Experiments and Single-Subject Designs

To understand this chapter:

- From Chapter 2, recall why participants are randomly assigned to conditions and the difference between a true and quasi-independent variable.
- From Chapter 3, recall the definition of a correlational design, and its issues of internal and external validity.
- From Chapter 18, understand how to interpret an interaction.

Then your goals in this chapter are to learn:

- The common types of quasi-experimental designs and their pitfalls.
- The types of single-subject designs.
- The pros and cons of conducting small *N* research.

In this chapter we focus on design issues as opposed to statistical procedures. First, we'll discuss quasi-experiments, learning additional ways that such research is commonly designed. Then we'll discuss an entirely different research approach, in which an experiment involves only one participant. Finally, we'll introduce research conducted on a grand scale, called program evaluation. None of these design topics is especially difficult, but be forewarned, there are a number of variations—each with its own name—so pay attention to the terminology.

UNDERSTANDING QUASI-EXPERIMENTS

Recall that in a true experiment, the researcher randomly assigns participants to the conditions of the independent variable, so it is the researcher who determines each individual's "score" on the *X* variable. Sometimes, however, the nature of the variable is such that participants cannot be randomly assigned to conditions. For example, let's say we think that personality type influences creative ability. We cannot randomly assign people to a certain personality, so, instead, we compare the creative abilities of a group of people already having one type of personality to a group having another type, and so on. Such a design is a quasi-experiment. As discussed in Chapter 2, the participants in a **quasi-experiment** are assigned to a particular condition because they have already experienced or currently exhibit that condition of the variable. The term *quasi* means "seemingly," so this design has the appearance of a true experiment. Because we do not truly manipulate the independent variable, a quasi-experiment involves a **quasi-independent variable**: We lay out the design and compare the scores between conditions as in a true experiment, but we only appear to administer the independent variable.

Quasi-experiments bear a remarkable resemblance to correlational designs. In both, participants have a score on the *X* variable because they have already experienced or exhibit that level of the variable. After all, whether we call it an experiment or not, the above example is equivalent to a study in which we merely approach a number of people, measure their personality and their creativity, and then look at the relationship between their scores. Thus, technically, a quasi-experiment is a correlational design that tests the hypothesis that a relationship exists.

The name "quasi-experiment," however, communicates two important differences from a correlational design. First, in a correlational design, participants determine the range of *X* scores obtained, and then the researcher examines the relationship across the full range. In a quasi-experiment, the researcher chooses a few, specific values of the *X* variable to examine. Thus, we might identify only a few personality types as the conditions of our quasi-independent variable, while in a truly correlational design, participants might demonstrate the full range of different personality types.

The second distinction is that a correlational design usually implies that there is little control of extraneous variables, while a quasi-experiment usually implies more control. Thus, our quasi-experiment would presumably maintain better control of researcher, environmental, and task variables than the correlational version. Ideally, such controls will yield a more reliable and internally valid study for describing the relationship.

Essentially, then, a quasi-experiment is a more controlled version of a correlational design. Because it is essentially a correlational design, however, a quasi-experiment still has two major limitations. First, its greater internal validity does not extend to extraneous subject variables. By not randomly assigning participants to conditions, we do not balance subject variables between and within conditions, so the conditions may be confounded by such variables. For example, people differing in personality type may also differ in intelligence, physiology, genetics, or history, any one of which may cause differences in creativity.

Second, we may not be able to identify the true temporal order in which the variables occur. For example, someone's personality may cause a certain creativity level to develop, but it's also possible that one's creativity level causes a certain personality to develop. Because of these restrictions, a quasi-experiment provides little confidence that differences in the independent variable cause differences in the dependent variable.

> *REMEMBER* A quasi-independent variable is confounded by subject variables, so it severely restricts internal validity for inferring the cause of a behavior.

Still, a quasi-experiment is a common and legitimate research approach. We simply accept that the results of a single study only suggest a causal variable and that, as usual, we build confidence in a conclusion only through replication.

Quasi-experiments generally occur in one of three situations: When the independent variable involves a subject variable, when it involves an uncontrollable environmental event, and when it involves the passage of time.

QUASI-INDEPENDENT VARIABLES INVOLVING SUBJECT VARIABLES

Researchers are studying a quasi-independent variable whenever they study a subject variable. Such variables include differences in individuals in terms of their anxiety, depression, self-esteem, attitudes, cognitive or physical characteristics, history and experiences, or social or work classifications. We "manipulate" such variables to the extent that we select the different types of participants that are present in the experiment. Thus, for example, Russel (1976) compared the conditions of males and females in terms of their ability to recognize sex-specific body odors (by smelling a person's well-worn t-shirt!). Burke, Chrisler, and Devlin (1989) studied brain organization by comparing left-handed and right-handed people on the dependent variable of creativity. In field research, researchers use quasi-independent variables when they examine, for example, factory workers whose jobs differ in level of responsibility and the resulting stress they produce (Martin & Wall, 1989). Quasi-independent variables also occur in animal research that compares the behaviors of different species or compares animals who differ in innate aggressiveness or dominance. In addition, a quasi-independent variable is involved any time the conditions compare "normal" to "abnormal" participants, as is often the case in clinical research.

Creating the Conditions of a Quasi-Independent Subject Variable

Identifying the participants for each condition requires first measuring individuals on the quasi-independent variable. Often pretesting is needed, either observing potential participants' overt behavior or administering a questionnaire that measures their characteristics. Using the scores from the pretest, we then operationally define each condition.

For example, let's say we hypothesize a relationship between a person's having low, medium, or high self-esteem and his or her willingness to take risks. From the research literature we can obtain any number of existing self-esteem tests, and one classic measure of risk-taking is the distance at which people stand from the target in a ring-toss game (Atkinson & Litwin, 1960). After administering the self-esteem test to a large pool of people, we use their scores to select participants for each condition. Because few people are likely to exhibit an identical level of self-esteem, however, we can define low self-esteem as a test score of between 0 and 10, medium self-esteem as a score between 45 and 55, and high self-esteem as a score between 90 and 100. The design for this study is shown in Table 20.1.

TABLE 20.1 Diagram of a One-Way Experiment with a Quasi-Independent Variable

	Self-esteem level		
	Low (0–10)	*Medium (45–55)*	*High (90–100)*
Risk-taking scores	*X*	*X*	*X*
	X	*X*	*X*
	X	*X*	*X*
	.	.	.
	.	.	.
	$\bar{X}$	$\bar{X}$	$\bar{X}$

Except for the absence of random assignment to conditions, this design is the same as in a true experiment. We face all of the usual concerns, such as ethics, standardized procedures, demand characteristics, reliable scoring, and so on. Also, we may combine the factor of self-esteem with other variables in a factorial design: We might also manipulate the number of confederates present when a person tosses rings. We may employ any combination of true and quasi-independent variables.

How effectively the conditions of a quasi-independent variable are manipulated hinges on the selection pretest. (Because this test is itself a measurement procedure, we have the usual design concerns to consider with it, such as scoring criteria, sensitivity, reliability, and demand characteristics.) Then, as always, we seek a valid manipulation of the independent variable, so for example, the above pretest must validly identify differences in self-esteem. Also, we seek a reliable, consistent manipulation, so the self-esteem scores should reliably reflect differences in self-esteem. Finally, we seek to maximize statistical power by creating a strong manipulation. Therefore, participants should have *very* distinctly low, medium, and high self-esteem scores so that the conditions are very different from one another. Then we should see large, significant differences in risk-taking.

The Problem of Regression Toward the Mean

There is a potential flaw in reliability that can occur whenever we seek to identify participants who are relatively extreme on a variable. Recall that any measurement technique may be unreliable to some extent, containing measurement error because of random distractions and flukes. Simply by chance, these influences may conspire in such a way that some participants obtain extreme scores: Some people will be particularly lucky or unlucky at guessing answers, some may feel particularly good while others might be having a bad day, or there may be quirks in the measurement procedure that cause some to score especially well or especially poorly. Such random, momentary influences will not always be present, however, and they do have a way of averaging out. Therefore, if we measure the same individuals again, their scores will tend to be less extreme, simply by chance. This time, the high scores aren't so high and the lows aren't so low, but tend more toward the middle. Because the mean falls in the middle of scores, another way to say this is that a participant's typical score will tend to be closer to the mean. This outcome is known as regression toward the mean. **Regression toward the mean** occurs when, because of inconsistent random factors, extreme scores tend to change in the direction of moving closer to the mean.

The problem with regression toward the mean is that it reduces the strength of a manipulation. For example, people identified by the pretest as having very high or very low self-esteem scores are likely to exhibit a more average level of self-esteem. Therefore, our three conditions may not actually differ in self-esteem as much as we think. If self-esteem does cause risk-taking, then with smaller differences between the levels of self-esteem, we may find smaller, possibly nonsignificant differences in risk taking.

An additional problem is that regression toward the mean also threatens internal validity, because what appears to be a change in scores due to the treatment may actually be nothing more than a change in random measurement error. For example, say that in a different study we test a counseling technique for raising a person's low self-esteem, using a pretest-posttest design. We measure and select people having low self-esteem, then apply the treatment, and then measure their self-esteem again. To some extent the peculiarities that produced very low self-esteem when we tested participants the first time will not be present the second time. Therefore, their second score will tend to be higher (closer to the mean), regardless of whether the treatment works or not.

We try to counteract regression toward the mean by employing multiple trials from the most reliable selection tests possible. Also, we can include a control group—another group that is measured at the same times as the experimental group but does not experience the treatment. The extent to which the control group's scores change will show the extent of extraneous influences, including that of regression toward the mean.

> *REMEMBER* Regression toward the mean is a change in an extreme score toward a less extreme score that occurs because random influences are not consistently present.

Interpreting the Quasi-Experiment

Even in a quasi-experiment, you can attempt to control extraneous subject variables. You can select participants from a limited population or match them on relevant variables (e.g., we might select only children and match them across the conditions on their ring-tossing ability). You may control other subject variables by balancing them (e.g., by selecting an equal number of males and females for each level of self-esteem). Recognize, however, that such controls do not eliminate the problem that a quasi-independent variable is still likely to be confounded by other extraneous subject variables.

Also recall that to minimize error variance, you want to eliminate differences *within* each group. In particular, the more that participants differ on the independent variable within a condition, the more that dependent scores will also differ. Therefore, you may narrowly define the range of selection scores that create each condition. Thus, in our self-esteem study, the goal is to select very similar people within each condition in terms of self-esteem, because then they should score consistently in terms of risk-taking. And, whenever there is lessened control and potentially large error variance, you can compensate by testing a relatively large N.

Finally, recall that any pretest may add to demand characteristics: It may alert participants to the variables under study and thus cause them to behave differently than they otherwise would. To counter this, recognize that in any correlational study, you can test the variables in the order that is least biasing to participants, capitalizing on the fact that one variable does not truly precede the other anyway. Thus, we might measure many people on the dependent variable first, and then give them the selection test to determine who will be placed in each condition when we analyze the data.

You analyze the results of quasi-experiments using the same procedures as in previous experiments. Compute the mean (or other summary measure) for each condition and then perform the *t*-test, ANOVA, or other inferential procedure that is appropriate for the study's design. However, unless you have created matched-groups or a pretest-posttest design, a quasi-independent variable is virtually always a between-subjects factor. Also, despite the problems of causality, it is again appropriate to compute the effect size of any significant quasi-independent variable.

As usual, the final step is to attempt to explain psychologically how and why the independent and dependent variables are related in nature. But tread softly around the issue of causality. In the example above, people who differ in self-esteem probably also differ on many other hidden variables. Therefore, we have little confidence that it is self-esteem itself that causes differences in risk-taking.

QUASI-INDEPENDENT VARIABLES INVOLVING ENVIRONMENTAL EVENTS: THE TIME-SERIES DESIGN

A second type of quasi-experiment arises when investigating the effect that an uncontrollable, environmental event has on behavior. Natural disasters (such as floods, hurricanes, and earthquakes) can dramatically affect an individual's mental health. Governments, schools, and industries institute programs that can influence a person's productivity and satisfaction. And societal events, such as wars, riots, and economic recessions, can alter individuals' expectations and attitudes.

Usually such variables cannot be validly studied in the laboratory (how do you create a war?). Instead, researchers study such events using the general quasi-experimental approach known as a time-series design. A **time-series design** is a repeated-measures design in which participants' behavior is measured prior to the occurrence of an event and again after it has occurred. Although this sounds like the typical pretest-posttest design, it is a quasi-experiment because participants cannot be randomly assigned to receive the treatment: We cannot randomly select those people who will experience an earthquake or who will have their school adopt a new program. We also have difficulty in creating control groups, and we cannot control the occurrence of the independent variable (in a city hit by a hurricane, not everybody experiences the same ferocity). We therefore have considerably less internal validity for concluding that the independent variable causes the dependent behavior, as well as less external validity for concluding that the same relationship is found with other participants and settings.

While there are numerous approaches to time-series designs, the four major types are discussed below. (See Campbell & Stanley, 1963, for the definitive brief text on such designs.)

One-Group Pretest-Posttest Designs

In the **one-group pretest-posttest design**, we obtain a single pretest measure on a group and then, after the event, obtain a single posttest measure of the group. For example, Nolen-Hoeksema and Morrow (1991) examined the mental stress of people before and after an earthquake. Or, Frank and Gilovich (1988) hypothesized that wearing black uniforms leads to more aggressive behavior, so they examined the number of penalty minutes incurred by a National Hockey League team before and after it changed to black uniforms.

Note that such designs provide extremely weak internal validity for inferring the causes of a behavioral change. The overwhelming problem is that they lack a control group. For example, without knowing the penalty scores of a control hockey team that is repeatedly measured, we have no idea whether the penalty scores might have changed in the experimental group, even if the uniforms hadn't been changed. Perhaps there was a simultaneous change in some other confounding factor that actually produced the increase in penalties. Maybe all teams became more aggressive, or the referees began calling more penalties than they had previously. Or perhaps the scores changed because of the players' ongoing history and maturation (maybe aggressiveness naturally increases as players become older and more experienced). Or maybe the results reflect mortality effects (with less aggressive players leaving the team between measurements). Or maybe the results reflect regression toward the mean (perhaps at the pretest, players were coincidentally experiencing very low penalty rates, and the increase at the posttest merely reflects natural fluctuations in scores).

When a study involves a simple one-group pretest-posttest design, there is no way to eliminate these possibilities. Therefore, because the conclusions from such a study are so weak, this design is typically used only when no alternative design is possible.

> *REMEMBER* In the one-group pretest-posttest design, the absence of a control group means that we cannot eliminate the possibility that extraneous variables caused the dependent scores to change.

Nonequivalent Control Group Designs

You may think that the solution to the problems with the one-group design is simply to add a control group. Implicitly, however, we always seek an *equivalent* control group. By equivalent we mean a control group that is similar to the experimental group in terms of subject variables and in terms of experiences between the pretest and the posttest. In true experiments, we attempt to obtain an equivalent control group by (1) randomly assigning participants to conditions so that we balance subject variables, and (2) keeping all experiences the same for both groups. Thus, the ideal would be to randomly select half of a team to change uniforms and the other half not to. Or select half of a city to experience an earthquake and the other half not. Then the experimental and control groups would have similar characteristics and similar experiences between the pretest and posttest. Ideally, the only difference between the groups would be that one group experiences the treatment, so any differences in their posttest scores could be attributed to it.

The problem, of course, is that we cannot create such an equivalent control group. In most cases, all members of the relevant participant pool automatically experience the treatment. The best we can do is to obtain a **nonequivalent control group**, a group that has different characteristics and different experiences during the study. For example, we might observe a hockey team that did not change to black uniforms during the same season as our experimental team. This would be a nonequivalent control group because different teams have players with different styles of play, different coaches, different game strategies, and different experiences during the season. Likewise, if we selected people who live in a different city as the control group for people who experience an earthquake, this too would be a nonequivalent group, because people living in another city may be intrinsically different and have different daily experiences.

Nonetheless, a nonequivalent control group is better than nothing. To analyze such results, however, we should not simply compare the posttest scores of the experimental and control groups. Any difference here is confounded by initial differences between the groups and by differences in experiences during the study. Instead, we can examine the *difference* between the pretest and posttest scores in each group. To illustrate, let's say the hockey teams produced the penalty data shown in Figure 20.1. Computing the difference for each group indicates the *relative* change that occurs from pretest to posttest. The experimental team showed an increase in penalty minutes from a mean of 4 to a mean of 10 minutes (a difference of 6), while the control team showed an increase from 10 to 12 (a difference of only 2). Regardless of the actual number of penalties in each group, the important finding is that, over the same time period, there was a larger increase for the team that changed uniforms.

To determine whether this difference is significant, we could first compute a pretest-posttest difference score for each player in the control group and in the experimental group. Then, because these form two independent samples of difference scores, we would perform the independent samples t-test, comparing the mean difference score for the control group with that of the experimental group. Alternatively, we could perform a two-way ANOVA on the raw penalty scores and examine the interaction. As Figure 20.1 shows, the relationship between pretest posttest scores and penalty minutes *depends* on whether we are talking about the control group or the experimental group.

FIGURE 20.1 Data for a Nonequivalent Control Group Design

These data show penalties for both experimental and control teams over the same pretest and posttest period.

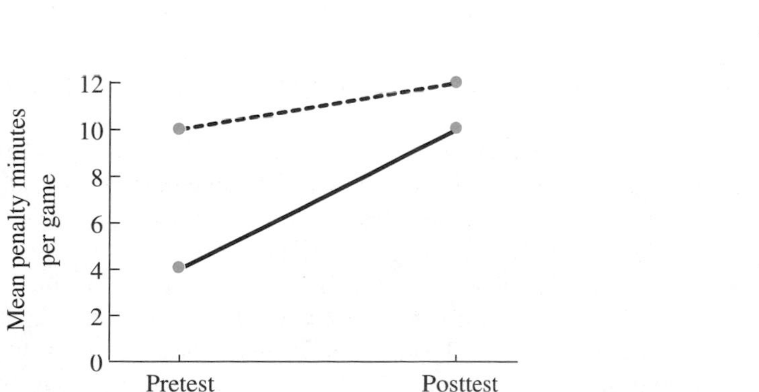

A nonequivalent control group design provides some degree of improvement in internal validity compared to the previous one-group design. The nonequivalent control group helps to eliminate potential confounding, but *only* from factors that are *common* to both groups. In the hockey study, for example, Figure 20.1 suggests that there was no confounding factor common to both teams that produced the increase in penalties. If any maturation, history, or environmental effects common to all hockey players had been responsible for the results, then the difference between pretest and posttest would be the same for both teams. Because the experimental group exhibited a larger change, something else was present for only that team which produced the change. However, it is still possible that this "something else" was not the change in uniforms. A nonequivalent groups design does not eliminate the possibility of a unique confounding or a random fluctuation that occurred only in the experimental group. Thus, it might have been some event specific to only the experimental team that actually brought about the increase in penalties (maybe a new coach was hired who actively promoted more aggressive play).

> *REMEMBER* A nonequivalent control group design eliminates only potential confounding by variables that are common to both the experimental and control groups.

Interrupted Time-Series Designs

Sometimes, we do not have access to a control group that is even approximately equivalent. For example, it is difficult to imagine the control group for survivors of an airplane crash or for those who have served as president of the United States. In such cases, we can test whether the pretest-to-posttest changes in scores would have occurred without the treatment by examining the scores of the experimental group at other times before and after the treatment. In an **interrupted time-series design**, we make observations at several spaced times prior to the occurrence of the independent variable and at several times after it. In fact, this was the approach taken by Frank and Gilovich (1988), who examined the penalty records for 10 years before and 6 years after the hockey team changed to black uniforms. Their results were similar to those shown in Figure 20.2. The researchers also incorporated the idea of a nonequivalent control group by comparing the team to the entire league. To do this, they transformed the team's yearly total penalty minutes to a z-score based on the average penalty time for the entire league. Recall that z-scores describe relative standing, so here a below-average score produces a negative z and an above-average score produces a positive z. As you can see in Figure 20.2, before the uniform change the team was consistently below the league average in penalties but after the change it was consistently above average.

From such a pattern we see two things. First, the many pretest and posttest observations demonstrate the normal random fluctuations in scores from year to year. These are not as large as the change from before to after the uniform change, so the apparent treatment effect cannot be dismissed as a random fluctuation. Second, because we see a long-term stable level of responding before the treatment and then a different long-term stable response after the treatment, it is unlikely that history, maturation, or environmental variables produced the observed change. These variables would be expected to operate over the entire 16-year period, producing similar changes at other points in time. Yet

FIGURE 20.2 Data for Interrupted Time-Series Design

Shown here are the yearly penalty records (in z-scores) of the hockey team before and after changing to black uniforms.

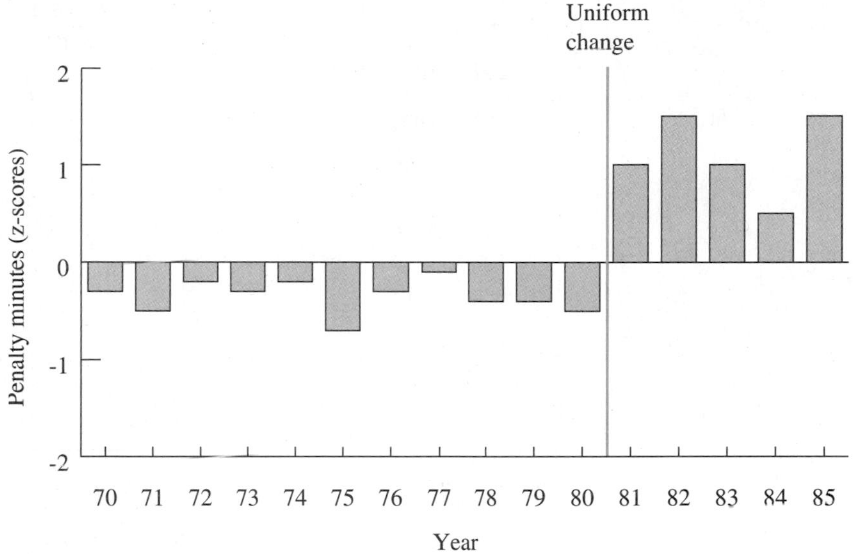

the change occurred only after the treatment was introduced. (Advanced statistical procedures are available for determining whether this is a significant change; see Cook and Campbell, 1979.)

Thus, the interrupted time-series design allows us to confidently conclude that the pretest-to posttest change in dependent scores did not result from a random fluctuation in scores or from a repeatedly occurring confounding variable. The one weakness in the design is that some variable may have coincidentally changed *once*, at the same time the treatment was introduced. For example, hockey teams change players yearly, and perhaps by chance more penalty-prone players were acquired during the same year as the uniform change. However, such an explanation would require a rather exceptional coincidence, considering all of the changes in the makeup and experiences of the team over this 16-year period. Therefore, we have substantial confidence that the change in behavior is due to the treatment. (Frank and Gilovich, 1988, provided additional confidence by also reporting a laboratory study demonstrating that wearing black does increase aggressiveness.)

> *REMEMBER* The numerous pretest and posttest observations of an interrupted time-series design reduce, but do not eliminate, the possibility that the treatment is confounded with some other event.

Multiple Time-Series Designs

To further increase confidence in the conclusions from a quasi-experiment, we can combine the interrupted time-series design and the nonequivalent control group design,

FIGURE 20.3 Data for a Multiple Time-Series Design

The yearly penalty record of hockey team before and after changing to black uniforms, and of a nonequivalent control team.

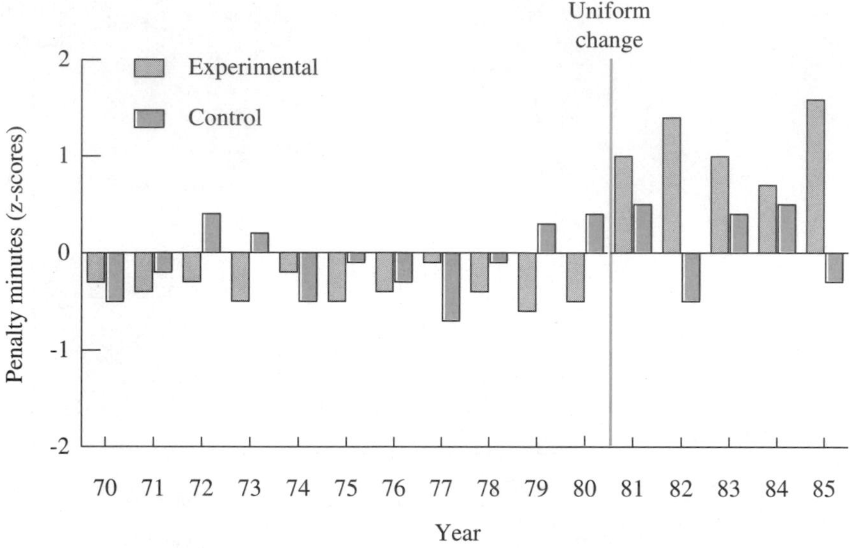

creating a **multiple time-series design**. Here, we observe an experimental group and a nonequivalent control group, and obtain several spaced pretest scores and several spaced posttest scores for each. Thus, for example, we might examine several years of penalty records both for the team that changes uniforms and for another team that does not, as shown in Figure 20.3 above. This figure shows the effect of the treatment in two ways. First, within the experimental group, the change in behavior occurs only after the treatment has been introduced, and otherwise there is one stable behavior before and a different stable behavior after the treatment. Second, the change from pretest to posttest scores in the experimental group is larger than that in the control group.

Although there may still be some factor that occurred simultaneously with the treatment, the fact that it does not produce the same results in the control group means that it is localized and specific to the experimental group. Further, the fact that throughout all these years it occurs only once and simultaneously with the treatment means that it would have to be an extreme coincidence. Together, therefore, these findings make it very unlikely that a confounding variable produced the change in the experimental group.

REMEMBER A multiple time-series design examines numerous pretest and posttest observations for both an experimental and nonequivalent control group.

THE QUASI-INDEPENDENT VARIABLE OF THE PASSAGE OF TIME

One important quasi-independent variable in psychology is the passage of time. The entire field of developmental psychology is built around the variable of age, focusing on how it relates to changes in social, emotional, and cognitive behavior. Researchers also study the passage of time in other settings, as when comparing experienced workers with inexperienced workers or following an individual's change in memory ability over the course of a day. These are quasi-independent variables, because we cannot randomly assign people to be a certain age or to experience only a certain amount of work. Therefore, such studies are essentially time-series designs, in that we sample participants' behavior at different points, before and after the passage of a certain amount of time. However, they differ from the previous time-series designs in one important way. In the previous designs, an environmental event was the variable of interest, while the accompanying passage of time between measurements allowed for potential confoundings from maturation and history. In the present designs, the passage of time—with the accompanying maturation and history—is the variable of interest, while environmental events are potential confoundings.

There are three general approaches to studying the passage of time: Longitudinal designs, cross-sectional designs, and cohort designs.

Longitudinal Designs

In a **longitudinal design**, we observe the effect of the passage of time by repeatedly measuring a group of participants. For example, let's say we want to study vocabulary development in a group of children. To do so, we'll test them yearly from ages four to eight (as in Gathercole, Willis, Emslie & Baddeley, 1992). As shown in Table 20.2, such a design is set up and analyzed in the same way as any other repeated-measures experiment. Collapsing vertically, any differences between the mean scores for the conditions will reflect changes in vocabulary skills as a function of age. As usual, this factor may be part of a multifactor design, in which other true or quasi-independent variables are examined.

We might also study briefer periods of time. For example, Nelson and Sutton (1990) examined white-collar workers over a nine-month period to determine how they coped

TABLE 20.2 Diagram of a One-Way Longitudinal Study, Showing Repeated Observations of Each Subject at Different Ages

Xs represent vocabulary scores.

	Age (in years)				
	4	*5*	*6*	*7*	*8*
Subject 1	*X*	*X*	*X*	*X*	*X*
Subject 2	*X*	*X*	*X*	*X*	*X*
Subject 3	*X*	*X*	*X*	*X*	*X*
"	.	.	.	.	.
"	.	.	.	.	.
	$\overline{X}$	$\overline{X}$	$\overline{X}$	$\overline{X}$	$\overline{X}$

with work-related stress. And here's an interesting twist: Gladue and Delaney (1990) investigated whether men and women become more attractive to one another as the closing time of a bar approaches. Using a repeated-measures design, they asked bar patrons to rate the attractiveness of the other patrons on several occasions during the evening. They found that attractiveness increased as time wore on. (Interestingly, the ratings were not positively correlated with alcohol consumption, so alcohol was not the reason for this increase!)

The overriding advantage of a longitudinal study is that, as a repeated-measures design, it keeps subject variables reasonably constant between the conditions. Thus, observing the same children as they age keeps constant such variables as their genetic makeup, their parents, the environments they're raised in, and so on.

There are, however, several disadvantages to longitudinal designs. First, just keeping in touch with participants over a lengthy period can be difficult, so these studies often involve a small N and the results may be biased by subject mortality. Second, the design is repeated measures and so successive conditions may be confounded by order effects (which cannot be counterbalanced). Third, and most important, a longitudinal study is inherently confounded by any extraneous variable that participants experience during the study. For example, an increase in a boy's vocabulary between ages four and five might appear to reflect normal development but may actually be due to his learning to read or watching certain television shows. Finally, a longitudinal study may not generalize well to future generations, because the society and culture are constantly changing. A study of language development in the 1950s, for example, may not generalize to children in the 1990s, because of such recent innovations as educational television and preschool education.

> *REMEMBER* Longitudinal designs are confounded by extraneous events that occur during the course of the study, and they may not generalize over time.

Cross-Sectional Designs

We can also study the passage of time using a **cross-sectional design**. This is a between-subjects quasi-experiment in which participants are observed at different ages or at different points in a temporal sequence. Thus, for example, we might select a cross-section of ages, testing the vocabulary of a group of four-year-olds, a different group of five-year-olds, and so on, as shown in Table 20.3. Basically, this is an example of the design we examined earlier in which the researcher selects participants for each condition using a subject variable—except that the variable here is based on time. (Again, this factor may be part of a multifactor design, which also examines other true or quasi-independent variables.)

The major advantage of a cross-sectional design is that the study can be conducted rather quickly and easily. The major disadvantage is that the conditions may differ in terms of many confounding variables. For example, the five-year-old participants will differ from the four-year-olds in genetic makeup, family environment, and other variables that may cause the differences in dependent scores.

> *REMEMBER* Cross-sectional designs involve a between-subjects comparison of different age groups. They may be confounded by any other variable that also distinguishes the groups.

TABLE 20.3 Diagram of a One-Way Cross-Sectional Study, Showing Observations of a Different Group of Subjects at Each Age

Xs represent vocabulary scores.

Age (in years)				
4	*5*	*6*	*7*	*8*
X	*X*	*X*	*X*	*X*
X	*X*	*X*	*X*	*X*
X	*X*	*X*	*X*	*X*
.	.	.	.	.
.	.	.	.	.
$\overline{X}$	$\overline{X}$	$\overline{X}$	$\overline{X}$	$\overline{X}$

In fact, a special confounding can occur in cross-sectional studies because of differences in subject history. It is called a cohort effect. **Cohort effects** occur when age differences are confounded by differences in subject history. The larger the differences in age, the greater the potential for cohort effects. For example, let's say we study memory ability as a function of age by testing people born in the United States in 1930, 1950, and 1970. These groups differ not only in age but also in that each group grew up during a different era. Therefore their backgrounds differ in terms of health and nutritional care, educational programs, and cultural experiences: One group reached adolescence during World War II, another during the birth of television, and the third when drugs and "disco" were common. Thus, any observed differences in memory ability may actually be due to these differences in history. Further, if each group's unique background influences performance, then the results will generalize poorly to other generations that have different backgrounds.

> *REMEMBER* Cohort effects are the confounding of age differences with generational history differences.

Because of the greater likelihood of potential confoundings, cross-sectional designs are generally considered to be less effective than longitudinal designs. However, especially if you match participants across conditions on relevant variables, such designs do provide an immediate comparison of individuals who differ in age or time-related experiences.

Cohort Designs

You cannot completely prevent cohort effects, but you can identify when they are present by employing a cohort design. A **cohort design** is a longitudinal study of several groups, each from a different generation. For example, let's say we repeatedly study the vocabulary development in one generation of children beginning when they were four years old in 1990 and in another generation of children beginning when they were four in 1994. As shown in Table 20.4, this design is set up and analyzed in the same way as any other two-way experiment.

TABLE 20.4 Diagram of a Two-Way Cohort Study

Xs represent vocabulary scores.

		Repeated measures over age						
		4	*5*	*6*	*7*	*8*		
1990 Subjects	*1*	X	X	X	X	X		
	2	X	X	X	X	X	$\overline{X}$ ←	Generation main effect
	3	.	.	.	.	.		
1994 Subjects	*1*	X	X	X	X	X		
	2	X	X	X	X	X	$\overline{X}$ ←	
	3	.	.	.	.	.		
Age main effect →		$\overline{X}$	$\overline{X}$	$\overline{X}$	$\overline{X}$	$\overline{X}$		

Collapsing over scores vertically produces the mean score for each age, and differences between the means show the main effect of age. This factor provides the longitudinal, developmental information. Collapsing scores horizontally produces the main effect mean of each generation. If there is no significant difference between these means, then there is no evidence that the scores differ on the basis of a child's generation. If there is a difference between the two groups, however, then cohort effects may be present and the developmental results are less likely to generalize to other generations. Likewise, the absence of a significant interaction between generation and age would suggest that changes in scores as a function of age are basically similar—parallel—regardless of each generation's history. A significant interaction, however, would indicate that the type of changes with age that we see *depends* on which generation we examine—in which case we have a cohort effect.

> REMEMBER A cohort design is the longitudinal study of several groups, each from a different generation.

To help you remember the names and procedures of all the preceding quasi-experimental designs, they are summarized in Table 20.5.

UNDERSTANDING SMALL *N* RESEARCH AND THE SINGLE-SUBJECT DESIGN

So far, we have focused on experiments involving *groups* of people or animals. However, there is an entirely different kind of experiment, in which only one participant is studied. A **single-subject design** is a repeated-measures experiment conducted on one participant. Typically, the experiment (with an *N* of 1) is then replicated on a few more participants, so this research is also known as **small *N* research**. A single-subject design is different from a case study discussed in Chapter 5. A case study is a descriptive

TABLE 20.5 Summary of Quasi-Experimental Designs

Type of design	*Procedure*
Designs involving subject variables	Create conditions on the basis of a subject characteristic
Time-series designs	
One-group pretest-posttest design	Measure one group once before and once after event.
Nonequivalent control group design	Conduct pretest and posttest on both an experimental and nonequivalent control group
Interrupted time-series design	Obtain repeated measurements of one group, both before and after event.
Multiple time-series design	Conduct interrupted time-series measurements on both an experimental and control group
Designs involving temporal variables	
Longitudinal design	Obtain repeated measures as a function of age or experience.
Cross-sectional design	Use a between-subjects design based on age or experience.
Cohort design	Examine repeated-measures factor based on age with between-subjects factor based on generation.

study. A single-subject design is an experiment that manipulates an independent variable. Before considering the particulars of such designs, let's discuss why we would want to use them.

The Argument for Small *N* Designs

Some researchers argue that there are three unacceptable flaws in experiments that study groups of participants (Sidman, 1960). The first pertains to error variance, the random differences between scores found within the conditions. Typically with group designs, we employ random assignment of participants and counterbalancing of extraneous variables. Because we therefore include the influence of fluctuating variables within each condition, the design itself produces much of the error variance. The inconsistency in scores then makes it difficult to see a relationship hidden in the data (so that we must rely on obtuse, inferential statistics). Further, researchers then ignore the differences in behavior reflected by error variance and the variables that cause them. We ignore differences in behavior between participants (*intersubject* differences), and different behaviors in the same participant from moment to moment (*intrasubject* differences). Yet, these variables are potentially important aspects of the behavior under study.

The second flaw is that, because of the variability in individual scores, we must compute the mean (or similar measures) in each condition. Yet a mean score may misrepresent the behavior of any and all individuals (how often does the mean score accurately describe your performance on an exam?). Then, incredibly, after using group means to describe a relationship, we turn around and generalize the findings to individuals! Psychology studies the laws of behavior as they apply to the individual, but in group designs we never examine a relationship in terms of the individual.

The third flaw involves the problem of demonstrating a consistent, reliable effect of the independent variable. Usually, we demonstrate a relationship only once in a particular study, typically testing participants only briefly under the various conditions. Then we rely on inferential statistics to conclude that the study is reliable. That is, if the results are significant, the relationship is unlikely to be due to random chance. Instead, it is likely to be caused by something that makes it reproducible, so a significant relationship is also described as a reliable relationship. We do not, however, have any *empirical* evidence that the relationship is reproducible. Other researchers may replicate a study, but their situation and participants inevitably will differ from ours. And in their replication, researchers also seldom demonstrate empirically that the relationship is reliable.

> *REMEMBER* Group designs ignore the causes of error variance, they rely on mean scores to describe individuals, and they do not empirically show that an effect is reliable.

Small N designs address these problems in the following ways:

1. They control subject variables and individual differences not by balancing them, but by keeping them constant. With only one participant, there can be no intersubject differences in scores. At any point, if we see an inconsistency in the participant's response, we know that some extraneous variable is responsible, so we can attempt to identify and understand it.
2. Our analysis of the data is usually accomplished by visual inspection. (That's right, we don't perform statistics when $N = 1$!) Instead, we look at a graph of the data to see whether there is a relationship between the independent and dependent variables. Because we rely on visual inspection, we accept that there is an effect only when it is obvious. The size of the effect then is the obvious amount that the participant's response changes between conditions.
3. To be sure that the effect of the independent variable is reliable, we perform the manipulation repeatedly on the same participant, or we perform a replication of the experiment on a few additional participants. Each replication is treated as a separate study, however, so we don't gloss over individuals by combining their results. Ultimately, then, because the relationship is based on individuals, we are studying reliable relationships in the psychology of individuals.

The Logic of Baseline Designs

If an experiment contains only one participant then, of course, it must be a repeated measures design. Typically, the study involves only two levels of the independent variable: A control condition with zero amount of the variable present and a treatment condition with some nonzero amount of the variable present. Under each condition, we measure the participant's behavior with a dependent variable that usually reflects the quantity of responding, such as the rate of responding over time or the magnitude of responses. For example, the old stand-by of counting the number of times a rat presses a lever under different conditions of reward fits this design.

Observing the participant under a control condition establishes a baseline. A **baseline** is the level of performance on the dependent variable when the independent variable is not present. It is used for comparison to the level of performance when the variable is present. To establish the baseline, the participant is observed for a substantial period of time, observing numerous responses. Once the participant has habituated to the procedure so that the baseline is stable, we have the typical response rate when the treatment is not present. Then we introduce the experimental treatment condition and establish the participant's response rate in this situation. If the response rate with the treatment is different from that without the treatment, we have demonstrated an effect of the independent variable.

Baseline designs are most often associated with animal research. Because of the flaws in group studies, baseline designs became the mainstay of B. F. Skinner and others who study instrumental conditioning. This approach is often referred to as the "experimental analysis of behavior" (and the *Journal of the Experimental Analysis of Behavior* is devoted to it). In a typical baseline experiment, a researcher might place a rat in a cage containing a lever and then establish the baseline rate of lever-pressing. In another experiment, a pigeon's baseline rate of pecking at a target might be established. A particular reward, punishment, or environmental stimulus is then introduced, and once responding is again stable, the response rate with the treatment is compared to the baseline rate. The procedure is then replicated on several other animals, with the results from all published as one research report.

Baseline designs are also performed with humans. For example, in applied studies of behavior modification, researchers establish baselines for anxiety attacks, phobias, psychotic episodes, eating disorders, and other dysfunctional behaviors. Then they introduce rewards, punishments, or other forms of treatment and observe the change in the frequency or magnitude of the behavior. Similar designs are also used in industrial settings to demonstrate the effects of various treatments on a worker's productivity, or in educational settings to study factors that improve a child's performance.

The fundamental logic of baseline designs is to compare the baseline response rate with the treatment response rate, but as the following sections show, there are two general design approaches we may take.

Reversal Designs

The simplest approach would be to test a participant first when the independent variable is not present, in order to obtain the baseline (call this condition A). Then we could observe the participant after the variable is present (call this condition B). This simple "AB" design could be used to show that rats, for example, will press a lever more often when food is dispensed as a reward than in a baseline condition when food is not dispensed.

By showing that responding is different with the treatment, we may be tempted to conclude that the treatment has an effect. If this is all we do, however, we are open to the rival hypothesis that it was some confounding factor that produced the change in responding. Maybe some confounding environmental stimulus led to the increased lever-pressing. Or, perhaps changes in the rat's ongoing history or maturation coincidentally caused the increased lever-pressing (or maybe the rat got bored and started pressing to entertain itself).

To demonstrate that it is the presence of the treatment and not some other variable that is controlling the participant's behavior, the strategy is to return the participant to the control condition after the treatment condition is over. If responding "reverses" to the baseline rate, we have evidence that the behavior is controlled by the treatment. This approach is called a reversal design. In a **reversal design**, the researcher repeatedly alternates between the baseline condition and the treatment condition. When we present the baseline phase, the treatment phase, and then the baseline phase again, the design is described as an ABA reversal design. For even more convincing evidence, we may reintroduce the treatment condition again, employing an ABAB design (or any extended sequence, such as an ABABAB design).

To see the effects of the manipulation, we graph the results, as shown in Figure 20.4. Going from testing under condition A to testing under condition B, we see that the introduction of food leads to increased responding. Then, after removing the reward and returning to condition A, the response "extinguishes," eventually returning to its original baseline rate. Reintroducing the reward reinstates the response rate, and so on. Because it is unlikely that a confounding variable would repeatedly and simultaneously change with each of the conditions, we are confident that the treatment caused the behavioral change. Then replicating this study on a few other participants further reduces the possibility that the behavioral change was due to a confounding variable that coincidentally changed with the treatment.

> ***REMEMBER*** A reversal design demonstrates the effect of a variable by repeatedly alternating between testing with and without the treatment condition.

Multiple-Baseline Designs

Recall that any repeated-measures design may introduce the problem of carry-over effects from one condition to the other. If the above reversal designs are to work, the carry-over effect of the treatment must be reversible. Many treatments, however, involve a permanent, irrevocable change. For example, once a rat has learned to respond to a stimulus, some learning may remain so that the animal's responding never returns to the original baseline rate. Further, some clinical treatments are not reversible for ethical reasons. For example, a researcher may believe that it is unethical to

FIGURE 20.4 Ideal Results from an ABAB Reversal Design

Here, lever-pressing rate is shown to be a function of the presence or absence of a food reinforcer.

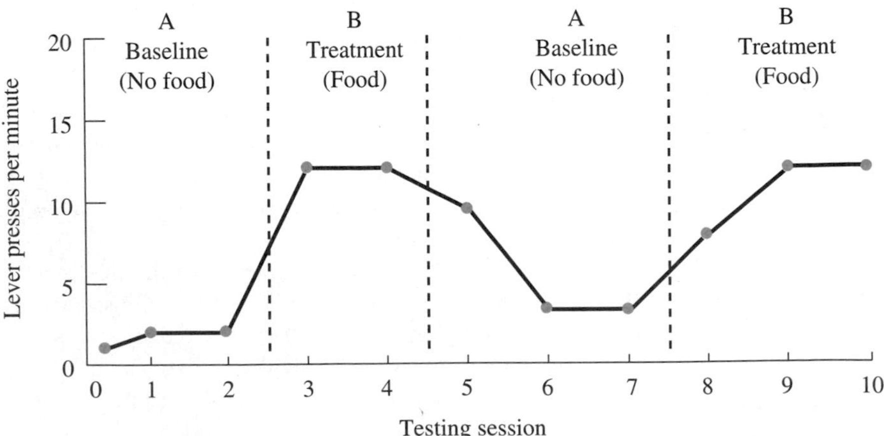

discontinue a treatment that reduces a person's phobic reactions, just for the sake of the research.

If we do not reverse the treatment in such situations, we do not eliminate the possibility that the change in behavior is due to maturation, to history, or to environmental effects that occurred coincidentally with the treatment. The solution is to employ a multiple-baseline design. A **multiple-baseline design** reduces the possibility of confounding factors by examining more than one baseline. The logic is that we eliminate potential confoundings by demonstrating that a behavior changes only when the treatment is introduced, regardless of when it is introduced. There are three general variations of the multiple-baseline design.

One approach is to establish **multiple baselines across participants**. Here we measure a baseline for several individuals on the same behavior, but we introduce the treatment for each at a different time. For example, the argument that maturation, history, or some other variable might cause a rat to increase lever-pressing relies on the idea that the variable changed at the precise moment that we introduced the treatment. To counter this argument, we might obtain the baselines for several rats, but for each we introduce the food reward at a different point in time. Let's say we obtain the data shown in Figure 20.5. We eliminate the argument that some other variable produced the results, because even though we start the treatment at a different time for each participant, the treatment still alters the behavior. An incredible coincidence would be required for a confounding variable to change simultaneously with the onset of treatment for each participant. (This approach is also used when replicating the previous reversal design with different participants, by varying the time at which the ABA conditions are instituted for each subject.)

A second approach is to collect **multiple baselines across behaviors**. Here, we measure a baseline for several behaviors from one participant and apply the treatment to each behavior at a different time. For example, say we are concerned with a child who is disruptive in school and we establish baselines for aggressive acts, for temper tantrums, and for attention-seeking behavior. We develop a treatment involving verbal feedback that should reduce all of these behaviors. The treatment is applied first to one behavior, later to the second behavior, and still later to the third. If the incidence of each type of behavior drops only when the treatment is introduced, it is implausible that an extraneous confounding variable coincidentally caused the change in each behavior.

Alternatively, we might hypothesize that the treatment will affect only one of the above behaviors. After introducing the treatment, we should find that the target behavior changes but that the other behaviors remain at their baseline rate. If so, we can be confident that it was not changes in some extraneous variable that produced the change, because it should have changed all of the behaviors.

The third approach is to establish **multiple baselines across situations**. Here, we establish baselines for one behavior on the same participant, but in different situations. For example, we might establish a baseline for a child's temper tantrums when at school and also when at home. Then at different points in time we apply a treatment phase to eliminate the tantrums. It is unlikely that an extraneous variable that decreases tantrums would coincidentally occur with the treatment at different times, and *both* at school and at home.

REMEMBER A multiple-baseline design shows the effect of a variable by demonstrating a change in the target behavior only when the treatment is introduced.

FIGURE 20.5 Idealized Data from a Multiple-Baseline Design Across Subjects

Note the different points in time at which a food reward was introduced.

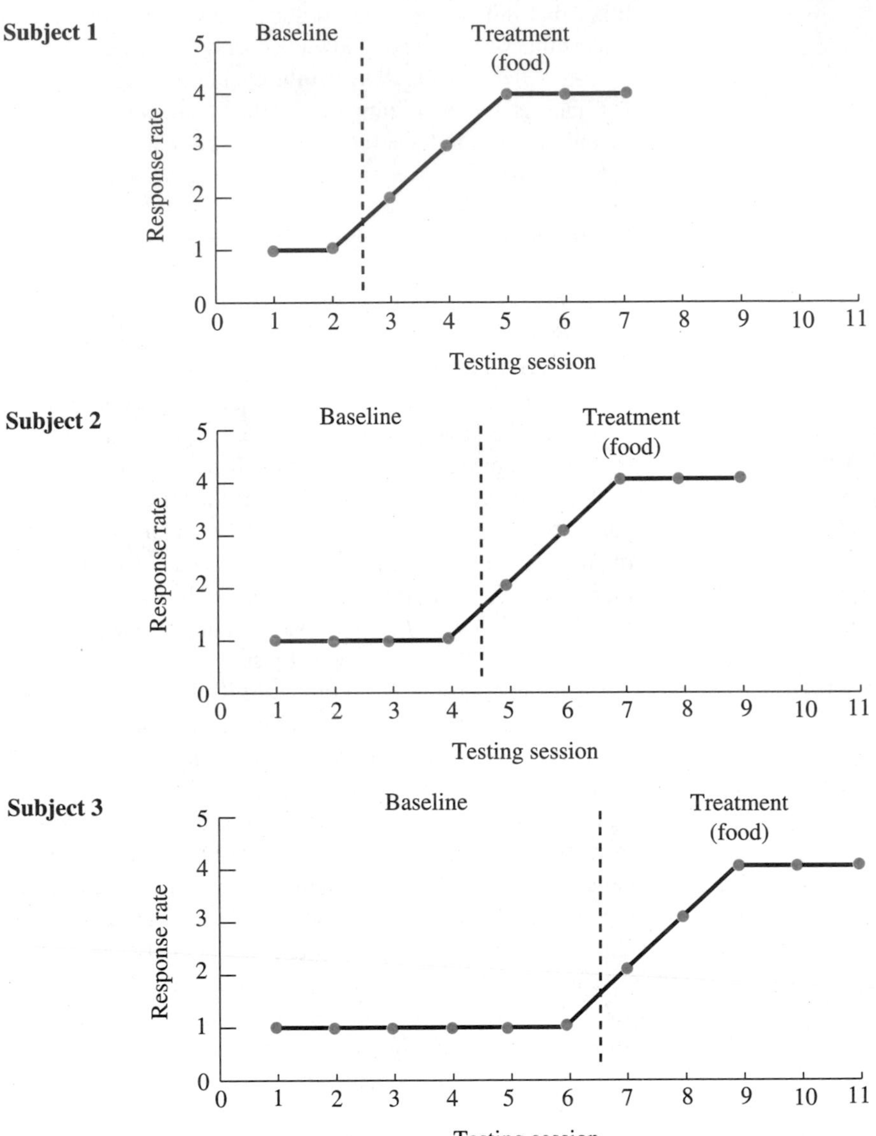

Choosing Between Single-Subject and Group Designs

There are many variations to the single-subject approach (see for example Barlow & Hernsen, 1984). We can, for example, study the effects of several levels of a variable: First we observe the control baseline condition (A), then observe one treatment level (B), then observe a different treatment level (C), and so on. To reverse the effects of each treatment, we insert a control condition between each treatment, producing an ABACAD design. We can even investigate an interaction effect by combining two treatments.

In addition to the baseline designs we've discussed, a single-subject design can also include the kinds of repeated-measures experiments described in previous chapters. We test only one participant, but still measure a typical dependent variable, compute a mean or other summary score for multiple trials per condition, and so on. This approach is common when studying someone with an extraordinary ability. For example, Ericsson & Polson (1988) performed experiments on a waiter with an astounding memory for customers' dinner orders. Likewise, this approach is used to study "split-brain" patients, in whom the two hemispheres of the brain are not connected (see Gazzaniga, 1983). Such designs are often found in clinical research, because the symptoms of each patient are so unique (e.g., Sacks, 1985).

The major advantage of all types of single-subject designs is that they allow us to examine a relationship between variables in a single individual. Further, sometimes they are necessary because the behavior of interest is found in an extremely small percentage of the population, and a researcher can find only a few participants to study. Sometimes, too, a design entails so much time and effort per participant that a small N is required.

The disadvantages of these designs stem first from the fact that they involve repeated measures. They become less feasible the more that carry-over effects occur. In addition, they are not commonly used for studying an interaction involving several levels of each factor, because they create impossibly complex schemes. Finally, any single-subject design is wiped out by mortality effects: Rats may die during a study, and humans require extreme patience and motivation for long-term participation. The solution to such problems may be a between-subjects group design. Between-subject designs can also be conducted more quickly, and they allow for the use of deception and other procedures that are not feasible with repeated measures.

As usual, selecting a design depends on the hypothesis and variables being studied, as well as all of the design concerns we've discussed. Of particular importance in selecting a design is finding a balance between internal and external validity. On the one hand, the potentially great control of variables in a single-subject design achieves a high degree of internal validity. On the other hand, a major drawback of single-subject research is its limited external validity for generalizing to other individuals and situations. Because results are tied to one or a few unique participants, the results may be very different from those we would find with other participants (remember that random sampling does not balance individual differences well in small samples). Likewise, results are tied to a highly controlled and individualized setting, so generalizability is limited to similar settings only. A group design, however, provides greater confidence that the conclusions generalize to other individuals and settings.

REMEMBER Single-subject designs provide substantial internal validity but have limited external validity.

PROGRAM EVALUATION

There is one other type of design that you should be familiar with, which incorporates a combination of the various procedures already discussed. Sometimes researchers conduct studies on community human services programs, such as programs for preventing or treating drug and alcohol abuse (e.g., Werch, Meers, & Hallan, 1992; McCusker et al., 1992) or educational programs like Project Head Start (e.g., McKey et al., 1985). **Program evaluation** refers to a variety of procedures for developing and evaluating social programs. This is the ultimate applied research because such programs are essentially grand experiments in social change (Campbell, 1969). They are based on the experimental principle that by providing some form of treatment to participants—to society—we'll see a corresponding change in behavior. Program evaluation provides feedback to administrators and service providers, as well as providing scientific information about the effect of a program as a quasi-independent variable.

Although this research encompasses many procedures (see Posavac & Carey, 1989), it usually consists of four basic phases:

1. *Need assessment*: Before designing a particular program, researchers identify the services that are needed and determine whether potential users of the program will use it.
2. *Program planning*: When designing the program, researchers apply findings from the literature that suggest the best methods to implement for the behavior and situation being addressed.
3. *Program monitoring*: Once the program is implemented, researchers monitor the program to ensure that it provides the intended services and that clients are using them.
4. *Outcome evaluation*: Eventually, researchers determine whether the program is having its intended effect, by using a time-series study that compares behaviors before and after the program's implementation.

The data in each of these phases comes from such sources as field surveys of the community; archival studies of hospital, school, and police records; and interviews, unobtrusive observations, and case studies of service providers and clients. On the one hand, such procedures suffer from all of the flaws we have discussed, especially because they are quasi-experiments. On the other hand, program evaluation serves a very real applied need and flawed data are generally considered better than no data (as long as the flaws are recognized and considered).

REMEMBER Program evaluation involves procedures for creating and evaluating social programs.

PUTTING IT ALL TOGETHER

The typical, true laboratory experiment involving groups of participants is common in psychology, because it's practical and efficient, it provides a relatively high degree of internal validity, and it provides some external validity for generalizing to other situations and subjects. Quasi-experiments and small *N* designs are also important approaches, however. Quasi-experiments greatly expand the types of influences on behavior that can be studied, and small *N* designs eliminate error variance, especially variability due to individual differences in participants that group designs incorporate.

Perhaps the most important thing to remember about quasi-experiments is that they often look like true experiments, especially when a quasi-independent subject variable is studied in a laboratory setting. When evaluating your own research or that designed by others, however, you won't see a red flag signaling the nature of the design. Instead, you must carefully examine whether random assignment to conditions truly occurs. If not, remember that the results only *suggest* the causes of the behavior under study.

CHAPTER SUMMARY

1. In a *quasi-experiment* involving a *quasi-independent variable*, participants cannot be randomly assigned to conditions. Instead, they are assigned to a condition based upon some inherent characteristic. Because the independent variable may be confounded by subject variables, quasi-experiments have much less internal validity than true experiments.
2. Effective manipulation of a quasi-independent variable involving a subject variable hinges on the selection of participants for each condition who are similar to each other but very different from those in other conditions.
3. *Regression toward the mean* occurs when, because of inconsistent random factors, extreme scores tend to change in the direction of coming closer to the mean.
4. A *time-series design* is a quasi-experimental repeated-measures design in which a behavior is sampled at different times.
5. A *one-group pretest-posttest* design has no control group.
6. In a *nonequivalent control group design*, the control group has different subject characteristics and different experiences during the study than the experimental group.
7. In an *interrupted time-series design* observations are made at several spaced times prior to the treatment and at several times after it.
8. In a *multiple time-series design*, both an experimental group and a nonequivalent control group are observed at several times before the treatment and several times after it.
9. In a *longitudinal design*, participants are repeatedly measured to observe the effect of the passage of time.
10. A *cross-sectional design* is a between-subjects experiment in which participants are observed at different ages or at different points in a temporal sequence.
11. *Cohort effects* occur when differences in age are confounded by differences in subject history.
12. A *cohort design* is a factorial design consisting of a longitudinal study of several groups, each from a different generation.
13. Group designs have been criticized because they (a) ignore the variables that cause error variance, (b) rely on mean scores to describe individuals, and (c) do not empirically show that an effect is reliable.

14. A *single-subject* or *small N design* involves a repeated-measures experiment conducted on one participant that is replicated with a few other participants. The advantages of this design are that it (a) keeps subject variables constant, (b) provides a clear indication of the effect size of a variable, and (c) empirically demonstrates the reliability of the effect.

15. A *baseline* is the level of performance on the dependent variable when the independent variable is not present. It is used as a comparison to performance when the variable is present.

16. A *reversal design* repeatedly alternates between the baseline and the treatment conditions.

17. When the influence of a manipulation cannot be reversed, a *multiple-baseline design* is employed, in which a baseline is established for one behavior from several participants, for several behaviors from one participant, or for one behavior from one participant in several situations.

18. Single-subject designs are appropriate when studying a detailed description of one participant, when the treatment does not produce large carry-over effects, and when many observations per participant are necessary. Otherwise, a group design may be preferable.

19. *Program evaluation* involves procedures for developing and evaluating applied social programs.

PRACTICE PROBLEMS

(Answers for odd-numbered problems are provided in Appendix D.)

1. (a) What is the difference between a true experiment and a quasi-experiment? (b) In what ways are quasi-experiments and correlational designs similar? (c) In what way(s) are they different?

2. (a) What three types of variables are studied as quasi-independent variables? (b) Why can't you confidently infer causality from a quasi-experiment?

3. (a) How do you design a quasi-independent variable in order to study a subject variable? (b) What is the goal when using the scores from a selection pretest to create conditions? (c) What bias may be produced by a pretest?

4. At the beginning of a gym class, you obtain the highest score on a physical-fitness test. During the semester, however, your scores get worse, while the initially unfit students tend to score higher. You conclude that the gym class helps unfit people but harms the most fit. (a) What rival explanation involving random factors might explain these results? (b) How would it cause the changes in scores? (c) How would you test your hypothesis?

5. (a) What is a one group pretest-posttest design? (b) What is missing from this design? (c) What extraneous variables may confound this design?

6. (a) What is a nonequivalent control group design? (b) What potential confounding variables does it eliminate? (c) What potential confounding variables are not eliminated?

7. (a) What is an interrupted time-series design? (b) What potential confounding variables does it eliminate? (c) What potential confoundings does it not eliminate?

8. (a) What is a multiple time-series design? (b) What potential confounding variables does it eliminate? (c) What potential confoundings does it not eliminate?

9. (a) What is a longitudinal design? (b) What is the major advantage of this design? (c) What is the major flaw in this design?

10. (a) What is a cross-sectional design? (b) What is its major strength? (c) What is its major weakness?

11. (a) What are cohort effects? (b) What is a cohort design? (c) What is the advantage of this design? (d) How do you determine whether cohort effects are present?

12. What is program evaluation?

13. (a) What is a reversal design? (b) What is the logic for eliminating confoundings in this design? (c) What is the major factor that prohibits its use?

14. (a) What is a multiple-baseline design involving one behavior from several participants? (b) What is the logic for eliminating confoundings in this design?

15. What is a multiple-baseline design (a) Involving several behaviors from one participant? (b) Involving one behavior for one participant in several situations?

16. Let's say you conduct a single-subject study of the influence of relaxation training on participants' momentary anxiety (measured using heart rate). (a) Should this study involve an ABA reversal design or a multiple-baseline design? (b) Describe the specific design to use.

17. (a) What are three advantages to using a single-subject design in question 16? (b) What are two disadvantages?

18. Foofy tests a group of sorority pledges' happiness at the beginning of the semester and again at the end of the semester after being admitted to the sorority. She finds that the latter scores are higher, and so she concludes that joining a sorority increases a woman's happiness. (a) What is the name of this design? (b) What major flaw is present? (c) What rival hypotheses can you suggest to explain her results? (d) What would you do to improve the study?

19. In a study, the maturity levels of students at four times during the college year are measured (factor A). You compare freshman from 10 years ago with present-day freshman (factor B). (a) What type of variable is factor A? (b) What type of variable is factor B? (c) What type of design have you created?

20. In question 19, you obtain the following cell means. What should you conclude about: (a) The presence of cohort effects? (b) How maturity changes during a college student's year?

	Time 1	*Time 2*	*Time 3*	*Time 4*
10 years ago	10	20	30	40
present-day	25	25	25	25

21. In two, nonequivalent control group designs, you obtain the following mean stress scores for people before and after an earthquake. (a) What should you conclude about whether the treatment caused the change in scores in the experimental group in Study A? (b) What other hypotheses are plausible? (c) What should you conclude in Study B? (d) What hypotheses are eliminated in Study B? (e) What possible confounding is still possible in Study B?

Study A	*Before*	*After*
Control	20	50
Experimental	30	60

Study B	*Before*	*After*
Control	60	70
Experimental	50	85

21

Nonparametric Statistical Procedures for Frequency Data and Ranked Data

To understand this chapter:

- From Chapter 5, recall the various observational and descriptive methods researchers use.
- From Chapter 6, recall the four types of measurement scales (nominal, ordinal, interval, and ratio).
- From Chapters 15 and 16, remember why we create within-subjects or between-subjects designs, the independent and the dependent samples *t*-test, and homogeneity of variance.
- From Chapters 17 and 19, recall the uses of the within-subjects and between-subjects ANOVA, *post hoc* tests, and eta squared.

Then your goals are to learn:

- When to use nonparametric statistics.
- The logic and use of the one-way chi square.
- The logic and use of the two-way chi square.
- The nonparametric procedures corresponding to the independent and dependent samples *t*-test, and to the one-way between-subjects and within-subjects ANOVA.

Recall that there are two general categories of inferential statistics, parametric and nonparametric. So far, we've performed only parametric procedures, but now we'll turn to nonparametric ones. Don't despair, however, because the designs and logic you'll see here are very similar to that of previous statistics: Essentially, we'll look at variations of procedures you've already learned, that as usual, can be applied to descriptive or experimental designs, from the laboratory or field. First we will discuss the most common nonparametric procedure, called chi square. Then we'll discuss the nonparametric procedures that are analogous to t-tests and ANOVAs, except that they are used with rank-ordered scores.

THE REASONS FOR USING NONPARAMETRIC PROCEDURES

Throughout our discussions, we have performed parametric inferential statistics in which we assumed that the samples represented raw score populations that were normally distributed, that the population variances were homogeneous (equal), and that the dependent scores reflect an interval or ratio scale. Sometimes we do not obtain data like this, however, and we cannot insist on using a parametric procedure when we seriously violate these assumptions. Instead, we look to the appropriate nonparametric procedure. Recall that nonparametric procedures do not assume a normal distribution or homogeneous variance, and the data may be nominal (categorical) or ordinal (rank ordered).

Nonparametric procedures are never our first choice. First, they are less powerful than parametric procedures. Second, recall that nominal and ordinal scales are less precise and sensitive ways of measuring differences in behavior, and even their measures of central tendency (the mode or median) are less precise and informative. But sometimes we must go this route. Understand that whether to use these statistics depends solely on the characteristics of the dependent variable. In the typical laboratory setting we are in control, so often we can choose a better, interval or ratio scale of measurement. Usually, the variable will also be normally distributed and have homogeneous variance. (When in doubt, check the literature to see how others have treated your variable.) If the scores are not normal or have heterogeneous variance, then we use nonparametric procedures. More commonly, nonparametric procedures are used in descriptive research or in field experiments, where we are forced to use a nominal or ordinal dependent variable. For example, in observational research we may be forced to rank order different behaviors, in surveys or field experiments we may need to count the frequency of different behaviors, or in archival research the best we can do may be to categorize behaviors.

> ***REMEMBER*** Nonparametric statistics are used when scores form very skewed or otherwise nonnormal distributions, when the population variance is heterogeneous, or when dependent scores are from an ordinal (ranked) or nominal (categorical) variable.

The reason for using nonparametric procedures when appropriate is this: Recall that the bottom line from previous hypothesis testing was that whenever we reject the null hypothesis, there is some probability we made a Type I error (rejecting H_0 when it is true). Implicit in this probability, however, is the proviso "assuming your data are normally distributed, have homogeneous variance, and so on." Parametric procedures are robust, meaning that they will tolerate *some* violation of their assumptions. If the data severely violate the rules, however, then the actual probability of a Type I error will be substantially *larger* than the alpha level we've set. Therefore, the probability of making a Type I error will be unacceptably large. Nonparametric procedures, however, provide accurate estimates of the probability of Type I errors with such data.

Nonparametric procedures are still inferential statistics used for deciding whether the differences between samples accurately represent differences in corresponding populations. Therefore, the concepts of H_0 and H_a, sampling distributions, Type I and Type II errors, alpha levels, critical values, and maximizing power all apply. We still ask the same questions that we asked in previous chapters, but here, the characteristics of the data simply dictate computing the answers differently.

CHI SQUARE PROCEDURES

In previous chapters, we've measured the *amount* of a dependent variable that a participant demonstrates. With nominal, or categorical, variables, however, we obtain scores that do not indicate an amount, but rather indicate the *category* into which a participant falls. Thus, we have categorical variables when we determine how many individuals answer yes, no, or maybe to a question; how many indicate that they are male or female; how many claim to vote Republican, Democratic, or Communist; how many say that they were or were not abused as children; and so on. In each case, we count the number or *frequency*, of participants who fall into each category.

For example, we might find that out of 100 participants, 40 say yes to a particular question and 60 say no. These numbers indicate how the *frequencies are distributed* across the categories of yes/no. As usual, we want to draw inferences about the population: If we were to ask everyone this question, can we infer that 40% of the population would say yes and 60% would say no? Or would the frequencies be distributed in a different manner? To make inferences about the frequencies that would be found in the population, we perform the procedure called chi square (pronounced "kigh square"). The **chi square** procedure is the nonparametric inferential procedure for testing whether the frequencies in each category in a sample represent certain frequencies in the population.

> *REMEMBER* Whenever you measure the number of participants that fall in different categories, use the chi square procedure for significance testing.

The symbol for the chi square statistic is χ^2. Theoretically, there is no limit to the number of categories, or levels, you may have in a variable and no limit to the number of variables you may have. A chi square design is described in the same way we described ANOVAs: When a study has only one variable, perform the one-way chi square; when a study has two variables, use the two-way chi square; and so on.

THE ONE-WAY CHI SQUARE: THE GOODNESS OF FIT TEST

The one-way chi square is used when data consist of the frequencies with which participants belong to the different categories of *one* variable. As usual, we're examining a relationship, but here it's a relationship between the different categories and the frequency with which individuals fall in each category. We ask, "As the categories change, do the frequencies with which participants fall into the categories change in a consistent fashion?"

Here is an example study that calls for a one-way chi square: Scientists believe that being right-handed or left-handed is related to brain organization and function. Interestingly, many of history's great geniuses were left-handed. To explore the relationship between the frequencies of left- and right-handedness in geniuses, say that, using an IQ test, we identify a random sample of 50 geniuses. Then we ask them whether they're left-handed or right-handed (ambidextrous is not an option). The total numbers of left-handers and right-handers are the frequencies in the two categories. The results are shown in the following diagram.

Handedness

Left-handers	*Right-handers*
$f_o = 10$	$f_o = 40$

$k = 2$
$N = \text{total } f_o = 50$

Each column contains the frequency with which participants fall in that category. We call this value the *observed frequency*, symbolized by f_o. The sum of the f_os from all categories must equal N, the total number of participants in the study. Notice that the symbol k again stands for the number of levels, or categories, in a one-way chi square. Here $k = 2$.

The results of the above study seem pretty straightforward: 10 of the 50 geniuses, or 20%, are left-handers, and 40 of them, or 80%, are right-handers. Because this is a random sample, we might conclude that the same distribution of 20% left-handers and 80% right-handers would occur in the population of all geniuses. But, of course, there is the usual problem: Sampling error. Maybe, by luck, the people in this sample are unrepresentative, so in the population of all geniuses, we would not find this distribution of right- and left-handers. Maybe these results poorly represent some *other* distribution. As usual, this is the null hypothesis, implying that we are being misled by sampling error.

What is that "other distribution" of frequencies that the sample poorly represents? To answer this, we create a *model* of the distribution of the frequencies we expect to find in the population when H_0 is *true*. Recall that H_0 always implies that the study did not work—that it failed to demonstrate the predicted relationship. Therefore, the H_0 model describes the distribution of frequencies in the population if there is not the predicted relationship.

You'll notice that the subtitle of this section is "The Goodness of Fit Test." This is because the one-way χ^2 procedure tests how "good" the "fit" is between your data and the H_0 model. Thus, goodness of fit is merely another way of determining whether sample data are likely to represent the distribution of frequencies in the population described by H_0.

Creating the Statistical Hypotheses for Chi Square

Usually, researchers test the H_0 that there is no difference between the frequencies in the categories in the population. This, of course, means that H_0 states that there is no relationship in the population. For the handedness study, we're learning about the population of geniuses, so say that for the moment we ignore that there are generally more right-handers than left-handers in the world. We might start the study by proposing no relationship in the population, which means that there is no difference between the frequencies of left- and right-handers. Thus, our H_0 model is that the frequencies of left-handed and right-handed geniuses in the population are equal. There is no conventional way to write these hypotheses in symbols, so simply write H_0 as:

H_0: all frequencies in the population are equal

This implies that if the observed frequencies, f_o, in the sample are not equal, it's because of sampling error.

The alternative hypothesis always implies that the study did demonstrate the predicted relationship. Here, H_a is:

H_a: all frequencies in the population are not equal

H_a implies that the observed frequencies represent different frequencies of left- and right-handers in the population of geniuses.

We can test only whether the sample frequencies are different from those described by H_0, so the one-way χ^2 tests only two-tailed hypotheses.

Computing the Expected Frequencies

To compute the χ^2 statistic, we translate the H_0 model into a specific expected frequency for each category. The **expected frequency** is the frequency we would expect in a category if the sample data perfectly represented the distribution of frequencies in the population described by the null hypothesis. The symbol for an expected frequency is f_e.

In our study, H_0 is that the frequencies of left- and right-handers are equal in the population of geniuses. If the sample perfectly represents this population, then out of our 50 participants, 25 should be right-handed and 25 should be left-handed. Thus, when H_0 is true, the expected frequency in each category is $f_e = 25$.

For future reference, notice that f_e is actually based on a probability. If the frequencies in the population are equal, then the probability of someone's being left-handed equals the probability of someone's being right-handed. With only two possible categories, the probability of someone's falling in either category is .5. Recall that probability is the same as relative frequency, so we expect .5 of all geniuses to be left-handed and .5 of all geniuses to be right-handed. Therefore, out of the 50 geniuses in the study, we expect to have a frequency of (.5)(50), or 25, in each category. *Thus, the expected frequency in a category is equal to the probability of someone's falling in that category multiplied times the N of the study.*

Whenever H_0 is that the frequencies in the categories are equal, the expected frequency for each category can be computed as the total N in the study divided by the number of categories. Thus,

THE COMPUTATIONAL FORMULA FOR EACH EXPECTED FREQUENCY, WHEN TESTING AN H_0 OF NO DIFFERENCE, IS

$$f_e \text{ in each category} = \frac{N}{k}$$

Thus, in the handedness study, with an N of 50 and $k = 2$,

$$f_e \text{ in each category} = \frac{50}{2} = 25$$

Sometimes f_e may contain a decimal. For example, if we included a third category, ambidextrous, then k would be 3, and for 50 subjects, each f_e would be 16.67.

As with any statistical test, we first check that the study meets the assumptions of the test.

Assumptions of the One-Way Chi Square

The assumptions of the one-way χ^2 are:

1. Participants are categorized along one variable having two or more categories, and the frequency (the number) in each category is counted.
2. Each participant is measured only once and can be in one and only one category. (You cannot have repeated measures with χ^2.)
3. Category membership is independent: The fact that a particular individual falls in one category does not influence the probability that another participant will fall in any particular category.
4. The computations are based on the responses of *all* participants in the study. That is, you would not count only the number of right-handers. In a different study, if you counted the number of people who agreed with some statement, you would also include the second category of those who disagreed with the statement.
5. So that the data meet certain theoretical rules, the f_e in any category should equal at least 5.

Computing Chi Square

If the sample perfectly represents the situation where there are no differences in handedness in the population, then we expect to find 25 geniuses in each category. In other words, when H_0 is true, the f_o "should" equal our f_e. Any difference between f_o and f_e is chalked up to sampling error. Of course, the greater the difference between the observed frequency and the expected frequency, the less likely it is that the difference is due to sampling error. Therefore, the greater the difference between f_o and f_e, the less likely it is that H_0 is true and that the sample represents an equal distribution of frequencies in the population.

The χ^2 is a way to measure the overall differences between f_o and f_e in the categories in a study. We compute an obtained χ^2, which we'll call χ^2_{obt}:

THE COMPUTATIONAL FORMULA FOR CHI SQUARE, χ^2_{obt}, IS

$$\chi^2_{obt} = \Sigma\left(\frac{(f_o - f_e)^2}{f_e}\right)$$

In English, you find the difference between f_o and f_e in each category, and square that difference. Then divide each squared difference by the f_e for that category. After doing this for all categories, sum the quantities, and the answer is χ^2_{obt}. (Note that because each difference is squared, χ^2 can never be a negative number.)

For the handedness study, we have these frequencies:

Handedness	
Left-handers	*Right-handers*
$f_o = 10$ $f_e = 25$	$f_o = 40$ $f_e = 25$

Filling in the formula gives

$$\chi^2_{obt} = \Sigma\left(\frac{(f_o - f_e)^2}{f_e}\right) = \left(\frac{(10 - 25)^2}{25}\right) + \left(\frac{(40 - 25)^2}{25}\right)$$

After subtracting,

$$\chi^2_{obt} = \left(\frac{(-15)^2}{25}\right) + \left(\frac{(15)^2}{25}\right)$$

Squaring then gives

$$\chi^2_{obt} = \left(\frac{225}{25}\right) + \left(\frac{225}{25}\right)$$

After dividing,

$$\chi^2_{obt} = 9 + 9 = 18.0$$

so $\chi^2_{obt} = 18.0$.

Interpreting Chi Squared

As always, to interpret an obtained statistic you must determine its location on the sampling distribution when H_0 is true. The H_0 sampling distribution of χ^2 contains all possible values of χ^2 when H_0 is true (the observed frequencies in a sample do represent the model described by H_0). Thus, for the handedness study, the χ^2-distribution is the distribution of all possible values of χ^2 when there are two categories and the frequencies in the two categories in the population are equal. You can envision the χ^2-distribution as shown in Figure 21.1.

Even though the χ^2-distribution is not at all normal, it is used in the same way as previous sampling distributions. When the data perfectly represent the H_0 model so that each f_o equals the corresponding f_e, χ^2 is zero. The larger the value of χ^2, the larger the differences between the expected and observed frequencies and the less frequently they occur when the observed frequencies do represent the H_0 model. Therefore, the larger the χ^2_{obt}, the less likely it is that our data represent the population frequencies described by H_0. With chi square, we again have two-tailed hypotheses but one region of rejection. If χ^2_{obt} is larger than the critical value, then it is in the region of rejection: It occurs less than 5% of the time when H_0 is true. Then χ^2 is significant. That is, the observed frequencies are so different from the expected frequencies that the observed frequencies are too unlikely to accept as representing the distribution of expected frequencies in the population described by H_0.

To determine if χ^2_{obt} is significant, you compare it to the critical value, symbolized by χ^2_{crit}. As with previous statistics, the χ^2-distribution changes shape as the degrees of freedom change, so to find the appropriate value of χ^2_{crit} for a study, you must first have the degrees of freedom.

> **In a one-way χ^2, the degrees of freedom equals $k - 1$, where k is the number of categories.**

To determine the critical value of χ^2, turn to Table 8 in Appendix C, entitled "Critical Values of Chi Square." Then, for the appropriate degrees of freedom ($k - 1$) and α, locate the critical value of χ^2. For the handedness study, $k = 2$, so $df = 1$, and with $\alpha = .05$, $\chi^2_{crit} = 3.84$. The χ^2_{obt} of 18.0 is larger than the χ^2_{crit} of 3.84, so these results are significant. Thus, reject the H_0 that each f_o represents an equal frequency in the population, and report these results as

$$\chi^2(1) = 18.0, p < .05$$

Notice our df of 1 is in parentheses.

FIGURE 21.1 Sampling Distribution of χ^2 When H_0 Is True

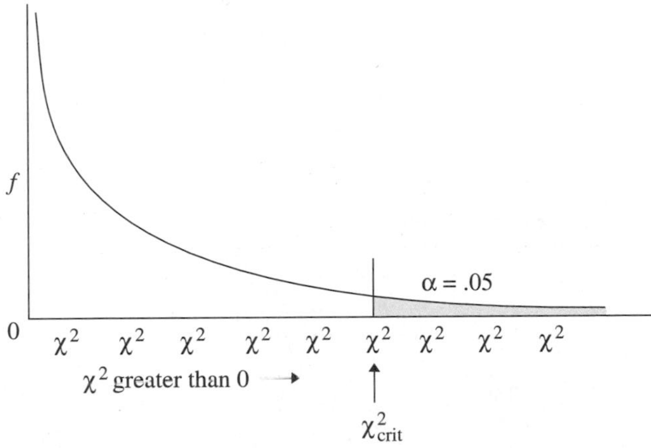

By rejecting H_0, we accept the H_a that the sample represents frequencies in the population that are not equal. In fact, as in our samples, we would expect to find about 20% left-handers and 80% right-handers in the population of geniuses. We conclude that we have evidence of a relationship between the categories of handedness and the frequency with which geniuses fall in each. Then as usual, we interpret this relationship psychologically, attempting to explain what aspects of being left-handed and a genius are related (this is a quasi-experiment, however, so we have no basis for inferring the causes of anything).

If χ^2_{obt} had not been significant, we would have failed to reject H_0. As usual, failing to reject H_0 does not prove that it is true. Therefore, we would *not* be able to say that the distribution of left-handed and right-handed geniuses in the population was equal. We would simply remain unconvinced that the distribution was unequal.

Other Uses of the Goodness of Fit Test

Instead of testing an H_0 that the frequencies in all categories are distributed equally, you can also test the goodness of fit to other H_0 models, which say that the frequencies are distributed in some other way. For example, above we assumed that geniuses are different from the everyday general population of people, in which only about 10% of the population is left-handed. Instead, we might test whether the distribution of handedness in our sample of geniuses fits this model of the distribution of handedness in the general population.

The null hypothesis is still that the data fit the H_0 model, so we can state H_0 as: 10% left-handed, 90% right-handed. For simplicity, we can write H_a as: Not H_0. This H_a implies that the observed frequencies represent a population that does not fit the H_0 model, so the population of geniuses is not 10% left-handed and 90% right-handed.

As usual, we compute the f_es based on H_0, but now the new model is that 10% of the population is left-handed and 90% is right-handed. If H_0 is true and the sample is perfectly representative of the general population, then left-handed geniuses should occur 10% of the time. For the 50 geniuses in the study, 10% is 5, so the expected frequency for left-handers is $f_e = 5$. Likewise, right-handed geniuses should occur 90% of the time, and because 90% of 50 is 45, the expected frequency for right-handers is $f_e = 45$. As usual, according to H_0, any differences between the observed and expected frequencies are due to sampling error in representing this model.

We should *not* perform two χ^2 procedures on the same data, but for the sake of illustration, we'll compare the previous handedness data and our new expected frequencies. We have:

Handedness

Left-handers	*Right-handers*
$f_o = 10$ $f_e = 5$	$f_o = 40$ $f_e = 45$

$k = 2$

Total $f_o = 50$

We compute χ^2 using the same formula we used in the previous section. Putting the above values in the formula gives

$$\chi^2_{obt} = \Sigma\left(\frac{(f_o - f_e)^2}{f_e}\right) = \left(\frac{(10 - 5)^2}{5}\right) + \left(\frac{(40 - 45)^2}{45}\right)$$

(Notice that we now have a different value of f_e in each fraction.) Working through the formula, we have

$$\chi^2_{obt} = 5.0 + .56$$

so, $\chi^2_{obt} = 5.56$.

With $\alpha = .05$ and $k = 2$, the critical value of χ^2 for $df = 1$ is again 3.84. Because the χ^2_{obt} of 5.56 is larger than the χ^2_{crit} of 3.84, these results fall in the region of rejection. Therefore, we reject H_0 and conclude that the observed frequencies are significantly different from what we would expect if handedness in the population of geniuses was distributed as it is in the general population. Instead, our best guess is that the population of geniuses would be distributed as in our sample data, with 20% left-handers and 80% right-handers.

If χ^2_{obt} had not been significant, we would have failed to reject H_0 and would simply remain unconvinced that handedness is distributed differently in geniuses than it is in the general population.

Graphing the Results in a One-Way Chi Square

As usual, creating a graph is a useful way to summarize data, especially if there are more than two categories. Label the Y axis with the frequencies and the X axis with the levels or categories, and then plot the f_o in each category. Figure 21.2 shows the results of the handedness study. Handedness is a nominal variable, and recall that when the X variable is a nominal variable, you should create a bar graph.

Notably, unlike ANOVA, the one-way chi square usually is not followed by *post hoc* comparisons. A significant χ^2_{obt} indicates that, across all categories of the variable, the frequencies are distributed in a manner that is significantly different from that described by H_0. Thus, use the observed frequency in each category to estimate the frequencies that would be found in the population. Likewise, do not compute such measures as eta squared

FIGURE 21.2 Frequencies of Left- and Right-Handed Geniuses

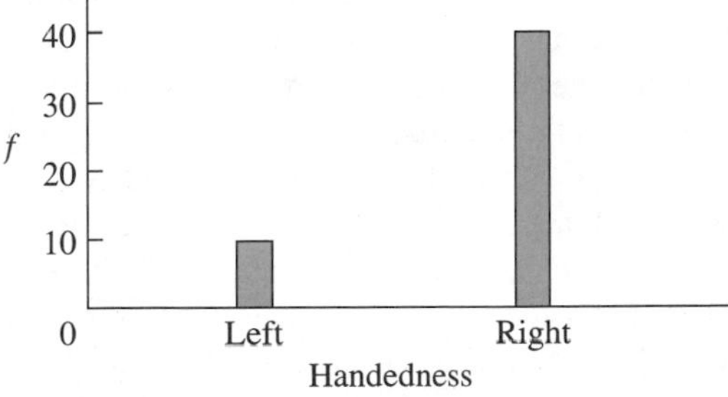

THE TWO-WAY CHI SQUARE: THE TEST OF INDEPENDENCE

The *two-way* chi square procedure is used when the data consist of the frequencies with which participants belong to the categories in each of *two* variables. This is similar to the two-way factorial design you saw in previous chapters. Depending on the number of categories in each variable, the design can be a 2 × 2, a 2 × 3, a 4 × 3, and so on. The procedures for computing χ^2 are the same regardless of the design.

The assumptions of the two-way chi square are the same as for the one-way chi square. (If an f_e is less than 5, do not compute χ^2. Instead, perform Fisher's exact test.[1])

Logic of the Two-Way Chi Square

Here is a study that calls for a two-way chi square: At one time, psychologists claimed to have identified two personality types: Type A and Type B. The Type A personality tends to be a very pressured, hostile individual who never seems to have enough time. The Type B personality tends not to be so time-pressured, being more relaxed and mellow. A controversy developed over whether people with Type A personalities are less healthy, especially when it comes to the big one—having heart attacks. Say that we enter this controversy by randomly selecting a sample of 80 people. Using the appropriate personality test, we determine how many are Type A and how many Type B. We then count the frequency with which Type A and Type B people have had heart attacks. We must also count the frequency with which Type A and Type B people have *not* had heart attacks (see item 4 in "Assumptions of the One-Way Chi Square"). Therefore, we have two categorical variables: Personality type (A or B) and health (heart attack or no heart attack). We can diagram this study as shown below.

		Personality type	
		Type A	*Type B*
Subject's health	*Heart attack*	f_o	f_o
	No heart attack	f_o	f_o

Although this looks like a two-way ANOVA, it is not analyzed like one. Instead of a series of tests for main effects and an interaction, *the two-way χ^2 procedure tests only what is essentially the interaction.* That is, it tests whether the distribution of the frequencies in the categories of one variable *depends* on which category of the other variable we examine. Because of this, the two-way χ^2 is called a **test of independence**: It determines whether the frequency of participants falling in a particular category of one variable is independent of the frequency of their falling into a particular category of the other variable. Thus, in our study, we will test whether the frequencies for having or not having a heart attack are independent of the frequencies for being Type A or Type B. Table 21.1 shows an ideal example of data when the two variables are independent. Here, the frequency of having or not having a heart attack does not depend on the frequency of being Type A or Type B. Thus, the two variables are independent.

[1]Described in S. Siegel and N. J. Castellan (1988), *Nonparametric Statistics for the Behavioral Sciences*, 2nd ed. (New York: McGraw Hill).

TABLE 21.1 Observed Frequencies When Personality Type and Heart Attacks Are Independent of Each Other

		Personality type	
		Type A	*Type B*
Participant's health	*Heart attack*	$f_o = 20$	$f_o = 20$
	No heart attack	$f_o = 20$	$f_o = 20$

Another way to view the two-way χ^2 is as a test of whether a correlation exists between the two variables. When the variables are independent, there is no correlation. Then, using the categories from one variable is no help in predicting the frequencies for the other variable. In Table 21.1, using Type A or Type B does not help in predicting how frequently people have or do not have a heart attack (and the categories of heart attack and no heart attack do not help to predict the frequency of each personality type).

On the other hand, Table 21.2 shows an ideal example of data when the two variables are not independent, but are dependent. Here, the frequency of a heart attack or no heart attack *depends* on personality type. Thus, a correlation exists, because whether people are Type A or Type B is a very good predictor of whether they have or have not had a heart attack (and vice versa).

The null hypothesis always says that there is zero correlation in the population, so the null hypothesis in the two-way χ^2 always says that the variables are *independent* in the population. If, in the sample data, the variables appear to be dependent and correlated, H_0 says that this is due to sampling error. The alternative hypothesis is that the variables are dependent (correlated).

Computing the Expected Frequencies in the Two-Way Chi Square

As usual, expected frequencies are based on the model of the population described by H_0, so here we compute the f_e in each category based on the idea that the variables are independent. To see how to do this, say that in the heart attack study, we obtained the data in Table 21.3.

TABLE 21.2 Observed Frequencies When Personality Type and Heart Attacks Are Dependent on Each Other

		Personality type	
		Type A	*Type B*
Participant's health	*Heart attack*	$f_o = 40$	$f_o = 0$
	No heart attack	$f_o = 0$	$f_o = 40$

TABLE 21.3 Frequencies as a Function of Personality Type and Subject's Health

		Personality type		
		Type A	*Type B*	
Participant's health	*Heart attack*	$f_o = 25$	$f_o = 10$	row total = 35
	No heart attack	$f_o = 5$	$f_o = 40$	row total = 45
		column total = 30	column total = 50	total = 80 $N = 80$

After recording the f_o for each cell, you should first compute the total of the observed frequencies in each column and in each row. Also, compute the total of all frequencies, which equals *N*. (To check your work, the sum of the row totals should equal the sum of the column totals, which equals *N*.)

Now compute the expected frequency for each cell when the variables are independent. As with the one-way χ^2, the expected frequency is based on the probability of a participant's being located in the cell. First let's compute the probability of a person being in the top left cell of heart attack and Type A. Out of a total of 80 people in the study, 35 reported having had a heart attack (the row total). Thus, the probability of someone reporting a heart attack is 35/80, or .438. Similarly, the probability of someone being Type A is 30 (the column total) out of 80, or 30/80, which is .375. We want to know the probability of someone being Type A *and* reporting a heart attack. To determine the probability of having two independent events occur simultaneously, we *multiply* the probability of one event times the probability of the other event. (A discussion of this multiplication rule for independent events is given in Part 4 of Appendix B.) Thus, the probability of a person reporting a heart attack *and* being Type A equals the probability of reporting a heart attack multiplied times the probability of being Type A, so it's (.438)(.375), which is .164. Thus, the probability that a person falls into the cell for a heart attack and Type A is .164, if the two variables are independent.

Because probability is the same as relative frequency, we expect .164 of our participants to fall in this cell if the variables are independent. This means that out of the 80 participants, we expect .164 times 80, or 13.125 people to be in this cell. Therefore, the expected frequency for this cell is $f_e = 13.125$.

Luckily, there is a shortcut formula for calculating each f_e. Above, we multiplied 35/80 times 30/80 and then multiplied the answer times 80. The 35 is the total f_o of the *row* that contains the cell, 30 is the total f_o of the *column* that contains the cell, and 80 is the total *N* of the study. Using these components, we can construct a formula.

THE COMPUTATIONAL FORMULA FOR COMPUTING THE EXPECTED FREQUENCY IN A CELL OF A TWO-WAY CHI SQUARE IS

$$f_e = \frac{(\text{cell's row total } f_o)(\text{cell's column total } f_o)}{N}$$

To find f_e for a cell, multiply the total observed frequencies for the row containing the cell times the total observed frequencies for the column containing the cell, and then divide by the N of the study.

Table 21.4 shows the finished diagram for our study, giving the computed f_e for each cell. To check your work, confirm that the sum of the f_es in each column or row equals the column or row total.

If H_0 is true and the variables are independent, then each observed frequency should equal each corresponding expected frequency. The larger the value of χ^2_{obt}, however, the larger the difference between f_o and f_e, so the less likely it is that the data represent variables that are independent.

Computing the Two-Way Chi Square

Compute the χ^2_{obt} for the two-way χ^2 using the same formula used in the one-way design, which is

$$\chi^2_{obt} = \Sigma\left(\frac{(f_o - f_e)^2}{f_e}\right)$$

Using the data in Table 21.4 from the heart attack study, we have

$$\chi^2_{obt} = \left(\frac{(25 - 13.125)^2}{13.125}\right) + \left(\frac{(10 - 21.875)^2}{21.875}\right) + \left(\frac{(5 - 16.875)^2}{16.875}\right) + \left(\frac{(40 - 28.125)^2}{28.125}\right)$$

As before, in the numerator of each fraction is the observed frequency minus the expected frequency for a cell, and in the denominator is the expected frequency for that cell. Solving each fraction gives

$$\chi^2_{obt} = 10.74 + 6.45 + 8.36 + 5.01$$

so $\chi^2_{obt} = 30.56$.

TABLE 21.4 Diagram Containing f_o and f_e for Each Cell

Each f_e equals the row total times the column total, divided by N.

		Personality type		
		Type A	*Type B*	
Participant's health	*Heart attack*	$f_o = 25$ $f_e = 13.125$ (35)(30)/80	$f_o = 10$ $f_e = 21.875$ (35)(50)/80	row total = 35
	No heart attack	$f_o = 5$ $f_e = 16.875$ (45)(30)/80	$f_o = 40$ $f_e = 28.125$ (45)(50)/80	row total = 45
		column total = 30	column total = 50	total = 80

Although this is a rather large value, such answers are possible. (If you get one, however, it's a good idea to triple-check your computations.)

To evaluate χ^2_{obt}, you need to find the appropriate χ^2_{crit}, so first determine the degrees of freedom.

The degrees of freedom in a two-way chi square are
df = (number of rows − 1)(number of columns − 1)

For our study, df is $(2 - 1)(2 - 1)$, or 1. Again find the critical value of χ^2 in Table 8 in Appendix C. At $\alpha = .05$ and $df = 1$, the χ^2_{crit} is 3.84.

The χ^2_{obt} of 30.56 is larger than the χ^2_{crit} of 3.84, so the obtained χ^2 is significant. When the two-way χ^2 is significant, the observed frequencies are too unlikely to accept as poorly representing frequencies from variables that are independent. Therefore, we reject H_0 that the variables are independent and accept the alternative hypothesis: We are confident that the sample represents frequencies from two variables that are dependent in the population. In other words, we conclude that there is a significant correlation such that the frequency of having or not having a heart attack depends on the frequency of being Type A or Type B (and vice versa). Report the results as:

$$\chi^2\,(1) = 30.56, p < .05$$

If χ^2_{obt} had not been larger than the critical value, we would not have rejected H_0. In that case, we could not say whether these variables are independent or not.

REMEMBER A significant two-way χ^2 indicates that the sample data are likely to represent two variables that are dependent (or correlated) in the population.

Additional Procedures in the Two-Way Chi Square

As usual, when you find a significant two-way χ^2_{obt}, you then want to understand the relationship in the data and interpret it psychologically. Therefore, as always, graph the data and describe the strength of the relationship.

Graphing the two-way chi square Graph the data in a two-way chi square in the same way that you graphed a two-way interaction in previous chapters, except that here you create a bar graph. Frequency is plotted along the Y axis, and one of the categorical variables is plotted along the X axis. The other categorical variable is indicated within the body of the graph. Figure 21.3 shows such a bar graph for the heart attack study. It is interpreted in the same way that you interpreted the table of frequencies: The frequency of people having or not having a heart attack depends on whether you are talking about Type A or Type B personalities.

Describing the relationship in a two-way chi square A significant two-way chi square indicates that there is a significant relationship between the two variables. Remember, however, that you don't keep the strength of a significant relationship secret. Instead, as usual, first think "correlation coefficient."

If you have performed a 2 × 2 chi square (and it is significant), describe the strength of the relationship by computing a new correlation coefficient known as the **phi coefficient**.

FIGURE 21.3 Frequency of Heart Attacks and Personality Type

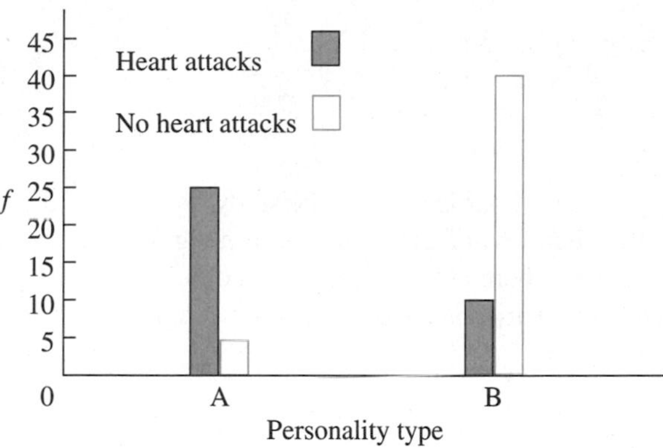

The symbol for the phi coefficient is ϕ and its value can be between 0 and +1. You can think of phi as comparing your actual data to the ideal situations when the variables are or are not perfectly dependent. A value of 0 would indicate that the data are not dependent, matching a pattern of independent frequencies such as back in Table 21.1. The larger the value of phi, however, the closer the data come to fitting an ideal pattern of dependent data, such as in Table 21.2.

THE COMPUTATIONAL FORMULA FOR THE PHI COEFFICIENT IS

$$\phi = \sqrt{\frac{\chi^2_{obt}}{N}}$$

N equals the total number of subjects in the study.

For our heart attack study, χ^2_{obt} was 30.56 and N was 80, so ϕ is

$$\phi = \sqrt{\frac{\chi^2_{obt}}{N}} = \sqrt{\frac{30.56}{80}} = \sqrt{.382} = .62$$

Thus, on a scale of 0 to +1, where +1 indicates that the variables are perfectly dependent, we found a correlation of .62 between the frequency of heart attacks and the frequency of personality types.

But remember that the best way to evaluate a relationship is to think "squared correlation coefficient" and compute the proportion of variance accounted for. If we did not take the square root in the above formula, we would have ϕ^2 (phi squared). This is analogous to r^2 or η^2, indicating how much more accurately we can predict scores by using the relationship. In the study above, $\phi^2 = .38$, so we are 38% more accurate in predicting the frequency of heart attacks/no heart attacks when we know personality type (or vice versa).

If you obtain a significant two-way chi square but it is *not* from a 2 × 2 design, then do *not* compute the phi coefficient. Instead, compute the **contingency coefficient**, symbolized by C.

THE COMPUTATIONAL FORMULA FOR THE CONTINGENCY COEFFICIENT, C, IS

$$C = \sqrt{\frac{\chi^2_{obt}}{N + \chi^2_{obt}}}$$

N is the number of participants in the study. Interpret C the same way you interpret ϕ.

Finally, when one of the variables in a significant two-way χ^2 contains more than two categories, there are advanced procedures available that are somewhat analogous to *post hoc* comparisons for determining which cells differ from which.[2]

NONPARAMETRIC PROCEDURES FOR RANKED DATA

In addition to chi square, there are several other nonparametric procedures you should be aware of. You perform these procedures when you have rank-ordered (ordinal) scores (where scores indicate 1st, 2nd, 3rd, and so on). You obtain ranked scores in a study for one of two reasons. First, sometimes participants' scores on the dependent variable are directly measured using ranked scores. Second, sometimes the dependent variable is initially measured using interval or ratio scores, but the data then violate the assumptions of parametric procedures by not being normally distributed or not having homogeneous variance. Therefore, you can transform the scores by assigning them ranks: The highest raw score is ranked 1, the next highest score is ranked 2, and so on. Then compute one of the following nonparametric inferential statistics to determine whether there are significant differences in the ranked scores for the different conditions of the independent variable.

The Logic of Nonparametric Procedures for Ranked Data

Instead of computing the mean of each condition in the experiment, as we did with t or F, with nonparametric procedures we add the ranked scores in each condition and then examine these sums of ranks. The symbol for a sum of ranks is ΣR. (So that we will have an accurate ΣR, we handle tied ranks as described in Chapter 10.) Then we compare the observed sum of ranks to an expected sum of ranks. To see the logic of this, say we have ranked the performance of a small college class, and then separated the participants into males and females as follows:

Female	*Male*
1	2
4	3
5	6
8	7
$\Sigma R = 18$	$\Sigma R = 18$

[2]Described in S. Siegel and N. J. Castellan (1988), *Nonparametric Statistics for the Behavioral Sciences*, 2nd ed. (New York: McGraw Hill).

Here there is no difference between the groups, with each group containing both high and low ranks. When the high and low ranks are distributed equally between the two groups, the sums of ranks are equal (here ΣR is 18 in each group). Because there is no difference between these two samples, they may represent the same population of ranks. That is, call it male or call it female, it's still one population that contains both low and high ranks. The null hypothesis always states that the populations from the conditions are the same, so here H_0 is that we have the same population of ranks. And notice, that when the data supports H_0, the sum of the ranks in each group equals 18. This is the expected sum of ranks if H_0 is true.

But say that the rankings had turned out as below:

Female	*Male*
1	5
2	6
3	7
4	8
$\Sigma R = 10$	$\Sigma R = 26$

The female group contains all of the low ranks, and the male group contains all of the high ranks. Because these samples are different, they may represent two different populations. The alternative hypothesis is always that the populations represented by the samples are different, so here H_a says that one population contains predominantly low ranks and the other contains predominantly high ranks. Notice that when the data support H_a, the sum of ranks in each sample is different from the expected sum of ranks: Here, each ΣR is not equal to 18.

As usual, the problem is that there is another reason that each observed sum of ranks may not equal the expected sum of ranks. It may be that H_0 is true and the groups represent one population, but that the data reflect sampling error. However, the larger the difference between the expected and observed sum of ranks, the less likely it is that this difference is due to sampling error, and the more likely it is that each sample represents a different distribution of ranks in the population.

Therefore, in each of the following procedures, we compute a statistic that measures the difference between the expected and the observed sum of ranks. If the statistic is a certain size, then we reject H_0 and accept H_a: We are confident that the reason the observed sum of ranks is different from the expected sum of ranks is that the samples represent different populations of ranks. (If the ranks reflect underlying interval or ratio scores, a significant difference in ranks indicates that the raw score populations are also different.)

Choosing a Nonparametric Procedure

Each of the major parametric procedures found in previous chapters has a corresponding nonparametric procedure for ranked data. Your first task is to know which nonparametric procedure to choose for the type of design you are testing. Table 21.5 shows the name of the nonparametric version of each parametric procedure we have discussed.

TABLE 21.5 Parametric Procedures and Their Nonparametric Counterparts

Type of design	*Parametric test*	*Nonparametric test*
Two independent samples	Independent samples t-test	Mann-Whitney U or Rank Sums test
Two dependent samples	Dependent samples t-test	Wilcoxon T test
Three or more independent samples	Between-subjects ANOVA (*Post hoc* test: protected t-test)	Kruskal-Wallis H test (*Post hoc* test: Rank Sums test)
Three or more dependent samples	Within-subjects ANOVA (*Post hoc* test: Tukey's *HSD*)	Friedman χ^2 test (*Post hoc* test: Nemenyi's test)

The steps in calculating these nonparametric tests are described in the following sections.

Tests for Two Independent Samples

There are two nonparametric procedures that are analogous to the t-test for two independent samples: The Mann-Whitney U test and the Rank Sums test. Both are used to test for significant differences between ranked scores measured under two, between-subjects conditions of an independent variable. Which test you use depends on the n in each condition.

The Mann-Whitney *U* test The Mann-Whitney test is appropriate when the n in each condition is equal to or less than 20 and you have two independent samples of ranks. For example, say we measure the reaction times of two groups of people to certain symbols. For one group, the symbols are printed in black ink, and for the other group the symbols are printed in red ink. We wish to know whether there is a significant difference between reaction times for each colored symbol. The raw score population of such reaction times tends to be highly positively skewed, however, so we cannot perform the t-test. Therefore, we convert the reaction time scores to ranks. Say that the n in each condition is 5 (but we can perform this procedure when the ns are not equal). Table 21.6 gives the reaction times (measured in milliseconds) and their corresponding ranks for this study.

To perform the Mann-Whitney U test, do the following:

1. *Assign ranks to all scores in the experiment.* As shown in Table 21.6, assign the rank of 1 to the lowest score in the experiment, regardless of which group it is in. Assign the rank of 2 to the second lowest score in the experiment, and so on.
2. *Compute the sum of the ranks for each group.* Compute ΣR for each group, and note its n, the number of scores in the group.
3. *Compute two versions of the Mann-Whitney U.* First, compute U_1 for Group 1, using the formula:

TABLE 21.6 Example Ranked Data from Two Independent Samples

Red symbols		*Black symbols*	
Reaction time	*Ranked score*	*Reaction time*	*Ranked score*
540	2	760	7
480	1	890	8
600	5	1105	10
590	3	595	4
605	6	940	9
	$\Sigma R = 17$		$\Sigma R = 38$
	$n = 5$		$n = 5$

$$U_1 = (n_1)(n_2) + \frac{n_1(n_1 + 1)}{2} - \Sigma R_1$$

where n_1 is the n of Group 1, n_2 is the n of the other group, and ΣR_1 is the sum of ranks from Group 1. Let's call the red symbol group Group 1, so filling in the above formula with the values from Table 21.6 gives

$$U_1 = (5)(5) + \frac{5(5 + 1)}{2} - 17 = 40 - 17 = 23.0$$

Now, compute U_2 for Group 2, using the formula

$$U_2 = (n_1)(n_2) + \frac{n_2(n_2 + 1)}{2} - \Sigma R_2$$

Here, the numerator of the fraction involves n_2 instead of n_1, and the sum of ranks from Group 2, ΣR_2. Call the black symbol group Group 2, so filling in the formula gives

$$U_2 = (5)(5) + \frac{5(5 + 1)}{2} - 38 = 40 - 38 = 2.0$$

4. *Determine the Mann-Whitney U_{obt}.* In a two-tailed test, the value of U_{obt} equals the *smaller* of U_1 or U_2. In the example, $U_1 = 23.0$ and $U_2 = 2.0$, so $U_{obt} = 2.0$. In a one-tailed test, you predict that one of the groups has the larger sum of ranks. The corresponding value of U_1 or U_2 from that group becomes U_{obt}.
5. *Find the critical value of U in Table 9 of Appendix C entitled "Critical Values of Mann-Whitney U."* Choose the appropriate part of the table for either a two-tailed or a one-tailed test. Then, locate U_{crit} using n_1 across the top of the table and n_2 along the left-hand side of the table. For the example, with a two-tailed test and $n_1 = 5$ and $n_2 = 5$, U_{crit} is 2.0.
6. *Compare U_{obt} to U_{crit}.* WATCH OUT! This is a biggie! Unlike any statistic we've discussed, the U_{obt} is significant if it is *equal to or less than* U_{crit}. (This is because the *smaller* the U_{obt}, the more likely it is that the group represents a distribution of ranks that is different from the distribution represented by the other group.)

REMEMBER The Mann-Whitney U_{obt} is significant if it is *less than or equal to* the critical value of U.

In the example, $U_{obt} = 2.0$ and $U_{crit} = 2.0$, so these results are significant. Therefore, conclude that the distribution of ranked scores represented by one sample is significantly different from the distribution represented by the other sample. Because the ranks reflect reaction time scores, also conclude that the populations of reaction time scores for participants who see red and black symbols are different. With $\alpha = .05$, the probability of a Type I error is $p < .05$.

7. *To describe the effect size, compute eta squared.* The only way to do this is to re-analyze your data using the following rank sums test.

The rank sums test The rank sums test is used to test two independent samples of ranks when the n in either condition is *greater* than 20. To illustrate how this statistic is calculated, however, we'll violate this rule and use the previous ranked scores from the reaction time study.

To perform the rank sums test, do the following:

1. *Assign ranks to the scores in the experiment.* As in Table 21.6, rank-order all scores in the experiment.
2. *Choose one group and compute the sum of the ranks.* Compute ΣR for one group, and note n, the number of scores in the group.
3. *Compute the expected sum of ranks, ΣR_{exp}, for the chosen group.* Use the formula

$$\Sigma R_{exp} = \frac{n(N + 1)}{2}$$

where n is the n of the chosen group and N is the total N of the study. Let's compute ΣR_{exp} for the red symbol group, which had $\Sigma R = 17$ and $n = 5$ (N is 10). Filling in the formula gives

$$\Sigma R_{exp} = \frac{n_1(N + 1)}{2} = \frac{5(10 + 1)}{2} = \frac{55}{2} = 27.5$$

Thus, $\Sigma R_{exp} = 27.5$.

4. *Compute the rank sums statistic, symbolized by z_{obt}. Use the formula*

$$z_{obt} = \frac{\Sigma R - \Sigma R_{exp}}{\sqrt{\frac{(n_1)(n_2)(N + 1)}{12}}}$$

where ΣR is the sum of the ranks for the chosen group, ΣR_{exp} is the expected sum of ranks for the chosen group, n_1 and n_2 are the ns of the two groups, and N is the total N of the study.

For the example,

$$z_{obt} = \frac{\Sigma R - \Sigma R_{exp}}{\sqrt{\frac{(n_1)(n_2)(N + 1)}{12}}} = \frac{17 - 27.5}{\sqrt{\frac{(5)(5)(10 + 1)}{12}}}$$

$$z_{obt} = \frac{-10.5}{\sqrt{22.92}} = \frac{-10.5}{4.79} = -2.19$$

Thus, $z_{obt} = -2.19$.

5. *Find the critical value of z in the z-tables (Table 1 in Appendix C).* At $\alpha = .05$, the two-tailed $z_{crit} = \pm 1.96$. (If you had predicted that the sum of ranks of the chosen group would be only greater than or only less than the expected sum of ranks, then you would use the one-tailed value of either $+1.645$ or -1.645.)
6. *Compare z_{obt} to z_{crit}.* If the absolute value of z_{obt} is larger than the corresponding z_{crit}, there is a significant difference between the two samples. In the example, $z_{obt} = -2.19$ and $z_{crit} = \pm 1.96$. Therefore, conclude that the distribution of ranked scores represented by one sample is significantly different from the distribution represented by the other sample. Because the ranks reflect reaction time scores, also conclude that the populations of reaction time scores for participants who see red and black symbols are different ($p < .05$).
7. *Describe a significant relationship using eta squared.* Here eta squared is analogous to r^2_{pb}, (discussed in Chapter 15). To compute eta squared, use the formula

$$\eta^2 = \frac{(z_{obt})^2}{N - 1}$$

where z_{obt} is computed in the above rank sums test and N is the total number of subjects.

In the example, z_{obt} is -2.19 and N is 10, so $(2.19)^2/9$ is .53. Thus, the color of the symbols accounts for approximately .53 of the variance, or differences, in the ranks. Because the ranks reflect reaction time scores, *approximately* 53% of the differences in reaction time scores are associated with the color of the symbol.

The Wilcoxon *T*-Test

The Wilcoxon test is analogous to the dependent samples *t*-test for ranked data. Recall that you have dependent samples when you match samples or when you have repeated measures. For example, say we perform a study similar to the above reaction time study, but this time we measure the reaction times of the *same* participants to both the red and black symbols. Table 21.7 gives the data we might obtain.

TABLE 21.7 Example Data for the Wilcoxon Test for Two Dependent Samples

Participant	*Reaction time to red symbols*	*Reaction time to black symbols*	*Difference, D*	*Ranked scores*	*R−*	*R+*
1	540	760	−220	6	6	
2	580	710	−130	4	4	
3	600	1105	−505	9	9	
4	680	880	−200	5	5	
5	430	500	−70	3	3	
6	740	990	−250	7	7	
7	600	1050	−450	8	8	
8	690	640	+50	2		2
9	605	595	+10	1		1
10	520	520	0			
			$N = 9$		$\Sigma R = 42$	$\Sigma R = 3$

To determine whether the two samples of scores represent different populations of ranks, compute the Wilcoxon T_{obt} as follows (this T_{obt} is not the t_{obt} from Chapters 15 and 16!):

1. *Determine the difference score, D, for each pair of scores.* For each pair of scores, subtract the score in one condition from the score in the other. It makes no difference which score is subtracted from which, but subtract the scores the same way for all pairs. Record the difference scores.
2. *Determine the N of the difference scores, but ignore all difference scores equal to zero.* The N is the total number of nonzero difference scores. In the above study, there is one difference of zero (for subject 10), so even though there were ten participants, $N = 9$.
3. *Assign ranks to the nonzero difference scores.* Ignore the sign (+ or −) of each difference. Assign the rank of 1 to the smallest difference, the rank of 2 to the second smallest difference, and so on. Record ranked scores in a column.
4. *Separate the ranks, using the sign of the difference scores.* Create two columns of ranks, labeled "$R-$" and "$R+$." The $R-$ column contains the ranks assigned to negative differences in step 3 above. The $R+$ column contains the ranks assigned to positive differences.
5. *Compute the sums of ranks for the positive and negative difference scores.* Compute ΣR for the column labeled "$R+$." Then compute ΣR for the column labeled "$R-$."
6. *Determine the Wilcoxon T_{obt}.* In the two-tailed test, the Wilcoxon T_{obt} is equal to the *smallest* ΣR *in step 5*. In the example, the smallest ΣR equals 3, so $T_{obt} = 3$. (In the one-tailed test, you predict whether most differences are positive or negative, depending on your experimental hypotheses, so you predict whether the $R+$ or $R-$ column contains the smaller ΣR. The ΣR predicted to be smallest is the value of T_{obt}. Thus, say we predicted that red symbols would produce the largest reaction time scores. Given the way we subtracted, we would predict that ΣR for the $R-$ column would be smaller, so T_{obt} would be 42.)
7. *Find the critical value of T in Table 10 of Appendix C, entitled "Critical Values of the Wilcoxon T."* Find T_{crit} for the appropriate alpha level and N, the number of nonzero difference scores. Above, $N = 9$, so for $\alpha = .05$, T_{crit} is 5.0.
8. *Compare T_{obt} to T_{crit}.* Again, watch out: The Wilcoxon T is significant if it is *equal to or less than* T_{crit}.

REMEMBER The Wilcoxon T_{obt} is significant if it is *less than or equal to* the critical value of T.

In the above example, for a two-tailed test, the T_{obt} of 3.0 is less than the T_{crit} of 6.0, so it is significant. Therefore, conclude that each sample represents a different distribution of ranks and thus a different population of reaction time scores ($p < .05$).

There is no recognized way to compute η^2 for this procedure.

The Kruskal-Wallis *H* Test

The Kruskal-Wallis H test is analogous to a one-way between-subjects ANOVA for ranked data. It assumes that the study involves one independent variable and that there are *three* or more independent samples with at least five participants in each sample. The null hypothesis is that all conditions represent the same distribution of ranks in the population.

As an example, consider a study that explores the relationship between the independent variable of a golfer's height and the dependent variable of the distance he or she hits the ball. We test three groups of novice golfers, classified on the factor of height as either short, medium, or tall. We measure the distance each person drives the ball in meters. However, say that based on the F_{max} test discussed in Chapter 19, we cannot assume that the distance scores have homogeneous variance, so we dare not use the parametric ANOVA. Instead, we rank the scores and perform the Kruskal-Wallis H test. These data are shown in Table 21.8.

To compute the Kruskal-Wallis H, do the following:

1. *Assign ranks, using all scores in the experiment.* Assign the rank of 1 to the lowest score in the experiment, the rank of 2 to the second lowest score, and so on.
2. *Compute the sum of the ranks in each condition.* Compute the sum of the ranks, ΣR, in each column. Also note the n in each condition.
3. *Compute the sum of squares between groups, SS_{bn}.* Use the formula

$$SS_{bn} = \frac{(\Sigma R_1)^2}{n_1} + \frac{(\Sigma R_2)^2}{n_2} + \ldots + \frac{(\Sigma R_k)^2}{n_k}$$

For each level, square the sum of the ranks and then divide that quantity by the n in the level. (There may be a different n in each level.) After doing this for all k levels in the factor, add the amounts together.

For the example, from Table 21.8

$$SS_{bn} = \frac{(21)^2}{5} + \frac{(35)^2}{5} + \frac{(64)^2}{5} = 88.2 + 245 + 819.2$$

so, $SS_{bn} = 1152.4$.

4. *Compute the H_{obt}.* Use the formula

$$H_{obt} = \left(\frac{12}{N(N + 1)}\right)(SS_{bn}) - 3(N + 1)$$

TABLE 21.8 Example Data for the Kruskal-Wallis H Test

Height						
Short		*Medium*		*Tall*		
Score	*Rank*	*Score*	*Rank*	*Score*	*Rank*	
10	2	24	3	68	14	
28	6	27	5	71	15	
26	4	35	7	57	10	
39	8	44	9	60	12	
6	1	58	11	62	13	
$\Sigma R_1 = 21$ $n_1 = 5$		$\Sigma R_2 = 35$ $n_2 = 5$		$\Sigma R_3 = 64$ $n_3 = 5$		$N = 15$

where N is the total N of the study. Divide 12 by $N(N + 1)$ and multiply the answer times the SS_{bn}. Then subtract $3(N + 1)$.

In the example,

$$H_{obt} = \left(\frac{12}{15(15 + 1)}\right)(1152.4) - 3(15 + 1) = (.05)(1152.4) - 48$$

$$H_{obt} = 57.62 - 48$$

Thus, the answer is $H_{obt} = 9.62$.

5. *Find the critical value of H in the χ^2 tables (Table 8 in Appendix C).* Values of H have the same sampling distribution as χ^2. The degrees of freedom are $df = k - 1$, where k is the number of levels in the factor.

   In our example, k is 3, so $df = 2$. In the χ^2 tables for $\alpha = .05$ and $df = 2$, χ^2_{crit} is 5.99.

6. *Compare the obtained value of H to the critical value of χ^2.* If H_{obt} is *larger* than the critical value found in the χ^2-tables, then H_{obt} is significant. For the golfing study, the H_{obt} of 9.62 is larger than the χ^2_{crit} of 5.99, so it is significant. This means that at least two of the conditions represent different populations of ranks. Because the distance each participant hit the ball underlies each rank, conclude that at least two of the populations of distances for short, medium, and tall golfers are not the same ($p < .05$).

7. *Perform post hoc comparisons using the rank sums test.* When H_{obt} is significant, determine which specific conditions differ by performing the rank sums test on every pair of conditions. This is analogous to Fisher's protected t-test, discussed in Chapter 17, and it is used regardless of the n in each group. To perform the procedure, treat each pair of conditions being compared as if they comprised the entire study, then follow the procedure described previously for the rank sums test.

   REMEMBER When performing the rank sums test as a *post hoc* test for H_{obt}, re-rank the scores using only those scores in the two conditions being compared.

   In the example, comparing the ranks of short and medium-height golfers produces a z_{obt} of 1.36, comparing short and tall golfers produces a z_{obt} of 2.62, and comparing medium-height and tall golfers produces a z_{obt} of 2.40. With $\alpha = .05$, z_{crit} is ± 1.96. Therefore, the scores of short and medium-height participants are not significantly different, but they both differ significantly from the scores in the tall condition. Conclude that the tall golfers represent one population of distances that is different from the population for short and medium-height golfers.

8. *Describe a significant relationship using eta squared.* Use the formula

$$\eta^2 = \frac{H_{obt}}{N - 1}$$

   where H_{obt} is the value computed in the Kruskal-Wallis test and N is the total number of subjects. In the example, $H_{obt} = 9.62$ and $N = 15$. Substituting into the above formula, $\eta^2 = 9.62/14$, or .69. Therefore, the variable of a player's height accounts for approximately .69 of the variance in these distance scores.

The Friedman Chi Square Test

The Friedman Chi Square test is analogous to a one-way, within-subjects ANOVA for ranks. It assumes that the study involves one factor, with either a matched groups or a repeated measures design with *three* or more conditions. If there are only three levels of the factor, there must be at least 10 scores per condition. If there are only four levels of the factor, there must be at least five scores per condition.

As an example, consider a study in which the scores we collect are already ranked. The three levels of our independent variable are the teaching styles of Dr. Highman, Dr. Shyman, and Dr. Whyman. We survey students who have taken courses from all three instructors, and each student rank-orders them. Table 21.9 gives the data we obtain.

To perform the Friedman χ^2 test, follow these steps:

1. *Assign ranks within the scores of each participant.* If the scores are not already ranks, assign the rank of 1 to the lowest score received by Participant 1, assign the rank of 2 to the second lowest score received by Participant 1, and so on. Repeat the process for each participant.
2. *Compute the sum of the ranks, ΣR, in each condition.* Find the sum of the ranks in each column.
3. *Compute the sum of squares between groups, SS_{bn}.* Use the formula

   $$SS_{bn} = (\Sigma R_1)^2 + (\Sigma R_2)^2 + \ldots + (\Sigma R_k)^2$$

   Square the sum of the ranks in each of the k conditions, and then add the squared sums together. In the example,

   $$SS_{bn} = (12)^2 + (23)^2 + (25)^2$$

   so $SS_{bn} = 1298$.

TABLE 21.9 Example Data for the Friedman Test

	Rankings for three instructors		
Participant	*Dr. Highman*	*Dr. Shyman*	*Dr. Whyman*
1	1	2	3
2	1	3	2
3	1	2	3
4	1	3	2
5	2	1	3
6	1	3	2
7	1	2	3
8	1	3	2
9	1	3	2
10	2	1	3
$N = 10$	$\Sigma R_1 = 12$	$\Sigma R_2 = 23$	$\Sigma R_3 = 25$

4. *Compute the Friedman χ^2 statistic.* Use the formula

$$\chi^2_{obt} = \left(\frac{12}{(k)(N)(k + 1)}\right)(SS_{bn}) - 3(N)(k + 1)$$

where N is the number of participants and k is the number of levels of the factor. First divide 12 by the quantity $(k)(N)(k + 1)$. Then multiply this number times SS_{bn}. Then subtract the quantity $3(N)(k + 1)$.

In the example,

$$\chi^2_{obt} = \left(\frac{12}{(3)(10)(3 + 1)}\right)(1298) - 3(10)(3 + 1)$$

$$\chi^2_{obt} = (.10)(1298) - 120 = 129.8 - 120$$

And the survey says: $\chi^2_{obt} = 9.80$.

5. *Find the critical value of χ^2.* Use the χ^2-tables (Table 8 in Appendix C). The degrees of freedom are $df = k - 1$, where k is the number of levels in the factor. For the example, $k = 3$, so for $df = 2$ and $\alpha = .05$, the critical value is 5.99.

6. *Compare χ^2_{obt} to the critical value of χ^2.* If χ^2_{obt} is larger than χ^2_{crit}, the results are significant. The above χ^2_{obt} of 9.80 is larger than the χ^2_{crit} of 5.99, so the results are significant. Thus, conclude that at least two of the samples represent different populations ($p < .05$).

7. *When the Friedman χ^2 is significant, perform post hoc comparisons using Nemenyi's procedure.* This procedure is analogous to Tukey's *HSD* procedure. Compute one value that is the *critical difference*. Any two conditions that differ by more than this critical difference are significantly different. To perform Nemenyi's procedure, follow these steps.

   (a) *Compute the critical difference.* Use the formula

   $$\text{Critical difference} = \sqrt{\left(\frac{k(k + 1)}{6(N)}\right)(\chi^2_{crit})}$$

   where k is the number of levels of the factor, N is the number of participants, and χ^2_{crit} is the critical value used to test the Friedman χ^2. Multiply k times $k + 1$, and then divide by $6(N)$. Multiply this number times χ^2_{crit}, and then find the square root.

   In the example, $\chi^2_{crit} = 5.99$, $k = 3$, and $N = 10$. Thus

   $$\text{Critical difference} = \sqrt{\left(\frac{k(k + 1)}{6(N)}\right)(\chi^2_{crit})} = \sqrt{\left(\frac{3(3 + 1)}{6(10)}\right)(5.99)}$$

   $$\text{Critical difference} = \sqrt{(.2)(5.99)} = \sqrt{1.198} = 1.09$$

   so the critical difference is ± 1.09.

   (b) *Compute the mean rank for each condition.* For each condition, divide the sum of ranks, ΣR, by the number of participants. In the example, the sums of ranks are 12, 23, and 25 in the three conditions, and N is 10. Therefore, the mean ranks are 1.2, 2.3, and 2.5 for Highman, Shyman, and Whyman, respectively.

(c) *Compute the differences between all pairs of mean ranks.* Subtract each mean rank from the other mean ranks. Any absolute difference between two mean ranks that is greater than the critical difference indicates that the two conditions differ significantly. In the example, the differences between the mean ranks for Dr. Highman and the other two instructors are 1.10 and 1.30, respectively, and the difference between Shyman and Whyman is .20. The critical difference is 1.09, so only Dr. Highman's ranking is significantly different from those of the other two instructors. Thus, conclude that if the entire population were to rank the three instructors, Dr. Highman would be ranked superior to the other two instructors.

8. *Describe a significant relationship using eta squared.* Use the formula

$$\eta^2 = \frac{\chi^2_{obt}}{(N)(k) - 1}$$

where χ^2_{obt} is the value computed in the Friedman χ^2 test, N is the number of participants, and k is the number of levels of the factor. For the example

$$\eta^2 = \frac{\chi^2_{obt}}{(N)(k) - 1} = \frac{9.80}{(10)(3) - 1} = \frac{9.80}{30 - 1} = .34$$

Thus, the instructor variable accounts for .34 of the variability, or differences, in these rankings.

PUTTING IT ALL TOGETHER

Congratulations. You are now familiar with the basic statistical and research procedures used in psychology and other behavioral sciences. Even if you someday go to graduate school, you'll find that there is little in the way of basics for you to learn.

CHAPTER SUMMARY

1. *Nonparametric procedures* are used when data do not meet the assumptions of parametric procedures.

2. *Chi square*, χ^2, is used with one or more categorical variables and the data are the frequencies with which participants fall in a category.

3. The *one-way* χ^2 is a goodness of fit test that determines whether the observed frequencies fit the model of the expected frequencies described by H_0. The larger the χ^2_{obt}, the larger the overall differences between the observed and expected frequencies. A significant χ^2_{obt} indicates that the observed frequencies are unlikely to represent the distribution of frequencies in the population described by H_0.

4. In the *two-way* χ^2, H_0 states that the frequencies in the categories of the two variables are independent. The χ^2_{obt} describes the differences between the observed frequencies and the expected frequencies when H_0 is true and the variables are independent. A significant χ^2 indicates that the observed frequencies are unlikely to represent variables that are independent in the population, so conclude that the two variables are dependent, or correlated.

5. In a significant 2 × 2 chi square, the strength of the relationship is described by the *phi correlation coefficient*, ϕ. In a significant two-way chi square that is not 2 × 2, the strength of the relationship is described by the *contingency coefficient*, *C*. The larger these coefficients are, the closer the variables are to being perfectly dependent, or correlated. Squaring ϕ or C gives the proportion of variance accounted for, which indicates how much more accurately the frequencies of category membership on one variable can be predicted by knowing participants' category membership on the other variable.

6. There are two nonparametric versions of the independent samples *t*-test for ranks. The *Mann-Whitney U test* is performed when the *n* in each condition is less than 20. The *rank sums test* is performed when the *n* in either condition is greater than 20.

7. The *Wilcoxon T test* is the nonparametric equivalent of the dependent samples *t*-test for ranks.

8. The *Kruskal-Wallis H test* is the nonparametric equivalent of the one-way, between-subjects ANOVA for ranks. The rank sums test is used as the *post hoc* test to determine which conditions differ.

9. The *Friedman* χ^2 *test* is the nonparametric equivalent of the one-way, within-subjects ANOVA for ranks. *Nemenyi's test* is the *post hoc* test to determine which conditions differ.

10. *Eta squared* describes the relationship found in experiments involving ranked data.

PRACTICE PROBLEMS

(Answers for odd-numbered problems are provided in Appendix D.)

1. What are the three major aspects of a design that determine the specific statistical procedures you should employ?
2. (a) What do all nonparametric procedures have in common with all parametric procedures? (b) What aspects of the data cause you to use nonparametric inferential procedures? (c) Why, if possible, should you design a study so that the data meet the assumptions of a parametric procedure? (d) Why shouldn't you use parametric procedures for data that clearly violate their assumptions?
3. (a) When do you use the chi square? (b) When do you use the one-way chi square? (c) When do you use the two-way chi square?
4. (a) What is the symbol for observed frequency? What does it mean? (b) What is the symbol for expected frequency? What does it mean?
5. (a) What does a significant one-way chi square indicate? (b) What does a significant two-way chi square indicate?
6. In the general population, the distribution of political party affiliation is 30% Republican, 55% Democratic, and 15% other. We wish to determine whether this distribution is also found among the elderly. In a sample of 100 senior citizens, we find 18 Republicans, 64 Democrats, and 18 other. (a) What are H_0 and H_a? (b) What is f_e for each group? (c) Compute χ^2_{obt}. (d) With $\alpha = .05$, what do you conclude about party affiliation in the population of senior citizens?

7. A survey finds that, given the choice, 34 females prefer males much taller than themselves, and 55 females prefer males only slightly taller than themselves. (a) What are H_0 and H_a for this survey? (b) With $\alpha = .05$, what would you conclude about the preference of females in the population? (c) Describe how you would graph these results.
8. In a study, Foofy counts the students who say they like Professor Demented and those who say they like Professor Randomsampler. She then performs a one-way χ^2 to determine if there is a significant difference between the frequency with which students like each professor. (a) Why is this approach incorrect? (b) How should she analyze the data?
9. (a) What is the phi coefficient, and when is it used? (b) What does the squared phi coefficient indicate? (c) What is the contingency coefficient, and when is it used? (d) What does the squared contingency coefficient indicate?
10. A study similar to the one in problem 6 determines the frequency of the different political party affiliations for male and female senior citizens.

		Affiliation		
		Republican	*Democrat*	*Other*
Gender	*Male*	18	43	14
	Female	39	23	18

(a) What are H_0 and H_a? (b) What is f_e in each cell? (c) Compute χ^2_{obt}. (d) With $\alpha = .05$, what should we conclude about gender and party affiliation in the population of senior citizens? (e) How consistent is this relationship?

11. The following data reflect the frequency with which people voted in the last election and were satisfied with the officials elected:

		Satisfied	
		yes	*no*
Vote	*yes*	48	35
	no	33	52

(a) What are H_0 and H_a? (b) What is f_e in each cell? (c) Compute χ^2_{obt}. (d) With $\alpha = .05$, what should we conclude about the correlation here? (e) How consistent is the relationship for these data?

12. What is the nonparametric version of each of the following? (a) A one-way, between-subjects ANOVA (b) An independent samples t-test ($n < 20$) (c) A dependent samples t-test (d) An independent samples t-test ($n > 20$) (e) A one-way, repeated measures ANOVA (f) Fisher's protected t-test (g) Tukey's HSD test
13. Select the statistical procedure that should be used to analyze the data from each of the following studies: (a) An investigation of the effects of a new pain reliever on rankings of the emotional content of words describing pain. A randomly selected group of participants is tested before and after administration of the drug. (b) An investigation of the effects of eight different colors of spaghetti sauce on tastiness scores. A different random sample of participants tastes each color of sauce, and then the tastiness scores are ranked. (c) An investigation of the effects of increasing amounts of alcohol consumption on reaction-time scores. The scores are ranked, and the same group of participants is tested after 1, 3, and 5

drinks. (d) An investigation of two levels of the variable of family income. In two random samples, we rank-order the percentage of participants' income spent on new clothing last year.

14. A study compares the maturity level of a group of students who have completed a research course to a group of students who have not. Maturity scores for college students tend to be skewed. For the following interval scores, answer the questions below.

No research	*Research*
43	51
52	58
65	72
23	81
31	92
36	64

(a) Do the groups differ significantly ($\alpha = .05$)? (b) What do you conclude about maturity scores you expect to find in the population of students who have taken research and in the population that hasn't?

15. We wish to compare the attitude scores of people when tested in the morning to their attitude scores when tested in the afternoon. From a morning and an afternoon attitude test, we obtain the following interval data, but they have significantly heterogeneous variance. With $\alpha = .05$, determine if there is a significant difference in scores as a function of testing times.

Morning	*Afternoon*
14	36
18	31
20	19
28	48
3	10
34	49
20	20
24	29

16. An investigator evaluated the effectiveness of activity therapy on three types of patients. She collected the following improvement ratings. (In the population, these data form highly skewed distributions.)

Depressed	*Manic*	*Schizophrenic*
16	7	13
11	9	6
12	6	10
20	4	15
21	8	9

(a) Which procedure should be used to analyze these data? Why? (b) What should she do first to the data? (c) If the results are significant, what should she do next? (d) Ultimately, what conclusions can be drawn from this study?

17. A therapist evaluates the progress of a sample of clients in a new treatment program after one month, after two months, and again after three months. Such progress data don't have homogeneous variance. (a) What statistical procedure should be used to analyze the data? Why? (b) What is the first thing the therapist must do? (c) If the results are significant, what should the therapist then do? (d) Ultimately, what will the therapist be able to identify?

18. What is the basic logic underlying the testing of H_0 in all nonparametric procedures for ranked data?

19. If a researcher summarizes scores using the mode and computes a chi square, how has he or she measured the dependent variable?

20. A research article specifies that a Wilcoxon test was performed. (a) What does this test indicate about the design and scores used in the study? (b) What would be indicated if a Freidman test had been performed?

21. To study nonverbal communication, you show participants a picture of a person smiling, frowning, or smirking. The different participants in each condition indicate whether the pictured person was either happy or sad. (a) Precisely describe the factor(s) and level(s) in this design and how you will analyze the results. (b) What flaws are built into the study in terms of the statistics you'll use and the scores you'll obtain? (c) How can you improve this study?

SUMMARY OF FORMULAS

A. Summary of chi square formulas

1. *The computational formula for chi square is*

$$\chi^2_{\text{obt}} = \Sigma\left(\frac{(f_o - f_e)^2}{f_e}\right)$$

where f_o is the observed frequency in a cell and f_o is the expected frequency in a cell.

(a) Computing expected frequency

(1) *In a one-way chi square,* the expected frequency in a category is equal to the probability of someone's falling in that category multiplied times the N of the study. *In testing an H_0 of no difference, the computational formula for each expected frequency is*

$$f_e \text{ in each category} = \frac{N}{k}$$

where N is the total N in the study and k is the number of categories.

(2) *In a two-way chi square, the computational formula for finding the expected frequency in each cell is*

$$f_e = \frac{(\text{cell's row total } f_o)(\text{cell's column total } f_o)}{N}$$

(b) *Critical values of* χ^2 *are found in Table 8 of Appendix C.*
(1) In a one-way chi square, the degrees of freedom are

$$df = k - 1$$

where k is the number of categories in the variable.
(2) In a two-way chi square, the degrees of freedom are

$$df = (\text{number of rows} - 1)(\text{number of columns} - 1)$$

2. *The computational formula for the phi coefficient is*

$$\phi = \sqrt{\frac{\chi^2_{obt}}{N}}$$

where N is the total number of participants in the study.

3. *The computational formula for the contingency coefficient, C, is*

$$C = \sqrt{\frac{\chi^2_{obt}}{N + \chi^2_{obt}}}$$

where N is the total number of participants in the study.

B. Summary of nonparametric formulas
1. *Formulas for two independent samples*

(a) *When N is equal to or less than 20, the computational formula for the Mann-Whitney* U *test for independent samples is*

$$U_1 = (n_1)(n_2) + \frac{n_1(n_1 + 1)}{2} - \Sigma R_1$$

and

$$U_2 = (n_1)(n_2) + \frac{n_2(n_2 + 1)}{2} - \Sigma R_2$$

n_1 and n_2 are the *ns* of the groups. After ranks are assigned based on all scores, ΣR_1 is the sum of ranks in Group 1, and ΣR_2 is the sum of ranks in Group 2.

In a two-tailed test, the value of U_{obt} equals the smaller of U_1 or U_2. In a one-tailed test, the value of U_1 or U_2 from the group predicted to have the largest sum of ranks is U_{obt}. Critical values of U are found in Table 9 in Appendix C. (U is significant if it is equal to or less than the critical value.)

(b) *When either* n *is greater than 20, the computational formula for the rank sums test for independent samples is*

$$z_{obt} = \frac{\Sigma R - \Sigma R_{exp}}{\sqrt{\frac{(n_1)(n_2)(N + 1)}{12}}}$$

n_1 and n_2 are the *n*s of the two groups, and *N* is the total *N*. After ranks are assigned based on all scores, ΣR is the sum of the ranks for the chosen group. ΣR_{exp} is the expected sum of ranks for the chosen group, found using the formula

$$\Sigma R_{exp} = \frac{n(N + 1)}{2}$$

where *n* is the *n* of the chosen group. Critical values of *z* are found in Table 1 of Appendix C.

(c) *Eta squared is computed using the formula*

$$\eta^2 = \frac{(z_{obt})^2}{N - 1}$$

2. *For dependent samples, the computational formula for the Wilcoxon* T *is*

$$T_{obt} = \Sigma R$$

After the difference scores are found and assigned ranks, in the two-tailed test, ΣR is the smaller of the sum of ranks for the positive difference scores or the sum of ranks for the negative difference scores. In the one-tailed test, ΣR is the sum of ranks that is predicted to be the smallest. Critical values of *T* are found in Table 10 of Appendix C, where *N* is the number of nonzero difference scores. (*T* is significant if it is equal to or less than the critical value.)

3. *For three or more independent samples, the computational formula for the Kruskal-Wallis* H *test is*

$$H_{obt} = \left(\frac{12}{N(N + 1)}\right)(SS_{bn}) - 3(N + 1)$$

where *N* is the number of participants in the study. After ranks are assigned using all scores, SS_{bn} is found using the formula

$$SS_{bn} = \frac{(\Sigma R_1)^2}{n_1} + \frac{(\Sigma R_2)^2}{n_2} + \ldots + \frac{(\Sigma R_k)^2}{n_k}$$

each *n* is the number of scores in a level, each ΣR is the sum of ranks for that level, and *k* is the number of levels of the factor.

Critical values of *H* are found in Table 8 of Appendix C, for $df = k - 1$, where *k* is the number of levels in the factor.

(a) When H_{obt} is significant, *post hoc* comparisons are performed using the rank sums test, regardless of the size of *n*.

(b) Eta squared is computed using the formula

$$\eta^2 = \frac{H_{obt}}{N - 1}$$

4. *For three or more dependent samples, the computational formula for the Friedman χ^2 test is*

$$\chi^2_{obt} = \left(\frac{12}{(k)(N)(k + 1)}\right)(SS_{bn}) - 3(N)(k + 1)$$

where N is the number of participants and k is the number of levels of the factor. After ranks are assigned within the scores of each subject, SS_{bn} is found using the formula

$$SS_{bn} = (\Sigma R_1)^2 + (\Sigma R_2)^2 + \ldots + (\Sigma R_k)^2$$

where each $(\Sigma R)^2$ is the squared sum of ranks for a level. Critical values of χ^2 are found in Table 8 of Appendix C, for $df = k - 1$, where k is the number of levels in the factor.

a. When the Friedman χ^2 is significant, *post hoc* comparisons are performed using Nemenyi's procedure.

(1) Compute the critical difference using the formula

$$\text{Critical difference} = \sqrt{\left(\frac{k(k + 1)}{6(N)}\right)(\chi^2_{crit})}$$

where k is the number of levels of the factor and N is the number of participants. χ^2_{crit} is the critical value of χ^2 for the appropriate α at $df = k - 1$.

(2) Compute the mean rank in each condition as $\Sigma R/n$.

(3) Any two mean ranks that differ by more than the critical difference are significantly different.

b. Eta squared is found using the formula

$$\eta^2 = \frac{\chi^2_{obt}}{(N)(k) - 1}$$

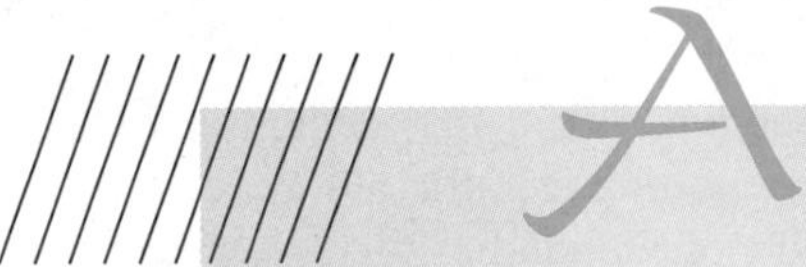

ORGANIZING AND COMMUNICATING RESEARCH USING APA FORMAT

To understand the information presented in this Appendix:

- From Chapter 1, recall that a research hypothesis must be testable, falsifiable, rational, and parsimonious.
- From Chapter 2, recall that research begins with broad constructs, is "whittled down" to precise operational definitions, and is then generalized back to the broad constructs.
- From Chapter 3, recall the requirements for designing the independent and dependent variable in an experiment.
- From Chapter 7, recall the uses and interpretation of the mean score in each condition, and how to graph an experiment's results.
- From Chapter 15, recall the independent samples *t*-test.

Then your goals are to learn:

- The basic components of the psychological literature and how to use it in a research project.
- How to organize the needed information when reporting a study.
- The parts of an APA-style report and the purpose of each.
- What information should and should not be reported in a research report.
- The style and tone that a research report should have.

Recall that science is a community activity in which we try to correct each other's errors by skeptically evaluating each study. Also recall that scientific facts are ultimately built through replication of findings. To allow critical evaluation and to build evidence for a hypothesis through replication, researchers share the results of their studies by publishing them and thus contributing to the research literature. This appendix discusses the literature and describes how a research article is created. By knowing the process an author uses in writing an article, you can read the literature more effectively. Also, as a psychology student, you'll probably be reporting your own study sooner or later.

Therefore, in the following sections, we'll first design an experiment and then see what goes into a manuscript for reporting it in the literature. The final section shows a sample APA-style research report of the study. Be forewarned that the study here is very simple, and does not incorporate all of the design or statistical issues discussed in this textbook. The idea is to show you the basics. Even though your particular study may be more elaborate, you create a report following the same logic and format.

AN EXAMPLE STUDY

Let's say that in a cognitive psychology course, you read the study by Bower, Karlin, and Dueck (1975). They studied short-term memory by presenting participants with 28 simple cartoons called "droodles." Each droodle is a meaningless geometric shape, but the neat thing is that a particular droodle will instantly become meaningful when an accompanying verbal interpretation is provided. Two examples of droodles are provided in Figure A.1. Some participants were told that droodle A shows a "midget playing a trombone in a telephone booth" and that droodle B shows "an early bird that caught a very strong worm." Other participants were not given any interpretation. Bower et al. found that those participants who had been given an interpretation for all droodles could recall (sketch) more of the droodles than those who had not. The authors concluded that the interpretations made the droodles more meaningful, allowing participants to integrate the droodles with their knowledge in memory. Then when participants tried to recall the droodles, this knowledge provided useful "retrieval cues."

Let's say that this strikes us as an interesting topic to investigate. In reading the study, we notice that the accompanying interpretations not only make the droodles meaningful but do so in a humorous way. By employing interpretations that can be seen as both humorous and meaningful, the droodle study may have been confounded: On the one hand, the interpretations may make the droodles meaningful and thus more

FIGURE A.1 Examples of "Droodles"

These droodles were accompanied by the following interpretations: (a) "A midget playing a trombone in a telephone booth," and (b) "An early bird that caught a very strong worm."

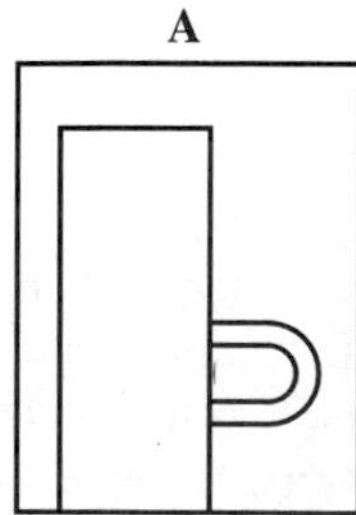

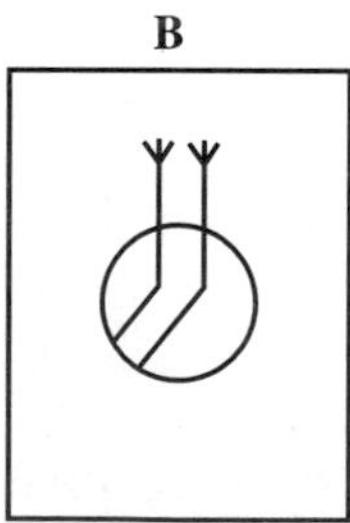

From G.H. Bower, M.B. Karlin & A. Dueck (1975), Comprehension and memory for pictures, *Memory and Cognition*, 3(2), 216–222. Reprinted by permission of the Psychonomic Society, Inc. and the author.

memorable, as the authors suggested. On the other hand, the interpretations may make the droodles humorous, and their humor makes the droodles more memorable. This second idea leads us to a rival hypothesis: When the contexts in which stimuli occur differ in humor, differences in memory for the stimuli are produced.

Our hypothesis suggests that greater humor causes improved memory, so a well-controlled, internally valid laboratory experiment to test the hypothesis is in order. In fact, the design of Bower et al. seems appropriate. They compared the effects of an interpretation versus the absence of an interpretation on memory for the droodles. We can investigate our hypothesis by presenting participants with humorous and nonhumorous interpretations. The prediction is that droodles with humorous interpretations will be better recalled.

Before proceeding, we need to be sure the study is rational, ethical, and practical. Rationally speaking, the idea that humor acts as a cue for recalling information seems to fit known memory processes (but we'll check), and understanding memory is a worthwhile psychological study. Ethically, asking participants to remember droodles does not appear to cause any major harm (but we'll check). And practically speaking, such a study seems doable and does not require inordinate time, expense, or hard-to-find participants or equipment.

As we begin creating the design for the study, we find that we don't know very much about this topic, so—to the literature!

THE RESEARCH LITERATURE

Previous research is the ultimate source for learning about a particular psychological topic. Published research provides background in the issues that pertain to your research question so that your hypothesis fits with existing conceptualizations of constructs and with the results of previous studies. The literature also suggests numerous ideas for interesting studies and describes established procedures that you can incorporate into a study.

Recall that defining the target population early in the design process helps to develop a more precise hypothesis as well as directing your attention to the most relevant portions of the literature. The Bower et al. experiment studied memory in normal adults. Therefore, we need to find past research on adults regarding (1) how the constructs of "humor" and "meaningfulness" are defined and how they may influence "memory," (2) whether humorous stimuli are better retained than nonhumorous stimuli, and (3) whether other studies using the Bower et al. design have identified flaws in it, have replicated it, or (heaven forbid) have already tested the effects of humorous and nonhumorous interpretations as we will do.

What Constitutes the Research Literature?

The term "research literature" does not mean books or newspapers found in the supermarket or popular magazines such as *Time* or *Psychology Today*. At best, these contain synopses of research articles that most likely omit necessary details. Therefore, go to professional books and psychological journals. Books provide useful background, but because of the time required to create them, even new books may be a few years behind the latest developments. Therefore, focus on journals for the most current developments in a research topic, because they are published once or more a year.

Not all professional journals, however, are of the same quality. In some, the primary requirement for publishing an article is that the author(s) pay the publication costs. These journals have less stringent requirements for quality research. Other journals are "refereed," meaning that each article has been reviewed by several psychologists who are knowledgeable about the topic being studied. To gauge the quality of a journal, check the section that describes its editorial policies. Also, look for journals published by professional organizations of psychologists such as the American Psychological Association or the Psychonomic Society. One function of these organizations is to disseminate quality research. However, although such journals tend to provide convincing studies, you still should approach each study with a critical eye.

Searching the Literature

To search the literature, begin with some topic in mind. Most journals are organized around a subarea of psychology (such as social, cognitive, abnormal) identified by the journal's title (such as *Cognitive Psychology* or the *Journal of Personality and Social Psychology*). By perusing such journals, you may come across articles that you wish to read. If you have a specific research idea, however, you can consult reference sources that search the literature in an organized manner. Eventually, the goal of any study is to explain "psychologically" the results and behaviors observed in the study. Therefore, this is the focus of the literature search.

Psychological Abstracts The *Psychological Abstracts* is a monthly publication that describes studies recently published in other psychology journals. Its index is organized using the variables and hypothetical constructs commonly studied by psychologists. When using this index, try to be specific, selecting terms you think would be used in the titles of relevant articles. For the droodle study, we would first look in the index of recent issues of *Psychological Abstracts* under such terms as humor, meaningfulness, and short-term memory. *Psychological Abstracts* provides a separate author index, so we would also look up the authors of the original droodle study to see if they have reported other similar research. For each article listed, an "abstract"—a brief synopsis—of the study is presented. By reading the abstract, you can determine whether to read the entire article.

Computerized literature searches Many college libraries provide a user-friendly computer program that searches the literature for you. Programs such as PsycLIT contain a large database covering years of research literature. You simply enter the names of your constructs and variables, at which point the computer provides abstracts and the references for studies filed under those terms. The computer will call up many unrelated references for a general term, so carefully cross-reference your terms to ensure a more selective search. For example, merely entering "memory" as a search term will produce hundreds of unrelated studies. But entering "memory and humor" will call up references more directly related to our study.

Bibliographies of research articles When you find an article on a topic of interest, notice that its accompanying bibliography contains references to related studies. By reading the published articles that it cites, and then reading the references in those articles, you can work backward in time and learn about research that came *before* the original study. Sometimes you will also see references to psychological conventions and meetings at which researchers orally present their research. For a copy of a particular presentation, contact the first author cited in the reference. (For assistance, the American Psychological Association provides a directory of its members' addresses.)

Social Science Citation Index When you find a relevant article that is several years old, you can search for more recent articles using the *Social Science Citation Index*. This publication identifies a research article by authors and date, and then lists articles published in a given year that have cited it. Thus, for example, we could look up the original 1975 droodle study in the index for 1995, 1996, and so on, to perhaps find more recent related articles.

Review articles The purpose of a review article is to survey and summarize a large body of theoretical and empirical literature dealing with a particular topic. Such articles provide a useful overview and they provide references to many specific studies. The title of a review article usually contains the word review and some books and journals, such as *Psychological Bulletin* and *Annual Review of Psychology*, specialize in review articles.

References on testing materials Often your design will employ a paper-and-pencil test to measure intelligence, personality, creativity, attitudes, emotions, motivations, and so on. Instead of creating your own test and being uncertain of its reliability or its content and construct validity, look for acceptable tests that already exist. To find them, consult reference books that describe common psychological tests. Such books usually have titles conveying the fact that they describe tests (such as *The Mental Measurements Yearbook* or *Measures of Personality and Social Psychological Attitudes*).

You can also use a computerized literature search to find research articles that have employed such tests. To be efficient, cross-reference the name of the attribute to be measured with the term "assessment." For example, if in a different study you wish to measure depression, using the terms "depression" and "assessment" will limit your search to studies that involve the measurement of depression.

Completing the Study

The literature search would provide no studies that would cause us to question the original Bower et al. (1975) procedure of presenting droodles as to-be-remembered stimuli. There are no ethical or practical concerns and because this procedure produced informative and powerful results (they were significant) we can adopt it for our study.

We would find many studies that replicate the finding that, for a variety of types of stimuli, the more meaningful a stimulus is, the better it is retained. For example, when learning a list of words, participants who use each word in a sentence will recall the words better than if they merely think of a rhyme for each word (see Lockhart & Craik, 1990).

Surprisingly, my literature search turned up little research that directly studied how and why humor improves memory (but see MacAninch, Austin & Derks, 1992, and Dixon, Willingham, Strano & Chandler, 1989). Numerous studies, however, show that more *distinctive* stimuli are better retained than less distinctive stimuli (e.g., Schmidt, 1985). For example, in a list of words, a word printed in a different style of print is retained better than the other words that are visually similar (Hunt & Elliott, 1980).

The literature will not always address an issue from exactly your perspective, so often you must generalize from previous findings and constructs to fit them to your hypothesis. Thus, in suggesting that humor influences memory, we can propose that a humorous interpretation makes a droodle more distinctive in memory, thereby making it more memorable. Essentially, then, we have proposed that humor is one component of the construct of distinctiveness—one way to make a stimulus distinctive.

Although the meaningfulness of a stimulus may seem to be the same thing as its distinctiveness, researchers do distinguish between the two concepts. Desrochers and Begg (1987), for example, suggest that distinctiveness is the extent to which unique cues are associated with the particular context in which the stimulus was encountered. In short, a distinctive event is notable and thus stands out in memory. Therefore, greater distinctiveness enhances access to the stimulus, allowing us to "find" it in memory. Meaningfulness, on the other hand, is the extent to which the components of the stimulus are organized and integrated. A meaningful event is tied together such that we know all of its "parts." Therefore, once we access a memory of a stimulus, greater meaningfulness enhances recall of the components of the stimulus (see also Einstein, McDaniel & Lackey, 1989).

Although the preceding discussion greatly simplifies the debate about the hypothetical constructs of distinctiveness and meaningfulness, for our study it boils down to this: On the one hand, the importance of the interpretations in the original droodle study may be that they were humorous, and thus made the droodles more distinctive and in turn more memorable. On the other hand, the importance of the interpretations may have nothing to do with the humor involved. Perhaps they simply made the droodles more meaningful and thus more memorable. Our task is to design a study that clearly shows the influence of humor, separate from the influence of meaningfulness.

By discussing the constructs of memory, humor, and meaningfulness in terms of droodles and their interpretations, we have begun to create operational definitions of them. Now we "whittle down" these constructs, completing the design by defining the specific variables and procedures.

Defining the variables Our independent variable involves changing the amount of humor attributed to the droodles by the interpretations. The challenge, however, is to manipulate the amount of humor in the interpretations while producing equally meaningful interpretations. If the droodles are not equally meaningful, then humor and meaningfulness will be confounded. Then we will be unable to tell whether more humor or more meaningfulness improves retention of a droodle.

What seems to make an interpretation in Bower et al. humorous is that it provides an unusual explanation involving unexpected objects, people, or animals. So, if we revise the original interpretations to provide common explanations involving predictable objects, people, and animals, they should be less humorous (and less distinctive) but just as meaningful as the originals. For example, from the humorous interpretation "This is a midget playing the trombone in a telephone booth," we can derive the less humorous interpretation "This is a telephone booth with a technician inside, repairing the broken door handle." In both cases the droodle features a telephone booth, so if a telephone booth is particularly meaningful and memorable, it is equally so in both the humorous and nonhumorous conditions. Also both interpretations involve the meaningful integration of a person, a telephone booth, and an object (either a trombone or a door handle).

Thus, we will have two conditions of the independent variable: In one we provide participants with nonhumorous interpretations (as defined above), and in the other we provide humorous interpretations (again, as defined above). If humor is an attribute that aids memory, then the humorous interpretations should produce better recall of the droodles. If humor is not psychologically important in this way, then there should be no difference in retention between the two conditions.

Now, we must consider all of the details involved in devising a reliable and valid study. First, if there is only one droodle per condition, participants might forget or remember it because of some hidden peculiarity in it. The original Bower et al. study presented 28 droodles per condition, so we'll use the same number of droodles. The easiest way to obtain the droodles is to use the ones from Bower et al. (To borrow stimuli not fully presented in an article, you can write to the first author of the article.) If instead we decided to create stimuli, then we must also attempt to control extraneous variables so that all stimuli are comparable. Thus, we would specify rules for creating stimuli so that they all have equal complexity and memorability: All are of equal size, all are drawn in black ink, all contain only two basic geometric shapes, and so on. Although Bower et al. handed each drawing of a droodle to participants, for better control we can present the droodles using a slide projector with an electronic timer or have participants sit at a computer-controlled video monitor.

Also for consistency, all interpretations will contain roughly the same number and type of words, and all will begin with the phrase "This is a . . . ." As in the original study, we will test participants one at a time, reading them the interpretation as they first view a droodle. We'll read all interpretations at the same speed and volume, with the same tone of voice and expressiveness. (To further ensure consistency, we might record the interpretations and time the playback to occur when participants view each droodle.) Likewise, the humorous interpretations should all be consistently humorous for a wide range of participants, the nonhumorous interpretations should be consistently nonhumorous, and, as a group, the humorous interpretations should be consistently more humorous than the nonhumorous interpretations.

The dependent variable is recall of the droodles, but we must also decide how to define and measure it. As in Bower et al. (1975), the participants will study each droodle for 10 seconds, so that all subjects have the same amount of study-time and the same retention period. Immediately after all droodles have been presented, participants will sketch all droodles on sheets of paper containing a grid of approximately 3-by-3-inch squares. They will place each droodle in a square, so that we can tell what shapes a participant believes go together to form one droodle.

Completing the design We—the researchers—could score the sketches as correct or incorrect ourselves, but our judgment may not be reliable. Instead, therefore, we'll enlist two other people as scorers who are "blind" to the purposes of the study. A response is correct if both scorers agree that it matches an original droodle.

We must also create clear and precise instructions for participants so that they know exactly what to do at each step, and so that we can control their extraneous behaviors. The instructions should be, to the extent possible, worded identically for all conditions, consistently read or recorded, of the same duration, and so on. Further, the researcher must attempt to behave identically when testing all participants, and the environment must be constant for them all.

We must also decide on the specific participants to be examined. Variables such as age, gender, and cultural background may influence what people consider humorous, and we want them all to see the droodles clearly and to understand the interpretations that accompany them. To keep such variables constant, we will randomly select as participants, Introductory Psychology students who are similar in age and background, with good eyesight, hearing, and English abilities. To avoid practice effects from showing the same people both conditions of droodles, we'll test a separate, independent sample of participants in each condition, and we'll balance gender. To have a powerful *N*, we'll test 40 people per condition, selecting 20 males and 20 females for each.

Finally, we must plan out the statistical analysis. Each person's score will be the number of correctly recalled droodles. These are ratio scores that can be assumed to meet the requirements of parametric statistics, so the inferential statistical procedure we'll use is the independent samples *t*-test. Because we predict that humor will improve recall scores, we have a one-tailed test.

Although there are other designs we might create, let's assume we conducted the above study and found that the average number of droodles recalled was 15.2 with nonhumorous interpretations and 20.5 with humorous interpretations. The *t*-test indicated a significant difference, so this is a believable relationship. Now we interpret these results psychologically, first inferring that humor does influence recall, and then working back to the broader hypothetical constructs of how humor and distinctiveness operate on memory.

To share the results with other researchers, we'll prepare a written report of the study following APA format.

ORGANIZATION OF A RESEARCH ARTICLE

Most psychological research articles follow the rules set down in the *Publication Manual of the American Psychological Association* (1994). Now in its fourth edition, this is the reference source for answering *any* question regarding the organization, content, and style of a research report. Although APA format may at first appear to be a very rigid, arbitrary set of rules, it is necessary. This format minimizes publishing costs by defining precisely the space and effort that a publication requires. It also specifies the information that any report should contain, how the information should be organized, and how it should be reported.

Especially for beginning researchers, APA format is a very useful organizational scheme. As a reader, you will learn where to look in an article to find certain information and how to understand the shorthand codes used to present it. As an author, you will see how to organize your paper, what to say, and how to say it. And as a researcher, you'll develop a framework for remembering the many design aspects of a study that must be considered. Asking yourself the question "What will I say in each section of a report of this study?" is a cue for remembering the issues that you must deal with.

The sections of an article describe the various aspects of a study in the order in which they logically occur. Recall that in Chapter 2 you saw that the flow of a study can be depicted using two funnels, as shown in Figure A.2. In conducting a study, you work from the general to the specific and then back to the general. As shown in Figure A.2, an APA-style report is organized following these same steps, with four major sections:

- The *Introduction* presents the hypothetical constructs as they are used in relevant literature from past research, develops the hypothesized relationship between the variables for the target population, and provides the specific predictions of the study.
- The *Method* section describes the specifics of the design and how the data were collected.
- The *Results* section reports the descriptive and inferential statistics performed and describes the statistical relationship found.
- The *Discussion* section interprets the results first in terms of the variables, and then generalizes to the broader relationship between the hypothetical constructs with which we began.

FIGURE A.2 The Parallels Between Research Activity and APA Format

The flow of a research study is from a general hypothesis to the specifics of the study, and then back to the general hypothesis. The APA format also follows this pattern.

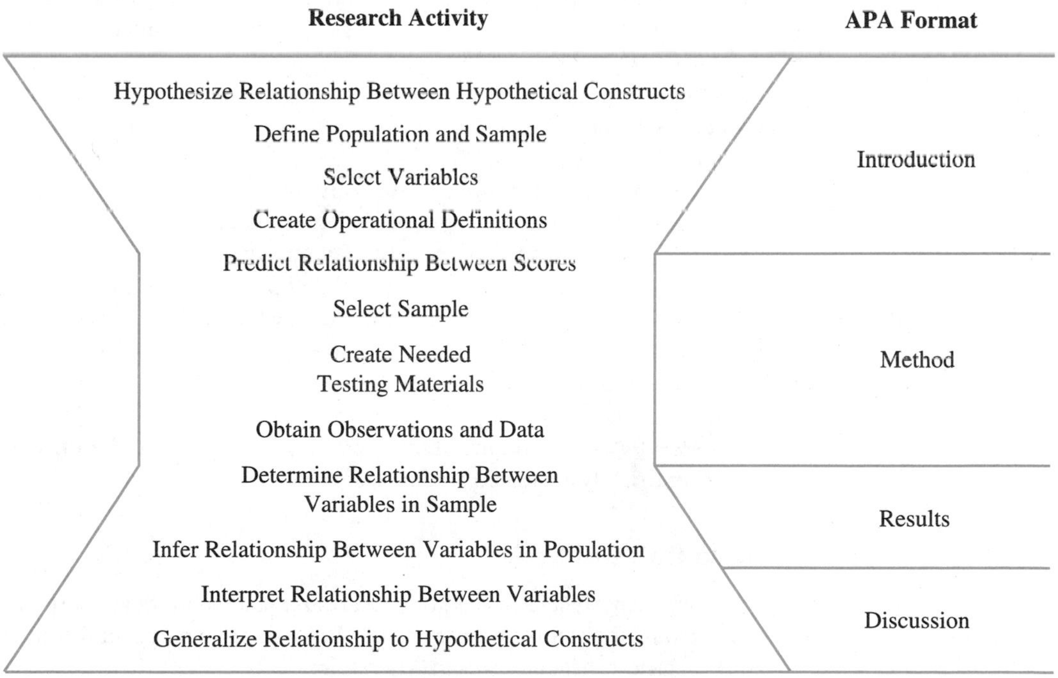

Research Activity	APA Format
Hypothesize Relationship Between Hypothetical Constructs Define Population and Sample Select Variables Create Operational Definitions	Introduction
Predict Relationship Between Scores Select Sample Create Needed Testing Materials Obtain Observations and Data	Method
Determine Relationship Between Variables in Sample Infer Relationship Between Variables in Population	Results
Interpret Relationship Between Variables Generalize Relationship to Hypothetical Constructs	Discussion

REMEMBER The organization of a research report follows the logical order of the steps performed in conducting the research.

The ultimate goal of APA format is precision in communication. At the same time, you need to conserve space and avoid overstatements or redundancy. Thus, strive to state each idea clearly, to say it once, and to report only the necessary information. To meet the goal of precise yet concise communication, both the author and the reader make certain implicit assumptions.

The Assumptions of the Author and Reader

The reader assumes that the author understands statistics and research methods, that the author has described any unusual or unexpected events, and that he or she is a reasonable, ethical, and competent researcher. Many things are left unsaid in a research article, because the reader can assume that commonly accepted procedures were used and that the details of the procedures are unimportant. Thus, for example, you need not state, "I compared the obtained statistic to the critical value" because all researchers know this must be done. Stating it would be redundant.

Likewise, an author assumes that the reader is a competent psychologist. Therefore, a report does not give readers a detailed background of the topic under study, because the assumption is that they already know something about it, or that they will read the references provided. The author also assumes that the reader understands statistics and research methods. *Do not teach statistics and design principles to the reader.* Do not say "Reliable data were important because . . ." or "A *t*-test was performed because . . . " The reader should already know why reliability is important and why a *t*-test is performed. Finally, always use common terminology (such as reliable, valid, confounding), but without providing definitions. The author assumes that the reader either understands them or will find out what they mean.

As the author, you should focus on providing readers with the information they cannot get elsewhere: *Your thoughts and actions as a researcher*. What conclusions did *you* draw from a previous article? What do *you* mean when using certain hypothetical constructs? What logic did *you* use in deriving a hypothesis or prediction? And what do *you* think a result indicates about the behavior under study? As the author, you are the expert, so give the reader the benefit of your wisdom. Your job is to describe clearly and concisely all of the important mental and physical activities you performed in creating, conducting, and interpreting the study. The goal is to provide readers the information necessary to (1) understand the study, (2) evaluate the study, and (3) perform a literal replication of the study.

REMEMBER A good research report allows the study to be fully understood, scientifically evaluated, and literally replicated.

Some Rules of Style

There are many specific rules for preparing a research article, so refer to the *Publication Manual* for complete instructions. Below are some general rules for preparing a research article that conforms to APA style:

1. A report describes a completed study, so it is written in the past tense ("I predicted that . . ."). The exception is to state in the present tense any conclusions that apply to present or future situations ("Humor influences recall by . . .").
2. Cite all sources from which you obtained information, using only the last names of the author(s) and the date. You may use the reference as the subject of a sentence: "Smith and Jones (1992) defined distinctiveness as . . ." Or you may state an idea and provide the reference in parentheses: "Distinctiveness is defined as . . . (Smith & Jones, 1992)." (In a parenthetical citation, "&" is used instead of "and.") When citing an article with three to six authors, include all names the first time you cite it. Thereafter, refer to it using only the first author and the Latin phrase et al. Thus, first we say "Bower, Karlin, and Dueck (1975)," but subsequently we say "Bower et al. (1975)." When citing an article with more than six authors, even the first time you cite it use only the first author and et al.
3. Refrain from directly quoting an article. Instead, paraphrase and summarize the idea, so that *you* tell the readers what they should understand about the idea. Also, address a study itself, not its authors. For example, the phrase "Bower et al." refers primarily to a reported experiment, not to the people who conducted it. Thus, we write "The results are reported *in* Bower et al. (1975)" instead of "The results are reported *by* Bower et al. (1975)."
4. To distinguish your study from other studies, refer to it as "this study" or "the present study." However, do not use these phrases in a way that attributes human actions to nonhuman sources, as in "This study attempted to demonstrate that. . . ." Instead, use "I" as the subject of these verbs. (Use "we" *only* if you have a co-author.)
5. Use accepted psychological terminology as much as possible. When you use a nonstandard term or you name a variable, define the word the first time you use it and then use that word consistently. In the droodle study, we'll define "humorous" and use only this term, rather than mixing in related terms such as "funny" or "entertaining." This prevents confusion about whether we mean something slightly different by "funny" or "entertaining." In addition, avoid using contractions or slang terms. A reader from a different part of the country or another country may not understand such terms.
6. Avoid abbreviations. They are justified only if (a) a term consists of several words, (b) it appears *very* frequently throughout the report, and (c) you are not using many different abbreviations. If you must abbreviate, do so by creating an acronym, using the first letter of each word of the term. Define the complete term the first time it is used, with its acronym in parentheses. Thus, you might say "Short-term Memory (STM) is . . . ." Then use *only* the acronym, *except* as the first word of a sentence: There, always use the complete term.
7. Use words for numbers between zero and nine, and digits for numbers that are 10 and larger. However, use digits for any size number if (a) you are writing a series of numbers in which at least one is 10 or larger, or (b) the number contains a decimal or refers to a statistical result or to a precise measurement (such as a specific score or the number of participants). Thus, you would say "The three conditions, with 5 individuals per condition . . . ." Also, never begin a sentence with a number expressed in digits.

8. In research published before 1994, you'll see the generic term "subjects" to refer to the individuals that researchers study. A post-1994 change in APA style now requires the use of less impersonal and more precise terms. The generic term to use is "participants," but where appropriate use more descriptive terms such as students, children, men, women, rats, or pigeons.
9. Finally, use precise wording. In the droodle study, we won't say that participants "saw" or "looked at" a droodle, or that they "forgot" a droodle, because we don't *know* that these events occurred. We know only that participants were presented a droodle or failed to recall it. In addition, APA rules stress that you avoid gender-biased language. Thus, refer to the gender of participants using the equivalent terms "male" and "female" and to individuals as "he" or "she." When possible, use neutral terms such as "Chairperson."

THE COMPONENTS OF AN APA-STYLE RESEARCH ARTICLE

These are the components of an APA-style manuscript in the order in which they occur:

Title page
Abstract page
Introduction
Methods
  Participants
  Materials or Apparatus
  Procedure
Results
Discussion
References
Tables and Figures

All parts are typed, double-spaced, and without "justifying" the righthand margin. The following sections examine each component in detail, using examples from a manuscript of the droodle study. (The complete manuscript is presented in the last section of this Appendix.) Throughout this discussion, compare the previous steps we went through when designing the study—and all that was said—to what is actually reported. Translating and summarizing your thoughts and activities are the keys to creating a research report.

The Title

A **title** allows readers to determine whether they want to read the article. It should clearly communicate the variables and relationship being studied, but it should consist of no more than 12 words. Titles often contain the phrase "as a function of." For example, "Helping Behavior as a Function of Self-Esteem" indicates that the researcher examined the relationship between participants' helping behavior and different amounts of their self-esteem. A title such as "Decreased Errors in Depth Perception as a Function of Increased Illumination Levels" provides the added information that the observed relationship is negative, such that greater illumination is associated with fewer errors. Because illumination level can be manipulated easily, this title probably describes an experiment in which "illumination level" was the independent variable and "errors," the dependent variable.

Titles also often begin with the phrase "Effect of," as in "Effect of Alcohol Consumption on Use of Sexist Language." The word effect means "influence." Such a title is a causal statement, implying that an experiment was conducted and that changes in the independent variable (amount of alcohol consumed) caused a change in the dependent variable (amount of sexist language used by participants). Note the difference between effect (usually a noun) and affect (usually a verb). If *X affects Y*, then there is an *effect* of *X* on *Y*. (Here's a trick for remembering this distinction: *Effect* means *end result*, and both begin with *e*. *Affect* means *alter* and both begin with *a*.)

The title you create should provide sufficient information for readers to determine whether the article is relevant to their literature search. Choose terms that are specific, and never use abbreviations or terms that need to be defined. Thus, for our study, we won't include "droodles" because most people won't know what they are. Instead, we might use the title "Effect of Humorous Interpretations on Immediate Recall of Nonsense Figures." This wording identifies the variables, specifying that we are studying short-term memory of drawings. Contrast this with such terrible titles as: "A Study of Humor and Memory" (of course it's a study!), or "When Does Memory Work Better?" (what does "work" mean?). Either of these would be useless for determining whether the article is relevant to a specific research topic.

In an APA-style manuscript, the title page is a separate page containing the title, your name, and the formal title of your college or university. A sample title page appears in Figure A.3 on the next page. The title page is page number 1, with the number placed in the upper right corner of the page, as it is on *all* other pages. (The minimum margin all around is 1 inch).

The title page also contains two other components. First, left of the page number is typed the "manuscript page header," consisting of the first two or three words from your title. The header appears on all subsequent pages, so if any pages become separated, the publisher can identify them as belonging to your manuscript. (This is another reason for all researchers not to use titles beginning "A Study of.") Second, on the first line below the header is typed the words "Running head:" followed by an abbreviated title. This running head will be printed at the top of each page in the published article. (On this page of your textbook, the running head is "The Components of an APA-Style Research Article.")

The Abstract

The title page is followed by the **abstract**, which is a brief summary of the study. The abstract describes the specific variables used, important subject characteristics, a brief description of the overall design, and the key general statistical relationship obtained. It also indicates the theoretical approach taken in interpreting the results, though often without giving the actual interpretation.

Although the abstract accompanies the article, it is also reproduced in *Psychological Abstracts*, so it must be able to stand alone, containing no abbreviations or uncommon terms (no "droodles"). It should include only details that answer the reader's question:

FIGURE A.3 Sample Title Page of a Research Manuscript

Notice the location and spacing of the various components.

Effect of Humorous 1

Running head: EFFECT OF HUMOROUS INTERPRETATIONS ON RECALL

Effect of Humorous Interpretations on
Immediate Recall of Nonsense Figures
Gary W. Heiman
Podunk University

"Is this article relevant to my literature search?" Most authors write the abstract after they have written the report, so they can summarize the key points easily. If you find it difficult to compress a lengthy paper into 100–120 words, think of the abstract as an elaboration of the title. Given the title, what else would you say to communicate the gist of the article?

REMEMBER The *title* describes the relationship under investigation. The *abstract* summarizes the report. Together, they allow readers to determine whether the article is relevant to their literature search.

The abstract for the droodle study appears in Figure A.4.

FIGURE A.4 Sample Abstract Page

The abstract page is page number 2, with a centered heading reading "Abstract." The abstract itself is one paragraph long. Note that the first line is not indented.

Effect of Humorous 2

Abstract

The effect of humor on the immediate recall of simple visual stimuli was investigated. Eighty college students (20 men and 20 women per condition) viewed 28 nonsensical line drawings that were each accompanied by either a humorous or nonhumorous verbal interpretation. Although the interpretations were comparable in the meaningfulness they conveyed, those participants presented with humorous interpretations correctly recalled significantly more drawings than those presented nonhumorous interpretations. The results suggest that a meaningful and humorous context provides additional retrieval cues beyond those cues provided by a meaningful yet nonhumorous context. The effect of the cues produced by humor is interpreted as creating a more distinctive and thus more accessible memory trace.

The Introduction

The **Introduction** should reproduce the logic you used to derive your hypothesis and to design the study to test it. It is the Introduction that shows the "whittling down" process, beginning with broad descriptions of behaviors and hypothetical constructs and translating them into the specific variables of a study. It then describes the predicted relationship between scores that will be measured using the operational definitions.

Researchers read an introduction with two goals in mind. First, a reader wants to understand the logic of the study. Thus, the author should introduce the hypothesis and the psychological explanations being tested, the general design (e.g., whether correlational or experimental), the reasons that certain operational definitions are employed, and why a particular result will support the predictions and hypothesis of the study. Both the purpose of the study and the population under study should be clear. (Unless specified, we assume that the relationship between variables applies to the broadest population.) Readers also evaluate the hypothesis of the study and its logic. Are the explanations circular pseudo-explanations? Are there rival hypotheses and extraneous variables to be considered?

Second, a reader looks for empirical evidence that supports the hypothesis. The Introduction is where virtually all references to past research occur, including those studies that do and do not support the hypothesis. Further, if the study is successful, the author will attempt to interpret and explain the findings "psychologically," so the Introduction also contains the conceptual and theoretical issues to be discussed later in the paper.

The reader assumes that, unless otherwise noted, a study cited in support of a hypothesis is reasonably convincing. Previous studies are reported very briefly, usually with the author merely citing them by name (rather than explaining them in detail) to indicate that they provide support. If discussed at all, previous studies are described in terms of the specific information the author judged to be important when deriving his or her hypotheses. The details of a study are provided only when (1) they are necessary for the reader to understand the author's comments about that study, or (2) they are necessary for showing support for the author's position. Therefore, the Introduction does not usually contain such details as the number of participants, the statistics used, or the specifics of the design employed in previous research.

A portion of the Introduction for our droodle study appears in Figure A.5. Although you know how the study turns out, the Introduction is written as if you do not, describing the process you went through *before* collecting the data.

In the droodle study, the Introduction begins within the larger context of the hypothetical constructs of meaningfulness and its influence on memory. Then we work logically from the broad ideas to the specific example of the droodle study. We immediately focus, however, on the perspective taken to study the hypothetical constructs. We orient the reader, providing the major relevant conclusions from past research and their references. We also identify when we are merely speculating. The purpose of the study is stated *early* in the Introduction—in this case, at the end of the first paragraph. In subsequent paragraphs, we further retrace our logic, defining what we mean by the constructs of meaningfulness and distinctiveness and explaining how humor might influence memory. Then we describe how we define and manipulate humor while keeping meaningfulness constant.

FIGURE A.5 Sample Portion of the Introduction

Note that the title is repeated, and that we do not label this section as the Introduction.

Effect of Humorous 3

Effect of Humorous Interpretations
on Immediate Recall of Nonsense Figures

Researchers have consistently demonstrated that retention of to-be-learned material improves when the material is presented in a context that leads to meaningful processing (Lockhart & Craik, 1990). In particular, Bower, Karlin, and Dueck (1975) presented college students with a series of "droodles," which are each a meaningless line drawing that can be made meaningful by presentation of an accompanying verbal interpretation. Those individuals who were provided the interpretations correctly recalled (sketched) significantly more of the droodles immediately following their presentation than did those individuals given no interpretations. However, each interpretation in Bower et al. (1975) defined a droodle in a humorous fashion, using unexpected and incongruent actors and actions. Thus, differences in the meaningfulness attributed to

Notice that you must make a connection between past research and the present study. Usually, after presenting previous findings, you can point out a question or flaw that has not been addressed. Once the background and important issues are discussed, you might say something like "However, this interpretation does not consider . . . ." or "However, this variable was not studied . . . ." Then, address the problem you have raised.

A good strategy is to logically lead up to a final paragraph that says something like "Therefore, in the present study . . . ." Then state the specific hypothesis and relationship to be studied, describe your general approach for defining and manipulating the variables, and specify your prediction. The details of how the data were collected are provided in the next section.

REMEMBER The Introduction presents all information that will be used to interpret the results: The conceptual and theoretical logic of the study, relevant past research, and the predictions of the study.

The Method

The **Method** section contains the information needed to understand, critique, and literally replicate the data-collection procedures. To collect data we need participants, testing materials and equipment, and a specific testing procedure and design. APA style requires that these categories be presented in three separate subsections, in this order: (1) Participants, (2) Materials or Apparatus, and (3) Procedure. The beginning of our Method section is shown in Figure A.6.

Participants Here, describe the participants you've enlisted so that other researchers can obtain comparable participants and look for uncontrolled subject variables. Thus, identify important subject characteristics (e.g., gender, age, school affiliation) and specify any criteria used when selecting participants. Always report the number of individuals tested. Because their motivation is important, describe any form of reimbursement that was used. Also, in this section or in a letter sent to the journal editor, an author must certify that the participants were treated in accordance with the ethical principles of the APA. (Requiring this statement, and calling this the "Participants" section instead of the "Subjects" section are other changes instituted by the APA in 1994.)

Materials or apparatus This section immediately follows the Participants section (see Figure A.6 below). Usually it is called Materials because most studies involve mainly testing materials such as stimulus objects, tests and printed material, slides,

FIGURE A.6 Sample Portion of the Method Section

Notice the placement of the headings, as well as the use of capital letters and underlining.

humorous interpretations should be more frequently recalled than those accompanied by nonhumorous interpretations.

Method

Participants

Forty female and 40 male undergraduate students from an introductory psychology course at Podunk University each received $3.00 for their voluntary participation. All were between 20 and 22 years of age (mean age = 20.7 years), were born in the United States, were raised in English speaking families, and had normal or corrected eyesight and hearing. Participants were randomly assigned to either the humorous or nonhumorous condition, with 20 males and 20 females in each condition.

Materials

The 28 droodles from Bower et al. (1975) were reproduced, each consisting of a black-ink line drawing involving two drawings, and so on. You may call this section Apparatus if testing mainly involves equipment such as computers, recording devices, and the like. (If extensive discussion is required, you may divide this section into two sections.) Regardless of its title, describe both the relevant materials and apparatus you prepared, but without explaining how they are used. Again organize the information according to the logical order in which the components occur: We'll present a droodle, give an interpretation, and then measure retention.

Supplies must be described clearly so that the reader can understand, evaluate, and reproduce them. Therefore, if supplies are purchased, indicate the manufacturer and model, or the edition or version. If materials are borrowed from previous research, briefly describe them and provide the citation. If you build equipment, describe it so a researcher can produce equipment that is similar in operation. If you create visual stimuli, describe the rules used to create them in terms of their dimensions, their color, and so on. If you create verbal stimuli, describe the rules used to select them, such as the length of words or sentences, their meaning and content, their difficulty level, and so on. For any paper-and-pencil tests, describe the number of questions, the format of each question, and the way in which participants indicate their responses. Also report information about the reliability and validity of a procedure. For example, the speed and error rates of equipment should be indicated because such rates affect reliability. With paper-and-pencil tests, either note their previously demonstrated validity and reliability or briefly report any procedures you performed to determine this.

Note that all physical dimensions are reported using the metric system, and that common units of measurement are abbreviated. Table A.1 provides the most common abbreviations used in psychological research. If you measure in nonmetric units, report the measurement both in nonmetric and in converted metric units.

Keep in mind that only *important* elements are reported. Readers know the necessary steps in designing a study, and generally understand why and how each component is chosen. (They've *had* this course!) Therefore, do not specify such things as how participants were randomly selected or how you determined their age. Likewise, do not describe obvious equipment (e.g., whether participants used a pencil or a pen to complete a questionnaire, or what furniture was present in the room where testing took place). Note a detail only if it (1) would not be expected by a reasonable researcher, or (2) would seriously influence the reliability or validity of the measurements.

TABLE A.1 Common Abbreviations Used in the APA Format

Notice that these abbreviations do not take periods.

Unit	*Symbol*	*Unit*	*Symbol*
centimeters	cm	meters	m
grams	g	milliliters	mL
hours	hr	millimeters	mm
kilograms	kg	minutes	min
liters	L	seconds	s

Procedure The **Procedure** section describes how you brought the participants, materials, and apparatus together to actually perform the study. A portion of the Procedure section for the droodle study is presented in Figure A.7

The best way to organize this section is to follow the temporal sequence that occurred in the study. The first thing we do is give participants their instructions, so first summarize these instructions. Then describe the tasks performed by the participants in the order they were performed. A useful strategy is to initially describe those aspects of the procedure that are common to all participants and then to distinguish one condition from another. (Always work from the general to the specific.)

Along the way, the various parts of the Method section should communicate the complete design of the study. If the design is not clear by this point, an optional *Design* section may be added. Here you describe the layout of the study in terms of the conditions, participants, and variables used (so that essentially a reader can diagram the study as we have diagrammed them in previous chapters.) But note that this section should be necessary only if you are describing a very complicated study, involving numerous groups or variables, or elaborate steps in testing.

In some instances, still other sections can be created—but, again, only if they're truly necessary. For example, if we took extensive steps to determine the reliability of a procedure and this element was central to the study, we might describe these steps in a *Reliability of Measures* section.

FIGURE A.7 Sample Portion of the Procedure Section

with a loop attached to the lower right side. The humorous interpretation was "This shows a midget playing a trombone in a telephone booth." The nonhumorous interpretation was "This shows a telephone booth with a technician inside fixing the broken door handle."

Response forms for recalling the droodles consisted of a grid of 3 by 3 in. (7.62 cm by 7.62 cm) squares printed on standard sheets of paper.

Procedure

Participants were tested individually and viewed all 28 droodles accompanied by either the humorous or nonhumorous interpretations. Participants were instructed to study each droodle during its presentation for later recall and were told that the accompanying interpretation would be helpful in remembering it. A timer in the slide projector presented each

REMEMBER The Participants section describes the important characteristics of subjects, the Materials or Apparatus section describes the characteristics of the testing materials and equipment, and the Procedure section describes the testing situation and design.

The Results

The **Results** section reports the statistical procedures you performed and the statistical outcomes you obtained. However, *don't* interpret the results here, only report them. A portion of the Results section for the droodle study appears in Figure A.8. Describe the results in the same order in which you perform the steps of the analysis. First, you must have some scores to analyze, so first describe how you operationally defined and tabulated each participant's score. (We described how each droodle was scored as correct or incorrect.) Also describe any transformations performed on the raw scores, such as if you converted each participant's number correct to a percentage. It is at this point that any information regarding the reliability of the data is noted (such as the inter-rater reliability of the scorers).

The analysis then involves first computing the descriptive statistics that summarize the scores and relationship. We usually report the mean and standard deviation for each condition. Note that the symbol for a mean is "M" and for a standard deviation is "SD." Write "The mean number of droodles," however, not "the M number of droodles."

FIGURE A.8 Sample Portion of the Results Section

Notice that the heading is centered.

instructed to recall the droodles in any order, sketching each droodle within one grid on the response sheet.

Results

Two assistants who were unaware of the purposes of the study scored the participants' sketches. A sketch was considered to indicate correct recall if both scorers agreed that it depicted a droodle. (On only 2% of the responses did the scorers disagree.) Each participant's score was then the total number of correctly recalled droodles.

The mean number of droodles correctly recalled was 20.50 in the humorous interpretation condition (<u>SD</u> = 3.25) and 15.20 in the nonhumorous interpretation condition (<u>SD</u> = 4.19). With an alpha level of .05, a one-tailed independent samples <u>t</u>-test indicated a significant difference between the conditions, <u>t</u>(78) = 6.32, <u>p</u> <.05. The relationship between amount of humor

The next step is to perform the appropriate inferential procedure, so next provide the formal name for the procedure and describe how it was applied to the data. Then report the results of the analysis. You must indicate the alpha level you employed, and the terms *significant* or *not significant* must appear for each result. For any statistic, as you did in previous chapters, report the symbol for the obtained value (e.g., t or F), the degrees of freedom in parentheses, the obtained value, and the probability of a Type I error. (Slightly different formatting rules are used when reporting a statistic whose symbol is from the Greek alphabet. See the *Publication Manual.*) Statistical symbols are to be italicized when published so they are underlined in your typed manuscript.

If the results of the inferential procedure are significant, you have convincing evidence of a relationship. At this point, you conduct any secondary analyses that describe the relationship. As in Chapter 11, for example, with a correlation, you also compute the regression line. Or, as in Chapter 17, with a significant F, you perform *post hoc* comparisons. Also, as in Chapter 15, you should indicate the *effect size* of each independent variable, reporting the proportion of variance accounted for by the relationship. (For the droodle study we could include that the squared Point-Biserial correlation coefficient, r^2_{pb}, is .34, indicating that the humor factor accounts for 34% of the differences in recall scores.) For each of these secondary analyses, again identify the procedure and then report the results.

Remember that for any inferential procedure, one result cannot be "more significant" than another, and there is no such thing as a "very significant" or "highly significant" result. Also, do not attribute human actions to statistical procedures, saying such things as "according to the ANOVA . . ." or "the t gave significance." Instead, say that "there were significant differences between the means" or that "there was a significant effect of the independent variable."

Figures Because of cost and space considerations, include graphs, tables, or other artwork only when the information is complicated enough that the reader will benefit from a visual presentation. Usually, graphs—called **"Figures"**—provide the clearest way to summarize the pattern in a relationship. However, the author should also have something to say about each figure, telling readers what they should see in it. (For illustrative purposes, a figure is included in the droodle study manuscript, although, technically, the relationship is so simple that a figure is unnecessary.)

Every figure needs to be numbered (even if there is only one). At the point in the narrative where readers should look at the figure, you should direct their attention to it, saying something like "As can be seen in Figure 1 . . . ." Refer to Figure A.9 for the wording as it appears in the manuscript of the droodle study. *Do not*, however, physically place the figure here (the journal publisher would do that). The graph is drawn on a separate page and placed after the references at the *end* of the manuscript. The figure from our sample manuscript appears in Figure A.10. There are many rules for preparing a figure, so check the *Publication Manual.* In general, the height of the Y axis should be about 60 to 75 percent of the length of the X axis. Fully label each axis, using the names of the variables and their amounts. Use black ink, because color is expensive to publish (and a reader might be color blind!). If the figure contains more than one line in the body of the graph, use different symbols for each (e.g., one solid line and one dashed). Then provide a "legend" or "key" to the symbols at the side of the figure. (For an example, see Figure 18.2 in Chapter 18.)

FIGURE A.9 Sample Portion of the Results Section Showing Reference to a Figure

t(78) = 6.32, p <.05. The relationship between amount of humor and recall scores can be seen in Figure 1. Although a positive relationship was obtained, the slope of this curve indicates that the rate of change in recall scores as a function of increased humor was not large.

Discussion

The results of the present study indicate that humorous interpretations lead to greater retention of droodles than do

Every figure has an explanatory title, called the "figure caption," which briefly identifies the variables and relationship depicted. (In the droodle study, the caption is "Mean number of droodles correctly recalled as a function of nonhumorous and humorous interpretations.") Instead of positioning each caption on the corresponding figure (the publisher also does that), place all captions on one **Figure Caption page**, another separate page at the end of the manuscript.

Tables Create a table of results rather than a figure when it is important for the reader to see the precise numerical values of means, percentages, and so on. A **table** is called for only when there are too many numbers to efficiently include in the narrative. All

FIGURE A.10 Sample Figure

The caption for the graph reads: "Mean number of droodles correctly recalled as a function of nonhumorous and humorous interpretations."

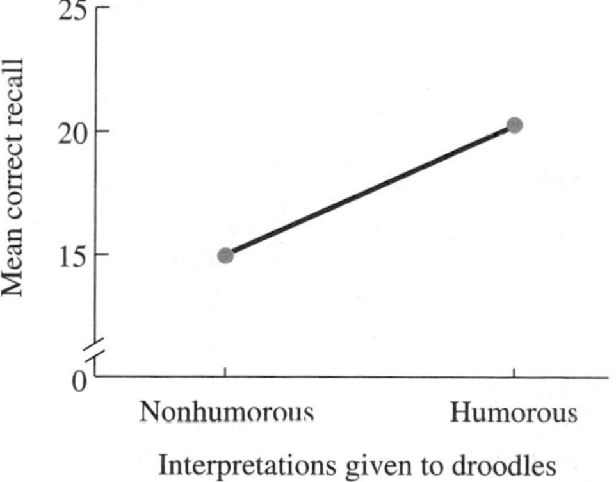

tables are numbered consecutively, and, as with figures, the reader is directed to them at appropriate points in the Results section. The actual tables are placed at the end of the manuscript. Note that you *do* place the title for each table on the table itself. (The *Publication Manual* provides detailed instructions for laying out a table.)

> *REMEMBER* The Results section summarizes the data and the relationship obtained, and reports the outcomes of the inferential procedures performed.

The Discussion

In the **Discussion** section you interpret the results and draw your conclusions. Here the questions originally posed in the Introduction are answered. The Discussion section begins at the point where you have *already* reported a significant relationship (do not report any statistics here), so your first question is, "Do the results confirm (1) the predictions, and thus (2) the hypothesis?" You can almost always answer this question by beginning the discussion with the phrase "The results of the present study . . . ." See Figure A.9 again.

As the term Discussion implies, however, do not merely state your conclusions—discuss them. Thus, after answering the original research question, your task is to use the answer to explain what you have learned about behavior. Recall from the funnel diagram back in Figure A.2 that, in the Discussion section, you work backward, from the specific to the general. Beginning with the narrowly defined relationship in the study, generalize to the relationship between the variables that might be found with other individuals or situations. Then generalize the findings based on the variables to the constructs you originally set out to study.

Begin by focusing on the descriptive statistics you've reported (inferential statistics are no help here). Your job is to become a psychologist again, translating the numbers and statistics into descriptions of behaviors and explanations of the variables that influence them. For example, the droodle study produced higher recall scores with humorous interpretations. This indicates better retention, a mental *behavior* that is different from what occurred when nonhumorous interpretations were given. Thus, based on the scores in each condition, as in Figure A.11, the discussion proposes how manipulating humor influences the memory system. Of course, we must not discount those droodles given nonhumorous interpretations, because they were frequently recalled, and the addition of humor had no great effect. Thus, as here, researchers factor into their conclusions what the direction and rate of change in scores indicate, what the consistency or inconsistency in the relationship indicates, and what the proportion of variance accounted for indicates. Recall that, in studies with more than two conditions, only some conditions may represent an actual relationship. In such cases also consider why and how this result occurs.

All explanations must rationally fit with previous findings and theoretical explanations. Thus, the goal is to provide an integrated and consistent explanation, answering the question, "Given our findings and past findings, what is the present state of knowledge about the behavior or construct?" In the droodle study, for example, we relate our findings to current explanations of the role of distinctiveness and meaningfulness in memory. Notice, however, the use of such words as *presumably*, *probably*, and *apparently*. Do not say "prove," or provide explanations as if they are fact. And although you may say *causes* or *influences*, the difficulty in identifying causal variables should lead you to use these words cautiously.

FIGURE A.11 Sample Portion of the Discussion Section

Effect of Humorous 8

nonhumorous interpretations. Because the meaningfulness of the droodles provided by the interpretations was presumably constant in both conditions, it appears that humor provides an additional source of retrieval cues. This conclusion is consistent with the proposal that humor increases the distinctiveness of a stimulus, thereby facilitating recall by increasing the accessibility of the stimulus in memory.

The improvement in recall produced by humor, however, was relatively small. This result may be due to the fact that all droodles were made meaningful, although sometimes by a nonhumorous interpretation. As in other research (Lockhart &

You must also consider any major flaws in the design that limit confidence in your conclusions. When looking for flaws, however, do not use the tired old argument that the data may not be representative or that a larger sample is needed. Your significant inferential statistics have eliminated these arguments. Instead, any questions you raise should be based on these important design issues:

1. *Reliability*: Did changing the interpretations given to the droodles consistently manipulate the variable of humor? Did we consistently and only measure a person's recall of the droodles?
2. *Internal validity*: Were the humorous and nonhumorous interpretations truly equal in meaningfulness? Perhaps the unusual elements in a humorous interpretation yield a broader, more wide-ranging meaning. If so, then by manipulating humor we also manipulated meaningfulness, and greater meaningfulness may have increased recall scores. If it is likely that this (or another) confounding occurred, then we have reduced internal validity for saying that amount of humor influenced recall.
3. *External validity*: Is it appropriate to generalize this relationship to other people and situations? Did participants have an unusual sense of humor, so that the results are unique? Droodles are simple visual stimuli, so can we generalize the effect of humor to verbal material or to complex visual material?
4. *Content and Construct validity*: Do the recall scores actually reflect "memory" for the droodles? Have we correctly defined "meaningfulness" and "humor"? In particular, we only speculated that humor produces "distinctiveness" but we have no empirical evidence of this.

Although you should attempt to prevent potential flaws when designing a study, they are sometimes unavoidable. Do not try to hide a major flaw. Instead, evaluate the study and either provide counterarguments to explain why a potential flaw does not seriously reduce your confidence in a conclusion, or qualify and limit your conclusions in light of the troubling flaw.

Researchers often conclude the Discussion section by pointing out the next steps to be taken in the research area. Remember, as the author you are the expert, so indicate what hypotheses should be tested next. (For example, we would note the lack of evidence that humor makes a stimulus more distinctive, and suggest that researchers attempt to confirm this hypothesis.)

> *REMEMBER* The Discussion section answers the questions posed by the study, interpreting the results in terms of what is now known about the underlying behavior.

The Reference Page

The final section following the Discussion is the **Reference** page(s). This lists alphabetically the complete references for all sources cited in the article. *Each source should be one that you have read.* If, for example, you learn about Jones' article from reading Smith's report, you should read Jones too (because Smith may be misleading). If you do not, then your reference to Jones should indicate that it is "as cited in" Smith.

As shown in Figure A.12, each reference is typed as a paragraph. (The published version will be converted to the normal hanging-indent format, as in the References in this book.) For a journal article, provide the last name and first initials of all authors, listed in the same order as they appear in the article. Next, give the year of publication, the article's title, and the title of the journal, its volume number, and the page numbers of the article. The *Publication Manual* provides slightly different rules for referencing books, book chapters, monthly magazines, and so on.

PUTTING IT ALL TOGETHER

Most research ideas come from the literature. Authors may suggest rival explanations for their results, or point out untested hypotheses. Also, no study is perfect, so you may find flaws that suggest a research hypothesis. Or you may discover two published studies that contradict each other, so you can design a study to resolve the debate. You may also literally replicate a study, but it is more interesting and informative to add a new twist or perspective to the design that simultaneously expands our knowledge. For example, in replicating the above droodle study, you might add a third, no-interpretation, control condition.

Developing your research ideas will be easier if you recognize that any published article makes the study *sound like* it was a smooth-running, perfectly planned, well-organized process. In reality, it wasn't. An article will not report all the people who volunteered for the study but never showed up. It will gloss over the difficulty the researcher had in finding an artist who could draw droodles or the hours it took to invent their interpretations. And it will not mention the number of prior attempts, using different stimuli or procedures, that failed to produce interpretable results. Thus,

FIGURE A.12 Sample Portion of Reference Page

Notice the punctuation and underlining.

Effect of Humorous 10

References

Bower, G. H., Karlin, M. B., & Dueck, A. (1975). Comprehension and memory for pictures. Memory and Cognition, 3, 216-220.

Desrochers, A., & Begg, I. (1987). A theoretical account of encoding and retrieval processes in the use of imagery-based mnemonic techniques: The special case of the keyword method. In M. A. McDaniel & M. Pressley (Eds.), Imagery and related mnemonic processes: Theories, individual differences, and applications (pp. 56-77). New York: Springer-Verlag.

Einstein, G. O., McDaniel, M. A., & Lackey, S. (1989). Bizarre imagery, interference, and distinctiveness. Journal of Experimental Psychology: Learning, Memory, and Cognition, 15, 137-146.

Hunt, R. R., & Elliott, J. M. (1980). The role of

although an article may give you the impression that the researcher was omniscient and that the study ran like clockwork, don't be fooled. Research is much more challenging—and more fun—than that.

CHAPTER SUMMARY

1. A research report is organized based on the sequence in which the various aspects of the study occur.
2. Most psychological articles follow the rules set down in the *Publication Manual of the American Psychological Association*, so that they conform to "APA format."
3. The *Psychological Abstracts* contains the abstracts of articles published in other psychological journals.
4. The *Social Science Citation Index* is used to identify recent articles that cite a previous article.

5. A research report should not present information that is redundant with a reader's knowledge or other reference sources. Rather, it should provide the necessary information for (a) understanding the study, (b) evaluating it, and (c) literally replicating it.

6. A report's *Title* should clearly communicate the variables and relationship being studied. The *Abstract* should summarize the report. Together, these elements allow readers to determine whether they want to read the article.

7. The *Introduction* presents all information that will be used to interpret the results. It reconstructs the logic and cites the literature the researcher used in working from the hypothetical constructs to the specific predicted relationship of the study.

8. The *Method* section provides the information needed to understand, critique, and literally replicate the data-collection procedures. It consists of three subsections: *Participants*, *Materials* or *Apparatus*, and *Procedure*.

9. The *Participants* section defines the subjects of a study in terms of their characteristics, their total number, and the reimbursement given.

10. The *Materials* or *Apparatus* section describes the characteristics of the stimuli, the response materials, and the equipment used to test participants. Information about the reliability and validity of the material and apparatus is also included here.

11. The *Procedure* section describes the situation(s) in which participants were tested. It should summarize the instructions given to participants and the task(s) performed by them, and complete the description of the design.

12. The *Results* section describes the statistical procedures performed and the outcomes obtained. In this order, it (a) describes the scores, (b) reports the descriptive statistics, (c) identifies each inferential procedure performed, and (d) reports the results of the procedure. Report primary analyses first, followed by secondary analyses.

13. In the *Discussion* section, the results are interpreted and the conclusions are drawn. The initial questions posed by the study are answered and the findings are related to past findings, providing an integrated description of a behavior or construct. It is here that major potential flaws in the study are discussed.

14. In an APA-style manuscript, after the Discussion section comes the *References*, the *Tables*, the *Figure Caption Page*, and the *Figures*.

PRACTICE PROBLEMS

(Answers for odd-numbered problems are provided in Appendix D.)

1. Why is it necessary to conduct a literature search?
2. How would you go about finding literature that investigates the possible connection between violence on television and heightened aggressiveness in adolescents?
3. You find an article related to the topic of television and aggressiveness in adolescents. How can you use this article to find relevant research that occurred prior to it?

4. The article you find was published in 1988. How would you use this article to find more recent related research?
5. You wish to measure the aggressiveness of a sample of adolescents. Other than creating your own test, what two approaches can you take to obtain a valid and reliable test of aggressiveness?
6. What information should you include in the title of a research article?
7. What information should you include in the abstract of a research article?
8. From another researcher's perspective, what is the purpose of reading your title and abstract?
9. (a) How do you create the manuscript page header that appears at the top of each page of a manuscript? (b) What is the running head, and how is it used?
10. (a) Summarize the information that is presented in the Introduction. (b) In general terms, what information about a study is not included here?
11. (a) What are the three standard components of the Method section? (b) What information is contained in each subsection? (c) In general terms, what information about a study is not reported in the Methods section?
12. (a) How is the Results section organized? (b) In general terms, what information about a study is not reported here?
13. (a) When should you include a figure in a research report? (b) When should you include a table in a research report?
14. (a) How is the Discussion section related to the Introduction? (b) In addition to drawing a conclusion about your specific prediction, what other issues are addressed in the Discussion section?
15. For each of the following statements, indicate two reasons that its format is incorrect: (a) "40 students will hear the music and be tested." (b) "Because the critical value is 2.45 and the obtained value is 24.7, the results are very significant." (c) "The mean scores in the respective conditions for men were 1.4, 3.0, 2.7, 6.9, 11.8, 14.77, 22.31, 25.6, 33.7, and 41.2. For girls, the mean scores were . . . ." (d) "To create the groups, the participants were split in half, with five individuals in each." (e) "The results were significant, indicating the null hypothesis should be rejected. Therefore, I conclude that the relationship demonstrates . . . ." (f) Title: "Type of Interpretation as a Function of Remembering Funny Droodles."
16. In question 15 above, revise each statement so that it employs correct APA format.
17. (a) What practical application might the droodle study have in the preparation of lectures or textbooks? (b) Why might your confidence in such a generalization be limited? (c) How would you use the literature to increase your confidence?

Sample APA-Style Research Report

Running head: EFFECT OF HUMOROUS INTERPRETATIONS ON RECALL

Effect of Humorous Interpretations on

Immediate Recall of Nonsense Figures

Gary W. Heiman

Podunk University

Abstract

The effect of humor on the immediate recall of simple visual stimuli was investigated. Eighty college students (20 men and 20 women per condition) viewed 28 nonsensical line drawings that were each accompanied by either a humorous or nonhumorous verbal interpretation. Although the interpretations were comparable in the meaningfulness they conveyed, those participants presented with humorous interpretations correctly recalled significantly more drawings than those presented nonhumorous interpretations. The results suggest that a meaningful and humorous context provides additional retrieval cues beyond those cues provided by a meaningful yet nonhumorous context. The effect of the cues produced by humor is interpreted as creating a more distinctive and thus more accessible memory trace.

Effect of Humorous Interpretations
on Immediate Recall of Nonsense Figures

Researchers have consistently demonstrated that retention of to-be-learned material improves when the material is presented in a context that leads to meaningful processing (Lockhart & Craik, 1990). In particular, Bower, Karlin, and Dueck (1975) presented college students with a series of "droodles," which are each a meaningless line drawing that can be made meaningful by presentation of an accompanying verbal interpretation. Those individuals who were provided the interpretations correctly recalled (sketched) significantly more of the droodles immediately following their presentation than did those individuals given no interpretations. However, each interpretation in Bower et al. (1975) defined a droodle in a humorous fashion, using unexpected and incongruent actors and actions. Thus, differences in the meaningfulness attributed to the droodles may have been confounded by differences in the humor associated with the droodles. The purpose of the present study was to investigate the effect of humorous interpretations when the meaningfulness of the droodles is kept constant.

Few studies can be found that directly examine how the humor associated with a stimulus influences recall of the stimulus. However, it is reasonable to speculate that the relevant dimension of humor may be that it is simply one type of context that makes a stimulus meaningful. Desrochers and Begg (1987) defined the meaningfulness of a stimulus as the extent to which the components of the stimulus are organized and integrated. Therefore, meaningfulness provides retrieval cues that enhance recall of the components of the stimulus, once the stimulus has been accessed in memory. From this perspective, either a humorous or a nonhumorous context should produce equivalent recall of stimuli, as long as both contexts provide an equivalent level of meaningful organization.

On the other hand, humor may play a different role than that of only providing a meaningful context. Because it provides an unusual and unexpected interpretation, humor may make a stimulus more distinctive in memory. The distinctiveness of a stimulus is defined as the number of novel attributes that it can be assigned (Schmidt, 1985). Research has shown that greater distinctiveness does improve retrieval (Hunt & Elliott, 1980). Desrochers and Begg (1987) and Einstein, McDaniel, and Lackey (1989) suggest that distinctiveness is created by unique cues that are associated with the particular context in which the stimulus was encountered. Therefore, distinctiveness enhances access to the overall memory trace for a stimulus. From this perspective, a humorous context should facilitate recall of a stimulus to a greater extent than a nonhumorous context, because, in addition to organizing the components of a stimulus through its meaning, humor provides additional retrieval cues that make the memory trace for the stimulus more distinctive and thus more accessible.

In this study, I tested the above proposals by determining

whether droodles accompanied by humorous interpretations are better retained than when they are accompanied by nonhumorous interpretations. For each of the humorous interpretations of Bower et al. (1975), I produced a non-humorous version that would provide an equally meaningful interpretation of the droodle. If humor adds retrieval cues over and above those produced by meaningful processing, then droodles accompanied by humorous interpretations should be more frequently recalled than those accompanied by nonhumorous interpretations.

Method

Participants

Forty female and 40 male undergraduate students from an introductory psychology course at Podunk University each received $3.00 for their voluntary participation. All were between 20 and 22 years of age (mean age = 20.7 years), were born in the United States, were raised in English speaking families, and had normal or corrected eyesight and hearing. Participants were randomly assigned to either the humorous or nonhumorous condition, with 20 males and 20 females in each condition.

Materials

The 28 droodles from Bower et al. (1975) were reproduced, each consisting of a black-ink line drawing involving two interconnected geometric shapes. Droodles were copied to film slides for presentation by a standard Kodak carousel projector (model 28-b).

For each humorous interpretation in Bower et al. (1975), a non-humorous version was created. Each interpretation consisted of a 10 to 14 word sentence, beginning with the phrase "This shows a. . . ." A humorous interpretation referred to an unusual action by unexpected people or animals using incongruent objects. A nonhumorous interpretation was derived by changing the humorous interpretation so that it described common actions by predictable actors using congruent objects. The meaning of each droodle was altered as little as possible, with only the humorous components being replaced with comparable, nonhumorous components. For example, one droodle consisted of a rectangle with a loop attached to the lower right side. The humorous interpretation was "This shows a midget playing a trombone in a telephone booth." The nonhumorous interpretation was "This shows a telephone booth with a technician inside fixing the broken door handle."

Response forms for recalling the droodles consisted of a grid of 3 by 3 in. (7.62 cm by 7.62 cm) squares printed on standard sheets of paper.

Procedure

Participants were tested individually and viewed all 28 droodles accompanied by either the humorous or nonhumorous interpretations. Participants were instructed to study each droodle during its presentation for later recall and were told that the accompanying interpretation would be helpful in remembering it. A timer in the slide projector presented each slide containing a droodle for 10 s, with approximately 2 s between slides. As each slide was presented, I recited the appropriate interpretation. The recall task began immediately after the final droodle was presented. Participants were instructed to recall the droodles in any order, sketching each droodle within one grid on the response sheet.

Results

Two assistants who were unaware of the purposes of the study scored the participants' sketches. A sketch was considered to indicate correct recall if both scorers agreed that it depicted a droodle. (On only 2% of the responses did the scorers disagree.) Each participant's score was then the total number of correctly recalled droodles.

The mean number of droodles correctly recalled was 20.50 in the humorous interpretation condition ($\underline{SD}$ = 3.25) and 15.20 in the nonhumorous interpretation condition ($\underline{SD}$ = 4.19). With an alpha level of .05, a one-tailed independent samples $\underline{t}$-test indicated a significant difference between the conditions, $\underline{t}(78) = 6.32$, $\underline{p} < .05$. The relationship between amount of humor and recall scores can be seen in Figure 1. Although a positive relationship was obtained, the slope of this curve indicates that the rate of change in recall scores as a function of increased humor was not large.

Discussion

The results of the present study indicate that humorous interpretations lead to greater retention of droodles than do nonhumorous interpretations. Because the meaningfulness of the droodles provided by the interpretations was presumably constant in both conditions, it appears that humor provides an additional source of retrieval cues. This conclusion is consistent with the proposal that humor increases the distinctiveness of a stimulus, thereby facilitating recall by increasing the accessibility of the stimulus in memory.

The improvement in recall produced by humor, however, was relatively small. This result may be due to the fact that all droodles were made meaningful, although sometimes by a nonhumorous interpretation. As in other research (Lockhart & Craik, 1990), the meaningful processing produced by a nonhumorous interpretation may have provided relatively effective retrieval cues. Then the additional retrieval cues produced by the distinctiveness of a humorous interpretation would only moderately improve the retrievability of the droodles. In addition, these results may have occurred because a nonhumorous interpretation given to such a simple visual stimulus produced a reasonably distinctive trace. Additional unique cues provided by a humorous interpretation would then only moderately increase a droodle's distinctiveness, resulting in only a moderate improvement in recall.

It is possible, of course, that humor added to the meaningfulness of a droodle, instead of to its distinctiveness. Desrochers and Begg (1987) suggested that increased meaningfulness results in increased organization of a stimulus in memory. Humor may have added to the meaningfulness of a droodle by providing additional ways to organize it, so that its components were better retrieved. Further research is needed to determine whether humor produces a more distinctive or a more meaningful stimulus, especially when the stimulus is more complex than a simple droodle.

References

Bower, G. H., Karlin, M. B., & Dueck, A. (1975). Comprehension and memory for pictures. Memory and Cognition, 3, 216-220.

Desrochers, A., & Begg, I. (1987). A theoretical account of encoding and retrieval processes in the use of imagery-based mnemonic techniques: The special case of the keyword method. In M. A. McDaniel & M. Pressley (Eds.), Imagery and related mnemonic processes: Theories, individual differences, and applications (pp. 56-77). New York: Springer-Verlag.

Einstein, G. O., McDaniel, M. A., & Lackey, S. (1989). Bizarre imagery, interference, and distinctiveness. Journal of Experimental Psychology: Learning, Memory, and Cognition, 15, 137-146.

Hunt, R. R., & Elliott, J. M. (1980). The role of nonsemantic information in memory: Orthographic distinctiveness effects on retention. Journal of Experimental Psychology: General, 109, 49-74.

Lockhart, R. S., & Craik, F. I. M. (1990). Levels of processing: A retrospective commentary on framework for memory research. Canadian Journal of Psychology, 44, 87-112.

Schmidt, S. R. (1985). Encoding and retrieval processes in the memory for conceptually distinctive events. Journal of Experimental Psychology: Learning, Memory, and Cognition, 11, 565-578.

Figure Caption

Figure 1. Mean number of droodles correctly recalled as a function of nonhumorous and humorous interpretations.

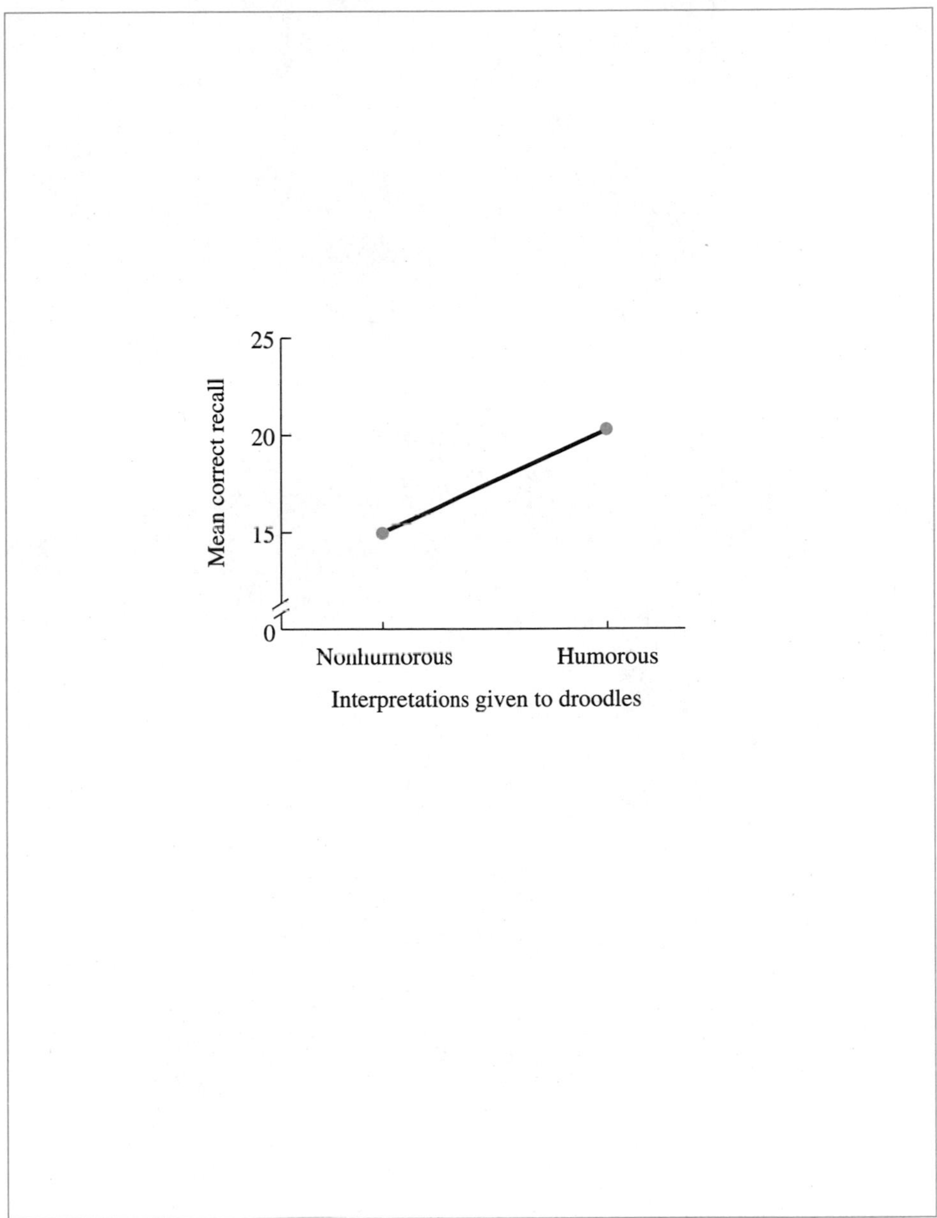
25
20
15
0
Mean correct recall
Nonhumorous
Humorous
Interpretations given to droodles

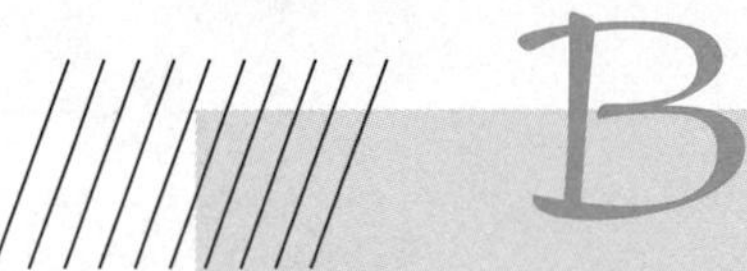

ADDITIONAL STATISTICAL FORMULAS

B.1: COMPUTING PERCENTILES

This part shows how to calculate either the percentile of a particular score, or the score that is at a particular percentile, as discussed in Chapter 6. To understand these formulas, however, you must first understand how to create grouped frequency distributions.

CREATING GROUPED FREQUENCY DISTRIBUTIONS

In a **grouped distribution** different scores are grouped together, and then the total *f*, *rel. f*, or *cf* of each group is reported. For example, say that we have the following scores:

3	4	4	18	4	28	26	41	5	40	4	6	5
18	22	3	17	12	26	4	20	8	15	38	36	

First determine the number of scores the data span. Here, the scores are between a low of 3 and a high of 41, spanning 39 different scores. You can count them on your fingers, or you can calculate the number of values spanned between any two scores using this formula:

$$\text{Number of values} = (\text{High score} - \text{Low score}) + 1$$

Thus, including the 41 and the 3, there is a span of 39 values between 41 and 3.

Next decide how many scores to put in each group, with the same range of scores in each group. To help, you can operate as if the sample contained a wider range of scores than was actually in the data. For example, we'll operate as if the above scores were from 0 to 44, spanning 45 scores. This allows us to create nine groups, each spanning 5 scores, resulting in the grouped distribution shown in Table B.1.

TABLE B.1 Grouped Distribution Showing *f*, *rel. f*, and *cf* for Each Group of Scores

The column on the left identifies the lowest and highest score in each class interval.

Scores	*f*	*rel. f*	*cf*
40–44	2	.08	25
35–39	2	.08	23
30–34	0	.00	21
25–29	3	.12	21
20–24	2	.08	18
15–19	4	.16	16
10–14	1	.04	12
5– 9	4	.16	11
0– 4	7	.28	7
Total:	25	1.00	

The group labeled "0–4" contains the scores 0, 1, 2, 3, and 4, while "5–9" contains scores 5 through 9, and so on. Each group is called a *class interval*, and the number of values spanned by an interval is called the *interval size*. Here, the interval size is 5, so each group spans five scores. (For example, using the above formula, $(4 - 0) + 1 = 5$.) Try to choose an interval size that is easy to work with (such as 2, 5, 10, or 20), rather than something unfriendly (like 17). Also, choose an interval size that results in between 8 and 18 intervals.

Notice several things about the layout of the score column in Table B.1. First, each interval is labeled with the low score on the left. Second, the low score in each interval is a whole-number multiple of the interval size of 5. Third, every class interval has the same size, including the highest and lowest intervals. (Even though the highest score in the data is only 41, we have the complete interval of 40–44.) Finally, the intervals are arranged so that higher scores are located toward the top of the column.

To complete the table, find the *f* for each class interval by summing the individual frequencies of all scores in the group. In our original data there were no scores of 0, 1, or 2, but there were two 3s and five 4s. Thus, the 0–4 interval has a total *f* of 7. For the 5–9 interval, there were two 5s, one 6, no 7s, one 8, and no 9s, so the 5–9 interval has a total *f* of 4. And so on.

Compute the relative frequency for each interval by dividing the *f* for the interval by *N*. Remember, *N* is the total number of raw scores (here 25), not the number of class intervals. Thus, for the 0–4 interval, *rel. f* equals 7/25 or .28.

Compute the cumulative frequency for each interval by counting the number of scores that are at or below the *highest* score in the interval. Begin with the lowest interval. There are 7 scores at 4 or below, so the *cf* for interval 0–4 is 7. Next, *f* is 4 for the scores between 5 and 9, and adding the 7 scores below the interval produces a *cf* of 11 for the interval 5–9. Likewise, the *cf* for each interval is the *f* for that interval plus the *cf* for the interval immediately below it.

Real Versus Apparent Limits

What if one of the scores in the above example were 4.6? This score seems too large to be in the 0–4 interval, but too small to be in the 5–9 interval. To allow for such scores, we consider the "real limits" of each interval. The upper and lower numbers of each interval seen in the score column of the frequency table are called the *apparent upper limit* and the *apparent lower limit*, respectively. As shown in Table B.2, however, the apparent limits for each interval imply corresponding real limits.

Note that (1) each real limit is halfway between the lower apparent limit of one interval and the upper apparent limit of the interval below it, and (2) the lower real limit of one interval is always the same number as the upper real limit of the interval below it. Thus, 4.5 is halfway between 4 and 5, so 4.5 is the lower real limit of the 5–9 interval and the upper real limit of the 0–4 interval. Also, the difference between the lower real limit and the upper real limit always equals the interval size $(9.5 - 4.5 = 5)$.

Real limits eliminate the gaps between intervals, so now a score such as 4.6 falls in the interval 5–9, because it falls between 4.5 and 9.5. If scores equal a real limit (such as two scores of 4.5), put half in the lower interval (between -0.5 and 4.5) and half in the upper interval (4.5–9.5). If one such score is left over, flip a coin to pick the interval.

TABLE B.2 Real and Apparent Limits

The apparent limits in the column on the left imply the real limits in the column on the right.

Apparent limits (lower-upper)	*imply*	*Real limits (lower-upper)*
40–44	→	39.5–44.5
35–39	→	34.5–39.5
30–34	→	29.5–34.5
25–29	→	24.5–29.5
20–24	→	19.5–24.5
15–19	→	14.5–19.5
10–14	→	9.5–14.5
5– 9	→	4.5– 9.5
0– 4	→	−0.5– 4.5

The principle of real limits also applies to ungrouped data. Implicitly, each individual score is actually a class interval with an interval size of 1. Thus, when the score column in an ungrouped distribution is labeled with 6, this is both the upper and the lower apparent limits. However, the lower real limit for this interval is 5.5, and the upper real limit is 6.5.

Graphing Grouped Distributions

Grouped distributions are graphed in the same way as ungrouped distributions, *except* that the X axis is labeled differently. To graph grouped simple frequency or grouped relative frequency, label the X axis using the *midpoint* of each class interval. To find the midpoint, multiply .5 times the interval size, and add the result to the lower real limit. Above, the interval size was 5, which multiplied times .5 is 2.5. For the 0–4 interval, the lower real limit was $-.5$. Adding 2.5 to $-.5$ yields 2. Thus, the score of 2 on the X axis identifies the class interval of 0–4. Similarly, for the 5–9 interval, 2.5 plus 4.5 is 7, so this interval is identified using 7.

As usual, for nominal or ordinal scores create a bar graph and for interval or ratio scores create a histogram or polygon. Figure B.1 presents a histogram and polygon for the grouped simple frequency distribution created back in Table B.1. The height of each data point or bar corresponds to the total simple frequency of all scores in the class interval. Plot a relative frequency distribution in the same way, except that the Y axis is labeled in increments between 0 and 1.

Figure B.2 presents the grouped cumulative frequency polygon for the preceding data. With a grouped cumulative frequency distribution, the X axis is labeled using the *upper real limit* of each interval. Thus, the 0–4 interval is represented at 4.5 on the X axis, and the 5–9 interval is at 9.5. Then each data point is the *cf* for a group.

FIGURE B.1 Grouped Frequency Polygon and Histogram

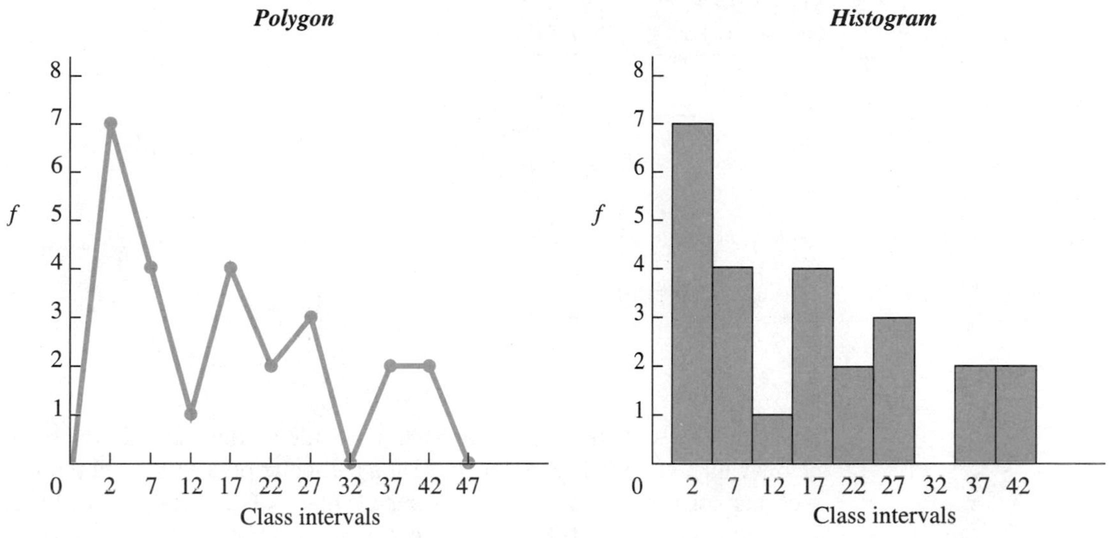

DETERMINING THE SCORE AT A GIVEN PERCENTILE

Percentiles are computed from a cumulative frequency distribution. As an example, say that we have collected scores that reflect the number of minutes required for a rat to find the end of a maze. The cumulative frequency distribution for these data is presented in Table B.3.

FIGURE B.2 Grouped Cumulative Frequency Polygon

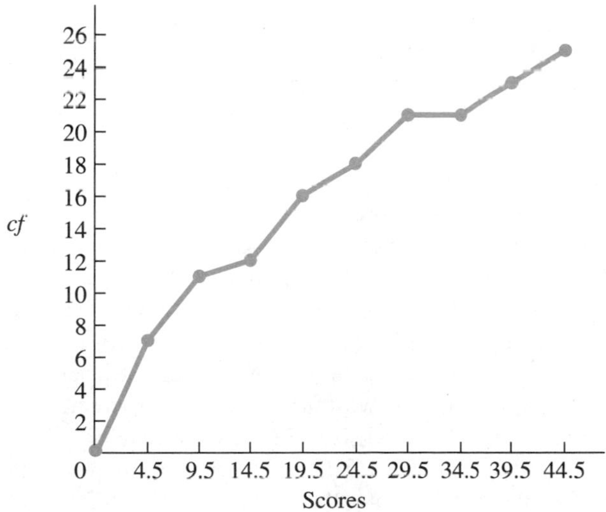

TABLE B.3 Cumulative Frequency Distribution of Maze-Running Times for Laboratory Rats

Score	*f*	*cf*
5	1	10
4	1	9
3	2	8
2	3	6
1	3	3
	$N = 10$	

Say that we seek the score at the 50th percentile, asking "50% of the scores are at or below which score?" To find the score at a particular percentile, find the score that has a *cf* that corresponds to that particular percentage of *N*. Here we are looking for the score with a *cf* that is 50% of *N*. Because *N* is 10, the score at the 50th percentile is the score having a *cf* of 5. The trouble is that no score has a *cf* of exactly 5. The score of 1 has a *cf* of only 3, and the score of 2 has a *cf* of 6. Obviously, the score that has a *cf* of 5 is somewhere between the scores of 1 and 2.

To compute a percentile, treat the scores as if they were from a continuous variable that allows decimals. Then look at the real limits. Looking at the real limits for the scores of 1 and 2, we see

Scores	*f*	*cf*
1.5–2.5	3	6
.5–1.5	3	3

Because the score we seek has a *cf* of 5, the score must be above 1.5 (which has a *cf* of only 3), so the score is in the interval 1.5–2.5. Thus, we'll proceed into this interval far enough beyond 1.5 to accumulate a *cf* of 5. We assume that the frequency in an interval is evenly spread throughout the interval, so that, for example, if we go to a score that is halfway between the upper and lower limits, we accumulate one-half of the frequency in the interval. Conversely, if we accumulate one-half of the frequency in an interval, we're at the score that is halfway between the upper and lower limits. Above, the score we seek has a *cf* of 5, so we want the score above 1.5 that increases the *cf* by 2. If we went to a score of 2.5, we would accumulate an additional *f* of 3, increasing the *cf* by 3, which is too much. We want an *f* of 2 out of the 3, so we want two-thirds of the total frequency in the interval. To accumulate two-thirds of the frequency in the interval, we go to the score that is two-thirds of the way between the lower and upper limits. To find the score that is two-thirds of the way between 1.5 and 2.5, multiply two-thirds, or .667, times the interval size of 1, which gives .667. Then, adding .667 to 1.5 takes us to the score of 2.17. Thus, 2.17 is two-thirds of the way through this interval, so it has a *cf* of 5. Therefore 2.17 is at the 50th percentile, so 50% of the rats completed the maze in 2.17 minutes or less.

Luckily, there's a formula that accomplishes all of the above at once.

THE FORMULA FOR FINDING THE SCORE AT A GIVEN PERCENTILE IS

$$\text{Score} = \text{LRL} + \left(\frac{\text{target } cf - cf \text{ below interval}}{f \text{ within interval}}\right)(\text{interval size})$$

In English, the formula says that you need these components.

1. target *cf*: The cumulative frequency of the score you seek. To find it, transform the percentile into a proportion and then multiply it by *N*. The interval containing this *cf* contains the score you seek. (In the example above, the target *cf* is 5.) You must find the target *cf* before you can find any other component.
2. LRL: The lower real limit of the interval that contains the score you seek. (In the example above, it is 1.5)
3. *cf* below interval: The cumulative frequency for the interval below the interval containing the score you seek. (In the example, it is 3.)
4. *f* within interval: The frequency in the interval that contains the score. (In the example, it is 3.)
5. Interval size: The interval size used to create the frequency distribution. (Above, it is 1.)

Table B.4 below shows where in the original cumulative frequency distribution you would find the components needed to determine the score at the 50th percentile. Putting these numbers into the formula for computing the score at a given percentile:

$$\text{Score} = \text{LRL} + \left(\frac{\text{target } cf - cf \text{ below interval}}{f \text{ within interval}}\right)(\text{interval size})$$

becomes

$$\text{Score} = 1.5 + \left(\frac{5 - 3}{3}\right)(1)$$

TABLE B.4 Cumulative Frequency Distribution Showing Components for Computing the Score at the 50th Percentile

	Score	f	cf	
	4.5–5.5	1	10	target *cf* of
	3.5–4.5	1	9	5 is in this
	2.5–3.5	2	8	interval
LRL →	1.5–2.5	3	6	*cf* below
	.5–1.5	3	3 ←	interval

interval size = 1   *f* within interval

First, deal with the fraction. After subtracting $5 - 3$, we have 2/3, which is .667. So

$$\text{Score} = 1.5 + (.667)(1)$$

After multiplying,

$$\text{Score} = 1.5 + .667$$

So finally,

$$\text{Score} = 2.17$$

Again, the score at the 50th percentile is these data is 2.17.

Although this example has an interval size of 1, you can use the above formula for any grouped distribution having any interval size.

Finding a Percentile for a Given Score

You can also work from the opposite direction when you have a score in mind and wish to determine its percentile. To find the percentile for a given score, find the *cf* of the score within the interval, plus the *cf* below the interval, and then determine the percent of scores that are at or below the score.

You can accomplish this using the following formula.

THE FORMULA FOR FINDING THE PERCENTILE OF A GIVEN SCORE IS

$$\text{Percentile} = \left(\frac{cf \text{ below interval} + \left(\frac{\text{score} - \text{LRL}}{\text{interval size}} \right) \left(\begin{matrix} f \text{ within} \\ \text{interval} \end{matrix} \right)}{N} \right) (100)$$

This formula requires the following components:

1. score: The score for which you are computing the percentile.
2. *cf* below interval: The cumulative frequency of the interval below the interval containing the score.
3. LRL: The lower real limit of the interval containing the score.
4. *f* within interval: The frequency in the interval containing the score.
5. interval size: The interval size used to create the grouped distribution.
6. *N*: The total number of scores in the sample.

Say that we want to find the percentile of the score of 4 in the previous rat data. The components for the formula are shown in Table B.5. Putting these numbers into the formula for finding the percentile of a given score gives:

$$\text{Percentile} = \left(\frac{8 + \left(\frac{4.0 - 3.5}{1} \right)(1)}{10} \right) (100)$$

TABLE B.5 Cumulative Frequency Distribution Showing Components for Computing the Percentile of a Given Score of 4

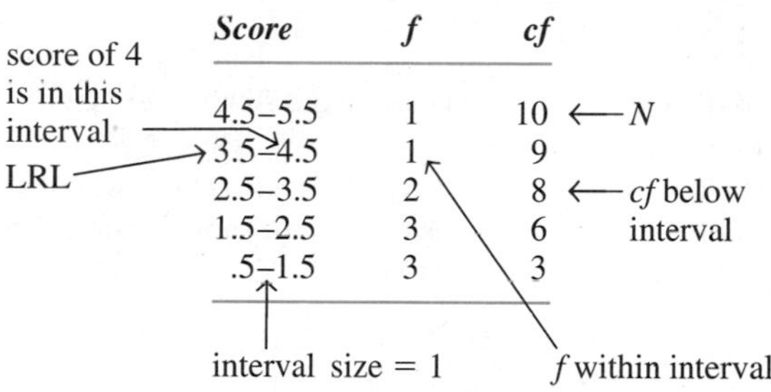

Score	*f*	*cf*
4.5–5.5	1	10
3.5–4.5	1	9
2.5–3.5	2	8
1.5–2.5	3	6
.5–1.5	3	3

Working on the fraction in the numerator first, 4.0 minus 3.5 is .5, which divided by 1 is still .5, so

$$\text{Percentile} = \left(\frac{8 + (.5)(1)}{10}\right)(100)$$

Multiplying .5 by 1 gives .5, so

$$\text{Percentile} = \left(\frac{8 + .5}{10}\right)(100)$$

After adding,

$$\text{Percentile} = \left(\frac{8.5}{10}\right)(100)$$

and after dividing,

$$\text{Percentile} = (.85)(100)$$

Finally, the answer is

$$\text{Percentile} = 85$$

Thus, the score of 4.0 in the above distribution is at the 85th percentile, so 85% of the rats completed the maze in 4 minutes or less.

PRACTICE PROBLEMS

(Answers for odd-numbered problems are provided in Appendix D.)

1. Organize the scores below in an ungrouped distribution showing simple frequency, cumulative frequency, and relative frequency.

49	52	47	52	52	47	49	47	50
51	50	49	50	50	50	53	51	49

(a) What is the percentile for the score of 51? (b) What score is at the 50th percentile?

2. A group of students received the following grades on a test of typing ability. Using an interval size of 5, group the scores and construct a table that shows simple, relative, and cumulative frequency.

76	66	80	82	76	80	84	86	80	86
85	87	74	90	92	87	91	94	94	91
94	93	57	82	76	76	82	90	87	91
66	80	57	66	74	76	80	84	94	66

(a) Find the score that corresponds to the 70th percentile. (b) What is the percentile for a score of 91?

3. Listed below are the weights (in pounds) of 28 high school students. Using an interval size of 4, group the scores and construct a table showing simple, relative and cumulative frequency. The lowest apparent limit is 100.

122	117	116	114	110	109	107
105	103	102	129	126	123	123
122	122	119	118	117	112	108
117	117	126	123	118	113	112

(a) What is the percentile for a student who weighs 117 pounds? (b) What weight corresponds to the 80th percentile?

B.2: COMPUTING THE SEMI-INTERQUARTILE RANGE

This part presents the procedures for computing the semi-interquartile range discussed in Chapter 8.

There are four *quartiles*, each referring to a quarter of a distribution: The first quartile contains the lowest 25% of the distribution, the second quartile contains the next 25% (between the 25th percentile and the median), and so on. The **semi-interquartile range** is one-half of the distance between the scores at the 25th and 75th percentiles.

THE FORMULA FOR COMPUTING THE SEMI-INTERQUARTILE RANGE IS

$$\frac{\text{Score at 75th percentile} - \text{score at 25th percentile}}{2}$$

First determine the scores at the 25th and 75th percentiles using the formula given in Part B.1 for finding the score at a given percentile. Then subtract the score at the 25th percentile from the score at the 75th percentile and divide by 2.

For example, say that in a set of data the score of 12 is at the 25th percentile and the score of 17 is at the 75th percentile. Using the above formula, the semi-interquartile range is (17 − 12)/2, which is 5/2, or 2.5. Essentially, we've determined that the average distance between the median and the scores at the 25th and 75th percentile is 2.5. In other words, the 25% of the distribution immediately below or above the median is, on average, within 2.5 points of the median.

B.3: PERFORMING LINEAR INTERPOLATION

This part presents the procedures for linear interpolation of z-scores as discussed in Chapter 9, and of values of t_{crit} discussed in Chapter 14.

INTERPOLATING FROM THE z-TABLES

You should interpolate to find an exact proportion that is not shown in the z-table or when dealing with a z-score that has three decimal places (carry all computations to four decimal places).

Finding an Unknown z-Score

Say that we seek a z-score that corresponds to a proportion of exactly .45 (.4500) of the curve between the mean and z. To interpolate, enter the z-tables and identify the two bracketing proportions—the closest proportions above and below the target proportion. Note their corresponding z-scores. For .4500, the bracketing proportions are .4505 at $z = 1.6500$ and .4495 at $z = 1.6400$. Arrange the values this way:

	Known proportion under curve	*Unknown z-score*
Upper bracket	.4505	1.6500
Target	.4500	?
Lower bracket	.4495	1.6400

Notice the labels. The "unknown" target z-score corresponds to the "known" target proportion of .4500. The known target proportion is bracketed by .4505 and .4495. Therefore, the unknown target z-score is bracketed by—falls between—1.6500 and 1.6400.

In fact, the target proportion of .4500 is halfway between .4495 and .4505. That is, the difference between the lower known proportion and the target proportion is one-half of the difference between the two known proportions. We assume that the z-score corresponding to .4500 is also halfway between the two bracketing z-scores of 1.6400 and 1.6500. The difference between the two bracketing z-scores is .010, and one-half of that is .005. Thus, to go to halfway between 1.6400 and 1.6500, we add .005 to 1.6400. Thus, a z-score of 1.6450 corresponds to .4500 of the curve between the mean and z.

The answer will not always be as obvious as in this example, so use the following steps.

Step 1. Determine the difference between the upper known bracket and the lower known bracket. In the above example, $.4505 - .4495 = .0010$. This is the total distance between the two known proportions.

Step 2. Determine the difference between the known target and the lower known bracket. Above, .4500 − .4495 = .0005.

Step 3. Form a fraction, with the answer from step 2 as the numerator and the answer from step 1 as the denominator. Above, the fraction is .0005/.0010, which equals .5. Thus, .4500 is one-half of the distance from .4495 to .4505.

Step 4. Find the difference between the two brackets in the unknown column. Above, 1.6600 − 1.6400 = .010. This is the total distance between the two z-scores that bracket the unknown target z-score.

Step 5. Multiply the proportion found in step 3 by the answer found in step 4. Above, (.5)(.010) = .005. This indicates that the unknown target z-score is .005 larger than the lower bracketing z-score.

Step 6. Add the answer in step 5 to the lower bracketing z-score. Above, .005 + 1.640 = 1.645. Thus, .4500 of the normal curve lies between the mean and $z = 1.645$.

Finding an Unknown Proportion

You can also apply the above steps to find an unknown proportion for a known three-decimal z-score. For example, say we seek the proportion between the mean and a z of 1.382. From the z-table, the upper and lower brackets around a z of 1.382 are 1.390 and 1.380. Arrange the z-scores and corresponding proportions as shown below.

	Known z-score	***Unknown proportion under curve***
Upper bracket	1.390	.4177
Target	1.382	?
Lower bracket	1.380	.4162

Here the z-scores are in the "known" column and the proportions are in the "unknown" column. To find the target proportion, use the above steps.

Step 1: 1.390 − 1.380 = .010

This is the total difference between the known bracketing z-scores.

Step 2: 1.382 − 1.380 = .002

This is the distance between the lower known bracketing z-score and the target z-score.

Step 3: $\frac{.002}{.010} = .20$

This is the proportion of the distance that the target z-score lies from the lower bracket. A z of 1.382 is .20 of the distance from 1.380 to 1.390.

Step 4: .4177 − .4162 = .0015

This indicates that .0015 is the total distance between the brackets of .4177 and .4162 in the unknown column.

The known target z-score is .20 of the distance from the lower bracketing z-score to the higher bracketing z-score. Therefore, the proportion we seek is .20 of the distance from the lower bracketing proportion to the upper bracketing proportion.

Step 5: $(0.20)(0.0015) = .0003$

Thus, .20 of the distance separating the bracketing proportions in the unknown column is .0003.

Step 6: $.4162 + .0003 = .4165$

Increasing the lower proportion in the unknown column by .0003 takes us to the point corresponding to .20 of the distance between the bracketing proportions. This point is .4165, which is the proportion that corresponds to the z-score of 1.382.

INTERPOLATING CRITICAL VALUES

Sometimes you must interpolate between the critical values in a table. Apply the same steps described previously, except now use degrees of freedom and critical values.

For example, say that we seek the critical value of t corresponding to 35 *df* (with $\alpha = .05$, two-tailed test). The t tables have values only for 30 *df* and 40 *df*, giving the following:

	Known df	*Unknown critical value*
Upper bracket	30	2.042
Target	35	?
Lower bracket	40	2.021

Because 35 *df* is halfway between 30 *df* and 40 *df*, the critical value for 35 *df* is halfway between 2.042 and 2.021. To find the unknown target critical value, follow the steps described for z-scores.

Step 1: $40 - 30 = 10$

This is the total distance between the known bracketing *df*s.

Step 2: $35 - 30 = 5$

Notice a change here: This is the distance between the *upper* bracketing *df* and the target *df*.

Step 3: $\frac{5}{10} = .50$

This is the proportion of the distance that the target *df* lies from the upper known bracket. The *df* of 35 is .50 of the distance from 30 to 40.

Step 4: $2.042 - 2.021 = .021$

This tells us that .021 is the total distance between the bracketing critical values of 2.042 and 2.021 in the unknown column.

Our *df* of 35 is .50 of the distance between the bracketing *df*s, so the critical value we seek is .50 of the distance between 2.042 and 2.021, or .50 of .021.

Step 5: (.050)(.021) = .0105

Thus, .50 of the distance separating the bracketing critical values is .0105. Because critical values decrease as *df* increases, and we are going from 30 *df* to 35 *df*, we subtract .0105 from the larger value, 2.042.

Step 6: 2.042 − .0105 = 2.0315

Thus, $t = 2.0315$ is the critical value associated with 35 *df* at $\alpha = .05$ for a two-tailed test.

The same logic can be applied to find critical values for any other statistical procedure.

PRACTICE PROBLEMS

(Answers for odd-numbered problems are provided in Appendix D.)

1. What is the *z*-score you must score above to be in the top 25% of scores?
2. Foofy obtains a *z*-score of 1.909. What proportion of scores are between her score and the mean?
3. For $\alpha = .05$, what is the two-tailed t_{crit} for $df = 50$?
4. For $\alpha = .05$, what is the two-tailed t_{crit} for $df = 55$?

B.4: ADDITIONAL FORMULAS FOR COMPUTING PROBABILITY

This part extends the discussion of computing probability found in Chapter 12.

THE MULTIPLICATION RULE

When computing the probability of complex events, sometimes we are "satisfied" only if several events occur. We use the multiplication rule when the word "and" links the events that must all occur for us to be satisfied. *The following multiplication rule can be used only with independent events.* (For dependent events, a different, more complex rule is needed.)

THE MULTIPLICATION RULE FOR INDEPENDENT EVENTS IS

$$p(\text{A and B}) = p(\text{A}) \times p(\text{B})$$

The multiplication rule states that the probability of several independent events is equal to the probabilities of the individual events *multiplied* together. (When you say "and," think "multiply.") Thus, the probability that we will be satisfied by having both A and B occur is equal to the probability of A multiplied times the probability of B. (If there were three events, then all three probabilities would be multiplied together, and so on.)

We may use the multiplication rule when describing a *series* of independent events. Say that we want to know the probability of obtaining 3 heads on 3 coin tosses. We can restate the problem as the probability of obtaining a head *and* then a head *and* then a head. Thus, by the multiplication rule, the probability of 3 heads is

$$p(\text{head}) \times p(\text{head}) \times p(\text{head}) = .5 \times .5 \times .5 = .125$$

The answer we obtain is the same one we would find by forming a fraction with the number of sequences of heads and tails that would satisfy us divided by the number of possible sequences we could obtain when tossing 3 coins. The multiplication rule simply provides a shorter, less complicated route.

We may also use the multiplication rule if we are satisfied when two or more independent events occur *simultaneously*. For example, the probability of drawing the king of hearts can be restated as the probability of drawing a king *and* a heart simultaneously. Because there are 4 kings, the probability of drawing a king is 4/52, or .0769. There are 13 hearts, so the probability of drawing a heart is 13/52, or .25. Thus, the probability of drawing a king and a heart is (.0769 × .25), so $p(\text{king and heart}) = .0192$. (As a check on this answer, there is 1 king of heart in 52 cards, so the probability of drawing the king of hearts is 1/52, which is again .0192. Amazing!)

THE ADDITION RULE

Sometimes we are satisfied by any *one* of a number of outcomes that may occur in one sample. We use the addition rule when the word "or" links the events that will satisfy us. For example, if we will be satisfied by either A *or* B, then we seek $p(\text{A or B})$. There are two versions of the addition rule, however, depending on whether we are describing mutually exclusive or mutually inclusive events. **Mutually exclusive events** are events that cannot occur together: The occurrence of one event prohibits, or excludes, the occurrence of another. Heads and tails, for example, are mutually exclusive on any *one* flip of a coin. Conversely, **mutually inclusive events** are events that can occur together. For example, drawing a king from a deck is mutually inclusive with drawing a heart, because we can draw the king of hearts.

THE ADDITION RULE FOR MUTUALLY EXCLUSIVE EVENTS IS

$$p(\text{A or B}) = p(\text{A}) + p(\text{B})$$

This formula says that the probability of being satisfied by having either A or B occur is equal to the probability of A plus the probability of B. (For mutually exclusive events, when you say "or," think "add.") For example, the probability of randomly drawing a queen or a king is found by adding the probability of a king (which is 4/52, or .0769) and the probability of a queen (which is also .0769). Thus, $p(\text{king or queen}) = .0769 + .0769 = .1538$. In essence, we have found that there are a total of 8 cards out of 52 that can satisfy us with a king or a queen, and 8/52 corresponds to a p of .1538.

When events are mutually *inclusive*, we may obtain A, we may obtain B, or we may obtain A and B simultaneously. To see how this can play havoc with the computations, say that we seek the probability of randomly drawing either a king *or* a heart in one draw. We *might* think that with 4 kings and 13 hearts, there are a total of 17 cards that will satisfy us—right? Wrong! If you count the cards in a deck that will satisfy us, you will find only 16, not 17. The problem is that we counted the king of hearts twice, once as a king and once as a heart. To correct this, we must subtract the "extra" king of hearts.

THE ADDITION RULE FOR MUTUALLY INCLUSIVE EVENTS IS

$$p(\text{A or B}) = p(\text{A}) + p(\text{B}) - [p(\text{A}) \times p(\text{B})]$$

This version of the addition rule states that the probability of obtaining any one of several mutually inclusive events is equal to the sum of the probabilities of the individual events *minus* the probability of those events' occurring simultaneously (minus the probability of A and B). We compute the probability of A and B using the multiplication rule, where $p(\text{A and B}) = p(\text{A}) \times p(\text{B})$. The probability of a king is 4/52, or .0769, and the probability of a heart is 13/52, or .25. Because we have counted the king of hearts both as a king and as a heart, we then subtract the extra king of hearts: the probability of a king and a heart is 1/52, or .0192. Therefore, altogether we have

$$p(\text{king or heart}) = .0769 + .25 - .0192 = .3077$$

This is equivalent to finding the ratio of the number of outcomes that satisfy the event to the total number of possible outcomes. There are 16 out of 52 outcomes that can satisfy king or heart, and 16/52 corresponds to $p = .3077$.

For fun, you can combine the addition and multiplication rules. For example, what is the probability of drawing either the jack of diamonds *or* the king of spades on one draw, *and* then drawing either the 5 *or* the 6 of diamonds on a second draw? In symbols, this is

$$p[(\text{A or B}) \text{ and } (\text{C or D})]$$

Because these events are all mutually exclusive, we have

$$p[(\text{A or B}) \text{ and } (\text{C or D})] = [p(\text{A}) + p(\text{B})] \times [p(\text{C}) + p(\text{D})]$$

If we are sampling with replacement, the answer is .00148. If we are sampling without replacement, the answer is .00151.

THE BINOMIAL EXPANSION

In Chapter 12 we computed the probability of obtaining 7 heads in 7 coin tosses. In doing so, we listed all of the possible sequences of heads and tails that might occur, and then we counted the number of outcomes that would satisfy us. We could use the same technique to determine the probability of obtaining 1 head in 3 coin tosses. If we list all of the possible sequences of head and tails we might obtain with 3 tosses, we have

head	head	head	tail	tail	tail
head	head	tail	tail	tail	head
head	tail	head	head	tail	tail
tail	head	head	tail	head	tail

There are a total of 8 possible combinations, 3 of which contain 1 head. Thus, the probability of obtaining 1 head in 3 tosses is 3/8, or .375.

Instead of listing all of the possible outcomes and then counting those that satisfy us, we can use a mathematical formula, called the *binomial expansion*, for computing such probabilities. A "binomial" situation exists when one of only two possible outcomes occurs on each occasion and the two outcomes are mutually exclusive. Then the binomial expansion can be used to compute the probability of obtaining a certain number of one of the outcomes in some total number of tries. In statistical terms, we find the probability of a certain *combination* of N events taken r at a time. Because either a head or a tail occurs on each toss of the coin, the binomial expansion can be used to determine the probability of obtaining some number of heads in a certain number of tosses. We'll use the symbol p_C to stand for the probability of the particular combination that satisfies us.

THE FORMULA FOR THE BINOMIAL EXPANSION IS

$$p_C = \left(\frac{N!}{r!(N - r)!}\right)(p^r)(q^{N-r})$$

N stands for the total number of tries or occasions, and r stands for the number of events that satisfy us. Thus, to find the probability of obtaining 1 head in 3 tries, $N = 3$ and $r = 1$. The symbol p stands for the probability of the desired event, and we raise it to the r power (multiply it times itself r times). Here heads is the desired event, so $p =$ .5. With $r = 1$, we have $.5^1$. The symbol q stands for the probability of the event that is not desired, and it is raised to the $N - r$ power. Tails is the undesirable event, so $q = .5$. Because $N - r$ equals 2, we have $.5^2$. Thus, filling in the binomial expansion, we have

$$p_C = \left(\frac{3!}{1!(3 - 1)!}\right)(.5^1)(.5^2)$$

The exclamation point (!) is the symbol for *factorial*, meaning that you multiply the number times all whole numbers less than it down to 1. Thus, 3! equals 3 times 2 times 1, for an answer of 6. The quantity 1! is (1)(1), or 1, and the quantity $(3 - 1)!$ is (2)!, which is (2)(1), or 2. Now the formula becomes

$$p_C = \left(\frac{6}{1(2)}\right)(.5^1)(.5^2)$$

Any number raised to the first power is that number, so $.5^1$ equals .5. (In a different problem, if you had to raise p or q to the zero power, the answer would be 1.) Because $.5^2$ is .25, we have

$$p_C = \left(\frac{6}{1(2)}\right)(.5)(.25)$$

Multiplying 1 times 2 yields 2, divided into 6 is 3. Multiplying .5 times .25 gives

$$p_C = 3(.125)$$

and $p_C = .375$. Thus, the probability of obtaining 1 head in 3 coin tosses is, as we found initially, equal to .375. Notice that this is not the probability of obtaining at least 1 head. Rather, it is the probability of obtaining precisely 1 head (and 2 tails) in 3 coin tosses.

We can also use the binomial expansion when we can classify various events as either "yes" or "no." For example, say we are playing with dice, and want to determine the probability of showing a two on 4 out of 6 rolls of one die. The desired event of a two is the "yes." We want it to happen 4 times, so $r = 4$. Its probability on any single roll is p, which equals 1/6, or .167. Showing any other number on the die is a "no," and the probability of any other number on 1 roll of the die is q, which equals 5/6, or .83. Thus, filling in the binomial expansion, we have

$$p_C = \left(\frac{N!}{r!(N - r)!}\right)(p^r)(q^{N-r}) = \left(\frac{6!}{4!(6 - 4)!}\right)(.167^4)(.83^2)$$

which becomes

$$p_C = \left(\frac{720}{24(2)}\right)(.00078)(.689)$$

which equals

$$p_C = \left(\frac{720}{48}\right)(.00054)$$

Thus,

$$p_C = 15(.00054) = .0081$$

so the probability of rolling a die 6 times and showing a two on 4 of the rolls equals .0081.

PRACTICE PROBLEMS

(Answers for odd-numbered problems are provided in Appendix D.)

1. (a) When you state a question in terms of the probability of this "and" that, what mathematical procedure do you employ? (b) When you state the question in terms of the probability of this "or" that, what do you do? (c) When you phrase a question using "or," what characteristics of the events must you consider?
2. Which of the following events are mutually inclusive and which are mutually exclusive? (a) Being male or female; (b) Being sunny and rainy; (c) Being tall and weighing a lot; (d) Being age 16 and being a registered voter.
3. Researchers have found that for every 100 people, 34 have an IQ above 116 and the rest are below 116, and 40 are introverted, 35 are extroverted, and 25 are in-between. (a) What is the probability of randomly selecting two people with an IQ above 116? (b) What is the probability of selecting either an introverted or an extroverted subject? (c) What is the probability of selecting someone with an IQ above 116 who is introverted? (d) What is the probability of selecting either someone with an IQ above 116 or someone who is

introverted? (e) What is the probability of selecting someone in-between introverted and extroverted, and then selecting either someone who has an IQ above 116 or who is introverted?

4. You want to determine the probability of obtaining three heads in a row in a coin toss. (a) Determine the probability by making a fraction of the total number of ways three heads can occur and the total number of possible combinations of heads and tails. (b) Determine the probability using the multiplication rule.
5. (a) When do you use the binomial expansion? (b) Out of 5 coin tosses, what is the probability of obtaining 4 heads? (c) What is the probability of obtaining 1 head in 5 coin tosses? (d) Why is the answer in (c) the same as in (b)?
6. When rolling dice, what is the probability of each of the following? (a) Getting a 4 or a 5 rolling one die; (b) Getting a 4 twice in a row rolling one die; (c) Getting a 5 on only one die when rolling two dice at once; (d) Getting three 1s in 5 rolls of one die.

B.5: THE TWO-WAY WITHIN-SUBJECTS ANALYSIS OF VARIANCE

Table B.6 shows example data from a 2 × 2 within-subjects design, with three scores per cell.

TABLE B.6 Data from 2 × 2 Within-Subjects Design

		Factor A		
Factor B		A_1	A_2	
B_1	Subject 1	8	18	
	Subject 2	12	17	
	Subject 3	13	15	
		$\bar{X} = 11$ $\Sigma X = 33$ $\Sigma X^2 = 377$ $n = 3$	$\bar{X} = 16.7$ $\Sigma X = 50$ $\Sigma X^2 = 838$ $n = 3$	$\bar{X} = 13.85$ $\Sigma X = 83$ $n = 6$
B_2	Subject 1	9	6	
	Subject 2	10	8	
	Subject 3	17	4	
		$\bar{X} = 12$ $\Sigma X = 36$ $\Sigma X^2 = 470$ $n = 3$	$\bar{X} = 6$ $\Sigma X = 18$ $\Sigma X^2 = 116$ $n = 3$	$\bar{X} = 9$ $\Sigma X = 54$ $n = 6$
		$\Sigma X = 69$ $n = 6$ $\bar{X} = 11.5$	$\Sigma X = 68$ $n = 6$ $\bar{X} = 11.33$	$\Sigma X_{total} = 137$ $\Sigma X^2_{total} = 1801$ $N = 12$ $k_A = 2$ $k_B = 2$

TABLE B.7 A × Subject Table After Collapsing Across Factor B

	Factor A		
	A_1	A_2	ΣX_{sub}
Subject 1	17	24	41
Subject 2	22	25	47
Subject 3	30	19	49

Step 1: In each cell, compute the sum of scores, ΣX, the sum of the squared scores, ΣX^2, n, and the mean (the interaction means). Determine k_A, the number of levels of factor A, and for each column, compute ΣX, n, and the mean (the main effect means of factor A). Determine k_B, the number of levels of factor B, and for each row, compute ΣX, n, and the mean (the main effect means of factor B).

Step 2: Also determine:

$$\Sigma X_{total} = 69 + 68 = 137$$

$$\Sigma X^2_{total} = 377 + 838 + 470 + 116 = 1801$$

$$N = 3 + 3 + 3 + 3 = 12$$

Then create a table in which you collapse across factor B (as in Table B.7 above), and create another table in which you collapse across factor A (as in Table B.8). Note: the ΣX_{sub} for each subject must be the same in each table.

Step 3: Compute the Correction term:

$$\text{Correction Term} = \left(\frac{(\Sigma X_{total})^2}{N}\right) = \frac{137^2}{12} = 1564.08$$

Step 4: As you perform the following calculations, create the Analysis of Variance Summary Table shown in Table B.9. (Note that the table has components due to each factor and due to the interaction of factors and subjects.)

TABLE B.8 B × Subject Table After Collapsing Across Factor A

	Factor B		
	B_1	B_2	ΣX_{sub}
Subject 1	26	15	41
Subject 2	29	18	47
Subject 3	28	21	49

TABLE B.9 Summary Table of Two-Way, Within-Subjects ANOVA

Source	*Sum of Squares*	*df*	*Mean Square*	*F*
Factor				
A	.09	1	.09	.004
B	70.09	1	70.09	52.70
A×B	102.07	1	102.07	23.52
Subjects				
A×S	44.66	2	22.33	
B×S	2.66	2	1.33	
A×B×S	8.68	2	4.34	
Total	236.92	11	(not computed)	

Step 5: Compute the total sum of squares:

$$SS_{tot} = \Sigma X^2_{total} - \text{Step 3}$$
$$SS_{tot} = 1801 - 1564.08 = 236.92$$

Step 6: Compute the sum of squares for the column factor A:

$$SS_A = \Sigma\left(\frac{(\text{sum of scores in each column})^2}{n \text{ of scores in the column}}\right) - \text{Step 3}$$

$$SS_A = \left(\frac{(69)^2}{6} + \frac{(68)^2}{6}\right) - 1564.08 = .09$$

Step 7: Compute the sum of squares for the row factor B:

$$SS_B = \Sigma\left(\frac{(\text{sum of scores in each row})^2}{n \text{ of scores in the row}}\right) - \text{Step 3}$$

$$SS_B = \left(\frac{(83)^2}{6} + \frac{(54)^2}{6}\right) - 1564.08 = 70.09$$

Step 8: Compute the total sum of squares between groups (not reported in Summary Table):

$$SS_{bn} = \Sigma\left(\frac{(\text{Sum of scores in each cell})^2}{n \text{ of scores in the cell}}\right) - \text{Step 3}$$

$$SS_{bn} = \left(\frac{(33)^2}{3} + \frac{(50)^2}{3} + \frac{(36)^2}{3} + \frac{(18)^2}{3}\right) - 1564.08$$

$$SS_{bn} = 172.25$$

Step 9: Compute the sum of squares for the A × B interaction:

$$SS_{A\times B} = SS_{bn} - SS_A - SS_B$$

$$SS_{A\times B} = \text{Step 8} - \text{Step 6} - \text{Step 7}$$

$$SS_{A\times B} = 172.25 - .09 - 70.09 = 102.07$$

Step 10: Compute the sum of squares for Subjects (not reported in Summary Table):

$$SS_{\text{subs}} = \frac{(\Sigma X_{\text{sub1}})^2 + (\Sigma X_{\text{sub2}})^2 \ldots + (\Sigma X_{\text{n}})^2}{(k_{\text{A}})(k_{\text{B}})} - \text{Step 3}$$

$$SS_{\text{subs}} = \frac{(41)^2 + (47)^2 + (49)^2}{(2)(2)} - 1564.08$$

$$SS_{\text{subs}} = 8.67$$

Step 11: Compute the sum of squares for the A × S interaction (Table B.7):

$$SS_{\text{A}\times\text{S}} = \Sigma\frac{(\text{sum of each A} \times \text{Subject score})^2}{k_{\text{B}}} - \text{Step 3} - SS_{\text{A}} - SS_{\text{subs}}$$

$$SS_{\text{A}\times\text{S}} = \Sigma\frac{(\text{sum of each A} \times \text{Subject score})^2}{k_{\text{B}}} - \text{Step 3} - \text{Step 6} - \text{Step 10}$$

$$SS_{\text{A}\times\text{S}} = \frac{(17)^2 + (24)^2 + (22)^2 + (25)^2 + (30)^2 + (19)^2}{2} - 1564.08 - .09 - 8.67$$

$$SS_{\text{A}\times\text{S}} = 44.66$$

Step 12: Compute the sum of squares for the B × S interaction (Table B.8):

$$SS_{\text{B}\times\text{S}} = \Sigma\frac{(\text{sum of each B} \times \text{Subject score})^2}{k_{\text{A}}} - \text{Step 3} - SS_{\text{B}} - SS_{\text{subs}}$$

$$SS_{\text{B}\times\text{S}} = \Sigma\frac{(\text{sum of each B} \times \text{Subject score})^2}{k_{\text{A}}} - \text{Step 3} - \text{Step 7} - \text{Step 10}$$

$$SS_{\text{B}\times\text{S}} = \frac{(26)^2 + (15)^2 + (29)^2 + (18)^2 + (28)^2 + (21)^2}{2} - 1564.08 - 70.09 - 8.67$$

$$SS_{\text{B}\times\text{S}} = 2.66$$

Step 13: Compute the sum of squares for the A × B × S interaction:

$$SS_{\text{A}\times\text{B}\times\text{S}} = SS_{\text{tot}} - SS_{\text{A}} - SS_{\text{B}} - SS_{\text{A}\times\text{B}} - SS_{\text{subs}} - SS_{\text{A}\times\text{S}} - SS_{\text{B}\times\text{S}}$$

$$SS_{\text{A}\times\text{B}\times\text{S}} = \text{Step 5} - \text{Step 6} - \text{Step 7} - \text{Step 9} - \text{Step 10} - \text{Step 11} - \text{Step 12}$$

$$SS_{\text{A}\times\text{B}\times\text{S}} = 236.92 - .09 - 70.09 - 102.07 - 8.67 - 44.66 - 2.66$$

$$SS_{\text{A}\times\text{B}\times\text{S}} = 8.68$$

Step 14: Compute the Degrees of Freedom

(a) Factor A:

$$df_{\text{A}} = k_{\text{A}} - 1$$
$$df_{\text{A}} = 2 - 1 = 2$$

(b) Factor B:

$$df_{\text{B}} = k_{\text{B}} - 1$$
$$df_{\text{B}} = 2 - 1 = 1$$

(c) A × B interaction:

$$df_{\text{A}\times\text{B}} = (df_{\text{A}})(df_{\text{B}})$$
$$df_{\text{A}\times\text{B}} = (1)(1) = 1$$

(d) Subjects:

$$df_{\text{S}} = \text{Number of subjects} - 1$$
$$df_{\text{S}} = 3 - 1 = 2$$

(e) A × Subjects interaction:

$$df_{\text{A}\times\text{S}} = (df_{\text{A}})(df_{\text{S}})$$
$$df_{\text{A}\times\text{S}} = (1)(2) = 2$$

(f) B × Subjects interaction:

$$df_{\text{B}\times\text{S}} = (df_{\text{B}})(df_{\text{S}})$$
$$df_{\text{B}\times\text{S}} = (1)(2) = 2$$

(g) A × B × Subjects interaction:

$$df_{\text{A}\times\text{B}\times\text{S}} = (df_{\text{A}})(df_{\text{B}})(df_{\text{S}})$$
$$df_{\text{A}\times\text{B}\times\text{S}} = (1)(1)(2) = 2$$

(h) Total:

$$df_{\text{tot}} = N - 1$$
$$df_{\text{tot}} = 12 - 1 = 11$$

Step 15: Compute the mean square for factor A:

$$MS_{\text{A}} = \frac{SS_{\text{A}}}{df_{\text{A}}}$$

$$MS_{\text{A}} = \frac{\text{Step 6}}{\text{Step 14.a}}$$

$$MS_{\text{A}} = \frac{.09}{1} = .09$$

Step 16: Compute the mean square for factor B:

$$MS_{\text{B}} = \frac{SS_{\text{B}}}{df_{\text{B}}}$$

$$MS_{\text{B}} = \frac{\text{Step 7}}{\text{Step 14.b}}$$

$$MS_{\text{B}} = \frac{70.09}{1} = 70.09$$

Step 17: Compute the mean square for the A × B interaction:

$$MS_{A \times B} = \frac{SS_{A \times B}}{df_{A \times B}}$$

$$MS_{A \times B} = \frac{\text{Step 9}}{\text{Step 14.c}}$$

$$MS_{A \times B} = \frac{102.07}{1} = 102.07$$

Step 18: Compute the mean square for the A × S interaction:

$$MS_{A \times S} = \frac{SS_{A \times S}}{df_{A \times S}}$$

$$MS_{A \times S} = \frac{\text{Step 11}}{\text{Step 14.e}}$$

$$MS_{A \times S} = \frac{44.66}{2} = 22.33$$

Step 19: Compute the mean square for the B × S interaction:

$$MS_{B \times S} = \frac{SS_{B \times S}}{df_{B \times S}}$$

$$MS_{B \times S} = \frac{\text{Step 12}}{\text{Step 14.f}}$$

$$MS_{B \times S} = \frac{2.66}{2} = 1.33$$

Step 20: Compute the mean square for the A × B × S interaction:

$$MS_{A \times B \times S} = \frac{SS_{A \times B \times S}}{df_{A \times B \times S}}$$

$$MS_{A \times B \times S} = \frac{\text{Step 13}}{\text{Step 14.g}}$$

$$MS_{A \times B \times S} = \frac{8.68}{2} = 4.34$$

Step 21: Compute the F for the main effect of factor A:

$$F_A = \frac{MS_A}{MS_{A \times S}}$$

$$F_A = \frac{\text{Step 15}}{\text{Step 18}}$$

$$F_A = \frac{.09}{22.33} = .004$$

Step 22: Compute the F for the main effect of factor B:

$$F_B = \frac{MS_B}{MS_{B \times S}}$$

$$F_B = \frac{\text{Step 16}}{\text{Step 19}}$$

$$F_B = \frac{70.09}{1.33} = 52.70$$

Step 23: Compute the F for the A × B interaction:

$$F_{A \times B} = \frac{MS_{A \times B}}{MS_{A \times B \times S}}$$

$$F_{A \times B} = \frac{\text{Step 17}}{\text{Step 20}}$$

$$F_{A \times B} = \frac{102.07}{4.34} = 23.52$$

Step 24: For each obtained F above, find the appropriate critical value in Table 5 in Appendix C, using as degrees of freedom:

- For Factor A: df_A is the df between and $df_{A \times S}$ is the df within. Above, for $\alpha = .05$ and $df_A = 1$, and $df_{A \times S} = 2$, F_{crit} is 18.51, so F_{obt} of .004 is not significant.
- For Factor B: df_B is the df between and $df_{B \times S}$ is the df within. Above, for $\alpha = .05$ and $df_B = 1$, and $df_{B \times S} = 2$, the F_{crit} is 18.51, so F_{obt} of 52.70 is significant.
- For A × B: $df_{A \times B}$ is the df between and $df_{A \times B \times S}$ is the df within. Above, for $\alpha = .05$ and $df_{A \times B} = 1$ and $df_{A \times B \times S} = 2$, the F_{crit} is 18.51, so F_{obt} of 23.52 is significant.

Step 25: For each significant F with more than two levels, compute Tukey's *HSD* using the formula:

$$HSD = (qk)\left(\sqrt{\frac{\text{denominator in } F \text{ ratio}}{n}}\right)$$

where "denominator in F ratio" is the *MS* used as the denominator when calculating the F_{obt}, and n is the number of scores that each mean being compared is based upon.

Step 26: For each significant F, compute η^2 using the formula:

$$\eta^2 = \frac{\text{Sum of squares for the effect}}{SS_{tot}}$$

where: "sum of squares for the effect" is the sum of squares used in calculating the numerator of the F_{obt}, whether SS_A, SS_B or $SS_{A \times B}$. The SS_{tot} is the total sum of squares in the ANOVA.

Step 27: Compute the confidence interval for the population μ represented by the mean of any level or cell using the formula:

$$\left(\sqrt{\frac{MS_{\text{wn}}}{n}}\right)(-t_{\text{crit}}) + \overline{X} \leq \mu \leq \left(\sqrt{\frac{MS_{\text{wn}}}{n}}\right)(+t_{\text{crit}}) + \overline{X}$$

where MS_{wn} is the denominator used in computing F_{obt}, $\overline{X}$ is the mean of the level or cell being described, n is the number of scores the mean is based upon, and t_{crit} is the two-tailed critical value from Table 2 in Appendix C, using as *df* the *df* used in computing the denominator of the F_{obt}.

PRACTICE PROBLEMS

(Answers to odd-numbered problems are in Appendix D)

1. You measure the dependent variable of participants' hypnotic suggestibility as a function of whether they meditate before being tested, and whether they were shown a film containing a low, medium, or high amount of fantasy. The same participants are tested under all conditions. Perform all appropriate statistical analyses, and determine what you should conclude about this study.

	Amount of fantasy		
	Low	*Medium*	*High*
Meditation	5 6 2 2 5	7 5 6 9 5	9 8 10 10 10
No meditation	10 10 9 10 10	2 5 4 3 2	5 6 5 7 6

2. You study whether alcohol affects performing a simple eye-hand coordination task and whether the time of year the alcohol is consumed affects performance. Each participant performed the task either before drinking 0 or 3 drinks and each was tested during the summer and during the winter. With $n = 3$ in each cell the following cell means were obtained.

		Drinks Prior to Task Performance	
		A_1: 0 drinks	*A_2: 3 drinks*
Time of Year	B1: summer	16	6
	B2: winter	11	12

Summary Table

Source	*Sum of Squares*	*df*	*Mean Square*	*F*
Factor				
A	90.75	______	______	______
B	6.75	______	______	______
A×B	47.50	______	______	______
Subjects				
A×S	8.00	______	______	
B×S	2.00	______	______	
A×B×S	4.25	______	______	
Total	236.92	______	(not computed)	

(a) Complete the ANOVA summary table. (b) With an $\alpha = .05$, what do you conclude about each F_{obt}. (c) Perform the appropriate *post hoc* comparisons. What do you conclude about the relationships in this study? (d) Determine the effect size where appropriate and interpret it.

B.6: THE TWO-WAY MIXED-DESIGN ANALYSIS OF VARIANCE

This design is diagrammed differently than previous designs. Arrange the data so that the row factor, Factor A, is the between-subjects factor, and factor B, the column factor, is the within-subjects factor. Table B.10 shows example data from a 2×2 mixed design.

Step 1: In each cell, compute the sum of scores, ΣX, the sum of the squared scores, ΣX^2, n, and the mean (the interaction means). Determine k_A, the number of levels of factor A, and for each level, compute ΣX, n, and the mean (the main effect means of factor A). Determine k_B, the number of levels of factor B, and for each column, compute ΣX, n, and the mean (the main effect means of factor B).

Also calculate the sum of the scores obtained by each participant, ΣX_{sub}.

Step 2: Determine:

$$\Sigma X_{total} = 54 + 36 = 90$$

$$\Sigma X^2_{total} = 83 + 525 + 116 + 114 = 838$$

$$N = 3 + 3 + 3 + 3 = 12$$

Step 3: Compute the Correction term:

$$\text{Correction Term} = \left(\frac{(\Sigma X_{total})^2}{N}\right) = \frac{90^2}{12} = 675$$

Step 4: As you perform the following calculations, create the Analysis of Variance Summary Table shown in Table B.11. (This table is organized differently from previous two-way summary tables. Here, the components of the between-subjects, Factor A are placed together, and then the components of the within-subjects factor and the interaction are placed together.)

TABLE B.10 Data from Two-Way Mixed Design

Factor A is the between-subjects factor and Factor B is the within-subjects factor.

Factor A		Factor B: B_1	B_2	ΣX_{sub}
A_1	Subject 1	3	10	13
	Subject 2	5	16	21
	Subject 3	7	13	20
		$\overline{X} = 5$ $\Sigma X = 15$ $\Sigma X^2 = 83$ $n = 3$	$\overline{X} = 13$ $\Sigma X = 39$ $\Sigma X^2 = 525$ $n = 3$	$\overline{X} = 9$ $\Sigma X = 54$ $n = 6$
A_2	Subject 1	8	7	15
	Subject 2	6	7	13
	Subject 3	4	4	8
		$\overline{X} = 6$ $\Sigma X = 18$ $\Sigma X^2 = 116$ $n = 3$	$\overline{X} = 6$ $\Sigma X = 18$ $\Sigma X^2 = 114$ $n = 3$	$\overline{X} = 6$ $\Sigma X = 36$ $n = 6$
		$\Sigma X = 33$ $n = 6$ $\overline{X} = 5.5$	$\Sigma X = 57$ $n = 6$ $\overline{X} = 12.5$	$\Sigma X_{total} = 90$ $\Sigma X^2_{total} = 838$ $N = 12$ $k_A = 2$ $k_B = 2$

TABLE B.11 Summary Table of Two-Way Mixed Design ANOVA

Source	*Sum of Squares*	*df*	*Mean Square*	*F*
Between Groups				
Factor A	27.00	1	27.00	3.38
Error Between	32.00	4	8.00	
Within Groups				
Factor B	48.00	1	48.00	24.00
A×B Interaction	48.00	1	48.00	24.00
Error Within	8.00	4	2.00	
Total	163.00	11		

Step 5: Compute the total Sum of Squares:

$$SS_{\text{tot}} = \Sigma X^2_{\text{total}} - \text{Step 3}$$

$$SS_{\text{tot}} = 838 - 675 = 163$$

Step 6: Compute the sum of squares for Subjects (not reported in summary table):

$$SS_{\text{subs}} = \frac{(\Sigma X_{\text{sub1}})^2 + (\Sigma X_{\text{sub2}})^2 \ldots + (\Sigma X_{\text{n}})^2}{k_{\text{B}}} - \text{Step 3}$$

$$SS_{\text{subs}} = \frac{(13)^2 + (21)^2 + (20)^2 + (15)^2 + (13)^2 + (8)^2}{2} - 675$$

$$SS_{\text{subs}} = 59$$

Step 7: Compute the sum of squares for the between-subjects, row factor A:

$$SS_{\text{A}} = \Sigma\left(\frac{(\text{sum of scores in each row})^2}{n \text{ of scores in the row}}\right) - \text{Step 3}$$

$$SS_{\text{A}} = \left(\frac{(54)^2}{6} + \frac{(36)^2}{6}\right) - 675 = 27$$

Step 8: Compute the sum of squares for the within-subjects, column factor B:

$$SS_{\text{B}} = \Sigma\left(\frac{(\text{sum of scores in each column})^2}{n \text{ of scores in the column}}\right) - \text{Step 3}$$

$$SS_{\text{B}} = \left(\frac{(33)^2}{6} + \frac{(57)^2}{6}\right) - 675 = 48$$

Step 9: Compute the sum of squares for error between subjects:

$$SS_{\text{e:bn}} = SS_{\text{subs}} - SS_{\text{A}}$$

$$SS_{\text{e:bn}} = \text{Step 6} - \text{Step 7}$$

$$SS_{\text{e:bn}} = 59 - 27 = 32$$

Step 10: Compute the total sum of squares between groups (not reported in summary table):

$$SS_{\text{bn}} = \Sigma\left(\frac{(\text{Sum of scores in each cell})^2}{n \text{ of scores in the cell}}\right) - \text{Step 3}$$

$$SS_{\text{bn}} = \left(\frac{(15)^2}{3} + \frac{(39)^2}{3} + \frac{(18)^2}{3} + \frac{(18)^2}{3}\right) - 675$$

$$SS_{\text{bn}} = 123$$

Step 11: Compute the sum of squares for the A $\times$ B interaction:

$$SS_{\text{A}\times\text{B}} = SS_{\text{bn}} - SS_{\text{A}} - SS_{\text{B}}$$

$$SS_{\text{A}\times\text{B}} = \text{Step 10} - \text{Step 7} - \text{Step 8}$$

$$SS_{\text{A}\times\text{B}} = 123 - 27 - 48 = 48$$

Step 12: Compute the sum of squares for error within subjects:

$$SS_{\text{e:wn}} = SS_{\text{tot}} - SS_{\text{subs}} - SS_{\text{B}} - SS_{\text{A}\times\text{B}}$$

$$SS_{\text{e:wn}} = \text{Step 5} - \text{Step 6} - \text{Step 8} - \text{Step 11}$$

$$SS_{\text{e:wn}} = 163 - 59 - 48 - 48 = 8$$

Step 13: Compute the Degrees of Freedom

(a) Factor A:

$$df_{\text{A}} = k_{\text{A}} - 1$$

$$df_{\text{A}} = 2 - 1 = 1$$

(b) Factor B:

$$df_{\text{B}} = k_{\text{B}} - 1$$

$$df_{\text{B}} = 2 - 1 = 1$$

(c) A × B interaction:

$$df_{\text{A}\times\text{B}} = (df_{\text{A}})(df_{\text{B}})$$

$$df_{\text{A}\times\text{B}} = (\text{Step 13.a})(\text{Step 13.b})$$

$$df_{\text{A}\times\text{B}} = (1)(1) = 1$$

(d) Error between groups:

$$df_{\text{e:bn}} = (k_{\text{A}})\,(n - 1)$$

$$df_{\text{e:bn}} = (2)\,(3 - 1) = 4$$

(e) Error within subjects:

$$df_{\text{e:wn}} = (k_{\text{B}} - 1)\,(k_{\text{A}})\,(n - 1)$$

$$df_{\text{e:wn}} = (2 - 1)\,(2)\,(3 - 1) = 4$$

(f) Total:

$$df_{\text{tot}} = N - 1$$

$$df_{\text{tot}} = 12 - 1 = 11$$

Step 14: Compute the mean square for factor A:

$$MS_{\text{A}} = \frac{SS_{\text{A}}}{df_{\text{A}}}$$

$$MS_{\text{A}} = \frac{\text{Step 7}}{\text{Step 13.a}}$$

$$MS_{\text{A}} = \frac{27}{1} = 27$$

Step 15: Compute the mean square for factor B:

$$MS_B = \frac{SS_B}{df_B}$$

$$MS_B = \frac{\text{Step 8}}{\text{Step 13.b}}$$

$$MS_B = \frac{48}{1} = 48$$

Step 16: Compute the mean square for the A × B interaction:

$$MS_{A \times B} = \frac{SS_{A \times B}}{df_{A \times B}}$$

$$MS_{A \times B} = \frac{\text{Step 11}}{\text{Step 13.c}}$$

$$MS_{A \times B} = \frac{48}{1} = 48$$

Step 17: Compute the mean square for error between groups:

$$MS_{e:bn} = \frac{SS_{e:bn}}{df_{e:bn}}$$

$$MS_{e:bn} = \frac{\text{Step 9}}{\text{Step 13.d}}$$

$$MS_{e:bn} = \frac{32}{4} = 8$$

Step 18: Compute the mean square for error within subjects:

$$MS_{e:wn} = \frac{SS_{e:wn}}{df_{e:wn}}$$

$$MS_{e:wn} = \frac{\text{Step 12}}{\text{Step 13.e}}$$

$$MS_{e:wn} = \frac{8}{4} = 2$$

Step 19: Compute the F for the main effect of A:

$$F_A = \frac{MS_A}{MS_{e:bn}}$$

$$F_A = \frac{\text{Step 14}}{\text{Step 17}}$$

$$F_A = = \frac{27}{8} = 3.38$$

Step 20: Compute the F for the main effect of B:

$$F_{\text{B}} = \frac{MS_{\text{B}}}{MS_{\text{e:wn}}}$$

$$F_{\text{B}} = \frac{\text{Step 15}}{\text{Step 18}}$$

$$F_{\text{B}} = = \frac{48}{2} = 24$$

Step 21: Compute the F for the A × B interaction:

$$F_{\text{A}\times\text{B}} = \frac{MS_{\text{A}\times\text{B}}}{MS_{\text{e:wn}}}$$

$$F_{\text{A}\times\text{B}} = \frac{\text{Step 16}}{\text{Step 18}}$$

$$F_{\text{A}\times\text{B}} = \frac{48}{2} = 24$$

Step 22: For each obtained F above, find the appropriate critical value in Table 5 in Appendix C, using as degrees of freedom:

- For Factor A: df_{A} is the *df* between and $df_{\text{e:bn}}$ is the *df* within. Above, for $\alpha = .05$ and $df_{\text{A}} = 1$, and $df_{\text{e:bn}} = 4$, the F_{crit} is 7.71, so the main effect of Factor A is not significant.
- For Factor B: df_{B} is the *df* between and $df_{\text{e:wn}}$ is the *df* within. Above, for $\alpha = .05$ and $df_{\text{B}} = 1$, and $df_{\text{e:wn}} = 4$ the F_{crit} is 7.71, so the main effect of Factor B is significant.
- For A × B: $df_{\text{A}\times\text{B}}$ is the *df* between and $df_{\text{e:wn}}$ is the *df* within. Above, for $\alpha = .05$ and $df_{\text{A}\times\text{B}} = 1$ and $df_{\text{e:wn}} = 4$, the F_{crit} is 7.71, so the interaction effect is significant.

Step 23: For each significant F with more than two levels, compute Tukey's *HSD* using the formula given in Step 25 in the previous section on the two-way within-subjects ANOVA.

Step 24: For each significant F, compute eta squared using the formula given in Step 26 in the previous section on the two-way within-subjects ANOVA.

Step 25: Compute the confidence interval for the population μ represented by the mean of any level or cell using the formula given in Step 27 in the previous section on the two-way within-subjects ANOVA.

PRACTICE PROBLEMS

(Answers to odd-numbered problems are in Appendix D.)

1. You measure the dependent variable of participants' hypnotic suggestibility as a function of whether they meditate before being tested, and whether they were shown a film containing a low, medium, or high amount of fantasy. The meditation factor is between subjects, the fantasy-level factor is repeated measures.

	Amount of fantasy		
	Low	*Medium*	*High*
Meditation	5 6 2 2 5	7 5 6 9 5	9 8 10 10 10
No meditation	10 10 9 10 10	2 5 4 3 2	5 6 5 7 6

(a) Perform all appropriate statistical analyses, and determine what you should conclude about this study.

2. A researcher studies the influence of four doses of a new drug to reduce depression in adult women who either have the AIDs virus (are HIV+) or do not have it (are HIV−). Dosage is a repeated measures factor and HIV status is a between-subjects factor. With $n = 3$ in each cell, the following overall mean mood improvement scores were obtained.

		Factor B: Dose of Antidepressant			
		B_1: control	*B_2: Low*	*B_3: Med.*	*B_4: High*
Factor A: HIV status	A_1: HIV−	4	5	13	17
	A_2: HIV+	3	6	12	19

Summary Table

Source	*Sum of Squares*	*df*	*Mean Square*	*F*
Between Groups				
Factor A	21.00	_____	_____	_____
Error Between	20.50	_____	_____	_____
Within Groups				
Factor B	67.75	_____	_____	_____
A×B Interaction	48.00	_____	_____	_____
Error Within	34.50	_____	_____	
Total	163.00	_____		

(a) Complete the ANOVA summary table. (b) With an $\alpha = .05$, what do you conclude about each F_{obt}? (c) Perform the appropriate *post hoc* comparisons. What do you conclude about the relationships in this study? (d) Determine the effect size where appropriate. What does it indicate about the observed effects?

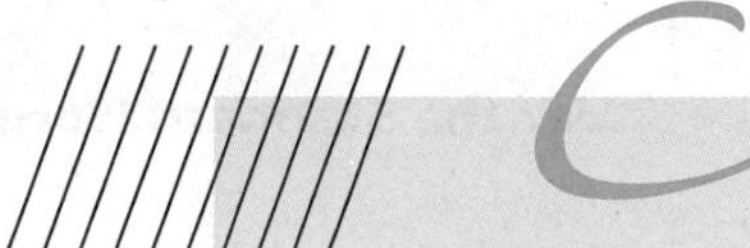

STATISTICAL TABLES

Table 1 Proportions of Area Under the Standard Normal Curve: The *z*-Tables

Column (A) lists *z*-score values. Column (B) lists the proportion of the area between the mean and the *z*-score value. Column (C) lists the proportion of the area beyond the *z*-score. *Note:* Because the normal distribution is symmetrical, areas for negative *z*-scores are the same as those for positive *z*-scores.

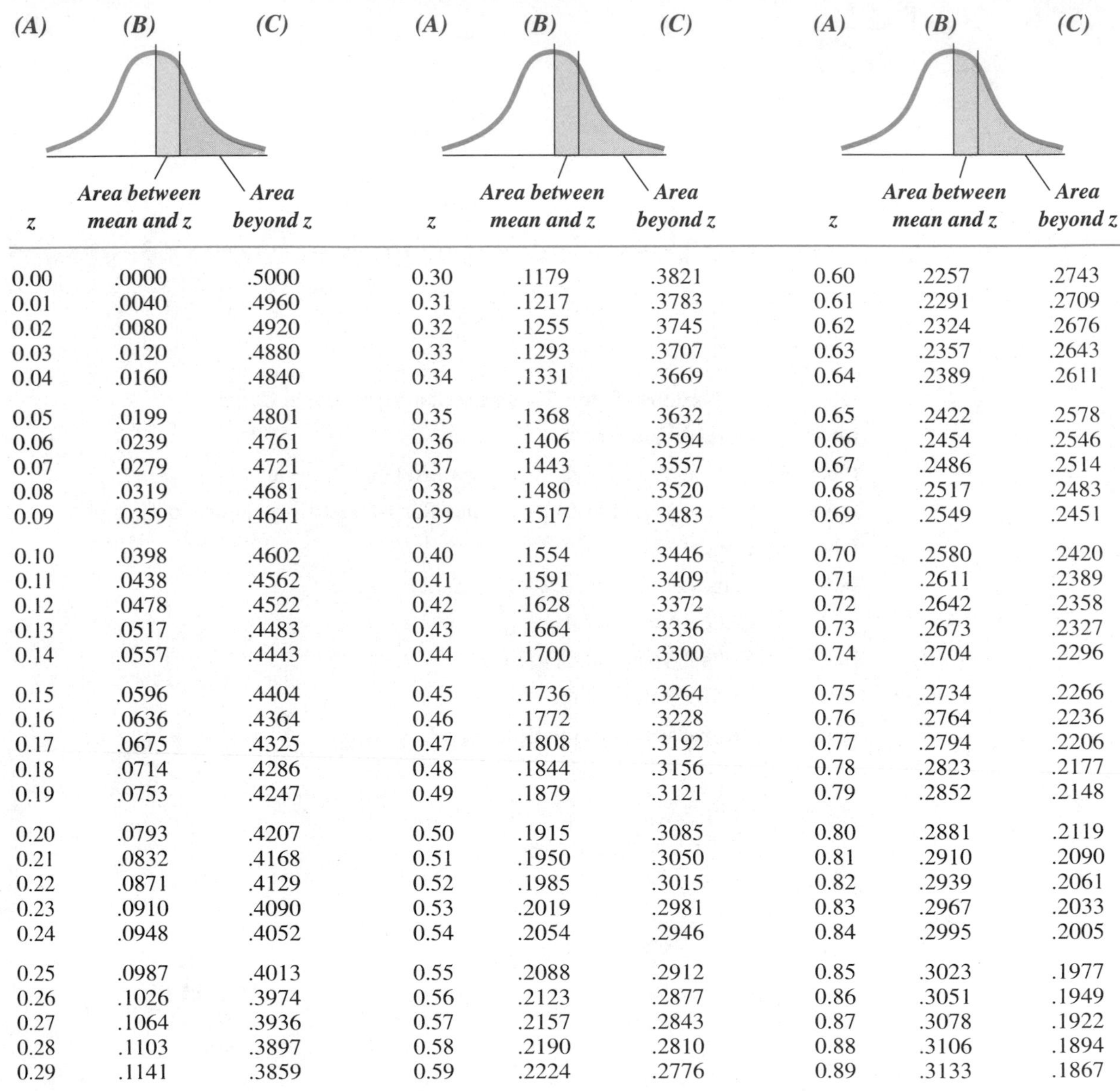

(A) *z*	(B) *Area between mean and z*	(C) *Area beyond z*	(A) *z*	(B) *Area between mean and z*	(C) *Area beyond z*	(A) *z*	(B) *Area between mean and z*	(C) *Area beyond z*
0.00	.0000	.5000	0.30	.1179	.3821	0.60	.2257	.2743
0.01	.0040	.4960	0.31	.1217	.3783	0.61	.2291	.2709
0.02	.0080	.4920	0.32	.1255	.3745	0.62	.2324	.2676
0.03	.0120	.4880	0.33	.1293	.3707	0.63	.2357	.2643
0.04	.0160	.4840	0.34	.1331	.3669	0.64	.2389	.2611
0.05	.0199	.4801	0.35	.1368	.3632	0.65	.2422	.2578
0.06	.0239	.4761	0.36	.1406	.3594	0.66	.2454	.2546
0.07	.0279	.4721	0.37	.1443	.3557	0.67	.2486	.2514
0.08	.0319	.4681	0.38	.1480	.3520	0.68	.2517	.2483
0.09	.0359	.4641	0.39	.1517	.3483	0.69	.2549	.2451
0.10	.0398	.4602	0.40	.1554	.3446	0.70	.2580	.2420
0.11	.0438	.4562	0.41	.1591	.3409	0.71	.2611	.2389
0.12	.0478	.4522	0.42	.1628	.3372	0.72	.2642	.2358
0.13	.0517	.4483	0.43	.1664	.3336	0.73	.2673	.2327
0.14	.0557	.4443	0.44	.1700	.3300	0.74	.2704	.2296
0.15	.0596	.4404	0.45	.1736	.3264	0.75	.2734	.2266
0.16	.0636	.4364	0.46	.1772	.3228	0.76	.2764	.2236
0.17	.0675	.4325	0.47	.1808	.3192	0.77	.2794	.2206
0.18	.0714	.4286	0.48	.1844	.3156	0.78	.2823	.2177
0.19	.0753	.4247	0.49	.1879	.3121	0.79	.2852	.2148
0.20	.0793	.4207	0.50	.1915	.3085	0.80	.2881	.2119
0.21	.0832	.4168	0.51	.1950	.3050	0.81	.2910	.2090
0.22	.0871	.4129	0.52	.1985	.3015	0.82	.2939	.2061
0.23	.0910	.4090	0.53	.2019	.2981	0.83	.2967	.2033
0.24	.0948	.4052	0.54	.2054	.2946	0.84	.2995	.2005
0.25	.0987	.4013	0.55	.2088	.2912	0.85	.3023	.1977
0.26	.1026	.3974	0.56	.2123	.2877	0.86	.3051	.1949
0.27	.1064	.3936	0.57	.2157	.2843	0.87	.3078	.1922
0.28	.1103	.3897	0.58	.2190	.2810	0.88	.3106	.1894
0.29	.1141	.3859	0.59	.2224	.2776	0.89	.3133	.1867

Table 1 (cont.) Proportions of Area Under the Standard Normal Curve: The z-Tables

(A) *z*	(B) *Area between mean and z*	(C) *Area beyond z*	(A) *z*	(B) *Area between mean and z*	(C) *Area beyond z*	(A) *z*	(B) *Area between mean and z*	(C) *Area beyond z*
0.90	.3159	.1841	1.25	.3944	.1056	1.60	.4452	.0548
0.91	.3186	.1814	1.26	.3962	.1038	1.61	.4463	.0537
0.92	.3212	.1788	1.27	.3980	.1020	1.62	.4474	.0526
0.93	.3238	.1762	1.28	.3997	.1003	1.63	.4484	.0516
0.94	.3264	.1736	1.29	.4015	.0985	1.64	.4495	.0505
0.95	.3289	.1711	1.30	.4032	.0968	1.65	.4505	.0495
0.96	.3315	.1685	1.31	.4049	.0951	1.66	.4515	.0485
0.97	.3340	.1660	1.32	.4066	.0934	1.67	.4525	.0475
0.98	.3365	.1635	1.33	.4082	.0918	1.68	.4535	.0465
0.99	.3389	.1611	1.34	.4099	.0901	1.69	.4545	.0455
1.00	.3413	.1587	1.35	.4115	.0885	1.70	.4554	.0446
1.01	.3438	.1562	1.36	.4131	.0869	1.71	.4564	.0436
1.02	.3461	.1539	1.37	.4147	.0853	1.72	.4573	.0427
1.03	.3485	.1515	1.38	.4162	.0838	1.73	.4582	.0418
1.04	.3508	.1492	1.39	.4177	.0823	1.74	.4591	.0409
1.05	.3531	.1469	1.40	.4192	.0808	1.75	.4599	.0401
1.06	.3554	.1446	1.41	.4207	.0793	1.76	.4608	.0392
1.07	.3577	.1423	1.42	.4222	.0778	1.77	.4616	.0384
1.08	.3599	.1401	1.43	.4236	.0764	1.78	.4625	.0375
1.09	.3621	.1379	1.44	.4251	.0749	1.79	.4633	.0367
1.10	.3643	.1357	1.45	.4265	.0735	1.80	.4641	.0359
1.11	.3665	.1335	1.46	.4279	.0721	1.81	.4649	.0351
1.12	.3686	.1314	1.47	.4292	.0708	1.82	.4656	.0344
1.13	.3708	.1292	1.48	.4306	.0694	1.83	.4664	.0336
1.14	.3729	.1271	1.49	.4319	.0681	1.84	.4671	.0329
1.15	.3749	.1251	1.50	.4332	.0668	1.85	.4678	.0322
1.16	.3770	.1230	1.51	.4345	.0655	1.86	.4686	.0314
1.17	.3790	.1210	1.52	.4357	.0643	1.87	.4693	.0307
1.18	.3810	.1190	1.53	.4370	.0630	1.88	.4699	.0301
1.19	.3830	.1170	1.54	.4382	.0618	1.89	.4706	.0294
1.20	.3849	.1151	1.55	.4394	.0606	1.90	.4713	.0287
1.21	.3869	.1131	1.56	.4406	.0594	1.91	.4719	.0281
1.22	.3888	.1112	1.57	.4418	.0582	1.92	.4726	.0274
1.23	.3907	.1093	1.58	.4429	.0571	1.93	.4732	.0268
1.24	.3925	.1075	1.59	.4441	.0559	1.94	.4738	.0262

Table 1 (cont.) Proportions of Area Under the Standard Normal Curve: The z-Tables

(A) z	(B) Area between mean and z	(C) Area beyond z	(A) z	(B) Area between mean and z	(C) Area beyond z	(A) z	(B) Area between mean and z	(C) Area beyond z
1.95	.4744	.0256	2.30	.4893	.0107	2.65	.4960	.0040
1.96	.4750	.0250	2.31	.4896	.0104	2.66	.4961	.0039
1.97	.4756	.0244	2.32	.4898	.0102	2.67	.4962	.0038
1.98	.4761	.0239	2.33	.4901	.0099	2.68	.4963	.0037
1.99	.4767	.0233	2.34	.4904	.0096	2.69	.4964	.0036
2.00	.4772	.0228	2.35	.4906	.0094	2.70	.4965	.0035
2.01	.4778	.0222	2.36	.4909	.0091	2.71	.4966	.0034
2.02	.4783	.0217	2.37	.4911	.0089	2.72	.4967	.0033
2.03	.4788	.0212	2.38	.4913	.0087	2.73	.4968	.0032
2.04	.4793	.0207	2.39	.4916	.0084	2.74	.4969	.0031
2.05	.4798	.0202	2.40	.4918	.0082	2.75	.4970	.0030
2.06	.4803	.0197	2.41	.4920	.0080	2.76	.4971	.0029
2.07	.4808	.0192	2.42	.4922	.0078	2.77	.4972	.0028
2.08	.4812	.0188	2.43	.4925	.0075	2.78	.4973	.0027
2.09	.4817	.0183	2.44	.4927	.0073	2.79	.4974	.0026
2.10	.4821	.0179	2.45	.4929	.0071	2.80	.4974	.0026
2.11	.4826	.0174	2.46	.4931	.0069	2.81	.4975	.0025
2.12	.4830	.0170	2.47	.4932	.0068	2.82	.4976	.0024
2.13	.4834	.0166	2.48	.4934	.0066	2.83	.4977	.0023
2.14	.4838	.0162	2.49	.4936	.0064	2.84	.4977	.0023
2.15	.4842	.0158	2.50	.4938	.0062	2.85	.4978	.0022
2.16	.4846	.0154	2.51	.4940	.0060	2.86	.4979	.0021
2.17	.4850	.0150	2.52	.4941	.0059	2.87	.4979	.0021
2.18	.4854	.0146	2.53	.4943	.0057	2.88	.4980	.0020
2.19	.4857	.0143	2.54	.4945	.0055	2.89	.4981	.0019
2.20	.4861	.0139	2.55	.4946	.0054	2.90	.4981	.0019
2.21	.4864	.0136	2.56	.4948	.0052	2.91	.4982	.0018
2.22	.4868	.0132	2.57	.4949	.0051	2.92	.4982	.0018
2.23	.4871	.0129	2.58	.4951	.0049	2.93	.4983	.0017
2.24	.4875	.0125	2.59	.4952	.0048	2.94	.4984	.0016
2.25	.4878	.0122	2.60	.4953	.0047	2.95	.4984	.0016
2.26	.4881	.0119	2.61	.4955	.0045	2.96	.4985	.0015
2.27	.4884	.0116	2.62	.4956	.0044	2.97	.4985	.0015
2.28	.4887	.0113	2.63	.4957	.0043	2.98	.4986	.0014
2.29	.4890	.0110	2.64	.4959	.0041	2.99	.4986	.0014

Table 1 (cont.) Proportions of Area Under the Standard Normal Curve: The z-Tables

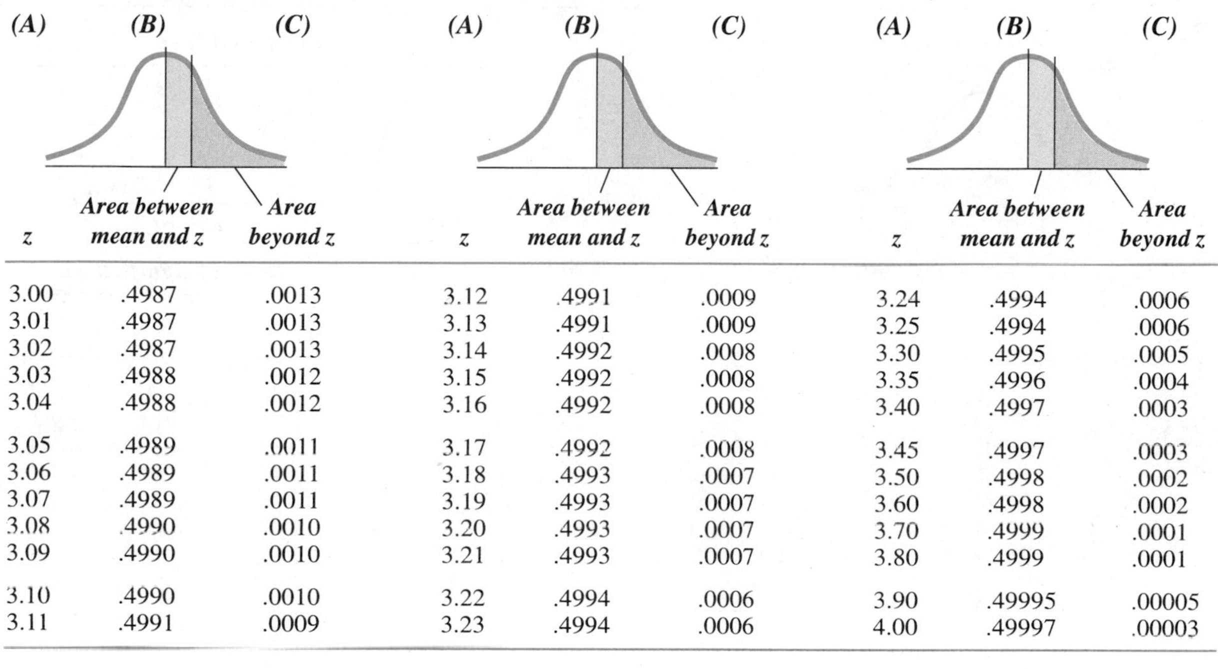

(A) z	(B) Area between mean and z	(C) Area beyond z	(A) z	(B) Area between mean and z	(C) Area beyond z	(A) z	(B) Area between mean and z	(C) Area beyond z
3.00	.4987	.0013	3.12	.4991	.0009	3.24	.4994	.0006
3.01	.4987	.0013	3.13	.4991	.0009	3.25	.4994	.0006
3.02	.4987	.0013	3.14	.4992	.0008	3.30	.4995	.0005
3.03	.4988	.0012	3.15	.4992	.0008	3.35	.4996	.0004
3.04	.4988	.0012	3.16	.4992	.0008	3.40	.4997	.0003
3.05	.4989	.0011	3.17	.4992	.0008	3.45	.4997	.0003
3.06	.4989	.0011	3.18	.4993	.0007	3.50	.4998	.0002
3.07	.4989	.0011	3.19	.4993	.0007	3.60	.4998	.0002
3.08	.4990	.0010	3.20	.4993	.0007	3.70	.4999	.0001
3.09	.4990	.0010	3.21	.4993	.0007	3.80	.4999	.0001
3.10	.4990	.0010	3.22	.4994	.0006	3.90	.49995	.00005
3.11	.4991	.0009	3.23	.4994	.0006	4.00	.49997	.00003

Table 2 Critical Values of t: The t-Tables

Note: Values of $-t_{crit}$ = values of $+t_{crit}$.

Two-Tailed Test

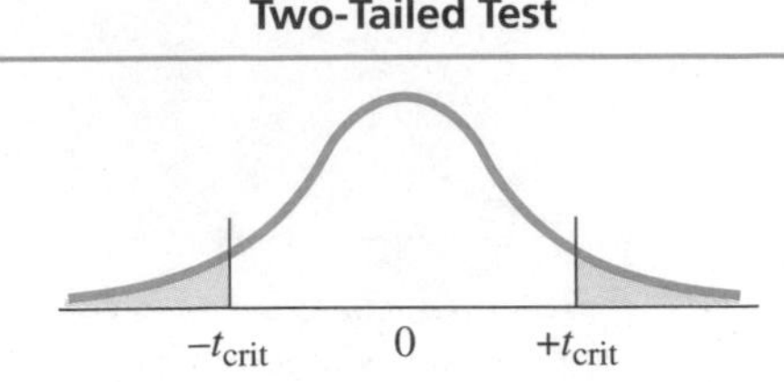

df	Level of significance $\alpha = .05$	Level of significance $\alpha = .01$
1	12.706	63.657
2	4.303	9.925
3	3.182	5.841
4	2.776	4.604
5	2.571	4.032
6	2.447	3.707
7	2.365	3.499
8	2.306	3.355
9	2.262	3.250
10	2.228	3.169
11	2.201	3.106
12	2.179	3.055
13	2.160	3.012
14	2.145	2.977
15	2.131	2.947
16	2.120	2.921
17	2.110	2.898
18	2.101	2.878
19	2.093	2.861
20	2.086	2.845
21	2.080	2.831
22	2.074	2.819
23	2.069	2.807
24	2.064	2.797
25	2.060	2.787
26	2.056	2.779
27	2.052	2.771
28	2.048	2.763
29	2.045	2.756
30	2.042	2.750
40	2.021	2.704
60	2.000	2.660
120	1.980	2.617
∞	1.960	2.576

One-Tailed Test

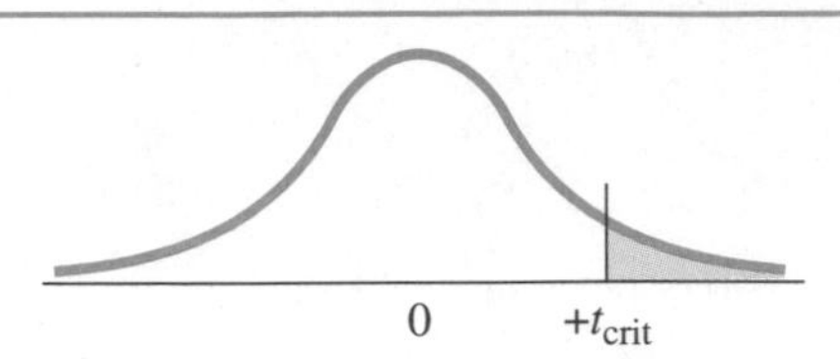

df	Level of significance $\alpha = .05$	Level of significance $\alpha = .01$
1	6.314	31.821
2	2.920	6.965
3	2.353	4.541
4	2.132	3.747
5	2.015	3.365
6	1.943	3.143
7	1.895	2.998
8	1.860	2.896
9	1.833	2.821
10	1.812	2.764
11	1.796	2.718
12	1.782	2.681
13	1.771	2.650
14	1.761	2.624
15	1.753	2.602
16	1.746	2.583
17	1.740	2.567
18	1.734	2.552
19	1.729	2.539
20	1.725	2.528
21	1.721	2.518
22	1.717	2.508
23	1.714	2.500
24	1.711	2.492
25	1.708	2.485
26	1.706	2.479
27	1.703	2.473
28	1.701	2.467
29	1.699	2.462
30	1.697	2.457
40	1.684	2.423
60	1.671	2.390
120	1.658	2.358
∞	1.645	2.326

From Table 12 of E. Pearson and H. Hartley, *Biometrika Tables for Statisticians*, Vol 1, 3rd ed. Cambridge: Cambridge University Press, 1966. Reprinted with permission of the Biometrika trustees.

Table 3 Critical Values of the Pearson Correlation Coefficient: The *r*-Tables

Two-Tailed Test

$-r_{crit}$ 0 $+r_{crit}$

Level of significance

df (*no. of pairs* − 2)	α = .05	α = .01
1	.997	.9999
2	.950	.990
3	.878	.959
4	.811	.917
5	.754	.874
6	.707	.834
7	.666	.798
8	.632	.765
9	.602	.735
10	.576	.708
11	.553	.684
12	.532	.661
13	.514	.641
14	.497	.623
15	.482	.606
16	.468	.590
17	.456	.575
18	.444	.561
19	.433	.549
20	.423	.537
21	.413	.526
22	.404	.515
23	.396	.505
24	.388	.496
25	.381	.487
26	.374	.479
27	.367	.471
28	.361	.463
29	.355	.456
30	.349	.449
35	.325	.418
40	.304	.393
45	.288	.372
50	.273	.354
60	.250	.325
70	.232	.302
80	.217	.283
90	.205	.267
100	.195	.254

One-Tailed Test

0 $+r_{crit}$

Level of significance

df (*no. of pairs* − 2)	α = .05	α = .01
1	.988	.9995
2	.900	.980
3	.805	.934
4	.729	.882
5	.669	.833
6	.622	.789
7	.582	.750
8	.549	.716
9	.521	.685
10	.497	.658
11	.476	.634
12	.458	.612
13	.441	.592
14	.426	.574
15	.412	.558
16	.400	.542
17	.389	.528
18	.378	.516
19	.369	.503
20	.360	.492
21	.352	.482
22	.344	.472
23	.337	.462
24	.330	.453
25	.323	.445
26	.317	.437
27	.311	.430
28	.306	.423
29	.301	.416
30	.296	.409
35	.275	.381
40	.257	.358
45	.243	.338
50	.231	.322
60	.211	.295
70	.195	.274
80	.183	.256
90	.173	.242
100	.164	.230

From Table IV of R. A. Fisher and F. Yates, *Statistical Tables for Biological, Agricultural and Medical Research*, 6th ed. London: Longman Group Ltd., 1974. Reprinted by permission of Addison-Wesley Longman, Ltd.

Table 4 Critical Values of the Spearman Rank-Order Correlation Coefficient: The r_s-Tables

Note: To interpolate the critical value for an N not given, find the critical values for the N above and below your N, add them together, and then divide the sum by 2.

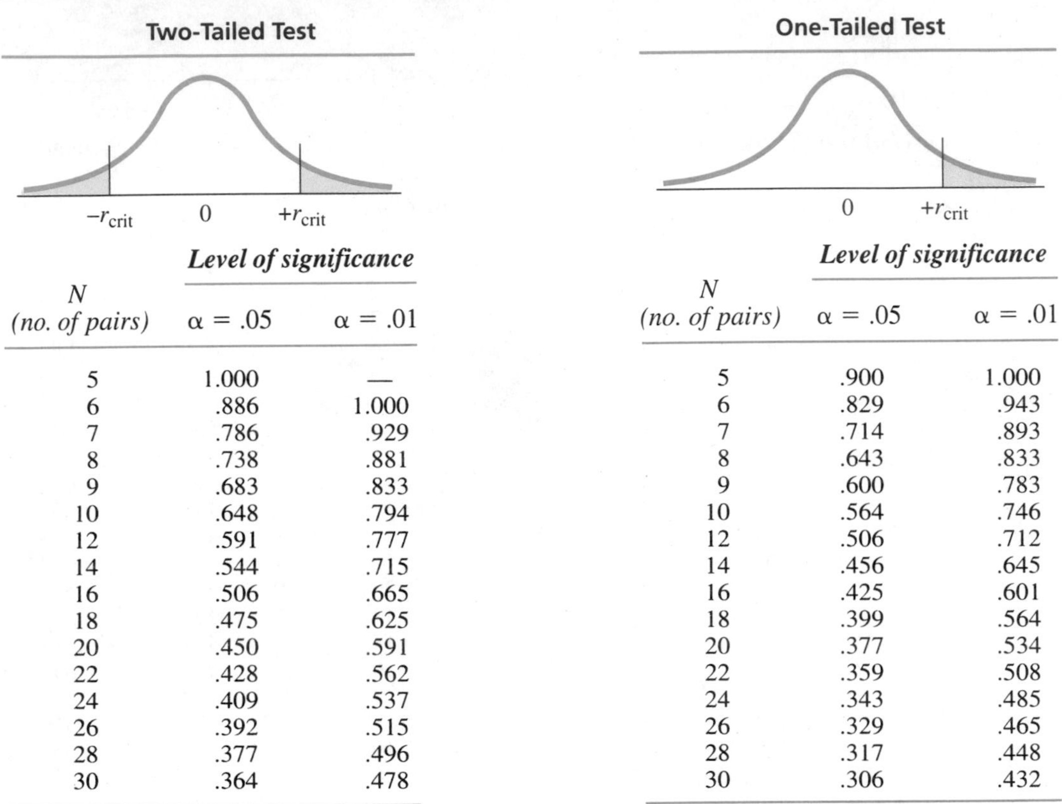

N (no. of pairs)	Two-Tailed Test, Level of significance $\alpha = .05$	Two-Tailed Test, Level of significance $\alpha = .01$	One-Tailed Test, Level of significance $\alpha = .05$	One-Tailed Test, Level of significance $\alpha = .01$
5	1.000	—	.900	1.000
6	.886	1.000	.829	.943
7	.786	.929	.714	.893
8	.738	.881	.643	.833
9	.683	.833	.600	.783
10	.648	.794	.564	.746
12	.591	.777	.506	.712
14	.544	.715	.456	.645
16	.506	.665	.425	.601
18	.475	.625	.399	.564
20	.450	.591	.377	.534
22	.428	.562	.359	.508
24	.409	.537	.343	.485
26	.392	.515	.329	.465
28	.377	.496	.317	.448
30	.364	.478	.306	.432

From E. G. Olds (1949), The 5 Percent Significance Levels of Sums of Squares of Rank Differences and a Correction, *Ann. Math. Statist.*, **20**, 117–118, and E. G. Olds (1938), Distribution of Sums of Squares of Rank Differences for Small Numbers of Individuals, *Ann. Math. Statist.*, **9**, 133–148. Reprinted by permission of the Institute of Mathematical Statistics.

Table 5 Critical Values of *F*: The *F*-Tables

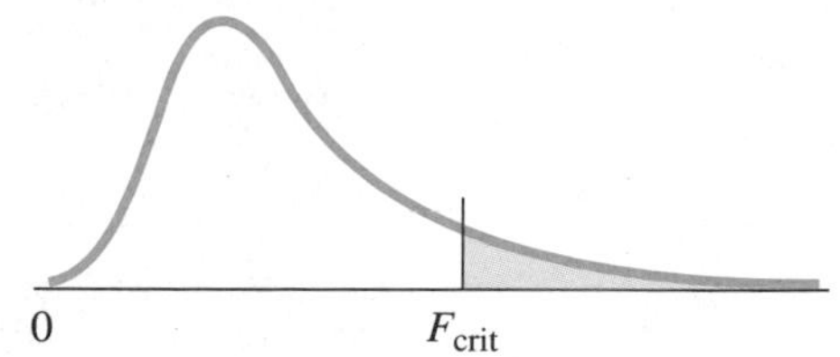

Critical values for α = .05 are in **dark numbers**.
Critical values for α = .01 are in light numbers.

Degrees of freedom within groups (degrees of freedom in denominator of F ratio)	α	*Degrees of freedom between groups (degrees of freedom in numerator of F ratio)* 1	2	3	4	5	6	7	8	9	10	11	12	14	16	20
1	**.05**	**161**	**200**	**216**	**225**	**230**	**234**	**237**	**239**	**241**	**242**	**243**	**244**	**245**	**246**	**248**
	.01	4,052	4,999	5,403	5,625	5,764	5,859	5,928	5,981	6,022	6,056	6,082	6,106	6,142	6,169	6,208
2	**.05**	**18.51**	**19.00**	**19.16**	**19.25**	**19.30**	**19.33**	**19.36**	**19.37**	**19.38**	**19.39**	**19.40**	**19.41**	**19.42**	**19.43**	**19.44**
	.01	98.49	99.00	99.17	99.25	99.30	99.33	99.34	99.36	99.38	99.40	99.41	99.42	99.43	99.44	99.45
3	**.05**	**10.13**	**9.55**	**9.28**	**9.12**	**9.01**	**8.94**	**8.88**	**8.84**	**8.81**	**8.78**	**8.76**	**8.74**	**8.71**	**8.69**	**8.66**
	.01	34.12	30.82	29.46	28.71	28.24	27.91	27.67	27.49	27.34	27.23	27.13	27.05	26.92	26.83	26.69
4	**.05**	**7.71**	**6.94**	**6.59**	**6.39**	**6.26**	**6.16**	**6.09**	**6.04**	**6.00**	**5.96**	**5.93**	**5.91**	**5.87**	**5.84**	**5.80**
	.01	21.20	18.00	16.69	15.98	15.52	15.21	14.98	14.80	14.66	14.54	14.45	14.37	14.24	14.15	14.02
5	**.05**	**6.61**	**5.79**	**5.41**	**5.19**	**5.05**	**4.95**	**4.88**	**4.82**	**4.78**	**4.74**	**4.70**	**4.68**	**4.64**	**4.60**	**4.56**
	.01	16.26	13.27	12.06	11.39	10.97	10.67	10.45	10.27	10.15	10.05	9.96	9.89	9.77	9.68	9.55
6	**.05**	**5.99**	**5.14**	**4.76**	**4.53**	**4.39**	**4.28**	**4.21**	**4.15**	**4.10**	**4.06**	**4.03**	**4.00**	**3.96**	**3.92**	**3.87**
	.01	13.74	10.92	9.78	9.15	8.75	8.47	8.26	8.10	7.98	7.87	7.79	7.72	7.60	7.52	7.39
7	**.05**	**5.59**	**4.47**	**4.35**	**4.12**	**3.97**	**3.87**	**3.79**	**3.73**	**3.68**	**3.63**	**3.60**	**3.57**	**3.52**	**3.49**	**3.44**
	.01	12.25	9.55	8.45	7.85	7.46	7.19	7.00	6.84	6.71	6.62	6.54	6.47	6.35	6.27	6.15
8	**.05**	**5.32**	**4.46**	**4.07**	**3.84**	**3.69**	**3.58**	**3.50**	**3.44**	**3.39**	**3.34**	**3.31**	**3.28**	**3.23**	**3.20**	**3.15**
	.01	11.26	8.65	7.59	7.01	6.63	6.37	6.19	6.03	5.91	5.82	5.74	5.67	5.56	5.48	5.36
9	**.05**	**5.12**	**4.26**	**3.86**	**3.63**	**3.48**	**3.37**	**3.29**	**3.23**	**3.18**	**3.13**	**3.10**	**3.07**	**3.02**	**2.98**	**2.93**
	.01	10.56	8.02	6.99	6.42	6.06	5.80	5.62	5.47	5.35	5.26	5.18	5.11	5.00	4.92	4.80
10	**.05**	**4.96**	**4.10**	**3.71**	**3.48**	**3.33**	**3.22**	**3.14**	**3.07**	**3.02**	**2.97**	**2.94**	**2.91**	**2.86**	**2.82**	**2.77**
	.01	10.04	7.56	6.55	5.99	5.64	5.39	5.21	5.06	4.95	4.85	4.78	4.71	4.60	4.52	4.41
11	**.05**	**4.84**	**3.98**	**3.59**	**3.36**	**3.20**	**3.09**	**3.01**	**2.95**	**2.90**	**2.86**	**2.82**	**2.79**	**2.74**	**2.70**	**2.65**
	.01	9.65	7.20	6.22	5.67	5.32	5.07	4.88	4.74	4.63	4.54	4.46	4.40	4.29	4.21	4.10
12	**.05**	**4.75**	**3.88**	**3.49**	**3.26**	**3.11**	**3.00**	**2.92**	**2.85**	**2.80**	**2.76**	**2.72**	**2.69**	**2.64**	**2.60**	**2.54**
	.01	9.33	6.93	5.95	5.41	5.06	4.82	4.65	4.50	4.39	4.30	4.22	4.16	4.05	3.98	3.86
13	**.05**	**4.67**	**3.80**	**3.41**	**3.18**	**3.02**	**2.92**	**2.84**	**2.77**	**2.72**	**2.67**	**2.63**	**2.60**	**2.55**	**2.51**	**2.46**
	.01	9.07	6.70	5.74	5.20	4.86	4.62	4.44	4.30	4.19	4.10	4.02	3.96	3.85	3.78	3.67
14	**.05**	**4.60**	**3.74**	**3.34**	**3.11**	**2.96**	**2.85**	**2.77**	**2.70**	**2.65**	**2.60**	**2.56**	**2.53**	**2.48**	**2.44**	**2.39**
	.01	8.86	6.51	5.56	5.03	4.69	4.46	4.28	4.14	4.03	3.94	3.86	3.80	3.70	3.62	3.51
15	**.05**	**4.54**	**3.68**	**3.29**	**3.06**	**2.90**	**2.79**	**2.70**	**2.64**	**2.59**	**2.55**	**2.51**	**2.48**	**2.43**	**2.39**	**2.33**
	.01	8.68	6.36	5.42	4.89	4.56	4.32	4.14	4.00	3.89	3.80	3.73	3.67	3.56	3.48	3.36
16	**.05**	**4.49**	**3.63**	**3.24**	**3.01**	**2.85**	**2.74**	**2.66**	**2.59**	**2.54**	**2.49**	**2.45**	**2.42**	**2.37**	**2.33**	**2.28**
	.01	8.53	6.23	5.29	4.77	4.44	4.20	4.03	3.89	3.78	3.69	3.61	3.55	3.45	3.37	3.25

Table 5 (cont.) Critical Values of *F*: The *F*-Tables

Degrees of freedom within groups (degrees of freedom in denominator of F ratio)	α	*Degrees of freedom between groups (degrees of freedom in numerator of F ratio)*														
		1	*2*	*3*	*4*	*5*	*6*	*7*	*8*	*9*	*10*	*11*	*12*	*14*	*16*	*20*
17	**.05**	**4.45**	**3.59**	**3.20**	**2.96**	**2.81**	**2.70**	**2.62**	**2.55**	**2.50**	**2.45**	**2.41**	**2.38**	**2.33**	**2.29**	**2.23**
	.01	8.40	6.11	5.18	4.67	4.34	4.10	3.93	3.79	3.68	3.59	3.52	3.45	3.35	3.27	3.16
18	**.05**	**4.41**	**3.55**	**3.16**	**2.93**	**2.77**	**2.66**	**2.58**	**2.51**	**2.46**	**2.41**	**2.37**	**2.34**	**2.29**	**2.25**	**2.19**
	.01	8.28	6.01	5.09	4.58	4.25	4.01	3.85	3.71	3.60	3.51	3.44	3.37	3.27	3.19	3.07
19	**.05**	**4.38**	**3.52**	**3.13**	**2.90**	**2.74**	**2.63**	**2.55**	**2.48**	**2.43**	**2.38**	**2.34**	**2.31**	**2.26**	**2.21**	**2.15**
	.01	8.18	5.93	5.01	4.50	4.17	3.94	3.77	3.63	3.52	3.43	3.36	3.30	3.19	3.12	3.00
20	**.05**	**4.35**	**3.49**	**3.10**	**2.87**	**2.71**	**2.60**	**2.52**	**2.45**	**2.40**	**2.35**	**2.31**	**2.28**	**2.23**	**2.18**	**2.12**
	.01	8.10	5.85	4.94	4.43	4.10	3.87	3.71	3.56	3.45	3.37	3.30	3.23	3.13	3.05	2.94
21	**.05**	**4.32**	**3.47**	**3.07**	**2.84**	**2.68**	**2.57**	**2.49**	**2.42**	**2.37**	**2.32**	**2.28**	**2.25**	**2.20**	**2.15**	**2.09**
	.01	8.02	5.78	4.87	4.37	4.04	3.81	3.65	3.51	3.40	3.31	3.24	3.17	3.07	2.99	2.88
22	**.05**	**4.30**	**3.44**	**3.05**	**2.82**	**2.66**	**2.55**	**2.47**	**2.40**	**2.35**	**2.30**	**2.26**	**2.23**	**2.18**	**2.13**	**2.07**
	.01	7.94	5.72	4.82	4.31	3.99	3.76	3.59	3.45	3.35	3.26	3.18	3.12	3.02	2.94	2.83
23	**.05**	**4.28**	**3.42**	**3.03**	**2.80**	**2.64**	**2.53**	**2.45**	**2.38**	**2.32**	**2.28**	**2.24**	**2.20**	**2.14**	**2.10**	**2.04**
	.01	7.88	5.66	4.76	4.26	3.94	3.71	3.54	3.41	3.30	3.21	3.14	3.07	2.97	2.89	2.78
24	**.05**	**4.26**	**3.40**	**3.01**	**2.78**	**2.62**	**2.51**	**2.43**	**2.36**	**2.30**	**2.26**	**2.22**	**2.18**	**2.13**	**2.09**	**2.02**
	.01	7.82	5.61	4.72	4.22	3.90	3.67	3.50	3.36	3.25	3.17	3.09	3.03	2.93	2.85	2.74
25	**.05**	**4.24**	**3.38**	**2.99**	**2.76**	**2.60**	**2.49**	**2.41**	**2.34**	**2.28**	**2.24**	**2.20**	**2.16**	**2.11**	**2.06**	**2.00**
	.01	7.77	5.57	4.68	4.18	3.86	3.63	3.46	3.32	3.21	3.13	3.05	2.99	2.89	2.81	2.70
26	**.05**	**4.22**	**3.37**	**2.98**	**2.74**	**2.59**	**2.47**	**2.39**	**2.32**	**2.27**	**2.22**	**2.18**	**2.15**	**2.10**	**2.05**	**1.99**
	.01	7.72	5.53	4.64	4.14	3.82	3.59	3.42	3.29	3.17	3.09	3.02	2.96	2.86	2.77	2.66
27	**.05**	**4.21**	**3.35**	**2.96**	**2.73**	**2.57**	**2.46**	**2.37**	**2.30**	**2.25**	**2.20**	**2.16**	**2.13**	**2.08**	**2.03**	**1.97**
	.01	7.68	5.49	4.60	4.11	3.79	3.56	3.39	3.26	3.14	3.06	2.98	2.93	2.83	2.74	2.63
28	**.05**	**4.20**	**3.34**	**2.95**	**2.71**	**2.56**	**2.44**	**2.36**	**2.29**	**2.24**	**2.19**	**2.15**	**2.12**	**2.06**	**2.02**	**1.96**
	.01	7.64	5.45	4.57	4.07	3.76	3.53	3.36	3.23	3.11	3.03	2.95	2.90	2.80	2.71	2.60
29	**.05**	**4.18**	**3.33**	**2.93**	**2.70**	**2.54**	**2.43**	**2.35**	**2.28**	**2.22**	**2.18**	**2.14**	**2.10**	**2.05**	**2.00**	**1.94**
	.01	7.60	5.42	4.54	4.04	3.73	3.50	3.33	3.20	3.08	3.00	2.92	2.87	2.77	2.68	2.57
30	**.05**	**4.17**	**3.32**	**2.92**	**2.69**	**2.53**	**2.42**	**2.34**	**2.27**	**2.21**	**2.16**	**2.12**	**2.09**	**2.04**	**1.99**	**1.93**
	.01	7.56	5.39	4.51	4.02	3.70	3.47	3.30	3.17	3.06	2.98	2.90	2.84	2.74	2.66	2.55
32	**.05**	**4.15**	**3.30**	**2.90**	**2.67**	**2.51**	**2.40**	**2.32**	**2.25**	**2.19**	**2.14**	**2.10**	**2.07**	**2.02**	**1.97**	**1.91**
	.01	7.50	5.34	4.46	3.97	3.66	3.42	3.25	3.12	3.01	2.94	2.86	2.80	2.70	2.62	2.51
34	**.05**	**4.13**	**3.28**	**2.88**	**2.65**	**2.49**	**2.38**	**2.30**	**2.23**	**2.17**	**2.12**	**2.08**	**2.05**	**2.00**	**1.95**	**1.89**
	.01	7.44	5.29	4.42	3.93	3.61	3.38	3.21	3.08	2.97	2.89	2.82	2.76	2.66	2.58	2.47
36	**.05**	**4.11**	**3.26**	**2.86**	**2.63**	**2.48**	**2.36**	**2.28**	**2.21**	**2.15**	**2.10**	**2.06**	**2.03**	**1.98**	**1.93**	**1.87**
	.01	7.39	5.25	4.38	3.89	3.58	3.35	3.18	3.04	2.94	2.86	2.78	2.72	2.62	2.54	2.43
38	**.05**	**4.10**	**3.25**	**2.85**	**2.62**	**2.46**	**2.35**	**2.26**	**2.19**	**2.14**	**2.09**	**2.05**	**2.02**	**1.96**	**1.92**	**1.85**
	.01	7.35	5.21	4.34	3.86	3.54	3.32	3.15	3.02	2.91	2.82	2.75	2.69	2.59	2.51	2.40
40	**.05**	**4.08**	**3.23**	**2.84**	**2.61**	**2.45**	**2.34**	**2.25**	**2.18**	**2.12**	**2.07**	**2.04**	**2.00**	**1.95**	**1.90**	**1.84**
	.01	7.31	5.18	4.31	3.83	3.51	3.29	3.12	2.99	2.88	2.80	2.73	2.66	2.56	2.49	2.37
42	**.05**	**4.07**	**3.22**	**2.83**	**2.59**	**2.44**	**2.32**	**2.24**	**2.17**	**2.11**	**2.06**	**2.02**	**1.99**	**1.94**	**1.89**	**1.82**
	.01	7.27	5.15	4.29	3.80	3.49	3.26	3.10	2.96	2.86	2.77	2.70	2.64	2.54	2.46	2.35

Table 5 (cont.) Critical Values of *F*: The *F*-Tables

Degrees of freedom within groups (degrees of freedom in denominator of F ratio)	α	*Degrees of freedom between groups (degrees of freedom in numerator of F ratio)* 1	2	3	4	5	6	7	8	9	10	11	12	14	16	20
44	**.05**	**4.06**	**3.21**	**2.82**	**2.58**	**2.43**	**2.31**	**2.23**	**2.16**	**2.10**	**2.05**	**2.01**	**1.98**	**1.92**	**1.88**	**1.81**
	.01	7.24	5.12	4.26	3.78	3.46	3.24	3.07	2.94	2.84	2.75	2.68	2.62	2.52	2.44	2.32
46	**.05**	**4.05**	**3.20**	**2.81**	**2.57**	**2.42**	**2.30**	**2.22**	**2.14**	**2.09**	**2.04**	**2.00**	**1.97**	**1.91**	**1.87**	**1.80**
	.01	7.21	5.10	4.24	3.76	3.44	3.22	3.05	2.92	2.82	2.73	2.66	2.60	2.50	2.42	2.30
48	**.05**	**4.04**	**3.19**	**2.80**	**2.56**	**2.41**	**2.30**	**2.21**	**2.14**	**2.08**	**2.03**	**1.99**	**1.96**	**1.90**	**1.86**	**1.79**
	.01	7.19	5.08	4.22	3.74	3.42	3.20	3.04	2.90	2.80	2.71	2.64	2.58	2.48	2.40	2.28
50	**.05**	**4.03**	**3.18**	**2.79**	**2.56**	**2.40**	**2.29**	**2.20**	**2.13**	**2.07**	**2.02**	**1.98**	**1.95**	**1.90**	**1.85**	**1.78**
	.01	7.17	5.06	4.20	3.72	3.41	3.18	3.02	2.88	2.78	2.70	2.62	2.56	2.46	2.39	2.26
55	**.05**	**4.02**	**3.17**	**2.78**	**2.54**	**2.38**	**2.27**	**2.18**	**2.11**	**2.05**	**2.00**	**1.97**	**1.93**	**1.88**	**1.83**	**1.76**
	.01	7.12	5.01	4.16	3.68	3.37	3.15	2.98	2.85	2.75	2.66	2.59	2.53	2.43	2.35	2.23
60	**.05**	**4.00**	**3.15**	**2.76**	**2.52**	**2.37**	**2.25**	**2.17**	**2.10**	**2.04**	**1.99**	**1.95**	**1.92**	**1.86**	**1.81**	**1.75**
	.01	7.08	4.98	4.13	3.65	3.34	3.12	2.95	2.82	2.72	2.63	2.56	2.50	2.40	2.32	2.20
65	**.05**	**3.99**	**3.14**	**2.75**	**2.51**	**2.36**	**2.24**	**2.15**	**2.08**	**2.02**	**1.98**	**1.94**	**1.90**	**1.85**	**1.80**	**1.73**
	.01	7.04	4.95	4.10	3.62	3.31	3.09	2.93	2.79	2.70	2.61	2.54	2.47	2.37	2.30	2.18
70	**.05**	**3.98**	**3.13**	**2.74**	**2.50**	**2.35**	**2.23**	**2.14**	**2.07**	**2.01**	**1.97**	**1.93**	**1.89**	**1.84**	**1.79**	**1.72**
	.01	7.01	4.92	4.08	3.60	3.29	3.07	2.91	2.77	2.67	2.59	2.51	2.45	2.35	2.28	2.15
80	**.05**	**3.96**	**3.11**	**2.72**	**2.48**	**2.33**	**2.21**	**2.12**	**2.05**	**1.99**	**1.95**	**1.91**	**1.88**	**1.82**	**1.77**	**1.70**
	.01	6.96	4.88	4.04	3.56	3.25	3.04	2.87	2.74	2.64	2.55	2.48	2.41	2.32	2.24	2.11
100	**.05**	**3.94**	**3.09**	**2.70**	**2.46**	**2.30**	**2.19**	**2.10**	**2.03**	**1.97**	**1.92**	**1.88**	**1.85**	**1.79**	**1.75**	**1.68**
	.01	6.90	4.82	3.98	3.51	3.20	2.99	2.82	2.69	2.59	2.51	2.43	2.36	2.26	2.19	2.06
125	**.05**	**3.92**	**3.07**	**2.68**	**2.44**	**2.29**	**2.17**	**2.08**	**2.01**	**1.95**	**1.90**	**1.86**	**1.83**	**1.77**	**1.72**	**1.65**
	.01	6.84	4.78	3.94	3.47	3.17	2.95	2.79	2.65	2.56	2.47	2.40	2.33	2.23	2.15	2.03
150	**.05**	**3.91**	**3.06**	**2.67**	**2.43**	**2.27**	**2.16**	**2.07**	**2.00**	**1.94**	**1.89**	**1.85**	**1.82**	**1.76**	**1.71**	**1.64**
	.01	6.81	4.75	3.91	3.44	3.14	2.92	2.76	2.62	2.53	2.44	2.37	2.30	2.20	2.12	2.00
200	**.05**	**3.89**	**3.04**	**2.65**	**2.41**	**2.26**	**2.14**	**2.05**	**1.98**	**1.92**	**1.87**	**1.83**	**1.80**	**1.74**	**1.69**	**1.62**
	.01	6.76	4.71	3.88	3.41	3.11	2.90	2.73	2.60	2.50	2.41	2.34	2.28	2.17	2.09	1.97
400	**.05**	**3.86**	**3.02**	**2.62**	**2.39**	**2.23**	**2.12**	**2.03**	**1.96**	**1.90**	**1.85**	**1.81**	**1.78**	**1.72**	**1.67**	**1.60**
	.01	6.70	4.66	3.83	3.36	3.06	2.85	2.69	2.55	2.46	2.37	2.29	2.23	2.12	2.04	1.92
1000	**.05**	**3.85**	**3.00**	**2.61**	**2.38**	**2.22**	**2.10**	**2.02**	**1.95**	**1.89**	**1.84**	**1.80**	**1.76**	**1.70**	**1.65**	**1.58**
	.01	6.66	4.62	3.80	3.34	3.04	2.82	2.66	2.53	2.43	2.34	2.26	2.20	2.09	2.01	1.89
∞	**.05**	**3.84**	**2.99**	**2.60**	**2.37**	**2.21**	**2.09**	**2.01**	**1.94**	**1.88**	**1.83**	**1.79**	**1.75**	**1.69**	**1.64**	**1.57**
	.01	6.64	4.60	3.78	3.32	3.02	2.80	2.64	2.51	2.41	2.32	2.24	2.18	2.07	1.99	1.87

Table 6 Values of Studentized Range Statistic, q_k

For a one-way ANOVA, or a comparison of the means from a main effect, the value of k is the number of means in the factor.

To compare the means from an interaction, find the appropriate design (or number of cell means) in the table below and obtain the adjusted value of k. Then use adjusted k as k to find the value of q_k.

Values of Adjusted k

Design of study	*Number of cell means in study*	*Adjusted value of k*
2 × 2	4	3
2 × 3	6	5
2 × 4	8	6
3 × 3	9	7
3 × 4	12	8
4 × 4	16	10
4 × 5	20	12

Values of q_k for $\alpha = .05$ are **dark numbers** and for $\alpha = .01$ are light numbers.

Degrees of freedom within groups (degrees of freedom in denominator of F ratio)	α	*k = number of means being compared*										
		2	*3*	*4*	*5*	*6*	*7*	*8*	*9*	*10*	*11*	*12*
1	**.05**	**18.00**	**27.00**	**32.80**	**37.10**	**40.40**	**43.10**	**45.40**	**47.40**	**49.10**	**50.60**	**52.00**
	.01	90.00	135.00	164.00	186.00	202.00	216.00	227.00	237.00	246.00	253.00	260.00
2	**.05**	**6.09**	**8.30**	**9.80**	**10.90**	**11.70**	**12.40**	**13.00**	**13.50**	**14.00**	**14.40**	**14.70**
	.01	14.00	19.00	22.30	24.70	26.60	28.20	29.50	30.70	31.70	32.60	33.40
3	**.05**	**4.50**	**5.91**	**6.82**	**7.50**	**8.04**	**8.48**	**8.85**	**9.18**	**9.46**	**9.72**	**9.95**
	.01	8.26	10.60	12.20	13.30	14.20	15.00	15.60	16.20	16.70	17.10	17.50
4	**.05**	**3.93**	**5.04**	**5.76**	**6.29**	**6.71**	**7.05**	**7.35**	**7.60**	**7.83**	**8.03**	**8.21**
	.01	6.51	8.12	9.17	9.96	10.60	11.10	11.50	11.90	12.30	12.60	12.80
5	**.05**	**3.64**	**4.60**	**5.22**	**5.67**	**6.03**	**6.33**	**6.58**	**6.80**	**6.99**	**7.17**	**7.32**
	.01	5.70	6.97	7.80	8.42	8.91	9.32	9.67	9.97	10.20	10.50	10.70
6	**.05**	**3.46**	**4.34**	**4.90**	**5.31**	**5.63**	**5.89**	**6.12**	**6.32**	**6.49**	**6.65**	**6.79**
	.01	5.24	6.33	7.03	7.56	7.97	8.32	8.61	8.87	9.10	9.30	9.49
7	**.05**	**3.34**	**4.16**	**4.69**	**5.06**	**5.36**	**5.61**	**5.82**	**6.00**	**6.16**	**6.30**	**6.43**
	.01	4.95	5.92	6.54	7.01	7.37	7.68	7.94	8.17	8.37	8.55	8.71
8	**.05**	**3.26**	**4.04**	**4.53**	**4.89**	**5.17**	**5.40**	**5.60**	**5.77**	**5.92**	**6.05**	**6.18**
	.01	4.74	5.63	6.20	6.63	6.96	7.24	7.47	7.68	7.87	8.03	8.18
9	**.05**	**3.20**	**3.95**	**4.42**	**4.76**	**5.02**	**5.24**	**5.43**	**5.60**	**5.74**	**5.87**	**5.98**
	.01	4.60	5.43	5.96	6.35	6.66	6.91	7.13	7.32	7.49	7.65	7.78

Table 6 (cont.) Values of Studentized Range Statistic, q_k

Degrees of freedom within groups (degrees of freedom in denominator of F ratio)	α	*k = number of means being compared*										
		2	*3*	*4*	*5*	*6*	*7*	*8*	*9*	*10*	*11*	*12*
10	**.05**	**3.15**	**3.88**	**4.33**	**4.65**	**4.91**	**5.12**	**5.30**	**5.46**	**5.60**	**5.72**	**5.83**
	.01	4.48	5.27	5.77	6.14	6.43	6.67	6.87	7.05	7.21	7.36	7.48
11	**.05**	**3.11**	**3.82**	**4.26**	**4.57**	**4.82**	**5.03**	**5.20**	**5.35**	**5.49**	**5.61**	**5.71**
	.01	4.39	5.14	5.62	5.97	6.25	6.48	6.67	6.84	6.99	7.13	7.26
12	**.05**	**3.08**	**3.77**	**4.20**	**4.51**	**4.75**	**4.95**	**5.12**	**5.27**	**5.40**	**5.51**	**5.62**
	.01	4.32	5.04	5.50	5.84	6.10	6.32	6.51	6.67	6.81	6.94	7.06
13	**.05**	**3.06**	**3.73**	**4.15**	**4.45**	**4.69**	**4.88**	**5.05**	**5.19**	**5.32**	**5.43**	**5.53**
	.01	4.26	4.96	5.40	5.73	5.98	6.19	6.37	6.53	6.67	6.79	6.90
14	**.05**	**3.03**	**3.70**	**4.11**	**4.41**	**4.64**	**4.83**	**4.99**	**5.13**	**5.25**	**5.36**	**5.46**
	.01	4.21	4.89	5.32	5.63	5.88	6.08	6.26	6.41	6.54	6.66	6.77
16	**.05**	**3.00**	**3.65**	**4.05**	**4.33**	**4.56**	**4.74**	**4.90**	**5.03**	**5.15**	**5.26**	**5.35**
	.01	4.13	4.78	5.19	5.49	5.72	5.92	6.08	6.22	6.35	6.46	6.56
18	**.05**	**2.97**	**3.61**	**4.00**	**4.28**	**4.49**	**4.67**	**4.82**	**4.96**	**5.07**	**5.17**	**5.27**
	.01	4.07	4.70	5.09	5.38	5.60	5.79	5.94	6.08	6.20	6.31	6.41
20	**.05**	**2.95**	**3.58**	**3.96**	**4.23**	**4.45**	**4.62**	**4.77**	**4.90**	**5.01**	**5.11**	**5.20**
	.01	4.02	4.64	5.02	5.29	5.51	5.69	5.84	5.97	6.09	6.19	6.29
24	**.05**	**2.92**	**3.53**	**3.90**	**4.17**	**4.37**	**4.54**	**4.68**	**4.81**	**4.92**	**5.01**	**5.10**
	.01	3.96	4.54	4.91	5.17	5.37	5.54	5.69	5.81	5.92	6.02	6.11
30	**.05**	**2.89**	**3.49**	**3.84**	**4.10**	**4.30**	**4.46**	**4.60**	**4.72**	**4.83**	**4.92**	**5.00**
	.01	3.89	4.45	4.80	5.05	5.24	5.40	5.54	5.56	5.76	5.85	5.93
40	**.05**	**2.86**	**3.44**	**3.79**	**4.04**	**4.23**	**4.39**	**4.52**	**4.63**	**4.74**	**4.82**	**4.91**
	.01	3.82	4.37	4.70	4.93	5.11	5.27	5.39	5.50	5.60	5.69	5.77
60	**.05**	**2.83**	**3.40**	**3.74**	**3.98**	**4.16**	**4.31**	**4.44**	**4.55**	**4.65**	**4.73**	**4.81**
	.01	3.76	4.28	4.60	4.82	4.99	5.13	5.25	5.36	5.45	5.53	5.60
120	**.05**	**2.80**	**3.36**	**3.69**	**3.92**	**4.10**	**4.24**	**4.36**	**4.48**	**4.56**	**4.64**	**4.72**
	.01	3.70	4.20	4.50	4.71	4.87	5.01	5.12	5.21	5.30	5.38	5.44
∞	**.05**	**2.77**	**3.31**	**3.63**	**3.86**	**4.03**	**4.17**	**4.29**	**4.39**	**4.47**	**4.55**	**4.62**
	.01	3.64	4.12	4.40	4.60	4.76	4.88	4.99	5.08	5.16	5.23	5.29

From B. J. Winer, *Statistical Principles in Experimental Design,* McGraw-Hill, 1962; abridged from H. L. Harter, D. S. Clemm, and E. H. Guthrie, The probability integrals of the range and of the studentized range, WADC Tech. Rep. 58–484, Vol. 2, 1959, Wright Air Development Center, Table II.2, pp. 243–281. Reproduced with permission of The McGraw-Hill Companies, Inc.

Table 7 Critical Values of the F_{max} Test

Critical values for α = .05 are **dark numbers** and for α = .01 are light numbers.

Note: n = number of scores in each condition or cell.

		k = number of samples in the study										
n − 1	α	2	3	4	5	6	7	8	9	10	11	12
4	**.05**	**9.60**	**15.50**	**20.60**	**25.20**	**29.50**	**33.60**	**37.50**	**41.40**	**44.60**	**48.00**	**51.40**
	.01	23.20	37.00	49.00	59.00	69.00	79.00	89.00	97.00	106.00	113.00	120.00
5	**.05**	**7.15**	**10.80**	**13.70**	**16.30**	**18.70**	**20.80**	**22.90**	**24.70**	**26.50**	**28.20**	**29.90**
	.01	14.90	22.00	28.00	33.00	38.00	42.00	46.00	50.00	54.00	57.00	60.00
6	**.05**	**5.82**	**8.38**	**10.40**	**12.10**	**13.70**	**15.00**	**16.30**	**17.50**	**18.60**	**19.70**	**20.70**
	.01	11.10	15.50	19.10	22.00	25.00	27.00	30.00	32.00	34.00	36.00	37.00
7	**.05**	**4.99**	**6.94**	**8.44**	**9.70**	**10.80**	**11.80**	**12.70**	**13.50**	**14.30**	**15.10**	**15.80**
	.01	8.89	12.10	14.50	16.50	18.40	20.00	22.00	23.00	24.00	26.00	27.00
8	**.05**	**4.43**	**6.00**	**7.18**	**8.12**	**9.03**	**9.78**	**10.50**	**11.10**	**11.70**	**12.20**	**12.70**
	.01	7.50	9.90	11.70	13.20	14.50	15.80	16.90	17.90	18.90	19.80	21.00
9	**.05**	**4.03**	**5.34**	**6.31**	**7.11**	**7.80**	**8.41**	**8.95**	**9.45**	**9.91**	**10.30**	**10.70**
	.01	6.54	8.50	9.90	11.10	12.10	13.10	13.90	14.70	15.30	16.00	16.60
10	**.05**	**3.72**	**4.85**	**5.67**	**6.34**	**6.92**	**7.42**	**7.87**	**8.28**	**8.66**	**9.01**	**9.34**
	.01	5.85	7.40	8.60	9.60	10.40	11.10	11.80	12.40	12.90	13.40	13.90
12	**.05**	**3.28**	**4.16**	**4.79**	**5.30**	**5.72**	**6.09**	**6.42**	**6.72**	**7.00**	**7.25**	**7.48**
	.01	4.91	6.10	6.90	7.60	8.20	8.70	9.10	9.50	9.90	10.20	10.60
15	**.05**	**2.86**	**3.54**	**4.01**	**4.37**	**4.68**	**4.95**	**5.19**	**5.40**	**5.59**	**5.77**	**5.93**
	.01	4.07	4.90	5.50	6.00	6.40	6.70	7.10	7.30	7.50	7.80	8.00
20	**.05**	**2.46**	**2.95**	**3.29**	**3.54**	**3.76**	**3.94**	**4.10**	**4.24**	**4.37**	**4.49**	**4.59**
	.01	3.32	3.80	4.30	4.60	4.90	5.10	5.30	5.50	5.60	5.80	5.90
30	**.05**	**2.07**	**2.40**	**2.61**	**2.78**	**2.91**	**3.02**	**3.12**	**3.21**	**3.29**	**3.36**	**3.39**
	.01	2.63	3.00	3.30	3.40	3.60	3.70	3.80	3.90	4.00	4.10	4.20
60	**.05**	**1.67**	**1.85**	**1.96**	**2.04**	**2.11**	**2.17**	**2.22**	**2.26**	**2.30**	**2.33**	**2.36**
	.01	1.96	2.20	2.30	2.40	2.40	2.50	2.50	2.60	2.60	2.70	2.70

Table 8 Critical Values of Chi Square: The χ^2-Tables

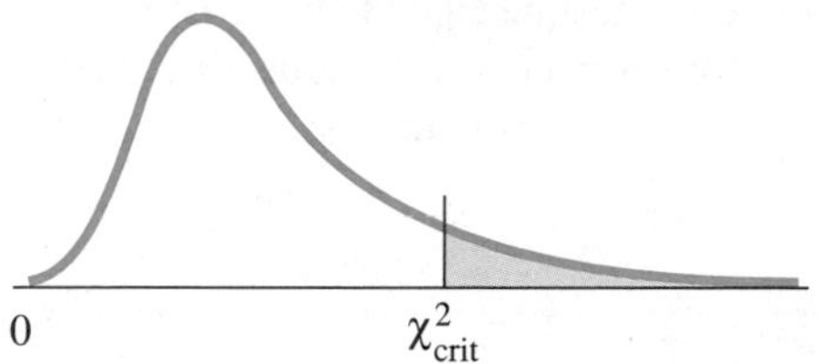

	Level of significance	
df	$\alpha = .05$	$\alpha = .01$
1	3.84	6.64
2	5.99	9.21
3	7.81	11.34
4	9.49	13.28
5	11.07	15.09
6	12.59	16.81
7	14.07	18.48
8	15.51	20.09
9	16.92	21.67
10	18.31	23.21
11	19.68	24.72
12	21.03	26.22
13	22.36	27.69
14	23.68	29.14
15	25.00	30.58
16	26.30	32.00
17	27.59	33.41
18	28.87	34.80
19	30.14	36.19
20	31.41	37.57
21	32.67	38.93
22	33.92	40.29
23	35.17	41.64
24	36.42	42.98
25	37.65	44.31
26	38.88	45.64
27	40.11	46.96
28	41.34	48.28
29	42.56	49.59
30	43.77	50.89
40	55.76	63.69
50	67.50	76.15
60	79.08	88.38
70	90.53	100.42

From Table IV of R. A. Fisher and F. Yates, *Statistical Tables for Biological, Agricultural and Medical Research*, 6th ed. London: Longman Group Ltd., 1974. Reprinted by permission of Addison-Wesley Longman, Ltd.

Table 9 Critical Values of the Mann-Whitney *U*

To be significant, the U_{obt} must be equal to or be *less than* the critical value. (Dashes in the table indicate that no decision is possible.) Critical values for $\alpha = .05$ are **dark numbers** and for $\alpha = .01$ are light numbers.

Two-Tailed Test

n_2 (no. of scores in Group 2)	α	n_1 (no. of scores in Group 1) 1	2	3	4	5	6	7	8	9
1	**.05**	**—**	**—**	**—**	**—**	**—**	**—**	**—**	**—**	**—**
	.01	—	—	—	—	—	—	—	—	—
2	**.05**	**—**	**—**	**—**	**—**	**—**	**—**	**—**	**0**	**0**
	.01	—	—	—	—	—	—	—	—	—
3	**.05**	**—**	**—**	**—**	**—**	**0**	**1**	**1**	**2**	**2**
	.01	—	—	—	—	—	—	—	—	0
4	**.05**	**—**	**—**	**—**	**0**	**1**	**2**	**3**	**4**	**4**
	.01	—	—	—	—	—	0	0	1	1
5	**.05**	**—**	**—**	**0**	**1**	**2**	**3**	**5**	**6**	**7**
	.01	—	—	—	—	0	1	1	2	3
6	**.05**	**—**	**—**	**1**	**2**	**3**	**5**	**6**	**8**	**10**
	.01	—	—	—	0	1	2	3	4	5
7	**.05**	**—**	**—**	**1**	**3**	**5**	**6**	**8**	**10**	**12**
	.01	—	—	—	0	1	3	4	6	7
8	**.05**	**—**	**0**	**2**	**4**	**6**	**8**	**10**	**13**	**15**
	.01	—	—	—	1	2	4	6	7	9
9	**.05**	**—**	**0**	**2**	**4**	**7**	**10**	**12**	**15**	**17**
	.01	—	—	0	1	3	5	7	9	11
10	**.05**	**—**	**0**	**3**	**5**	**8**	**11**	**14**	**17**	**20**
	.01	—	—	0	2	4	6	9	11	13
11	**.05**	**—**	**0**	**3**	**6**	**9**	**13**	**16**	**19**	**23**
	.01	—	—	0	2	5	7	10	13	16
12	**.05**	**—**	**1**	**4**	**7**	**11**	**14**	**18**	**22**	**26**
	.01	—	—	1	3	6	9	12	15	18
13	**.05**	**—**	**1**	**4**	**8**	**12**	**16**	**20**	**24**	**28**
	.01	—	—	1	3	7	10	13	17	20
14	**.05**	**—**	**1**	**5**	**9**	**13**	**17**	**22**	**26**	**31**
	.01	—	—	1	4	7	11	15	18	22
15	**.05**	**—**	**1**	**5**	**10**	**14**	**19**	**24**	**29**	**34**
	.01	—	—	2	5	8	12	16	20	24
16	**.05**	**—**	**1**	**6**	**11**	**15**	**21**	**26**	**31**	**37**
	.01	—	—	2	5	9	13	18	22	27
17	**.05**	**—**	**2**	**6**	**11**	**17**	**22**	**28**	**34**	**39**
	.01	—	—	2	6	10	15	19	24	29
18	**.05**	**—**	**2**	**7**	**12**	**18**	**24**	**30**	**36**	**42**
	.01	-	—	2	6	11	16	21	26	31
19	**.05**	**—**	**2**	**7**	**13**	**19**	**25**	**32**	**38**	**45**
	.01	—	0	3	7	12	17	22	28	33
20	**.05**	**—**	**2**	**8**	**13**	**20**	**27**	**34**	**41**	**48**
	.01	—	0	3	8	13	18	24	30	36

Table 9 (cont.) Critical Values of the Mann-Whitney *U*

n_1 (no. of scores in Group 1)										
10	*11*	*12*	*13*	*14*	*15*	*16*	*17*	*18*	*19*	*20*
—	—	—	—	—	—	—	—	—	—	—
—	—	—	—	—	—	—	—	—	—	—
0	**0**	**1**	**1**	**1**	**1**	**1**	**2**	**2**	**2**	**2**
—	—	—	—	—	—	—	—	—	0	0
3	**3**	**4**	**4**	**5**	**5**	**6**	**6**	**7**	**7**	**8**
0	0	1	1	1	2	2	2	2	3	3
5	**6**	**7**	**8**	**9**	**10**	**11**	**11**	**12**	**13**	**13**
2	2	3	3	4	5	5	6	6	7	8
8	**9**	**11**	**12**	**13**	**14**	**15**	**17**	**18**	**19**	**20**
4	5	6	7	7	8	9	10	11	12	13
11	**13**	**14**	**16**	**17**	**19**	**21**	**22**	**24**	**25**	**27**
6	7	9	10	11	12	13	15	16	17	18
14	**16**	**18**	**20**	**22**	**24**	**26**	**28**	**30**	**32**	**34**
9	10	12	13	15	16	18	19	21	22	24
17	**19**	**22**	**24**	**26**	**29**	**31**	**34**	**36**	**38**	**41**
11	13	15	17	18	20	22	24	26	28	30
20	**23**	**26**	**28**	**31**	**34**	**37**	**39**	**42**	**45**	**48**
13	16	18	20	22	24	27	29	31	33	36
23	**26**	**29**	**33**	**36**	**39**	**42**	**45**	**48**	**52**	**55**
16	18	21	24	26	29	31	34	37	39	42
26	**30**	**33**	**37**	**40**	**44**	**47**	**51**	**55**	**58**	**62**
18	21	24	27	30	33	36	39	42	45	48
29	**33**	**37**	**41**	**45**	**49**	**53**	**57**	**61**	**65**	**69**
21	24	27	31	34	37	41	44	47	51	54
33	**37**	**41**	**45**	**50**	**54**	**59**	**63**	**67**	**72**	**76**
24	27	31	34	38	42	45	49	53	56	60
36	**40**	**45**	**50**	**55**	**59**	**64**	**67**	**74**	**78**	**83**
26	30	34	38	42	46	50	54	58	63	67
39	**44**	**49**	**54**	**59**	**64**	**70**	**75**	**80**	**85**	**90**
29	33	37	42	46	51	55	60	64	69	73
42	**47**	**53**	**59**	**64**	**70**	**75**	**81**	**86**	**92**	**98**
31	36	41	45	50	55	60	65	70	74	79
45	**51**	**57**	**63**	**67**	**75**	**81**	**87**	**93**	**99**	**105**
34	39	44	49	54	60	65	70	75	81	86
48	**55**	**61**	**67**	**74**	**80**	**86**	**93**	**99**	**106**	**112**
37	42	47	53	58	64	70	75	81	87	92
52	**58**	**65**	**72**	**78**	**85**	**92**	**99**	**106**	**113**	**119**
39	45	51	56	63	69	74	81	87	93	99
55	**62**	**69**	**76**	**83**	**90**	**98**	**105**	**112**	**119**	**127**
42	48	54	60	67	73	79	86	92	99	105

Table 9 (cont.) Critical Values of the Mann-Whitney *U*

One-Tailed Test

n_2 (no. of scores in Group 2)	α	n_1 (no. of scores in Group 1) 1	2	3	4	5	6	7	8	9
1	**.05**	**—**	**—**	**—**	**—**	**—**	**—**	**—**	**—**	**—**
	.01	—	—	—	—	—	—	—	—	—
2	**.05**	**—**	**—**	**—**	**—**	**0**	**0**	**0**	**1**	**1**
	.01	—	—	—	—	—	—	—	—	—
3	**.05**	**—**	**—**	**0**	**0**	**1**	**2**	**2**	**3**	**3**
	.01	—	—	—	—	—	—	0	0	1
4	**.05**	**—**	**—**	**0**	**1**	**2**	**3**	**4**	**5**	**6**
	.01	—	—	—	—	0	1	1	2	3
5	**.05**	**—**	**0**	**1**	**2**	**4**	**5**	**6**	**8**	**9**
	.01	—	—	—	0	1	2	3	4	5
6	**.05**	**—**	**0**	**2**	**3**	**5**	**7**	**8**	**10**	**12**
	.01	—	—	—	1	2	3	4	6	7
7	**.05**	**—**	**0**	**2**	**4**	**6**	**8**	**11**	**13**	**15**
	.01	—	—	0	1	3	4	6	7	9
8	**.05**	**—**	**1**	**3**	**5**	**8**	**10**	**13**	**15**	**18**
	.01	—	—	0	2	4	6	7	9	11
9	**.05**	**—**	**1**	**3**	**6**	**9**	**12**	**15**	**18**	**21**
	.01	—	—	1	3	5	7	9	11	14
10	**.05**	**—**	**1**	**4**	**7**	**11**	**14**	**17**	**20**	**24**
	.01	—	—	1	3	6	8	11	13	16
11	**.05**	**—**	**1**	**5**	**8**	**12**	**16**	**19**	**23**	**27**
	.01	—	—	1	4	7	9	12	15	18
12	**.05**	**—**	**2**	**5**	**9**	**13**	**17**	**21**	**26**	**30**
	.01	—	—	2	5	8	11	14	17	21
13	**.05**	**—**	**2**	**6**	**10**	**15**	**19**	**24**	**28**	**33**
	.01	—	0	2	5	9	12	16	20	23
14	**.05**	**—**	**2**	**7**	**11**	**16**	**21**	**26**	**31**	**36**
	.01	—	0	2	6	10	13	17	22	26
15	**.05**	**—**	**3**	**7**	**12**	**18**	**23**	**28**	**33**	**39**
	.01	—	0	3	7	11	15	19	24	28
16	**.05**	**—**	**3**	**8**	**14**	**19**	**25**	**30**	**36**	**42**
	.01	—	0	3	7	12	16	21	26	31
17	**.05**	**—**	**3**	**9**	**15**	**20**	**26**	**33**	**39**	**45**
	.01	—	0	4	8	13	18	23	28	33
18	**.05**	**—**	**4**	**9**	**16**	**22**	**28**	**35**	**41**	**48**
	.01	—	0	4	9	14	19	24	30	36
19	**.05**	**0**	**4**	**10**	**17**	**23**	**30**	**37**	**44**	**51**
	.01	—	1	4	9	15	20	26	32	38
20	**.05**	**0**	**4**	**11**	**18**	**25**	**32**	**39**	**47**	**54**
	.01	—	1	5	10	16	22	28	34	40

n_1 *(no. of scores in Group 1)*										
10	*11*	*12*	*13*	*14*	*15*	*16*	*17*	*18*	*19*	*20*
—	**—**	**—**	**—**	**—**	**—**	**—**	**—**	**—**	**0**	**0**
—	—	—	—	—	—	—	—	—	—	—
1	**1**	**2**	**2**	**2**	**3**	**3**	**3**	**4**	**4**	**4**
—	—	—	0	0	0	0	0	0	1	1
4	**5**	**5**	**6**	**7**	**7**	**8**	**9**	**9**	**10**	**11**
1	1	2	2	2	3	3	4	4	4	5
7	**8**	**9**	**10**	**11**	**12**	**14**	**15**	**16**	**17**	**18**
3	4	5	5	6	7	7	8	9	9	10
11	**12**	**13**	**15**	**16**	**18**	**19**	**20**	**22**	**23**	**25**
6	7	8	9	10	11	12	13	14	15	16
14	**16**	**17**	**19**	**21**	**23**	**25**	**26**	**28**	**30**	**32**
8	9	11	12	13	15	16	18	19	20	22
17	**19**	**21**	**24**	**26**	**28**	**30**	**33**	**35**	**37**	**39**
11	12	14	16	17	19	21	23	24	26	28
20	**23**	**26**	**28**	**31**	**33**	**36**	**39**	**41**	**44**	**47**
13	15	17	20	22	24	26	28	30	32	34
24	**27**	**30**	**33**	**36**	**39**	**42**	**45**	**48**	**51**	**54**
16	18	21	23	26	28	31	33	36	38	40
27	**31**	**34**	**37**	**41**	**44**	**48**	**51**	**55**	**58**	**62**
19	22	24	27	30	33	36	38	41	44	47
31	**34**	**38**	**42**	**46**	**50**	**54**	**57**	**61**	**65**	**69**
22	25	28	31	34	37	41	44	47	50	53
34	**38**	**42**	**47**	**51**	**55**	**60**	**64**	**68**	**72**	**77**
24	28	31	35	38	42	46	49	53	56	60
37	**42**	**47**	**51**	**56**	**61**	**65**	**70**	**75**	**80**	**84**
27	31	35	39	43	47	51	55	59	63	67
41	**46**	**51**	**56**	**61**	**66**	**71**	**77**	**82**	**87**	**92**
30	34	38	43	47	51	56	60	65	69	73
44	**50**	**55**	**61**	**66**	**72**	**77**	**83**	**88**	**94**	**100**
33	37	42	47	51	56	61	66	70	75	80
48	**54**	**60**	**65**	**71**	**77**	**83**	**89**	**95**	**101**	**107**
36	41	46	51	56	61	66	71	76	82	87
51	**57**	**64**	**70**	**77**	**83**	**89**	**96**	**102**	**109**	**115**
38	44	49	55	60	66	71	77	82	88	93
55	**61**	**68**	**75**	**82**	**88**	**95**	**102**	**109**	**116**	**123**
41	47	53	59	65	70	76	82	88	94	100
58	**65**	**72**	**80**	**87**	**94**	**101**	**109**	**116**	**123**	**130**
44	50	56	63	69	75	82	88	94	101	107
62	**69**	**77**	**84**	**92**	**100**	**107**	**115**	**123**	**130**	**138**
47	53	60	67	73	80	87	93	100	107	114

From the *Bulletin of the Institute of Educational Research*, 1, No. 2, Indiana University, with permission of the publishers.

Table 10 Critical Values of the Wilcoxon T

To be significant, the T_{obt} must be equal to or *less than* the critical value. (Dashes in the table indicate that no decision is possible.) In the table, N is the number of nonzero differences that occurred when T_{obt} was calculated.

Two-Tailed Test

N	$\alpha = .05$	$\alpha = .01$	N	$\alpha = .05$	$\alpha = .01$
5	—	—	28	116	91
6	0	—	29	126	100
7	2	—	30	137	109
8	3	0	31	147	118
9	5	1	32	159	128
10	8	3	33	170	138
11	10	5	34	182	148
12	13	7	35	195	159
13	17	9	36	208	171
14	21	12	37	221	182
15	25	15	38	235	194
16	29	19	39	249	207
17	34	23	40	264	220
18	40	27	41	279	233
19	46	32	42	294	247
20	52	37	43	310	261
21	58	42	44	327	276
22	65	48	45	343	291
23	73	54	46	361	307
24	81	61	47	378	322
25	89	68	48	396	339
26	98	75	49	415	355
27	107	83	50	434	373

Table 10 (cont.) Critical Values of the Wilcoxon *T*

One-Tailed Test

N	α = *.05*	α = *.01*	*N*	α = *.05*	α = *.01*
5	0	—	28	130	101
6	2	—	29	140	110
7	3	0	30	151	120
8	5	1	31	163	130
9	8	3	32	175	140
10	10	5	33	187	151
11	13	7	34	200	162
12	17	9	35	213	173
13	21	12	36	227	185
14	25	15	37	241	198
15	30	19	38	256	211
16	35	23	39	271	224
17	41	27	40	286	238
18	47	32	41	302	252
19	53	37	42	319	266
20	60	43	43	336	281
21	67	49	44	353	296
22	75	55	45	371	312
23	83	62	46	389	328
24	91	69	47	407	345
25	100	76	48	426	362
26	110	84	49	446	379
27	119	92	50	466	397

From F. Wilcoxon and R. A. Wilcox, *Some Rapid Approximate Statistical Procedures, Revised Edition* (Pearl River, NY: Lederle Laboratories, 1964).

ANSWERS TO ODD-NUMBERED PRACTICE PROBLEMS

This appendix provides answers to the odd-numbered questions in the "Practice Problems" section found at the end of each chapter.

Chapter 1

1. To conduct their own research and to understand that of others.
3. The researcher will use statistics to organize, summarize, and communicate the IQ scores, and to draw conclusions about what the scores indicate about intelligence.
5. (a) Scientists are uncertain, open-minded, skeptical, cautious, and ethical.
   (b) To protect science from misleading information.
7. To describe, explain, predict, and control behavior.
9. A hypothesis should be: *testable* (a test of the hypothesis is possible); *falsifiable* (the hypothesis potentially may be shown to be false); *precise* (the hypothesis involves specific terms and applies to a specific situation); *rational* (the hypothesis fits what is known about nature); *parsimonious* (the hypothesis is as simple an explanation as possible).
11. A causal hypothesis postulates a particular cause of a behavior. A descriptive hypothesis postulates particular characteristics of a behavior or provides a goal of observations.
13. Because the design may contain flaws that determine our level of confidence in its conclusions.
15. Replication is the process of repeatedly conducting studies to build confidence in a hypothesis. It eliminates coincidental influences in a study that may mislead researchers.
17. Ask what empirical, objective, systematic, and controlled evidence is there? How is the research flawed, has it been replicated, and what level of confidence is there for the conclusions?
19. The researcher's, because your observations are unlikely to be objective, systematic, and controlled.
21. It is unscientific to require faith in the events and explanations being studied.
23. (a) To confirm it, you must find people who don't wake up and then die. If they're dead, however, you can't determine what they were dreaming about.
    (b) By looking for disconfirmation: Finding people who dreamt they hit the bottom of the cliff but did not die would disconfirm this hypothesis.
    (c) The data are not very objective or empirical: Maybe participants didn't really dream of falling off a cliff, or maybe they woke up briefly prior to hitting bottom, but they forgot.

Chapter 2

1. A relationship exists when certain scores on one variable are associated with certain scores on the other variable, and as the scores on one variable change, the scores on the other variable also tend to consistently change.
3. (a) A population consists of all the members of a specific group for which a law of nature applies, and a sample is a relatively small subset of the population.
   (b) The behaviors or scores observed in the sample are used to infer the scores or behaviors that would be found in the population if it could be studied.
5. A data point is a dot plotted on a graph to represent a pair of X and Y scores.
7. (a) A representative sample occurs because, by chance, the scores selected have the same characteristics as those in the population.
   (b) An unrepresentative sample occurs because, by chance, the scores do not have the same characteristics as those in the population.

9. Inferential statistics are used to decide whether the sample data represent a particular relationship in the population.
11. (a) The independent variable is the overall variable assumed to be the causal variable; conditions are the amounts or categories of the independent variable under which participants are tested.
    (b) The dependent variable measures participants' behavior and is assumed to be influenced by the independent variable.
13. With a true independent variable, participants can be randomly assigned to any condition. With a quasi-independent variable, participants must be assigned to a particular condition because of an inherent characteristic.
15. Study A is an experiment, because the researcher manipulates the amount of alcohol participants consume. Study B is a correlational study, because the researcher merely measures, without manipulating, the amount of alcohol participants consume.
17. Parts (a), (d), and (e) describe experiments; parts (b) and (c) describe quasi-independent variables.
19. Samples A and D.
21. A relationship is in Study A and Study C because as the scores on one variable change, the scores on the other variable tend to consistently change.
23. Because each relationship suggests there is something about the way nature operates so that as the amount of the X variable changes, the amount of the Y variable also consistently changes.
25. (a) Participants must select from the possible response choices provided (e.g., multiple choice or yes-no questions).
    (b) Participants rate their reaction to a statement using a rating scale (e.g., indicating the extent they agree with the statement).
    (c) Participants indicate their judgments by sorting stimuli into different groups.
    (d) Participants describe their feelings or thoughts.

Chapter 3

1. (a) Reliability is the degree to which measurements are reproducible and do not contain error.
   (b) Content validity is the degree to which a measurement reflects the intended variable or behavior.
   (c) Construct validity is the degree to which a measurement reflects the intended hypothetical construct.
   (d) Internal validity is the degree to which there are no unintended variables reflected in the relationship observed in the study.
   (e) External validity is the degree to which results generalize to other individuals and settings.
3. Whether the experimental situation generalizes to natural settings and natural behaviors.
5. (a) The advantage is that experimental methods tend to have high internal validity. The disadvantage is that they tend to have reduced external validity.
   (b) We have greater confidence in identifying the causes of the behavior, but less confidence that a natural behavior is being observed.
7. (a) With a true independent variable, participants can be randomly assigned to any condition. With a quasi-independent variable, subjects must be assigned to a particular condition because of an inherent characteristic.
   (b) A quasi-independent variable is likely to be confounded by extraneous subject variables, so there are many differences between the conditions that may cause differences in dependent scores.
9. (a) A confounding variable.
   (b) Keep your mood constant throughout the study or balance it by testing each type of model while you are in both moods.
11. (a) A quasi-experiment.
    (b) We have little confidence, because participants' differences in amount of wine consumed may be confounded by other differences (e.g., their attitudes toward health and fitness, their diet, their medical history).
13. (a) Amount of aspirin is a confounding variable.
    (b) Because reduced heart disease may be caused by greater amounts of aspirin—instead of more red wine—this confounding reduces internal validity for concluding that red wine reduces heart disease.
15. Conduct either a quasi-experiment (conditions of the independent variable are the amount of participants' inherent fear of disease and dependent scores reflect number of partners) or a correlational study (measuring each person's fear and number of partners). Either design is flawed because we cannot conclude that greater fear causes fewer sexual partners. Confounding variables may actually determine the number of partners (e.g., subjects' age, attractiveness, marital status).
17. (a) A correlational design.
    (b) No, this type of design has little internal validity for inferring causality.

(c) The age and personality of those who drive brightly colored cars (young, liberal, less cautious) may make them more accident-prone compared with those who drive muted-colored cars (older, conservative, cautious) who are less accident-prone.

(d) A true experiment, randomly assign participants to conditions of the independent variable where they are given a car of a certain color to drive for a time, during which we measure their accident rates.

Chapter 4

1. (a) It is a measurement, in addition to the dependent variable, that determines whether the treatment had its intended effect.

(b) A manipulation check is performed during the study, while a pilot study is performed prior to the study to validate stimuli and debug procedures.

3. (a) Practice effects are the influence on trials resulting from practicing trials.

(b) Carry-over effects are the influence on trials from experiencing previous trials.

(c) A response set is a bias toward responding in a particular way because of previous responses made.

5. A sensitive measure produces different scores for small differences in behavior.

7. (a) Automation is using electronic or mechanical devices to present stimuli and/or to measure and record responses.

(b) It increases reliability.

(c) It may heighten reactivity and, over time, instrumentation effects may actually decrease reliability.

9. (a) Experimental realism is the extent to which a measurement task engages participants. We seek it so that participants ignore demand characteristics.

(b) Ecological validity occurs if the task reflects natural behaviors; experimental realism occurs if the task actively engages participants.

11. Research ethics deal with balancing a researcher's right to study a behavior with the right of participants to be protected from abuse.

13. Obtain informed consent and provide a debriefing.

15. (a) In role playing, subjects pretend they are in a particular situation.

(b) It minimizes risks to participants.

(c) Participants may not provide realistic or accurate responses.

17. (a) You cannot read the lists consistently for all participants and differences may influence their memory.

(b) You expect the conditions to influence memory in a certain way, and inconsistency in your pronunciation, voice, inflection, and so on can make a list more or less memorable and produce the expected results.

(c) Through automation: Play a tape recording of the lists to all participants.

(d) Instrumentation effects may occur, with the recording becoming distorted and unclear with use. (Several copies of the original recording are needed.)

19. (a) Recalling only one list may not reliably reflect participants' typical memory performance in a condition.

(b) Provide practice trials.

(c) Provide several lists per condition, and compute a summary score from these multiple trials.

(d) Order effects may result from multiple trials.

(e) Counterbalance the order of the lists presented to different participants in each condition.

21. (a) To improve reliability, because only one film might be particularly arousing or not arousing.

(b) Order effects due to the order in which films are presented.

(c) Counterbalance so that different participants in a condition view the films in different orders.

23. By adding to reactivity and by communicating experimenter expectancies that influence participants and thus decrease internal validity, and that reduce generalizability to other situations and thus reduce external validity.

25. (a) Participants may respond to demand characteristics and provide the expected response, even though their views remain the same.

(b) Subjects may be hesitant to reveal their private feelings so they may all behave in the same, neutral fashion, producing no differences between conditions.

(c) Employ a manipulation check (e.g., a questionnaire) to directly assess participants' sexist views.

Chapter 5

1. That the researcher observes and describes participants or their behavior, usually in natural settings, without manipulating or controlling any variables.

3. (a) Behaviors are observed in a natural setting that is not influenced by the researcher or by participants reacting to demand characteristics.
   (b) The results may be biased by researcher expectations, we may not have random sampling, and we obtain qualitative data that may lack precision, accuracy, and sensitivity.
5. (a) A case study involves a description of one individual, organization, or event.
   (b) It provides an in-depth, complete description of a subject or event.
   (c) A single case provides little external validity for generalizing to other cases.
7. (a) Simple random sampling is randomly selecting subjects from a list of the population. Systematic random sampling is randomly selecting a starting point and then selecting every *n*th participant on the list.
   (b) Stratified random sampling is proportionate random selection from each important subgroup in the population.
   (c) Cluster sampling is randomly selecting certain groups and then observing all subjects in each group.
9. (a) Quota sampling proportionately samples from the population but relies on convenience sampling to obtain each proportion.
   (b) Snowball sampling is identifying potential subjects through previously tested subjects.
11. (a) When mailing is the best method for reaching participants, a lengthy survey is required, and the researcher is not in a hurry to obtain the data.
   (b) When the appropriate sample can be reached by telephone, the survey is not lengthy, and it is necessary to obtain the data rapidly.
   (c) So that the sample is representative.
13. (a) Open-ended questions provide a wide range of responses and allow us to discover many relevant variables. Their disadvantage is that they involve subjective scoring, which may decrease reliability and validity.
   (b) Closed-ended questions can be objectively and reliably scored. Their disadvantage is that they may provide limited information, because some variables or response choices are not included.
15. (a) They are differently worded versions of the same questionnaire.
   (b) That they are comparable in terms of reliability, validity, demand characteristics, and so on.
   (c) They are necessary when repeatedly testing the same participants.
17. (a) This is the bias in a study's results that arises from the particular people who will participate in a study.
   (b) Volunteer participants differ from the general population in social status, intelligence, and need for social approval. In addition, volunteers may be motivated by strong feelings about the issues raised in a study.
19. (a) The judgments and their scoring may not be reliable.
   (b) Train the judges to use explicit rules for evaluating and assigning scores for aggressiveness.
   (c) Employ multiple raters who are blind to the study's goals.
   (d) Determine that there is high inter-rater reliability.

Chapter 6

1. (a) A transformation is a mathematical procedure for systematically converting a set of scores into different scores.
   (b) To make scores easier to work with and to make different kinds of scores comparable.
3. Consider what you wish to know about the data, the specific design employed, and the scale of measurement used to measure the scores.
5. (a) N is the total number of scores in a sample.
   (b) f is frequency, the number of times a score occurs.
   (c) *rel.* f is relative frequency, the proportion of time certain scores occur.
   (d) cf is cumulative frequency, the number of times scores at or below a certain score occur.
7. (a) Bar graph.
   (b) Polygon.
   (c) Bar graph.
   (d) Histogram.
9. It means that the score is either a high or low extreme score relative to the middle scores in the distribution, and that it occurs with a relatively low frequency.
11. (a) Simple frequency is the number of times a score occurs in a sample; relative frequency is the proportion of time the score occurs.
   (b) Cumulative frequency is the number of times that scores at or below the score occur; percentile is the percent of the time that scores at or below the score occur.
13. (a) Two decimal places.
   (b) Four decimal places.
   (c) Two decimal places.
   (d) Two decimal places.

15. (a) 13.75
(b) 10.04
(c) 10.05
(d) .08
(e) 1.00

17. (a) The middle IQ score has the highest frequency in a symmetrical distribution; the higher and lower scores have lower frequencies, and the highest and lowest scores have relatively very low frequency.
(b) Agility scores form a symmetrical distribution containing two distinct "humps" where there are two scores that occur more frequently than the surrounding scores.
(c) Memory scores form an asymmetrical distribution in which there are some very infrequent, extremely low scores, but there are not corresponding infrequent high scores.

19. The *N* should be 10; the scores should be listed in descending order; a row containing the score of 6 should be included; the *cf* for the score of 1 should be 1; and the *cf* for all other scores should be corrected so that at the score of 7 the *cf* is 10.

21. (a) 70, 72, 60, 85, 45.
(b) The 20th percentile, because .20 of the scores are below 60.
(c) .30; there's .50 of the curve to the left of 70, and .20 to the left of 60, so .50 − .20 = .30.
(d) The 75th percentile; with .25 of the scores between 80 and 70, and .50 of the scores below 70, in total .75 or 75% of the scores are below 80.

23.

Score	*f*	*rel. f*	*cf*
16	5	.33	15
15	1	.07	10
14	0	.00	9
13	2	.13	9
12	3	.20	7
11	4	.27	4

Chapter 7

1. (a) A statistic describes a characteristic of a sample of scores. A parameter describes a characteristic of a population of scores.
(b) The symbols for statistics are letters from the English alphabet. The symbols for population parameters are letters from the Greek alphabet.

3. (a) The mode is the most frequently occurring score. It is used with nominal scores.
(b) The median is the score at the 50th percentile (50% of the scores are at or below the median). It is used with ordinal scores or with highly skewed interval or ratio scores.
(c) The mean is the average of the scores. It is used with symmetrical, unimodal distributions of interval or ratio scores.

5. The mean, because psychological variables usually form a normal distribution, and such distributions are best summarized by the mean.

7. (a) $\Sigma X = 638, N = 11, \overline{X} = 58$.
(b) The mode is 58.

9. (a) The low grade produced a negatively skewed distribution.
(b) Compute her median grade.

11. (a) 18 is the mode.
(b) This must be a positively skewed distribution because the mean is larger than the mode, so there are a few extreme high scores that pull the mean upwards.

13. (a) Mean.
(b) Median (these ratio scores are skewed).
(c) Mode (this is a nominal variable).
(d) Median (this is an ordinal variable).

15. (a) The subject scoring +5.
(b) The subject scoring −10.
(c) −10, +5, −2, +1.

17. μ is the symbol for the population mean. It is estimated from the mean of a random sample taken from the population.

19. Mean errors do not change until there has been 5 hours of sleep deprivation. Mean errors then increase as a function of increasing sleep deprivation.

21. (a) Income on *Y* axis, age on *X* axis; line graph; find median income per age group (income is skewed).
(b) Positive votes on *Y* axis, presence or absence of a wildlife refuge on *X* axis; bar graph; find mean number of votes per *X* group.
(c) Running speed on *Y* axis, amount of carbohydrates consumed on *X* axis; line graph; find mean running speed per amount if normally distributed.
(d) Alcohol abuse on *Y* axis, ethnic group on *X* axis; bar graph; find mean rate of alcohol abuse per group.

23. (a) The mean score on the test decreases as amount of sunlight increases.
(b) Raw scores tend to decrease as amount of sunlight increases.

(c) Assuming the data passes the inferential test, then the populations of scores and their μs tend to decrease as the conditions increase.
(d) These data provide evidence of a relationship in nature.

Chapter 8

1. Perform squaring and taking a square root first, then multiplication and division, then addition and subtraction.
3. (a) The larger the variability, the more the scores differ or are spread out.
(b) The smaller the variability, the more consistent their scores and thus their behavior.
5. (a) They both communicate how much the scores are spread out around the mean.
(b) The standard deviation, because it can be more directly interpreted as the "average" distance the scores are from the mean.
7. (a) All are forms of the standard deviation, communicating the "average" amount scores differ from the mean.
(b) S_X is a sample's standard deviation; s_X is an estimate of the population's standard deviation based on a sample, and σ_X is the population's true standard deviation.
9. This S_X^2 indicates no differences between the scores and the mean, so all subjects obtained the same score, equal to the mean.
11. (a) $\overline{X} = 5$ so the typical creativity score, μ, is expected to be 5.
(b) $s_X^2 = (668 - 500)/19 = 8.84$.
(c) $s_X = 2.97$.
(d) Between 2.03 (5.00 − 2.97) and 7.97 (5.00 + 2.97).
13. Error variance is the variance in each condition, reflecting the error when using the mean of a condition to predict scores in that condition.
15. (a) The sample tends to be normally distributed, so we'd expect the population to be normal.
(b) Because $\overline{X} = 1297/17 = 76.29$, we expect the typical score, μ, to be 76.29.
(c) $s_X^2 = (99{,}223 - 98{,}953.47)/16 = 16.85$.
(d) $s_X = 4.10$.
(e) Between 72.19 (76.29 − 4.10) and 80.39 (76.29 + 4.10).
17. (a) Pluto, because his S_X is smaller.
(b) Pluto, because his scores are closer to 60.
19. Predict the $\overline{X}$ of 65 for each student. Error is measured as variance, so $S_X^2 = 6^2 = 36$.
21. (a) Compute the mean and standard deviation of each condition.
(b) Experiment 1: As the conditions change, dependent scores tend to increase from around 11.75 to 33.5 to 4.5. Experiment 2: As the conditions change, dependent scores tend to decrease from 13.33 to 8.33 to 5.67.
(c) Experiment 1.
(d) In Experiment 1, for conditions 1, 2 and 3 the S_X = 1.48, 2.29, 2.06, respectively. In Experiment 2 for conditions 1, 2 and 3, the S_X = 3.68, 2.05, and 2.49, respectively.
(e) Experiment 2, because it produces larger values of S_X, which, when squared, indicate larger error variance.
(f) With more consistency and less variability, the independent variable in Experiment 1 apparently has a greater influence in determining scores. Greater variability in Experiment 2 suggests other variables were influencing scores.

Chapter 9

1. A z-score indicates the distance, measured in standard deviation units, that a score is above or below the mean.
3. A z-distribution is the distribution that results when a distribution of raw scores is transformed into z-scores.
5. Because z-scores standardize or equate different distributions so that they can be compared and graphed on the same set of axes.
7. (a) He should consider the size of each class's standard deviation.
(b) Small. A small S_X will give him a large positive z-score, placing him far above the mean.
(c) Large. With a large S_X, he will have a small negative z and still be close to the mean.
9. (a) It is the model of a perfect normal z-distribution.
(b) It is used as a model of any normal distribution of raw scores after they have been transformed to z-scores.
(c) The raw scores should be at least approximately normally distributed, the scores should be from a continuous interval or ratio variable, and the sample should be large.
11. $\Sigma X = 103$, $\Sigma X^2 = 931$, and $N = 12$, so $S_X = 1.98$ and $\overline{X} = 8.58$.
(a) For $X = 10$, $z = (10 - 8.58)/1.98 = +.72$.
(b) For $X = 6$, $z = (6 - 8.58)/1.98 = -1.30$.
13. (a) $z = +1.0$
(b) $z = -2.8$
(c) $z = -.70$
(d) $z = -2.0$

15. (a) .4706
(b) .0107
(c) .3944 + .4970 = .8914
(d) .0250 + .0250 = .05

17. From the z-table, the 25th percentile is at approximately $z = -.67$. The cutoff score is then $X = (-.67)(10) + 75 = 68.3$.

19. For City A, her salary has a z of $(27{,}000 - 50{,}000)/15{,}000 = -1.53$. For City B, her salary has a z of $(12{,}000 - 14{,}000)/1000 = -2.0$. City A is the better offer, because her income will be closer to the average cost of living in that city.

21. Convert $\overline{X}$ to a z-score. First, compute $\sigma_{\overline{X}} = 6/\sqrt{50} = .849$. Then $z = (18 - 19.4)/.849 = -1.65$. From the z-tables, .0495 of the curve is below this score. Out of 1,000 samples, you would expect $(.0495)(1000) = 49.5$ sample means to be below 18.

Chapter 10

1. (a) In experiments, the researcher manipulates one variable and measures participants' responses on another variable; in correlational studies, the researcher merely measures participants' responses on two variables.
(b) In experiments, the researcher computes the mean of the dependent scores (Y scores) for each condition of the independent variable (each X score); in correlational studies, the researcher examines the relationship over all X–Y pairs simultaneously by computing a correlation coefficient.

3. In correlational research, we don't necessarily know which variable occurred first, nor have we controlled or eliminated any confounding variables that might cause scores to change.

5. (a) A scatterplot is a graph of the individual data points formed from a set of X–Y pairs.
(b) A regression line is the summary straight line that best fits through the scatterplot.

7. (a) As the X scores increase, the Y scores tend to increase.
(b) As the X scores increase, the Y scores tend to decrease.
(c) As the X scores increase, the Y scores do not tend to only increase or only decrease.

9. (a) The scatterplot has a circular or horizontal elliptical shape.
(b) The variability in Y at each X is equal to the overall variability across all Y scores in the data.
(c) The Y scores are not at all relatively close to the regression line.
(d) Knowing X does not improve accuracy in predicting Y.

11. He is incorrectly drawing the inference that more people cause fewer bears. It may not be the number of people that affects the bear population so much as the number of hunters, or the amount of pesticides used, or the noise level associated with more people.

13. (a) Convergent validity is the extent to which a procedure correlates with other procedures that have accepted validity. Discriminant validity is the extent a procedure is not correlated with procedures that measure other unintended variables or constructs.
(b) Convergent and discriminant validity correlates the scores from two procedures or tests. Criterion validity correlates the scores from one procedure with an observable behavior.
(c) Concurrent validity is the extent a procedure correlates with a present behavior; predictive validity is the extent a procedure correlates with a future behavior.

15. (a) r.
(b) r_S.
(c) r_{pb}.
(d) r_s (after the liquid-consumed scores are transformed to rank-order scores).

17. (a) ρ stands for the Pearson correlation coefficient in the population.
(b) It is estimated from a sample's r.
(c) It indicates the strength and type of linear relationship found between all pairs of X and Y scores in the population.

19. Disagree. The problem is a restricted range, because exceptionally smart subjects produce a small range of IQ scores and grade averages. With an unrestricted range, there may be a much larger r.

21. First compute r_{pb}. For those with degrees, $\overline{Y}_2 = 8.6$; for those without degrees, $\overline{Y}_1 = 5.2$; $S_Y = 3.208$, $p = .50$, and $q = .50$. $r_{pb} = (1.06)(.50) = .53$. Looking at those without college degrees and then at those with degrees, this is a positive linear relationship with an intermediate degree of association.

23. To answer this question, compute r. $\Sigma X = 38$, $\Sigma X^2 = 212$, $(\Sigma X)^2 = 1444$, $\Sigma Y = 68$, $\Sigma Y^2 = 552$, $(\Sigma Y)^2 = 4624$, $\Sigma XY = 317$, and $N = 9$. $r = (2853 - 2584)/\sqrt{(464)(344)} = +.67$. This is a positive linear relationship of intermediate strength, so a nurse's "burnout" score will allow reasonably accurate prediction of the individual's absenteeism.

25. Compute r_s: $\Sigma D^2 = 312$; $r_s = 1 - (1872/990) = -.89$. There is a strong negative relationship between these variables, so that the most dominant tend to weigh the most, and less dominant weigh less.

Chapter 11

1. It is the line that summarizes a scatterplot by, on average, passing through the center of the Y scores at each X.
3. (a) Y' is the predicted Y score for a given X.
   (b) It is computed from the linear regression equation.
5. (a) The Y-intercept indicates the value of Y when the regression line crosses the Y axis.
   (b) The slope indicates the direction of the regression line and the degree to which it slants.
7. (a) It is the standard error of the estimate.
   (b) It is a standard deviation, indicating the "average" amount that the Y scores at each X deviate from their corresponding values of Y'.
   (c) It indicates the "average" amount that the actual scores differ from the Y' scores, so it is the "average" error.
9. (a) $S_{Y'}$ is inversely related to the absolute value of r.
   (b) $S_{Y'}$ is at its maximum (equal to S_Y) when $r = 0$, because with no relationship the amount that the Y scores deviate from the Y' scores equals the overall spread in the data.
   (c) $S_{Y'}$ is at its minimum (equal to 0) when $r = \pm 1.0$, because with a perfect relationship there are no differences between Y scores and the corresponding Y' scores.
11. (a) r^2 is called the coefficient of determination, or the proportion of variance in Y that is accounted for by the relationship with X.
   (b) r^2 can be interpreted as the proportional improvement when using the relationship with X to predict Y scores, compared to using the overall mean of Y to predict Y scores.
13. (a) Foofy. The positive r indicates that the higher the statistics grade, the higher the test score.
   (b) The relationship does not account for 83% of the variance, and $S_{Y'}$ is large, so predictions will not be very accurate, as with Bubbles and Foofy.
15. (a) He should use multiple correlation and multiple regression procedures, simultaneously considering both a subject's concentration and visualization abilities when predicting memory ability.
   (b) With a multiple R of +.67, R^2 is .45: He is 45% more accurate in predicting memory ability by considering both concentration and visualization abilities than if these predictors are not considered.
17. (a) The standard error of the estimate most directly communicates how much better (or worse) than predicted she is likely to perform.
   (b) Its predictive validity.
19. (a) Compute r: $\Sigma X = 45$, $\Sigma X^2 = 259$, $(\Sigma X)^2 = 2025$, $\Sigma Y = 89$, $\Sigma Y^2 = 887$, $(\Sigma Y)^2 = 7921$, $\Sigma XY = 460$, and $N = 10$, so $r = (4600 - 4005)/\sqrt{(565)(949)} = +.81$.
   (b) $b = (4600 - 4005)/565 = +1.05$ and $a = 8.9 - (1.05)(4.5) = 4.18$, so $Y' = (+1.05)X + 4.18$.
   (c) Using the completed regression equation, for participants with an attraction score of 9, the predicted anxiety score is $Y' = (+1.05)9 + 4.18 = 13.63$.
   (d) Compute $S_{Y'}$; $S_Y = 3.081$, so, $S_{Y'} = 3.081\sqrt{1 - .81^2} = 1.81$. Our "average error" is 1.81 when we use Y' to predict each anxiety score.
21. Square each coefficient: $.20^2 = .04$. Knowing students' class rankings or knowing a student's gender each results in 4% more accuracy in predicting studiousness scores.

Chapter 12

1. (a) It is the expected relative frequency of the event.
   (b) It is based on the event's relative frequency in the population.
3. (a) $p = 1/6 = .167$.
   (b) $p = 13/52 = .25$.
   (c) $p = 1/4 = .25$.
   (d) $p = 0$: After selecting the ace the first time, it would no longer be in the deck to select again.
5. No. The sex of a child is an independent event, so previous children do not influence the probability that a child will be a boy or girl.
7. The p of a hurricane here is $160/200 = .80$. The uncle is looking at an unrepresentative sample, consisting of the past 13 years. Poindexter uses the gambler's fallacy, failing to realize that this p is based on the long run, and in the next few years there may not be a hurricane.
9. (a) Dependent: You're less likely to golf in rain, snow, and so on.
   (b) The answer depends on the amount of money you have. If you're poor, they're probably dependent—after buying a car, you're less able to buy shoes. If you're rich, the events are independent.
   (c) Dependent: Weight loss depends on calories consumed.
   (d) Independent: Your chances of winning are the same, whether you use the same or different numbers.
11. (a) $z = (27 - 43)/8 = -2.0$; $p = .0228$.
   (b) $z = (51 - 43)/8 = +1.0$; $p = .1587$.

(c) $z = (42 - 43)/8 = -.13$; $z = (44 - 43)/8 = +.13$; $p = .0517 + .0517 = .1034$.
(d) $z = (33 - 43)/8 = -1.25$; $z = (49 - 43)/8 = +.75$; $p = .1056 + .2266 = .3322$.

13. Transform 24 to z: $\sigma_{\bar{X}} = 12/\sqrt{30} = 2.19$; $z = (24 - 18)/2.19 = +2.74$; $p = .0031$.
15. Either the sample is unrepresentative of the population because of the luck of who was selected for the sample, or the sample represents some other population.
17. No. With a $z = +2.74$, this mean falls beyond the critical value of $+1.96$. It is too unlikely to be accepted as representing this population.
19. (a) The $\bar{X} = 321/9 = 35.67$; $\sigma_{\bar{X}} = 5/\sqrt{9} = 1.67$. Then $z = (35.67 - 30)/1.67 = +3.40$. With a critical value of ± 1.96, conclude that the football players do not represent this population.
(b) Football players, as represented by your sample, form a population different from non-football players, having a μ of about 35.67.
21. (a) For Fred's sample we'd estimate $\mu = 26$, and for Ethel's, $\mu = 18$.
(b) The population with $\mu = 26$ is most likely to produce a sample with $\bar{X} = 26$, and the population with $\mu = 18$ is most likely to produce a sample with $\bar{X} = 18$.

Chapter 13

1. (a) Sampling error occurs by luck in drawing a sample so that a sample statistic is different from the population parameter it represents.
(b) A sample may poorly represent one population because of sampling error, or it may represent some other population.
3. (a) In a real relationship, there is something in nature that ties different X scores to different Y scores to form a relationship. Sampling error can produce scores that by chance pair up to produce the appearance of a relationship.
(b) No, only that there is a relationship even if we have not correctly identified the variables.
5. (a) They are more powerful than nonparametric procedures.
(b) Parametric procedures are robust, which means that violating their assumptions somewhat does not result in a large error in accurately determining the probability of a Type I error.
7. A one-tailed test is used when you predict the direction in which the scores will change. A two-tailed test is used when you predict a relationship but do not predict the direction in which scores will change.
9. (a) Changing the independent variable from a week other than finals week to finals week increases the dependent variable of amount of pizza consumed; the experiment will not demonstrate an increase.
(b) Changing the independent variable from not performing breathing exercises to performing them changes the dependent variable of blood pressure; the experiment will not demonstrate a change.
(c) Changing the independent variable by increasing hormone levels changes the dependent variable of pain sensitivity; the experiment will not demonstrate a change.
(d) Changing the independent variable by increasing amount of light will decrease the dependent variable of frequency of dreams; the experiment will not demonstrate a decrease.
11. (a) α stands for the criterion probability; it determines the size of the region of rejection and is the theoretical probability of a Type I error.
(b) By defining the size of the region of rejection, the smaller the α the larger the absolute value of z_{crit}, and so the larger z_{obt} must be to be significant.
13. (a) A one-tailed test: *beneficial* implies only higher scores.
(b) H_0: $\mu \leq 50$, H_a: $\mu > 50$.
(c) $\sigma_{\bar{X}} = 12/\sqrt{49} = 1.71$; $z_{obt} = (54.63 - 50)/1.71 = +2.71$.
(d) $z_{crit} = +1.645$.
(e) Because z_{obt} is beyond z_{crit}, his results are significant: He has evidence of a relationship in the population, so changing from the condition of no music to the condition of music results in test scores' changing from a μ of 50 to a μ of around 54.63.
15. (a) The probability of a Type I error is $p < .05$. The error would be concluding that music influences scores, when really it does not. (b) The probability of a Type II error is 0 (he rejected H_0). It would be concluding that music does not influence scores, when really it does.
17. Because confirming the hypothesis rests on not rejecting H_0, in which case the researcher is still confronted by H_0 and H_a—that there may or may not be a difference.
19. (a) Power is the probability of rejecting a false H_0 (the probability of not making a Type II error).
(b) So that if they fail to reject H_0, they are confident that they would have rejected it if it were false.
(c) In a one-tailed test the critical value is smaller than in a two-tailed test; so the obtained value is

more likely to be larger than the critical value, and thus is more likely to be significant.

21. She is correct about it being easier to reject H_0, but she is incorrect about Type I errors. For a given α, the total size of the region of rejection is the same regardless of whether a one- or a two-tailed test is used. Because α is the probability of making a Type I error, this error is equally likely using either type of test.

23. (a) She is correct; that is what $p < .0001$ indicates.
(b) She is incorrect. In both studies, the researchers decided the results were too unlikely to reflect sampling error from the H_0 population; they merely defined *too unlikely* differently.
(c) The probability of a Type I error is less in Study B.

Chapter 14

1. (a) The *t*-test and the *z*-test.
(b) Compute *z* if the true standard deviation of the raw score population (σ_X) is known; compute *t* if σ_X must be estimated by s_X.

3. (a) $s_{\overline{X}}$ is the estimated standard error of the mean; $\sigma_{\overline{X}}$ is the true standard error of the mean.
(b) Both are used as a standard deviation to locate a sample mean on the sampling distribution of means.

5. The *df* of 80 are .33 of the distance between the *df* at 60 and 120, so the target t_{crit} is .33 of the distance from 2.000 to 1.980: $2.000 - 1.980 = .020$, so $(.020)(.33) = .0066$, and thus $2.000 - .0066$ equals the target t_{crit} of 1.993.

7. (a) Power is the probability of rejecting H_0 when it really is false (the probability of not making a Type II error).
(b) Because we don't know if we were likely to reject H_0, even if it really was false.
(c) They initially design the study to maximize power.

9. (a) H_0: $\mu = 68.5$; H_a: $\mu \neq 68.5$
(b) $s_X^2 = 130.5$; $s_{\overline{X}} = \sqrt{130.5/10} = 3.61$; $t_{obt} = (78.5 - 68.5)/3.61 = +2.77$
(c) With $df = 9$, $t_{crit} = \pm 2.262$.
(d) Using this book rather than other books produces a significant improvement in exam scores: $t_{obt}(9) = 2.77, p < .05$.
(e) $(3.61)(-2.262) + 78.5 \leq \mu \leq (3.61)(+2.262) + 78.5 = 70.33 \leq \mu \leq 86.67$

11. (a) H_0: $\mu = 50$; H_a: $\mu \neq 50$.
(b) $t_{obt} = (53.25 - 50)/8.44 = +.39$.
(c) For $df = 7$, $t_{crit} = \pm 2.365$.
(d) $t(7) = +.39, p > .05$.
(e) The results are not significant, so do not compute the confidence interval.
(f) She has no evidence that strong arguments change people's attitudes toward this issue.

13. Disagree. Everything Poindexter said was meaningless, because he failed to first perform significance testing to eliminate the possibility that his correlation was merely a fluke resulting from sampling error.

15. (a) H_0: $\rho = 0$; H_a: $\rho \neq 0$.
(b) With $df = 70$, $r_{crit} = \pm .232$.
(c) $r(70) = +.38, p < .05$.
(d) The correlation is significant, so he should conclude that the relationship exists in the population, and he should estimate that ρ is approximately +.38.
(e) The regression equation and r^2 should be computed.

17. (a) r_{pb}
(b) H_0: $\rho_{pb} = 0$; H_a: $\rho_{pb} \neq 0$.
(c) For $df = 40$, $r_{crit} = \pm .304$.
(d) The r_{pb} is significant, so she expects the ρ_{pb} to be approximately .33. She should expect it to be a positive relationship only if left-handers are assigned a lower score than right-handers on the variable of handedness.
(e) $(r_{pb})^2 = .11$. The relationship accounts for only 11% of the variance in personality scores, so the results are not very useful.

19. (a) Math majors are likely to produce a restricted range of scores—a ceiling effect—when measured on mathematical ability, resulting in a small *r*.
(b) With only three puns, he may have an insensitive procedure that produces a restricted range—perhaps a floor effect if no one thinks any of the puns are funny.
(c) His *N* should be considerably larger.

21. Compute the appropriate correlation coefficient, determine whether it is significant, and if it is, perform linear regression, determine the proportion of variance accounted for by the relationship, estimate the value of ρ, and interpret the relationship in psychological terms.

Chapter 15

1. (a) The independent samples *t*-test and the dependent samples *t*-test.
(b) Whether the design involves independent samples or not.

3. By identifying subject variables that are strongly correlated with the independent or dependent variable.
5. (a) It improves internal validity by eliminating a potential confounding. It improves external validity by including a wider variety of participants.
   (b) It results in fluctuating variables within a condition that can increase error variance and decrease the strength of the relationship and power.
7. A pretest may alert subjects to the variables being studied and communicate demand characteristics.
9. Collapsing across a variable means to combine scores from the different amounts or categories of that variable.
11. She should graph the results, compute the appropriate confidence interval, and compute the effect size.
13. (a) H_0: $\mu_1 - \mu_2 = 0$; $H_a = \mu_1 - \mu_2 \neq 0$.
   (b) $s^2_{pool} = 23.695$; $s_{\bar{X}_1-\bar{X}_2} = 1.78$; $t_{obt} = (43 - 39)/1.78 = +2.25$.
   (c) With $df = (15 - 1) + (15 - 1) = 28$, $t_{crit} = \pm 2.048$.
   (d) The results are significant: in the population, hot baths (with μ about 43) produce different relaxation scores than cold baths (with μ about 39).
   (e) $(1.78)(-2.048) + 4 \leq \mu_1 - \mu_2 \leq (1.78)(+2.048) + 4 = .35 \leq \mu_1 - \mu_2 \leq 7.65$
   (f) $r^2_{pb} = (2.25)^2/[(2.25)^2 + 28] = .15$, so bath temperatures do not have a very large effect.
   (g) Label the X axis bath temperature; label the Y axis mean relaxation score; plot the data point for cold baths at a Y of 39 and for hot baths at a Y of 43; connect the data points with a straight line.
15. (a) She should retain H_0, because in her one-tailed test the signs of t_{obt} and t_{crit} are different.
   (b) She probably did not subtract the sample means in the same way that she subtracted the μs in her hypotheses.
17. (a) With $df = 10$, $t_{crit} = \pm 2.228$, so $t(10) = +1.38$, $p > .05$.
   (b) The results are not significant. There is no evidence that changing the type of background music affects the irritability of air-traffic controllers.
   (c) No other statistics need to be computed.
   (d) The experiment is likely to have insufficient power.
   (e) He could use larger samples, design and give the test so as to reduce variability in each group, and select and define the type of music so as to obtain larger differences between the means.
   (f) It would increase the likelihood of rejecting H_0 if the relationship exists.

Chapter 16

1. (a) The independent samples t-test and the dependent samples t-test.
   (b) Whether the scientist employed a between-subjects or a within-subjects design, respectively.
3. The within-subjects design, because the dependent samples t-test is intrinsically more powerful.
5. (a) They reduce external validity when the participants in the samples are unrepresentative of the population.
   (b) They reduce internal validity when participants differ between conditions so that a confounding exists.
7. They reduce internal validity because differences in responses between conditions may result from these factors instead of the conditions of the independent variable. They reduce external validity because the behaviors and participants do not represent those found in other situations where these factors are different or not present.
9. Scores may be influenced by participants' history and maturation, by demand characteristics resulting from experiencing all conditions, by subject mortality, and by order effects between conditions.
11. (a) They improve internal validity by eliminating a potential confounding from one order. They improve external validity by including many orders, as might be found in other situations.
   (b) They include the influence of another fluctuating variable within a condition that can increase error variance and decrease power.
13. She should graph the results, compute the appropriate confidence interval, and compute the effect size.
15. (a) It is impossible to test people when they are male and again when they are female.
   (b) Match each male with a corresponding female on relevant subject variables.
17. (a) Over 6 weeks of testing, a third of the participants will hear the messages in order 1, 2, 3; a third in order 2, 3, 1; and a third in order 3, 1, 2.
   (b) Mortality, maturation, and history effects.
   (c) Carryover effects
   (d) Use a between-subjects design.
19. (a) H_0: $\mu_D \leq 0$; H_a: $\mu_D > 0$.
   (b) $t_{obt} = (2.63 - 0)/.75 = +3.51$.
   (c) With $df = 7$, $t_{crit} = +1.895$, so $t(7) = +3.51$, $p < .05$
   (d) $.86 \leq \mu_D \leq 4.40$.
   (e) The $\bar{X}$ of 15.5; the $\bar{X}$ of 18.13.
   (f) $r^2_{pb} = (3.51)^2/[(3.51)^2 + 7] = .64$; they are on average about 64% more accurate.

(g) Participants exposed to high amounts of sunshine exhibit a significantly higher well-being score than do participants exposed to low amounts, with the μ of the difference scores between .86 and 4.40.

21. (a) H_0: $\mu_D = 0$; H_a: $\mu_D \neq 0$.
(b) $t_{obt} = (3.0 - 0)/1.592 = +1.88$.
(c) With $df = 7$, $t_{crit} = \pm 2.365$, $p < .05$.
(d The results are not significant. We have no evidence of a relationship.
(e) We cannot determine effect size with a nonsignificant result, because we do not know if a relationship exists or not.

Chapter 17

1. (a) Analysis of variance.
(b) A study that contains one independent variable.
(c) An independent variable.
(d) A condition of the independent variable.
(e) All samples are independent.
(f) All samples are related, because either a repeated measures or a matched samples design is used.

3. To test the hypothesis adequately, to demonstrate a nonlinear relationship, and to obtain the maximum information from a study.

5. (a) It is the overall probability of making a Type I error after comparing all possible pairs of means in an experiment.
(b) Multiple *t*-tests will result in an experiment-wise error rate larger than *alpha*. After performing ANOVA and then *post hoc* tests, the experiment-wise error rate will equal *alpha*.

7. (a) H_0: $\mu_1 = \mu_2 = \mu_3 = \mu_4$
(b) H_a: not all μs are equal.
(c) H_0 maintains that all μs represented by the levels are the same; H_a maintains that not all μs represented by the levels are the same.

9. (a) The MS_{bn} is less than the MS_{wn}; either term is a poor estimate of σ^2_{error} and H_0 is assumed to be true.
(b) He made a computational error because F_{obt} cannot be a negative number.

11. (a) It is necessary when F_{obt} is significant and k is greater than 2. The F_{obt} indicates only that two or more sample means differ significantly; therefore, *post hoc* tests are used to determine which specific levels produced significant differences.
(b) It is not necessary when F_{obt} is not significant, because we are not convinced there are any differences to be found. It's also not necessary when $k = 2$, because there is only one possible difference between means in the study.

13. It describes the effect size in the sample—the proportion of variance in dependent scores accounted for by changing the levels of the independent variable.

15. (a)

Source	*Sum of squares*	*df*	*Mean square*	*F*
Between	134.80	3	44.93	17.08
Within	42.00	16	2.63	
Total	176.80	19		

(b) With $df = 3$ and 16, $F_{crit} = 3.24$, so F_{obt} is significant, $p < .05$.
(c) For $k = 4$ and $df_{wn} = 16$, $q_k = 4.05$, so $HSD = (4.05)(\sqrt{2.63/5}) = 2.94$; $\overline{X}_4 = 4.4$, $\overline{X}_6 = 10.8$, $\overline{X}_8 = 9.40$, $\overline{X}_{10} = 5.8$.
(d) This is an inverted U-shaped function, in which only ages 4 and 10 and ages 6 and 8 do not differ significantly.
(e) Because $\eta^2 = 134.8/176.8 = .76$; this relationship accounts for 76% of the variance, so it is a very important relationship.
(f) Label the *X* axis with the independent variable of age and the *Y* axis with the mean creativity score. Plot the mean score for each condition, and connect adjacent data points with straight lines.

17. (a) H_0: $\mu_1 = \mu_2 = \mu_3 = \mu_4$; H_a: not all μs are equal.
(b)

Source	*Sum of Squares*	*df*	*Mean Square*	*F*
Between Groups	47.69	3	15.90	9.19
Within Groups	20.75	12	1.73	
Total	68.44	15		

(c) for $df = 3, 12$, $F_{crit} = 3.49$.
(d) $F(3,12) = 9.19$, $p < .05$.
(e) $\overline{X}_1 = 2.0$, $\overline{X}_2 = 3.0$, $\overline{X}_3 = 5.75$, $\overline{X}_4 = 6.0$. $q_k = 4.20$, $HSD = (4.20)(\sqrt{1.73/4}) = 2.76$, significant differences are between negligible and moderate, negligible and severe, and minimal and severe.
(f) Increasing stress levels significantly increases infection rate, although only differences in nonadjacent stress conditions produced significant differences in number of infections.

(g) $\eta^2 = 47.69/68.44 = .70$; changing stress levels accounts for 70% of the variance in infection scores in the sample data.
(h) $(\sqrt{1.73/4})(-2.179) + 6.0 \leq \mu \leq (\sqrt{1.73/4})(+2.179) + 6.0 = 4.57 \leq \mu \leq 7.43$.

Chapter 18

1. (a) The study contains two independent variables.
(b) All levels of one factor are combined with all levels of the other factor.
(c) The combination of a level of factor A with a level of factor B.
(d) A two-factor design where all cells contain independent samples.
3. Incorrect: Two factors were combined, one with 3 levels and one with 2 levels, so that 6 cells were created.
5. (a) A main effect mean is based on scores in a level of one factor while collapsing across the other factor. A cell mean is the mean of scores from a particular combination of a level of A with a level of B without collapsing.
(b) That changing the levels of the factor produced a significant difference somewhere among the level means.
(c) That changing the levels of both factors produced a significant difference somewhere among the cell means such that an interaction is formed.
(d) Because the interaction indicates that the effect of either factor depends on the level of the other factor that is present, so overall conclusions about a main effect of A or B are inaccurate.
7. (a) H_0 is that the μs represented by the level means from factor A are all equal; H_a is that not all μs are equal.
(b) H_0 is that the μs represented by the level means from factor B are equal; H_a is that they are not all equal.
(c) H_0 is that the μs represented by the cell means do not form an interaction, H_a is that they do form an interaction.
9. Study 1: For A, means are 7 and 9; for B, means are 3 and 13. Apparently there are effects for A and B but not for A × B. Study 2: For A, means are 7.5 and 7.5; for B, means are 7.5 and 7.5. There is no effect for A or B, but there is an effect for A × B. Study 3: For A, means are 8 and 8; for B, means are 11 and 5. There is no effect for A, but there are effects for B and A × B.
11. Perform Tukey's *post hoc* comparisons on each main effect and the interaction, graph each main effect and interaction and compute its η^2; where appropriate, compute confidence intervals for the μ represented by a cell or level mean.
13. (a) Maximize the differences between the levels of the main effects or the cells of the interaction, minimize the variability of scores in each cell, and maximize n.
(b) The power of the *post hoc* comparisons.
15. Only the main effect for difficulty level is significant.
17. (a) A confounded comparison is when two cells differ along more than one factor. It occurs when comparing cell means that are diagonally positioned in a design's diagram.
(b) An unconfounded comparison is when two cells differ along only one factor. It occurs when comparing cell means within the same column or row.
(c) Because we cannot determine which of the factors produced the significant difference, if there is one.
19. (a) Because as the amount of reward increases, performance does not first increase and then decrease under every level of practice as indicated by the main effect means.
(b) Because increasing the amount of practice does not increase performance under every level of reward as indicated by the main effect means.
21. (a)

Source	*Sum of Squares*	*df*	*Mean Square*	*F*
Between groups				
Factor A	7.20	1	7.20	1.19
Factor B	115.20	1	115.20	19.04
Interaction	105.80	1	105.80	17.49
Within groups	96.80	16	6.05	
Total	325.00			

For each factor, $df = 1$ and 16, so $F_{crit} = 4.49$: factor B and the interaction are significant, $p < .05$.
(b) For factor A, $\overline{X}_1 = 8.9$, $\overline{X}_2 = 10.1$; for factor B, $\overline{X}_1 = 11.9$, $\overline{X}_2 = 7.1$; for the interaction, $\overline{X}_{A_1B_1} = 9.0$, $\overline{X}_{A_1B_2} = 8.8$, $\overline{X}_{A_2B_1} = 14.8$, $\overline{X}_{A_2B_2} = 5.4$.
(c) Because factor A is not significant and factor B contains only two levels, such tests are unnecessary for the main effects. For A × B, adjusted

APPENDIX D

$k = 3$, so $q_k = 3.65$, $HSD = (3.65)(\sqrt{6.05/5}) = 4.02$: the only significant differences are between males and females tested early, and between females tested early and females tested late.

(d) We can conclude only that a relationship exists between gender and test scores when testing is done early in the day, and that early and late testing produce a relationship with test scores for females, $p < .05$.

(e) For B, $\eta^2 = 115.2/325 = .35$; for A × B, $\eta^2 = 105.8/325 = .33$.

23. (a)

Source	Sum of Squares	df	Mean Square	F
Between				
Factor A	8.42	1	8.42	2.43
Factor B	76.79	3	25.60	7.40
Interaction	23.71	3	7.90	2.28
Within	110.72	32	3.46	
Total	219.64	39		

(b) For 1 and 32 *df*, $F_{crit} = 4.15$; for 3 and 32 *df*, $F_{crit} = 2.90$. Only the F_{obt} for factor B is significant.

(c) For factor B, $n = 10$, $df_{wn} = 32$, $q_k = 3.83$, and $HSD = 2.25$.

(d) Except for between B_1 and B_2, changing each level of B results in a significant increase in scores.

(e) For audience size, $\eta^2 = .35$, so it is a reasonably important variable in determining subjects' anxiety. Because the self-confidence factor was not significant, we are not sure if it even relates to anxiety scores, so we do not determine η^2.

Chapter 19

1. (a) She may be able to use either a *t*-test or a one-way ANOVA.

(b) She can use a two-way between-subjects ANOVA; a two-way within-subjects ANOVA; or a two-way mixed design ANOVA.

(c) The procedure to use depends on whether: both factors are tested using independent samples; both factors are tested using dependent samples (usually with repeated measures); or one factor involves independent samples and one involves dependent samples.

3. (a) That the same participant's weight gain was repeatedly measured over time.

(b) That as weight increased, their mood decreased.

5. (a) H_0: $\mu_1 = \mu_2 = \mu_3$; H_a: not all μs are equal.

(b) $SS_{tot} = 477 - 392.04$; $SS_A = 445.125 - 392.04$; and $SS_{subs} = 1205/3 - 392.04$.

Source	Sum of Squares	df	Mean Square	F
Factor A	53.08	2	26.54	16.69
Subjects	9.63	7		
A × Subjects	22.25	14	1.59	
Total	84.96	23		

(c) With $df_A = 2$ and $df_{A \times subs} = 14$, the F_{crit} is 3.74. The F_{obt} is significant.

(d) The $q_k = 3.70$ and $HSD = 1.65$. The means for zero, one, and two hours are 2.13, 4.25, and 5.75, respectively. Significant differences occurred between zero and one hour and between zero and two hours, but not between one and two hours.

(e) Eta squared (η^2) = 53.08/84.96 = .62.

(f) The variable of amount of practice is important in determining performance scores, but although 1 or 2 hours of practice significantly improved performance compared to no practice, 2 hours was not significantly better than 1 hour.

7. (a) Use F_{max} to determine whether to assume homogeneity of variance when performing the *t*-test or ANOVA.

(b) $F_{max} = 43.68/9.50 = 4.598$. With $k = 3$ and $n - 1 = 15$, the critical value is 3.54. These variances differ significantly, so do not assume homogeneity.

(c) Technically, do not perform ANOVA. Instead perform a nonparametric procedure.

9. (a) Two-way, mixed design.

(b) One *F* for Factor A indicating whether, overall, fantasy level produced a significant difference; one *F* for Factor B indicating whether, overall, meditation level produced a significant difference; and one *F* for the interaction indicating whether, the influence of meditation level depends on fantasy level (or vice versa).

(c) the means are

	low	*med*	*high*	
Meditation	4.0	6.4	9.4	6.6
No Meditation	9.8	3.2	5.8	6.3
	6.9	4.8	7.6	

The main effect of fantasy level and the interaction are probably significant but the main effect of meditation is probably not.

(d) Based on the interaction, the relationship between increasing fantasy levels and hypnotic suggestibility is essentially a positive linear relationship for subjects who meditate, but it is a nonlinear, U-shaped relationship for those who do not meditate.

11. (a) It is the effect of one factor at one level of a second factor within the interaction.

(b) There appears to be a positive linear relationship where increasing fantasy levels produce increased suggestibility.

13. (a) The main effect means for boys = 15 and girls = 10; for candy = 11.5 and money = 13.5; and for morning = 12.5 and afternoon = 12.5. The means for the two-way interactions are:

	Boys	*Girls*
Morning	$\overline{X} = 15$	$\overline{X} = 10$
Afternoon	$\overline{X} = 15$	$\overline{X} = 10$

	Candy	*Money*
Morning	$\overline{X} = 9$	$\overline{X} = 16$
Afternoon	$\overline{X} = 14$	$\overline{X} = 11$

	Boys	*Girls*
candy	$\overline{X} = 15$	$\overline{X} = 8$
money	$\overline{X} = 15$	$\overline{X} = 12$

The three-way interaction is given in the problem.

(b) Only in the three-way interaction are there significant differences.

(c) Using money as a reward significantly improves performance for boys only in the morning but decreases their performance in the afternoon.

15. (a) A two-way mixed design.

(b) F_B will indicate the overall difference between depression for the four doses, F_A will indicate the overall difference between scores for positive-negative status, and $F_{A \times B}$ will indicate whether the difference between scores for HIV+ and HIV− depends on drug dose (and vice versa).

(c) For the four doses, the means are 3.5, 5.5, 12.5 and 18. For HIV−, the mean is 9.75, and for HIV+, the mean is 10.

(d) Probably only the main effect for dose and the interaction are significant.

(e) From the interaction, only the low and high drug doses produced an increase in mood scores of HIV+ participants relative to those who are HIV−.

Chapter 20

1. (a) In a true experiment, participants are randomly assigned to conditions; in a quasi-experiment, participants are assigned to a condition because of an inherent characteristic or past experience.

(b) In both, participants' score on the X variable is determined by an inherent characteristic or past experience.

(c) Quasi-experiments tend to be more controlled, and only some levels of the X variable are investigated.

3. (a) For each condition, select participants who demonstrate the desired level of the quasi-independent variable.

(b) To select participants for a condition who are very similar on the independent variable but very different from those in other conditions.

(c) It may alert participants to the variables and hypotheses, producing demand characteristics.

5. (a) The design where a group is measured once before and once after a treatment.

(b) A control group.

(c) Participants' history and maturation, subject mortality, confounding environmental effects, experimenter effects, or regression toward the mean.

7. (a) The design where the experimental group is measured at several spaced times prior to the treatment and at several times after it.
   (b) Confounding variables that continuously fluctuate and thus would cause scores to change prior to or after the treatment.
   (c) Confounding variables that change only with the onset of the treatment.
9. (a) The design where one group of participants is repeatedly measured to observe the effect of the passage of time.
   (b) Because it is a repeated measures design, it eliminates confounding due to subject variables by keeping them constant between conditions.
   (c) Any changing environmental variables may confound the independent variable.
11. (a) Cohort effects are confoundings that occur when differences in age are confounded by differences in subject history.
   (b) The design consisting of a longitudinal study of several groups, each from a different generation.
   (c) It allows us to determine whether cohort effects are present.
   (d) They are present if there are significant differences between the generations tested, or if there is a significant interaction between age and generation.
13. (a) A design that alternates testing a subject between the baseline condition and the experimental condition.
   (b) It is unlikely that a confounding variable would repeatedly and coincidentally change at the same time that we change the conditions.
   (c) Carry-over effects that cannot be reversed.
15. (a) A design that measures a baseline for several behaviors from a participant but applies the treatment to each behavior at a different point in time.
   (b) A design that measures baselines for one behavior on the same participant but in different situations, and then institutes the treatment in each situation at different points in time.
17. (a) It keeps subject variables constant, it shows the effect size of the variable, and it empirically demonstrates the reliability of the effect.
   (b) It has the problems of any repeated measures design, and external validity is limited.
19. (a) Longitudinal; repeated measures.
   (b) Cohort; between subjects.
   (c) Cohort design.
21. (a) The before-after difference of 30 occurred for both the control and experimental groups, so there is little evidence for any influence of the earthquake.
   (b) Some experience common to both groups increased stress scores.
   (c) The before-after difference for the control group is 10, but for the experimental group the difference is 35, so there is evidence for influence by the earthquake.
   (d) No confounding variable common to both groups could be responsible for the experimental group's data.
   (e) A confounding variable unique to the experimental group could have produced these data.

Chapter 21

1. Whether the study is an experiment or descriptive study, whether the data meet the requirements of parametric or nonparametric procedures, and whether the design is between-subjects or within-subjects.
3. (a) When the data consist of the frequency that participants fall in each category of one or more variables.
   (b) When categorizing participants along only one variable.
   (c) When simultaneously categorizing participants along two variables.
5. (a) That the sample frequencies are unlikely to represent the distribution of frequencies in the population described by H_0.
   (b) That category membership on one variable depends on, or is correlated with, category membership on the other variable.
7. (a) H_0: the frequencies of females preferring slightly taller or much taller men are equal in the population; H_a: the frequencies of females preferring slightly taller or much taller men are not equal in the population.
   (b) With $N = 89$, $f_e = 89/2 = 44.5$ for each group. The $\chi^2 = 2.48 + 2.48 = 4.96$. With $df = 1$, $\chi^2_{crit} = 3.84$, so the results are significant. Conclude in the population of females, about 55/89, or 62%, prefer slightly taller males, and about 38% prefer much taller males, $p < .05$.
   (c) Label the Y axis "frequency" and the X axis with each type of female. For each, draw a bar graph to the height of their frequencies.
9. (a) It is a correlation coefficient used to describe the strength of the relationship in a significant two-way χ^2 that involves a 2×2 design.

(b) Phi squared is the proportion of variance accounted for by the relationship between the two variables.
(c) It is a correlation coefficient used to describe the strength of the relationship in a significant two-way χ^2 that does not involve a 2×2 design.
(d) The squared contingency coefficient indicates the proportion of variance accounted for by the relationship between the two variables.

11. (a) H_0: in the population, the frequency with which people did or did not vote is independent of the frequency of whether they were satisfied with the outcome and vice versa. H_a: the frequencies are dependent.
(b) For voters, satisfied $f_e = 40.02$, dissatisfied $f_e = 42.98$; for nonvoters, satisfied $f_e = 40.98$, dissatisfied $f_e = 44.02$.
(c) $\chi^2_{obt} = 6.07$.
(d) $\chi^2_{crit} = 3.84$. The results are significant; therefore, in the population, whether people are satisfied with the election results depends on whether they voted, $p < .05$.
(e) The phi coefficient = .19, so it is not a very consistent relationship.

13. (a) The Wilcoxon T test
(b) The Kruskal-Wallis H test
(c) The Friedman χ^2 test
(d) The Mann-Whitney U test or the rank sums test, depending on the size of n.

15. Sum of positive ranks = 1, sum of negative ranks = 27, $T_{obt} = 1$. $N = 7$, so $T_{crit} = 2$. The two samples of ranks, and the underlying interval scores, differ significantly, $p < .05$.

17. (a) The Friedman χ^2 test. Skewed data violate the assumptions for a parametric procedure, and there are three levels of one factor in a repeated-measures design.
(b) She must assign ranks to the three scores for each participant.
(c) She should perform Nemenyi's procedure.
(d) She will be able to determine which amounts of time spent in the program result in significant differences in progress scores.

19. The variable was measured using a nominal (categorical) variable.

21. (a) One factor is the pictured facial expression having three categories. Categorizing each picture as happy or sad produces a two-way chi square design.
(b) Measuring perceived happiness as either happy or sad is not a very sensitive or powerful dependent variable, nor is the nonparametric chi square powerful.
(c) Change the dependent variable to an interval or ratio scale that is more sensitive and that meets the requirements of a parametric procedure.

Appendix A

1. To learn how constructs and behaviors are conceptualized, and how previous studies have been conducted.

3. By reading the references cited in the article and then the references cited in those articles.

5. Search previous research articles or search in reference manuals that describe tests.

7. A brief overview of the hypothesis, methods, results, and conclusions of the study.

9. (a) It is the first two or three words of the article's title.
(b) It is a different, brief title typed at the bottom of the manuscript's title page that will appear at the top of each page of the published article.

11. (a) The Participants, Material/Apparatus, and Procedure sections.
(b) In Participants describe the number and type of participants tested, in Material/Apparatus describe the stimuli and equipment used in testing, and in Procedure describe the steps involved in testing.
(c) The results and conclusions of the study.

13. (a) When the reader should see the overall pattern in the obtained relationship.
(b) When the reader should see the precise values of means, etc., and there are too many to include in the narrative.

15. (a) The sentence begins with the number 40; it is written in future tense. (And "hear" is inappropriate.)
(b) This teaches statistics; a result is never "very" significant.
(c) So many numbers should be presented in a table or figure; the words "men" and "girls" imply different ages and may be construed as sexist.
(d) "Split in half" cannot be what the author truly means, implying that participants were sliced up; the number of participants should be in digits, not words.
(e) This teaches statistics; conclusions are not reported in the Results, nor are statistics given in the Discussion.
(f) The name of each variable will not be understood; the order of the variables reverses the

independent and dependent variables, "funny" is inappropriate.

17. (a) Perhaps lectures and texts should be humorous because then the information will be more distinctive and memorable.
    (b) The nature of the stimuli and setting in this study is very different from that of lectures and textbooks, so external validity may be poor.
    (c) By a search directed at the effects of humor on memory for complex material presented in lectures and texts.

Appendix B

Part B.1: Computing Percentiles

1.

Score	*f*	*rel. f*	*cf*
53	1	.05	18
52	3	.17	17
51	2	.11	14
50	5	.28	12
49	4	.22	7
48	0	.00	3
47	3	.17	3

(a) The score of 51 is at the 72nd percentile.
(b) The score at the 50th percentile is:

$$\text{Score} = 49.5 + \left(\frac{9-7}{5}\right)(1) = 49.90$$

3.

Score	*f*	*rel. f*	*cf*
128 − 131	1	.04	28
124 − 127	2	.07	27
120 − 123	6	.21	25
116 − 119	8	.29	19
112 − 115	4	.14	11
108 − 111	3	.11	7
104 − 107	2	.07	4
100 − 103	2	.07	2

(a) $$\text{percentile} = \left(\frac{11 + \frac{117 - 115.5}{4}(8)}{28}\right)(100) =$$

$$\left(\frac{11 + .375(8)}{28}\right)(100) = \left(\frac{11+3}{28}\right)(100) =$$

50th percentile

(b) $$\text{Score} = 119.5 + \left(\frac{22.4 - 19}{6}\right)(4) = 119.5 + .567(4)$$

$$\text{Score} = 119.5 + 2.268 = 121.768 = 121.77$$

Part B.3: Performing Linear Interpolation

1. The target z-score is between $z = .670$ at .2514 of the curve and $z = .680$ at .2483. With .2500 at .0014/.0031 of the distance between .2514 and .2483, the corresponding z-score is .00452 above .67, at .67452.
3. The *df* of 50 is bracketed by $df = 40$ with $t_{crit} = 2.021$, and $df = 60$ with $t_{crit} = 2.000$. Because 50 is at .5 of the distance between 40 and 60, the target t_{crit} is .5 of the .021 between the brackets, which is 2.0105.

Part B.4: Additional Formulas for Computing Probability

1. (a) With "and" we multiply the individual probabilities times each other.
   (b) With "or" we add the individual probabilities together.
   (c) We must consider whether the events are mutually inclusive or mutually exclusive.
3. (a) The probability is the same as that of first selecting one such person and then selecting another. The $p(\text{above } 116) = .34$, so $p(\text{two people above } 116) = (.34)(.34) = .1156$.
   (b) These are mutually exclusive events, so with $p(\text{introverted}) = .40$ and $p(\text{extroverted}) = .35$, $p(\text{introverted or extroverted}) = .40 + .35 = .75$.
   (c) With $p(\text{above } 116) = .34$ and $p(\text{introverted}) = .40$, $p(\text{above 116 and introverted}) = (.34)(.40) = .136$.
   (d) These are mutually inclusive events, so with $p(\text{above } 116) = .34$, $p(\text{introverted}) = .40$, and $p(\text{above 116 and introverted}) = .136$, $p(\text{above 116 or introverted}) = (.34 + .40) - .136 = .604$.
   (e) The $p(\text{in-between}) = .25$, and from part (d), $p(\text{above 116 or introverted}) = .604$. Therefore, $p(\text{in-between and then above 116 or introverted}) = (.25)(.604) = .151$.
5. (a) We use the binomial expansion when we want to find the probability of obtaining a sequence of events in which each event involves one of two mutually exclusive possibilities.
   (b) $p(4 \text{ heads}) = (5!/4!(1))\ (.5^4)(.5^1) = (120/24)(.03125) = .15625$.
   (c) $p(1 \text{ head}) = (5!/(1!(4!))\ (.5^1)(.5^4) = (120/24)(.03125) = .15625$.
   (d) Obtaining 1 head out of 5 tosses is equivalent to obtaining 4 tails; since $p(4 \text{ tails}) = p(4 \text{ heads})$, the answer in (c) is the same as in (b).

Part B.5: The Two-way Within-Subjects Analysis of Variance

1.

Factor B: Amount of Fantasy

Factor A		B_1: low	B_2: med	B_3: high	
	Sub 1	5	7	9	
	Sub 2	6	5	8	
	Sub 3	2	6	10	
	Sub 4	2	9	10	
	Sub 5	5	5	10	
A_1: Meditation		$\overline{X} = 4$ $\Sigma X = 20$ $\Sigma X^2 = 94$ $n = 5$	$\overline{X} = 6.4$ $\Sigma X = 32$ $\Sigma X^2 = 216$ $n = 5$	$\overline{X} = 9.4$ $\Sigma X = 47$ $\Sigma X^2 = 445$ $n = 5$	$\overline{X} = 6.6$ $\Sigma X = 99$ $n = 15$
	Sub 1	10	2	5	
	Sub 2	10	5	6	
	Sub 3	9	4	5	
	Sub 4	10	3	7	
	Sub 5	10	2	6	
A_2: No Meditation		$\overline{X} = 9.8$ $\Sigma X = 49$ $\Sigma X^2 = 481$ $n = 5$	$\overline{X} = 3.2$ $\Sigma X = 16$ $\Sigma X^2 = 58$ $n = 5$	$\overline{X} = 5.8$ $\Sigma X = 29$ $\Sigma X^2 = 171$ $n = 5$	$\overline{X} = 6.09$ $\Sigma X = 94$ $n = 15$
		$\Sigma X = 69$ $\overline{X} = 6.9$ $n = 10$	$\Sigma X = 48$ $\overline{X} = 4.8$ $n = 10$	$\Sigma X = 76$ $\overline{X} = 7.6$ $n = 10$	$\Sigma X_{total} = 193$ $\Sigma X^2_{total} = 1465$

$N = 30$ $k_A = 2$ $k_B = 3$

A × Subject Table After Collapsing Across Factor B:

Factor A

	A_1	A_2	A_2	ΣX_{Sub}
Subject 1	15	9	14	38
Subject 2	16	10	14	40
Subject 3	11	10	15	36
Subject 4	12	12	17	41
Subject 5	15	7	16	38

B × Subject Table After Collapsing Across Factor A

Factor B

	B_1	B_2	ΣX_{Sub}
Subject 1	21	17	38
Subject 2	19	21	40
Subject 3	18	18	36
Subject 4	21	20	41
Subject 5	20	18	38

$$\text{Correction Term} = \left(\frac{(\Sigma X_{\text{total}})^2}{N}\right) = \frac{193^2}{30} = 1241.63$$

$$SS_{\text{tot}} = 1465 - 1241.63 = 223.37$$

$$SS_{\text{A}} = \left(\frac{(69)^2 + (68)^2 + (68)^2}{10}\right) - 1241.63 = 42.47$$

$$SS_{\text{B}} = \left(\frac{(99)^2 + (94)^2}{15}\right) - 1241.63 = .84$$

$$SS_{\text{bn}} = \left(\frac{(33)^2 + (50)^2 + (36)^2 + (18)^2 + (36)^2 + (18)^2}{5}\right)$$

$$- 1241.63 = 184.57$$

$$SS_{\text{A}\times\text{B}} = 184.57 - 42.47 - .84 = 141.26$$

$$SS_{\text{subs}} = \frac{(38)^2 + (40)^2 + (36)^2 + (41)^2 + (38)^2}{(2)(3)} - 1241.63$$

$$SS_{\text{subs}} = 2.54$$

$$SS_{\text{A}\times\text{S}} = \frac{(15)^2 + (9)^2 + (14)^2 + (16)^2 + (10)^2 + (14)^2 + (11)^2 + (10)^2}{2} +$$

$$\frac{(15)^2 + (12)^2 + (12)^2 + (17)^2 + (15)^2 + (7)^2 + (16)^2}{2} -$$

$$1241.63 - 42.47 - 2.54$$

$$SS_{\text{A}\times\text{S}} = 16.86$$

$$SS_{\text{B}\times\text{S}} = \frac{(21)^2 + (17)^2 + (19)^2 + (21)^2 + (18)^2 + (18)^2 + (21)^2 + (20)^2}{2} +$$

$$\frac{(20)^2 + (18)^2}{3} - 1241.63 - .84 - 2.54 = 3.32$$

$$SS_{\text{A}\times\text{B}\times\text{S}} = 223.37 - 42.47 - .84 - 141.26 -$$
$$2.54 - 16.86 - 3.32 = 16.08$$

Source	*Sum of Squares*	*df*	*Mean Square*	*F*
Factor				
A	42.47	2	21.35	10.12
B	.84	1	.84	1.01
A×B	141.26	2	70.63	35.14
Subjects				
A×S	16.86	8	2.11	
B×S	3.32	4	.83	
A×B×S	16.08	8	2.01	
Total	223.37	29		

For $\alpha = .05$ and $df_{\text{A}} = 2$, and $df_{\text{A}\times\text{S}} = 8$, $F_{\text{crit}} = 4.46$, so A is significant. For $df_{\text{B}} = 1$, and $df_{\text{B}\times\text{S}} = 4$, $F_{\text{crit}} = 7.71$, so B is not significant. For $df_{\text{A}\times\text{B}} = 2$ and $df_{\text{A}\times\text{B}\times\text{S}} = 8$, $F_{\text{crit}} = 4.46$, so A × B is significant.

For Factor A, $k = 3$ and $df = 8$, so $q_k = 4.04$. $HSD = (4.04)(\sqrt{2.11/10}) = (4.04)(0.46) = 1.86$. All levels differ significantly. For the interaction, $k = 5$ and $df = 8$, so $q_k = 4.89$. $HSD = (4.89)(\sqrt{2.01/5}) = (4.89)(0.63) = 3.1$. Meditation and no meditation differ at each level of fantasy. With meditation, low versus high differ; with no meditation, low versus medium and low versus high differ.

For Factor A, $\eta^2 = 42.47 / 223.37 = .19$. For A × B, $\eta^2 = 141.26 / 223.37 = .63$.

Part B.6: The Two-way Mixed Design Analysis of Variance

1. (a)

Level of Fantasy: Factor A

Factor A Meditation		B_1: low	B_2: med	B_3: high	ΣX_{sub}
	Sub 1	5	7	9	21
	Sub 2	6	5	8	19
	Sub 3	2	6	10	18
	Sub 4	2	9	10	21
	Sub 5	5	5	10	20
A_1 Meditation		$X = 4$ $\Sigma X = 20$ $\Sigma X^2 = 94$ $n = 5$	$\overline{X} = 6.4$ $\Sigma X = 32$ $\Sigma X^2 = 216$ $n = 5$	$\overline{X} = 9.4$ $\Sigma X = 47$ $\Sigma X^2 = 445$ $n = 5$	$\overline{X} = 6.6$ $\Sigma X = 99$ $n = 15$
	Sub 1	10	2	5	17
	Sub 2	10	5	6	21
	Sub 3	9	4	5	18
	Sub 4	10	3	7	20
	Sub 5	10	2	6	18
A_2 No Meditation		$\overline{X} = 9.8$ $\Sigma X = 49$ $\Sigma X^2 = 481$ $n = 5$	$\overline{X} = 3.2$ $\Sigma X = 16$ $\Sigma X^2 = 58$ $n = 5$	$\overline{X} = 5.8$ $\Sigma X = 29$ $\Sigma X^2 = 171$ $n = 5$	$\overline{X} = 6.09$ $\Sigma X = 94$ $n = 15$
		$\Sigma X = 69$ $\overline{X} = 6.9$ $n = 10$ $N = 30$	$\Sigma X = 48$ $\overline{X} = 4.8$ $n = 10$ $k_A = 2$	$\Sigma X = 76$ $\overline{X} = 7.6$ $n = 10$ $k_B = 3$	$\Sigma X_{total} = 193$ $\Sigma X^2_{total} = 1465$

$$\text{Correction Term} = \left(\frac{193^2}{30}\right) = 1241.63$$

$SS_{tot} = \Sigma X^2\text{total} - \text{Step 3} = 1465 - 1241.63 = 223.37$

$SS_{\text{subs}} =$

$$\frac{(21)^2 + (19)^2 + (18)^2 + (21)^2 + (20)^2 + (17)^2 + (21)^2}{3} +$$

$$\frac{(18)^2 + (20)^2 + (18)^2}{3} - 1241.63 = 6.70$$

$$SS_A = \left(\frac{(99)^2 + (94)^2}{15}\right) - 1241.63 = .84$$

$$SS_B = \left(\frac{(69)^2 + (48)^2 + (76)^2}{10}\right) - 1241.63 = 42.47$$

$$SS_{\text{e:bn}} = 6.70 - .84 = 5.86$$

$$SS_{\text{bn}} = \left(\frac{(20)^2 + (32)^2 + (47)^2 + (49)^2 + (16)^2 + (29)^2}{5}\right) -$$

$$1241.63 = 184.57$$

$$SS_{A\times B} = 184.57 - .84 - 42.47 = 141.26$$

$$SS_{\text{e:wn}} = 223.37 - 6.70 - 42.47 - 141.26 = 32.94$$

Source	*Sum of Squares*	*df*	*Mean Square*	*F*
Between Groups				
A (Meditation)	.84	1	.84	1.15
Error Between	5.86	8	.73	
Within Groups				
B (Fantasy)	42.47	2	21.24	10.31
A×B	141.26	2	70.63	34.29
Error Within	32.94	16	2.06	
Total	223.37	29		

For $\alpha = .05$ and $df_A = 1$, and $df_{\text{e:bn}} = 8$, $F_{\text{crit}} = 5.32$, so A is not significant. For $df_B = 2$, and $df_{\text{e:wn}} = 16$, $F_{\text{crit}} = 3.63$, so B is significant. For $df_{A\times B} = 2$ and $df_{\text{e:wn}} = 16$, $F_{\text{crit}} = 3.63$, so A × B is significant.

For the B main effect, $k = 3$ and the $df = 12$, so $q_k = 3.65$. $HSD = (3.65)(\sqrt{2.06/10}) = (3.65)(0.45) = 1.66$. Only low versus medium and medium versus high differ significantly. For the interaction, $k = 5$ and $df = 16$, so $q_k = 4.33$. $HSD = (4.33)(\sqrt{2.06/5}) = (4.33)(0.64) = 2.78$. There is a significant difference between Meditation versus No meditation for each fantasy level. With Meditation, only low and medium fantasy don't differ. With No meditation, only medium and high fantasy don't differ.

For the fantasy main effect, $\eta^2 = 42.47/223.37 = .19$. For the interaction, $\eta^2 = 141.26/223.37 = .63$.

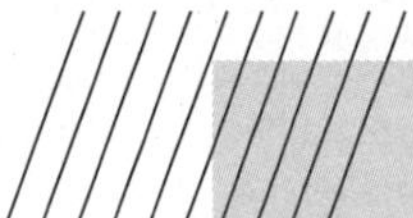

GLOSSARY

Alpha The Greek letter α, which symbolizes the criterion, the size of the region of rejection of a sampling distribution, and the theoretical probability of making a Type I error

Alternate forms Different versions of the same questionnaire

Alternative hypothesis The statistical hypothesis that describes the population parameters that the sample data represent if the predicted relationship does exist; symbolized by H_a

Analysis of variance The parametric procedure for determining whether significant differences exist in an experiment that involves two or more sample means; abbreviated ANOVA

ANOVA Abbreviation of *Analysis of variance*

Archival research Research for which written records constitute the source of data

Bar graph A graph in which a free-standing vertical bar is centered over each score on the X axis; used with nominal or ordinal scores for the independent variable, which represent discrete categories

Barnum statements Questions or statements that are so global and vague that everyone would agree with them or select the same response for them

Baseline The level of performance on the dependent variable when the independent variable is not present, used as a comparison to the level of performance when the independent variable is present

Beta The Greek letter β, which symbolizes the theoretical probability of making a Type II error

Between-subjects ANOVA The type of ANOVA that is performed when a study involves between-subjects factors

Between-subjects design A design in which all factors are between subjects factors

Between-subjects factor The type of factor created when an independent variable is studied using independent samples in all conditions

Bimodal distribution A symmetrical frequency polygon with two distinct humps; each hump represents high frequency scores and the center scores of each hump technically have the same frequency

Carry-over effects The influence that a participant's experience of a trial has on his or her performance of subsequent trials

Case study An in-depth description of one participant, organization, or event

Causal hypothesis A hypothesis that tentatively explains a particular influence on, or cause of, a behavior

Ceiling effects A restriction of range that occurs when a task is too easy, causing most scores to be near the highest possible score

Cell In a multi-factor ANOVA, the combination of one level of one factor with one level of the other factor(s)

Central limit theorem A statistical principle that defines the mean, standard deviation, and shape of a theoretical sampling distribution

χ^2-distribution The sampling distribution of all possible values of χ^2 that occur when the samples represent the distribution of frequencies described by the null hypothesis

Chi square procedure The nonparametric inferential procedure for testing whether the frequencies of category membership in the sample represent the predicted frequencies in the population; used with nominal data

Closed-ended question In a questionnaire or interview, a question accompanied by several answers from which a participant must select his or her response

Classification variable See *Qualitative variable*

Cluster sampling A sampling technique in which certain groups are randomly selected and all participants in each group are observed

Coefficient of alienation The proportion of variance not accounted for by a relationship; computed by subtracting the squared correlation coefficient from 1

Coefficient of determination The proportion of variance accounted for by a relationship; computed by squaring the correlation coefficient

Cohort design A factorial design consisting of a longitudinal study of several groups, each from a different generation

Cohort effects The situation that occurs when age differences are confounded by differences in subject history

Collapsing across a variable To combine scores from the different amounts or categories of that variable

Complete counterbalancing Testing different participants with different orders so that all possible orders of conditions or trials occur in a study

Complete factorial design A design in which all levels of one factor are combined with all levels of the other factors

Conceptual replication The repeated test or confirmation of a hypothesis using a design different from that of the original study

Concurrent validity The extent to which a procedure correlates with the present behavior of participants

Condition An amount or category of the independent variable that creates the specific situation under which participants' scores on the dependent variable are measured

Confederates People enlisted by a researcher to act as other participants or "accidental" passers-by, thus creating a social situation to which "real" participants then respond

Confidence interval A statistically defined range of values of the population parameter, one of which the sample statistic is likely to represent

Confidence interval for a single population μ A statistically defined range of values of μ, one of which is likely to be represented by the sample mean

Confidence interval for the difference between two μs A statistically defined range of differences between two population μs, one of which is likely to be represented by the difference between the two sample means

Confounded comparison In ANOVA, a comparison of two cells that differ along more than one factor

Confounded variables See *Confounding*

Confounding A situation that occurs when an extraneous variable systematically changes along with the variable hypothesized to be a causal variable

Construct validity The extent to which a measurement reflects the hypothetical construct of interest

Content analysis A scoring procedure for open-ended questions in which the researcher counts specific words or themes in a participant's responses

Content validity The extent to which a measurement reflects the variable or behavior of interest

Contingency coefficient The statistic that describes the strength of the relationship in a two-way chi square when there are more than two categories of either variable; symbolized by C

Continuous scale A measurement scale that allows for fractional amounts of the variable being measured

Control The elimination of unintended, extraneous factors that might influence the behavior being studied

Control group A group of participants who are measured on the dependent variable but receive zero amount of the independent variable, thus providing a baseline for comparison to the experimental group

Convenience sampling A sampling approach in which the researcher studies the participants who are conveniently available

Convergent validity The extent to which the scores obtained from one procedure are positively correlated with the scores obtained from another procedure that is already accepted as valid

Correlation coefficient A number, computed from the pairs of X scores and Y scores in a set of data, that summarizes and describes the type of relationship present and the strength of that relationship

Correlational design A design in which scores on two or more variables are measured to determine whether they form the predicted relationship

Counterbalancing The process of systematically changing the order of trials for different participants in a balanced way, so as to counter the biasing influence of any one order

Criterion The probability that provides the basis for deciding whether a sample is too unlikely to have occurred by chance and thus is unrepresentative of a particular population

Criterion validity The extent to which the scores obtained from a procedure correlate with an observable behavior

Criterion variable The variable in a relationship whose unknown scores are predicted through use of the known scores on the predictor variable

Critical value The value of the sample statistic that marks the edge of the region of rejection in a sampling distribution; values that fall beyond it fall in the region of rejection

Cross-sectional design A quasi-experimental between-subjects design in which participants are observed at different ages or at different points in a temporal sequence

Cumulative frequency The frequency of those scores at or below a particular score; symbolized by cf

Cumulative frequency distribution A distribution of scores organized to show the frequency of the scores at or below each score

Curvilinear relationship See *Nonlinear relationship*

Data point A dot plotted on a graph to represent a pair of X and Y scores

Debriefing The procedure by which researchers inform participants about all aspects of a study after they have participated in it, in order to remove any negative consequences of the procedure

Degree of association See *Strength of a relationship*

Degrees of freedom The number of scores in a sample that are free to vary, and thus the number that is used to calculate an estimate of the population variability; symbolized by *df*

Demand characteristics Cues within the research context that guide or bias a participant's behavior

Dependent events Events for which the probability of one event is influenced by the occurrence of the other

Dependent samples Samples created by matching each subject in one sample with a subject in the other sample or by repeatedly measuring the same subject under all conditions; also called *Related samples*

Dependent samples *t*-test The statistical procedure that is appropriate for significance testing when the scores meet the requirements of a parametric test, the design involves matched groups or repeated measures, and there are only two conditions of the independent variable

Dependent variable In an experiment, the variable that is measured under each condition of the independent variable

Descriptive design See *Descriptive research*

Descriptive hypothesis A hypothesis that tentatively describes a behavior in terms of its characteristics or the situation in which it occurs

Descriptive methods The research methods used to test descriptive hypotheses

Descriptive research The observation and description of a behavior, the situation it occurs in, or the individuals exhibiting it.

Descriptive statistics Mathematical procedures for organizing, summarizing, and describing the important characteristics of a sample of data

Design The specific manner in which a research study is conducted

Deviation The distance that separates a score from the mean and thus indicates how much the score differs from the mean

Dichotomous variable A discrete variable that has only two possible amounts or categories

Discrete scale A measurement scale that allows for measurement only in whole-number amounts

Discriminant validity The extent to which the scores obtained from one procedure are not correlated with the scores obtained from another procedure that measures other variables or constructs

Distribution An organized set of data

Double-barreled questions Questions that have more than one component

Double-blind procedure The procedure in which both the researcher who interacts with the participants and the participants themselves are unaware of the treatment being presented

Ecological validity The extent to which an experimental situation can be generalized to natural settings and behaviors

Effect size The proportion of variance accounted for in an experiment, which indicates how consistently differences in the dependent scores are influenced by changes in the independent variable

Empirical knowledge Knowledge obtained through observation of events

Empirical probability distribution A probability distribution based on observations of the relative frequency of events

Environmental variables The aspects of the environment that can influence scores

Error variance The variability in Y scores at each X score, or the inherent variability within a population, estimated in ANOVA by the mean square within groups

Estimated standard error of the mean An estimate of the standard deviation of the sampling distribution of means, used in calculating the single-sample *t*-test; symbolized by $s_{\overline{X}}$

Eta The correlation coefficient used to describe a linear or nonlinear relationship containing two or more levels of a factor; symbolized by η

Eta squared The proportion of variance in the dependent variable that is accounted for by changing the levels of a factor; the effect size of a factor in a sample; symbolized by η^2

***Ex post facto* research** Research conducted after a phenomenon has occurred

Expected frequency In chi square, the frequency expected in a category if the sample data perfectly represent the distribution of frequencies in the population described by the null hypothesis; symbolized by f_e

Experiment A design in which one variable is actively changed or manipulated and scores on another variable are measured to determine whether there is a relationship

Experiment-wise error rate The probability of making a Type I error when comparing all means in an experiment

Experimental group(s) Those participants who receive a nonzero amount of the independent variable and are then measured on the dependent variable

Experimental hypotheses Two statements made before a study is begun, describing the predicted relationship that the study may or may not demonstrate

Experimental methods The research methods used to test causal hypotheses

Experimental realism The extent to which the experimental task engages participants psychologically, such that they become less concerned with demand characteristics

Experimenter expectancies Subtle cues provided by the experimenter about the responses that participants should give in a particular condition

External validity The extent to which results generalize to other participants and other situations

Extraneous variables Variables that may influence the results of a study but are not the variables of interest

Extreme scores The scores that are relatively far above and below the middle score of a distribution

***F*-distribution** The sampling distribution of all possible values of F that occur when the null hypothesis is true and all conditions represent one population μ

***F*-ratio** In ANOVA, the ratio of the mean square between groups to the mean square within groups

***F* statistic** See *F-ratio*

Face validity The extent to which a measurement procedure appears to measure what it is intended to measure

Factor In ANOVA, an independent variable

Factorial design See *Complete factorial design*

Field experiment An experiment conducted in a natural setting

Field survey A procedure in which participants complete a questionnaire or interview in a natural setting

Fisher's protected *t*-test The *post hoc* procedure performed with ANOVA to compare means from a factor in which all levels do not have equal *n*

Floor effects A restriction of range that occurs when a task is too difficult, causing most or all scores to be near the lowest possible score

Forced-choice procedure A measure in which participants must select from a limited set of choices, such as a multiple-choice test

Frequency The number of times each score occurs within a set of data; also called simple frequency; symbolized by *f*

Frequency polygon A graph that shows interval or ratio scores (*X* axis) and their frequencies (*Y* axis), using data points connected by straight lines

Friedman Chi Square test The one-way, within-subjects ANOVA for ranked scores, performed when there are more than two levels of one factor

Grouped distribution A distribution formed by combining different scores to make small groups whose total frequencies, relative frequencies, or cumulative frequencies can then be manageably reported

Habituation The process by which participants are familiarized with a procedure before actual data collection is commenced, in order to reduce reactivity

Hawthorne effect A bias in participants' behavior—usually an improvement in performance—that results from the special treatment and interest shown by a researcher

Heterogeneity of variance A characteristic of populations when they do not have the same variance

Heteroscedasticity An unequal spread of *Y* scores around the regression line (that is, around the values of Y')

Histogram A graph similar to a bar graph but with adjacent bars touching, used to plot the frequency distribution of a small range of interval or ratio scores

Homogeneity of variance A characteristic of populations when they have the same variance

Homogeneity of variance test A test performed before a *t*-test or ANOVA is conducted to determine whether populations can be assumed to have homogeneity of variance; also called the F_{max} test

Homoscedasticity An equal spread of *Y* scores around the regression line (that is, around the values of Y')

Human Subjects Review Committee A committee at colleges and research institutions that is charged with the responsibility of reviewing all prospective research procedures to ensure the ethical and safe treatment of participants

Hypothesis A formally-stated expectation about a behavior that defines the purpose and goals of a research study

Hypothetical construct An abstract concept used in a particular theoretical manner to relate different behaviors according to their underlying features or causes

Incomplete factorial design An ANOVA design in which not all levels of all factors are combined

Independent events Events for which the probability of one event is not influenced by the occurrence of the other

Independent samples Samples created by selecting each participant for one sample without regard to the participants selected for any other sample

Independent samples *t*-test The statistical procedure used for significance testing that is appropriate when the scores meet the requirements of a parametric test, the design involves independent samples, and there are only two conditions of the independent variable

Independent variable In an experiment, the variable that is hypothesized to cause a change in the dependent variable and is systematically changed or manipulated by the researcher; also called a *factor*

Individual differences Variations in individuals' traits, backgrounds, genetic make-up, and other characteristics that make individuals different from one another and produce different responses to the same situation thus influencing the strength of a relationship

Inferential statistics Mathematical procedures for deciding whether a sample relationship represents a relationship that actually exists in the population

Informed consent The procedure by which researchers inform participants about a study prior to their participation in it, and obtain participants' explicit consent to participate

Instrumentation effects Changes in measurement procedures that occur through use of equipment over time, making the measurements less reliable

Inter-rater reliability The extent to which raters agree on the scores they assign to a participant's behavior

Interaction effect The effect produced by the concurrent manipulation of two independent variables such that the influence of changing the levels of one factor depends on which level of the other factor is present

Internal validity The extent to which the observed relationship reflects the relationship between the variables in a study

Interrupted time-series design A quasi-experimental repeated-measures design in which observations are made at several spaced times before and then after a treatment

Interval estimation An estimation technique where the value of a population parameter is assumed to lie within a specified interval

Interval scale A measurement scale in which each score indicates an actual amount, an equal unit of measurement separates consecutive scores, zero is not a true zero value, and negative scores are possible

Intervening variable An internal subject characteristic that is influenced by the independent variable and, in turn, influences the dependent variable

Kruskal-Wallis *H*-test The nonparametric version of the one-way between-subjects ANOVA for ranked scores

Leading questions Questions that are so loaded with social desirability or experimenter expectancies that there is one obvious response

Level In ANOVA, each condition of the factor (independent variable); also called *treatment*

Likert-type questions A measure in which participants rate statements, such as when using a scale of 1 to 5 where 1 indicates "strongly agree" and 5 indicates "strongly disagree"

Line graph A graph in which X scores from an interval or ratio variable are plotted by connecting adjacent data points with straight lines; used when the independent variable implies a continuous, ordered amount

Linear regression The procedure for predicting participants' scores on one variable based on the linear relationship with participants' scores on another variable

Linear regression equation The equation that defines the straight line summarizing a linear relationship by describing the value of Y' at each X

Linear regression line The best-fitting straight line that summarizes the scatterplot of a linear relationship by, on average, passing through the center of all Y scores

Linear relationship A correlation between the X scores and Y scores in a set of data in which the Y scores tend to change in only one direction as the X scores increase, forming a slanted straight regression line on a scatterplot

Literal replication The precise duplication of the specific design and results of a previous study

Longitudinal design A quasi-experimental design in which a researcher repeatedly measures a group of participants in order to observe the effect of the passage of time

Main effect In a multi-factor ANOVA, the effect on the dependent scores of changing the levels of one factor while collapsing over other factors in the study

Mann-Whitney *U* test The nonparametric version of the independent samples t-test for ranked scores when n is less than or equal to 20

Margin of error The confidence interval that is computed when estimating the population's responses to a field survey

Matched-groups design A research design in which each participant in one condition is matched with a subject in every other condition along an extraneous subject variable

Mean The average of a group of scores, interpreted as the score around which the scores in a distribution tend to be clustered

Mean square In ANOVA, an estimated population variance, symbolized by MS

Mean square between groups In ANOVA, the variability in scores that occurs between the levels in a factor or the cells in an interaction

Mean square within groups In ANOVA, the variability in scores that occurs in the conditions, or cells; also known as the *error term*

Measure of central tendency A score that summarizes the location of a distribution on a variable by indicating where the center of the distribution tends to be located

Measurement variables The aspects of the stimuli presented or the measurement procedure employed that may influence scores

Measures of variability Measures that summarize and describe the extent to which scores in a distribution differ from one another

Median The score located at the 50th percentile; symbolized by Mdn; also called the median score

Meta-analysis Statistical procedures for combining, testing, and describing the results from different studies

Mixed design A design with a mix of within-subjects and between-subjects factors

Modal score See *Mode*

Mode The most frequently occurring score in a sample; also called the *modal score*

Model A generalized, hypothetical description that, by analogy, explains the process underlying a set of common behaviors

Multi-factor experiment An experiment in which the researcher examines several independent variables and their interactions

Multiple-baseline design A design in which a baseline is established for one behavior from several participants, for several behaviors from one participant, or for one behavior from one participant in several situations

Multiple correlation and regression Statistical procedures performed when multiple predictor (*X*) variables are used to predict one criterion (*Y*) variable

Multiple time-series design A quasi-experimental repeated-measures design in which an experimental group and a nonequivalent control group are observed at several spaced times before and then after a treatment

Multivariate statistics The inferential statistical procedures used when a study involves multiple dependent variables

Naturalistic observation The unobtrusive observation of participants' behaviors in an unstructured fashion

Negative linear relationship A linear relationship in which the *Y* scores tend to decrease as the *X* scores increase

Negatively skewed distribution A frequency polygon with low frequency, extreme low scores but without corresponding low frequency, extreme high ones, so that its only pronounced tail is in the direction of the lower scores

Nemenyi's procedure The *post hoc* procedure performed with the Friedman χ^2-test

Nominal scale A measurement scale in which each score identifies a quality or category and does not indicate an amount

Nonequivalent control group In a quasi-experiment, a control group whose subject characteristics and experiences are different from those of the experimental group

Nonexperimental methods See *Descriptive methods*

Nonlinear relationship A relationship in which the Y scores change their direction of change as the X scores change, forming a curved regression line; also called a *curvilinear relationship*

Nonparametric statistics Inferential procedures that do not require stringent assumptions about the parameters of the raw score population represented by the sample data; usually used with scores most appropriately described by the median or the mode

Nonprobability sampling Any sampling technique in which every potential participant in the population does not have an equal likelihood of being selected for a study

Nonsignificant Describes results that are considered likely to result from sampling error when the predicted relationship does not exist; it indicates failure to reject the null hypothesis

Normal curve model The most common model of how nature operates; it is based on the normal curve and describes a normal distribution of a population of scores

Normal distribution A frequency distribution forming a bell-shaped curve that is symmetrical about the mean

Null hypothesis The statistical hypothesis that describes the population parameters the sample data represent if the predicted relationship does not exist; symbolized by H_0

Observational research Research where participants are observed in an unobtrusive manner; see also *Naturalistic observation*; *Systematic naturalistic observation*; and *Participant observation*

Observed frequency In chi square, the frequency with which subjects fall in a category of a variable; symbolized by f_o

One-group pretest-posttest design A quasi-experimental pretest-posttest design for which there is no control group

One-tailed test The test used to evaluate a statistical hypothesis that predicts that scores will only increase or only decrease

One-way ANOVA The analysis of variance performed when an experiment has only one independent variable

One-way chi square The chi square procedure performed when a study examines category membership along one variable

One-way design A research design involving the manipulation of one independent variable

Open-ended question In a questionnaire or interview, a question for which the participant determines both the alternatives to choose from and the response

Operational definition The definition of a construct or variable in terms of the operations used to measure it

Order effects The influence on a particular trial that arises from its position in a sequence of trials

Ordinal scale A measurement scale in which scores indicate rank order or a relative amount

Parameter See *Population parameter*

Parametric statistics Inferential procedures that require certain assumptions about the parameters of the raw score population represented by the sample data; usually used with scores most appropriately described by the mean

Partial counterbalancing Balancing order effects by testing different participants using only some of the possible orders

Participants The individuals in a sample

Participant observation The unobtrusive observation of a group in which the researcher is an active member

Pearson correlation coefficient The correlation coefficient that describes the strength and type of a linear relationship between two interval or ratio variables; symbolized by r

Percent A proportion multiplied times 100

Percentile A cumulative percentage; the percentage of all scores in the sample that are at or below a particular score

Phi coefficient The statistic that describes the strength of the relationship in a two-way chi square when there are only two categories for each variable; symbolized by ϕ

Placebo An inactive substance that provides the demand characteristics of a manipulation while presenting zero amount of the independent variable

Planned comparisons In ANOVA, statistical procedures for comparing only some conditions in an experiment

Point-biserial correlation coefficient The correlation coefficient that describes the strength of the linear relationship between scores from one continuous interval or ratio variable and one dichotomous variable; symbolized by r_{pb}

Point estimation The estimation procedure where the value of the population parameter is assumed to equal the value of the corresponding sample statistic

Pooled variance The weighted average of the sample variances in a two-sample experiment; symbolized by s^2_{pool}

Population The large group of all possible scores that would be obtained if the behavior of every individual of interest in a particular situation could be measured

Population parameter A number that describes a characteristic of a population of scores, symbolized by a letter from the Greek alphabet; also called a *parameter*

Positive linear relationship A linear relationship in which the Y scores tend to increase as the X scores increase

Positively skewed distribution A frequency polygon with low frequency, extreme high scores but without corresponding low frequency, extreme low ones, so that its only pronounced tail is in the direction of the higher scores

***Post hoc* comparisons** In ANOVA, statistical procedures used to compare all possible pairs of conditions to determine which ones differ significantly from each other

Power The probability that a statistical test will allow the rejection of a false null hypothesis

Practice effects The influence on performance that arises from practicing a task

Predicted Y score In linear regression, the best description and prediction of the Y scores at a particular X, based on the linear relationship summarized by the regression line; symbolized by Y'

Prediction A statement as to how a behavior will be manifested in a research situation, describing the specific results that will be found

Predictive validity The extent to which a procedure allows for accurate predictions about a participant's future behavior

Predictor variable The variable for which known scores in a relationship are used to predict unknown scores on another variable

Pretest A measure used to identify and select potential participants, prior to conducting a study

Pretest-posttest design A research design in which participants are measured before and after a treatment

Probability A mathematical statement indicating the likelihood that an event will occur when a particular population is randomly sampled; symbolized by p

Probability distribution The probability of every possible event in a population, derived from the relative frequency of every possible event in that population

Probability sampling Any sampling technique in which every potential participant in the population has an equal likelihood of being selected for a study

Program evaluation The procedures undertaken to evaluate the goals, activities, and outcomes of social programs

Proportion A decimal number between 0 and 1 that indicates a fraction of a total

Proportion of the area under the curve The proportion of the total area beneath the normal curve at certain scores, which represents the relative frequency of those scores

Proportion of variance accounted for The proportion of the error in predicting scores that is eliminated when, instead of using the mean of Y, the relationship with the X variable is used to predict Y scores; the proportional improvement in predicting Y scores thus achieved

Pseudo-explanation A circular statement that explains an event by renaming it

Psychological Abstracts A monthly publication that describes studies recently published in psychology journals

Publication Manual of the American Psychological Association The reference source regarding the organization, content, and style of a research manuscript

Qualitative variable A variable that reflects a quality or category

Quantitative variable A variable that reflects a quantity or amount

Quasi-experiment A study in which participants cannot be randomly assigned to conditions but, instead, are assigned to a particular condition on the basis of some inherent characteristic

Quasi-independent variable The independent variable in a quasi-experiment

Quota sampling A sampling technique in which, using convenience sampling, the sample has the same percentage of each subgroup as is found in the population

Random assignment A method of selecting a sample for an experiment such that the condition each participant experiences is determined in a random and unbiased manner

Random sampling A method of selecting samples whereby all members of the population have the same chance of being selected for a sample and all samples have the same chance of being selected

Randomization The creation of different random orders of trials under which different participants are tested

Range The difference between the highest and lowest scores in a set of data

Rank sums test The nonparametric version of the independent samples t-test for ranked scores when n is greater than 20; also the *post hoc* procedure performed with the Kruskal-Wallis H-test

Ratio scale A measurement scale in which each score indicates an actual amount, an equal unit of measurement separates consecutive scores, zero means zero amount, and negative scores are not possible

Reactivity The bias in responses that occurs when participants know they are being observed

Reaction time The amount of time a participant takes to respond to a stimulus

Rectangular distribution A symmetrical frequency polygon shaped like a rectangle; it has no discernible tails because extreme scores do not have relatively low frequencies

Region of rejection That portion of a sampling distribution that contains values considered too unlikely to occur by chance, found in the tail or tails of the distribution

Regression line The line drawn through the long dimension of a scatterplot that best fits the center of the scatterplot and thus visually summarizes the scatterplot and indicates the type of relationship that is present

Regression toward the mean The tendency of extreme scores to become less extreme; this occurs because random influences are not consistently present

Relationship A correlation between two variables so that a change in one variable is accompanied by a consistent change in the other variable

Relative frequency The proportion of time a score occurs in a distribution, equal to the proportion of the total number of scores that the score's simple frequency represents; symbolized by *rel. f*

Relative frequency distribution A distribution of scores, organized to show the proportion of time each score occurs in the data

Relative standing A description of a particular score derived from a systematic evaluation of the score using the characteristics of the sample or population in which it occurs

Reliability The extent to which a measurement is consistent, can be reproduced, and avoids error

Repeated-measures design A design in which each participant is measured repeatedly under all conditions of an independent variable

Replication The process of repeatedly conducting studies that test and confirm a hypothesis so that confidence in its truth can be developed

Representative sample A sample whose characteristics and behaviors accurately reflect those of the population from which it is drawn

Research design The way in which a study is laid out so as to demonstrate a relationship

Researcher variables The behaviors and characteristics of the researcher that may influence the reactions of subjects

Response scale The number and type of choices provided for each question in a questionnaire or interview

Response set A bias toward responding in a particular way because of previous responses made

Restriction of range Improper limitation of the range of scores obtained on a variable

Reversal design A design in which the researcher alternates between the baseline condition and the treatment condition

Robust procedure A procedure that alters the probability of a Type I error only a negligible amount, even if the assumptions of the procedure are not perfectly met; describes parametric procedures

Sample A relatively small subset of a population that is selected to represent or stand in for the population; a subset of the complete group of scores found in any particular situation

Sample standard deviation The square root of the sample variance

Sample statistic A number that describes a characteristic of a sample of scores, symbolized by a letter from the English alphabet; also called a statistic

Sample variance The average of the squared deviations of the scores around the mean

Sampling distribution of a correlation coefficient A frequency distribution showing all possible values of the coefficient that occur when samples of a particular size are drawn from a population whose correlation coefficient is zero

Sampling distribution of differences between the means A frequency distribution showing all possible differences between two means that occur when two independent samples of a particular size are drawn from the population of scores described by the null hypothesis

Sampling distribution of mean differences A frequency distribution showing all possible mean differences that occur when the difference scores from two dependent samples of a particular size are drawn from the population of difference scores described by the null hypothesis

Sampling distribution of means A frequency distribution showing all possible sample means that occur when samples of a particular size are drawn from the raw score population described by the null hypothesis

Sampling error The difference, due to random chance, between a sample statistic and the population parameter it represents

Sampling with replacement A sampling procedure in which a previously selected sample is returned to the population before additional samples are selected

Sampling without replacement A sampling procedure in which previously selected samples are not returned to the population before additional samples are selected

Scatterplot A graph of the individual data points from a set of X–Y pairs

Scientific method The assumptions, attitudes, goals, and procedures for creating and answering questions about nature in a scientific manner

Self-report A measure in which participants describe their feelings or thoughts

Semi-interquartile range The average distance between the median and the scores at the 25th and 75th percentiles (the quartiles), used to describe highly skewed distributions

Significant Describes results that are considered too unlikely to result from chance sampling error if the predicted relationship does not exist; it indicates rejection of the null hypothesis

Simple frequency distribution A distribution of scores, organized to show the number of times each score occurs in a set of data

Simple main effect The effect of one factor at one level of a second factor

Simple random sampling A sampling technique in which participants are randomly selected from a list of the members of the population

Single-blind procedure The procedure in which participants are unaware of the treatment they are receiving

Single-sample *t*-test The parametric procedure used to test the null hypothesis for a single-sample experiment when the standard deviation of the raw score population must be estimated

Single-subject design A repeated-measures experiment conducted on one participant

Skewed distribution A frequency polygon similar in shape to a normal distribution except that it is not symmetrical and it has only one pronounced tail

Slope A number that indicates how much a linear regression line slants and in which direction it slants; symbolized by b

Snowball sampling A sampling technique in which the researcher contacts potential participants who have been identified by previously tested participants

Social desirability The demand characteristic that causes participants to provide what they consider to be the socially acceptable response

Social Science Citation Index A reference source that identifies a given research article by authors and date, and then lists subsequent articles that have cited it

Sorting task A measure in which participants sort stimuli into different groups

Spearman rank-order correlation coefficient The correlation coefficient that describes the linear relationship between pairs of ranked scores; symbolized by r_s

Split-half reliability The consistency with which participants' scores on some trials match their scores on other trials

Squared correlation coefficient The proportion of total variance in Y scores that is systematically associated with changing X scores

Squared sum of *X* A result calculated by adding all scores and then squaring their sum; symbolized by $(\Sigma X)^2$

Standard deviation The statistic that communicates the average of the deviations of the scores from the mean in a set of data, computed by obtaining the square root of the variance; see also *Sample standard deviation*

Standard error of the difference The estimated standard deviation of the sampling distribution of differences between the means of independent samples in a two-sample experiment; symbolized by $s_{\bar{X}_1 - \bar{X}_2}$

Standard error of the estimate A standard deviation that indicates the amount that actual Y scores in a sample differ from their corresponding Y′ scores; symbolized as $S_{Y'}$

Standard error of the mean The standard deviation of the sampling distribution of means; used in the *z*-test (symbolized by $\sigma_{\bar{X}}$) and in the single-sample *t*-test (symbolized by $s_{\bar{X}}$)

Standard error of the mean difference The standard deviation of the sampling distribution of mean differences between dependent samples in a two-sample experiment; symbolized by $s_{\bar{D}}$

Standard normal curve A theoretical perfect normal curve, which serves as a model of the perfect normal z-distribution

Standard scores See *z-score*

Statistic See *Sample statistic*

Statistical hypotheses Two statements (H_0 and H_a) that describe the population parameters the sample statistics will represent if the predicted relationship exists or does not exist

Stratified random sampling A sampling technique involving the identification of important subgroups in

GLOSSARY

the population, followed by the proportionate random selection of participants from each subgroup

Strength of a relationship The extent to which one value of Y within a relationship is consistently associated with one and only one value of X; also called the degree of association

Strong manipulation Manipulation of the independent variable in such a way that participants' behavior is greatly differentiated, thus producing large differences in dependent scores between the conditions

Structured interview An interview in which participants are asked a specific set of predetermined questions in a controlled manner

Subject history The bias that arises due to participants' experiences that influence repeated measures

Subject maturation The bias that arises due to the changes that occur as an individual grows older and more mature that influence repeated measures

Subject mortality effects The bias that arises when participants fail to show up for a study or discontinue their participation before the study is completed

Subject variables Inherent, personal characteristics that distinguish one participant from another

Subjects See *Participants*

Sum of squares The sum of the squared deviations of a set of scores around a statistic

Sum of the deviations around the mean The sum of all differences between the scores and the mean; symbolized as $\Sigma(X - \overline{X})$

Sum of X The sum of the scores in a sample; symbolized by ΣX

Sum of the squared Xs The sum after squaring each score in a sample; symbolized by ΣX^2

Systematic naturalistic observation The unobtrusive observation of a particular behavior or situation in a structured fashion

Systematic random sampling A sampling technique in which every nth participant is selected from a list of the members of the population

***t*-distribution** The sampling distribution of all possible values of t that occur when samples of a particular size represent the raw score population(s) described by the null hypothesis

***t*-test for dependent samples** See *Dependent samples t-test*

***t*-test for independent samples** See *Independent samples t-test*

***t*-test** See *Dependent samples t-test*; *Independent samples t-test*

Tail (of a distribution) The far-left or far-right portion of a normal distribution, containing relatively low frequency, extreme scores

Test-retest reliability The consistency with which participants obtain the same overall score when tested at different times

Theoretical probability distribution A probability distribution based on a theoretical model of the relative frequencies of events in a population

Theory A logically organized set of proposals that defines, explains, organizes, and interrelates knowledge about many behaviors

Three-way design A three-way design involving three main effects, three two-way interactions, and one three-way interaction

Three-way interaction The interaction of three factors such that the two-way interaction between two factors changes as the levels of the third factor change

Tied rank The situation that occurs when two subjects in a sample receive the same rank-order score on a variable

Time-series design A quasi-experimental repeated-measures design in which participants' behavior is sampled before and then after the occurrence of an event

Total area under the curve The area beneath the normal curve, which represents the total frequency of all scores

Transformation A systematic procedure for converting a set of scores into a different but equivalent set of scores

Treatments The conditions of the independent variable; also called levels

Treatment variance In ANOVA, the variability between scores from different populations that would be created by the different levels of a factor

Trial A single complete instance of testing in an experimental series

True experiment A study in which the researcher actively changes or manipulates the independent variable, and in which participants can be randomly assigned to conditions

True independent variable The independent variable in a true experiment

Tukey's *HSD* test The *post hoc* procedure performed with ANOVA to compare means from a factor in which all levels have equal n

Two-tailed test The test used to evaluate a statistical hypothesis that predicts a relationship, but not whether scores will increase or decrease

Two-way between-subjects design A design in which an independent sample of participants is tested under each condition of two independent variables

Two-way chi square procedure The chi square procedure performed in testing whether, in the population, frequency of category membership on one variable is independent of frequency of category membership on another variable

Two-way design A design involving the manipulation of two independent variables

Two-way interaction The interaction of two factors such that the relationship between one factor and the dependent scores is different for and depends on each level of the other factor; also called two-way interaction effect

Two-way mixed design A design that involves one within-subjects factor and one between-subjects factor

Two-way within-subjects design A design in which matched groups or the same repeatedly-measured participants are tested in all conditions of two independent variables

Type I error A statistical error in which a large amount of sampling error causes rejection of the null hypothesis even though the null hypothesis is true (that is, when the predicted relationship does not exist)

Type II error A statistical error in which the closeness of the sample statistic to the population parameter described by the null hypothesis causes the null hypothesis to be retained even though it is false (that is, when the predicted relationship does exist)

Type of relationship The form of the pattern between the *X* scores and the *Y* scores in a set of data, determined by the overall direction in which the *Y* scores change as the *X* scores change

Unconfounded comparisons In an ANOVA, comparisons of cell means that differ along only one factor

Ungrouped distribution A distribution showing the frequency, relative frequency, or cumulative frequency of each individual score in the data

Unimodal distribution A distribution whose frequency polygon has only one hump and thus has only one score that qualifies as the mode

Univariate statistics Statistics that involve one dependent variable

Unobtrusive measures Procedures by which participants' behavior is measured without their being aware that measurements are being made

Unstructured interview An interview in which the questions are not rigidly predetermined, thus allowing for substantial discussion and interaction between participant and interviewer

Validity The extent to which a procedure measures what it is intended to measure

Variable Any measurable aspect of a behavior or influence on behavior that may change so that, when measured, can produce two or more different scores

Variance A measure of the variability of the scores in a set of data, computed as the average of the squared deviations of the scores around the mean; see also *Sample variance*; *Error variance*

Volunteer bias The bias that arises from the fact that a given sample contains only those participants who are willing to participate in the study

Wilcoxon *t*-test The nonparametric version of the dependent samples *t*-test for ranked scores

Within-subjects ANOVA The ANOVA performed when a study involves within-subjects factors

Within-subjects design A design in which all factors are within-subjects factors

Within-subjects factor The type of factor created when an independent variable is studied using dependent samples in all conditions, either because participants are matched or repeatedly measured

***Y*-intercept** The value of Y at the point where the linear regression line intercepts the *Y* axis; symbolized by *a*

Y prime The value of Y that falls on the regression line above any X; symbolized as Y'

***z*-distribution** The distribution produced by transforming all raw scores in a distribution into *z*-scores

***z*-score** The statistic that describes the location of a raw score in terms of its distance from the mean when measured in standard deviation units; symbolized by *z*; also known as a standard score because it allows comparison of scores on different kinds of variables by equating, or standardizing, the distributions

***z*-table** The table that gives the proportion of the total area under the standard normal curve for any two-decimal *z*-score

***z*-test** The parametric procedure used to test the null hypothesis for a single-sample experiment when the true standard deviation of the raw score population is known

REFERENCES

American Psychological Association. (1994). *Publication manual of the American Psychological Association* (4th ed.). Washington, DC: Author.

American Psychological Association. (1992). Ethical principles of psychologists and code of conduct. *American Psychologist, 47*, 1597–1611.

Anderson, C. A., & Anderson, D. C. (1984). Ambient temperature and violent crime: Tests of the linear and curvilinear hypotheses. *Journal of Personality and Social Psychology, 46*, 91–97.

Anderson, P. (1983). Decision making by objection and the Cuban missile crisis. *Administrative Science Quarterly, 28*, 201–222.

Asch, S. E. (1951). Effects of group pressure upon the modification and distortion of judgment. In H. Guetzknow (Ed.), *Groups, leadership, & men*. Pittsburgh: Carnegie.

Atkinson, J. W., & Litwin, G. H. (1960). Achievement motive and test anxiety conceived as a motive to approach success and to avoid failure. *Journal of Abnormal and Social Psychology, 60*, 52–63.

Atkinson, R. C., & Shiffrin, R. (1968). Human memory: A proposed system and its control processes. In K. Spence & J. Spence (Eds.), *The psychology of learning and motivation* (Vol. 2). New York: Academic Press.

Bandura, A., Ross, D., & Ross, S. A. (1961). Transmission of aggression through imitation of aggressive models. *Journal of Abnormal and Social Psychology, 63*, 575–582.

Barefoot, J. C., Hoople, H., & McClay, D. (1972). Avoidance of an act which would violate personal space. *Psychonomic Science, 28*, 205–206.

Barlow, D. H., & Hernsen, M. (1984). *Single case experimental designs: Strategies for studying behavior change* (2nd ed.). New York: Pergamon Press.

Bell, P. A. (1980). Effects of heat, noise, and provocation on retaliatory evaluative behavior. *Journal of Social Psychology, 110*, 97–100.

Bell, P. A., & Baron, R. A. (1976). Aggression and heat: The mediating role of negative affect. *Journal of Applied Social Psychology, 6*, 18–30.

Blakemore, J. E., LaRue, A. A., & Olejnik, A. B. (1979). Sex-appropriate toy preference and the ability to conceptualize toys as sex-role related. *Developmental Psychology, 15*, 339–340.

Boesch-Acherman, H., & Boesch, C. (1993). Tool use in wild chimpanzees: New light from dark forests. *Current Directions in Psychological Science, 2*, 18–21.

Bower, G. H., Karlin, M. B., & Dueck, A. (1975). Comprehension and memory for pictures. *Memory and Cognition, 3*(2), 216–220.

Bower, G. H., & Hilgard, E. R. (1981). *Theories of learning*. Englewood Cliffs, NJ: Prentice Hall.

Bramel, D., & Friend, R. (1981). Hawthorne, the myth of the docile worker, and class bias in psychology. *American Psychologist, 36*, 867–878.

Burke, B. F., Chrisler, J. C., & Devlin, A. S. (1989). The creative thinking, environmental frustration, and self-concept of left- and right-handers. *Creativity Research Journal, 2*(4), 279–285.

Campbell, D. T. (1969). Reforms as experiments. *American Psychologist, 24*, 409–429.

Campbell, D. T., & Stanley, J. C. (1963). *Experimental and quasi-experimental designs for research*. Boston: Houghton Mifflin.

Cann, D. R., & Donderi, D. C. (1986). Jungian personality typology and the recall of everyday and archetypal dreams. *Journal of Personality and Social Psychology, 50*, 1021–1030.

Carlson, M., Marcus-Newhall, A., and Miller, N. (1990). Effects of situational aggression cues: A quantitative review. *Journal of Personality and Social Psychology, 58*, 622–633.

Christensen, L. (1988). Deception in psychological research: When is it justified? *Personality and Social Psychology Bulletin, 14*, 664–675.

Cicchetti, D. V. (1972). Extension of multiple range tests to interaction tables in the analysis of variance. *Psychological Bulletin, 77*, 405–408.

Connors, J. G., & Alpher, V. S. (1989). Alcohol themes within country-western songs. *International Journal of the Addictions, 24*, 445–451.

Cook, T. D., & Campbell, D. T. (1979). *Quasi-experimentation: Design and analysis issues for field settings*. Chicago: Rand McNally.

Craik, F. I. M., & Lockhart, R. S. (1972). Levels of processing: A framework for memory research. *Journal of Verbal Learning and Verbal Behavior, 11*, 671–684.

Crusco, A. H., & Wetzel, C. G. (1984). The Midas touch: The effect of interpersonal touch on restaurant tipping. *Personality and Social Psychology Bulletin, 10*, 512–517.

Cunningham, M. R., Shaffer, D. R., Barbee, A. P., Wolff, P. L., & Kelley, D. J. (1990). Separate processes in the relation of elation and depression to helping: Social versus personal concerns. *Journal of Abnormal and Social Psychology, 26*, 13–33.

Darley, J. M., & Latané, B. (1968). Bystander intervention in emergencies: Diffusion of responsibility. *Journal of Personality and Social Psychology, 8*, 377–383.

Desrochers, A., & Begg, I. (1987). A theoretical account of encoding and retrieval processes in the use of imagery-based mnemonic techniques: The special case of the keyword method. In M. A. McDaniel & M. Pressley (Eds.), *Imagery and related mnemonic processes: Theories, individual differences, and applications* (pp. 56–77). New York: Springer Verlag.

Dixon, P. N., Willingham, W., Strano, D. A., & Chandler, C. K. (1989). Sense of humor as a mediator during incidental learning of humor-related material. *Psychological Reports, 64*, 851–855.

Eagly, A. H., Ashmore, R. D., MaKijani, M. G., & Longo, L. C. (1991). What is beautiful is good but . . . : A meta-analytic review of research on the physical attractiveness stereotype. *Psychological Bulletin, 110*, 109–128.

Ebert, P. D., & Hyde, J. S. (1976). Selection of agonistic behavior in wild female *Mus musculus. Behavior Genetics, 6*, 291–304.

Einstein, G. O., McDaniel, M. A., & Lackey, S. (1989). Bizarre imagery, interference, and distinctiveness. *Journal of Experimental Psychology: Learning, Memory, and Cognition, 15*, 137–146.

Ekman, P., & O'Sullivan, M. (1991). Who can catch a liar? *American Psychologist, 46*, 913–920.

Ellsworth, P. C., Carlsmith, J. M., & Henson, A. (1972). The stare as stimulus to flight in human subjects: A series of field experiments. *Journal of Personality and Social Psychology, 21*, 302–311.

Erdley, C. A., & D'Agostino, P. R. (1988). Cognitive and affective components of automatic priming effects. *Journal of Personality and Social Psychology, 54*, 741–747.

Ericsson, K. A., & Polson, P. G. (1988). An experimental analysis of the mechanisms of a memory skill. *Journal of Experimental Psychology: Learning, Memory, and Cognition, 74*, 476–484.

Faustman, W., & White, P. (1989). Diagnostic and psychopharmacological treatment characteristics of 536 inpatients with posttraumatic stress disorder. *The Journal of Nervous and Mental Disease, 177*, 154–159.

Firth, H., & Britton, P. (1989). "Burnout," absence and turnover amongst British nursing staff. *Journal of Occupational Psychology, 62*, 55–59.

Fisher, J. L., & Harris, M. B. (1973). Effect of note taking and review on recall. *Journal of Educational Psychology, 65*, 321–325.

Flowers, J. H., Warner, J. L., & Polansky, M. L. (1979). Response and encoding factors in "ignoring" irrelevant information. *Memory and Cognition, 7*, 86–94.

Fornell, C. (1992). A national customer satisfaction barometer: The Swedish experience. *Journal of Marketing, 56*, 6–21.

Frank, M. G., & Gilovich, T. (1988). The dark side of self- and social perception: Black uniforms and aggression in professional sports. *Journal of Personality and Social Psychology, 54*, 74–85.

Fromkin, V. A. (1980). *Errors in linguistic performance.* New York: Academic Press.

Gathercole, S. E., Willis, C. S., Emslie, H., & Baddeley, A. D. (1992). Phonological memory and vocabulary development during the early school years: A longitudinal study. *Developmental Psychology, 28*, 887–898.

Gazzaniga, M. S. (1983). Right-hemisphere language following brain bisection. *American Psychologist, 38*, 525–537.

George, J. M., Reed, T. F., Ballard, K. A., Colin, J., & Fielding, J. (1993). Contact with AIDS patients as a source of work-related distress: Effects of organizational and social support. *Academy of Management Journal, 36*, 157–171.

Gladue, B. A., & Delaney, H. J. (1990). Gender differences in perception of attractiveness of men and women in bars. *Personality and Social Psychology Bulletin, 16*, 378–391.

Goodall, J. (1986). *The chimpanzees of Gombe: Patterns of behavior*. Cambridge, MA: Belknap Press.

Goodall, J. (1990). *Through a window: My thirty years with the chimpanzees of Gombe*. Boston: Houghton Mifflin.

Greenspoon, J. (1955). The reinforcing effect of two spoken sounds on the frequency of two responses. *American Journal of Psychology, 68*, 409–416.

Haney, C., Banks, W. C., & Zimbardo, P. G. (1973). Interpersonal dynamics in a simulated prison. *International Journal of Criminology and Penology, 1*, 69–97.

Harrison, L., & Gfroerer, J. (1992). The intersection of drug use and criminal behavior: Results from the national household survey on drug abuse. *Crime and Delinquency, 38*, 422–443.

Hayduk, L. A. (1983). Personal space: Where we now stand. *Psychological Bulletin, 94*, 293–335.

Heslin, R., & Boss, D. (1980). Nonverbal intimacy in airport arrival and departure. *Personality and Social Psychology Bulletin, 6*, 248–252.

Hinkle, P. E., Wiersma, W., & Jurs, S. G. (1994). *Applied statistics for the behavioral sciences* (3rd ed.). Boston: Houghton Mifflin.

Hockey, G. R. F. (1970). Effect of loud noise on attentional selectivity. *Quarterly Journal of Experimental Psychology, 22*, 28–36.

Holsti, O. (1969). *Content analysis for the social sciences and humanities*. Reading, MA: Addison-Wesley.

Horne, J. A. (1978). A review of the biological effects of total sleep deprivation in man. *Biological Psychology, 7*, 55–102.

Hunt, R. R., & Elliott, J. M. (1980). The role of nonsemantic information in memory: Orthographic distinctiveness effects on retention. *Journal of Experimental Psychology: General 109*, 49–74.

Isen, A. M., & Levin, P. F. (1972). Effect of feeling good on helping: Cookies and kindness. *Journal of Personality and Social Psychology, 21*, 384–388.

Jenson, A. R. (1993). Why is reaction time correlated with psychometric *g*? *Current Directions in Psychological Science, 2*(2), 53–56.

Kanuk, L., & Berenson, C. (1975). Mail surveys and response rates: A literature review. *Journal of Marketing Research, 12*, 440–453.

Klein, G. S. (1964). Semantic power measured through the interference of words with color-naming. *American Journal of Psychology, 77*, 576–588.

Koocher, G. P. (1977). Bathroom behavior and human dignity. *Journal of Personality and Social Psychology, 35*, 120–121.

Krippendorf, K. (1980). *Content analysis: An introduction to its methodology*. Beverly Hills, CA: Sage.

Lavender, A. (1987). The effects of nurses changing from uniforms to everyday clothes on a psychiatric rehabilitation ward. *British Journal of Medical Psychology, 60*(2), 189–199.

Lavrakas, P. J. (1993). *Telephone survey methods* (2nd ed.). Thousand Oaks, CA: Sage.

Lockhart, R. S., & Craik, F. I. M. (1990). Levels of processing: A retrospective commentary on the framework for memory research. *Canadian Journal of Psychology, 44*, 87–112.

Loftus, E. F. (1975). Leading questions and the eyewitness report. *Cognitive Psychology, 7*, 560–572.

Martin, R., & Wall, T. D. (1989). Attentional demand and cost responsibility as a stressor in shopfloor jobs. *Academy of Management Journal, 32*, 69–86.

Martinez, R., & Dukes, R. L. (1987). Race, gender, and self-esteem among youth. *Hispanic Journal of Behavioral Science, 9*, 427–443.

Martorano, J. (1991). Case study: The use of the CEEG in treating premenstrual syndrome: An opportunity for treatment innovation. *Integrative Psychiatry, 7*, 63–64.

Mathews, K. E., Jr., & Cannon, L. K. (1975). Environmental noise level as a determinant of helping behavior. *Journal of Personality and Social Psychology, 32*, 571–577.

May, J. L., & Hamilton, P. A. (1980). Effects of musically evoked affect on women's interpersonal attraction toward and perceptual judgments of physical attractiveness in men. *Motivation and Emotion, 4*(3), 217–228.

McAninch, C. B., Austin, J. L., & Derks, P. L. (1992). Effect of caption meaning on memory for nonsense figures. *Current Psychology: Research and Reviews, 11*, 315–323.

McCusker, J., Stoddard, A. M., Zapka, J. G., Morrison, C. S., Zorn, M., & Lewis, B. F. (1992). AIDS education for drug abusers: Evaluation of short-term effectiveness. *American Journal of Public Health, 82*, 533–540.

McKey, R., Cordelli, L., Ganson, H., Barrett, B., McCorkey, C., & Plantz, M. (1985). *The impact of Head Start on children, families, and communities: Final report of the Head Start evaluation, synthesis, and utilization project* (No. OHDS 85-31193). Washington, DC: U.S. Government Printing Office.

Meyer, D. E., & Schvanveldt, R. W. (1971). Facilitation in recognizing pairs of words: Evidence of a dependence between retrieval operations. *Journal of Experimental Psychology, 90*, 227–234.

Middlemist, R. D., Knowles, E. S., & Matter, C. F. (1976). Personal space invasions in the lavatory: Suggestive evidence for arousal. *Journal of Personality and Social Psychology, 33*, 541–546.

Middlemist, R. D., Knowles, E. S., & Matter, C. F. (1977). What to do and what to report: A reply to Koocher. *Journal of Personality and Social Psychology, 35*, 122–124.

Milgram, S. (1963). Behavioral study of obedience. *Journal of Abnormal and Social Psychology, 67*, 371–378.

Nelson, D. L., & Sutton, C. (1990). Chronic work stress and coping: A longitudinal study and suggested new directions. *Academy of Management Journal, 33*, 859–869.

Neri, D. F., Shappell, S. A., & DeJohn, C. A. (1992). Simulated sustained flight operations and performance: I. Effects of fatigue. *Military Psychology, 4*(3), 137–155.

Nolen-Hoeksema, S., & Morrow, J. (1991). A prospective study of depression and posttraumatic stress symptoms after a natural disaster: The 1989 Loma Prieta earthquake. *Journal of Personality and Social Psychology, 61*, 115–121.

Orne, M. T. (1962). On the social psychology of the psychological experiment: With particular reference to demand characteristics and their implications. *American Psychologist, 17*, 776–783.

Patterson, F. G. (1978). The gestures of a gorilla: Language acquisition in another pongid. *Brain and Language, 5*, 72–97.

Phesterson, G. I., Kiesler, S. B., & Goldberg, P. A. (1971). Evaluation of the performance of women as a function of their sex, achievement, and personal history. *Journal of Personality and Social Psychology, 19*, 114–118.

Posavac, E. J., & Carey, R. G. (1989). *Program evaluation* (3rd ed.). Englewood Cliffs, NJ: Prentice Hall.

Pritchard, R. D., Dunnette, M. D., & Jorgenson, D. O. (1972). Effect of perceptions of equity and inequity on worker performance and satisfaction [Monograph]. *Journal of Applied Psychology, 56*, 75–94.

Robinson, J. P., Shaver, P. R., & Wrightsman, L. S. (1991). *Measures of personality and social psychological attitudes* (Vol. 1). San Diego, CA: Academic Press.

Roethlisberger, F. J., & Dickson, W. J. (1939). *Management and the worker*. Cambridge, MA: Harvard University Press.

Roper Organization. (1992). *Unusual personal experiences: An analysis of the data from three national surveys*. Las Vegas, NV: Bigelow Holding.

Rosenhan, D. L. (1973). On being sane in insane places. *Science, 179*, 250–258.

Rosenthal, R. (1976). *Experimenter effects in behavioral research*. New York: Ervington.

Rosenthal, R., & Rosnow, R. L. (1975). *The volunteer subject*. New York: Wiley.

Runco, M. A., & Albert, R. S. (1986). The threshold theory regarding creativity and intelligence: An empirical test with gifted and nongifted children. *The Creative Child and Adult Quarterly, 11*, 212–218.

Russel, M. J. (1976). Human olfactory communication. *Nature, 260*, 250–252.

Sacks, O. (1985). *The man who mistook his wife for a hat and other clinical tales*. New York: Simon & Schuster.

Schachter, S. (1968). Obesity and eating. *Science, 161*, 751–756.

Schachter, S. (1971). Some extraordinary facts about obese humans and rats. *American Psychologist, 26*, 129–144.

Schachter, S., Goldman, R., & Gordon, A. (1968). Effects of fear, food deprivation and obesity on eating. *Journal of Personality and Social Psychology, 10*, 91–97.

Schmidt, S. R. (1985). Encoding and retrieval processes in the memory for conceptually distinctive events. *Journal of Experimental Psychology: Learning, Memory, and Cognition, 11*, 565–578.

Schwarz, N., & Clore, G. L. (1983). Mood, misattribution, and judgments of well-being: Informative and directive functions of affective states. *Journal of Personality and Social Psychology, 45*, 513–523.

Shaffer, R. M., & Hendrick, C. (1975). Intervention in the library: The effect of increased responsibility on bystander willingness to prevent theft. *Journal of Personality and Social Psychology, 5*, 303–319.

Shah, I. (1970). *Tales of the Dervishes*. New York: Dutton.

Sherman, S. J., & Gorkin, L. (1980). Attitude bolstering when behavior is inconsistent with central attitudes. *Journal of Experimental Social Psychology, 16*, 388–403.

Silverman, L. H., Ross, D. L., Adler, J. M., & Lustig, D. A. (1978). Simple research paradigm for demonstrating subliminal psychodynamic activation: Effects of Oedipal stimuli on dart-throwing accuracy in college males. *Journal of Abnormal Psychology, 87*, 341–357.

Silverman, L. H., & Weinberger, J. (1985). Mommy and I are one: Implications for psychotherapy. *American Psychologist, 40*, 1296–1308.

Skinner, B. F. (1938). *The behavior of organisms: An experimental analysis*. New York: Appleton-Century-Crofts.

Stagray, J. R., & Truitt, L. (1992). Monaural listening therapy for auditory disorders: Opinions and a case study. *Canadian Journal of Rehabilitation, 6*, 45–49.

Sternberg, S. (1969). Memory scanning: Mental processes revealed by reaction time experiments. *American Scientist, 57*, 421–457.

Stevens, S. S. (1975). *Psychophysics: Introduction to its perceptual, neural, and social prospects*. New York: Wiley.

Strack, F., Martin, L. L., & Stepper, S. (1988). Inhibiting and facilitating conditions of the human smile: A nonobtrusive test of the facial feedback hypothesis. *Journal of Personality and Social Psychology, 5*, 768–777.

Stroop, J. R. (1935). Studies of interference in serial verbal reactions. *Journal of Experimental Psychology, 18*, 643–662.

Suedfield, P., Ballard, E. J., Baker-Brown, G., & Borrie, R. A. (1986). Flow of consciousness in restricted environmental stimulation. *Imagination, Cognition, and Personality, 5*, 219–230.

Taylor, S. P., & Leonard, K. F. (1983). Alcohol and human physical aggression. *Aggression, 2*, 77–101.

Tokunaga, K., Fukushima, M., Kemnitz, J., & Bray, G. (1986). Comparison of ventromedial and paraventricular lesions in rats that become obese. *American Journal of Physiology, 251*, R1221–R1227.

Tronick, E. Z. (1989). Emotions and emotional communication in infants. *American Psychologist, 44*, 112–119.

Werch, C. E., Meers, B. W., & Hallan, J. B. (1992). An analytic review of 73 college-based drug abuse prevention programs. *Health Values, 16*(5), 38–45.

Zimbardo, P. G., Cohen, A. R., Weisenberg, M., Dworskin, L., & Firestone, I. (1966). Control of pain motivation by cognitive dissonance. *Science, 151*, 217–219.

INDEX

List of Symbols

N	number of scores in the data
f	frequency
cf	cumulative frequency
rel. f	*relative frequency*
X	scores
Y	scores
k	constant
ΣX	sum of X
Mdn	median
$\bar{X}$	sample mean of Xs
$X - \bar{X}$	deviation
μ	mu; population mean
ΣX^2	sum of squared Xs
$(\Sigma X)^2$	squared sum of Xs
S_X	sample standard deviation
S_X^2	sample variance
σ_X	population standard deviation
σ_X^2	population variance
s_X	estimated population standard deviation
s_X^2	estimated population variance
df	degrees of freedom
$\pm$	plus or minus
z	z-score
$\sigma_{\bar{X}}$	standard error of the mean
$\bar{Y}$	sample mean of Ys
ΣY	sum of Ys
ΣY^2	sum of squared Ys
$(\Sigma Y)^2$	squared sum of Ys
ΣXY	sum of cross products of X and Y
D	difference score
r	Pearson correlation coefficient
r_s	Spearman correlation coefficient
r_{pb}	point-biserial correlation coefficient
ρ	rho; population correlation coefficient
Y'	Y prime; predicted value of Y
$S_{Y'}^2$	variance of the Y scores around Y'
$S_{Y'}$	standard error of the estimate
b	slope of the regression line
a	Y-intercept of the regression line
r^2	coefficient of determination
$1 - r^2$	coefficient of alienation
p	probability
$p(\text{A})$	probability of event A
$>$	greater than
$<$	less than
$\geq$	greater than or equal to
$\leq$	less than or equal to
$\neq$	not equal to
H_a	alternative hypothesis
H_0	null hypothesis
z_{obt}	obtained value of z-test
z_{crit}	critical value of z-test
α	alpha; theoretical probability of a Type I error
β	beta; theoretical probability of a Type II error
$1 - \beta$	power